Lecture Notes in Physics

AF856256

The Editorial Policy for Proceedings

The series Lecture Notes in Physics reports new developments in physical research and teaching – quickly, informally, and at a high level. The proceedings to be considered for publication in this series should be limited to only a few areas of research, and these should be closely related to each other. The contributions should be of a high standard and should avoid lengthy redraftings of papers already published or about to be published elsewhere. As a whole, the proceedings should aim for a balanced presentation of the theme of the conference including a description of the techniques used and enough motivation for a broad readership. It should not be assumed that the published proceedings must reflect the conference in its entirety. (A listing or abstracts of papers presented at the meeting but not included in the proceedings could be added as an appendix.)
When applying for publication in the series Lecture Notes in Physics the volume's editor(s) should submit sufficient material to enable the series editors and their referees to make a fairly accurate evaluation (e.g. a complete list of speakers and titles of papers to be presented and abstracts). If, based on this information, the proceedings are (tentatively) accepted, the volume's editor(s), whose name(s) will appear on the title pages, should select the papers suitable for publication and have them refereed (as for a journal) when appropriate. As a rule discussions will not be accepted. The series editors and Springer-Verlag will normally not interfere with the detailed editing except in fairly obvious cases or on technical matters.
Final acceptance is expressed by the series editor in charge, in consultation with Springer-Verlag only after receiving the complete manuscript. It might help to send a copy of the authors' manuscripts in advance to the editor in charge to discuss possible revisions with him. As a general rule, the series editor will confirm his tentative acceptance if the final manuscript corresponds to the original concept discussed, if the quality of the contribution meets the requirements of the series, and if the final size of the manuscript does not greatly exceed the number of pages originally agreed upon.
The manuscript should be forwarded to Springer-Verlag shortly after the meeting. In cases of extreme delay (more than six months after the conference) the series editors will check once more the timeliness of the papers. Therefore, the volume's editor(s) should establish strict deadlines, or collect the articles during the conference and have them revised on the spot. If a delay is unavoidable, one should encourage the authors to update their contributions if appropriate. The editors of proceedings are strongly advised to inform contributors about these points at an early stage.
The final manuscript should contain a table of contents and an informative introduction accessible also to readers not particularly familiar with the topic of the conference. The contributions should be in English. The volume's editor(s) should check the contributions for the correct use of language. At Springer-Verlag only the prefaces will be checked by a copy-editor for language and style. Grave linguistic or technical shortcomings may lead to the rejection of contributions by the series editors.
A conference report should not exceed a total of 500 pages. Keeping the size within this bound should be achieved by a stricter selection of articles and not by imposing an upper limit to the length of the individual papers.
Editors receive jointly 30 complimentary copies of their book. They are entitled to purchase further copies of their book at a reduced rate. As a rule no reprints of individual contributions can be supplied. No royalty is paid on Lecture Notes in Physics volumes. Commitment to publish is made by letter of interest rather than by signing a formal contract. Springer-Verlag secures the copyright for each volume.

The Production Process

The books are hardbound, and the publisher will select quality paper appropriate to the needs of the author(s). Publication time is about ten weeks. More than twenty years of experience guarantee authors the best possible service. To reach the goal of rapid publication at a low price the technique of photographic reproduction from a camera-ready manuscript was chosen. This process shifts the main responsibility for the technical quality considerably from the publisher to the authors. We therefore urge all authors and editors of proceedings to observe very carefully the essentials for the preparation of camera-ready manuscripts, which we will supply on request. This applies especially to the quality of figures and halftones submitted for publication. In addition, it might be useful to look at some of the volumes already published. As a special service, we offer free of charge LATEX and TEX macro packages to format the text according to Springer-Verlag's quality requirements. We strongly recommend that you make use of this offer, since the result will be a book of considerably improved technical quality. To avoid mistakes and time-consuming correspondence during the production period the conference editors should request special instructions from the publisher well before the beginning of the conference. Manuscripts not meeting the technical standard of the series will have to be returned for improvement.

For further information please contact Springer-Verlag, Physics Editorial Department V, Tiergartenstrasse 17, W-6900 Heidelberg, FRG

M. Napolitano F. Sabetta (Eds.)

Thirteenth International Conference on Numerical Methods in Fluid Dynamics

Proceedings of the Conference
Held at the Consiglio Nazionale delle Ricerche
Rome, Italy, 6-10 July 1992

Springer-Verlag
Berlin Heidelberg GmbH

Editors

M. Napolitano
Istituto di Macchine ed Energetica
Politecnico di Bari
Via Re David 200, I-70125 Bari, Italy

F. Sabetta
Dipartimento di Meccanica ed Aeronautica
Università di Roma "La Sapienza"
Via Eudossiana 18, I-00184 Roma, Italy

ISBN 978-3-662-14949-2 ISBN 978-3-540-47551-4 (eBook)
DOI 10.1007/978-3-540-47551-4

Originally published by Springer-Verlag Berlin Heidelberg New York in 1993

58/3140-543210 - Printed on acid-free paper

EDITORS' PREFACE

The present volume of Lecture Notes in Physics constitutes the Proceedings of the Thirteenth International Conference on Numerical Methods in Fluid Dynamics, held at the Consiglio Nazionale delle Ricerche, Rome, Italy on 6–10 July 1992, with the following format: a plenary session in the mornings and two parallel sessions in the afternoons. Each plenary session started with an invited lecture on the subject of the session, the five topics, chosen by the Conference International Committee, being: Multidimensional Upwinding; Turbulent Flows; Domain Decomposition Methods; Unstructured Grids; Flow Visualization. Finally, one session was devoted to a Workshop on Cell-Vertex Schemes, sponsored by the U.S. Department of Defense.

Ninety-eight contributed papers were accepted, out of 266 abstracts submitted from all over the world, by four paper selection committees, chaired by M. Holt, K. Oshima, M. Pandolfi and V. Rusanov. Seven papers were withdrawn before or after the conference and two have not been presented at the conference by any of the authors.

The remaining papers presented in this volume have been divided into seven parts: the first four broadly relate to the first four invited lectures, the fifth contains papers concerning supersonic and hypersonic flows, the sixth contains all other papers and the last contains the contributions to the workshop on cell-vertex schemes.

The conference was attended by 191 people from 22 countries. Its success owes much to the efforts of the International Organizing Committee, to our associates A. Angelini and V. Casalino of the Local Organizing Committee, to the staff of Progress Promozione Congressi and, last but not least, to the unique charm of Rome.

Rome
September 1992

M. Napolitano
F. Sabetta

Acknowledgements

The Organizing Committee wishes to express its gratitude to the following organizations for their generous financial support of the conference and this volume:

Alenia
CIRA S.p.A.
Consiglio Nazionale delle Ricerche
ENEL
IBM Semea S.p.A.
Politecnico di Bari
United Technologies Research Center
Università di Roma "La Sapienza"
U.S.A. Department of Defense

CONTENTS

INVITED LECTURES

UPWIND AND CHARACTERISTIC TYPE METHODS

TRANSITIONAL AND TURBULENT FLOWS

DOMAIN DECOMPOSITION METHODS

UNSTRUCTURED AND ADAPTIVE GRIDS

SUPERSONIC AND HYPERSONIC FLOWS

APPLICATIONS AND SPECIAL TOPICS

WORKSHOP ON CELL-VERTEX SCHEMES

PROGRESS IN MULTI-DIMENSIONAL UPWIND DIFFERENCING

Bram van Leer

University of Michigan, Ann Arbor, MI

1 Introduction

CFD algorithms for the coming generation of massively parallel computers will have to be extremely robust. They will most likely be implemented on adaptive unstructured grids, and will be used for ambitious simulations of steady and unsteady three-dimensional flows. In such a complex environment there is little place left for hand-tuning parameters that regulate accuracy, stability and convergence of the computations. A typical algorithm will make very intensive use of local data, with a minimum of message passing.

Algorithms of this nature exist already in CFD: they are the upwind-differencing schemes, computationally intensive but unsurpassed in their combination of accuracy and robustness. While these favorable properties are explainable for one-dimensional methods, it is a stroke of luck that upwind schemes work as well as they do for two- and three-dimensional flow. Their design is commonly based on one-dimensional physics, namely, the solution of the one-dimensional Riemann problem that describes the interaction of two fluid cells by finite-amplitude waves moving normal to their interface. The inadequacy of this technique clearly shows up when the numerical solution contains shock or shear waves not aligned with the grid, for instance, by a loss of resolution.

The need to incorporate genuinely multi-dimensional physics in upwind algorithms was recognized as early as 1983 by Phil Roe [1]. A study of discrete multi-dimensional wave models by Roe followed in 1985 (ICASE Report 85-18, also [2]), but it took until 1991 [3] before any algorithms based on such wave models became truly successful. Important contributions to this development were made by Herman Deconinck and collaborators [3, 4] at the Von Kármán Institute in Brussels. The new upwind schemes are formulated on unstructured grids with data in the vertices of triangular or tetrahedral cells.

While genuinely multi-dimensional methods were slowly developing, partial successes were booked by putting some multi-dimensional information into the Riemann solvers used in conventional upwind schemes. In particular, it became the fashion to obtain a plausible wave-propagation angle from the data, rather than accepting the angle dictated by the grid geometry. The earliest work of this kind is due to Steve Davis [5]; it recently was picked up by a number of authors: Levy, Powell and Van Leer [6], [7], Dadone and Grossman [8, 9], Obayashi and Goorjian [10], Tamura and Fujii [11]. Roughly speaking, they apply Riemann solvers in several, physically appealing, directions; I shall refer to their work as the *multi-directional* approach.

Related, but closer to the genuinely multi-dimensional approach is the work of Rumsey, Van Leer and Roe [12, 13, 14, 15] and Parpia and Michalek [16, 17]. These authors independently developed almost identical multi-dimensional wave models based on minimizing wave strengths. These wave models requires only two input states, just

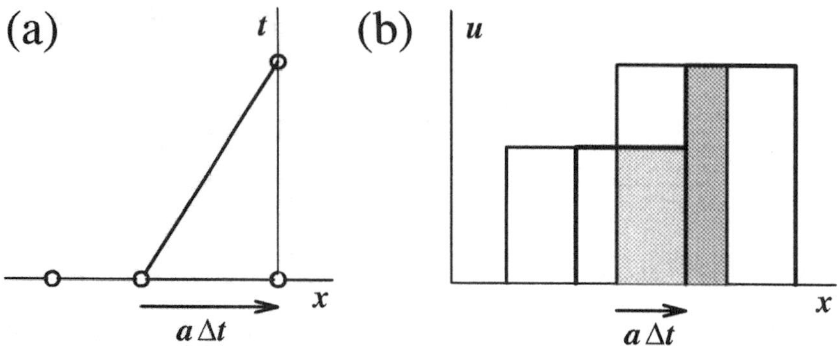

Figure 1: Two views on scalar upwind differencing: (a) nodal-point interpretation; (b) finite-volume interpretation.

as a regular Riemann solver.

In support of these quasi-multi-dimensional approaches, aimed at putting better physics into interface fluxes, some authors have dedicated efforts to improving the interpolation or reconstruction step that precedes the flux calculation. On a structured grid the reconstruction of a non-oscillatory distribution of flow variables from their cell-averages usually is done dimension by dimension; a fully multi-dimensional reconstrucion is indispensable in achieving higher accuracy. Barth and Frederickson [18] indicated how to reconstruct a smooth function up to arbitrarily high order on an unstructured triangulation; Abgrall [19] showed how to implement truly multi-dimensional limiting of higher derivatives.

In this lecture I shall review a decade of efforts toward multi-dimensional upwind-differencing, with the accent on the very latest developments. The discussion is limited to the multi-dimensional physics that goes into these methods; multi-dimensional reconstruction will not be further mentioned. For a somewhat different emphasis or point of view the reader is referred to three excellent other reviews of multi-dimensional methods [20, 21, 4] that have been presented in the past year.

2 Two views of one-dimensional upwinding

In order to appreciate the problems surrounding multi-dimensional upwinding it is useful to consider the principles of one-dimensional upwinding. The reader is assumed to be familiar with the theory of conservative upwind schemes; as a tutorial Roe's [22] review article is recommended.

Upwind differencing is a way of differencing convection terms. For the scalar convection equation

$$u_t + cu_x = 0, \tag{1}$$

the simplest upwind-difference scheme, of first-order accuracy, reads

$$\frac{u_i^{n+1} - u_i^n}{\Delta t} + c\frac{u_i^n - u_{i-1}^n}{\Delta x} \quad = \quad 0, \quad c \geq 0; \tag{2}$$

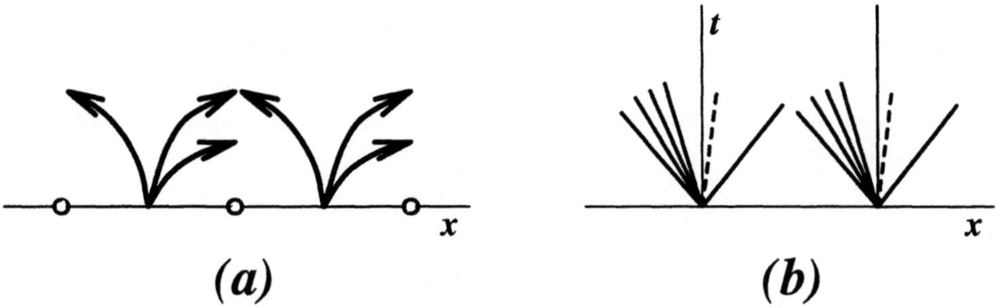

Figure 2: Two approaches to upwind differencing for the Euler equations: (a) fluctuation approach; (b) finite-volume approach.

$$\frac{u_i^{n+1}-u_i^n}{\Delta t}+c\frac{u_{i+1}^n-u_i^n}{\Delta x} = 0, \quad c<0. \tag{3}$$

Scheme (2,3) can be regarded as a formula for updating, from t^n to t^{n+1}, either the *nodal-point value* of u in x_i, or the *cell average* of u in cell i. These two view-points are illustrated in Figures 1a and 1b. The distinction is significant, because it leads to distinct methods for more complex equations. In the development of schemes for the one-dimensional Euler equations, the first view-point has led to the concept of *fluctuation splitting*, due to Roe [23, 22]; the second view-point is that of Godunov [24] and has led to the *projection/evolution* or *reconstruction/evolution* concept of finite-volume schemes, due to Van Leer [25, 26, 27]. Below I shall review the formulas pertinent to each approach.

2.1 Fluctuation splitting

Assume the system

$$U_t + F(U)_x = 0 \tag{4}$$

represents the Euler equations in conservation form, i.e., $U = (\rho, \rho u, \rho E)^T$ is the vector of conserved state quantities and $F(U) = (\rho u, \rho u^2 + p, \rho u H)^T$ is the vector of their fluxes. The equation shows that any local imbalance of the fluxes causes the local solution to change in time. Such a local imbalance is called a *fluctuation* by Roe [28, 1]. If source terms are present, their value must be included in the fluctuation [22].

Define the matrix $A(U)$ as the derivative of $F(U)$ with respect to U, so that

$$dF(U) = A(U)dU. \tag{5}$$

It is essential for the technique of fluctuation splitting that this differential relation be replaced by an exact finite-difference analogue, namely,

$$\Delta F = \hat{A}\Delta U, \tag{6}$$

where Δ indicates a difference between neighboring nodal points. Roe [29] has indicated how to construct a mean value $\hat{A}$ of A such that Eq. (6) holds exactly for arbitrary pairs of state vectors. For a calorically perfect gas a suitable mean value can easily be obtained by introducing the the parameter vector $w \equiv \sqrt{\rho}\,(1, u, H)^T$. Since both $U(w)$ and $F(w)$ are quadratic in the components of w, it follows that Eq. (6) is satisfied by $\hat{A} \equiv A\,(U(\bar{w}))$, where $\bar{w}$ is the *algebraic* average of w.

Fluctuation splitting requires that the matrix $\hat{A}$ be split into its positive and negative parts, i.e.,

$$\hat{A} = \hat{A}^{+} + \hat{A}^{-}, \tag{7}$$

so that

$$\Delta F = \hat{A}^{+}\Delta U + \hat{A}^{-}\Delta U. \tag{8}$$

A popular name for this procedure is "flux-difference splitting"; the term "fluctuation splitting" is preferable because it includes source-term splitting. The first term on the right-hand side combines disturbances that propagate forward; in consequence, this term is used to update the right nodal point. The second term combines backward-moving disturbances and is used to update the left nodal point. This concept is illustrated in Figure 2a. Conservation is ensured because the two terms add up to a perfect flux difference. The first-order update formula becomes

$$U_i^{n+1} = U_i^n - \frac{\Delta x}{\Delta t}\left(\left(\hat{A}^{+}\Delta U\right)_{i-\frac{1}{2}} + \left(\hat{A}^{-}\Delta U\right)_{i+\frac{1}{2}}\right). \tag{9}$$

In practice it often pays to abandon the matrix notation and expand ΔU and ΔF in terms of the individual disturbances. This yields

$$\Delta U = \sum_{k=1}^{3} \alpha_k R_k, \tag{10}$$

$$\Delta F = \sum_{k=1}^{3} \lambda_k \alpha_k R_k, \tag{11}$$

where λ_k is an eigenvalue of $\hat{A}$, R_k is the corresponding eigenvector, and α_k is the wave strength; note that Eqs. (6) and (10) imply Eq. (11). By considering that each fluctuation may move forward or backward through the grid, we recover the splitting formula (8):

$$\begin{aligned} \Delta F &= \sum_{\lambda_k<0} \lambda_k \alpha_k R_k + \sum_{\lambda_k>0} \lambda_k \alpha_k R_k \\ &\equiv \sum_{k=1}^{3} \lambda_k^{+} \alpha_k R_k + \sum_{k=1}^{3} \lambda_k^{-} \alpha_k R_k \\ &= \hat{A}^{+}\Delta U + \hat{A}^{-}\Delta U. \end{aligned} \tag{12}$$

2.2 Finite-volume approach

In the finite-volume approach the focus is on the numerical flux function $F\,(U_L, U_R)$, a recipe for computing the interface fluxes from the states U_L and U_R on the left and right sides of the interface. The generic formula for updating cell averages of the conserved quantities is

$$U_i^{n+1} = U_i^n - \frac{\Delta t}{\Delta x}\left(F_{i+\frac{1}{2}}^n - F_{i-\frac{1}{2}}^n\right). \tag{13}$$

In Godunov's first-order scheme the interface flux is taken from the solution at $t > 0$ of Riemann's initial-value problem with input data

$$U(x,0) = U_L, \quad x > 0, \tag{14}$$
$$U(x,0) = U_R, \quad x < 0; \tag{15}$$

this is illustrated in Figure 2b.

For many applications it is not necessary to use the exact solution to this problem, hence the activity in the research area of "approximate Riemann solvers" [23, 30, 31]. Adopting Roe's [23] approximate solution, which is the exact solution of the locally linearized equation

$$U_t + \hat{A}U_x = 0, \tag{16}$$

we find three equivalent formulas for the interface flux:

$$F(U_L, U_R) = F_L + \hat{A}^- \Delta U, \tag{17}$$
$$F(U_L, U_R) = F_R - \hat{A}^+ \Delta U, \tag{18}$$
$$F(U_L, U_R) = \frac{1}{2}(F_L + F_R) - |\hat{A}|\Delta U, \tag{19}$$

where

$$|\hat{A}| = \hat{A}^+ - \hat{A}^-. \tag{20}$$

In practice the formula (19) is preferred because of its symmetry; the expanded form is

$$F(U_L, U_R) = \frac{1}{2}(F_L + F_R) - \frac{1}{2}\sum_{k=1}^{3} |\lambda_k| \alpha_k R_k. \tag{21}$$

Inserting the flux (19) into the finite-volume scheme (13) yields an scheme that, with the help of the identity (6), reduces precisely to the fluctuation scheme (9). Yet, there exists an important difference between Eqs. (19,13) and Eq. (9): in the latter the matrix $\hat{A}$ *must* satisfy the identity (6) in order to maintain conservation, while in Eq. (19) the matrix $|\hat{A}|$ may be derived from *any* average $\hat{A}$ without endangering conservation. The flux formula (19), due to Van Leer [32, 33], preceded the fluctuation approach of Roe [23], based on (6), by a decade.

3 Intermezzo: how good is one-dimensional upwinding?

To appreciate the superior accuracy and robustness of upwind differencing in one dimension, consider the numerical results shown in Figure 3 and 4, taken from [34] and [35], respectively. In Figure 3a the exact and discrete Mach-number distributions for choked flow through a converging-diverging channel are superimposed. First-order fluctuation splitting was used, including source-term splitting [22, 36] and a special splitting near the sonic point [34]. Although the update formula is only first-order accurate, it can be shown that the scheme yields second-order accurate *steady* solutions. In fact, in the steady state the scheme reduces to the two-point box scheme on all meshes except near a sonic point and inside a shock structure, where it becomes a three-point scheme. This yields the smooth transition through the sonic point and the crisp shock transition in the displayed results. Figure 3b shows the residual-convergence histories for three increasingly powerfull marching techniques: global time-stepping, local time-stepping and characteristic time-stepping [35]; these look uneventful. In Figure 4a a shockless transonic solution is reached from initial values containing 7 shocks and 8 sonic points; again, the residual-convergence history in Figure 4b for local time-stepping shows nothing unusual.

It is this type of performance we wish to preserve when extending upwind differencing to higher dimensions.

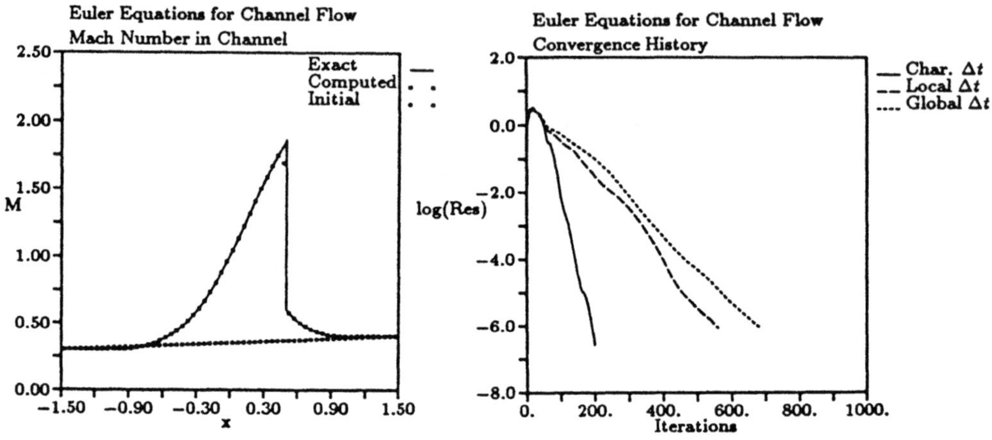

Figure 3: Choked flow through a converging-diverging channel, computed with a fluctuation scheme. (a) Initial and final Mach-number distributions; (b) residual-convergence histories for global, local and characteristic time-stepping.

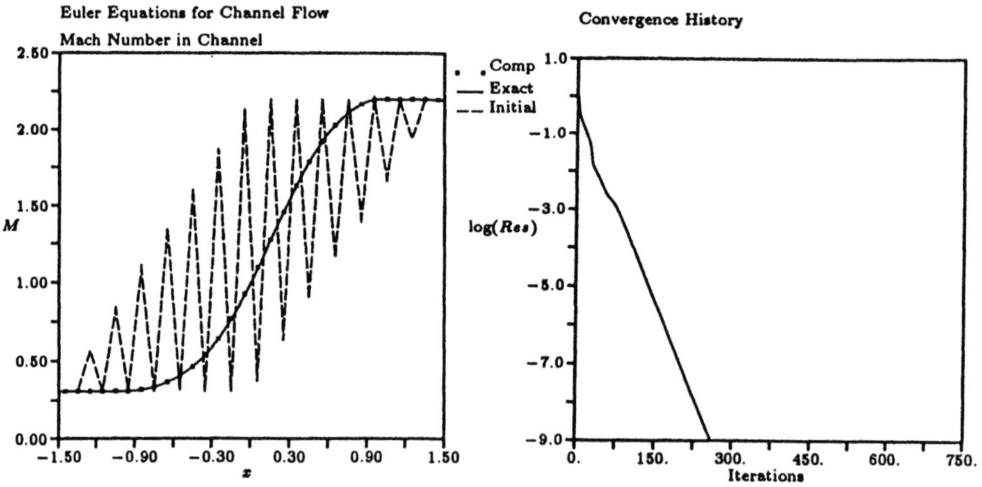

Figure 4: Transonic flow in a converging-diverging channel, computed with a fluctuation scheme. (a) Initial and final Mach-number distributions; (b) residual-convergence history for local time-stepping.

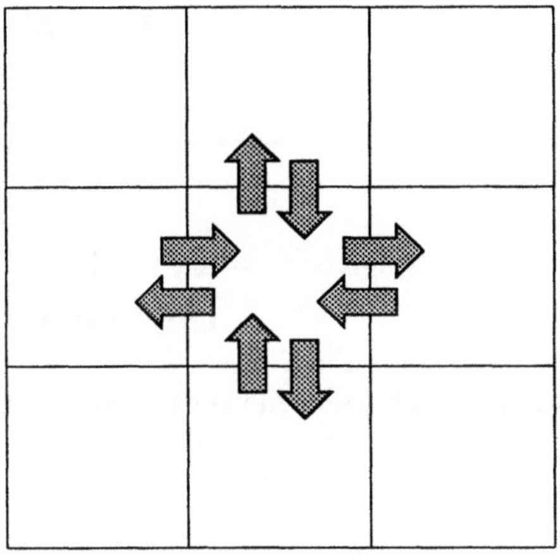

Figure 5: Extending the finite-volume method to two dimensions by solving one-dimensional Riemann problems at all cell faces. The arrows symbolize the exchange of information between cells in the direction normal to their interface.

4 Multi-dimensional extension of the finite-volume method

The standard way to extend upwind differencing to the multi-dimensional Euler equations is still the same as indicated by Godunov et al. [37] in 1961. For first-order accuracy, initial values are assumed to be constant in each cell, just as in one dimension; fluxes at cell interfaces again follow from solving one-dimensional Riemann problems of the type (14,15), with x now measuring distance along the normal to the interface. This is illustrated by Figure 5.

It is the projection of the true initial values onto cellwise constant distributions (or linear [25, 26] or quadratic [25, 38, 39] or even higher-order distributions [40]) that creates discontinuities at the interfaces. This leads us to introducing plane wave fronts parallel to the interface, and selecting, out of all possible directions, the interface normal as the direction for wave propagation. If the solution contains only shock and/or shear waves aligned or nearly aligned with the grid, this choice happens to be the correct one, and high resolution of such waves can be achieved in the steady state, just as in one dimension. If, however, such waves are far from aligned with the grid, they get misrepresented by the upwind scheme as pairs of grid-aligned waves, as shown in Figure 6 for a shear wave. Thus, a grid-oblique stationary wave may be represented by several grid-aligned running waves, leading to higher numerical dissipation and a considerable loss of resolution.

Another purely numerical artifact caused by grid-aligned upwinding is the presence of pressure disturbances across a grid-oblique shear layer. First observed by Venkatakrishnan [41], the explanation was provided by Rumsey et al. [12]; this phenomenon is further discussed in Section 4.2.

From the above critique one should not conclude that in higher dimensions the standard upwind methods are inferior to other methods; the loss of accuracy just is much more obvious for upwind methods.

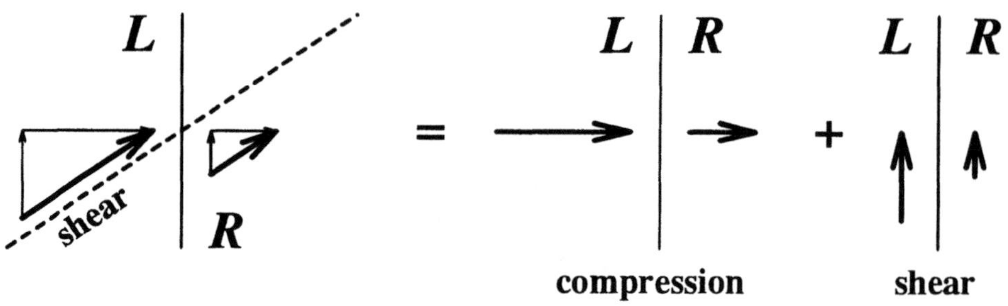

Figure 6: Misinterpretation of a grid-oblique shear wave by grid-aligned upwinding.

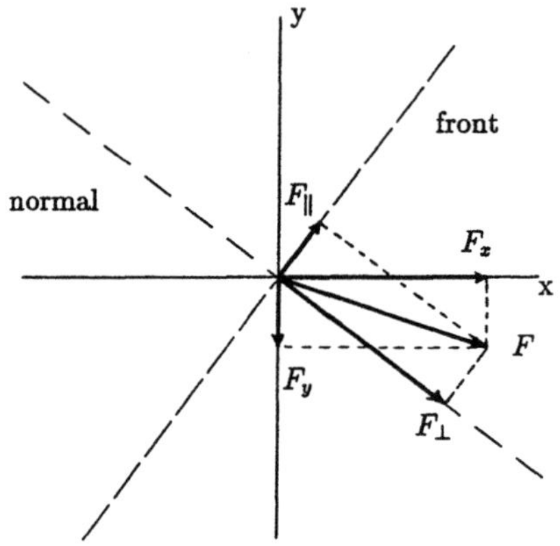

Figure 7: Fluxes in a frame aligned with a wave front oblique to the grid lines.

4.1 Multi-directional methods

The smearing of oblique shock waves in numerical solutions has received considerable attention, and a proportionally large research effort has been spent in mending this weakness. The prevailing idea is to solve the Riemann problem in a direction more appropriate than the grid direction. One immediate consequence of leaving the grid-aligned frame is that solving one Riemann problem no longer suffices. Figure 7 shows that, in two dimensions, both flux vectors in the rotated frame are needed for the construction of the fluxes normal to the interface.

Consider, for example, Figure 8, showing a rotated coordinate system aligned with level lines representing a shock front in a discrete solution. It makes sense to solve a one-dimensional Riemann problem in the direction normal to the front, i.e. using the flow-velocity components in that direction; this yields the flux in the normal direction. The input states for the Riemann solver are $U_{L\perp} = U_L$ and $U_{R\perp} = U_R$. The flux tangential to the shock should be obtained from state values located at $L_{\|}$ and $R_{\|}$; using U_L and U_R once more would completely destroy the effect of the rotation [7, 14]). These values could be approximated by

$$U_{L\|} = U_{R\|} = \frac{1}{2}\left(U_L + U_R\right); \tag{22}$$

this, however, implies central differencing along the shock and leads to odd-even decoupling in that direction [6, 7, 42].

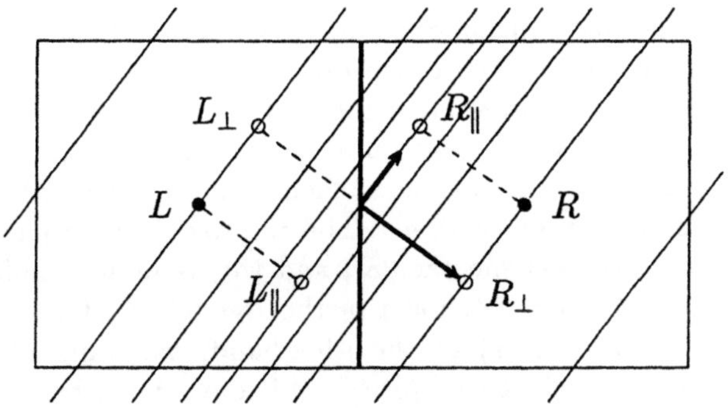

Figure 8: A simple multi-directional flux formula.

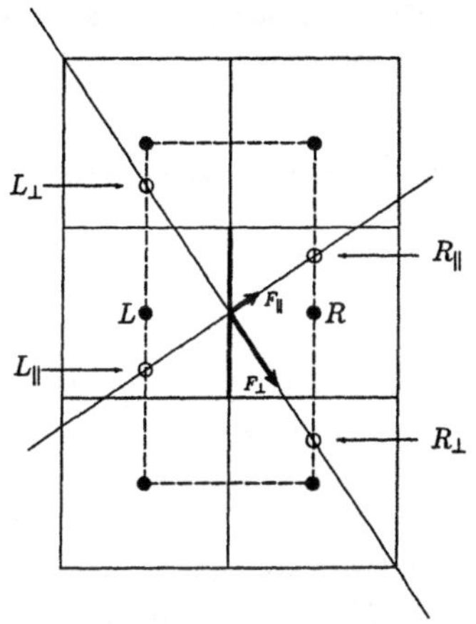

Figure 9: Input states for the Riemann problems in the flux computation according to Levy et al.

In the work of Davis [5, 43], dating back as far as 1983, the computation of the tangent flux actually is more complicated than that of the normal flux. The more recent work of Levy et al. [6, 7, 42] and Dadone and Grossman [8, 9] is more mature in that the fluxes are treated without distinction. Figure 9 shows how pairs of input states to the two Riemann problems, $(U_{L\perp}, U_{R\perp})$ and $(U_{L\|}, U_{R\|})$, are selected according to Levy et al. In their first-order method the input states in the rotated frame are obtained by linear interpolation between neighboring states in a ring of cells surrounding the interface; Dadone and Grossman simply take the value in the nearest cell, which apparently adds to the robustness of the method. Another, wider ring of cells is needed for achieving second-order accuracy.

Various choices can be made for the rotation angle of the frame in which the Riemann problems are solved. A sensitive quantity is the direction of the velocity-difference vector, $\vec{V}_R - \vec{V}_L$, which was adopted by Davis and also is crucial to the approach of Rumsey and Parpia (see Section 4.2). Levy et al. use the direction of the velocity-magnitude gradient $\vec{\nabla}|\vec{V}|$, which can detect both shock and shear waves, while Dadone and Grossman use the pressure gradient $\vec{\nabla}p$, which only detects shocks.

For a more detailed description of the multi-directional approach the reader may be referred to reference [9] in these proceedings.

After a decade of multi-directional methods, what benefits have been demonstrated? Surely, these methods yield impressive results when applied to first-order schemes: shock and shear waves not aligned with the grid are represented as if computed with a higher-order method. The improvement brought to higher-order schemes, though, is a lot less spectacular, and this is understandable. On the one hand, there is not much room left for a further reduction of wave spread (more for shear waves than for shock waves); on the other hand, loss of monotonicity may occur, against which there are no effective limiters, and convergence to a steady state suffers under the strong nonlinearity of the methods.

In my opinion, the multi-directional approach has had a clear impact on computational fluid dynamics. Although complete multi-directional methods will survive only if the problem of ensuring robustness can be solved, I expect that elements of such methods may find their way into standard, direction-split codes, to help resolve flow features arising in specific flow problems.

4.2 Minimum-strength wave models

In the work of Rumsey, Roe and Van Leer [14] and Parpia and Michalek [17], the orientation of the cell interface is de-emphasized. The spatial discretization is no longer regarded as generating a discontinuity along the interfaces; instead, an attempt is made to find out what waves are actually propagating near the interface. This, of course, requires data spanning a multi-dimensional part of space; if only the two states U_L and U_R are to be used, a theoretical conjecture must make up for the missing information.

In the basic wave model of Rumsey et al. a special set of 4 waves is used to match the state difference $U_R - U_L$; for uniqueness, the sum of the wave strengths is minimized. Three of these waves follow from solving a one-dimensional Riemann problem in the direction of the velocity difference $\Delta\vec{V}$, the fourth wave is a shear wave normal to the other three. This choice of waves makes sense from a kinematic point of view, as illustrated by Figure 10. It shows that a velocity difference $\Delta\vec{V}$ can be explained by an acoustic wave traveling in the direction of $\Delta\vec{V}$ as well as a shear wave traveling in the normal direction. Which explanation is the more likely one may be determined by also considering the pressure difference $p_R - p_L$: a large value favors the acoustic explanation, while a small value favors the shear explanation. The minimization procedure takes the full state difference $U_R - U_L$ and comes up with a plausible explanation in terms of all four waves. The method of Parpia and Michalek differs only in the choice of the functional that is minimized. Figure 11 shows the configuration of the plane waves crossing the interface. In practice both methods include a fifth wave, a weak shear wave, which corrects for the difference between the true direction of $\Delta\vec{V}$ and the direction actually used; the latter may have been held over from a previous iteration ("frozen"), for improvement of convergence.

The word "plausible" used above indicates that the minimization procedure only makes an educated guess: it is possible to compose a set of initial values that is totally misinterpreted. Consider, for instance, the head-on collision of two gases that have equal, negligible pressures. In reality two strong shocks are formed, moving into the

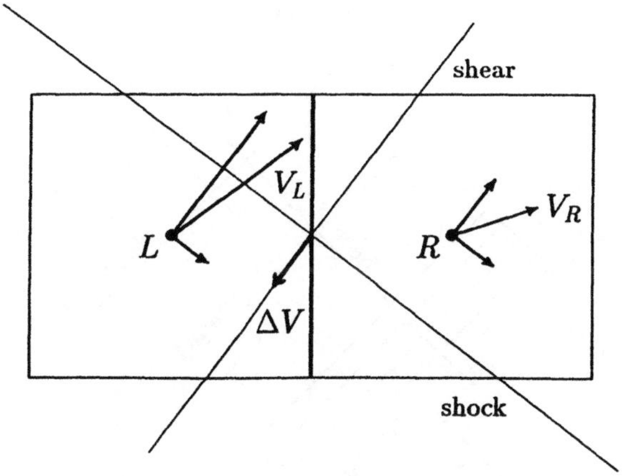

Figure 10: Shock or shear?

gases. The procedure sees as input a velocity difference not accompanied by a pressure difference, hence calls for a single shear wave, as if the gases *avoided* collision!

The flux formula based on the above wave model is worth some discussion. Assuming the system

$$U_t + F(U)_x + G(U)_y = 0, \tag{23}$$

with flux Jacobians $A(U)$ and $B(U)$, represents the two-dimensional Euler equations, we may again write ΔU as a sum:

$$\Delta U = \sum_{k=1}^{5} \alpha_k R_k. \tag{24}$$

The vector R_k is now an eigenvector of the matrix

$$\hat{A} \cos \theta_k + \hat{B} \sin \theta_k, \tag{25}$$

where θ_k indicates the propagation angle of the k-th wave; the matrices $\hat{A}$ and $\hat{B}$ are standard Roe-averages. The upwind-biased interface flux is defined by

$$F(U_L, U_R) = \frac{1}{2}(F_L + F_R) - \sum_{k=1}^{5} |\lambda_k \cos(\theta_k - \theta_{\text{normal}})| \, \alpha_k R_k, \tag{26}$$

i.e. still by formula (21), but with the wave speeds λ_k projected onto the interface normal. Although this formula seems trivial, it must be pointed out that there no longer exists a relation between ΔF and ΔU like (6).

In numerical practice minimum-strength wave models appear to bring the same benefits and problems as multi-directional methods: great improvements in shock and shear resolution for first-order methods, much smaller improvements for second-order methods, and possible loss of monotonicity and convergence.

To illustrate the performance of this class of methods, consider Figures 12a and 12b. Both show pressure plots for steady viscous flow over a NACA 0012 airfoil at 3° angle of attack and Reynolds number 5000, computed on a 129 × 49 O-grid by Rumsey [12, 14]. Under these conditions the flow separates from the upper surface,

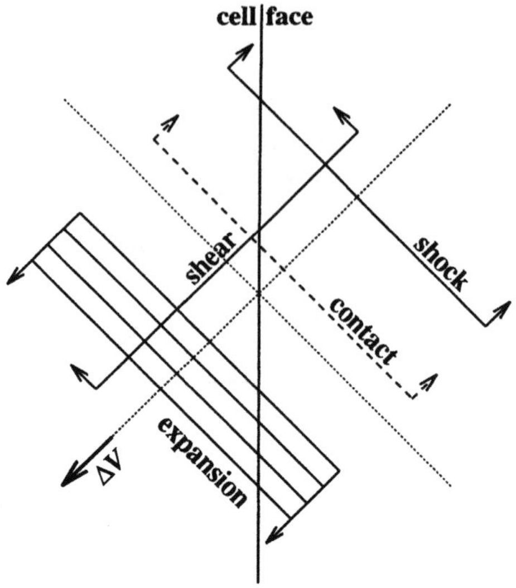

Figure 11: Plane waves crossing a cell face according to the model of Rumsey et al.

producing a detached shear layer oblique to the grid. For the results of Figure 12a a second-order MUSCL-type scheme [26, 44] was used, with Roe's [23] standard grid-aligned Riemann solver. The Riemann solver misinterprets the oblique shear as an grid-aligned shear plus an acoustic wave (see Figure 6); the latter causes a pressure rise or drop at the interface. Correspondingly, the steady solution shows pressure fluctuations across the shear layer, so that its presence can actually be detected in pressure plots. A grid-refinement study shows that the disturbances scale with the mesh size. This phenomenon was first observed by Venkatakrishnan [41] and correctly explained by Roe; in fact, it motivated the work of Rumsey, Van Leer and Roe. As seen from Figure 12b, the minimum-strength wave model properly recognizes the oblique shear layer and generates clean pressure contours.

The same method gives an unexpected improvement in the representation of inviscid stagnating flow. The explanation is found in Figure 13, showing the turning of the flow near a stagnation point S as represented by the discrete velocities in the three cells marked 1, 2 and 3.. A grid-aligned Riemann solver interprets the velocity difference between vertical neighbors 1 and 2 as a compression ($V_{y1} > V_{y2}$), and the velocity difference between horizontal neighbors 2 and 3 as an expansion ($V_{x2} < V_{x3}$); this leads to pressure variations of the order of ΔV. The wave model detects only very small pressure changes ($\Delta p \sim \rho\Delta(V^2)$) and therefore explains both velocity differences by shear waves. Although this still is not the right explanation, the result is a decrease in numerical entropy production. The effect is rather large for first-order methods, as can be judged from Figure 14 showing entropy contours for inviscid flow over a NACA 0012 airfoil at $M = 0.3$, $\alpha = 1°$, on a sequence of O-grids. The reduced entropy levels lead directly to reduced numerical drag levels, as Figure 15a shows. For second-order schemes the effect, as usual, is less dramatic; the drag values are given in Figure 15b.

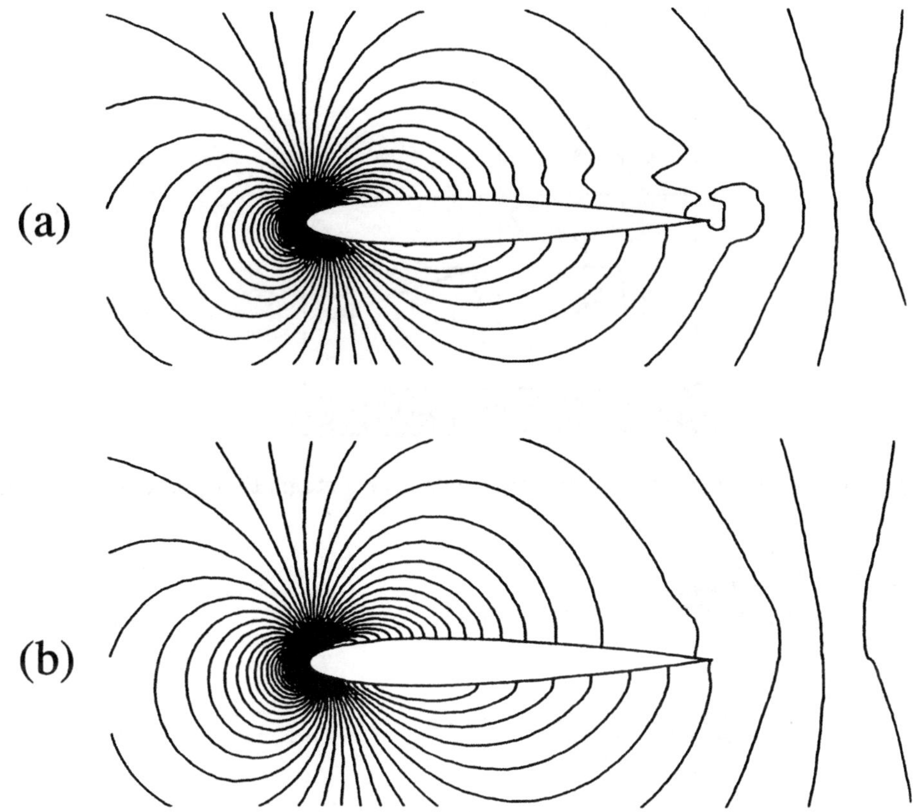

Figure 12: Viscous separating flow over a NACA 0012 airfoil at $M = 0.5$, $\alpha = 3°$ and $Re = 5000$. Pressure contours on a 129×49 C-grid, obtained with a second-order upwind scheme incorporating (a) Roe's grid-aligned Riemann solver; (b) the five-wave model of Rumsey et al.

5 Multi-dimensional fluctuation approach

The fluctuation approach to upwind differencing lends itself better to extension into higher dimensions than the finite-volume approach. Recall that a fluctuation is a local flux imbalance causing a non-zero time derivative of the local solution. For the one-dimensional Euler equations (4) the quantity $-\Delta F$ equals the residual evaluated on a one-dimensional mesh:

$$\int_{\text{mesh}} U_t dx = -\int_{\text{mesh}} F_x dx = -\Delta F. \tag{27}$$

This suggests extension of the fluctuation approach beyond one dimension by regarding each multi-dimensional mesh residual as the sum of a finite number of waves (say, m), moving in all possible directions. Thus we discretize the two-dimensional Euler equations as

$$\int\int_{\text{mesh}} U_t dx dy = -\oint (F dy - G dx) = \sum_{k=1}^{m} \lambda_k \alpha_k R_k, \tag{28}$$

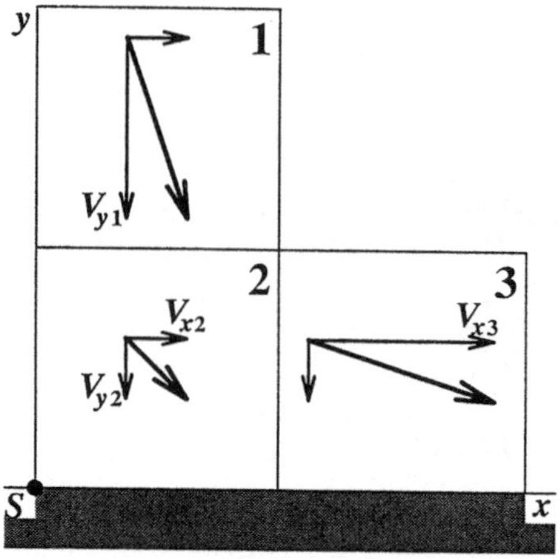

Figure 13: Turning of the flow in three cells near a stagnation point S at a wall.

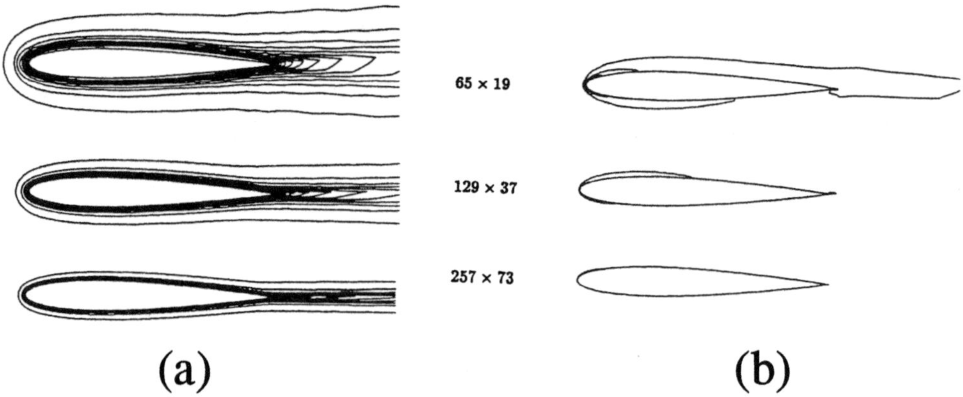

Figure 14: Entropy contours for inviscid flow over a NACA 0012 airfoil at $M = 0.3$, $\alpha = 1°$, generated on a sequence of O-grids with a first-order scheme incorporating (a) Roe's grid-aligned Riemann solver; (b) the five-wave model of Rumsey et al.

where the matrices $\hat{A}$ and $\hat{B}$ are multi-dimensional averages that remain to be defined. Since the fluctuation approach is a nodal-point approach, and we wish to develop only schemes of maximum compactness, we shall use a grid of triangular meshes, with data given in the nodal points. For the computation of the residual on such meshes it suffices to apply the trapezoidal integration rule on each side of the triangle. The fluctuations resulting from residual decomposition must be sent to the triangle's vertices according to some distribution scheme that approximates the convection equation.

It follows that, for the construction of a genuinely multi-dimensional upwind-differencing scheme, three components are needed:

1. A reliable multi-dimensional wave model for representing the residual;

2. A way to ensure conservation, i.e. a multi-dimensional extension of Roe's matrix average;

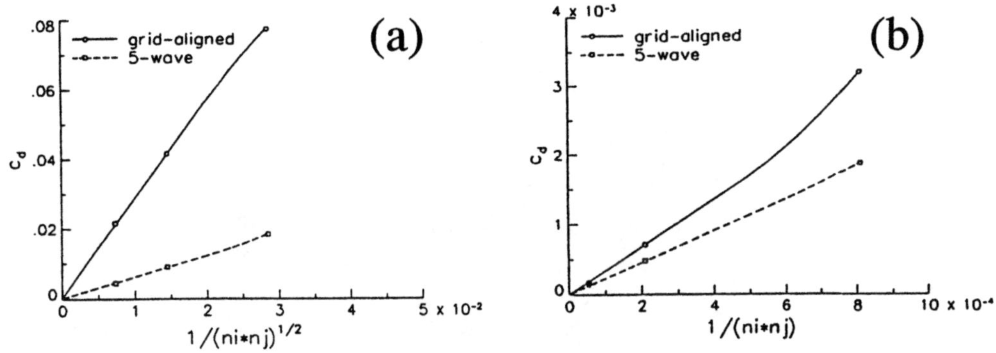

Figure 15: Grid-convergence study of the drag coefficient based on (a) the first-order solutions of Figure 14; (b) the corresponding second-order solutions.

3. A multi-dimensional convection scheme for advancing the waves.

Each of these will be discussed in a separate subsection.

5.1 Multi-dimensional wave models

The modeling of a local Euler residual by a finite number of waves was launched as a research subject by Roe [2]; his first paper, however, gave no specific instructions as to how the model would be used in a numerical integration of the Euler equations. This is not surprising, given that the other problems - multi-dimensional conservation and advection - had not yet been addressed.

The latest version of Roe's wave model calls for four acoustic waves, running along the principal strain axes of the local fluid element, a shear wave making a 45° angle with the acoustic waves, and an entropy wave running in the direction of the entropy gradient; see Figure 16. Thus, $m = 6$ in Eq. (28). These six waves are defined by two independent angles and six strengths; therefore, eight independent pieces of information need to be supplied per triangular mesh. This information is available in the form of the gradient of the state vector; its mesh value is computed with the trapezoidal rule from the following boundary integrals:

$$\widehat{U_x} \equiv \frac{1}{Area}\int\int_{\text{mesh}} U_x dxdy = \frac{1}{Area}\oint_{\text{mesh}} U dy; \tag{29}$$

$$\widehat{U_y} \equiv \frac{1}{Area}\int\int_{\text{mesh}} U_y dxdy = -\frac{1}{Area}\oint_{\text{mesh}} U dx. \tag{30}$$

A detailed discussion of this wave model, including the three-dimensional case, can be found in Roe's contribution to the present volume [45]; numerical results obtained with this model are presented in the contribution by Catalano et al. [46].

This section would not be complete without a discussion of the work of Hirsch and collaborators [47, 48, 49]. Their multi-dimensional approach is based on diagonalizing the Euler equations, i.e. changing these into a system of convection equations, by a transformation of state variables. The transformation itself depends on the local gradient of the solution, making the diagonalization essentially nonlinear. For certain data the transformation does not exist, in which case it is chosen so as to minimize

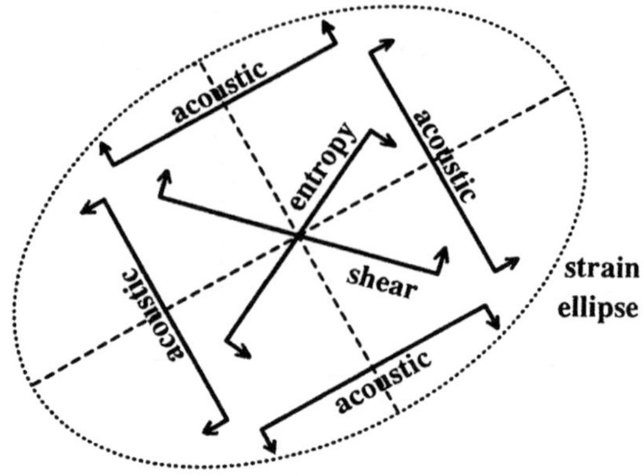

Figure 16: Roe's two-dimensional six-wave model. The acoustic waves run parallel to the principal strain axes (dashed); the strain ellips (dotted) shows the kinematic deformation of a circular fluid element.

the off-diagonal terms. The update scheme, though, can be made identical to a fluctuation-based scheme: decomposition of the residual along certain eigenvectors, followed by convection of the components [50]. In two dimensions the diagonalization is equivalent to using one particular four-wave model; clearly, the fluctuation approach offers much more flexibility.

5.2 Multi-dimensional conservation

The multi-dimensional extension of Roe's averaging of the flux Jacobian was independently discovered by Roe and Struijs, and is presented in a joint paper [51]. This very recent (1991) addition to the multi-dimensional toolbox applies exclusively to triangular meshes in two dimensions and tetrahedral meshes in three dimensions. The following description and explanation of the two-dimensional averaging apply to the special case of a calorically perfect gas.

To begin with, assume that the parameter vector $w = \sqrt{\rho}(1, u, v, H)$ is distibuted linearly over a mesh triangle with vertices labeled 1, 2 and 3. Denote the average of w over the triangle by $\bar{w}$; we then have

$$\bar{w} = \frac{1}{3}(w_1 + w_2 + w_3). \tag{31}$$

As before, $U(w)$ and $F(w)$, and also $G(U)$, are quadratic in the components of w, so that the Jacobian matrices U_w, F_w and G_w are linear in w, and therefore also in x and y. Considering that $U_x = U_w w_x$, $U_y = U_w w_y$, etc., where w_x and w_y are constant over the entire triangle, we conclude that $\vec{\nabla} U$, $\vec{\nabla} F$ and $\vec{\nabla} G$ also vary linearly over the triangle. Using the definition of the mesh-averaged gradient $\widehat{\vec{\nabla} U}$ given in Eqs. (29), (30), and similar definitions of $\widehat{\vec{\nabla} F}$ and $\widehat{\vec{\nabla} G}$, we easily derive the relations,

$$\widehat{\vec{\nabla} F} \equiv A(U(\bar{w}))\widehat{\vec{\nabla} U}, \tag{32}$$
$$\widehat{\vec{\nabla} G} \equiv B(U(\bar{w}))\widehat{\vec{\nabla} U}, \tag{33}$$

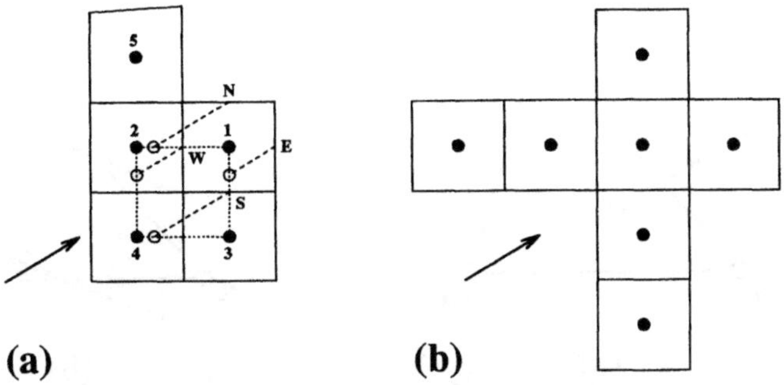

Figure 17: Stencils of two-dimensional upwind convection schemes; case $a > b > 0$. (a) Sidilkover's second-order scheme. The fluxes for cell 1 nominally are compurted by linear interpolation between upstream pairs of data, but the fluxes at the North and South faces must be limited to prevent numerical oscillations. The limiters are based on the ratios $a(u_1 - u_2)/[b(u_5 - u_2)]$ and $a(u_3 - u_4)/[b(u_2 - u_4)]$, respectively. (b) Standard second- or third-order grid-aligned scheme.

which are direct extensions of the one-dimensional relation (6). The extension to three-dimensional averaging is self-evident.

5.3 Multi-dimensional convection

The pursuit of multi-dimensional convection schemes has kept a number of authors busy over the past three years. In two dimensions the basic equation to be solved is

$$u_t + au_x + bu_y = 0, \tag{34}$$

where a and b are constant velocity components, or, in vector notation,

$$u_t + \vec{a} \cdot \vec{\nabla} u = 0 \tag{35}$$

The first significant work was that of Sidilkover [52], who, among other things, showed how a second-order upwind scheme, with residual computed on a square mesh, can be made non-oscillatory by standard limiters without undue spreading of the stencil. The domain of dependence for this algorithm is shown in Figure 17a, for the case $a > b > 0$; note how compact this is in comparison to the stencil of a standard second-order upwind scheme, shown in Figure 17b [27]. He also coined the name "N-scheme" for the first-order scheme that, on a cartesian grid, takes its data from the upwind triangle fitting the convection path most tightly (N stands for narrow). For example, for point 1 in Figure 17a it would be triangle (124). This scheme, as shown in [53], is optimal in the sense that, among all schemes with upwind triangular domain of dependence, it combines the smallest truncation error with the largest stable time-step. The three-dimensional extension is also described in [53].

While the triangles in Sidilkover's work were still considered subdivisions of squares, they become autonomous in later work by other authors. A major step in the development of two-dimensional convection schemes was the realization that there are two types of triangles [54]: those with one inflow side and those with two inflow sides. This is illustrated in Figure 18. If there is only one inflow side, the fluctuation ap-

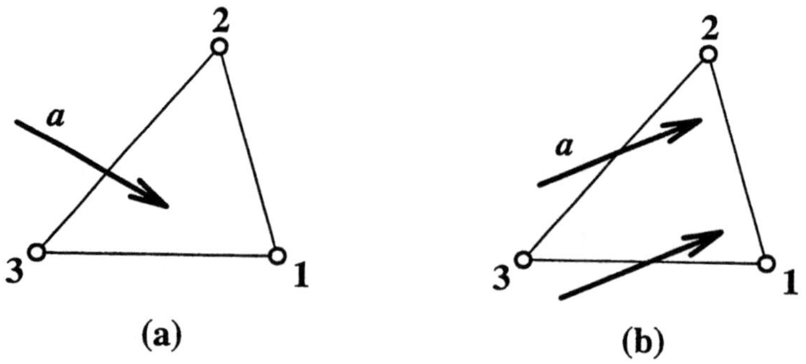

Figure 18: Two kinds of triangles: (a) with one inflow side; (b) with two inflow sides.

proach dictates that the entire residual be used to update the opposite node. This is the unique "single-target" form of the scheme, similar to the one-dimensional upwind scheme 2. If, however, there are two inflow sides, it may be argued that the residual be distributed over the two nodal points defining the third side. This is the "dual-target" form of the particular scheme; each choice of distribution weights defines a new scheme. The spreading of the residual information over two points implies a potential loss of resolution, inherent to multi-dimensional numerical convection; there is no one-dimensional analogue of this effect.

In the development of multi-dimensional convection schemes, three design criteria play a decisive role. According to these, it is desirable for a scheme to be

1. **linear:** for a given grid geometry and flow angle the solution depends linearly on the data. This promotes convergence to a steady numerical solution. It is well known that the presence of nonlinear devices in the scheme, such as limiters [44] and frame rotation (see Section 4.2) can slow down or even halt the convergence process;

2. **linearity preserving (LP):** data of the form

$$u(x,y) = bx - ay, \tag{36}$$

which is a steady solution of Eq. 34 are not changed by the scheme. This promotes the accuracy of the scheme. It can be shown [54] that LP schemes yield second-order-accurate steady solutions of Eq. 34;

3. **positive:** the scheme has positive coefficients. This is sufficient for preventing numerical oscillations.

From one-dimensional finite-difference theory we know - and have known so for a long time - that the above conditions are mutually exclusive. There is a famous theorem by Godunov [24] which says that no linear convection-diffusion scheme with positive coefficients can be more than first-order accurate. With reference to our design criteria for multi-dimensional convection schemes this theorem reads:

There are no linear positive LP schemes.

Again, nonlinearity is essential for the design of accurate, non-oscillatory schemes.

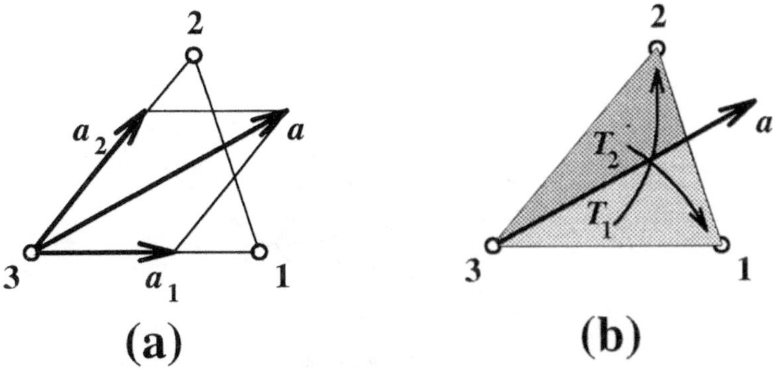

Figure 19: Dual-target form of convection schemes: (a) N-scheme; (b) LDA-scheme.

Among the various upwind convection schemes proposed in recent years, three schemes stand out; these are discussed below. They all are as compact as can be, requiring data on only one triangle for the approximation of the convection equation. A small miracle is that even positivity can be achieved *without leaving the triangle.* Of course, each nodal point is a vertex of a number of triangles and may receive fluctuations from several of these; programming therefore must be triangle-based. Some results of numerical experiments are presented in Section 6.

The N-scheme: the optimal linear positive scheme

The name of this scheme suggests equivalence to Sidilkover's N-scheme, but it actually is more general. Sidilkover's scheme is just the single-target form, common to all compact schemes; fluctuations from triangles requiring a dual-target scheme are ignored in the update. The dual-target form of the current N-scheme uses distribution weights proportional to the components of the convection speed along the two inflow sides, as depicted in Figure 19a. This makes the scheme optimal in the sense of having the largest stability range for the time-step [54]. It is also linear and positive, and therefore can be no more than first-order accurate.

The NN-scheme: the optimal nonlinear positive LP scheme

This scheme is a nonlinear variant of the N-scheme. hence the second N. The nonlinear procedure included in this scheme has absolutely nothing in common with the TVD-enforcing limiters included in one-dimensional convection schemes. It is based on the observation that in the convection equation (35) the component of the convection velocity $\vec{a}$ perpendicular to the solution gradient $\vec{\nabla} u$, has no effect on u_t. We therefore are allowed to replace $\vec{a}$ by any velocity that has the same component parallel to $\vec{\nabla} u$, as shown in Figure 20. This component, indicated by $\vec{a}_w$, is the velocity at which the level lines of u normal normal to themselves, i.e. the wave speed of the local distribution of u. This wave speed is the smallest of all admissible convection speeds; it actually vanishes with the residual. We may now adopt the following strategy: if both $\vec{a}$ and $\vec{a}_w$ call for a dual-target scheme, we replace $\vec{a}$ by $\vec{a}_w$ in the N-scheme; in all other cases the scheme becomes or remains a single-target scheme. In the case of

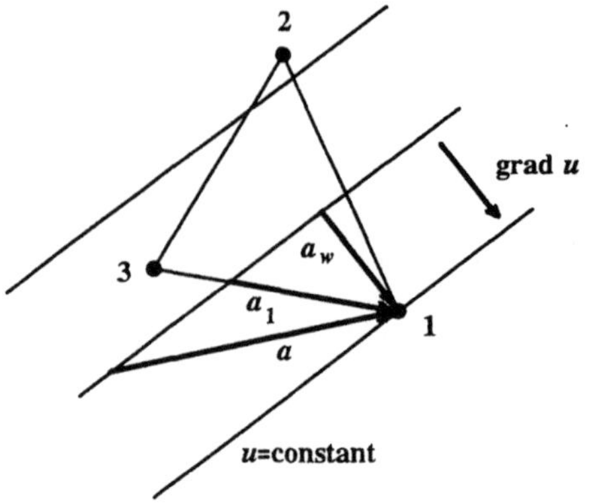

Figure 20: The NN-scheme: nonlinear single-target form. The convection speed $\vec{a}$ calls for a dual-target scheme but is replaced by $\vec{a}_1$, which calls for a single-target scheme. The wave speed $\vec{a}_w$ is the component of $\vec{a}$ and $\vec{a}_1$ parallel to $\vec{\nabla} u$.

Figure 20, $\vec{a}$ is replaced by $\vec{a}_1$, the nearest admissible speed yielding a single-target scheme. The resulting scheme does not change any nodal value if the residual vanishes, hence is LP, and maximizes the allowable time step.

Numerical results indicate that the accuracy of the NN-scheme lies between first- and second-order; see further Section 6.

The LDA scheme: a non-positive linear LP scheme

This scheme is one of several low-diffusion schemes, designed for a low truncation error. In the dual-target form of the scheme the distribution weights are inversely proportional to the areas of the triangles cut from the mesh triangle by a streamline through the inflow vertex; see Figure 19b. This scheme is not positive, but very accurate: on a uniform grid it achieves third order accuracy, as demonstrated in Section 6.

The above schemes have served as the basis for convection-diffusion schemes in a study by Tomaich and Roe [55]. Since the diffusion operator can not be approximated on a single triangle, their schemes are formulated with reference to a central nodal point. Numerical solutions of the Smith-Hutton [56] test problem demonstrate that these schemes rival the best exsisting convection-diffusion schemes in accuracy. In addition, their way of discretizing the Laplacean is directly applicable to any of the disipative terms included in the Navier-Stokes equations; thus, the basis for genuinely multi-dimensional Navier-Stokes codes has been laid.

6 Numerical results

To support some of the statements made about the new, compact convection schemes I first show how these schemes fared in a comparative grid-refinement study by Jens Müller [57]. The problem is that of convection of a Gaussian distribution over a semi-circle; Inflow is at $y = 0$, $x < 0$, outflow at $y = 0$, $x > 0$. Four kinds of grids were used, of which three examples are displayed in Figure 21. Grids α and β derive from

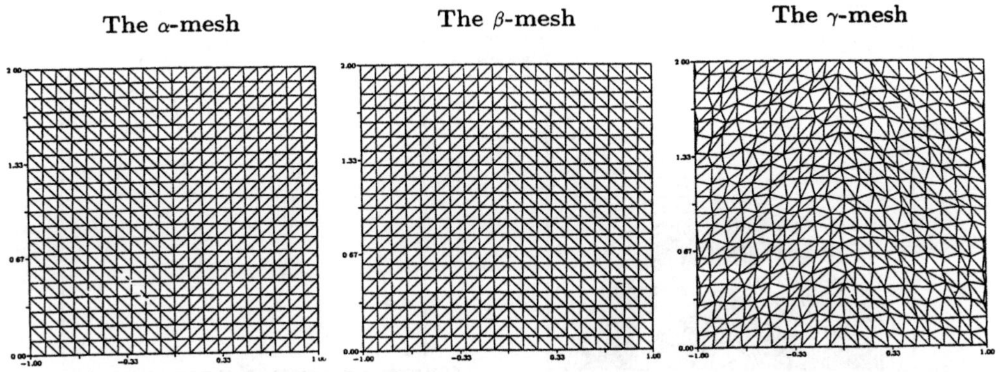

Figure 21: Three grids used in the circular-convection experiments.

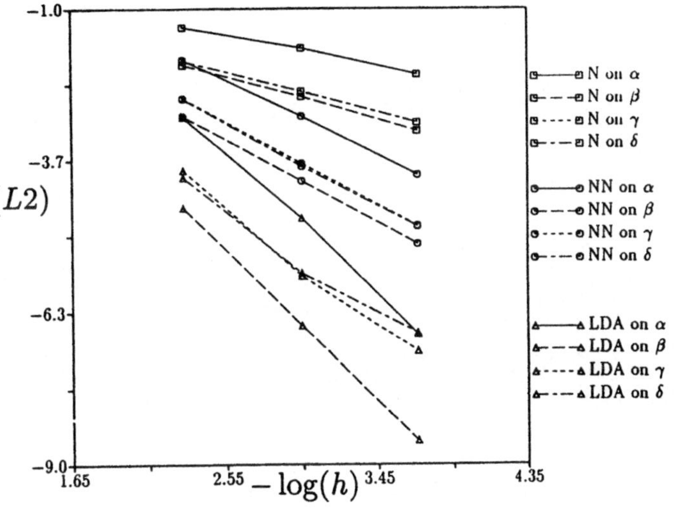

Figure 22: Grid-convergence of L_2-error for convection of a Gaussian over a semicircle by various schemes on various grids.

a uniform cartesian grid by adding diagonals. in grid α those diagonals are chosen that are least aligned with the convection direction, in β those most aligned. Grid γ is a irregular perturbation to β, while δ (not shown) is a minor perturbation to γ. Figure 22 shows the convergence of the L_2-error produced by the N-, NN- and LDA-schemes on the different grids. From the slope of the graphs of log(error) versus log(mesh-width) the following conclusions can be drawn:

1. The N-scheme is somewhat less than first-order accurate;
2. The NN-scheme is closer to being second-order accurate than first-order accurate;
3. The LDA-scheme is third-order accurate on a regular grid, second-order or less on a perturbed grid;

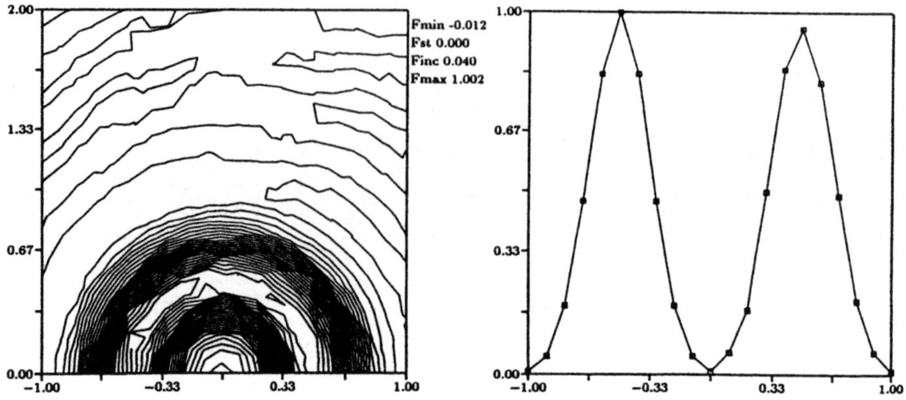

Figure 23: Contours (left) and cut along the x-axis (right) of the solution obtained with the LDA-scheme on a 20×20 β-grid.

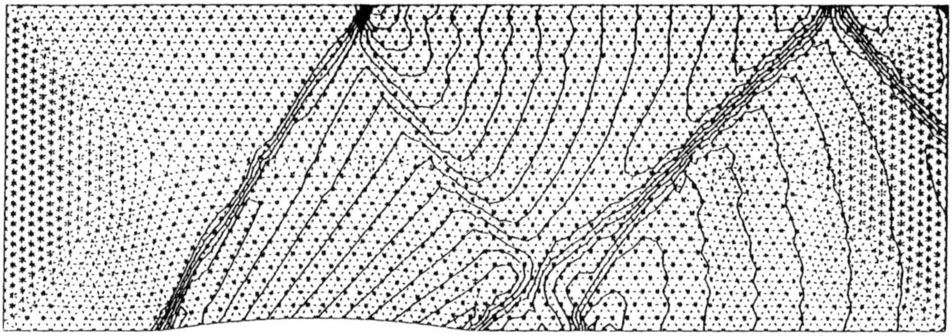

Figure 24: Inviscid flow at Mach 1.4 through an inlet, computed by a genuinely multi-dimensional upwind Euler code. Shown are Mach-number contours and the unstructured grid used.

4. All schemes decrease their error when diagonals are aligned with the flow.

Most surprising is the achievement of third-order accuracy on regular grids, considering the limited amount of information going into these compact schemes. Figure 23 gives an idea of this high accuracy by showing solution contours and a cut at $y = 0$ obtained with the LDA-scheme on the very coarse β-grid of Figure 21 (Gaussian represented on 10 meshes). Similar results obtained with three-dimensional extensions of the schemes can be found in Deconinck's comprehensive review paper [4].

While the search for compact convection schemes continues, several authors are trying to put together the ingredients listed in Section 5, producing a genuinely multi-dimensional upwind Euler code. Advanced numerical results can be found in the present volume in the contribution by Catalano et al. An earlier successful calculation of supersonic inlet flow by Struijs et al. [3] produced the Mach-contours shown in Figure 24; superimposed is the moderately irregular triangulation. The results

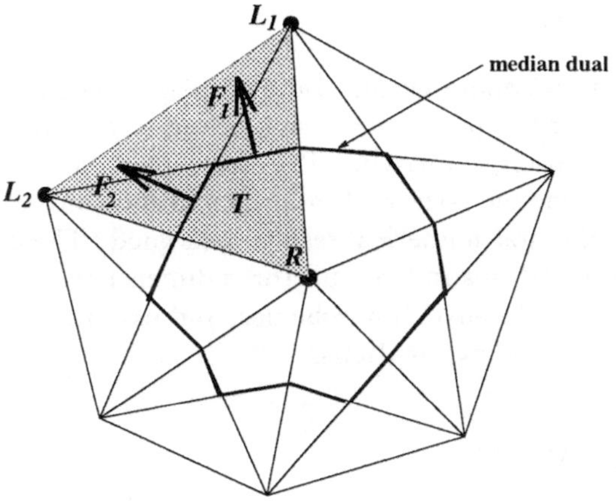

Figure 25: The finite-volume scheme of Barth, Powell and Parpia, based on a triangular grid and its median dual. The flux F_1 across one median element located in triangle T is computed using wave data from triangle T and fluxes from the vertices L_1 and R. For the flux F_2 across the other element in the same triangle the same wave data are used, but the fluxes are taken from L_2 and R.

demonstrate the excellent shock-capturing ability of the NN-scheme.

7 Finite-volume schemes revisited

From Section 5 the reader might get the impression that genuinely multi-dimensional schemes can only be formulated on triangular and tetrahedral meshes, and that they are incompatible with the finite-volume formulation. If this were true, it would mean a serious restriction on the use of such schemes within the CFD community, for it is not at all clear that unstructured triangular or tetrahedral grids are the way of the future. An alternative, for instance, is offered by adaptive cartesian grids [58]. The emphasis on triangles in Section 5 arises from the experience that the numerical building blocks, e.g. Roe's matrix averaging, take their simplest form on such meshes. Therefore, in developing a multi-dimensional scheme of a different format it would be good practice to start with the wave decomposition of residuals on an underlying triangular grid.

As an example of such practice, consider the two-dimensional finite-volume Euler scheme of Powell, Barth and Parpia [59], illustrated in Figure 25. The wave model indeed is applied to data on triangular meshes; for the update, though, a finite-volume scheme is chosen. Cell faces are formed from the medians of the triangles, yielding the so-called median-dual grid. Across each median element of the cell contour a flux is computed using an equation of the form (26), where L and R denote the vertices of the triangle side bisected by the median, and the sum includes all waves identified in the triangle. Note the difference with the scheme of Rumsey et al., where the wave model would be based solely on $U_R - U_L$. The resulting nonlinear scheme, applied to a scalar convection equation, is LP and positive and appears to be more accurate than the NN-scheme.

8 Conclusions

The state of the art in genuinely multi-dimensional upwind differencing has made dramatic advances over the past three years, owing to a shift from the finite-volume approach to the flctuation approach. The basic ingredients for multi-dimensional Euler codes, i.e. wave model, conservation principle and convection scheme, are ready for integration, and the first numerical results look good. The coming years will yield many more Euler applications in two and three dimensions, further improvements in wave models and compact convection schemes, and extension of the approach to the modeling of the Navier-Stokes equations.

Acknowledgement

In preparing this lecture I received much-needed help from Jens Müller, Tim Tomaich, Phil Roe and Ken Powell, for which I thank them profoundly.

References

[1] P. L. Roe, "Upwind schemes using various formulations of the Euler equations," in *Numerical Methods for the Euler equations of fluid dynamics*, pp. 14–31, 1985.

[2] P. L. Roe, "Discrete models for the numerical analysis of time-dependent multidimensional gas-dynamics," *Journal of Computational Physics*, vol. 63, 1986.

[3] R. Struijs, H. Deconinck, P. DePalma, P. Roe, and K. Powell, "Progress on multi-dimensional Euler solvers on unstructured grids," in *AIAA 10th Computational Fluid Dynamics Conference*, 1991.

[4] H. Deconinck, R. Struijs, G. Bourgois, H. Paillère, and P. L. Roe, "Multidimensional upwind methods for unstruuctured grids," in *Unstructured Grid Methods for Advection Dominated Flows*, 1992.

[5] S. F. Davis, "A rotationally-biased upwind difference scheme for the Euler equations," *Journal of Computational Physics*, vol. 56, 1984.

[6] D. Levy, K. G. Powell, and B. van Leer, "Implementation of a grid-independent upwind scheme for the Euler equations," in *AIAA 9th Computational Fluid Dynamics Conference*, 1989.

[7] D. Levy, *Use of a Rotated Riemann Solver for the Two-Dimensional Euler Equations*. PhD thesis, The University of Michigan, 1990.

[8] A. Dadone and B. Grossman, "A rotated upwind scheme for the Euler equations," AIAA Paper 91-0635, 1991.

[9] A. Dadone and B. Grossman, "A rotated upwind scheme for the Euler equations," in *Proceedings of the 13th International Conference on Numerical Methods in Fluid Dynamics*, 1992. To appear.

[10] S. Obayashi and P. M. Goorjian, "Improvements and applications of a streamwise upwind algorithm," in *AIAA 9th Computational Fluid Dynamics Conference*, 1989.

[11] Y. Tamura and K. Fujii, "A multi-dimensional upwind scheme for the Euler equations on unstructured grids," in *Fourth International Symposium on Computational FLuid Dynamics*, 1991.

[12] C. L. Rumsey, B. van Leer, and P. L. Roe, "A grid-independent approximate Riemann solver with applications to the Euler and Navier-Stokes equations," AIAA Paper 91-0239, 1991.

[13] C. L. Rumsey, B. van Leer, and P. L. Roe, "Effect of a multi-dimensional flux function on the monotonicity of Euler and Navier-Stokes computations," in *AIAA 10th Computational Fluid Dynamics Conference*, 1991.

[14] C. L. Rumsey, *Development of a grid-independent approximate Riemann solver*. PhD thesis, University of Michigan, 1991.

[15] C. L. Rumsey, B. van Leer, and P. L. Roe, "A grid-independent approximate Riemann solver for the Euler and Navier-Stokes equations," *JCP*, 1992. To appear.

[16] I. H. Parpia, "A planar oblique wave model for the Euler equations," in *AIAA 10th Computational Fluid Dynamics Conference*, 1991.

[17] I. H. Parpia and D. J. Michalek, "A nearly-monotone genuinely multi-dimensional scheme for the Euler equations," AIAA Paper 92-035, 1992.

[18] T. J. Barth and P. O. Frederickson, "Higher order solution of the Euler equations on unstructured grids using quadratic reconstruction," AIAA Paper 90-0013, 1990.

[19] R. Abgrall, "Design of an essentially non-oscillatory reconstruction procedure on finite-element type meshes." ICASE Report 91-84, 1991.

[20] P. L. Roe, "Beyond the Riemann problem," in *Algorithmic Trends in Computational Fluid Dynamics for the 90s*, 1992.

[21] H. Deconinck, "Beyond the Riemann problem ii," in *Algorithmic Trends in Computational Fluid Dynamics for the 90s*, 1992.

[22] P. L. Roe, "Characteristic-based schemes for the Euler equations," *Annual Review of Fluid Mechanics*, vol. 18, pp. 337–365, 1986.

[23] P. L. Roe, "The use of the Riemann problem in finite-difference schemes," *Lecture Notes in Physics*, vol. 141, 1980.

[24] S. K. Godunov, "A finite-difference method for the numerical computation and discontinuous solutions of the equations of fluid dynamics," *Matematicheskii Sbornik*, vol. 47, pp. 271–306, 1959.

[25] B. van Leer, "Towards the ultimate conservative difference scheme. IV. A new approach to numerical convection," *Journal of Computational Physics*, vol. 23, 1977.

[26] B. van Leer, "Towards the ultimate conservative difference scheme. V. A second-order sequel to Godunov's method," *Journal of Computational Physics*, vol. 32, 1979.

[27] B. van Leer, "Upwind-difference methods for aerodynamic problems governed by the Euler equations," in *Large-Scale Computations in Fluid Mechanics, Lectures in Applied Mathematics*, vol. 22, 1985.

[28] P. L. Roe, "Fluctuations and signals, a framework for numerical evolution problems," in *Numerical Methods in Fluid Dynamics*, pp. 219–257, 1982.

[29] P. L. Roe, "Approximate Riemann solvers, parameter vectors and difference schemes," *Journal of Computational Physics*, vol. 43, 1981.

[30] S. Osher and F. Solomon, "Upwind schemes for hyperbolic systems of conservation laws," *Mathematics and Computation*, vol. 38, 1982.

[31] A. Harten, P. D. Lax, and B. van Leer, "Upstream differencing and Godunov-type schemes for hyperbolic conservation laws," *SIAM Review*, vol. 25, 1983.

[32] B. van Leer, *A Choice of Difference Schemes for Ideal Compressible Flow.* PhD thesis, Leiden University, 1970.

[33] B. van Leer, "Towards the ultimate conservative difference scheme. III. Upstream-centered finite-difference schemes for ideal compressible flow," *Journal of Computational Physics*, vol. 23, 1977.

[34] B. van Leer, W. T. Lee, and K. G. Powell, "Sonic-point capturing," in *AIAA 9th Computational Fluid Dynamics Conference*, 1989.

[35] B. van Leer, W. T. Lee, and P. L. Roe, "Characteristic time-stepping or local preconditioning of the Euler equations," in *AIAA 10th Computational Fluid Dynamics Conference*, 1991.

[36] B. van Leer, "On the relation between the upwind-differencing schemes of Godunov, Engquist-Osher and Roe," *SIAM Journal on Scientific and Statistical Computing*, vol. 5, 1984.

[37] S. K. Godunov, A. W. Zabrodyn, and G. P. Prokopov, "A difference scheme for two-dimensional unsteady problems of gas dynamics and computation of flow with a detached shock wave," *Zhur. Vych. Mat. i Mat. Fyz.*, vol. 1, p. 1020, 1961.

[38] P. R. Woodward and P. Colella, "The numerical simulation of two-dimensional fluid flow with strong shocks," *Journal of Computational Physics*, vol. 54, pp. 115–173, 1984.

[39] P. Colella and P. R. Woodward, "The piecewise-parabolic method (PPM) for gas-dynamical simulations," *Journal of Computational Physics*, vol. 54, pp. 174–201, 1984.

[40] S. K. Chakravarthy, A. Harten, and S. Osher, "Essentially non-oscillatory shock-capturing schemes of arbitrarily-high accuracy," AIAA Paper 86-0339, 1986.

[41] V. Venkatakrishnan, "Newton solution of inviscid and viscous problems," AIAA Paper 88-0413, 1988.

[42] D. Levy, K. G. Powell, and B. van Leer, "Use of a rotated Riemann solver for the two-dimensional Euler equations," *Journal of Computational Physics*, 1991. Submitted.

[43] S. F. Davis, "Shock capturing." ICASE Report 85-25, 1985.

[44] W. K. Anderson, J. L. Thomas, and B. van Leer, "A comparison of finite volume flux vector splittings for the Euler equations," *AIAA Journal*, vol. 24, 1985.

[45] P. L. Roe and L. Beard, "An improved wave model for multidimensional upwinding of the Euler equations," in *Proceedings of the 13th International Conference on Numerical Fluid Dynamics*, 1992. To appear.

[46] L. A. Catalano, P. DePalma, and G. Pascazio, "A multidimensional solution-adaptive multigrid solver for the Euler equations," in *Proceedings of the 13th International Conference on Numerical Methods in Fluid Dynamics*, 1992. To appear.

[47] C. Hirsch, C. Lacor, and H. Deconinck, "Convection algorithm based on a diagonalization procedure for the multidimensional Euler equations," in *AIAA 8th Computational Fluid Dynamics Conference*, 1987.

[48] C. Hirsch and C. Lacor, "Upwind algorithms based on a diagonalization of the multidimensional Euler equations," AIAA Paper 89-1958, 1989.

[49] P. van Ransbeek, C. Lacor, and C. Hirsch, "A multidimensional cell-centered upwind algorithm based on a diagonalization of the Euler equations," in *Proceedings of the 12th International Conference on Numerical Methods in Fluid Dynamics*, 1990.

[50] K. G. Powell and B. van Leer, "A genuinely multi-dimensional upwind cell-vertex scheme for the Euler equations," AIAA Paper 89-0095, 1989.

[51] P. L. Roe, R. Struijs, and H. Deconinck, "A conservative linearisation of the multidimensional Euler equations," *Journal of Computational Physics*, 1992. To appear.

[52] D. Sidilkover, *Numerical Solution to Steady-State Problems with Discontinuities*. PhD thesis, Weizmann Institute of Science, 1989.

[53] P. L. Roe and D. Sidilkover, "Optimum positive linear schemes for advection in two and three dimensions." Submitted to *Journal of Computational Physics*, 1991.

[54] H. Deconinck, R. Struijs, K. Powell, and P. Roe, "Multi-dimensional schemes for scalar advection," in *AIAA 10th Computational Fluid Dynamics Conference*, 1991.

[55] G. T. Tomaich and P. L. Roe, "Compact schemes for advection-diffusion schemes on unstructured grids." Presented at the 23rd Annual Modeling and Simulation Conference, 1992.

[56] R. M. Smith and A. G. Hutton, "The numerical treatment of advection: A performance comparison of current methods," *Numerical Heat Transfer*, vol. 5, 1982.

[57] J.-D. Müller and P. L. Roe, "Experiments on the accuracy of some advection schemes on unstructured and partly structured grids." Presented at the 23rd Anuual Modeling and Simulation Conference, 1992.

[58] D. De Zeeuw and K. G. Powell, "An adaptively-refined carterian mesh solver for the Euler equations." To appear in *Journal of Computational Physics*, 1992.

[59] K. G. Powell, T. J. Barth, and I. F. Parpia, "A solution scheme for the Euler equations based on a multi-dimensional wave model." Extended abstract for the AIAA 31st Aerospace Sciences Meeting and Exhibit, 1992.

THE IMPACT OF TURBULENCE MODELLING ON THE NUMERICAL PREDICTIONS OF FLOWS.

Minh Hieu HA
Institut de Mécanique des Fluides de Toulouse
I.N.P.T.- U.R.A. C.N.R.S.005 - Av.du Pr. Camille Soula, 31400 Toulouse-France
and C.E.R.F.A.C.S., 42 Av. Gustave Coriolis, 31057 Toulouse Cedex, France

I.- INTRODUCTION.

A century has already passed since REYNOLDS [1] et BOUSSINESQ [2], after their long and patient observations on flows in Irwell and Ariège rivers, built the foundations of modern prediction methods for turbulence in fluids. Since then, despite important scientific, financial and human efforts devoted to a better understanding of the physics of turbulence, in order to obtain the suitable modelling of such phenomena, we have to admit that the turbulence of the Irwell and Ariège rivers have not yet revealed all their secrets and that numerical simulation has not yet become a daily optimization tool for engineers. This fact is worth analyzing and the state of the art is the main objective of this contribution.

One of the old dreams of engineers, is to be able to predict physical phenomena without the systematic use of experiments. To satisfy this wish, research scientists endlessly try to put the physical phenomena in "*equations*", to analytically or numerically solve these equations and to compare the results obtained with experimental data used as a validation reference. When the validation comparison seems to be correct, these equations then receive the label of "*mathematical models*". The study of turbulence is not an exception to this general approach.

Indeed, for the last thirty years, and in a more general context than the turbulence problem, the work tools of engineers have undergone enormous developments. Instead of the catalogues containing ready-made formulae and ready-drawn diagrams, engineers need more and more detailed knowledge, before efficiently optimizing their "product". Then, the engineer has to turn towards the basic mechanisms of Physics, by studying the models governing these physical processes, and also by numerically solving the latter.

The first "*oil shock*" in 1973 deeply modified engineering methodologies: an engineer could no longer be content with global methods and "*coefficient theories*". On the contrary, he had more and more to confront the elaborate optimization problems of complex "*systems*", this operation being widely helped by increasingly efficient computers, with decreasing computing costs.

Fluid mechanics is a branch of physics which has always been a favourite field to apply the progress obtained on the numerical resolution of partial differential equations, and then the use of super-computers. This state is due to a fact: Fluids Mechanics is rich with *non-linearities.*

These omnipresent non-linearities are not without important consequences on the problems studied, both on the *physical* aspects as on the *numerical* ones. Even in the usual fluids such as air and water, where the fluid behaviour law is *linear* (newtonian behaviour law), in addition to the *convective non-linearity*, we have to add three others: *i)*: the first can come from the **density variations** reinforcing the coupling of the transport equations by the convection process (for any origin of these density variations), *ii)* the second one can issue from important **source terms** such as the buoyancy effects, and *iii)* last, but not least, they can derive from the **behaviour law** itself. In this way, the non-linear rheological laws, so important in industrial applications, can play a major role in the flow stability. Owing to this fact, very few flows can remain for a long time in a laminar regime, after their commencement: they then become turbulent with all the consequences on the flow structures and on the transfer processes.

In any event, turbulent flows are much more complex than their laminar homologues, because of the multi-characteristic scale problem. For this reason, turbulent flows are not suitable to analytical solutions, and often a numerical approach becomes necessary.

But this fact is not the only reason for the use of a numerical approach in Fluid Mechanics.

The seventies were marked by the extraordinary development of numerical prediction methods for turbulent flows. The ampleness of this progress can be appreciated when we compare the results obtained between the first Stanford Conference on Turbulent Boundary Layers in 1969 [3] and the Stanford Conference on Complex Turbulent Flows in 1980-81 [4]. From simple thin shear layers we come rapidly to different complex turbulent flows, including recirculating separated flows, or shock-wave/boundary layer interactions [5], with very strong density variations, in complex geometry configurations, in heterogeneous medium, and even in reactive flows...Whatever the flow configuration to be studied, the work done in the 70's is usually within a very classical framework: **incompressible** fluids and **steady mean flows**. In fact, the statistical turbulence models performed in the 70's, give rise to satisfactory results for the numerical prediction of *thin* shear flows (jets, wakes, mixing layers...) and generally, in the *fully developed* turbulence zones of the far downstream fields, where self-similarity properties are observed for mean and turbulence quantities. However, the theoretical prediction seems to be less satisfactory in the zones of **transition** and in the **beginning of the turbulence development**. In the same way, if some numerical simulations of separated flows, including large recirculating zones, seem to be globally correct [6] (Fig.1), the theoretical prediction of wall momentum, heat and mass transfers, still remain to be perfected.

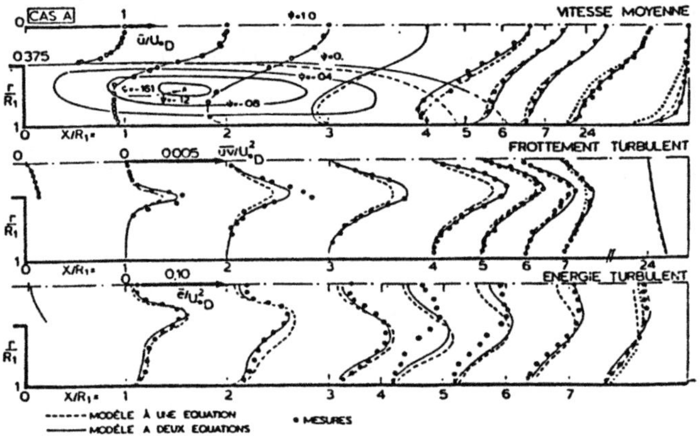

Fig.1.- *Turbulent flow in a sudden enlargement of a pipe flow: Comparison experiment-numerical prediction* [6], 1977.

The 80's are more marked by the modelling efforts of some complex effects of turbulence. By "complex", we mean the effects of **volume forces**, of **strong curvature**, the effects of **natural or forced unsteadiness**, the effects of **compressibility** or the **variable density** by mass heterogeneity, etc...

Due to the requirements imposed by the large diversity of industrial flow problems, one of the challenges in the last few years in the modelling of turbulence, was to correctly take into account the complex effects previously cited.

To underline these efforts compared with the previous state, it is worth to come back to the main characteristics of turbulence modelling in incompressible flows.

2.- THE CONTEXT OF STATISTICAL MODELLING.

Efforts devoted to the development of statistical turbulence models are usually within a context limited by two considerations: the first concerns the *physics* of the phenomenon, and the other is rather related to the *pragmatic* character of the search for tools destined for the prediction of complex industrial flows.

Almost all turbulent flows simultaneously contain both **organized** and **disorganized** characters. Their relative proportions depend mainly on the geometric patterns of the flow domain, on the flow Reynolds number, on initial conditions and on boundary conditions.

Thus, it seems that the mixing layers contain more *organized structures* than any other flows submitted to the same initial conditions. But this fact is rather superficial: the use of adequate measurement and data processing techniques or high performance visualization methods, shows that an axisymmetric jet even at a high Reynolds number, contains similar ordered structures which are almost imperceptible to the naked eye. In the same way, a mixing layer at a relatively low Reynolds number, and apparently unturbulent, may contain some *disordered structures* at small scales which visualization techniques reveal with difficulty.

However, it is too simplistic to state that the organized and disorganized parts of the flow often play different roles in the transport processes. A distinction between these two types of structures seems to be recommended, although this fact is not a *sine qua non* condition: the rather good results of the prediction of variable density mixing layers, obtained by SHIH et LUMLEY [7], constitutes a remarkable counter-example. Indeed, by using second order closures *ignoring* the presence of organized structures, these authors successfully modelled one of ROSHKO's variable density mixing layers where unsteady organized structures uncontestably exist, placed in a prominent position by the famous pictures of BROWN et ROSKHO [8] since 1974.

Nowadays, no-one seriously supposes that a "*universally valid*" model can be developed with a one-point closure context, which is able to describle *all* complex turbulent flows for industrial applications. Disappointments met by the numerical prediction of the *spreading rate of the axisymmetric round jet* is one of the classical examples illustrating the weakness of statistical one-point closure modelling. But this restrictive remark does not mean that statistical turbulence models have to be "*thrown away*". On the contrary, one must be perplexed in the face of the quality of the results obtained with the previous models in a large variety of flows as different as the plane jet and the wall boundary layer, in spite of the rather rough validity of some hypotheses on which the modelling closures are built. Figure 2 shows a remarkable example [9-10] of the behaviour of turbulence models in a three-dimensional configuration: a curved channel. The good quality of the prediction attests a well built base of statistical modelling, in spite of some weakneses which we will try to identify in this paper. Let us give back some "*charisma*" to these models: although qualified as *phenomenological* (perhaps by reason of their belonging to the same family of the *mixing-length theory*), the construction of these models is often based on mathematical considerations, more rigourous than they appear. The relative success of some one-point closure turbulence models on the one hand stimulates us to advance in this field, and on the other hand forces us to objectively examine here the *strength* and *weakness* of the models in order to improve their ability for the prediction of complex turbulent flows presenting some industrial interest. Let us however remain lucid in recognizing that it would probably be impossible to calculate the flows where the amplification or the suppression of some *particular frequencies* constitutes a basic mechanism for the physical phenomenon we try to model.

3.- THE GENERAL LANDSCAPE OF ONE-POINT CLOSURE TURBULENCE MODELLING.

For about twenty years, important efforts have been spent to develop various statistical turbulence models, using **one-point closures** (because different correlations and mean quantities are linked between them by relationships expressed at the *same* point in space and at the *same* time); sometimes these closure schemes even come from volume or surface integrals then taking into account in fact some *space correlations*, i.e.*closures in two-points*, and even *space-time correlations*.

In spite of the variety of the one-point closure models presented in the literature, we can distinguish two general groups: those built on the *eddy viscosity concept* and those determining the *second moment* (i.e. the turbulent momentum, heat and mass fluxes) from partial derivative equations (transport equations) governing their balance and their evolution in time and in space. We respectively designate these two groups, the E.V.M. (Eddy-Viscosity Models) and the R.S.T.M. (Reynolds Stress Transport Models).

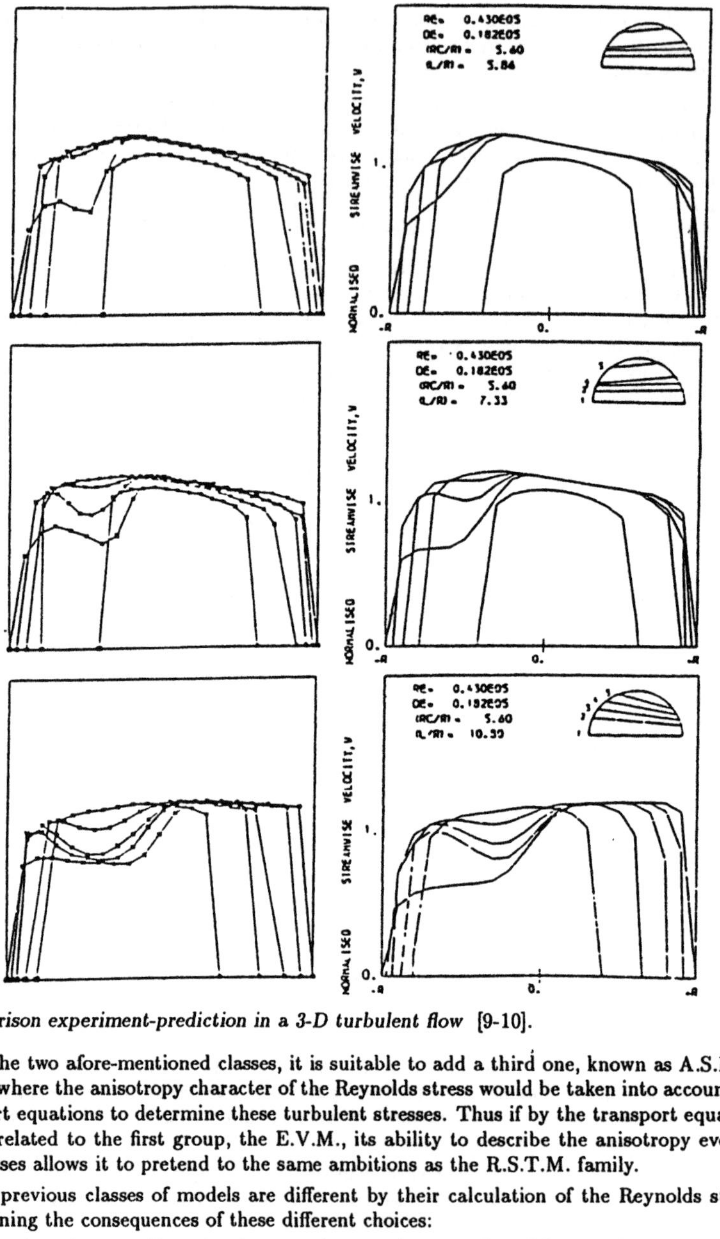

Fig.2.- *Comparison experiment-prediction in a 3-D turbulent flow* [9-10].

Between the two afore-mentioned classes, it is suitable to add a third one, known as A.S.M. (Algebraic Stress Model) where the anisotropy character of the Reynolds stress would be taken into account without the use of transport equations to determine these turbulent stresses. Thus if by the transport equation number, the A.S.M. is related to the first group, the E.V.M., its ability to describe the anisotropy evolution of the turbulent stresses allows it to pretend to the same ambitions as the R.S.T.M. family.

Since the previous classes of models are different by their calculation of the Reynolds stresses, let us start by examining the consequences of these different choices:

Every time there is a *non-linearity* character in a mathematical model, a statistical process applied to this model leads to the appearance of *new unknown correlations.*

For example, for the momentum conservative equation, in incompressible flows, the particle derivative becomes after an *ensemble avaraging*:

$$\frac{\partial \bar{U}_i}{\partial t} + \frac{\partial(\bar{U}_i \bar{U}_k)}{\partial x_k} + \frac{\partial(\overline{u_i u_k})}{\partial x_k} = ... \tag{1}$$

The different classes of turbulence models are distinguished by the way the new unknown correlations issuing from *convective non-linearity* are expressed.

The first class of models is based on a *formal* analogy with the *linear* newtonian behaviour law:

$$-\rho\left[\overline{u_i u_k}-\frac{2}{3}\delta_{ik}\bar{k}\right]=\mu_t\left[\frac{\partial \bar{U}_i}{\partial x_k}+\frac{\partial \bar{U}_k}{\partial x_i}-\frac{2}{3}\delta_{ik}\frac{\partial \bar{U}_l}{\partial x_l}\right] \tag{2}$$

In this *behaviour law*, the eddy-viscosity which is no longer an *intrinsic property* of the fluid, but becomes a *character of the flow itself*, and must be calculated by the mean of a *length scale* and a *velocity scale*: both of them must be characteristic of the turbulent structures governing the turbulent transfer process, inside the flows. The way these two turbulence scales are calculated, is used to distinguish the different levels of modelling of the eddy-viscosity concept family. However, we can point out that through the numerical approximations solving the equations (1) and (2), some *numerical linearization* (then artificial) is introduced, eventually decreasing some non-linearity properties of the physical phenomena.

An alternative to this behaviour law of gradient type consists in considering the transport equations and the balance relatives to each second order moment representing the turbulent fluxes, whence the name of *second order closure modelling*.

For the turbulent fluxes of momentum, exact transport equations can be derived as:

$$\begin{aligned}\rho\left[\frac{\partial(\overline{u_i u_j})}{\partial t}+\bar{U}_k\frac{\partial(\overline{u_i u_j})}{\partial x_k}\right]=&-\underbrace{\rho\frac{\partial}{\partial x_k}[\overline{u_i u_j u_k}]}_{Diffusion\ by\ fluctuating\ velocity}-\underbrace{\frac{\partial}{\partial x_k}[\delta_{ik}\overline{u_j p}+\delta_{jk}\overline{u_i p}]}_{Diffusion\ by\ fluctuating\ pressure}\\&+\underbrace{\frac{\partial}{\partial x_k}\left[\mu\overline{\left(u_j\left(\frac{\partial u_i}{\partial x_k}+\frac{\partial u_k}{\partial x_i}\right)+u_i\left(\frac{\partial u_j}{\partial x_k}+\frac{\partial u_k}{\partial x_j}\right)\right)}\right]}_{Viscous\ external\ power\ by\ fluctuating\ motion}\\&-\underbrace{\rho\left[\overline{u_j u_k}\frac{\partial \bar{U}_i}{\partial x_k}+\overline{u_i u_k}\frac{\partial \bar{U}_j}{\partial x_k}\right]}_{Direct\ production}+\underbrace{\overline{p\left(\frac{\partial u_i}{\partial x_j}+\frac{\partial u_j}{\partial x_i}\right)}}_{Redistribution}\\&-\underbrace{\mu\left[\overline{\left(\frac{\partial u_i}{\partial x_k}+\frac{\partial u_k}{\partial x_i}\right)\frac{\partial u_j}{\partial x_k}}+\overline{\left(\frac{\partial u_j}{\partial x_k}+\frac{\partial u_k}{\partial x_j}\right)\frac{\partial u_i}{\partial x_k}}\right]}_{Viscous\ internal\ power\ by\ fluctuating\ motion}\end{aligned} \tag{3}$$

Using a more compact form, we can write:

$$\frac{\partial(\overline{u_i u_j})}{\partial t}+\bar{U}_k\frac{\partial(\overline{u_i u_j})}{\partial x_k}=D_{ij}+P_{ij}+\phi_{ij}-DES_{ij} \tag{4}$$

where D_{ij}, P_{ij}, ϕ_{ij} and DES_{ij} respectively designate the tensors of *diffusion* (or transport), of *production*, of *redistribution* and of *destruction* for the second moment $\overline{u_i u_j}$. When additional complex effects on turbulence, such as those of rotation, of the strong curvatures, or of the buoyancy forces... are present, they must be included in the terms ϕ_{ij} et DES_{ij}.

As we will certainly see during this Conference, today, the model (k,ϵ) of JONES and LAUNDER [11] is probably the most frequently used, not only in subjects of a theoretical nature, but also in practical engineering problems. Even if, in most cases, this model has often yielded to satisfactory results, we must not forget the following points:

• From the fundamental point of view, to adopt the eddy-viscosity concept (2) amounts to giving turbulent stresses, which have a convective origin, a diffusive character with a gradient type, leading to a rather *linear* nature. Consequently, we have to expect that the turbulent effects *stabilize* the unsteadiness mechanisms issuing from the non-linearity of the Reynolds equations (1).

• The use of the eddy-viscosity represented by the ratio (k^2/ϵ) leads to a *nearly constant* value for the turbulent diffusivity in a large part of the flow fields. Indeed, the source terms (production and destruction) of the turbulence kinetic energy on the one hand, and those of the turbulence dissipation rate on the other hand, are nearly proportional. From this fact, a numerical simulation with a constant viscosity (equivalent to a false laminar regime) often leads to similar results (except in viscous sub-layers), but with much lower computing costs. In other respects, this eddy-viscosity in (k^2/ϵ) form, is also the origin of the difficulty to correctly calculate the diffusivity of the streamwise momentum in recirculating flows.

Another weakness of the eddy-viscosity concept is the representation of the three normal turbulent stresses, by their half-sum - the turbulence kinetic energy - which is equivalent to concede a *similar evolution* of these three components, even if the turbulence anisotropy of flows can be implicitly included in this modelling. Consequently, in flows where these three normal components vary in different ways (internal zones of wall flows, curved channels, dissymmetrical channels, shock waves, expansion zones, rapid distorsion, recirculating zones...), the eddy-viscosity concept models, even with two transport equations, do not yield very satisfactory results.

The eddy-viscosity concept also has important consequences from the numerical point of view when the partial differental equations are solved. Indeed, the acceptance of the eddy viscosity value, usually much higher than the molecular viscosity, actually leads to relatively weaker Reynolds numbers (local convection and diffusion ratio). This fact enables numerical schemes, usually submitted to low local Reynolds number criteria, to be easily extended to turbulent flows, although the real face-value of the Reynolds number is much higher.

Finally, in most of the numerical predictions performed so far, very few *fully implicit* numerical schemes are used to solve the transport equations for k and for ϵ. Consequently, in practical numerical solutions, even when the so-called implicit methods are used, there actually exists a time-step limit for time-dependent approaches. In the same way, the non-linearity of source terms nearly always needs a local time linearization.

The two-equation turbulence models are widely used in practical flow configurations, including very complex situations. In an excellent review,[12], RODI presented and discussed the numerical results of various patterns such as: boundary layers with or without pressure gradients, including eventually transition and suction, curved walls, plane or axisymmetrical channels, free shear layers, with or without natural convection, separated flows, flows with swirls, hydraulic applications, turbomachines, thermal exchangers... The EUROMECH 180 Congress at Karlsruhe in 1984, made a review of progress with models of the (k, ϵ) type [13].

By emphasizing the existence of some special cases (wall jet [14][15] or dissymmetrical channel [16][17]) where the turbulent shear stress does not cancel out with the mean velocity gradient, HANJALÍC and LAUNDER [18] replaced the constitutive law (2) by the solution of two of the transport equations (3), the contraction of which leads to the transport equation of the turbulence kinetic energy k.

Compared with the turbulence kinetic energy transport equation, the tensorial equation grows richer with a new term: the *pressure-strain correlation* [19].

In incompressible cases, this term does not concern the turbulence kinetic energy, since the fluctuating motion is divergence free, but it plays an important role in the transport equations relative to each shear or normal stress component. In compressible flows, this term adds a compressibility effect which would not be negligible through the shock waves. More generally, the role of this term is to make the turbulence state more isotropic, by sharing the turbulence energy between its components, hence its name of **redistribution**.

The closure of this redistribution term is obtained by the integration of a Poisson equation relative to the fluctuating pressure. This integration leads to two distinct mechanisms: *i)* the first, of a *linear nature*, is named **rapid**, since it results from the action of the mean flow on itself (through the mean velocity gradients), *ii)* and the second, of a *quadratic nature*, is qualified **slow term**, since it ensures a **return to isotropy** with a time-scale borrowed from the *decaying rate of turbulence* ($\tau = k/\epsilon$), nevertheless remaining proportional to the *anisotropy degree* of the flow. Thus, in contrast with the (k, ϵ) which has only one characteristic turbulence time-scale, the second order closures introduce in fact, *two* different time-scales: the first borrowed from the mean motion, and the other deduced from the fluctuating motion. Let us point out that the spectral methods for turbulence modelling, proposed by the Ecole Centrale de Lyon, during the 70's, concern an infinite number of time-scales.

The integration of the Poisson equation for the fluctuating pressure, necessary for the modelling of the pressure-strain correlation, in addition to the volume integrals, also contains surface integrals taking into account the boundary of the flow field. This second part, usually negligible in free flows, again becomes important in the regions close to the solid wall. Physically speaking, this term represents the "*mirror*" effects or the "*echo*" effects resulting from the fluctuation of pressure. Concretely, this term acts in the opposite way of the mean term of redistribution, since it searches to enforce the anisotropic character of the normal turbulence stresses in the very near-wall regions, with a strong production of the streamwise normal stress component, and a severe destruction of the transverse normal stress component. In practical applications, this term has not yet given very satisfactory results, and further local arrangements must be included to improve the quality of the results [20][21]. Several scientists have tried to neglect this term, and take into account the "echo" effect, by adjusting the constants of the model, but this "*computer optimization*" makes the calculated results of the external zones unreliable. Finally, we have to point out that the effects of this contribution are important, not only near the solid wall, but also at the source of the anisotropy mechanisms, due to the fluctuating pressure, leading to a maximum velocity under a free surface.

Because of the absence of any rigorous theory for the modelling of this "*mirror effect*" reflecting the fluctuating pressure (an inviscid mechanism), HANJALÍC and LAUNDER [22] proposed a closure scheme deduced from the modelling of the volume integral of the redistribution, i.e. including the rapid part and the quadratic part, associated with a scale ratio l/y (y being the normal distance to the wall) in order to cancel this effect far from the wall. Since the scale l is characteristic of the eddies ensuring the turbulent diffusion, it is of the same order as the shear layer thickness, which is in turn proportional to y, leading to a practically uniform effect of this corrective term close to the walls. Very often, the scale l is replaced by the quantity $k^{3/2}/\epsilon$, but this hypothesis is not logical as the last quantity represents a diffusion property whilst the reflecting of the fluctuating pressure is essentially an inviscid mechanism.

Nevertheless, the adoption of the previous closure often leads to important errors owing to some uncertainty in the computation of the turbulence dissipation rate ϵ. This weakness appears mainly in the simulation of curved channels and in stratified flows, and it gives rise to erroneous predictions of the ratio of normal turbulence stresses $\overline{u^2}/\overline{v^2}$. Finally, to end with this wall correction term, let us point out that according to LAUNDER [23], the damping effect of the normal stress $\overline{v^2}$ has some memory effect, even after the disappearance of the wall, as in the wake following a flat plate, hence the difficulty to calculate with accuracy in this zone.

After modelling, the transport equation of the Reynolds stresses takes the following general form, for incompressible cases:

$$\begin{aligned}
\frac{\partial(\overline{u_iu_j})}{\partial t} + \bar{U}_k\frac{\partial(\overline{u_iu_j})}{\partial x_k} = &-\underbrace{\left[\overline{u_ju_k}\frac{\partial \bar{U}_i}{\partial x_k} + \overline{u_iu_k}\frac{\partial \bar{U}_j}{\partial x_k}\right]}_{\textit{Production } P_{ij}} + \underbrace{\frac{\partial}{\partial x_k}\left[C_s^1\frac{\bar{k}}{\bar{\epsilon}}\overline{u_ku_l}\frac{\partial}{\partial x_l}\overline{u_iu_l}\right]}_{\textit{Turbulent Diffusion}} \\
&+\underbrace{\alpha\left[P_{ij}-\frac{1}{3}\delta_{ij}P_{kk}\right]+\beta\left[Q_{ij}-\frac{1}{3}\delta_{ij}P_{kk}\right]+\gamma\bar{k}\left[\frac{\partial\overline{U_i}}{\partial x_j}+\frac{\partial\overline{U_j}}{\partial x_i}\right]}_{\textit{Linear Redistribution}} \\
&-\underbrace{C_2\frac{\bar{\epsilon}}{\bar{k}}\left[\overline{u_iu_j}-\frac{\delta_{ij}}{3}\overline{u_ku_k}\right]}_{\textit{Quadratic Redistribution}} - \underbrace{\frac{2}{3}\bar{\epsilon}\left[(1-f_s)\delta_{ij}+\frac{\overline{u_iu_j}}{\frac{2}{3}\bar{k}}f_s\right]}_{\textit{Destruction}} \\
&+\underbrace{\left[0.125\frac{\bar{\epsilon}}{\bar{k}}(\overline{u_iu_j}-\frac{2}{3}\delta_{ij}\bar{k})+0.15(P_{ij}-D_{ij})\right]\frac{\bar{k}^{3/2}}{\bar{\epsilon}y}}_{\textit{Wall Redistribution}}
\end{aligned} \tag{5}$$

with

$$\alpha = -\frac{C_1+8}{11} \qquad \beta = -\frac{8C_1-2}{11} \qquad \gamma = -\frac{30C_1-2}{55}$$

$$P_{ij} = -\left(\overline{u}_iu_k\frac{\partial\overline{U}_j}{\partial x_k}+\overline{u}_ju_k\frac{\partial\overline{U}_i}{\partial x_k}\right) \qquad Q_{ij} = -\left(\overline{u}_iu_k\frac{\partial\overline{U}_k}{\partial x_j}+\overline{u}_ju_k\frac{\partial\overline{U}_k}{\partial x_i}\right)$$

$$P_{kk} = \overline{u}_m u_l \frac{\partial \overline{U}_m}{\partial x_l} \tag{6}$$

We can then state that as for all two equation turbulence models, based on the eddy viscosity concept, it is necessary to determine a time-scale representing the decay of the turbulence kinetic energy, by the use of the transport equation for the dissipation rate of the kinetic energy of the fluctuating motion. The modelling of this equation can be straightforwardly deduced from the Navier-Stokes equations, leading to this form [24]:

$$\begin{aligned} \frac{\partial \overline{\epsilon}}{\partial t} + \bar{U}_k \frac{\partial \overline{\epsilon}}{\partial x_k} = &+ \frac{\partial}{\partial x_k}\Big[\Big(\nu\delta_{kl} + C_\epsilon \frac{\bar{k}}{\bar{\epsilon}} \overline{u_l u_k}\Big)\frac{\partial \bar{\epsilon}}{\partial x_l}\Big] \\ &- C_\epsilon^1 \frac{\bar{\epsilon}}{\bar{k}} \overline{u_i u_k} \frac{\partial \bar{U}_i}{\partial x_k} - C_\epsilon^2 \frac{\bar{\epsilon}}{\bar{k}} \Big[\overline{\epsilon} - 2\nu\Big(\frac{\partial \sqrt{k}}{\partial x_n}\Big)^2\Big] \\ &+ C_\epsilon^3 \nu \frac{\bar{k}}{\bar{\epsilon}} \overline{u_l u_k} \frac{\partial^2 \bar{U}_i}{\partial x_j \partial x_l} \frac{\partial^2 \bar{U}_j}{\partial x_i \partial x_l} \end{aligned} \tag{7}$$

with $f_\epsilon = 1.0 - 0.22exp\Big[-\Big(\frac{Re_t{}^2}{6}\Big)\Big]$

A comment should be made for the second order modelling. In this context, the momentum transport equation, according to (1), would have a very small linear diffusion term, in comparison with the non-linear convective term. In fact, only the viscous stresses having a gradient type behaviour contribute to ensure a constantly dissipative diffusion. Consequently, one can expect that this second order modelling simulates the non-linear characteristics of the flows more faithfully than the turbulence models using the eddy viscosity concept. Then, second order modelling can allow a local transfer of energy from the fluctuating motion to the mean motion, a physical aspect observed in some flows containing unsteady organized structures. This fact would explain the relative success of second order closure modelling used to predict flows where a strong presence of unsteady organized structures exists.

The framework of this paper does not allow us to discuss in detail all the hypotheses used for the closure schemes, as well as the basic physical ideas contained. We restrict this discussion by presenting one example of the comparison between the behaviour of the two previous turbulence model families. The example selected is the turbulent flow in an S channel, with a circular section, experimentally investigated by STEVENS & FRY [25] and numerically predicted by JONES & MANNERS [26] (Figure 3). The results obtained with an eddy viscosity two-equation model, of the (k, ϵ) type, lead to a velocity profile far from the experimental evidences, (and in the opposite way) whilst second order modelling correctly reproduces the specificity of the problem.

Such behaviour differences seem to justify the use of second order modelling in complex flows. However despite second order modelling being proposed for more than fifteen years, it is not very often applied to turbulent complex flows, where it is perfectly adapted. Paradoxically, because of the numerical complexity for solving the equations when the second order closure is used, this kind of model has so far only been applied to the *thin shear layers* there where the (relative) simplicity of the flow configurations, should allow the eddy-viscosity concept turbulence models to be sufficient. Thus, very often, we feel that the relatively few improvements brought by second order modelling do not justify the complexity of its implementation.

If second order modelling is not often used in complex flow configurations as much as it merits, it is not owing to the computing cost as we might expect (because of the higher number of equations to be solved), but rather because of the mathematical structure of the equations contained in the model. We have already mentioned that the main advantage of the eddy viscosity concept models is the reinforcing of the linear terms in the convection-diffusion equations. It is then easy to keep low local Reynolds numbers, from a numerical point of view. This advantage disappears in the second order modelling: here the local Reynolds number recovers its face-value (convection/molecular diffusion), and the Reynolds stresses act as *the source terms*, and finally the momentum transport equation has a hyperbolic character like the Euler equations. However, the solution of Euler equations is not familiar to scientists working in incompressible flows. We have then to turn to the field of high speed aerodynamics, where already many high performance numerical schemes exist (see other papers in this Conference) well suited to Euler equations. Unfortunately, the major disavantage of the classical numerical schemes for Euler resolution is to add some artificial numerical viscosity to stabilize the numerical behaviour. It is then important to check on the numerical viscosity in order to keep it much lower than the effects of the turbulent diffusion represented by the divergence of the Reynolds stress tensor.

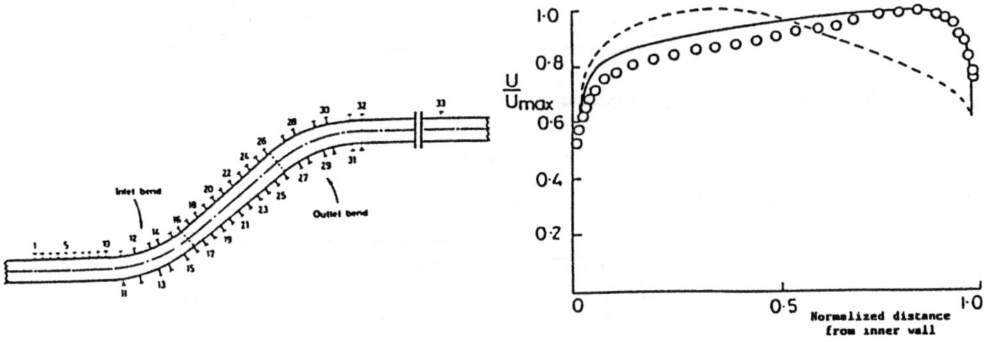

Fig.3.- *Comparison between the behaviours of different levels of modelling.*

From the numerical point of view, the solution of equations (5) and (7) needs special care for the source terms, because the modelled turbulence equations are not conservative. The numerical treatment used for the momentum, mass and energy conservation equations can be extended easily to the conservative terms (convection and diffusion) of the turbulence transport equations without major changes, but the success of the solution of equations (5) and (7) depends strongly on the treatment of source terms, particularly when implicit schemes are used [27], [28] and [29].

As already mentioned, one of the weaknesses of the eddy viscosity concept models is their inefficiency to represent the internal exchanges between different normal Reynolds stresses, i.e. the components of the turbulence kinetic energy. These exchanges are contained in the source terms of equation (5), and more particularly in the **redistribution** term. Not wishing to continue to second order modelling, several authors prefer to adopt an intermediate way: i.e. **Algebraic Stress Model** (A.S.M.).

Indeed, we can notice the conservative form of transport terms in the equation (5) governing the evolution of the turbulent stresses: because of the conservative property of convective and diffusive terms of this equation, it is often more convenient to group these transport terms as:

$$\begin{aligned} T_{ij} &= \left[\frac{\partial(\overline{u_i u_j})}{\partial t} + \overline{U}_k \frac{\partial(\overline{u_i u_j})}{\partial x_k} - D_{ij}\right] \\ &= \frac{\overline{u_i u_j}}{\overline{u_k u_k}} T_{mm} = \frac{\overline{u_i u_j}}{\overline{u_k u_k}}\left[P_{mm} - 2\bar{\epsilon}\right] \end{aligned} \tag{8}$$

It is from this form that the algebraic stress models are built.

This approximation of T_{ij}, which is perfectly justified, is due more to numerical simplicity considerations than to that of physical evidence. The right hand side of equation (8) neither contains differential operators applied to turbulent stresses, nor individual sources and sinks relative to each stress (for example no P_{ij}). Then, if the approximations adopted for the redistribution terms Φ_{ij} (see the previous definition) do not contain the differences of the turbulent stresses, the replacing of T_{ij} by the R.H.S. of (8) reduces a partial differential equation set (5) to an algebraic equation set. In such an approach, *only* the *trace* of the turbulent stress tensor (otherwise the turbulence kinetic energy $\bar{k} = 0.5\overline{u_i u_i}$) will be determinated by a transport equation. As for the two-equation models, these algebraic models must be achieved by another scale equation, usually the transport equation for the turbulence kinetic energy dissipation rate $\bar{\epsilon}$.

Let us remember that the general form of the turbulence stresses (5), in mean steady situations, can be rewritten as:

$$\underbrace{\bar{U}_k \frac{\partial(\overline{u_i u_j})}{\partial x_k}}_{C_{ij}} = -\underbrace{\left[\overline{u_j u_k}\frac{\partial \bar{U}_i}{\partial x_k} + \overline{u_i u_k}\frac{\partial \bar{U}_j}{\partial x_k}\right]}_{P_{ij}} + \underbrace{\overline{\frac{p}{\rho}\left(\frac{\partial u_i}{\partial x_j} + \frac{\partial u_j}{\partial x_i}\right)}}_{\Phi_{ij}}$$
$$\underbrace{-\frac{\partial}{\partial x_k}\left[\overline{u_i u_j u_k} + \frac{\overline{p u_j}}{\rho}\delta_{ik} + \frac{\overline{p u_i}}{\rho}\delta_{jk} - \nu\frac{\partial \overline{u_i u_j}}{\partial x_k}\right]}_{D_{ij}} \underbrace{- 2\nu\overline{\frac{\partial u_i}{\partial x_k}\frac{\partial u_j}{\partial x_k}}}_{\epsilon_{ij}} \tag{9}$$

a relationship which is symbolically written as:

$$C_{ij} \;-\; D_{ij} \;=\; P_{ij} \;+\; Phi_{ij} \;-\; \epsilon_{ij} \tag{10}$$

Morever, the exact form of turbulence kinetic energy is written as:

$$\underbrace{\bar{U}_j \frac{\partial(\bar{k})}{\partial x_j}}_{C_k} = \underbrace{-\overline{u_i u_j}\frac{\partial \bar{U}_i}{\partial x_j} + \overline{u_i u_k}\frac{\partial \bar{U}_j}{\partial x_k}}_{P_k} + \underbrace{\frac{\partial}{\partial x_j}\overline{\left[u_j\left(\frac{p}{\rho} + \frac{1}{2}u_i u_i\right)\right]}}_{D_k} \underbrace{-\nu\overline{\frac{\partial u_i}{\partial x_j}\frac{\partial u_i}{\partial x_j}}}_{\epsilon_{ij}} \tag{11}$$

a relationship recasted as:

$$C_k \;-\; D_k \;=\; P_k \;-\; \epsilon \tag{12}$$

Using (8) to link the transport terms (convective and diffusive) of the Reynolds stresses to those of turbulence kinetic energy $\bar{k}$, we then obtain:

$$C_{ij} - D_{ij} = \frac{\overline{u_i u_j}}{\bar{k}}(C_k - D_k) \tag{13}$$

which is the baseground of the algebraic stress models (A.S.M.). Indeed, using (10), (11) and (12), we can obtain:

$$P_{ij} + \Phi_{ij} - \epsilon_{ij} = \frac{\overline{u_i u_j}}{\bar{k}}(P_k - \bar{\epsilon}) \tag{14}$$

By inverting this relationship, we can finally obtain the following forms for the Reynolds stresses:

$$\overline{u_i u_j} = \frac{2}{3}\delta_{ij}\bar{k} + \frac{\bar{k}}{\bar{\epsilon}} \frac{(1 - C_1)(P_{ij} - \frac{2}{3}\delta_{ij}P_k) + \Phi_{ijw,1} + \Phi_{ijw,2}}{C_2 + \frac{P_k}{\bar{\epsilon}} - 1} \tag{15}$$

with: P_{ij} and P_k respectively the turbulent stress and turbulent kinetic energy production terms. Φ_{ij} represents the classical pressure-strain correlation, and $\bar{\epsilon}$ and $\bar{\epsilon}_{ij}$ represent destruction terms for the turbulence kinetic energy and the turbulent stresses respectively. Finally the two terms: $\Phi_{ijw,1}$ and $\Phi_{ijw,2}$ represent the wall effects of the pressure-strain correlation.

It is difficult to conclude clearly on the algebraic stress model performance. Applied to various turbulent flows (boundary layers, wakes, with adverse pressure gradient...) the A.S.M. did not mark a definite superiority compared with the eddy-viscosity concept models like the ($\bar{k}$, $\bar{\epsilon}$) type. On the other hand, in some flows including the curvature effects, for example in the case of a jet along a curved wall, RODI and SCHEUERER [30] have obtained clearly better results with the algebraic stress model compared to those given by the eddy-viscosity model.

We present here just an example of this comparison between the predictions obtained by an A.S.M. and by a two-equation model of $\bar{k}, \bar{\epsilon}$ type, in a boundary layer along a curved wall, using the experimental data of GILLIS *et al.* [31] from STANFORD.

DAVIDSON L. [32] has also shown that the use of an algebraic stress model can lead to much better results, than those of a $\bar{k}, \bar{\epsilon}$ model, predicting the separation over an airfoil.

Anyhow, we can point out that the algebraic stress models give rise to results less satisfactory than those of the second order modelling and sometimes numerical difficulties can arise, more particularly about the numerical convergence problem, for recirculating flows or for three-dimensional configurations.

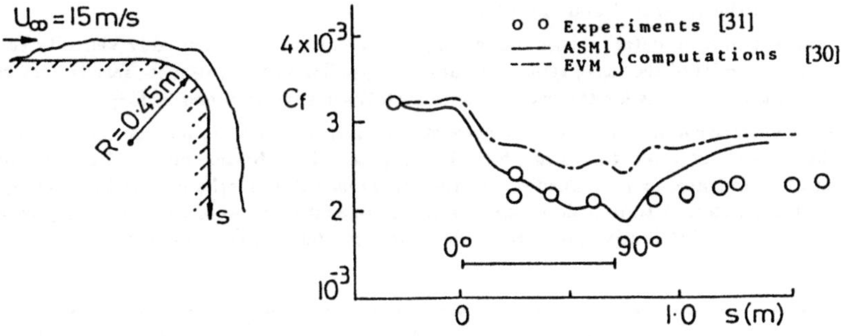

Figure 4.- *Skin friction development along a convex corner.*

4.- CURRENT REQUIREMENTS AND RESEARCH PROSPECTS

The development of new scientific and industrial programmes in general, and in aeronautics in particular, reveal urgent needs which we shall try to resume, albeit non-exhaustively.

4.1.- The unsteady flows.

As we have already mentioned, nearly all real flows contain *simultaneously* both ordered and disordered characters. However, it is advisable to distinguish two types of unsteady flows: those governed by a **natural unsteadiness** and those submitted to some forcing by the boundary conditions, or a **forced unsteadiness.**

The *naturally unsteady flows* are characterized by a strong presence of ordered structures, which are also called, under some conditions, the *coherent structures*. Very often, the presence of these structures in "industrial flows", is the origin of the rather bad prediction results obtained from classical statistical turbulence models. Some intermediate ways are proposed to fulfil this gap, in particular by Large Eddy Simulation (L.E.S.) completed by Sub-Grid Modelling (S.G.M.). The work done by research centres such as I.M.G. [33], O.N.E.R.A.[34], N.A.S.A.[35], Stanford [36]... lead to excellent simulations in more or less academic configurations .

Another approach, *semi-deterministic modelling*, uses the same methodology, but it tries to take into account all the past knowledges concerning classical statistical modelling [37]. This way consists in splitting any instantaneous physical quantity into a *time-dependent ensemble averaging part* and an *incoherent part*. The ensemble-average, a function of time and space, composes the *time-mean value* and the *coherent unsteadiness part*: this ensemble-average will be directly determined by the solution of the unsteady averaged Navier-Stokes equations (ensemble averaging sense). On the other hand, the *effects* of the incoherent part are taken into account by the solution of the closure equations, in the same way as for classical statistical modelling. From a practical point of view, this approach is related to the traditional way of classical turbulence modelling, it means that we will again have two modelling levels: the eddy-viscosity concept level and second order modelling. In fact, the so-called *semi-deterministic* modelling is different from the classical one, in the real sense of the fluctuating parts which currently *only* contain presently the *apparently random* characters. Consequently, and concretely, the role of the turbulence is then notably reduced in flows where the transition occurs or where the turbulence is not fully developed.

Let us consider the example of a backward-facing step, at high Reynolds numbers. The 1981 Stanford Congress contributions showed that nearly all the predictions obtained with the classical (k, ϵ) model, lead to a too short reattachment length, compared with the experimental evidence. A more elaborate analysis of the physical phenomenon reveals a very particular behaviour of turbulence in the recirculating zone, where the turbulent shear stress is unusually low compared with the level of normal turbulent stresses, the sum of which is used for the calculation of the eddy viscosity $\sim (k^2/\epsilon)$. The special character of turbulence in this recirculating zone is actually the consequence of the organized structures, issuing from instabilities of the Kelvin-Helmholtz type, in the mixing shear layer created at the step edge, the physical properties of which depend on those of the separated boundary-layer.

In the context of semi-deterministic modelling, with a closure using the eddy viscosity concept based on a two-equation model of the (k, ϵ) type, the numerical prediction leads to more satisfactory results, both for the reattachment length as for the wall pressure evolutions, (figures 4 and 5) [38].

The *forced unsteady flows* also give rise to new studies during the last decade, especially in periodic mean motions. The presence of a new time scale, fixed by the external conditions, may lead to some *amplification* phenomenon of turbulence, or in opposition, to the *relaminarization* mechanism of turbulence, in function of the ratio of the external time-scale compared to the relaxation-time of turbulence represented by the quadratic part of the redistributive pressure-strain correlation term in equation (5).

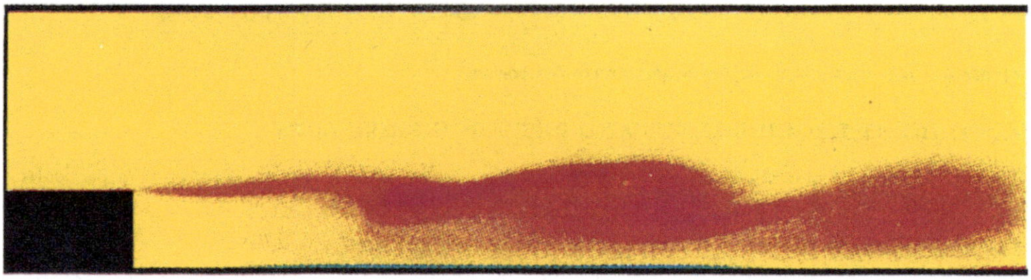

Figure 5.- *Iso-vorticity contours on the backward facing step predicted by a semi-deterministic model.*

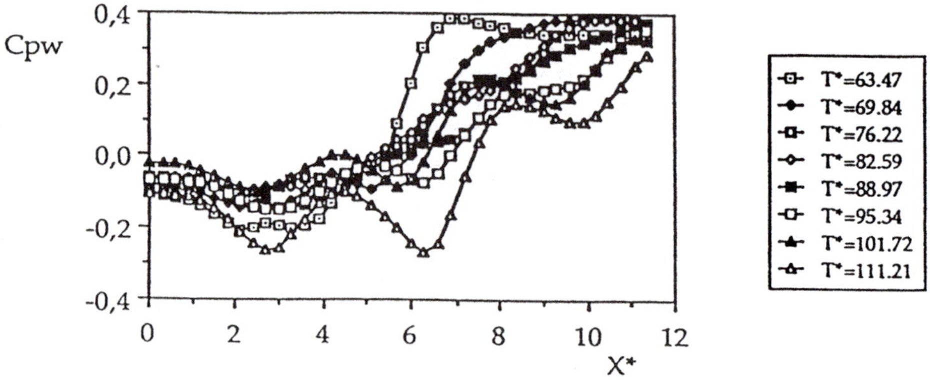

Figure 6.- *Wall-pressure evolution during a cycle.*

So far, few theoretical studies are devoted to unsteady turbulent flows, or even in periodic situations. According to LAUNDER [39], it is usual to admit four frequency categories:

- *Quasi-steady flows*: where the changes of the mean flow are so slow that their influence on the turbulent structures is negligible. Consequently, the classical turbulence models, developed in a stationary and ergodicity context, should be used without change.

- *Slow unsteady flows*: where the time variations of mean quantities, and more particularly, the mean momentum, are significant, but the time-scales remain intact. Then the classical turbulence can be used again.
- *Intermediate unsteady flows*: where the unsteadiness effects notably modify the turbulent structures via the *turbulent shear stress*. One can point out that the classical "universal" laws, such as the logarithmic velocity profile in the near-wall, are no longer valid. This flow category probably needs more modelling efforts before obtaining correct predictions.
- *High unsteady flows*: where the turbulence kinetic energy seems to be *"frozen"* over major regions of shear flow (but shear stress does not). It is noticed that the mean flow seems to become *independent of the turbulence* level. Again, turbulence models must be able to correctly describe this fact.

The importance of this aspect has led several research groups to undertake studies in order to obtain more knowledge for the improved modelling of unsteady turbulent flows. If in laminar flows, the unsteady character is represented by the square of the Stokes parameter $\omega L^2/\nu$, where ω is the angular frequency of a periodic motion and L is a characteristic length scale, in turbulent flow, it is tempting to replace ν by the eddy-viscosity built with the mean friction velocity U_τ (case of the wall boundary layers) and the same length scale L. The new unsteadiness parameter would be $\omega L/U_\tau$. According to some experimental investigations, see for example [40], the value of 5 of this parameter seems to be typical of the "intermediate" unsteady flows and the value of 50 corresponds to the high frequency behaviour. Using more details on the unsteady boundary layer, namely by direct simulations, and by second order turbulence modelling, the author and his colleagues tried to take into account the ratio of the *unsteadiness frequency* and the inverse of the local turbulent turnover time-scale, i.e. the quantity $\omega k/\epsilon$. Since the ratio k/ϵ increases as one proceeds from the pipe wall to the centre, it clarifies why the *"freezing"* behaviour occurs first in the core region.

Again, in this kind of unsteady flows, second order turbulence modelling often shows its superiority when predicting the turbulence characteristics, compared to the eddy-viscosity concept models. For example, in the oscillating boundary layer on a flat plate problem, the direct simulation of P. SPALART [41] showed a *relaminarization* in the near wall, a phenomenon which seems be correctly simulated only when second order closures, including the highly anisotropy terms, are used (figure 7) [42].

More specific studies are undertaken for unsteady boundary layers, by COUSTEIX's group at CERT-ONERA (see for example the reference [43]), and the successful application of the second order modelling in intermediate unsteady flows by the UMIST group of Pr. LAUNDER and LECHZINER [21].

In conclusion, the turbulence modelling of the unsteady flows could be done only when strong collaborative works exist, between fine experiments, for example [44], direct simulation studies, for example [45], and modelling efforts.

4.2.- Variable density turbulent flows.

If progress in turbulence modelling has incontestably been very important for the past twenty years, the turbulence models are usually obtained in incompressible situations, where, besides the relative simplicity of the equations to be closed, the divergence free of the mean and fluctuating velocities enables clear explanations of the physical processes, namely those of the *redistribution* due to the pressure-strain correlation. This simplicity does not exist in situations where the density varies in space and in time. Although the averaging equations remain identical in all cases, in practice, it is worth distinguishing various origins of these density variations in order to identify some specific simplifications in each case:

- Density variations by **compressibility** effects of *high speed* flows, where the coupling of equations governing the motion is due to the *mean density* changes, which in turn, come from the thermodynamical aspects.
- Density variations by **thermal** effects even in slow speed flows, situations where the thermodynamic aspects are still important, but the compressibility effects are negligible..
- Density variations by **mass heterogeneity** issuing from the monophasic mixing of different gases of different densities, even in isothermal and/or isobar situations.

- Density variations by global **compression**, when the whole room volume is subjected to important time variations. This state leads to some studies concerning *compressed turbulence* [46][47].

In some practical industrial applications, for example in the piston engine, usually several origins of density variations are *simultaneously* present, this fact substantially increases the difficulty of the problem to be solved.

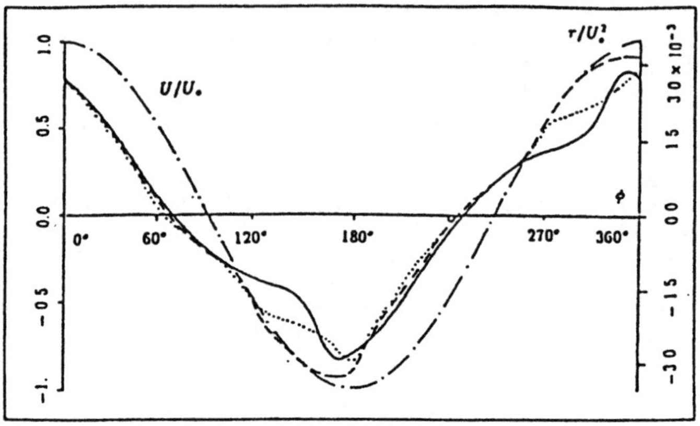

Figure 7.- *Skin friction during a period, for the boundary layer on an oscillating flat plate.*

The first problem to be examined is the higher number of unknown correlations to be handled: this fact is due to the fluctuating density. To limit the "inflation" of the unknowns to be modelled, several variable regroupings are proposed. Let us remember that for variable density turbulent flow, we have:

$$\begin{aligned}\overline{\rho U_i U_k} \;=\; & \underbrace{\bar{\rho}\bar{U}_i\bar{U}_k}_{(1)} + \underbrace{\overline{\rho' u'_i}\bar{U}_k}_{(2)} + \underbrace{\overline{\rho' u'_k}\bar{U}_i}_{(3)} \\ & + \underbrace{\bar{\rho}\overline{u'_i u'_k}}_{(4)} + \underbrace{\overline{\rho' u'_i u'_k}}_{(5)}\end{aligned} \tag{16}$$

In the incompressible case:

$$\overline{\rho U_i U_k} = \underbrace{\bar{\rho}\bar{U}_i\bar{U}_k}_{(1)} + \underbrace{\bar{\rho}\overline{u'_i u'_k}}_{(4)} \tag{17}$$

For physical reasons, but also for commodity, compared to the incompressible case, FAVRE [48] and [49], proposed a specific regrouping, leading to the so-called *mass weighted averaging*:

$$\begin{aligned}\overline{\rho U_i U_k} \;=\; & (1) + (2) + (3) + \frac{\overline{\rho' u'_i}.\overline{\rho' u'_k}}{\bar{\rho}} \rightarrow \bar{\rho}\tilde{U}_i\tilde{U}_k \\ & (4) + (5) \quad - \frac{\overline{\rho' u'_i}.\overline{\rho' u'_k}}{\bar{\rho}} \rightarrow \overline{\rho u''_i u''_k}\end{aligned} \tag{18}$$

Using other physical and numerical arguments, HA MINH-LAUNDER-Mc INNES [50] tried to distinguish the convective velocity and that of the transportable momentum:

$$\overline{\rho U_i U_k} = \overline{G_i U_k} = \underbrace{\bar{G}_i.\bar{U}_k}_{(1)+(2)} + \underbrace{\overline{g_i u_k}}_{(3)+(4)+(5)}$$

$$\overline{G}_i = \overline{\rho U_i} = \bar{\rho}\overline{U}_i + \overline{\rho' u_i'} \quad \text{and} \quad g_i' = \rho'\overline{U}_i + \bar{\rho}u_i' + \rho' u_i' - \overline{\rho' u_i'} \tag{19}$$

Desiring, on the contrary, to emphasize the specific role of density variations, and more particularly in the case of isothermal and isobaric mixing of different gases, CHASSAING [51][52], has developed a new approach:

$$\overline{\rho U_i U_k} = \underbrace{\underbrace{\bar{\rho}\overline{U}_i\overline{U}_k}_{(1)}}_{A_{ik}} + \underbrace{\underbrace{\overline{\rho u_i' u_k'}}_{(4)+(5)}}_{B_{ik}} + \underbrace{\underbrace{\overline{\rho' u_i'}\overline{U}_k}_{(2)} \quad \underbrace{\overline{\rho' u_k'}\overline{U}_i}_{(3)}}_{C_{ik}} \tag{20}$$

avoiding then the modelling of the term $\overline{u}_i = -\dfrac{\overline{\rho u_i'}}{\bar{\rho}}$ which appears in the FAVRE splitting. In the case of mixing two different density gases, the mass fluxes can be derived as:

$$\frac{1}{\rho} = \frac{C}{\rho_1} + \frac{1-C}{\rho_2} \Rightarrow \rho = a\,\rho\,C + b$$

$$\bar{\rho} = a(\bar{\rho}\bar{C} + \overline{\rho\gamma'}) + b \quad \text{and} \quad \rho' = a(\rho'\bar{C} + \rho\gamma' - \overline{\rho\gamma'}) \tag{21}$$

Consequently:

$$\overline{\rho' f'}\ (\equiv \overline{\rho f'}) = \frac{a}{1-a\bar{C}}\overline{\rho\gamma' f'} \Rightarrow \overline{\rho f'} = \alpha(\bar{C})\overline{\rho\gamma' f'}$$

$$\overline{\rho u_i'} = \alpha(\bar{C})\overline{\rho\gamma' u_i'} \quad \text{and} \quad \overline{\rho\gamma'} = \alpha(\bar{C})\overline{\rho\gamma'^2} \tag{22}$$

Finally, let us point out a new formulation of statistical equations, based on mass (and not on volume) integration, which is developed by C. REY [53], to handle subsonic compressible turbulent flows.

In any event, we can show that the more the decomposition is complete, the more the information is important, but complete decomposition means a higher number of equations to be solved, including a higher number of *empirical constants* to be "calibrated" leading to more uncertainty in modelling. Besides, don't forget that more equations to be solved means more numerical errors in the results. For all these reasons, and in spite of some improvements expected about the turbulence modelling, it seems to us that the FAVRE mass weighted averaging is presently the more reasonnable approach for the numerical simulation of complex compressible turbulent flows.

Finally, for the complex compressible flows, such as the shock/boundary layer interaction, the analysis of results, in particular about the lambda shock, imply the necessity for the use of second order turbulence modelling [5][54]. Indeed, figure 8 shows that a prediction with an eddy viscosity model of (k, ϵ) type, does not allow the natural capture of the second shock (the turbulent energy calculated by the (k, ϵ) model being too high), as obtained with second order modelling. The main reason for this fact is that in such a flow, the two normal turbulent stresses $\overline{u^2}$ and $\overline{v^2}$ play two different roles: the streamwise normal turbulent stress represents a general level of turbulence, and its kinetic energy, while the transverse component ensures a diffusion role. In recirculating flows resulting from the shock/boundary layer interaction, these two quantities are quite different in level and in evolution: only second order modelling would correctly simulate this behaviour.

4.3.- The wall transfers.

One of the major difficulties appearing in the indirect numerical simulation of turbulent flows (i.e. with the use of turbulence models) consists of the wall transfer prediction. Indeed, the strong gradients of physical quantities (mean and fluctuating) in the vicinity of solid walls, often require an extremely fine grid in this zone, leading to very high computing costs. In a schematic way, it is possible to distinguish two approaches for the calculation of this zone:

- Either by a *computation up to the true wall* with the handling of the viscous effects in this zone by including the "*damping low Reynolds functions*" in the standard turbulence models which were usually established in the fully developed turbulence context. The role of these low-Reynolds effect laws is to restore the molecular viscosity effects in the diffusion mechanism.

- Or by the use of the "*wall laws*", algebraically obtained by a local analytical integration near the wall of turbulence transport equations, under COUETTE's hypothesis. This approach, much cheaper than the previous one, is unfortunately valid only for fully developed turbulent boundary layers, without pressure gradients or with moderate favourable pressure gradients.

Today, several studies have shown that the "*low Reynolds damping law*" approach would be extended to cases with favourable or adverse pressure gradients, provided that these gradients are not too strong in the positive case (fig.9a). However, the use of the "*wall laws*" would only be possible for moderate favourable pressure gradients. (fig.9b), as shown in the thesis work of F. HANINE [55].

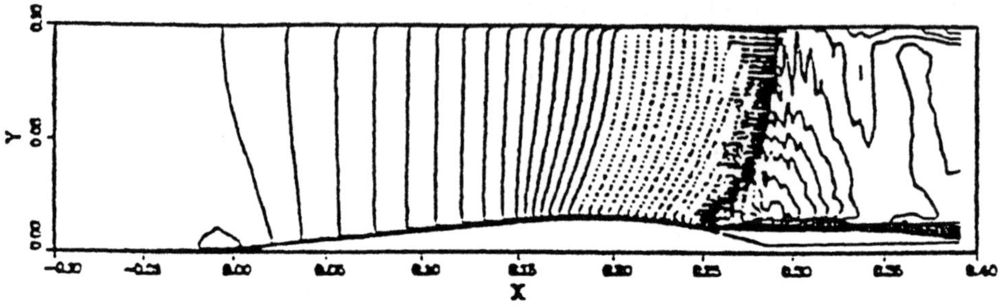

Figure 8a.- *Shock/Boundary Layer Interaction predicted by (k, ϵ) model* [5].

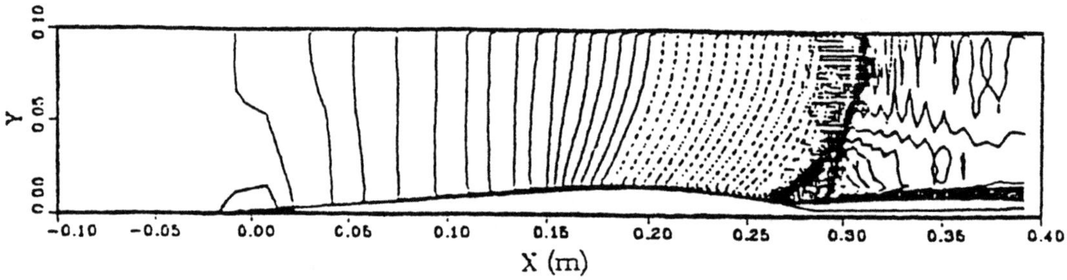

Figure 8b.- *Shock/Boundary Layer Interaction predicted by second-order modelling* [5].

5.- CONCLUDING REMARKS.

This review shows that we currently dispose of a fairly varied panoply of turbulence models to numerically simulate many turbulent flows from the more simple to the most complex configurations. We must first recognize that for twenty years, the development of turbulence modelling was possible owing to good numerical tools developed by numerical scientists for our benefit. The role of the physicist should then be, on the one hand, to understand the essential part of his physical problem to be solved, and, on the other hand, to have good knowkedge about the basic ideas on which the various numerical schemes proposed are built, in order that the numerical way chosen is in perfect coherence with the physical aspects to be simulated.

However, there remain important needs in numerics before we are able to solve many complex flows presenting practical industrial interests. For example, the high accuracy numerical schemes developed to handle the discontinuities in flows such as shock waves, where very strong streamwise gradients exist, do not seem to be reliable in flow situations which are dominated by very strong *transverse gradients*, for instance in the case of co-axial supersonic jets containing two very different density fluids (hydrogen and oxygen for example), a configuration which often exists in the cryotechnic rocket engines, including combustion.

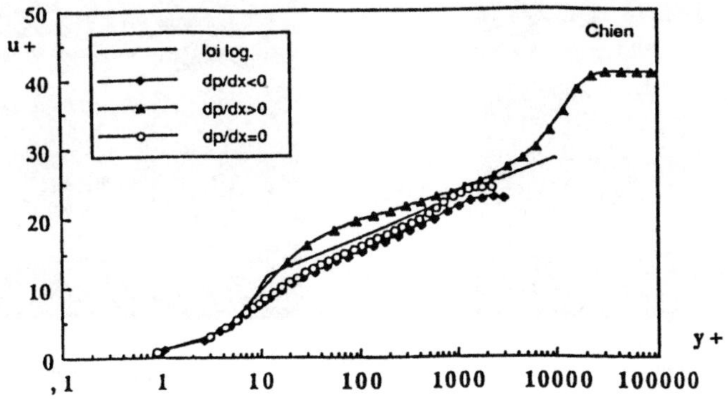

Figure 9a.- *Results obtained with the use of "low Reynolds damping functions"*.

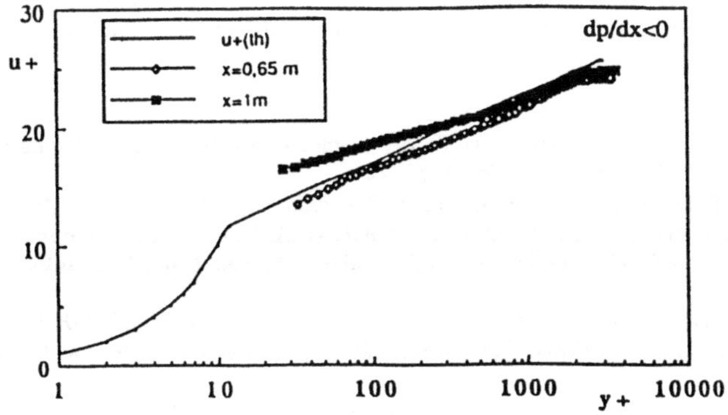

Figure 9b.- *Results obtained with the use of "wall fuctions"*.

As a conclusion, let us present in figures 9a and 9b, the views of variable density axisymmetric jets, obtained respectively by CHASSAING [56] and BINDER *et al.* [57]. These pictures reveal the existence of natural "*lateral jets*" that no current turbulence model can predict. Thus, in order to obtain, in the near future, available tools to numerically simulate this kind of "*mysterious*" physical phenemena, important efforts must be devoted to turbulence studies both from an experimental point of view, as for direct simulations, and theoretical modelling.

Figure 10a.- *Lateral jet* [56]

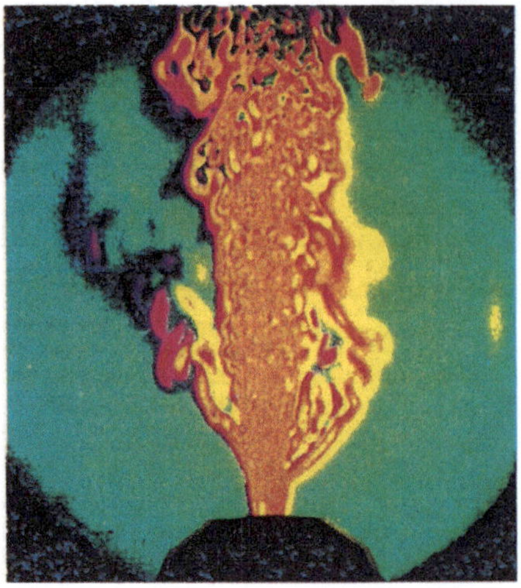

Figure 10b.- *Lateral jet* [57]

ACKNOWLEDGMENTS.

This contribution came from several DRET grants, groups VI and VII, and the NASA-UC Davis Contract NCC2-186. The author found much joy and friendship working with Pr B.E. LAUNDER, Pr R.W. MacCORMACK, Dr M.W. RUBESIN, Dr J.R. VIEGAS, Pr P. CHASSAING, and Pr D.D. VANDROMME. He also wishes to thank Dominique BENNETT and Christine NGUYEN-TRISTANI for their very kind help.

RÉFÉRENCES.

[1] REYNOLDS O. (1894) "*On the Dynamical Theory of Incompressible Viscous Fluids and the Determination of the Criterion*", Phil. Trans. Roy. Soc.A186, 123.

[2] BOUSSINESQ J. (1877) "*Théorie des Écoulements Tourbillonnaires*", C.R.A.S., T.23.

[3] KLINE S.J., MORKOVIN M.V., SOVRAN G., COCKRELL D.J. (1968) "*Computation of Turbulent Boundary Layers - 1968 - AFOSR-IFO-STANFORD Conference*", Vol. I and II, Stanford University, California USA 94305.

[4] KLINE S.J., CANTWELL B.J., LILLEY G.M. (1982) "*The 1980-1981 AFOSR-HTTM-STANFORD Conference on Complex Turbulent Flows: Comparison of Computation and Experiment.*", Vol. I, II and III, Stanford University, 1982.

[5] HA MINH H., RUBESIN M.W., VIEGAS J.R. and VANDROMME D.D. (1985) "*On the Use of Second Order Closure Modelling for the Prediction of Turbulent Boundary Layer / Shock Wave Interactions: Physical and Numerical Aspects.*", Numerical Methods in Fluid Mechanics I, Ed. by Koichi OSHIMA, Invited Lectures of the International Symposium on Computational Fluid Dynamics - Tokyo.

[6] HA MINH H. and HEBRARD P. (1977) "*Theoretical and Experimental Study of Flow Separation after a Sudden Increase in the Cross-Section of a Circular Pipe.*", Journal of Hydraulic Research, no 1, 15, pp. 1-18.

[7] SHIH T.H. (1984) "*Second Order Modeling of Scalar Turbulent Flow*" Ph.D. Thesis, Ithaca NY, Cornell.

[8] BROWN and ROSKHO (1974) "*On Density Effects and Large Structure in Turbulent Mixing Layers*", J. Fluid Mech., Vol.64, part 4, pp 775-816.

[9] IACOVIDES H, and LAUNDER B.E. (1984) "*The Computation of Momentum and Heat Transport in*

Turbulent Flow Around Pipe Bends", Proc. 1st UK National Heat Transfer Conf., IChemE Symposium 86, 1097, 1984.

[10] IACOVIDES H, (1986) "*Momentum and Heat Transport in Flow Through 180° Bends of Circular Cross-Section*", PhD Thesis, Faculty of Technology, Univ. Manchester.

[11] JONES W.P. and LAUNDER B.E. (1973) "*The Prediction of Laminarization with a Two-Model of Turbulence*", Int. Heat Mass Transferts, No 5, pp 301-314.

[12] RODI W. (1984) "*Examples of Turbulence Model Applications*", Turbulence models and their applications. CEA-EDF-INRIA École d'été d'analyse numérique. Eyrolles 1984.

[13] RODI W. (1986) "*Turbulence Modelling for Incompressible Flows*", P.C.H. Vol.7, No 5/6, pp 297-324.

[14] MATHIEU J. (1961) "*Contribution à l'étude aérothermique d'un jet plan évoluant en présence d'une paroi.* " P.S.T. du Ministère de l'Air, No 374.

[15] BÉGUIER C. (1971) "*Étude du jet plan dissymétrique en régime turbulent incompressible*", Thèse de Doctorat d'État. Université de Provence, I.M.S.T.

[16] ESKINAZI S., YEH H. (1956) "*An Investigation on Fully Developed Turbulent Flows in a Curved Channel*", Jr. Aero. Sc. Vol 23, No 1, pp 23-25.

[17] HANJALÍC K. and LAUNDER B.E. (1972) "*Fully Developed Asymetric Flow in a Plane Channel*", J.F.M., No 51, p. 301.

[18] HANJALÍC K. and LAUNDER B.E. (1972) "*A Reynolds Stress Model of Turbulence and its application to thin shear flows.*", J.F.M., No 52, part 4, pp 609-638.

[19] LAUNDER B.E., REECE G.J., RODI W (1975) "*Progress in the Development of a Reynolds Stress Turbulence Closure.*", J.F.M., Vol. 68-3, pp.537-566.

[20] VANDROMME D.D., HA MINH H., VIEGAS J.R., RUBESIN M.W., KOLLMANN W (1983) "*Second Order Closure for the Calculation of Compressible Wall Bounded Flows with an Implicit Navier-Stokes Solver*", Fourth T.S.F., Karlsruhe, Sept. 1983.

[21] LAUNDER B.E. and TSELEPIDAKIS D.P. (1991) "*Directions in Second-Moment Modelling of Near-Wall Turbulence.*", A.I.A.A. 29th Aerospace Sciences Meeting, RENO, January 1991.

[22] HANJALÍC K. and LAUNDER B.E. (1976) "*Contribution towards a Reynolds Stress Closure for Low Reynolds Number Turbulence.*", J.F.M., No 74, part 4, pp 593-610.

[23] LAUNDER B.E. (1991) "*An Introduction to Single-Point Closure Methodology.*", Introduction to the Modeling of Turbulence, V.K.I. Lecture Series 1991-02, March 18-21.

[24] HA MINH H. (1991) "*Physique et Modélisation de la Turbulence en écoulements de Fluides.*", École Printemps de Mécanique des Fluides Numérique, AUSSOIS 1991.

[25] STEVENS S.J. and FRY P. (1973) "*Measurements of the Boundary-Layer Growth in Annular Diffusers*, J. Aircraft, 10, 73.

[26] JONES W.P. and MANNERS A. (1989) "*The Calculation of the Flow through a two-dimensional faired diffuser*", Turbulent Shear Flow 6, (ed. J.C. André *et al.*), 18-31, Springer, Heidelberg, 1989.

[27] VANDROMME D.D. and HA MINH H. (1986) "*About the Coupling of Turbulence Closure Models with Averaged Navier-Stokes Equations*", Journal of Computational Physics, 65, No 2, pp 386-409, August.

[28] HA MINH H., VANDROMME D.D. and KOLLMANN W. (1983) "*Implicit Treatment of Multi-Equation Turbulence Models for Compressible Flows*", NASA Rep.NCC2-186 (A.R.C.) 1983.

[29] VANDROMME D.D. and LEBRET Y. (1991) "*Influence du Nombre de Mach Convectif sur Taux d'Ouverture*", Rapport DRET 89-193.

[30] RODI W. and SCHEUERER G. (1983) "*Calculation of Curved Shear Layers with Two-Equation Turbulence Models*", Phys. of Fluids 26(6), pp. 1422-1436.

[31] GILLIS J.C., J.P. JOHNSTON, W.M. KAYS and R.J. MOFFAT (1980) Stanford University Department of Mechanical Engineering Thermosciences Division, Report No.HMT-31.

[32] DAVIDSON L. (1991) "*Prediction Stall of a Two-Dimensional Airfoil using an Algebraic Reynolds Stress Model*", CERFACS TR/RF/91/52.

[33] MÉTAIS O. and LESIEUR M. (1991) "*Spectral Large-Eddy Simulation of Isotropic and Stably-Stratified Turbulence*", Submitted to Journal of Fl. Mech, revised version 14 February 1991.

[34] DESCHAMPS V. and DANG T.K. (1987) "*Evaluation of Subgrid-Scale Models for Large-Eddy Simulation of Transitional Channel Flows*", 6th Symposium on T.S.F., pp 20.1.1-7., Toulouse, September.

[35] MOIN P. and KIM J.J. (1982) "*Numerical Investigation of Turbulent Channel Flow*", Journal of Fluid Mech., 118, pp.341-377.
[36] FERZIGER J.H. (1985) "*Large-Eddy Simulation: Its Role in Turbulence Reasearch*", Theoretical Approches to Turbulence, Ed. by D.L. Dwoyer, M.Y. Hussaini and R.G. Voigt, Springer-Verlag.
[37] HA MINH H. (1989) "*Une Démarche de Modélisation de la Turbulence dans les Moteurs d'Aviation,*" Rapport D.R.E.T.
[38] KOURTA A. and H. HA MINH (1991) "*Modified turbulence model to predict unsteady separated flows.* 7th Int. Conf. on Numerical Methods in Laminar and Turbulent Flows, Stanford, California, USA, July 15-19, 1991.
[39] LAUNDER B.E. (1985) "*Progress and Prospects in Phenomenological Turbulence Models*", Theoretical Approches to Turbulence, Ed. by D.L. Dwoyer, M.Y. Hussaini and R.G. Voigt, Springer-Verlag.
[40] RAMAPRIAN B.R. and TU S.W. (1982) "*Study of Periodic Turbulent Pipe Flow*", IIHR report No 238, Institute of Hydraulic Research, Univ. of Iowa, Iowa City.
[41] SPALART P.R. and BALDWIN B.S.. (1987) "*Direct Simulation of a Turbulent Oscillating Boundary-Layer*", NASA Tech.Mem.No 89460.
[42] HA MINH H., VIEGAS J.R., RUBESIN M.W., VANDROMME D.D. and SPALART P.R.(1989) "*Physical Analysis and Second Order Modelling of an Unsteady Turbulent Flow: the Oscillating Boundary Layer on a Flat Plate*", 7th Symposium on T.S.F., Stanford University, Aug.21-23.
[43] HOUDEVILLE R., BONNET J.L. and COUSTEIX (1989) "*Experimental Study of a Flat Plate Oscillating in Pitch*", 7th Symposium on T.S.F., Stanford University, Aug.21-23.
[44] MAESTRI R.D., TARDU S. and G. BINDER (1991) "*Wall Shear-Stress Measurements in Unsteady Turbulent Flows in Diverging Channels*", 8th Symposium on T.S.F., Munich, Sept.9-11.
[45] RIDA S. and DANG T.K. (1991) "*Direct Simulation of Turbulent Pulsed Plane Channel Flows*", 8th Symposium on T.S.F., Munich, Sept.9-11.
[46] WU C.T., FERZIGER J.H. and CHAPMAN D.R. (1985) "*Simulation and Modelling of Compressed Turbulence*" Department of Mechanical Engineering; Stanford University.
[47] NUGLISH H.J., J. BORÉE and G.CHARNAY (1990) "*Measurements in a Compressed Air-Jet as Support to Support to the Modelling of Compressed Turbulence*" 5th Intern. Symposium on Application of Laser Techniques to Fluid Mechanics, Lisbon (Portugal) 9-12 Juillet 1990.
[48] FAVRE A.J.A. (1965) "*Équations Statistiques des Gaz Turbulents Compressibles*", Journal de Mécanique, Vol. 4, No 3, No 4.
[49] FAVRE A.J.A. (1991) "*Formulation of the Statistical Equations of Turbulent Flows with Variable Density*", Studies in Turbulence, Dedicated to J.L. Lumley on the Occasion of His Sixtieth Birthday, Springer-Verlag.
[50] HA MINH H., LAUNDER B.E.and Mc INNES J. (1980) "*A New Analysis of the Modelling of Variable Density Turbulence*", Physical Review, Turbulence Modelling, Thirty Third Meeting of the Americal Physical Society, Ithaca, USA, Nov. 1980.
[51] CHASSAING P. (1985) "*Une Alternative à la Formulation des Équations du Mouvement Turbulent d'un Fluide à Masse Volumique Variable*", Jl. Mécanique Théorique te Appliquée, Vol. 4, No 3, pp.375-389.
[52] CHASSAING P. (1992) "*Incidences du Traitement Statistique des Équations ouvertes sur la Modélisation au Second Ordre des Écoulements Turbulents à Masse Volumique Variable*", Deuxièmes Journées d'études sur les Écoulements à Masse Volumique Variable, Orléans, Mai 1992.
[53] REY C.(1992) "*Équations Statistiques des Écoulements Turbulents Compressibles Subsoniques*", Deuxièmes Journées d'études sur les Écoulements à Masse Volumique Variable, Orléans, Mai 1992.
[54] LECHZINER M.A. (1992) "*Euroval- A European Initiative in Validating C.F.D. Algorithms for Turbulent Aeronautical Flows*", 5th Biennial Colloquium on Computational Fluid Dynamics, U.M.I.S.T., Manchester, 27-28 May, 1992.
[55] HANINE F.(1992) "*Physique, Modélisation et Simulation des Couches Limites Turbulentes Compressibles*", Thèse de Doctorat de l'Institut National Polytechnique de Toulouse, soutenue le 2 Juillet 1992.
[56] CHASSAING P. (1991) Communication privée.
[57] RIVA R., BINDER G., TARDU S. and FAVRE-MARINET M. "*Visualizations of Hellium Jets in Air*", Turbulence 89: Organized Structures and Turbulence in Fluid Mechanics, Kluwer Academic Publishers, 1990, pp.113-124.

DOMAIN DECOMPOSITION METHODS WITH APPLICATIONS TO FLUID DYNAMICS

Yu. A. Kuznetsov

Institute of Numerical Mathematics
Academy of Sciences of Russia
Leninski prospect 32-a, Moscow 117334, Russia

AMS(MOS) subject classifications. 65F10, 65N30, 65N55

Keywords. overlapping and nonoverlapping domain decomposition, convection-diffusion equation, partitioning into subdomains, multiplicative and additive preconditioners

Abstract

In this presentation, a brief review of domain decomposition methods with emphasis on the applications to solving elliptic problems arising from the Navier-Stokes equations via operator splitting methods is given. The singularly perturbed convection-diffusion equation is chosen as a model problem. We consider both overlapping (multiplicative and additive Schwarz) and nonoverlapping (Neumann-Dirichlet and Neumann-Neumann) domain decomposition methods. Some convergence results for particular cases are presented.

1 Introduction

Domain decomposition methods have become increasingly important for the numerical solution of partial differential equations with many applications in natural science and engineering. Originally, the term "domain decomposition" probably appeared in [15]. Underlying the surge of interest is the need for studying complex mechanical and physical phenomena by coupling various mathematical models. Efficient numerical algorithms are needed for solving large scale algebraic problems arising from finite difference, finite element, and finite volume approximations of partial differential equations on various meshes including composite ones. In addition, numerical software for parallel computers is required.

Actually, we could speak about two topics: domain decomposition methods and domain decomposition methodology. The latter is much wider than the the former. It includes coupling of different mathematical models, mesh construction, and approximation via subdomains as well as block data partitionings to solve algebraic problems on parallel computers.

The goal of this presentation is to give a brief review of domain decomposition methods with emphasis on the application to solving elliptic problems arising from the Navier-Stokes equations via operator splitting methods [4]. The presentation consists of two main parts. In the first part, we formulate the problem under consideration and describe overlapping domain decomposition methods, in particular, multiplicative and additive Schwarz methods. In the second part, we consider several nonoverlapping domain decomposition methods based on subdomain iteration procedures with Neumann-Dirichlet boundary condition and some other substructuring algorithms.

To describe domain decomposition algorithms, we consider as a model differential problem the following singularly perturbed convection-diffusion equation:

$$\begin{aligned} \mathcal{L}_\omega u \equiv \mathcal{L}u + \omega^2 u \equiv -\nu\Delta u + (\vec{b}\circ\nabla)\,u + \omega^2 u &= g_\omega \quad \text{in} \quad \Omega, \\ u &= 0 \quad \text{on} \quad \partial\Omega, \end{aligned} \tag{1}$$

arising from approximations of parabolic problems:

$$\begin{aligned} \frac{\partial u}{\partial t} + \mathcal{L}u &= f && \text{in} \quad (t_0+\Delta t)\times\Omega, \\ u &= 0 && \text{on} \quad (t_0+\Delta t)\times\partial\Omega, \\ u(t_0,x) &= u_0(x) && \text{in} \quad \Omega, \end{aligned} \tag{2}$$

either by fully implicit or by Crank-Nicolson schemes. Here $\vec{b}$ is a given vector-function such that $\operatorname{div}\vec{b} = 0$ and $\omega^2 = 1/\Delta t$, where Δt is a considerably small parameter. The other given functions as well as the boundary $\partial\Omega$ are assumed to be at least piece-wise smooth.

Let Ω_h be a mesh partitioning of Ω. Ω_h consists of elementary cells e_h^i (triangles, rectangles, etc.) which are unified within subdomains (superelements, clusters) $\Omega_h^{(l)}$, $l = 1,\ldots,m$. Thus, the original mesh domain Ω_h is decomposed (partitioned, divided) into overlapping/nonoverlapping subdomains $\Omega_h^{(l)}$.

We denote a mesh problem arising from finite difference, finite volume, or finite element methods (or some combination of these methods) by

$$\begin{aligned} \mathcal{L}_{\omega,h} u^h &= g_\omega^h \quad \text{in} \quad \Omega_h, \\ u^h &= 0 \quad \text{on} \quad \partial\Omega_h. \end{aligned} \tag{3}$$

The algebraic representation of this mesh system sometimes can be more convenient:

$$Au \equiv A_\omega u = g_\omega, \; g_\omega \in R^N. \tag{4}$$

To define a matrix A, we can use the assembling procedure

$$A \equiv A_\omega = \left\{a_\omega^{(i)}\right\}, \tag{5}$$

where $a_\omega^{(i)}$ are stiffness matrices associated with mesh cells e_h^i. For every subdomain $\Omega_h^{(l)}$, we also can define stiffness matrices

$$A_\omega^{(l)} = \left\{a_\omega^{(i)}\right\}, \tag{6}$$

taking into account those cells e_i which belong to $\Omega_h^{(l)}$. If subdomains do not overlap, then

$$A_\omega = \left\{A_\omega^{(l)}\right\}. \tag{7}$$

It is clear that, in this case, with suitable permutation matrices P_k we have

$$A_\omega = \sum_{k=1}^{m} P_k \begin{bmatrix} A_\omega^{(k)} & 0 \\ 0 & 0 \end{bmatrix} P_k^T. \tag{8}$$

All these notations will be used to define various domain decomposition preconditioners.

2 Overlapping Domain Decomposition Methods

To define this set of methods, we assume that every mesh node of Ω_h belongs to the interior of at least one subdomain $\Omega_h^{(k)}$. For every subdomain $\Omega_h^{(k)}$, we divide the set of mesh nodes into two groups: the first collects all nodes from the interior of $\Omega_h^{(k)}$, while the second comprises those of nodes which belong to $\Gamma_h^{(k)} = \partial\Omega_h^{(k)} \setminus \partial\Omega_h$. After that, the matrix $A_\omega^{(k)}$ can be represented in the following 2×2 block form:

$$A_\omega^{(k)} = \begin{bmatrix} A_{11}^{(k)} & A_{1\Gamma}^{(k)} \\ A_{\Gamma 1}^{(k)} & A_{\Gamma\Gamma}^{(k)} \end{bmatrix}. \tag{9}$$

We also introduce the matrices

$$A_k = P_k \begin{bmatrix} A_{11}^{(k)} & 0 & 0 \\ 0 & 0 & 0 \\ 0 & 0 & 0 \end{bmatrix} P_k^T \tag{10}$$

and

$$H_k \equiv A_k^+ = P_k \begin{bmatrix} \left[A_{11}^{(k)}\right]^{-1} & 0 & 0 \\ 0 & 0 & 0 \\ 0 & 0 & 0 \end{bmatrix} P_k^T, \tag{11}$$

which will be used to define the overlapping domain decomposition preconditioners.

2.1 Multiplicative Schwarz methods

The iterative procedure

$$\begin{aligned} \mathcal{L}_{\omega,h} u^h_{l+\frac{k}{m}} &= g^h_\omega && \text{in } \Omega_h^{(k)}, \\ u^h_{l+\frac{k}{m}} &= u^h_{l+\frac{k-1}{m}} && \text{on } \Gamma_h^{(k)}, \\ u^h_{l+\frac{k}{m}} &= 0 && \text{on } \partial\Omega_h, \\ u^h_{l+\frac{k}{m}} &= u^h_{l+\frac{k-1}{m}} && \text{in } \Omega_h \setminus \Omega_h^{(k)}, \\ k &= 1, \ldots, m \end{aligned} \tag{12}$$

is said to be a multiplicative Schwarz method. Using matrix notation, we have

$$u^{l+\frac{k}{m}} = u^{l+\frac{k-1}{m}} - H_k(Au^{l+\frac{k-1}{m}} - g) \tag{13}$$

or, in more compact form,

$$u^{l+1} - u = T_{MS}(u^l - u), \tag{14}$$

where u is the solution vector of system (4),

$$T_{MS} = T_m \cdots T_1,$$

and

$$T_k = E - H_k A.$$

It can be easily shown that the matrices T_k are projectors. The convergence of the multiplicative Schwarz method is proved at least for two important cases, when the matrix A is either symmetric positive definite or an M-matrix. The first case appears from diffusion equations ($\vec{b} \equiv 0$). The second case appears from convection-diffusion equations under special assumptions about the discretization method (for instance, the usage of the upwinding technique).

The multiplicative Schwarz methods were studied by S. Sobolev, S. Mikhlin, I. Babuška, J.-L. Lions, M. Dryja, and O. Widlund. The most complete results were obtained by P.-L. Lions in [8]–[10].

On the basis of the references mentioned above, the following conclusion about the rate of convergence of multiplicative Schwarz methods can be made. Assume that a mesh node $x \in \Omega_h^{(k)}$ and denote the distance between x and $\partial\Omega_h^{(k)}$ by

$$d(x;\partial\Omega_h^{(k)}) = \min_{y\in\partial\Omega_h^{(k)}} |x - y|. \tag{15}$$

If for any mesh node $x \in \Omega_h$ a subdomain $\Omega_h^{(k)}$ exists, such that

$$d(x;\partial\Omega_h^{(k)}) \geq \frac{c_0}{\omega}, \tag{16}$$

where c_0 is a positive constant independent of Ω_h, the estimation

$$\rho(T_{MS}) \leq q \tag{17}$$

holds, where $q < 1$ is a positive constant independent of Ω_h. Here $\rho(T_{MS})$ is the spectral radius of T_{MS}.

In the case when the convergence of the multiplicative Schwarz method cannot be proved (for instance, for mesh systems arising from approximations of the Navier-Stokes or Euler equations), one should couple it with the GMRES-procedure. Remember also that block Gauss-Zeidel methods applied to system (4) are particular cases of the multiplicative Schwarz methods.

2.2 Additive Schwarz Methods

The matrix

$$H_{AS} = \sum_k H_k \tag{18}$$

is said to be an additive Schwarz preconditioner based on the partitioning of Ω_h into overlapping subdomains $\Omega_h^{(k)}$. The convergence of the stationary preconditioned Richardson method with the iteration matrix $T_{AS} = E - H_{AS}A$ with the preconditioner H_{AS} from (18) can be proved only for some special cases.

Assume that every mesh node of Ω_h belongs to the interior of only one subdomain $\Omega_h^{(k)}$, i.e., cannot belong to the interior of two different subdomains. Then this method is convergent if the matrix A is either symmetric positive definite or an M-matrix. In this particular case the method mentioned above is fully equivalent to the block Jacobi method [19].

The additive Schwarz method was introduced by A. Matsokin and S. Nepomnyaschikh. The theory of the additive domain decomposition methods was developed by J. Bramble, X.-C. Cia, M. Dryja, P.-L. Lions, J. Pasciak, O. Widlund, and J .Xu [5], [6], [8]–[12].

2.3 New Overlapping Domain Decomposition Algorithms

Partition the domain Ω_h into nonoverlapping subdomains $\widehat{\Omega}_h^{(k)}$ and extend them to subdomains $\Omega_h^{(k)}$ such that

$$d(\widehat{\Gamma}_k;\Gamma_k) = \min_{\substack{x\in\widehat{\Gamma}_k \\ y\in\Gamma_k}} |x-y| > 0, \tag{19}$$

where $\widehat{\Gamma}_k = \partial\widehat{\Omega}_h^{(k)} \setminus \partial\Omega_h$. Define diagonal matrices $\widehat{Q}_k$ with nonnegative diagonal elements, such that

$$\sum \widehat{Q}_k = E \tag{20}$$

and for any $x\overline{\in}\widehat{\Omega}_h^{(k)}$ the corresponding diagonal element of $\widehat{Q}_k$ is equal to zero. Define the overlapping domain decomposition preconditioner by

$$\widehat{H} = \sum_{k=1}^{m} \widehat{Q}_k H_k. \tag{21}$$

Under the constructions made for the iterative procedure

$$u^{k+1} = u^k - \widehat{H}(Au^k - f), \tag{22}$$

the following convergence result can be proved [17]. Assume that two independent of Ω_h positive constants c_1 and c_2 exist such that the inequality

$$d(\widehat{\Gamma}_k;\Gamma_k) \geq \frac{c_1}{\omega}\ln(c_2\cdot\omega^\alpha) \tag{23}$$

is valid for any $1 \leq k \leq m$ and $\alpha \geq 0$. Then the estimation

$$\left\| E - \widehat{H}A \right\| \leq c_3\frac{1}{\omega^\alpha} \equiv c_3(\Delta t)^\alpha \tag{24}$$

holds with some positive constant c_3, where c_3 is independent of Ω_h and $\|\cdot\|$ is the Euclidian norm. It is clear that similar estimations can also be established for some other important norms.

From inequalities (23) and (24), it immediately follows that a prescribed accuracy ε for the solution vector can be reached even within one iteration step if the overlappings between subdomains are sufficiently large. For instance, to reach accuracy $O(\Delta t^\alpha)$, we have to choose $d(\widehat{\Gamma}_k;\Gamma_k) = O\left(\alpha\sqrt{\Delta t}\times\ln\frac{1}{\Delta t}\right)$ if $\Delta t \ll 1$. Otherwise, we have to use several steps of the iterative procedure, or even accelerate it by GMRES if the value in the right hand side of (24) is not sufficiently small.

We describe two examples of practical applications of the above domain decomposition methods.

Example 1. Splitting into subproblems via interface boundaries.

Denote the interface boundary between subdomains $\widehat{\Omega}_h^{(k)}$ by

$$\widehat{\Gamma}_h = \bigcup_k \widehat{\Gamma}_h^{(k)}, \tag{25}$$

and extend this boundary to a subdomain G_h such that the following inequality

$$d(\widehat{\Gamma}_h; \partial G_h \setminus \partial\Omega_h) \equiv \min_{\substack{x\in\widehat{\Gamma}_h \\ y\in \partial G_h\setminus\partial\Omega_h}} |x-y| \geq \frac{c_4}{\omega}\ln(c_5\omega^\alpha) \tag{26}$$

holds with some positive constants c_4, c_5, and α, which are independent of Ω_h. Then for given $\varepsilon = (\Delta t)^\beta$, $\beta > 0$, the constants c_4, c_5, and the value of α can be chosen for the estimation

$$\left\|Q_\Gamma(H_G g - A^{-1}g)\right\| \leq \varepsilon \tag{27}$$

to be valid. Here H_G is the matrix from (11) related to the subdomain G_h, and Q_Γ is a diagonal matrix with the diagonal elements equal to 1 if they correspond to the mesh nodes belonging to $\widehat{\Gamma}_h$, and zero otherwise. After the calculation of the components of the solution vector for the boundary $\widehat{\Gamma}_h$, the original mesh system will be divided into m independent subproblems with the Dirichlet boundary condition on the interface $\widehat{\Gamma}_h$.

Two other methods to divide the original global mesh system into independent of subsystems via explicit schemes were proposed and developed by C. Dawson, T. Dupont [7], [10], [11], and H. Blum, S. Lisky, and R. Rannacher [2].

Example 2. Coupling of additive and multiplicative domain decomposition algorithms.

Let us partition Ω_h into square-like subdomains $\widehat{\Omega}_h^{(k)}$, $k = 1, \ldots, m$ similar to the black-red partitioning of mesh nodes in the SOR-methods (this partitioning is also known as the chess-like one). We unify the black colour subdomains $\widehat{\Omega}_h^{(1)}, \ldots, \widehat{\Omega}_h^{(s)}$ into the subdomain $G_h^{(1)}$, and the others into the subdomain $G_h^{(2)}$ (see Fig. 1). Then, we embed every subdomain $\widehat{\Omega}_h^{(k)}$ into larger subdomains $\Omega_h^{(k)}$, $k = 1, \ldots, s$ such that the inequalities (23) and estimation (24) with appropriate values of c_1, c_2, c_3, and α are valid (see Fig.2).

The computational procedure consists of two steps. At the first step, we solve subproblems from the subdomains $\Omega_h^{(k)}$, and then restrict the solutions onto the subdomains $\widehat{\Omega}_h^{(k)}$, $k = 1, \ldots, s$. At the second step, we solve the subproblems for the square-like subdomains $\widehat{\Omega}_h^{(k)}$, $k > s$ from $G_h^{(2)}$ with the Dirichlet boundary values having been calculated within the first step. A numerical investigation of the algorithm of this kind was done by G. Meurant [11].

There is one more interesting possibility of application of the overlapping domain decomposition approach described above to the unsteady convection-diffusion problems via schemes with the local time stepping.

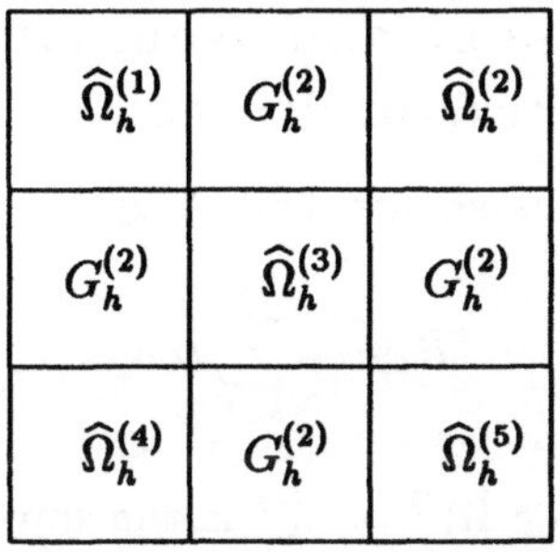

Figure 1: A piece of the chess partitioning

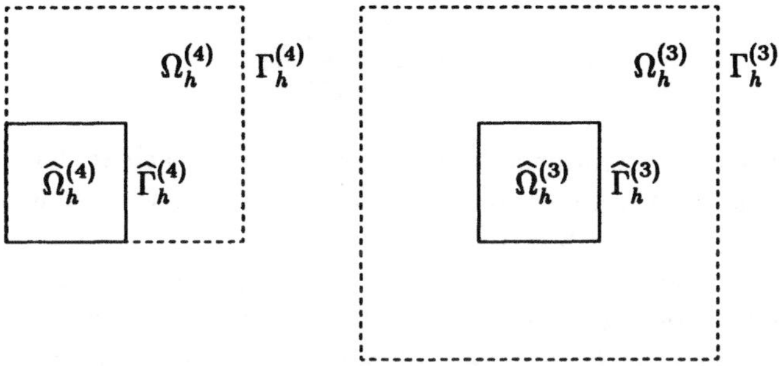

Figure 2: Examples of subdomains $\widehat{\Omega}_h^{(k)}$ and $\Omega_h^{(k)}$

3 Nonoverlapping Domain Decomposition Methods

In this section we assume that $\widehat{\Omega}_h^{(k)} = \Omega_h^{(k)}$, i.e., Ω_h is divided into nonoverlapping subdomains $\widehat{\Omega}_h^{(k)}$. Let us define two sets of matrices. The first one is defined by

$$\widetilde{H}_k = P_k D_k \begin{bmatrix} \left[A_\omega^{(k)}\right]^+ & 0 \\ 0 & 0 \end{bmatrix} D_k P_k^T, \tag{28}$$

where $\left[A_\omega^{(k)}\right]^+$ is the generalized inverse for $A_\omega^{(k)}$ (simple inverse, if $A_\omega^{(k)}$ are nonsingular), and D_k are diagonal matrices such that $\sum_k D_k = E$.

To define the second set of matrices, we have to introduce the Schur complements of $A_\omega^{(k)}$ by

$$S_k = A_{\Gamma\Gamma}^{(k)} - A_{\Gamma 1}^{(k)} \left[A_{11}^{(k)}\right]^{-1} A_{\Gamma 1}^{(k)}. \tag{29}$$

Then we define the required set of matrices by

$$\widetilde{R}_k = P_k^{(s)} D_k^{(s)} S_k^+ D_k^{(s)} \left[P_k^{(s)}\right]^T, \tag{30}$$

where $P_k^{(s)}$ and $D_k^{(s)}$ are the Schur complements of P_k and D_k correspondingly.

3.1 Element-by-Element and Neumann-Neumann Domain Decomposition Preconditioners

Element-by-element preconditioners are defined by

$$H_{EE} = \sum_{k=1}^{m} \widetilde{H}_k, \tag{31}$$

with $\widetilde{H}_k$ from (28). The choice $\widehat{\Omega}_h^{(k)} = e_h^{(k)}$ is the simplest. Numerous applications of element-by-element preconditioners can be found in papers by T. Tezduyar and his co-authors [10]–[12].

The Neumann-Neumann domain decomposition preconditioners were introduced and developed by R. Glowinski, P. Le Tallec, M. Wheller, and others [8]–[12]. These preconditioners can be defined by

$$R_{NN} = \sum_{k=1}^{m} \widetilde{R}_k. \tag{32}$$

Originally, they were introduced in a different way. It is clear that the preconditioner (32) is equivalent to the preconditioner (28) applied to the reduced system with the Schur complement matrix $S = \{S_k\}$.

The theoretical and numerical results concerning preconditioners (28) and (32) are obtained only for the symmetric positive definite case. In any case, the application of these preconditioners to convection-diffusion problems seems to be very attractive.

3.2 Neumann-Dirichlet Domain Decomposition Preconditioners

Divide Ω_h into two nonoverlapping domains $\Omega_h^{(1)}$ and $\Omega_h^{(2)}$ with the common boundary Γ_h, i.e., $\Gamma_h^{(1)} = \Gamma_h^{(2)} = \Gamma_h$ and define the matrix $A_\omega^{(2)}$ using the different ordering of vector components by

$$A_\omega^{(2)} = \begin{bmatrix} A_{\Gamma\Gamma}^{(2)} & A_{\Gamma 2}^{(2)} \\ A_{2\Gamma}^{(2)} & A_{22}^{(2)} \end{bmatrix}. \tag{33}$$

Then the matrix

$$H_{ND} = B_{ND}^{-1}, \tag{34}$$

where

$$B_{ND} = \begin{bmatrix} A_{11}^{(1)} & A_{1\Gamma}^{(1)} & 0 \\ A_{\Gamma 1}^{(1)} & A_{\Gamma\Gamma}^{(1)} + A_{\Gamma 2}^{(2)}\left[A_{22}^{(2)}\right]^{-1} A_{2\Gamma}^{(2)} & A_{\Gamma 2}^{(2)} \\ 0 & A_{2\Gamma}^{(2)} & A_{22}^{(2)} \end{bmatrix} \tag{35}$$

is said to be the Neumann-Dirichlet domain decomposition preconditioner for the matrix A. Preconditioners of this kind, based on the sequential solution of subproblems with alternating Dirichlet and Neumann boundary conditions, were introduced by P. Bjorstad, J. Bramble, M. Dryja, A. Matsokin, J. Pasciak, A. Schatz, O. Widlund, and other numerical analysts [1], [2], [8]–[12], [18]. The numerical technique was developed in [16].

All theoretical and numerical results were obtained from symmetric positive definite matrices. The methods involved can be used to solve Poisson equations for the pressure as well as to construct preconditioners for the Stokes problem. The application of the Neumann-Dirichlet domain decomposition technique to the Stokes problem was considered by J. Bramble, J. Pasciak, and A. Quarteroni [9]–[11].

4 Comparisons and Conclusions

It is almost impossible to give acceptable comparisons of the domain decomposition methods which were described above. These methods are very popular but we have very few information about their computational and communicational properties from numerical experiments and applications.

The only thing that can be said definitely is: the multiplicative domain decomposition methods have better convergent characteristics in comparison with additive methods (like block Gauss-Zeidel methods in comparison with block Jacobi ones). At the same time, the multiplicative methods are not well suited to parallel computers. Moreover, they require the use of more accurate inner solvers (especially for subproblems with Dirichlet boundary conditions), contrary to the corresponding additive domain decomposition algorithms.

We did not even give here a more or less complete overview of domain decomposition algorithms which can be applied, after some modifications, to computational fluid dymnamics. We only looked through those of them which can (probably) be used to solve applied problems on parallel computers. For instance, we did not consider here an interesting approach for the separation of the global problem (3) into several subproblems with nonoverlapping subdomains. It is based on specific explicit schemes on interface boundaries, developed in papers [2], [7]. We also did not consider algorithms with subdomain local time stepping [14].

References

[1] Bjostard, P.E., and Widlund, O.B. *To overlap or not to overlap: A note on a domain decomposition method for elliptic problems.* SIAM J. Sci. Stat. Comput., 10(1989), pp. 1053–1061

[2] Blum, H., Lisky, S., and Rannacher, R. *A domain decomposition algorithm for parabolic problems.* Preprint 92-08, Interdisziplinäres Zentrum für Wissenschaftliches Rechnen, Universität Heidelberg, 1992

[3] Bramble, J.H., Pasciak, J.E., and Schatz, A.H. *An iterative method for elliptic problems on regions partitioned into substructures.* Math. Comp., 46(1986), pp. 361–369

[4] Bristeau, M.O., Glowinski, R., and Periaux, J. *Numerical methods for the Navier-Stokes equations. Applications to the simulation of compressible and incompressibel viscous flows.* Computer Physics Report, North Holland, Amsterdam, 6(1987), pp. 73–187

[5] Cai, X.-C. *Additive Schwarz algorithms for parabolic convection-diffusion equations.* Numer. Math., 60(1991), pp. 41–62

[6] Cai, X.-C., and Widlund, O.B. *Domain decomposition algorithms for indefinite elliptic problems.* SIAM J. Sci. Stat. Comput., 13(1992), pp. 243–258

[7] Dawson, C.N., and Dupont, T.F. *Excplicit/implicit, conservative, Galerkin domain decomposition procedures for parabolic problems.* Report TR90-26, Rice University, Houston, 1990

[8] *Domain Decomposition Methods for Partial Differential Equations.* Proceedings of the 1st Int. Symp., Eds. R.Glowinski, G.H. Golub, G.A. Meurant, and J. Periaux, SIAM,Philadelphia, 1988

[9] *Domain Decomposition Methods for Partial Differential Equations.* Proceedings of the 2nd Int. Symp., Eds. T.F. Chan, R.Glowinski, G.A. Meurant, J. Periaux, and O.B. Widlund, SIAM,Philadelphia, 1989

[10] *Domain Decomposition Methods for Partial Differential Equations.* Proceedings of the 3rd Int. Symp., Eds. T.F. Chan, R.Glowinski, J. Periaux, and O.B. Widlund, SIAM,Philadelphia, 1990

[11] *Domain Decomposition Methods for Partial Differential Equations.* Proceedings of the 4th Int. Symp., Eds. R.Glowinski, Yu.A.Kuznetsov, G.A. Meurant, J. Periaux, and O.B. Widlund, SIAM,Philadelphia, 1991

[12] *Domain Decomposition Methods for Partial Differential Equations.* Proceedings of the 5th Int. Symp., Eds. T.F. Chan, D. Keyes, G.A. Meurant, J. Scroggs, and R.Voigt, SIAM,Philadelphia, 1991

[13] Dryja, M., and Widlund, O.B. *Multilevel additive methods for elliptic finite element problems.* In Parallel Algorithms for Partial Differential Equations, Proceedings of the 6th GAMM-Seminar, Kiev, January 19–21, 1990, W.Hackbusch, ed., Vieweg&Son, Braunschweig, Germany, 1991

[14] Ewing, R.E., Lazarov, R.D., Pasciak, J.E., and Vassilevski, P.S. *Finite element methods for parabolic problems with time steps variable in space.* Tech. Rep. No 1989-05, Inst. Sci. Comp. at Univ. of Wyoming, 1989

[15] Glowinski, R., Periaux, J., and Dihn, Q.V. *Domain decomposition methods for nonlinear problems in fluid dynamics.* INRIA Repports de Recherche, No. 147, Rocquencourt, France, 1982

[16] Kuznetsov, Yu.A. *Multilevel domain decomposition methods.* Appl. Numer. Math., 6(1989/1990), pp. 186–193

[17] Kuznetsov, Yu.A. *Domain Decomposition Methods for Unsteady Convection-Diffusion Problems.* In Computing Methods in Applied Sciences and Engineering, Proceedings of the 9th Int. Symp., Paris, France, January 29–February 2, 1990, SIAM, Philadelphia, 1990, pp. 211–227

[18] Marchuk, G.I., Kuznetsov, Yu.A., and Matsokin, A.M. *Fictitious domain and domain decomposition methods.* Soviet. J. Numer. Anal. and Math. Modelling, 1(1986), pp.3–35

[19] Varga, R.S. *Matrix iterative analysis.* Prentice Hall, Englewood Cliffs, 1962

UNSTRUCTURED MESH ALGORITHMS FOR AERODYNAMIC CALCULATIONS

D. J. Mavriplis

Institute for Computer Applications in Science and Engineering
NASA Langley Research Center
Hampton, Virginia, USA

1. INTRODUCTION

Over the last decade, much attention has been devoted to the development and use of unstructured mesh methodologies within the research community. This enthusiasm however, has not always been shared by the applications and industrial community. The promise of easily enabling the discretization of complex geometries has been counterbalanced by questions of accuracy and efficiency. Furthermore, the dearth of results concerning viscous flow calculations using unstructured meshes has produced skepticism concerning the value of unstructured mesh techniques for practical aerodynamic calculations.

There is no doubt that block-structured techniques have proved extremely effective in discretizing very complex geometries. However, unstructured grid techniques offer additional inherent advantages which may not at first appear evident. The possibility of easily performing adaptive meshing is perhaps the largest advantage of unstructured grid methods. In fact, the implementation of adaptive meshing techniques for structured meshes has generally been found to incur unstructured-mesh type overheads [1]. Furthermore, although unstructured grid data-sets are irregular, they are homogeneous (as opposed to block structured grids where differentiation between block boundaries and interiors must be made). One of the consequences of this property is that unstructured-mesh type solvers are relatively easily parallelizable. While unstructured mesh solvers always incur additional memory and CPU-time overheads due to the random nature of their data-sets, large gains in efficiency can be obtained by careful choices of data-structures, and by resorting to more efficient implicit or multi-level solution procedures. When combined with adaptive meshing and parallelization, these can result in truly competitive solution procedures.

In the following sections, a brief outline of some of the various approaches currently in use in unstructured mesh solution strategies is given, and the various advantages and trade-offs of each method are discussed. This is followed by a set of illustrative example solutions taken from the author's own work, which include two-dimensional viscous flows and three-dimensional inviscid flow solutions on sequential and parallel machine architectures.

2. DISCRETIZATIONS

2.1. Vertex Based and Cell-Centered Schemes

The first choice which arises in the context of unstructured mesh discretizations is the issue of cell-centered versus vertex-based schemes. Unlike the situation for structured grids, where the differences between these two types of schemes consist principally of different boundary condition treatments, the situation for unstructured meshes is quite different. Whereas a hexahedral structured mesh contains the same number of cells as vertices (asymptotically neglecting boundary effects), an unstructured tetrahedral mesh with N vertices contains α N tetrahedral cells, where α is usually between 5 or 6 (there are twice as many triangles as vertices in two dimensions). Thus a cell centered scheme for unstructured meshes requires the solution of 5 to 6 times more unknowns than a vertex based scheme operating on the same grid. Therefore, a cell-centered scheme can be expected to incur substantially higher memory and CPU overheads on a given grid than a vertex scheme.

On the other hand, the solution of a larger number of unknowns would suggest that higher accuracy may be achieved on the same grid using a cell-centered scheme. If one visualizes an unstructured mesh as a simple graph (i.e. a collection of vertices joined together by a set of edges or links), then the vertex scheme is seen to operate on the original graph of the grid, and the cell-centered scheme on a dual graph, i.e. the dual obtained by placing a vertex at the center of each tetrahedron, and associating a link with each triangular face of the tetrahedra, thus joining neighboring cell centers. The original graph thus contains N vertices and $(\alpha + 1)N$ links, whereas the dual graph contains αN vertices and $2\alpha N$ links. In the original graph, the degree of each vertex (number of incident links) is variable, but averages out to $2(\alpha + 1)$. In the dual graph, the degree of each vertex is fixed and equal to 4. By comparison, the degree of each vertex in a hexahedral mesh is 6, for both cell centered and vertex schemes.

Thus, although the vertex scheme contains 5 to 6 times less unknowns on a given grid than the cell-centered scheme, these vertices are more tightly coupled than those of the cell-centered scheme. This in turn suggests that, although there are less vertices, the discretization at each vertex may be more accurate than in the cell-centered scheme.

Practical evidence indicates that on a given grid, for inviscid flows, cell-centered schemes appear to yield somewhat higher accuracy than vertex schemes. The crucial question is thus whether this perceived increase in accuracy is sufficient to overcome the substantial memory and CPU overheads incurred by the cell centered schemes. Unfortunately, few direct comparisons have been made between vertex and cell-centered unstructured schemes, and these have usually been hindered by the use of different discretization schemes and/or different grids. This is an area which should be further investigated in the future. Furthermore, the above discussion illustrates the dangers of comparing vertex and cell-centered unstructured schemes with each other or with structured grid solvers based on the number of unknowns, without regard for the amount of connectivity between the unknowns.

2.2. Central Difference and Upwind Schemes

The same benefits, trade-offs and controversies exist concerning the use of central-difference schemes (with additional artificial dissipation) and upwind schemes on unstructured meshes as in the structured mesh context. The equivalent of a central-difference discretization can be formulated on an unstructured mesh as a Galerkin finite-element discretization where the variables are stored at the vertices of the mesh and the fluxes are assumed to vary piecewise linearly over the cells of the mesh [2,3]. This results in a nearest neighbor stencil which is non-dissipative. Additional dissipative terms are thus constructed as a blend of a Laplacian and a biharmonic operator, which correspond to the second and fourth differences employed in the structured mesh context for damping out oscillations in the vicinity of shocks, and in smooth regions of the flow, respectively. These schemes are simple to construct and relatively inexpensive to compute. Furthermore, they are easily linearizable for use with implicit schemes [4]. The explicit control over the amount of dissipation in the scheme can be viewed either as an advantage (additional control) or as a disadvantage (additional input parameters). Additional improvements to central difference schemes are possible, such as the use of matrix valued dissipation [5], which attempts to scale the dissipative terms among the various wave components of the governing equations.

Upwind schemes are more complex and expensive than simple Galerkin finite-element schemes, but offer the possibility of capturing shocks with higher resolution. The amount of dissipation is automatically determined by the scheme and split appropriately between the various wave components of the governing equations. The most successful upwind schemes for unstructured meshes have been those based on flux differencing [6,7,8]. The introduction of multi-dimensional reconstruction for the extension to second-order schemes [8] has put these methods on a more solid theoretical foundation. On the other hand, one-dimensional Rieman solvers are still required, although much research is currently devoted to developing truly multi-dimensional upwind schemes [9]. The use of limiters with such schemes, which is required in the presence of shock waves, has often been found to inhibit convergence to steady-state.

Higher order schemes have also been investigated by a number of researchers (see for example [10,11]). Such schemes offer the possibility of resolving complex flows in a more efficient manner using a more accurate (and expensive) representation of the data on a smaller number of mesh points. Although few if any such schemes are routinely used today, they will probably play an important role in the future for the solution of high Reynolds number viscous flows which presently require the use of tens of millions of grid points. It is interesting to note that in the structures field, unstructured higher order discretizations are the method of choice.

In the context of unstructured meshes, the increased cost of upwind or higher order methods must be weighed against the cost of retaining an inexpensive discretization and making use of adaptive meshing techniques, which constitute one of the main advantages of unstructured meshes. The combination of h-refinement (adaptive meshing) and P refinement (higher-order methods) should also be further pursued since this has been shown to enable exponential convergence [12].

3. SOLUTION TECHNIQUES

Once the governing equations have been discretized in space, they form a large set of coupled ordinary differential equations which must be integrated in time. The main interest in this paper relates to the solution of steady-state problems. In this case, time accuracy of the integration may be sacrificed in the interest of accelerating the convergence to steady-state. This may include the use of a low-accuracy time integration scheme, the use of large time steps, and lumping of the mass matrix for finite-element schemes. Furthermore, many of the convergence acceleration techniques developed for structured meshes carry over in a straight forward manner to unstructured meshes. These include the use of local time stepping, enthalpy damping for inviscid flows, and implicit residual averaging [13,14] (which must be implemented using a Jacobi iteration rather than a tridiagonal solver).

However, for large problems, the use of simple explicit schemes inevitably results in very slow convergence rates. Many of the solution algorithms employed for structured grids exploit the structure of the grid (e.g ADI schemes) or the limited bandwidth of the resulting Jacobian matrix, and thus are not applicable to unstructured meshes. The lack of efficient steady-state solution algorithms for unstructured mesh problems has been one of the main impedements towards greater use of unstructured mesh strategies. For large stiff problems, implicit methods based on sparse matrix technology or multi-level approaches modified for use on unstructured data-sets are required.

The ultimate implicit method is the direct solver, or Newton's method. If an exact linearization of the Jacobian is employed, and the resulting matrix is inverted at each time-step using sparse matrix techniques, quadratic convergence can be obtained, resulting in convergence to machine zero in under ten iterations. Direct methods have been demonstrated for both structured grids and unstructured grids [15,16,17]. Although these are among the most robust methods available, their operation count and storage requirements for unstructured meshes are non-optimal and are thus seldom employed in practice.

For a second-order method, the exact linearization results in a stencil which includes nearest neighbors as well as second to nearest neighbors. Thus, the storage requirements for the resulting sparse matrix become excessive, particularly in three dimensions. By employing a linearization of the corresponding first-order scheme, a nearest neighbor stencil is obtained, and the memory requirements for storing the corresponding sparse matrix are reduced substantially. However, this mismatch between the implicit and explicit operators ensures that quadratic convergence rates cannot be achieved. Furthermore, the exact inversion of the implicit matrix at each time-step is no longer necessary, since the implicit matrix itself is an inaccurate representation of the explicit operator. Thus, the use of iterative implicit methods, in which the implicit system of equations is only approximately solved at each time-step is more appropriate.

A large variety of iterative implicit methods have been developed. These may consist of a single iteration scheme, or a preconditioning operation followed for example by a GMRES (generalized minimum residual) technique. These implicit methods may be divided into methods which operate pointwise (such as Jacobi and Gauss-Siedel methods), and those which require storing the entire implicit matrix (such as LU factorization schemes).

Jacobi and Gauss-Siedel approaches have been employed successfully both as iteration schemes and as preconditioners for GMRES [4,7,18,19,20]. Improved efficiency over explicit schemes with minimal memory overheads have been demonstrated for a variety of problems. However, for large problems and very stiff equation sets, such as those encountered in the solution of high Reynolds number viscous flows, the local nature of these methods results in a degradation of the convergence rate.

Methods which operate on the entire implicit matrix such as LU factorization are of a more global nature; such methods promote the rapid transmittal of information across the entire domain. As such, these methods are more robust and their convergence degrades less significantly for very large and stiff problems. An incomplete LU factorization employed as a preconditioner followed by a GMRES technique has been found to yield one of the most competitive solution strategies for high-Reynolds number viscous two-dimensional flows [4]. However, such methods require the storage of the entire implicit matrix. While this is feasible in two-dimensions, for three dimensional calculations this matrix alone requires of the order of 300 N words of storage for a vertex scheme, where N represents the number of mesh vertices.

An interesting alternative approach which is not entirely local, but which obviates the need to store the implicit Jacobian matrix is based on the use of additional data-structures called linelets or snakes [21,22]. By joining series of neighboring points together in the unstructured mesh, a set of lines can be identified and employed to mimick the alternating direction implicit (ADI) type algorithms commonly employed on structured meshes. Such methods may be viewed as a compromise between the low storage requirements of point-wise methods and the global nature of LU factorization methods, and thus their performance can be expected to fall somewhere in this region. However, for problems where the stiffness is strongly directional, such as for high Reynolds number boundary layers, this approach may be capable of resolving much of the stiffness.

An entirely different approach involves the use of multi-level or multigrid strategies. These are hierarchical methods which make use of a sequence of coarser grids to accelerate the solution on a fine grid. The advantages of time stepping on coarse meshes are twofold: first, the permissible time-step is much larger, since it is proportional to the cell size, and secondly, the work is much less because of the smaller number of grid points. Generally, a simple explicit scheme is employed on each grid of the sequence. The process begins by performing a time-step on the finest grid of the sequence, and then interpolating the flow variables and residuals up to the next coarser grid of the sequence. On this grid, a correction equation is formulated, which consists of the governing flow equations augmented by a forcing function which represents the fine grid solution. This correction equation is time-stepped and the resulting flow variables and residuals are interpolated up to the next coarser grid, where the process is repeated recursively until the coarsest grid of the sequence is reached. The computed corrections are then recursively interpolated back down to the finest grid where they are employed to update the solution. This entire procedure constitutes one multigrid cycle. These cycles are repeated until convergence is obtained.

The use of a multigrid method with unstructured meshes presents an additional challenge. Consistent coarse tetrahedral grids can no longer be formed by simply grouping together neighboring sets of tetrahedra. An alternative would be to generate the fine mesh by repeatedly subdividing an initial coarse mesh in some manner. However, generally poor topological control of the fine mesh results from such a procedure. A strategy which has proven successful involves the use of independent non-nested coarse and fine grids. This approach provides great flexibility in determining the configuration of the coarsest and finest meshes. Coarse meshes may be designed to optimize the speed of convergence, whereas fine meshes may be constructed based on solution accuracy considerations. In general, beginning from a fine grid, a coarser level is constructed which contains roughly half the resolution in each coordinate direction throughout the domain (about 1/8 the number of vertices in three dimensions, or 1/4 in two dimensions). This process is repeated until the coarsest grid capable of representing the geometry topology is obtained. In the context of adaptive meshing, new finer meshes may be added to the multigrid sequence, using any given adaptive refinement technique, since no relation is assumed between the various meshes of the sequence. The key to the success of such a strategy lies in the ability to efficiently transfer variables, residuals and corrections back and forth between unrelated unstructured meshes. This may be performed using linear interpolation. For each vertex of a given grid, the tetrahedron which contains this vertex on the grid to which variables are to be interpolated is determined. The variable at this node is then linearly distributed to the four vertices of the enclosing tetrahedron (three vertices of the enclosing triangle in two dimensions). The main difficulty lies in efficiently determining the enclosing cell for each grid point. A naive search over all cells would lead to an $O(N^2)$ complexity algorithm, where N is the total number of grid points, and would be more expensive than the flow solution itself. Thus, an efficient search strategy such as a graph-traversal algorithm or a quad-tree approach is required.

This particular unstructured multigrid approach has been shown to be very effective [3,23,24,25]. Near grid independent convergence rates can be obtained while incurring minimal memory overheads. However, the need to manually generate a complete sequence of grids is viewed as tedious in a production environment, and several efforts at automating this process have been developed. On approach agglomerates or fuses together neighboring cells to form coarse super-cells which are generally not tetrahedral but polyhedral [26,27]. A second approach constructs triangular or tetrahedral grids by filtering a portion of the fine grid points and retriangulating the remaining points [28,29]. All of these approaches involve various tradeoffs. However, they all make use of extra geometrical constructs (i.e. coarse grids) to solve what is essentially an algebraic matrix inversion problem. This may be viewed as an inconvenience. However, unstructured multigrid methods are probably the most efficient solution methods available presently in terms of CPU and memory overhead for steady-state solutions.

4. GRID GENERATION AND ADAPTIVITY

Although one of the main motivations for the use of unstructured meshes has been the added flexibility they offer for dealing with complex geometries, grid generation remains a pacing item for unstructured mesh computations, especially in three dimensions. To be sure, part of the problem is associated with the lack of standard and flexible geometry definition standards and interfaces to current CAD systems employed in the industrial design process. However, much difficulty still rests with the grid generation algorithms themselves.

Unstructured mesh generation algorithms have traditionally been divided into advancing front type methods, and Delaunay triangulation methods, although this classification is somewhat arbitrary. In fact, these two approaches are not mutually exclusive. The advancing front algorithm begins with a surface mesh on the geometry which it then marches out into the flow field by adding points ahead of the front and joining them up to form tetrahedra with existing front faces, until the entire region has been discretized [30,31]. This algorithm essentially represents a point placement strategy. The reconnection strategy employed to form tetrahedral elements is somewhat arbitrary and resorts to checking the validity of each proposed tetrahedral cell by examining possible intersections with neighboring cells. The advantages of such a method are that it guarantees the integrity of the boundaries. This is evident since the geometry surface-grid constitutes the original front, and the method is of the "greedy type"; i.e. it never undoes what has already been constructed. The resulting placement of the mesh points is generally very satisfactory and very smooth element variations can be ensured. On the other hand, robustness is not guaranteed, due to the somewhat heuristic nature of the reconnection strategy. Sophisticated dynamically varying data-structures are required to accelerate the spatial searching routines employed in such an approach.

A Delaunay triangulation represents one of the most fundamental data-structures in computational geometry [32]. Given a set of points in a three dimensional volume (or in a two dimensional plane), the Delaunay triangulation of these points constitutes a set of non-overlapping tetrahedra (or triangles in 2-D), the union of which define the convex hull of the points, and for which a number of properties can be proved. Various algorithms exist for constructing the Delaunay triangulation [33]. A commonly employed algorithm in the mesh generation context, known as Bowyer's [34] or Watson's [35] algorithm, is based on the empty circumsphere criterion. This property states that the circumsphere of any tetrahedron cannot contain any other vertices of the mesh. Thus, if an initial triangulation is assumed to exist, new mesh points can be introduced one at a time and triangulated into the mesh by first locating all tetrahedra whose circumsphere is intersected by the newly introduced point, removing all such elements, and forming new tetrahedra by joining the new point up to the faces of the cavity which was created by the removal of the intersected elements.

Since proofs exist for the validity of Delaunay triangulation algorithms, robustness can potentially be built into a mesh generator by making use of such algorithms. However, the Delaunay triangulation is only valid within the convex hull of its defining points. Thus for geometries other than the convex hull, such as a body inside a flowfield, the integrity of the geometry cannot be guaranteed. One approach to this problem is to triangulate the entire set of mesh points, and then to attempt to reconstruct a prescribed surface grid on the geometry by swapping edges and faces of the mesh [36]. Another approach ensures that the placement of points in the mesh is such that the geometry integrity is observed [37].

Since a Delaunay triangulation merely describes a connectivity pattern for a set of points, arbitrary point placement strategies can be employed. In general, the point placement strategies employed are somewhat heuristic. Point sets may be predetermined by sets of overlapping structured grids, or generated incrementally by subdividing elements deemed to be too large [38]. These strategies have generally resulted in less than optimal point distributions, and the resulting meshes are generally less smooth than those obtained with the advancing front method.

On the other hand, Delaunay triangulation mesh generation strategies based on Bowyer's algorithm have proved to be extremely efficient and rather simple to implement. They obviate the need for the complex data-structures required in the advancing front technique, and do not perform tedious intersection checking. Perhaps an additional reason for such efficiency is the fact that Bowyer type algorithms generate the mesh one point at a time, whereas advancing front type algorithms proceed one tetrahedron at a time, and a typical unstructured mesh contains 5 to 6 times more tetrahedra than vertices.

As mentioned previously, adaptive meshing represents one of the principal advantages of the use of unstructured meshes. Similarly to the original mesh generation process, mesh adaptation requires the introduction (or removal) of new points at appropriate locations, and the reconnection of these points to neighboring points of the mesh. The simplest implementation of adaptive mesh refinement is to subdivide existing tetrahedra into smaller cells by introducing new points midway along the tetrahedra forming edges, and reconnecting these points according to a predetermined set of rules. This strategy is very efficient and simple to implement but can lead to distorted cells and vertices of very high degree after numerous refinement levels.

Bowyer's algorithm for Delaunay triangulation is ideally suited for adaptive meshing. Assuming the flow has been solved on a mesh which constitutes a Delaunay triangulation, the solution can be examined to determine regions of high discretization errors or large flow gradients where additional grid points are required. Each new grid point is then inserted into the mesh and locally retriangulated using Bowyer's algorithm, thus resulting in the Delaunay triangulation of the new augmented point set.

The advancing front technique makes use of a field function for determining the desired size of the mesh elements throughout the flow field. A simple adaptive remeshing strategy consists of replacing this initial field function by a function derived from the computed flow solution, and completely regenerating the entire mesh. This approach contains more flexibility for generating the new adaptive mesh, but is more expensive since a non-local mesh restructuring is performed, and may be impractical for transient type problems where many remeshings are required. Alternatively, individual regions of the mesh can be cut out, thereby defining new fronts to be advanced, and local mesh patches regenerated. (For an implementation of this procedure for the removal of distorted elements see [39]).

The main issue which needs to be addressed for all mesh generation and adaptation strategies is that of robustness. This can only be achieved through less heuristic and more theoretically sound approaches to the problem. New developments in computational geometry should enable the formulation of fool-proof algorithms. For example, the existence and construction methods for a constrained Delaunay triangulation in an arbitrary non-convex two-dimensional domain are now well known [40]. A constrained Delaunay triangulation is one which contains certain predefined edges in the final triangulation. Thus, the triangulation of a given set of points in an arbitrary two-dimensional domain with initial geometry boundaries is a relatively easily solved problem. What is required is the extension of such proofs to three dimensions, as well as solutions to various optimization problems, such as: what is the optimal distribution of points ? the optimal triangulation of these points (it need not necessarily be a Delaunay triangulation)? how does such a triangulation interact with the solver? and how can one construct such a triangulation in an efficient and rigorous manner?

These issues become even more important for the solution of high-Reynolds number viscous flows. In three dimensions, few if any practical high-Reynolds number viscous flow solutions have been demonstrated. In two-dimensions, turbulent viscous flow solution capabilities have only emerged in the last several years. Aside from the turbulence modeling issues, which will be described in the results section of this paper, the main difficulty for solving such flows on unstructured meshes relates to the requirement of generating very highly stretched triangular cells (or tetrahedral elements in three dimensions) which are required in order to efficiently resolve the thin shear layers which occur in such flows. This represents a somewhat non-standard application of unstructured meshes, since highly stretched triangular elements have traditionally been considered detrimental to numerical accuracy, and as such have been avoided. However, it has been shown that, while triangular elements with one large angle (~ 180 degrees) are detrimental, elements with small angles, such as a highly elongated right angle triangle are acceptable [41]. This interplay between numerical behavior and optimal triangulation (for a given numerical method) is important for accurately and efficiently resolving complex flows, and should increasingly result in a tighter coupling between the grid generation and flow solution processes.

The current method employed by the author for generating highly stretched two-dimensional triangulations consists of constructing a Delaunay triangulation of the set of grid points in a mapped space rather than in physical space [42,43]. The set of grid points is determined by generating a stretched structured grid over each individual component of the geometry and considering the union of the points defined by these overlapping grids. The locally mapped space is defined by the amount of local grid stretching desired, which in turn is dictated by the aspect ratios of the underlying structured grid cells. In this mapped space, the mesh points appear locally isotropic, and a regular Delaunay triangulation is constructed. The projection of this triangulation back into the physical space produces the desired stretching. Although this method does not guarantee the formation of non-obtuse triangles, the combination of the particular point distribution and reconnection strategy tends to produce nearly right-angle triangles in regions of high stretching.

Improvements to this strategy can be sought by drawing on more rigorous computational geometry algorithms which provide bounds on the angles of the generated triangles, or combined with point placement optimization techniques. These techniques will become necessary for the generation of suitable stretched meshes in three dimensions for viscous flow calculations.

Another alternative to the generation of highly stretched triangular or tetrahedral meshes for the resolution of viscous flows is the use of hybrid meshes, where a thin layer of quadrilaterals in two-dimensions, or prizms in three-dimensions [44,45] are employed in the boundary layer regions. These methods by-pass the difficulties associated with the generation and solution of flows on highly stretched triangular and tetrahedral elements. On the other hand, the resulting grid becomes structured in one of the directions, and part of the generality of the unstructured approach is lost. Such trade-offs must of course be weighed in terms of the complexity of the geometry.

5. TWO-DIMENSIONAL RESULTS

5.1. An Inviscid Case

In order to demonstrate the potential effectiveness of an unstructured mesh strategy, the solution of a steady-state inviscid internal flow, which incorporates adaptive meshing in conjunction with an unstructured multigrid algorithm is demonstrated. The basic discretization employed consists of a Galerkin finite-element approach with added artificial dissipation. The unstructured multigrid scheme makes use of a sequence of pre-generated unrelated coarse meshes, and mesh adaptation is achieved by introducing new points in regions of high density gradients and restructuring the mesh locally using Bowyer's algorithm.

The geometry consists of a two-dimensional turbine blade cascade which has been the subject of an experimental and computational investigation at the occasion of a VKI lecture series [46]. A total of seven meshes were used in the multigrid algorithm, with the last three meshes generated adaptively. The coarsest mesh of the sequence contains only 51 points, while the finest mesh, depicted in Figure 1, contains 9362 points. Extensive mesh refinement can be seen to occur in the neighborhood of shocks, and in other regions of high gradients. The inlet flow incidence is 30 degrees, and the average inlet Mach number is 0.27. The flow is turned 96 degrees by the blades, and the average exit isentropic Mach number is 1.3. At these conditions, the flow becomes supersonic as it passes through the cascade, and a complex oblique shock wave pattern is formed. These are evident from the computed Mach contours depicted in Figure 2. All shocks are well resolved, including some of the weaker reflected shocks, which non-adapted mesh computations often have difficulty resolving. Details of the flow in the rounded trailing edge region of the blade, where the flow separates inviscidly and forms a small recirculation region, are also well reproduced. Once the first four globally generated meshes were constructed, the entire flow solution - adaptive mesh enrichment cycle was performed three times, executing 25 multigrid cycles at each stage. This entire operation required 40 CPU seconds on a single processor of a Cray-YMP supercomputer. The residuals on the finest mesh were reduced by two and a half orders of magnitude, which should be adequate for engineering calculations.

5.2. Viscous Flows

The main difficulties involved in computing high-Reynolds-number viscous flows relate to the grid generation and turbulence modeling requirements. Since the grid generation issues have been previously discussed, the turbulence modeling issues will be briefly addressed in this section.

The most common turbulence models employed for aerodynamic flows are of the algebraic type. Such models typically require information concerning the distance of each point from the wall. Turbulence length scales are determined by scanning appropriate flow variables along specified streamwise stations. In the context of unstructured meshes, such information is not readily available and hence, the implementation of algebraic turbulence models on such meshes introduces additional complexities. A particular approach adopted by the author [47] consists of generating a set of background turbulence mesh stations. These are constructed by generating a hyperbolic structured mesh about each geometry component, based on the boundary-point distribution of the original unstructured mesh, and extracting the normal lines of the mesh. When performing adaptive meshing, new turbulence mesh stations must be constructed for each new adaptively generated boundary point, as illustrated in Figure 3. Each time the turbulence model is executed, the flow variables are interpolated onto the normal turbulence stations, the turbulence model is executed on each station, and the resulting eddy viscosity is interpolated back to the unstructured mesh. The method employed for interpolating variables back and forth between the unstructured mesh and the turbulence mesh stations is similar to that previously described for the unstructured multigrid algorithm.

Figures 4 through 7 illustrate a calculation which makes use of these various techniques to compute a complicated two-dimensional viscous flow over a high-lift multi-element airfoil. The final mesh employed is depicted in Figure 4, and contains a total of 48,691 points. This mesh was obtained using the stretched Delaunay triangulation technique previously described, followed by two levels of adaptive refinement. The height of the smallest cells at the wall is of the order of $2 x 10^{-5}$ chords and cell aspect ratios up to 500:1 are observed. The computed Mach number contours for this case are depicted in Figure 5. The freestream Mach number is 0.1995, the chord Reynolds number is 1.187 million, and the corrected incidence is 16.02 degrees. At these conditions, the flow remains entirely subcritical. Compressibility effects are nevertheless important due to the large suction peaks generated about each airfoil. For example, in the suction peak on the upper surface of the leading-edge slat, the local Mach number achieves a value of 0.77. The computed surface pressure coefficients are compared with experimental wind tunnel data [48] in Figure 6, and good overall agreement is observed, including the prediction of the height of the suction peaks. This case provides a good illustration of the importance of adaptive meshing in practical aerodynamic calculations. Adequate resolution of the strong suction peak on the upper surface of the slat can only be achieved with a very fine mesh resolution in this region. Failure to adequately capture this large suction peak results in the generation of numerical entropy which is then convected downstream, thus contaminating the solution in the downstream regions, and degenerating the global accuracy of the solution. Because these suction peaks are very localized, they are efficiently resolved with adaptive techniques. In order to obtain a similar resolution using global mesh refinement, of the order of 200,000 mesh points would be required, greatly increasing the cost of the computation. The convergence history for this case, as measured by the density residuals and the total lift coefficient versus the number of multigrid cycles, is depicted in Figure 7. A total of 400 multigrid cycles were executed, which required roughly 35 minutes of single processor CRAY-YMP time, and 14 Mwords of memory.

The discrepancy between the computed and experimental pressure coefficients on the trailing edge flap is due to a separated flow condition which is not reproduced by the algebraic turbulence model. These results strongly indicate the need for more sophisticated turbulence modeling. The use of single or multiple field-equation models appears to be the most appropriate choice for turbulent unstructured mesh computations. Such models can be discretized in a straight-forward manner on unstructured meshes. However, the task is now to ensure that such models adequately represent the flow physics, and that they can be solved in an efficient and robust manner. The implementation of a standard high-Reynolds-number $k-\varepsilon$ turbulence model with low-Reynolds-number modifications proposed by Speziale, Abid and Anderson [49], is demonstrated in the next example. The main effort was focused on devising a technique for efficiently solving the two turbulence equations in the context of the unstructured multigrid strategy [50]. The four flow equations and the two turbulence equations are solved as a loosely coupled system. The flow equations are solved explicitly, and the turbulence equations point-implicitly, using a time-step limit which ensures stability and positivity of k and ε. In the context of the unstructured multigrid algorithm, the turbulence eddy viscosity is assumed constant on all but the finest grid level where it is recomputed at each time-step. The transonic flow over a two-element airfoil configuration has been computed using this implementation of the model. For this case, the freestream Mach number is 0.5, the incidence is 7.5 degrees, and the Reynolds number is 4.5 million. Figures 8 and 9 depict the mesh and the solution obtained with the $k-\varepsilon$ turbulence model. Four meshes were employed in the multigrid sequence, with the finest mesh containing a total of 28,871 points. The convergence rates of the various equations for this case are plotted in Figure 10. As can be seen, the turbulence equations and flow equations converge at approximately the same rates. All flow variables and turbulence quantities are initialized with freestream values, and convergence to steady-state is achieved in several hundred multigrid cycles. The computed flow field exhibits regions of transonic flow with a small region of separated flow at the foot of the shock. These features are well reproduced by the turbulence model. Future efforts should concentrate on computationally predicting flows with large regions of separation, such as that inferred by Figure 6, and on developing models which better represent the flow physics.

6. THREE DIMENSIONAL RESULTS

Due to the limitations of present day supercomputers, and the difficulties associated with generating highly stretched tetrahedral meshes, three-dimensional computations have generally been confined to inviscid flows. Most of the techniques described in the context of two-dimensional inviscid flows extend readily to three dimensions. In particular, the unstructured multigrid algorithm and the adaptive meshing strategy have been found to be particular effective for three-dimensional computations [23]. As an example, an adaptive multigrid calculation of transonic flow about an ONERA M6 wing is illustrated in Figures 11 through 13. The final mesh, depicted in Figure 11, contains a total of 174,412 points and just over 1 million tetrahedral volumes. This represents the fourth mesh in the multigrid sequence and the second adaptive refinement level. Mesh refinement was based on the undivided gradient of density. The freestream Mach number and incidence for this case are 0.84 and 3.06 degrees respectively. The well known double shock pattern for this case is reproduced in the computed Mach contours of the solution in Figure 12. The leading edge expansion and shocks are well resolved due to the extensive mesh refinement in these regions. A globally refined mesh of this resolution would result in roughly 600,000 points and would thus require 3 to 4 times more computational resources. The multigrid convergence rate for this case is depicted in Figure 13, where 50 cycles were performed on the original grid, prior to adaptation, 50 cycles on the first adapted mesh, and 100 cycles on the finest adapted mesh. On this final mesh, the residuals were reduced by 5 orders of magnitude over 100 cycles, requiring a total of 35 CRAY-YMP single CPU minutes and 22 MW of memory.

6.1. Parallel Computing Results

As mentioned previously, due to their homogeneous (although random) nature, unstructured mesh data-sets are particularly well suited for parallel processing. An unstructured mesh solver typically consists of a single (indirect addressed) loop over all interior mesh edges, and another similar loop over all boundary elements. On a vector machine, each loop may be split into groups (colors) such that within each group, no recurrences occur. Each group can then be vectorized. A simple parallelization strategy for a shared memory machine is to further split each group into n subgroups, where n is the number of available processors. Each subgroup can then be vectorized and run in parallel on its associated processor. Because the original number of groups is not large (usually 20 to 30), the vector lengths within each subgroup are still long enough to obtain the full vector speedup of the machine, for a moderate number of processors. For more massively parallel distributed-memory scalar machines, the entire mesh must be subdivided and each resulting partition associated with a single processor. On each processor, the single scalar interior and boundary loops are then executed, with inter-processor communication occurring at the beginning and end of each loop. The mesh partitioning strategy must ensure good load balancing on all processors while minimizing the amount of inter-processor communication required.

6.2. CRAY-YMP-8 Results

Figure 14 illustrates an unstructured mesh generated over a three-dimensional aircraft configuration. This mesh contains a total of 106,064 points and 575,986 tetrahedra. This represents the second finest mesh employed in the multigrid sequence. The finest mesh, which is not shown due to printing resolution limitations, contains a total of 804,056 points and approximately 4.5 million tetrahedra. This is believed to be the largest unstructured grid problem attempted to date. The inviscid flow was solved on this mesh using all eight processors running in parallel on the CRAY-YMP supercomputer. A total of 4 meshes were used in the multigrid sequence. The convergence rate for this case is depicted in Figure 16. In 100 multigrid cycles, the residuals were reduced by almost 6 orders of magnitude. This run required a total of 16 minutes wall clock time running in dedicated mode on the 8 processor CRAY-YMP, including the time to read in all the grid files, write out the solution, and monitor the convergence by summing and printing out the average residual throughout the flow field at each multigrid cycle. The total memory requirements for this job were 94 million words. The ratio of CPU time to wall clock time was 7.7 on 8 processors, and the average speed of calculation was 750 Mflops, as measured by the CRAY hardware performance monitor [51]. For this case, the freestream Mach number is 0.768 and the incidence is 1.116 degrees. The computed Mach contours are shown in Figure 15, where good resolution of the shock on the wing is observed, due to the large number of mesh points employed.

6.3. Intel Touchstone Delta Results

The implementation of the unstructured multigrid Euler solver on the Intel Touchstone Delta distributed memory scalar multiprocessor machine, has been pursued using a set of software primitives designed to ease the porting of scientific codes to parallel machines [52]. The present implementation was undertaken as part of a more general project aimed at designing and constructing such primitives with experience gained from various implementations. The net effect of the use of such primitives is to relieve the programmer of most of the low level machine dependent software programming tasks. The mesh was partitioned using a spectral partioning algorithm which had previously been shown to produce good load balancing and minimize inter-processor communication [53]. The flow over the aircraft configuration previously described using the 804,056 vertex mesh was recomputed on the Intel Touchstone Delta machine using both an explicit single grid unstructured euler solver, and the unstructured multigrid euler solver. The single grid solver achieved a computational rate of 1.5 gigaflops on 512 processors, whereas the multigrid solver, using a V-cycle strategy achieved a rate of 1.2 gigaflops on the same number of processors. This represents a computational efficiency of 50% to 60%. These numbers are based on a single processor speed of approximately 5 Mflops, which corresponds to the computational rate achieved for a series of small meshes which were run on a single processor. The computational efficiencies are seen to vary with the particular solution strategy employed, and were also observed to vary with the size of the mesh. On the other hand, the CRAY-YMP-8 results were found to be relatively insensitive to the solution algorithm or the problem size. This is presumably due to the large bandwidth and shared-memory architecture of the machine. However, on the Intel Touchstone Delta, the multigrid strategy is still the method of choice, since in spite of its slightly lower computational efficiency, the numerical efficiency (convergence rate) achieved by this approach is approximately an order of magnitude greater than that of the simple single-grid explicit scheme. For this case, convergence to steady-state could be achieved in approximately 10 minutes of wall clock time using 512 processors.

7. CONCLUSION

This paper has illustrated the application of unstructured mesh techniques to various types of aerodynamic flows, and emphasized the advantages which can be obtained for complex geometries using adaptive meshing and parallelization. In two dimensions, a viscous flow solution capability has been demonstrated, while in three dimensions, efficient Euler solutions are possible. The main problems associated with three-dimensional viscous solutions are related to the development of reliable grid generation strategies, particularly with regards to the generation of highly stretched tetrahedral elements for capturing thin viscous layers. Turbulence modeling is also a limiting factor, although this difficulty is not particular to the field of unstructured meshes. Future work should also concentrate on more complete parallelization of the entire solution process, including items such as grid generation, partitioning, and adaptive meshing.

ACKNOWLEDGEMENTS

Many of the results shown here have been made possible due to the generous assistance of a large number of people. Among those, Daryl Bonhaus is acknowledged for his work in helping validate the 2-D viscous flow solver. The help of Shahyar Pirzadeh and Clyde Gumbert in providing assistance with the generation of three-dimensional unstructured grids using the VGRID program is acknowledged. Rob Vermeland and CRAY Research Inc. are thanked for providing dedicated time on the CRAY-YMP-8 machine, and the National Aerodynamic Simulation Facility (NAS), for providing the computational facilities which have made most of this work possible. Finally, Raja Das, Joel Saltz and Ravi Ponnusamy are acknowledged for actually having done most of the work on the distributed memory parallel implementation.

REFERENCES

1. Dannenhoffer, J. F., "Grid Adaptation for Complex Two-Dimensional Transonic Flows", *Sc.D Thesis, Department of Aeronautics and Astronautics, Massachusetts Institute of Technology,* August 1987.

2. Jameson, A., Baker, T. J., and Weatherill, N. P., "Calculation of Inviscid Transonic Flow over a Complete Aircraft", *AIAA paper 86-0103,* January, 1986.

3. Mavriplis, D. J., "Turbulent Flow Calculations Using Unstructured and Adaptive Meshes", *Int. J. Numer. Methods Fluids, Vol. 13, No. 9,* November 1991, pp. 1131-1152

4. Venkatakrishnan, V., and Mavriplis, D., "Implicit Solvers for Unstructured Meshes", *AIAA paper 91-1537CP, Proceedings of the 10th AIAA Computational Fluid Dynamics Conference,* Honolulu, Hawaii, June 1991

5. Turkel, E., and Vatsa, V. N., "Effect of Artificial Viscosity on Three Dimensional Flow Solutions", *AIAA paper 90-1444* June 1990.

6. Stoufflet, B., Periaux, J., Fezoui, F., and Dervieux, A., "Numerical Simulation of 3-D Hypersonic Euler Flows Around Space Vehicles Using Adapted Finite Elements", *AIAA paper 87-0560* January 1987.

7. Batina, J. T., "Implicit Flux-Split Euler Schemes for Unsteady Aerodynamic Analysis Involving Unstructured Dynamic Meshes", *AIAA paper 90-0936,* April 1990.

8. Barth, T. J., and Jespersen, D. C., "The Design and Application of Upwind Schemes on Unstructured Meshes" *AIAA paper 89-0366* January, 1989.

9. Struijs, R., Deconinck, H., de Palma, P., Roe, P., and Powell, K. G., "Progress on Multidimensional Upwind Euler Solvers for Unstructured Grids", *AIAA paper 91-1550CP, Proceedings of the 10th AIAA Computational Fluid Dynamics Conference,* Honolulu, Hawaii, June 1991

10. Barth, T. J. and Frederickson, P. O., "Higher Order Solution of the Euler Equations on Unstructured Grids Using Quadratic Reconstruction", *AIAA paper 90-0013,* January 1990.

11. Halt, D. W., and Agarwal, R. K., "A Compact Higher-Order Euler Solver for Unstructured Grids with Curved Boundaries" *AIAA paper 92-2696,* June 1992.

12. Oden, J. T., Demkowicz, L., Liszka, T., and Rachowicz, W., "h-p Adaptive Finite Element Methods for Compressible and Incompressible Flows", *Proceedings of the Symposium on Computational Technology on Flight Vehicles, Eds. A. K. Noor, S. L. Venneri, Permagon Press, pp. 523-534.* Washington, D.C., November 1990

13. Jameson, A., "Transonic Flow Calculations" *Princeton University Report MAE 1751,* 1984

14. Mavriplis, D. J., and Jameson, A., "Multigrid Solution of the Navier-Stokes Equations on Triangular Meshes", *AIAA Journal, Vol 28, No. 8, pp. 1415-1425,* August 1990.

15. Venkatakrishnan, V., "Newton Solution of Inviscid and Viscous Problems", *AIAA paper 88-0413,* January 1988.

16. Orkwis, P. D., and McRae, D. S., "A Newton's Method Solver for the Navier Stokes Equations", *AIAA paper 90-1524,* June 1990.

17. Venkatakrishnan, V. and Barth, T.J., "Application of Direct Solvers to Unstructured Meshes for the Euler and Navier-Stokes Equations Using Upwind Schemes", *AIAA Paper 89-0364,* January, 1989.

18. Struijs, R., Vankeirsbilck, P., Deconinck, H., "An Adaptive Grid Polygonal Finite-Volume Method for the Compressible Flow Equations", *AIAA paper 89-1959CP* June 1989.

19. Fezoui, F., Stoufflet, B., "A Class of Implicit Upwind Schemes for Euler Simulations with Unstructured Meshes", *Journal of Computational Physics,* Vol 84, No. 1, September 1989, pp. 174-206.

20. Whitaker, D. L., Slack, D. C., Walters, R. W., "Solution Algorithms for the Two-Dimensional Euler Equations on Unstructured Meshes", *AIAA paper 90-0697* January 1990.

21. Hassan, O. Morgan, K., Peraire, J., "An Implicit Finite-Element Method for High Speed Flows", *AIAA paper 90-0402,* January 1990.

22. Lohner, R., Martin, D., "An Implicit Linelet-Based Solver for Incompressible Flows", *AIAA paper 92-0668* January 1992.

23. Mavriplis, D. J., "Three Dimensional Unstructured Multigrid for the Euler Equations", *Proc. of the AIAA 10th Comp. Fluid Dyn. Conf.,* AIAA paper 91-1549, June, 1991.

24. Leclercq, M. P., "Resolution des Equations d'Euler par des Methodes Multigrilles Conditions aux Limites en Regime Hypersonique", *Ph.D Thesis, Applied Math, Universite'de Saint-Etienne,* April, 1990.

25. Peraire, J., Peiro, J., and Morgan K., "A 3D Finite-Element Multigrid Solver for the Euler Equations" *AIAA paper 92-0449* January 1992.

26. Lallemand, M. H., Dervieux, A., A Multigrid Finite-Element Method for Solving the Two-Dimensional Euler Equations", *Proceedings of the Third Copper Mountain Conference on Multigrid Methods, Lecture Notes in Pure and Applied Mathematics,* Ed S. F. McCormick, Marcel Dekker Inc., April 1987, pp. 337-363.

27. Smith, W. A., "Multigrid Solution of Transonic Flow on Unstructured Grids", *Recent Advances and Applications in Computational Fluid Dynamics, Proceedings of the ASME Winter Annual Meeting,* Ed. O. Baysal, November 1990.

28. Morano, E., Guillard, H., Dervieux, A., Leclercq, M. P., Stoufflet, B., "Faster Relaxations for Non-Structured MG with Voronoi Coarsening", *Proceedings of the First European Computational Fluid Dynamics Conference, Eds. Ch. Hirsch,* Brussels, September 1992.

29. Vassberg, J., "A Fast Implicit Unstructured Mesh Euler Method", *AIAA paper 92-2693,* June 1992.

30. Peraire, J., Vahdati, M., Morgan, K., and Zienkiewicz, O. C., "Adaptive Remeshing for Compressible Flow Computations", *J. Comp. Phys.,* Vol 72, October, 1987, pp. 449-466

31. Parikh, P., Pirzadeh, S., and Lohner, R., "A Package for 3-D Unstructured Grid Generation, Finite-Element Flow Solution and Flow-Field Visualization" *NASA CR-182090* September 1990.

32. Aurenhammer, F., "Voronoi Diagrams - A survey of a Fundamental Geometric Data Structure" *ACM Comput. Surveys,* Vol 23, 1991, pp. 345-406

33. Preparata, F. P., and Shamos, M. I., *Computational Geometry, An Introduction,* Texts and Monographs in Computer Science, Springer-Verlag, 1985.

34. Bowyer, A., "Computing Dirichlet Tessalations", *The Computer Journal,* Vol. 24, No. 2, 1981, pp. 162-166

35. Watson, D. F., "Computing the n-dimensional Delaunay Tessalation with Application to Voronoi Polytopes", *The Computer Journal, Vol 24, No. 2,* pp. 167-172, 1981.

36. George, P. L., Hecht, F., and Saltel, E., "Fully Automatic Mesh Generator for 3D Domains of any Shape", *Impact of Computing in Science and Engineering, Vol 2, No. 3,* pp. 187-218, 1990.

37. Baker, T. J., "Three Dimensional Mesh Generation by Triangulation of Arbitrary Point Sets", *Proc. of the AIAA 8th Comp. Fluid Dyn. Conf.,* AIAA paper 87-1124, June, 1987.

38. Holmes, D. G., and Snyder, D. D., "The Generation of Unstructured Meshes Using Delaunay Triangulation" Numerical Grid Generation in Computational Fluid Mechanics *Proc. of the Second International Conference on Numerical Grid Generation in Computational Fluid Dynamics, Miami, December 1988,* Eds. S. Sengupta, J. Hauser, P. R. Eisman, and J. F. Thompson, Pineridge Press Ltd., 1988.

39. Pirzadeh, S., "Recent Progress in Unstructured Grid Generation", *AIAA paper 92-0445* January, 1992.

40. Chew, L. P., "Constrained Delaunay Triangulations", *Algorithmica, Vol 4,* pp. 97-108, 1989.

41. Babuska, I., and Aziz, A. K., "On the Angle Condition in the Finite Element Method", *SIAM Journal of Numerical Analysis,* Vol 13, No. 2, 1976.

42. Mavriplis, D. J., "Adaptive Mesh Generation for Viscous Flows Using Delaunay Triangulation" *Journal of Comp. Physics, Vol. 90, No. 2,*

43. Mavriplis, D. J., "Unstructured and Adaptive Mesh Generation for High-Reynolds Number Viscous Flows", *Proc. of the Third International Conference on Numerical Grid Generation in Computational Fluid Dynamics and Related Fields,* Eds. A. S. Arcilla, J. Hauser, P. R. Eiseman, J. F. Thompson, North Holland, June 1991.

44. Nakahashi, N., "FDM-FEM Zonal Approach for Viscous Flow Computations Over Multiple Bodies", *AIAA paper 87-0604,* January, 1987.

45. Kallinderis, Y., and Ward, S., "Prismatic Grid Generation with an Efficient Algebraic Method for Aircraft Configurations", *AIAA paper 92-2721* June 1992.

46. Sieverding, C. H., "Experimental Data on Two Transonic Turbine Blade Sections and Comparisons with Various Theoretical Methods", *Transonic Flows in Turbomachinery,* VKI Lecture Series 59, 1973.

47. Mavriplis, D. J., "Algebraic Turbulence Modeling for Unstructured and Adaptive Meshes", *AIAA Paper 90-1653,* June, 1990.

48. Wigton, L. B., *Private Communication,* The Boeing Company

49. Speziale, C. G., Abid, R., and Anderson, E. C., "A Critical Evaluation of Two-Equation Models for Near Wall Turbulence", *ICASE Report 90-46, NASA CR 182068, AIAA paper 90-1481,* June, 1990

50. Mavriplis, D. J., "Multigrid Solution of Compressible Turbulent Flow on Unstructured Meshes Using a Two-Equation Model", *AIAA Paper 91-0237,* January 1991.

51. UNICOS Performance Utilities Reference Manual, *SR-2040 6.0* Cray Research Inc.

52. Das, R., Mavriplis, D. J., Saltz, J., Ponnusamy, R., "The Design and Implementation of a Parallel Unstructured Euler Solver Using Software Primitives", *AIAA paper 92-0562,* January 1992.

53. Simon, H., "Partitioning of Unstructured Mesh Problems for Parallel Processing", *Proceedings of the Conference on Parallel Methods on Large Scale Structural Analysis and Physics Applications,* Permagon Press, 1991.

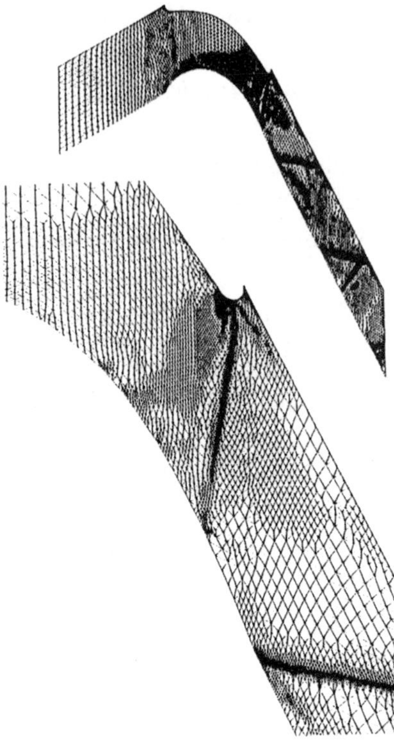

Figure 1: Adaptive Mesh Employed for Computing Transonic Inviscid Flow Through a Periodic Turbine Blade Cascade Geometry; Number of Nodes = 9362

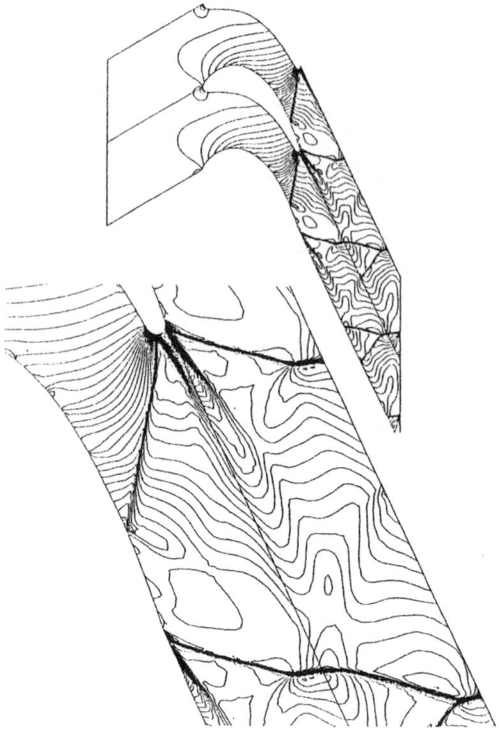

Figure 2: Computed Mach Contours for Flow Through a Periodic Turbine Blade Cascade Geometry

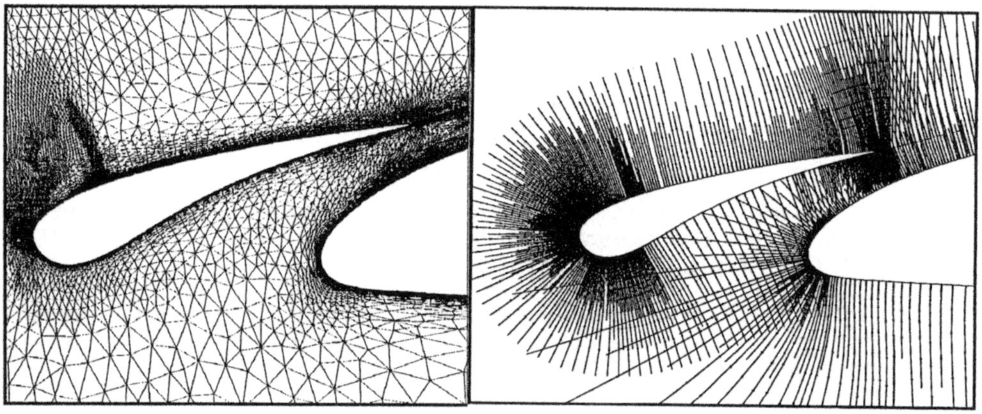

Figure 3: Illustration of Turbulence Mesh Stations Employed in Algebraic Model for an Adaptively Generated Mesh

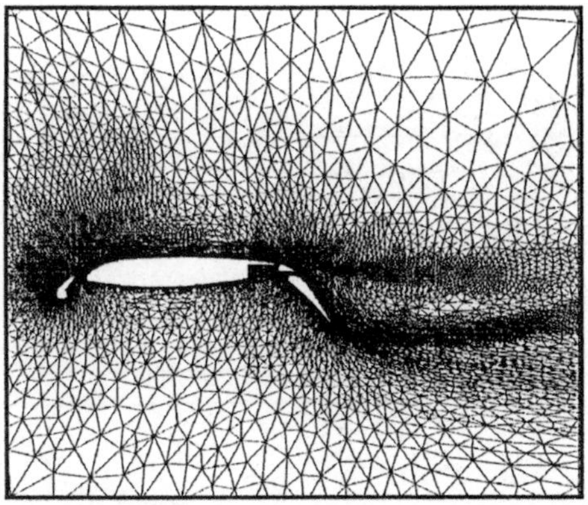

Figure 4: Adaptively Generated Unstructured Mesh about Four-Element Airfoil; Number of Nodes = 48,691

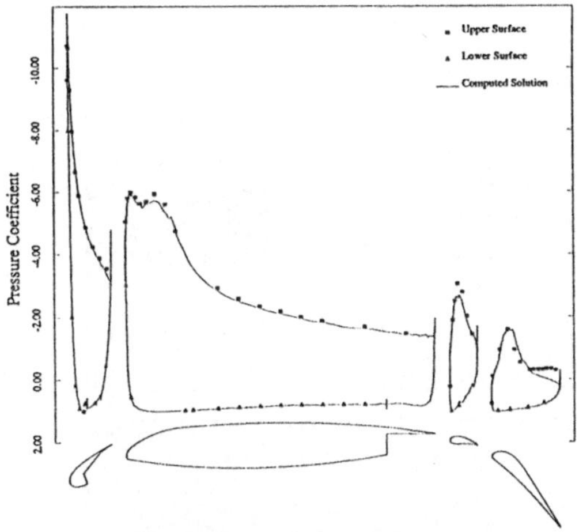

Figure 6: Comparison of Computed Surface Pressure Distribution with Experimental Wind-Tunnel Data for Flow Over Four-Element Airfoil Configuration; Mach = 0.1995, Reynolds Number = 1.187 million, Incidence = 16.02 degrees

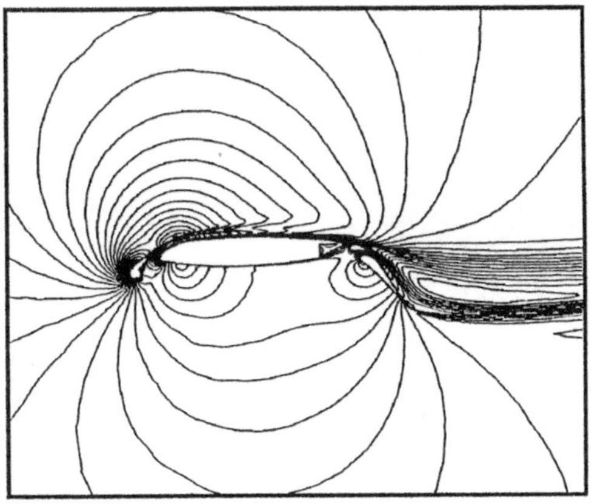

Figure 5: Computed Mach Contours for Flow over Four-Element Airfoil; Mach = 0.1995, Reynolds Number = 1.187 million, Incidence = 16.02 degrees

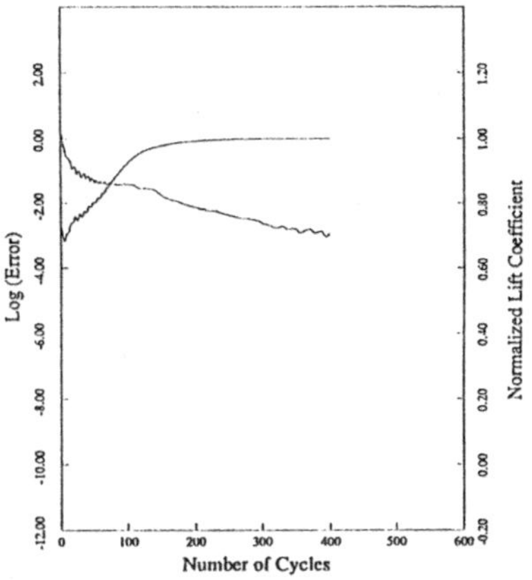

Figure 7: Convergence as Measured by the Computed Lift Coefficient and the Density Residuals Versus the Number of Multigrid Cycles for Flow Past a Four-Element Airfoil

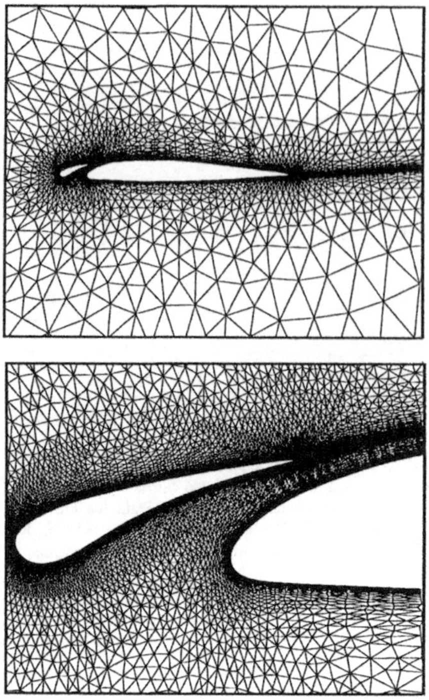

Figure 8: Global View of Coarse Unstructured Mesh and Close-Up View of Fine Unstructured Mesh Employed for Computing Flow Past a Two-Element Airfoil (Coarse Mesh Points = 7272, Fine Mesh Points = 28871)

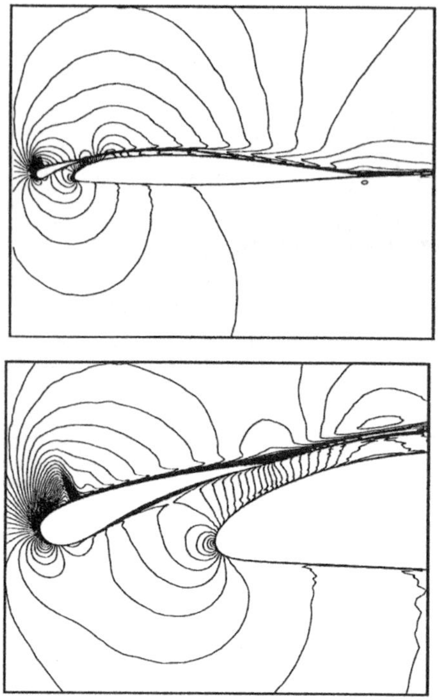

Figure 9: Computed Mach Contours Using Low-Reynolds Number Modification for Turbulence Equations for Supercritical Flow over a Two-Element Airfoil (Mach = 0.5, Re = 4.5 million, Incidence = 7.5 degrees)

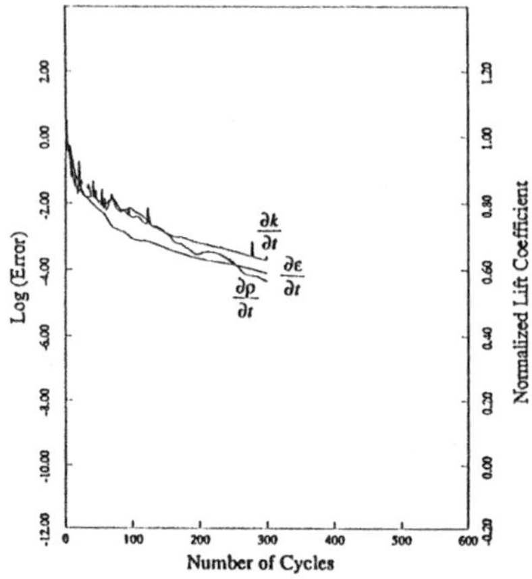

Figure 10: Multigrid Convergence Rate of the Density Equation and the Two Turbulence Equations Using Low-Reynolds Number Modifications for Flow Over Two-Element Airfoil (Mach = 0.5, Re = 4.5 million, Incidence = 7.5 degrees)

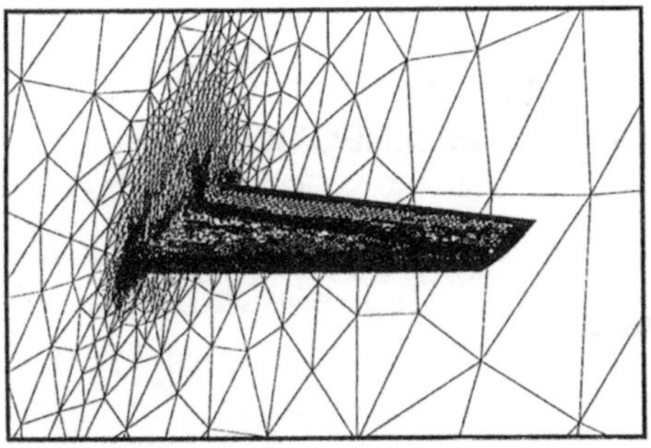

Figure 11: Finest Adapted Mesh Generated About ONERA M6 Wing (Number of Nodes = 173,412 Number of Tetrahedra = 1,013,718)

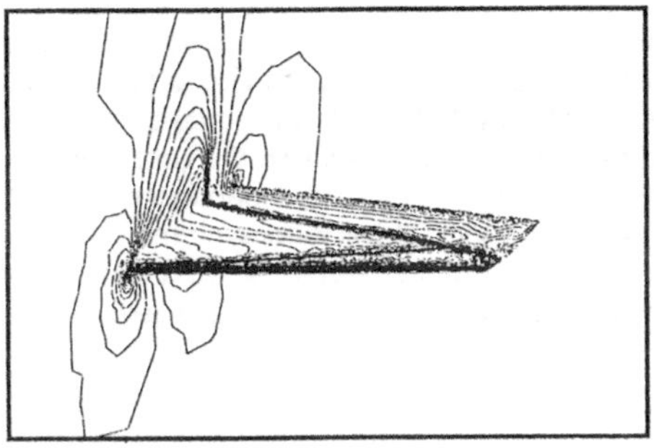

Figure 12: Computed Mach Contours on the Adaptively Generated Mesh About the ONERA M6 Wing (Mach = 0.84, Incidence = 3.06 degrees)

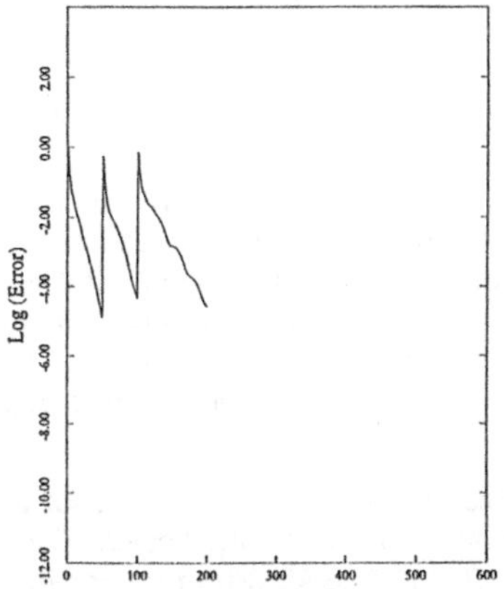

Figure 13: Convergence Rate of the Unstructured Multigrid Algorithm on the Adaptively Generated Sequence of Meshes about the ONERA M6 Wing as Measured by the Average Density Residuals Versus the Number of Multigrid Cycles

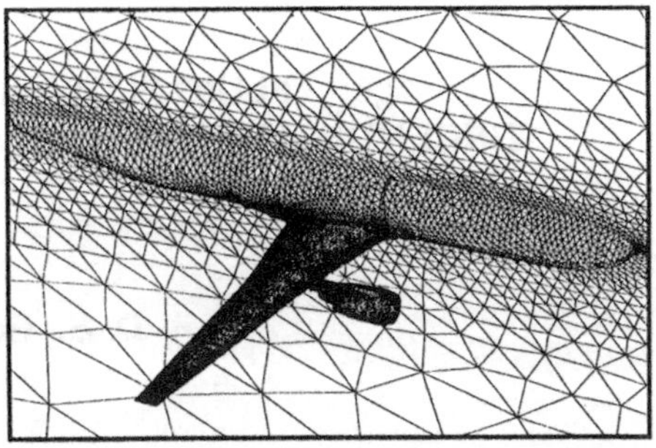

Figure 14: Coarse Unstructured Mesh about an Aircraft Configuration with Single Nacelle; Number of Points = 106,064, Number of Tetrahedra = 575,986 (Finest Mesh Not Shown)

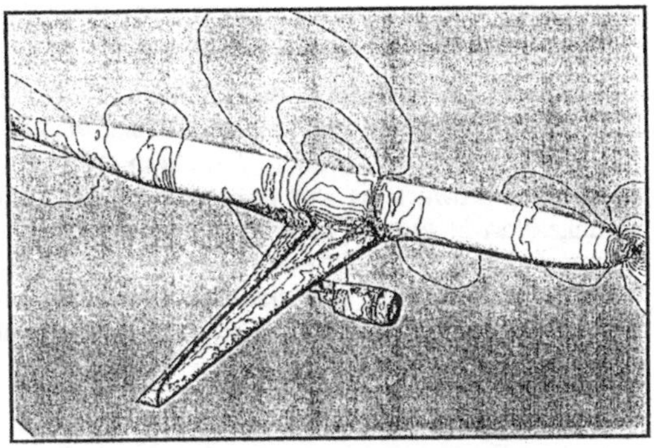

Figure 15: Mach Contours for Flow over Aircraft Configuration Computed on Fine Mesh of 804,056 Vertices and 4.5 million Tetrahedra (Mach = 0.768, Incidence = 1.116 degrees)

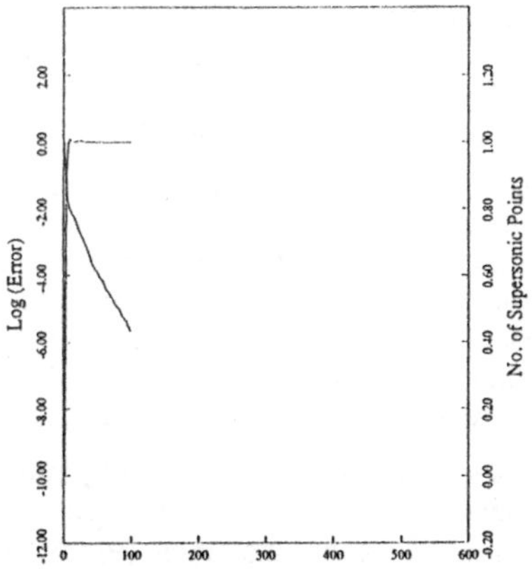

Figure 16: Multigrid Convergence Rate on Finest Mesh of the Multigrid Sequence for Transonic Flow over Aircraft-with-Nacelle Configuration

FLUID FLOW VISUALIZATION: RECENT DEVELOPMENTS AND FUTURE DIRECTIONS

D. E. Edwards

United Technologies Research Center
East Hartford, Connecticut 06108

Abstract

Visualization is playing an important role in the various stages of a computational simulation. The goal in using visualization is to assist existing scientific procedures by providing new insight through visual representation. Trends in scientific visualization will depend on advancements in computer hardware as well as trends in engineering disciplines such as CFD. Current trends in visualization include application of advanced visualization techniques, data management, data compression, feature extraction, graphical user-interfaces, migration to low-end systems, network computing, portability and visual programming environments. Future directions for visualization will be in the areas of virtual reality, automated feature identification and visual languages.

1 Introduction

As we enter the 1990's, the scientific community has witnessed a tremendous growth and availability of large scale computing in the engineering sciences. Many of today's advanced computer simulations create results containing billions of pieces of information. The sheer size of this amount of data results in an exceedingly difficult and time consuming process to interrogate and interpret the information and knowledge contained within the scientific data. This problem was recognized in a 1987 report of the ACM SIGGRAPH Panel on Graphics, Image Processing and Workstations [27], submitted to the National Science Foundation. The report described the need for advanced visualization technology to assist the scientist in the analysis of technical data. Visualization is used to probe the data to locate and identify physical phenomena or to identify limitations in the data creation process.

The term *Scientific Visualization* is used to describe the application of modern interactive computer graphics in the analysis of scientific data. It transforms the symbolic representation (number) into the geometric (image), enabling researchers to observe their simulations. The field of scientific visualization encompasses and unifies the fields of computer graphics, image processing, computer vision, computer-aided design, signal processing and user-interfaces. The goal in using visualization is to assist existing scientific procedures by providing new insight through visual representation.

Progress was made during the 1980's in developing scientific visualization procedures for the field of Computational Fluid Dynamics (CFD). The NASA Ames Workstation Application Office developed several visualization tools (PLOT3D [8], SURF [29], RIP [32], GAS [30]) for examining two- and three-dimensional, steady (time independent) or unsteady (time dependent) fluid dynamic data. In addition, two commercial software companies, Intelligent Light and Wavefront, developed generalized animation turnkey packages that can be used to visualize fluid dynamic data [12,13,16]. The software tools described above had two major limitations. The first limitation was that these tools were difficult to use without specific training. The person using these tools required a detailed understanding of computer graphics. The second limitation was that the process to create the visualization was slow, requiring a substantial amount of user time.

Several efforts are underway to develop systems that attempt to address the limitations of the earlier visualization systems. These systems include NASA Ames' FAST [3], MIT's VISUAL3 [15,17,18], Intelligent Light's FIELDVIEW [22,23], Wavefront's Data Visualizer [5] and AVS Inc.'s (formerly Stardent) AVS [35]. These systems take advantage of the increase in computational and graphics performance in new computer hardware.

This paper describes recent developments and current trends occurring in the field of scientific visualization for fluid dynamic applications. The software tools mentioned above will be discussed with an indication of how these tools are addressing the current trends. Future directions for areas of research in scientific visualization will also be discussed.

2 Recent Developments and Current Trends

Current trends in scientific visualization are dependent on trends in computer hardware. In the past five years there have been significant improvements in workstation computer hardware that have resulted in computational performance increasing by a factor of two every 12-18 months. As we move into the 1990's, the distinction between the different workstations and even between workstations and PCs will blur. Originally the differences were defined by display resolution, performance, memory, operating system and price. Relative rather than absolute performance and price range will differentiate low-, medium- and high-end workstations and PCs. The absolute performance of the high-end workstations of today will become the absolute performance of midrange workstations in 18 months and the absolute performance of low-end workstations in about three years. The same type of migration for computer graphics performance will occur but on a longer cycle, perhaps every 24-30 months [26]. This indicates that computational procedures currently on high-end workstations can be migrated to low-end systems within three years. It also indicates that a significantly larger number of end-users on low-end systems will utilize these procedures. In addition, high-speed fiber-optic networks, with bandwidths two orders of magnitude higher than today's Ethernet networks, will be available in the next several years.

Trends in scientific visualization are also dependent on the engineering discipline (in this case, CFD) where the technology is being applied. Research in CFD is currently focused on the investigation of three-dimensional, time-dependent analyses of complex configurations. An example of this is shown in the work of Dorney [10] who has examined the effect of hot streaks in a 3D rotor-stator interaction (temperature field is shown in Fig. 1) using the time-dependent Reynolds averaged Navier Stokes equations. The 3D simulation consisted of approximately 410,000 grid points, 14,000 time steps (2,000 time steps per cycle) and required 420 hours of CPU time on a Cray 2 supercomputer to obtain a converged solution.

The grid that will make up the volume in the CFD simulations in the 1990's will be either multi-blocked structured, unstructured or a combination of the two. Between one and ten million node points will be used in the simulation; if the problem being analyzed is unsteady, thousands of time steps can be required. For example, a million node point problem with 10 pieces of information at each node and 1000 time steps results in a data set containing 10 gigawords. Scientific visualization will be essential in analyzing these large CFD simulations. Visualization techniques as shown in Fig. 2 can be applied in the different stages of a computational simulation including pre-processing (grid generation), solvers and post-processing (examination of fluid flow results).

When considering the current trends in visualization, one must also examine the state-of-the-art visualization techniques presently available. The following software was reviewed: FAST (Fig. 3), VISUAL3 (Fig. 4) FIELDVIEW (Fig. 5), Data Visualizer (Fig. 6) and AVS (Fig. 7). For completeness, a brief description of each of these visualization software environments is given in the Appendix.

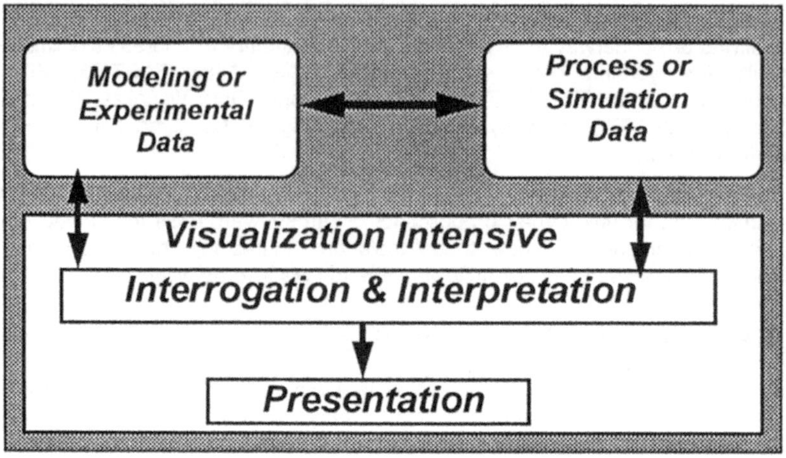

Fig. 2. Scientific Visualization in a Computational Simulation

Given the above information, I see the following trends for scientific visualization for CFD:

- **Advanced Visualization Techniques** - computer graphics procedures such as animation, hidden surface, light sources, transparency, and shading models will be utilized in the display of scalar and vector fields. Stereographics will also be applied.
- **Data Management** - procedures will be developed to efficiently access volumetric (unstructured or structured) time dependent data.
- **Data Compression** - procedures to compress volumetric data will be developed. Techniques may consist of subsampling of data, removing redundant data or using analytical representations of the volume. For example, wavelet theory is used in data compression of visual images.
- **Feature Extraction** - improved techniques to extract information from volume of data in terms of contours, contour lines, vectors, iso-surfaces, streaklines, streamlines, particle paths and cutting planes. Procedures will be develop for 3D and 4D data.
- **Graphical User-Interface (GUI)** - interactive graphical user-interfaces which are easy to learn and use are being developed. Designers work with end-users to determine functionality of interface.
- **Migration to Low-End Systems** - visualization procedures that are currently available on high-end graphic workstations will be ported to low-end workstations, PCs and X-Terminals.
- **Network Computing** - procedures will be developed to distribute scientific visualization analysis over a network of PCs, workstations, mini-supercomputers and supercomputers.
- **Portability** - visualization procedures will be developed to be portable over a range of UNIX based workstations using X-Windowing system and a standard graphics library (PEX, PHIGS, GL).
- **Visual Programming Environments** - visual programming environments will be applied by software developers to design visualization applications.

TREND	FAST	VISUAL3	FIELDVIEW	Data Visualizer	AVS
Advanced Visualization Techniques	Yes	Yes	Yes	Yes	Yes
Data Management	No	No	No	No	No
Data Compression	No	No	No	No	No
Feature Extraction	Yes	Yes	Yes	Yes	Yes
Graphical User-Interface	Yes	No	Yes	Yes	Yes
Migration to Low-End	No	No	Yes	Yes	Yes
Network Computing	Yes	No	No	No	Yes
Portability	No	Yes	Yes	Yes	Yes
Visual Programming Environment	No	No	No	No	Yes

Table 1 - Current trends in scientific visualization.

Table 1 shows that none of the visualization software reviewed in this paper address all the trends listed above. Current procedures are not addressing data management. With the trend in CFD to perform large scale simulations resulting in gigawords of information, managing the data is becoming extremely important. In addition to data management, these large scale simulations will also require the scientific visualization community to examine data compression techniques. While techniques exists in visualization procedures today to subsample the data in a crude fashion (for example, taking every other point in each direction in a structured grid) more sophisticated techniques are required. Data compression will also be important for the transmission of images to remote sites.

The migration of scientific visualization procedures to low-end systems is an obvious trend since technology advancements in computer hardware occur every 18 months. As high speed networks become available, network computing (for example, an application distributed over a network) will become more common [31]. In many ways, over the next several years the network will become the computer.

Finally, the use of visual programming environments is a current trend. While AVS is the only visualization environment out of the five reviewed to use visual programming, other visual programming environments such as SGI Explorer, IBM Data Explorer and Taravisual apE [9,11] will be available in 1992. These visual programming environments will be useful for software developers especially for the rapid prototyping of visualization software in the early phase of the design cycle.

While each of the visualization software reviewed for this paper does not address all of the current trends, it is noted that these tools are extremely powerful and can assist the engineer in analyzing CFD data.

3 Future Directions

The 1990's promises continuing dramatic advances in computer hardware that will be exploited in the utilization of scientific visualization. I see the following directions for future research in visualization:

- Virtual Reality
- Automated Feature Identification
- Visual Languages

3.1 Virtual Reality

Virtual Reality (VR) will offer the engineer the opportunity to interact with the computer in a three-dimensional environment in a variety of new and unique ways [4,6,14,21,25]. Two VR metaphors are currently in use [33]. One is an immersive VR metaphor where the user is immersed in a virtual world through a head mounted display. The other is a desktop VR metaphor where the user views the virtual world through a "window" on the desk similar to a sailor viewing the physical world through a periscope on a submarine.

Several research thrusts into immersive VR are currently underway. One example is the development of a virtual environment at NASA Ames by Bryson and Levit [7] for exploring computer simulated 3D unsteady flowfields. This environment (Fig. 8) has been created using 1) a boom-mounted head position sensitive CRT display system for viewing, 2) a VPL data glove for injecting tracers into the virtual flowfield and 3) a SGI 4D/380 VGX multiprocessor graphics workstation for computation and rendering.

Research in VR is also being conducted at IBM's T. J. Watson Research Center by Ling [2,24] to explore ways in which multiple users can interact in a virtual world (Fig. 9). Using datagloves and a network of IBM RS/6000 workstations, the users can not only point in 3D space but can also grasp and move virtual objects whose presence can be sensed through tactical feedback mechanisms. Tuori [34] of Alias is also conducting research in the use of immersive VR for design evaluation with current CAD and CAID software. He is developing a VR viewing tool based on a boom mounted stereoscopic display connected to an SGI Skywriter to complement the Alias design system.

An example of a desktop VR application is SimGraphics' Assembly Modeler [33] which was developed as part of the Air Force's Automated Airframe Assembly Program. The Assembly Modeler is an interactive assembly model simulation that is used to concurrently design and plan the parts, tools and manufacturing processes necessary to construct various airplane assemblies. This application is used to manipulate, test and verify the assembly of components in a virtual space. Users select parts or subassemblies to visualize part-to-part clearances and fits.

Today's research in virtual reality demonstrates both the potential of virtual reality and how far we must go before the hardware and software can provide useful environments for practical problem solving. Issues to address in VR include sophisticated, easy to use, high resolution head-mount displays, voice controls, input devices for tracking, pointing, drawing and shaping, and VR standards.

3.2 Automated Feature Identification

Automated feature detection and identification procedures need to be developed for the examination of large data sets, especially those for 3D time dependent analyses. As the performance of our computers increases, the size of our engineering problems will also increase resulting in massive data sets to be interpreted. Today, visualization procedures are used interactively to identify physical phenomena contained within the data. The knowledge required to identify the various phenomena (physical and numerical) contained within various types of fluid dynamic flows should be built into a knowledge base that could be used by an expert system to automatically locate, detect and identify phenomena in fluid dynamic data. The benefits of developing these automated feature identification systems are 1) automation procedures reduce time of engineer required to analyze data, 2) experience and knowledge of experts is made available to less experienced users, and 3) experience is accumulated.

3.3 Visual Languages

Visual languages [1,19,20] will provide an alternative to textural languages for the development of scientific visualization applications. The visual programming languages developed must provide the necessary constructs and predefined functions so that the user of this language can apply it to complex problems in scientific visualization. Data flow [1] is currently a popular computational model for visual programming languages. Data flow provides a view of computation which shows data flowing from one function to another, being transformed as it goes. It is noted that AVS, while a useful visualization programming environment, is not a visual language. Specifically AVS cannot create new modules while staying in the visual portion of AVS. Instead the new module is created using a textual language (C or FOR-TRAN). The benefit of developing a visual programming language for scientific visualization will be the increased ease in developing scientific visualization applications.

4 Concluding Remarks

Scientific visualization is playing an important role in the various stages of a computational simulation. The goal in using visualization is to assist existing scientific procedures by providing new insight through visual representation. Trends in scientific visualization will depend on advancements in computer hardware as well as trends in engineering disciplines such as CFD. Current trends in scientific visualization include application of advanced visualization techniques, data management, data compression, feature extraction, graphical user-interfaces, migration to low end systems, network computing, portability and visual programming environments. Future directions for scientific visualization will be in the areas of virtual reality, automated feature identification and visual languages.

Acknowledgements

The work in this paper was supported by United Technologies Research Center under the Corporate research program. The author expresses his gratitude to Prof. Mike Giles of MIT, Mr. David Kamins of AVS Inc., Mr. Steve Legensky of Intelligent Light, Mr. Martin Tuori of Alias, Mr. Mike Wilson of Wavefront and Dr. Val Watson of NASA Ames for their assistance in providing information for this publication.

5. Appendix

5.1 FAST

FAST (Flow Analysis Software Toolkit) is a general purpose visualization environment for CFD applications which has been developed by NASA Ames and Sterling Software. This toolkit has evolved into a collection of programs (modules) that communicate via UNIX sockets with a hub module which manages a pool of shared memory. The FAST Central module (Fig. 3) is the main module from the user's perspective since the graphical data from the other modules is managed and interactively viewed using this module. Other modules are called or closed as they are needed from the Central module. The modules include:

- Animation - used to create and record keyframe animation sequences to videotape.
- NAS File Input - reads in PLOT3D grid, solution and function files. Also reads in ARCGraph files.
- CFD Calculator - user can define scalar or vector function or select CFD built-in functions.
- SURFER - used to examine vector and scalar data on computational surfaces. Surfer can sweep through surfaces in a given computational direction.
- Titler - used to create annotation on images.
- Isoev - used to create iso-surfaces and cutting planes. Also creates vector field deformation surfaces on either iso-surface or cutting plane.
- Tracer - used to create particle traces, which can be animated after calculation.
- ColorMap - controls color map editing and background color.

The SGI 4D220/GTX was used as the development platform for this software. The SGI GL graphics library was used in developing the graphical procedures. The NASA Panel Library was used to develop the user-interface. This software is currently operational on SGI workstations.

5.2 VISUAL3

VISUAL3 is a visualization environment for the examination of 3D volumetric scientific data which is currently under development at MIT. The volume can be represented with either structured or unstructured grids. The data in the volume can be steady or time dependent. An interesting aspect of VISUAL3 is the dimensional windowing approach (Fig. 4). The three main plotting windows are:

- 3D Window - displays data on three-dimensional surfaces, either from the bounding domain of the volume or from cuts from a cutting plane or iso-surfaces. Vector fields can be displayed as either tufts or streamlines.
- 2D Window - displays data on a mapped domain or cutting surface. This window is used to seed particles for particle paths and initiate many of the probes and points in 3D space.
- 1D Window - displays data generated by probe functions in 2D window.

Using VISUAL3, information contained in the volume can be extracted and displayed in terms of vector clouds, cutting planes, iso-surfaces, vector tufts, pathlines, and tubes (pathlines with circular cross-sectional area based on local crossflow divergence). In addition, VISUAL3 has a series of probes to locate information at points on a surface, in the viscous boundary-layer and along or normal to a pathline.

VISUAL3 was developed using Stardent's XFDI extension to the X windowing system. Most of VISUAL3 is an X application with a few non-standard internal calls to perform Gouraud shading for surfaces and 3D support for drawing lines and triangles with hidden surface removal. The XFDI model is similar to the PEX standard with the exception that PEX currently does not support immediate mode capability to the server.

5.3 FIELDVIEW

FIELDVIEW is an end-user visualization environment that is designed for CFD applications. The software tool is being marketed by Intelligent Light and has been jointly developed with United Technologies Research Center (the prototype was developed in the UTRC Visualization Lab [28]). The visualization system's graphical user interface has been constructed using the Open Software Foundation's Motif and X. The 3D graphical procedures are developed using Intelligent Light's IVIEW-DORE, a portable enhanced version of the DORE graphics library. The layout of FIELDVIEW's graphical user-interface is shown in Fig. 5. The main panel across the top of the screen consists of several pull down menus which open various panels for:

- File Input - reads in PLOT3D grid, solution and function files.

- Function Calculator - user can select PLOT3D CFD built-in functions.
- Computational Surface - used to examine vector and scalar data on computational surfaces. Can sweep through surfaces in a given computational direction. Point probes and 2D line plots are available.
- Iso-surface - used to create iso-surfaces and cutting planes.
- Streamlines - controls interactive seeding of particle rakes and integration of current vector function. Vector pathlines can be animated after calculation.
- Titles - used to create annotation on 3D display.
- Colormap -controls color specifications (object color, color maps for scalar function and background color).
- Viewer - controls viewing position of objects in 3D display.

FIELDVIEW is designed specifically to be portable to low-end workstations. The software can be utilized on systems with either true color (24 bit displays) or pseudo color (8 bit displays). The software is currently operational on workstations from HP, IBM, SGI and Sun. In addition the software can be operated from color X-Terminals.

5.4 Data Visualizer

The Data Visualizer is an end-user visualization environment developed and marketed by Wavefront for the analysis of 3D scalar and vector data in a heterogeneous hardware environment. The graphics procedures are developed using a custom internal portable graphics layer. The Data Visualizer also uses a custom User Interface Management System (Fig. 6). Visualization techniques were developed in a tool oriented approach where the following tools are used:

- File Input - reads in data that can be either structured or unstructured. The PLOT3D format is available in this system.
- Data Probe - provide an interpolation lookup of data located in the 3D space.
- Cutting Plane - offers a variety of cutting planes for examining both scalar and vector data.
- Iso-surface - creates a 3D contour surface within the volume of the specified scalar value.
- Particle System - continuously launches particles from a user-defined area to trace vector values.
- Particle Trace - launches a particle to create a streak line through the selected vector field.
- Motion - used to created animated image sequences.
- Color Editor - provides control over scalar color maps.
- Volume Rendering - uses ray casting to produce a rendered picture of scalar data.

This software is available on a range of workstations including HP, IBM, SGI and Sun.

5.5 AVS

AVS (Application Visualization System) is a visual programming environment for both end-users and software developers that is marketed by AVS Inc. (formerly Stardent). AVS allows users to create their own applications by visually constructing flow graphs (called networks in AVS) comprised of icons that represent various processing modules. Existing applications (such as CFD analyses or CFD data) can be directly integrated into the AVS environment where the results from the application can be visualized (Fig. 7). AVS allows for the applications to be distributed over a computer network. The basic building blocks of AVS are modules, which perform special program functions such as reading in data, analyzing data and rendering the data.

In many ways, the AVS environment can be though of as a visual representation of a library of modules. AVS has a collection of modules to visualize volumetric data in terms of contours, vector fields, iso-surfaces, cutting planes, particle traces and voxel images. It also has modules for data probes and line plotting.

The AVS visualization environment is currently available on systems from Cray, Convex, DEC, HP, IBM, SGI and Sun.

References

1 Agerwalak, T. and Arvind, A., "Data Flow systems," *IEEE Computer* Vol. 15, pp. 10-13, 1982.

2 Appino, P. A., Lewis, J. B., Koved, L., Ling, D. T., Rabenhorst, D. A., and Codella, C. F., "An Architecture for Virtual Worlds," IBM Research Report RC 16446, July, 1991.

3 Bancroft, G. V., Merrit, F. J., Plessel, T. C., Kelaita, P. G., McCabe, R. K., and Globus, A., "FAST: A Multi-Processed Environment for Visualization of Computational Fluid Dynamics," *Proceeding of IEEE Visualization '90,* San Francisco, CA, October, 1990.

4 Blattner, M. M. and Grinstein, G., "Multimedia Environments for Scientists" (Panel Session), *Proceeding of IEEE Visualization '91,* San Diego, CA, October, 1991.

5 Brittain, D. L., Aller, J., Wilson, M., and Wang, S. C., "Design of an End-User Data Visualization System," *Proceeding of IEEE Visualization '90,* San Francisco, CA, October, 1990.

6 Brooks, F. P., Ouh-Young, M., Batter, J. J., and Kilpatrick, P. J., "Project GROPEHaptic Displays for Scientific Visualization," *SIGGRAPH Computer Graphics,* Vol. 24, No. 4, August, 1990.

7 Bryson, S., and Levit, C., "The Virtual Windtunnel: An Environment for the Exploration of Three-Dimensional Unsteady Flows," *Proceeding of IEEE Visualization '91,* San Diego, CA, October, 1991.

8 Buning, P. G. and Steger, J. L., "Graphics and Flow Visualization in Computational Fluid Dynamics," *Proceeding of IEEE Visualization 90,* October, 1990.

9 Csuri, C. A., Dyer, S., Faust, J., and Marshall, R., "A Flexible Integrated Graphics Environment for Supercomputers and Workstations," *Science and Engineering on Cray Supercomputers: Proceeding of the Third International Symposium,* Cray Research Inc., September, 1987.

10 Dorney, D.J., Davis. R. L., Edwards, D. E. and Madavan, N.K., "Unsteady Analysis of Hot Streak Migration in a Turbine Stage," AIAA Paper 90-2354, also to appear in the AIAA *Journal of Propulsion and Power.*

11 Dyer, S., "A Dataflow Toolkit for Visualization," *IEEE Computer Graphics & Applications,* July, 1990.

12 Edwards, D. E., "Application of Scientific Visualization to Fluid Dynamic Problems," AIAA Paper 88-3351, July, 1988.

13 Edwards, D. E., "Three Dimensional Visualization of Fluid Dynamic Problems," AIAA Paper 89-0136, January, 1989.

14 Fisher, S. S., McGreevy, M., Humphries, J., and Robinett, W., "Virtual Environment Display Systems," *Proceedings of the 1986 Workshop on Interactive 3-D Graphics,* Chapel Hill, NC, 1986.

15 Giles, M. and Haimes, R., "Advanced Interactive Visualization for CFD," *Computing Systems in Engineering,* Vol. 1, No. 1, pp. 51-62, 1990.

16 Haber, R. B., "Visualization in Engineering Mechanics: Techniques, Systems and Issues," Presented at Visualization Techniques in the Physical Sciences, SIGGRAPH, August, 1988.

17 Haimes, R. and Giles, M, "VISUAL3: Interactive Unsteady Unstructured 3D Visualization," AIAA Paper 91-0794, January, 1991.

18 Haimes, R. and Darmofal, D., "Visualization in Computational Fluid Dynamics: A Case Study," *Proceeding of IEEE Visualization '91,* San Diego, CA, October, 1991.

19 Hils, D., "DataVis: A Visual Programming Language for Scientific Visualization," *Proceeding of 1991 ACM Computer Science Conference*, San Antonio, Texas, March 1991.

20 Hirakawa, M., Monden, N., Yoshimoto, I., Tanaka, M. and Ichikawa, T., " HIVISUAL, A language supporting visual interaction in programming," *Visual Languages*, editors S. Chang, T. Ichikawa, and P. Ligomenides, Plenum Press, New York, 1986.

21 Iwata, H., "Artificial Reality with Force-Feedback: Development of Desktop Virtual Space with Compact Master Manipulator," *SIGGRAPH Computer Graphics,* Vol. 24, No. 4, August, 1990.

22 Legensky, S. M., "Interactive Investigation of Fluid Mechanics Data Sets," *Proceeding of IEEE Visualization '90,* San Francisco, CA, October, 1990.

23 Legensky, S. M., "Advanced Visualization on Desktop Workstations," *Proceeding of IEEE Visualization '91,* San Diego, CA, October, 1991.

24 Ling, D. T., "Beyond Visualization - Virtual Worlds for Data Understanding," IBM Research Report RC 15479, February, 1990.

25 Mackinlay, J. D., Card, S. K., and Robertson, G. G., "Rapid Controlled Movement Through a Virtual 3D Workspace," *SIGGRAPH Computer Graphics,* Vol. 24, No. 4, August, 1990.

26 Machover, C., "The Business of Computer Graphics," *IEEE Computer Graphics & Applications,* January, 1991.

27 McCormick, B. H., DeFanti, T. A., and Brown, M. D., "Visualization in Scientific Computing," NSF Workshop on Graphics, Image Processing and Workstations, July, 1987.

28 Miller, S. L., and Edwards, D. E., "VISA: An Interactive Graphics System for Scientific Visualization," AIAA Paper 91-0795, January, 1991.

29 Plessel, T., *SURF Users Manual,* NASA Ames Research Center, Code RFW, 1988.

30 Plessel, T., *GAS Users Manual,* NASA Ames Research Center, Code RFW, 1988.

31 Robertson, D. W., Jacobson, V. L., Johnston, W. E., Loken, S. C., Theil, E. H., and Tierney, B. L., "Distributed Visualization using Workstations, Supercomputers and High Speed Networks," *Proceeding of IEEE Visualization '91,* San Diego, CA, October, 1991.

32 Rogers, S., "Distributed Interactive Graphics Applications in Computational Fluid Dynamics," *International Journal of Supercomputing Applications*, Vol. 1, No. 4, Winter 1987.

33 Tice, S., Private Communication, September, 1991.

34 Tuori, M., "Immersive Simulation- Applications In Design," Presented at The International Virtual Reality Symposium, Nagoya, Japan, October, 1991.

35 Upson, C., Faulhaber, T., Kamins, D., Laidlaw, D., Schlegel, D., Vroom, J., Gurwitz, R., and van Dam, A., "The Application Visualization System: A Computer Environment for Scientific Visualization," *IEEE Computer Graphics & Applications,* July, 1989.

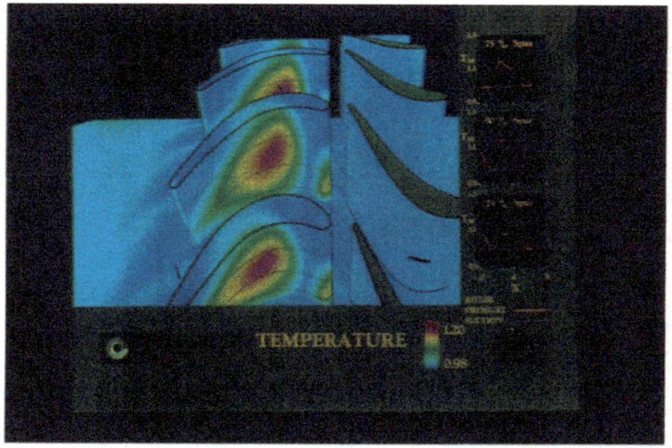

Fig. 1. 3D Hot Streak in a Rotor- Stator Interaction. Temperature Field.

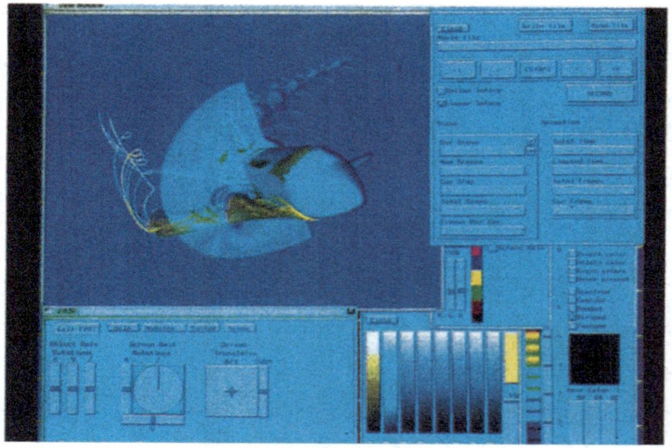

Fig. 3. FAST Central Module. Shuttle displayed in 3D window with streamlines.

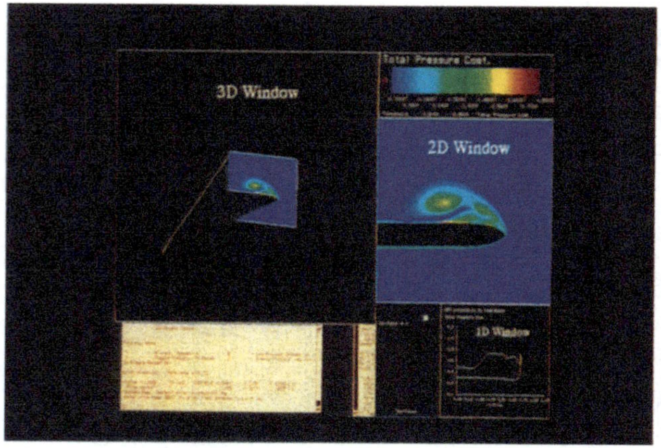

Fig. 4. VISUAL3 Window Layout. Cutting plane example.

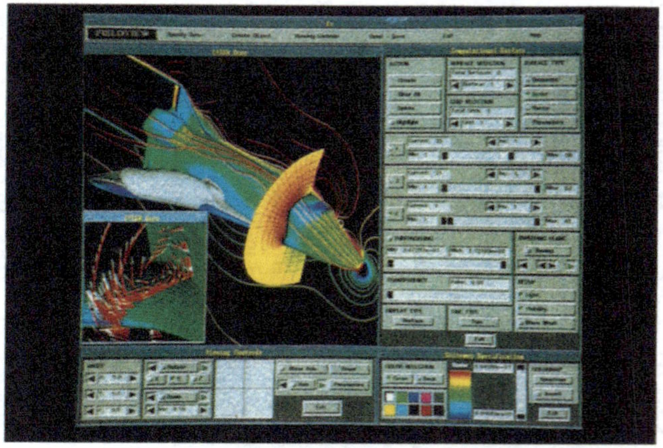

Fig. 5. FIELDVIEW graphical interface. Density contours with streamlines and velocity vectors.

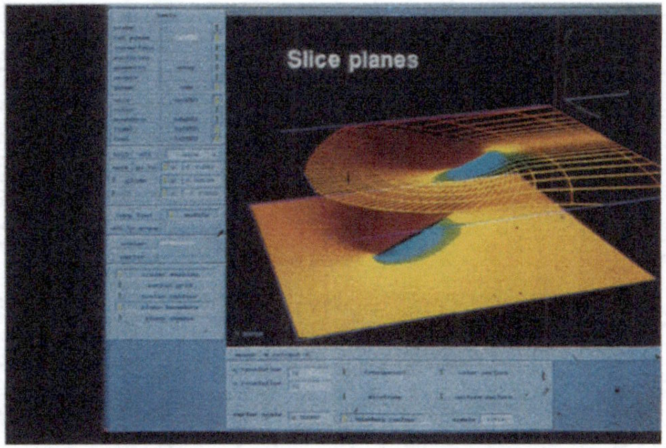

Fig. 6. Data Visualizer Graphical User-Interface. Cutting planes example.

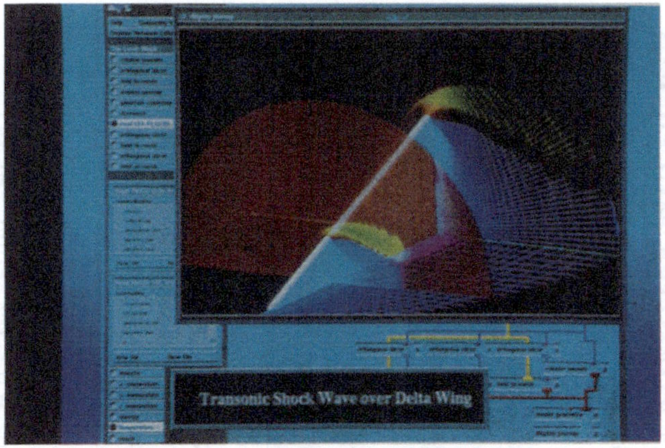

Fig. 7. AVS Graphical User-Interface. Image of transonic shock wave over delta wing.

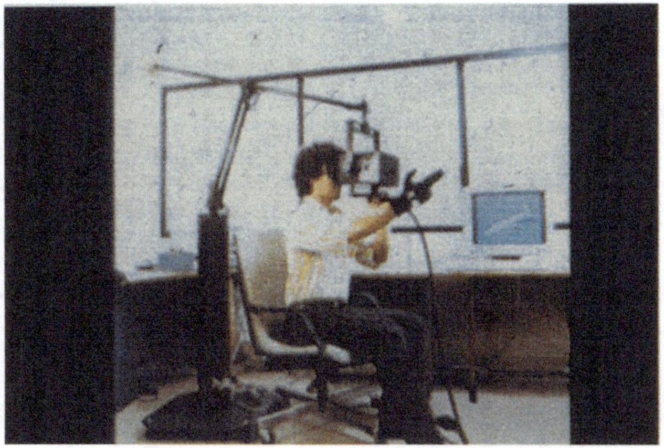

Fig. 8. NASA Ames Virtual Wind Tunnel. Flow particles are seeded using VR environment.

Fig. 9. Virtual Reality from IBM T. J. Watson Research Center.

A MULTI-DIMENSIONAL SOLUTION ADAPTIVE MULTIGRID SOLVER FOR THE EULER EQUATIONS

L. A. Catalano, P. De Palma, G. Pascazio

Politecnico di Bari, via Re David 200, 70125 Bari, Italy

Introduction

In the last few years, the development of numerical methods for solving the multi-dimensional Euler equations with improved shock-capturing properties has been an important research topic in computational fluid dynamics.

Two different approaches are under investigation. The first *ad hoc* one is based on the application of rotated 1-D Riemann solvers, see, e.g., the pioneering work of Davis [1]. The second, more general, *genuinely multi-dimensional* approach consists in solving an equivalent set of scalar wave equations with solution-dependent propagation directions, as recently proposed by Deconinck *et al.* [2] and Roe [3]. To date, results using the latter methodology, of interest here, have been obtained in conjunction with the Fluctuation Splitting [4] space discretization, see [5-7].

In this paper, some major improvements are reported, concerning both the wave modelling and a proper treatment of solid wall and subsonic inlet and outlet boundary conditions. Furthermore, the effectiveness of the explicit multi-stage multigrid approach of [7] is demonstrated also for subsonic and transonic flow problems. Finally, the present improved genuinely multi-dimensional multigrid method is combined with a solution-adaptive-grid strategy to further enhance its accuracy and efficiency.

The multigrid Euler solver

The present numerical method is based on simple wave solutions of the Euler equations [3]. This approach consists in selecting a number N of waves (acoustic, entropy, shear), each one having strength α and propagation direction $\mathbf{n}$, so that the primitive variable gradient can be decomposed as

$$\nabla \tilde{u} = \sum_{k=1}^{N} \nabla \tilde{u}^k = \sum_{k=1}^{N} \alpha^k \tilde{r}^k \mathbf{n}^k . \tag{1}$$

In eq. (1) $\tilde{r}^k$ is the right eigenvector of the Jacobian $(\tilde{A} n_x + \tilde{B} n_y)$ with corresponding eigenvalue λ^k. For linear initial data, each simple wave evolves in time according to

$$u^k(\mathbf{x}-\mathbf{x}_0, t) = u_0^k + \nabla u^k \cdot [(\mathbf{x}-\mathbf{x}_0) - \lambda^k \mathbf{n}^k t]. \tag{2}$$

The global variation of the conservative variables is obtained by adding all of the wave contributions. Using eq. (1), one gets

$$u(\mathbf{x}-\mathbf{x}_0, t) = u_0 + \sum_{k=1}^{N} \alpha^k r^k [(\mathbf{x}-\mathbf{x}_0)\cdot \mathbf{n}^k - \lambda^k t], \tag{3}$$

which, after differentiation in time, becomes

$$u_t = -\sum_{k=1}^{N} \alpha^k r^k \lambda^k = -(F_x + G_y). \tag{4}$$

Two fundamental issues need to be addressed, namely, the wave modelling and the numerical discretization. An analytical solution of eq. (1) can be obtained when employing four orthogonal acoustic, one entropy and one shear waves [3]. Such a solution provides the intensities of all waves as well as the directions of propagation of the acoustic and entropy waves, whereas the direction of the shear is somewhat arbitrary. All present models basically differ on the choice of such a direction. In [3], Roe chose such a direction perpendicular to the streamline, but the resulting model experiences difficulties in resolving oblique shocks. A strong improvement has been obtained by employing the direction of the pressure gradient [5]. However, two problems are still open: 1) the pressure gradient is not defined for pure shear flows; 2) numerical disturbances in smooth flow regions inhibit convergence to machine zero. In this paper, it is shown that the choice of the direction of the velocity vector easily overcomes such difficulties.

Concerning the numerical discretization, a cell-vertex structured quadrilateral grid is employed, each cell being subdivided into two triangles. Using eq. (4), the global fluctuation over a triangle S, defined as $\Phi_T = \int_S u_t \, dS$, is split into the wave contributions, as follows:

$$\Phi_T = -\int_S \sum_{k=1}^{N} \alpha^k r^k \lambda^k \, dS = -S \sum_{k=1}^{N} \bar{\alpha}^k \bar{r}^k \bar{\lambda}^k = \sum_{k=1}^{N} \Phi_T^k. \tag{5}$$

In eq. (5) the cell-averaged values $\bar{\alpha}^k$, $\bar{r}^k$, $\bar{\lambda}^k$ can be calculated analytically, if the parameter vector $z = \sqrt{\rho}(1, u, v, H)$ is assumed to vary linearly over each triangle [4,7]. Each wave contribution is then split among the vertices, according to the Fluctuation Splitting first-order-accurate N-scheme [4].

Characteristic subsonic boundary conditions are imposed: total enthalpy, entropy and flow angle are specified at the inlet boundaries, while the pressure is specified at the outlet boundaries. One row of auxiliary cells is used at walls, the state at the mirror-image nodes being calculated by imposing impermeability and Isentropic Simple Radial Equilibrium [8].

Concerning the efficiency of the numerical method, the classical FAS multigrid V-cycle [9] is employed, the relaxation being replaced by the optimally-smoothing Runge-Kutta scheme of [7]. Standard full-weighting collection and bi-linear prolongation operators are used.

Results on non-adaptive grid

The flow through the GAMM channel is calculated for two different values of the inlet Mach number, using a 128×64 uniform non-orthogonal diamond grid. The iso-Mach lines for the subsonic case ($M_i = 0.6$) are shown in fig. 1a. The solution is symmetric except for a small shift at the lower wall, due to entropy generated numerically in the re-compression region. The present solutions are considerably more accurate than the results of [6], thanks to the present choice of the propagation direction for the shear and to the improved boundary conditions. Furthermore, convergence to machine zero is obtained here, see fig. 1b, where the single- and the multi-grid convergence histories are presented. The logarithm of the L^1-norm of the residual of the mass conservation equation is plotted versus the work, one *work unit* being defined as *one single-stage residual* calculation on the *finest grid*. One pre- and one post-application of the optimal three-stage smoother of [7] are used. The reduction in the convergence rate, seen at $R \approx -2.5$, is due to the well-known multigrid alignment phenomenon. Semi-coarsening should reduce such a problem.

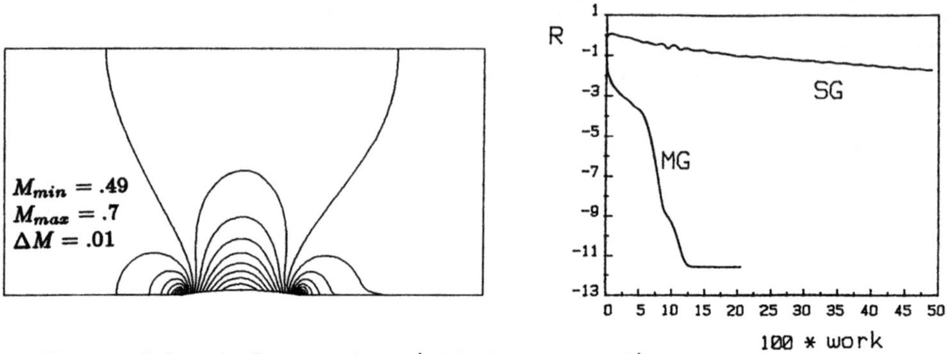

Fig. 1: Subsonic flow results. a) Mach contours b) convergence histories

Similar considerations apply to the solution of the transonic flow case (M_i = 0.83) shown in fig. 2a. A further decrease in the MG convergence rate is experienced in this case (see fig. 2b), since no extra-relaxations are applied in the shock region.

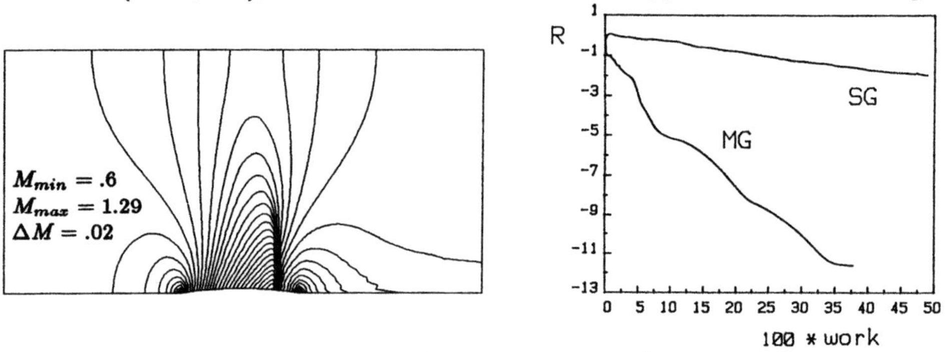

Fig. 2: Transonic flow results. a) Mach contours b) convergence histories

Adaptive grid strategy

One degree of freedom exists when subdividing each quadrilateral cell into two triangles. The positive linear scheme with minimum cross-diffusion is produced when choosing the diagonal more aligned with the direction of the wave having maximum intensity [10], or, in the presence of shocks, with the direction perpendicular to the pressure gradient. Since the CFL number, defined here as $\nu = max(\nu_x, \nu_y)$, is based on the fastest wave, two different sets of optimal predictor coefficients and Courant number have to be used, depending wether the diagonal is optimal or not for its propagation direction (see [7] for details). In the first case, the optimal coefficients are found to be constant ($c_1 = .1481$, $c_2 = .4$, $\nu = 1.5$), whereas, in the second one, the directional splitting first-order-accurate stencil is recovered. The optimal coefficients c_1, c_2 and ν depend on the advection speed component ratio R and are provided by the interpolation function $c_k\nu = \delta_k/(1+\gamma_k R)$, with $\delta_1 = .222$, $\gamma_1 = .64$, $\delta_2 = .6$, $\gamma_2 = .557$, $\delta_3 = 1.5$, $\gamma_3 = .548$, $c_3 = 1$.

Finally, a local solution-adaptive refinement strategy is employed in conjunction with the quad-tree data-structure provided by Hemker *et al.* [11]. Each quadrilateral cell is subdivided into four *kid*-cells, if needed. The choice is based on the value either of the static pressure gradient multiplied by the area, or of the divergence of the velocity. An hystogram composed by one hundred classes is used to determine the threshold value, the percentage of cells to be refined being assigned by the user. The conservative variables at *green* points, marked with a cross in fig. 3, are updated by linear interpolation only, after visiting the coarse grid.

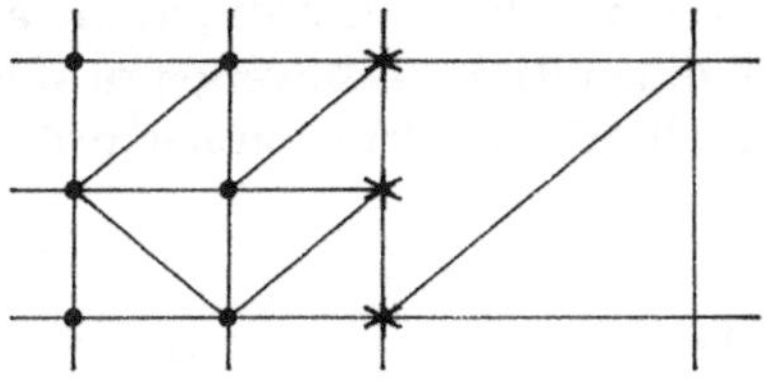

Fig. 3: Green points

Results on adaptive grid

A shock reflection problem with incident angle of 29° and inlet Mach number equal to 2.9 has been considered as a suitable test for the solution adaptive strategy described above. Figs. 4 and 5 show the Mach contours obtained when using the 192×64 standard diamond grid and optimal diagonal grid, respectively. A strong improvement in the resolution of the shocks is achieved, only a small amount of extra work being needed: see fig. 6, which shows the convergence histories for the optimal diagonal grid (dashed line) and for the standard diamond grid (dotted line). The low convergence rate in the initial adaptive stage is due to the dynamic adaptation of the diagonals.

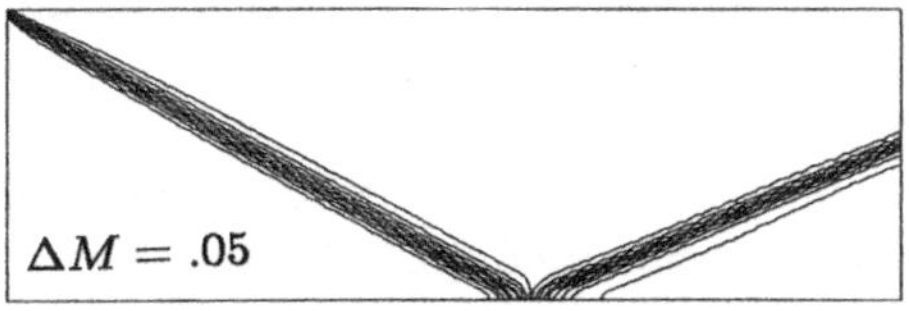

Fig. 4: Iso-Mach — diamond grid

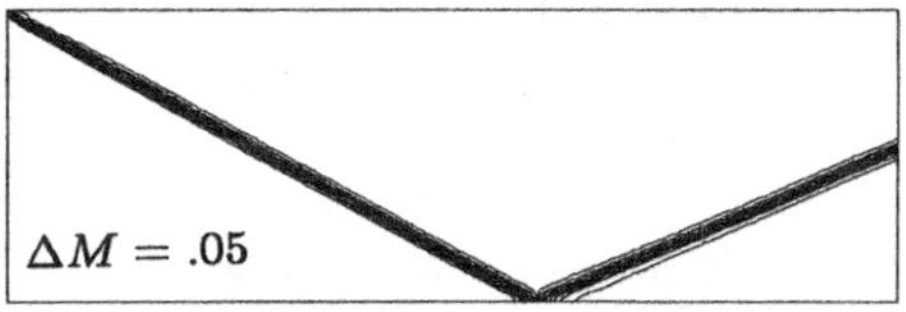

Fig. 5: Iso-Mach — adapted diagonals

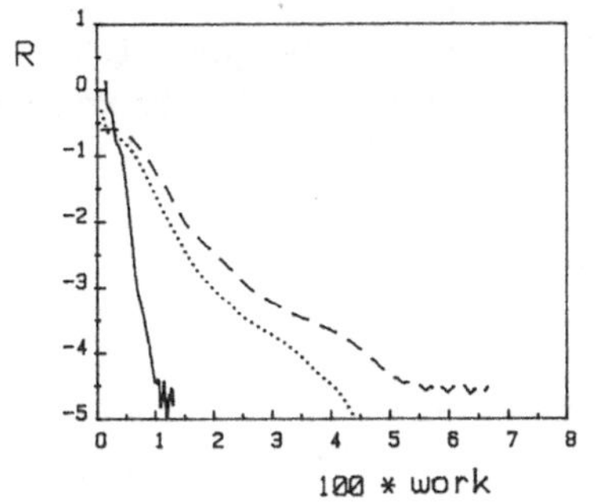

Fig. 6: Convergence histories

A considerable efficiency gain is finally obtained without loss of accuracy, when using local refinement. The composite grid shown in fig. 8, obtained, starting from a 12×4 uniform grid, after four nested local

Fig. 7: Iso-Mach — locally refined grid

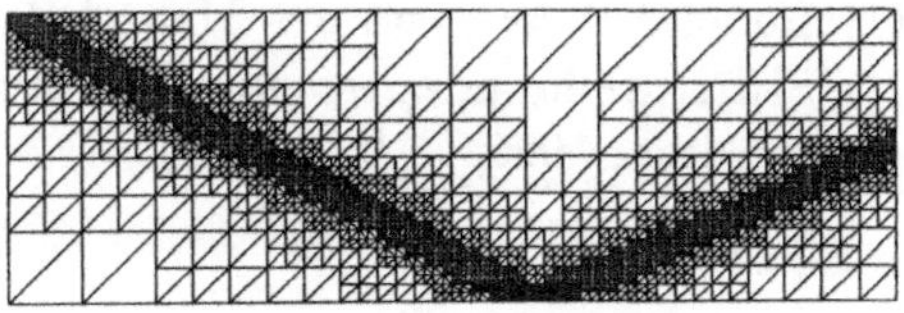

Fig. 8: Composite grid

refinements (2,226 cells), produces a solution (fig. 7) quite similar to the one of fig. 5 (12,288 cells), at a much lower cost: see the solid line in fig. 6.

Conclusions

The encouraging results, obtained using the simple wave decomposition approach in conjunction with the fluctuation splitting space discretization, stimulate further work to improve such a methodology. Some difficulties, as convergence stagnation, still need to be investigated. The explicit multigrid acceleration has proven effective in all cases considered so far.

Acknowledgements

The present research was carried out as part of the BRITE/EURAM contract AERO-2037-C in the Area 5 (aeronautics). The authors thank prof. M. Napolitano for his continuous guidance and prof. A. Dadone for some stimulating discussions.

References

[1] Davis S. F., "A rotationally biased upwind scheme for the Euler equations", *J. Comp. Phys.*, 56, 1984, pp. 65-92.

[2] Deconinck H., Hirsch C., Peuteman J., "Characteristic Decomposition Methods for the Multidimensional Euler Equations", *Lecture Notes in Physics*, 264, Springer Verlag, 1986, pp. 216-221.

[3] Roe P. L., "Discrete models for the numerical analysis of time-dependent multidimensional gas dynamics", *J. Comp. Phys.*, 63, 1986, pp. 458-476.

[4] Struijs R., Roe P. L., Deconinck H., "Fluctuation splitting schemes for the 2D Euler equations", *VKI-LS* 1991-01, von Karman Institute, Belgium, 1991.

[5] De Palma P., Deconinck H., Struijs R., "Investigation of Roe's 2D wave decomposition models for the Euler equations", $VKI - TN$ 172, Belgium, 1990.

[6] Struijs R., Deconinck H., De Palma P., Roe P. L., Powell K. G., "Progress on Multidimensional Upwind Euler Solver for Unstructured Grids", AIAA Paper 91-1550-CP, Honolulu, 1991.

[7] Catalano L. A., De Palma P., Napolitano M., "Explicit multigrid smoothing for multidimensional upwinding of the Euler equations", *NNFM*, 35, Vieweg, Braunschweig, 1992, pp. 69-78.

[8] Dadone A., "A numerical technique to compute Euler flows at impermeable boundaries based on physical considerations", 4th Int. Conf. on Hyperbolic Problems, Taormina, Italy, 1992.

[9] Brandt A., "Guide to multigrid development", *Lecture Notes in Mathematics* 960, Springer Verlag, Berlin, 1982, pp. 220-312.

[10] Pascazio G., Deconinck H., "Multidimensional upwind scheme for the Euler equations on quadrilateral grids", $VKI - PR$ 1991-15, Belgium, 1991.

[11] Hemker P. W., van der Maarel H. T. M., Everaars C. T. H., "BASIS: A Data Structure for adaptive multigrid computations", Report NM-R9014, CWI, Amsterdam, 1990.

A MULTI-DIMENSIONAL UPWIND SCHEME FOR THE EULER EQUATIONS

A. Dadone

Istituto di Macchine ed Energetica
Politecnico di Bari
70125 Bari, Italy

B. Grossman

Dept. Aerospace and Ocean Eng.
Virginia Polytechnic Inst. & State Univ.
Blacksburg, Virginia 24061, USA

INTRODUCTION

Among the numerical methods available for inviscid, compressible flows, upwind methods have the property of approximately mimicking physical wave-propagation phenomena. When dealing with one-dimensional flow problems, the choice of appropriate upwind directions is straightforward, since characteristic information may only be propagated forward or backward. For two or three-dimensional flow problems, the choice is more difficult, with most schemes using directions which are normal to the cell faces. Such grid dependent schemes do not always account for the domain of dependence of each grid point. The consequences are particularly evident when computing shocks oblique with respect to the computational mesh. In such a case, the domain of dependence is violated due to the fact that in the vicinity of the shock, the velocity component in one of the splitting directions may be subsonic, and accordingly, numerical information emanating downstream of the shock may reach the upstream region, causing excessive smearing of the oblique shock.

The authors have presented the rudiments of a rotated upwind scheme in Ref. 1, where flux-difference splitting has been applied along two orthogonal directions for each cell face. The directions were determined on the basis of pressure gradient data using procedures similar to those in Ref. 2. Rotated extrapolations and flux limiting were used to generate nominal second-order accuracy. Flux evaluations were based on characteristic variables and a form of characteristic boundary conditions was implemented. The method was evaluated on the basis of an oblique shock reflecting off a flat plate. The objective of the present effort is to develop a multi-dimensional upwind scheme able to satisfy the domain of dependence of each mesh cell by extending the previously outlined technique to general, non-Cartesian coordinate systems.

ROTATED FORMULATION

The first step of the procedure is the determination for each cell face A–B in Fig. 1, of the pressure gradient direction, $\hat{n}_\perp = \pm\nabla p/|\nabla p|$. The sign is selected such that $\hat{n}_\perp \cdot \hat{n} \geq 0$ with $\hat{n}$ being normal to cell face A–B. In the vicinity of a shock wave, $\hat{n}_\perp$ will be the direction normal to the shock. We also define a direction $\hat{n}_\parallel$ perpendicular to $\hat{n}_\perp$ and oriented such that $\hat{n}_\parallel \cdot \hat{n} \geq 0$. The details of the procedure follow from concepts presented in Ref. 2 and are described in Ref. 1. We define α to be the angle between $\hat{n}_\perp$ and $\hat{n}$ so that $-\pi/2 \leq \alpha \leq +\pi/2$.

Next, the flux $\boldsymbol{F}$ through cell face A–B is decomposed into two components, $\boldsymbol{F}_\perp$ in the direction of $\hat{n}_\perp$ and $\boldsymbol{F}_\parallel$ in the direction of $\hat{n}_\parallel$, such that

$$\boldsymbol{F}\Delta s = \boldsymbol{F}_\perp \Delta s \cos(\alpha) + \boldsymbol{F}_\parallel \Delta s \sin|\alpha|\,,$$

with Δs being the length of A–B. $\boldsymbol{F}_\perp$ and $\boldsymbol{F}_\parallel$ will be computed with different stencils.

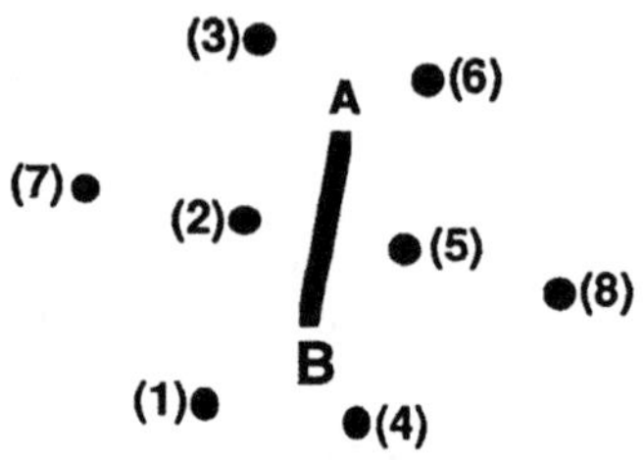

Fig. 1. *Cell face A–B and surrounding cell centers.*

The procedure for computing the first-order stencils may be illustrated by considering cell face A–B and the 6 neighboring cell centers, labeled (1) – (6) in Fig. 1. For the situation where $\hat{n}_\parallel$ is oriented inside the quadrant formed by connecting the cell centers (3)–(5)–(1), $\boldsymbol{F}_\parallel$ will be computed using the data of (2) for the left state and (5) for the right state. Correspondingly, $\boldsymbol{F}_\perp$ is computed using the data (3) and (4) if $\alpha < 0$ and using (1) and (6) if $\alpha > 0$.

For the opposite case where $\hat{n}_\parallel$ is oriented outside the quadrant formed by (3)–(5)–(1), the roles of $\boldsymbol{F}_\parallel$ and $\boldsymbol{F}_\perp$ are reversed. Namely $\boldsymbol{F}_\perp$ is computed from (2) and (5) and $\boldsymbol{F}_\parallel$ is computed from (3) and (4) for $\alpha > 0$ and from (1) and (6) for $\alpha < 0$.

Nominal second-order extrapolations have been determined for $\boldsymbol{F}_\parallel$ and $\boldsymbol{F}_\perp$. Let us label the first-order stencil cells as (V) and (W), located respectively on the left and right sides of A–B. The first step of our procedure is to estimate the value of the vector of characteristic variables, $\boldsymbol{Q}$ at the left cell face. We denote this estimate as $\boldsymbol{Q}^-$ and obtain its value by a standard mesh-oriented extrapolation as $\boldsymbol{Q}^- = \boldsymbol{Q}_2 + (\boldsymbol{Q}_2 - \boldsymbol{Q}_7)/2$. Then forward and backward differences are evaluated using $\boldsymbol{Q}^-$ and the first-order stencil values $\boldsymbol{Q}_V$ and $\boldsymbol{Q}_W$ as $\nabla\boldsymbol{Q}^- = 2(\boldsymbol{Q}^- - \boldsymbol{Q}_V)$ and $\Delta\boldsymbol{Q}^- = \boldsymbol{Q}_W - \boldsymbol{Q}_V$. Using a limiter on these differences will produce a limited backward difference $\bar{\nabla}\boldsymbol{Q}^-$ which can be used to produce a formally second-order left state $\boldsymbol{Q}_L$ as $\boldsymbol{Q}_L = \boldsymbol{Q}_V + \bar{\nabla}\boldsymbol{Q}^-/2$.

A similar procedure is used for the right cell face. An estimate $\boldsymbol{Q}^+$ is obtained using a standard mesh oriented extrapolation as $\boldsymbol{Q}^+ = \boldsymbol{Q}_5 - (\boldsymbol{Q}_8 - \boldsymbol{Q}_5)/2$. Then forward and backward differences are evaluated using $\boldsymbol{Q}^+$ and the first-order stencil values $\boldsymbol{Q}_V$ and $\boldsymbol{Q}_W$ as $\nabla\boldsymbol{Q}^+ = \boldsymbol{Q}_W - \boldsymbol{Q}_V$ and $\Delta\boldsymbol{Q}^+ = 2(\boldsymbol{Q}_W - \boldsymbol{Q}^+)$. The final right state is $\boldsymbol{Q}_R = \boldsymbol{Q}_W - \bar{\Delta}\boldsymbol{Q}^+/2$, where $\bar{\Delta}\boldsymbol{Q}^+$ is the limited right forward difference.

In our calculations we have used the Chakravarthy-Osher limiter such that

$$\bar{\nabla}\boldsymbol{Q}^- = \nabla\boldsymbol{Q}^- \max[0, \min(\Delta\boldsymbol{Q}^-/\nabla\boldsymbol{Q}^-, 2)]\,,$$
$$\bar{\Delta}\boldsymbol{Q}^+ = \Delta\boldsymbol{Q}^+ \max[0, \min(\nabla\boldsymbol{Q}^+/\Delta\boldsymbol{Q}^+, 2)]\,.$$

We utilize Roe's approximate Riemann solver for $\boldsymbol{F}_\parallel$ and $\boldsymbol{F}_\perp$. The boundary conditions have been determined using the improved reflection procedures of Ref. 3.

Further details of the basic numerical scheme appear in Ref. 1 including a description of the *classical* upwind scheme used in the following comparative results.

RESULTS and DISCUSSION

Our first goal was to test the ability of the new scheme to treat oblique shocks on arbitrary grids, but in a controlled setting. We utilized the same shock reflection from a flat plate described in Ref. 1, but with a different grid. Instead of the uniform Cartesian grid, we adapted a polar grid which subjects the shock to continuously changing angles between the shock and the cell faces. We considered the flat-plate reflection of an incoming shock penetrating a $M_\infty = 2.9$ free stream at an angle of 29°, computed with a 32 by 16 *polar* grid. The computed isobar patterns and pressure distributions, (not shown here because of space limitations), have reaffirmed the previous uniform grid results in Ref. 1. Namely, that the rotated scheme computed the shock with only 2 cell centers located within the shock transition region compared to classical results which contained at least 4 cell centers.

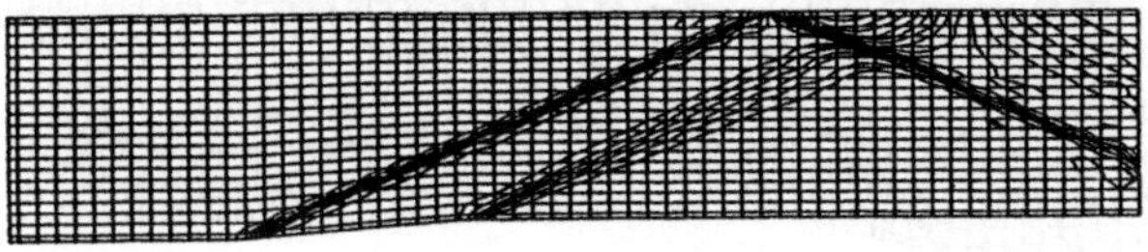

Fig. 2a. *Isobars with rotated formulation, 5°-ramp problem, M=3.0.*

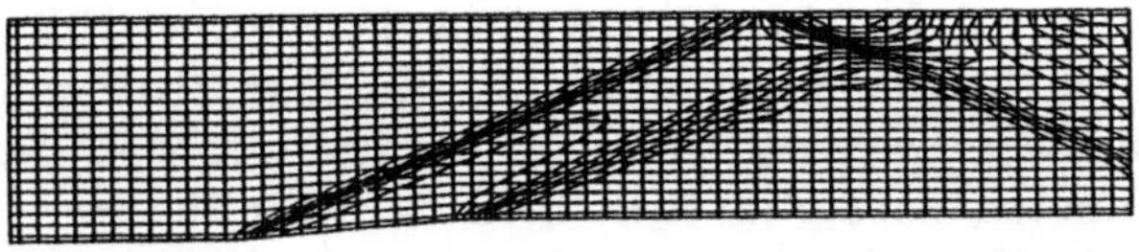

Fig. 2b. *Isobars with classical formulation, 5°-ramp problem, M=3.0.*

The first test case presented is the $M = 3.0$ flow through a channel with a 5° compression ramp followed by a 5° expansion corner. Computed isobars for a 61 by 21 grid for the rotated scheme are presented in Fig. 2a and for the classical scheme in Fig. 2b. The pressure distributions along the upper wall and along a line located at 62.5% of the channel height are shown in Figs. 3a and 3b respectively.

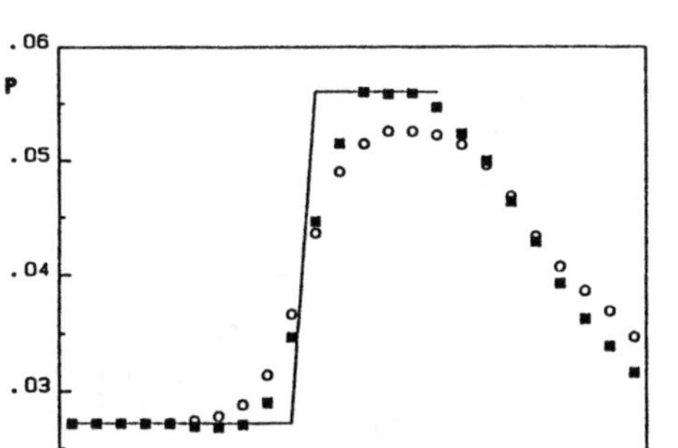

Fig. 3a. *Pressure distribution on upper wall, 5°-ramp problem, M=3.0.*

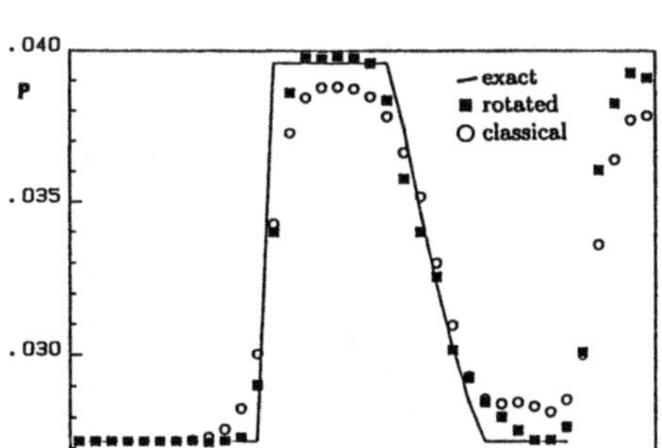

Fig. 3b. *Pressure distribution at 62.5% channel height.*

The rotated and classical results are compared to the exact solution. These results reaffirm the findings that the rotated formulation gives improved shock resolution. In addition, the rotated scheme gives the correct post-shock pressure levels, which indicates an improved accuracy over the classical scheme.

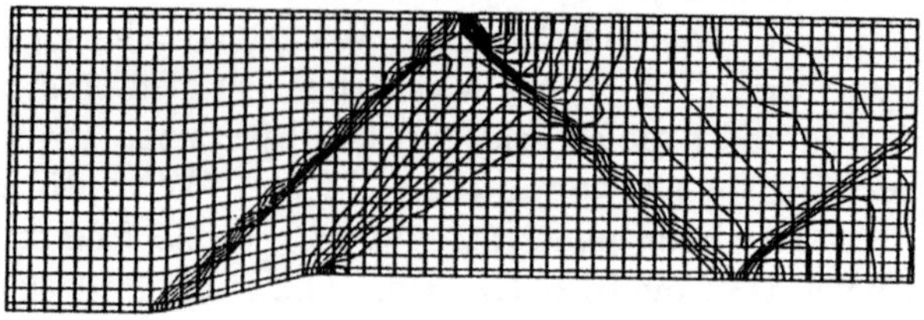

Fig. 4a. *Isobars with rotated formulation, 15°-ramp problem, M=2.0.*

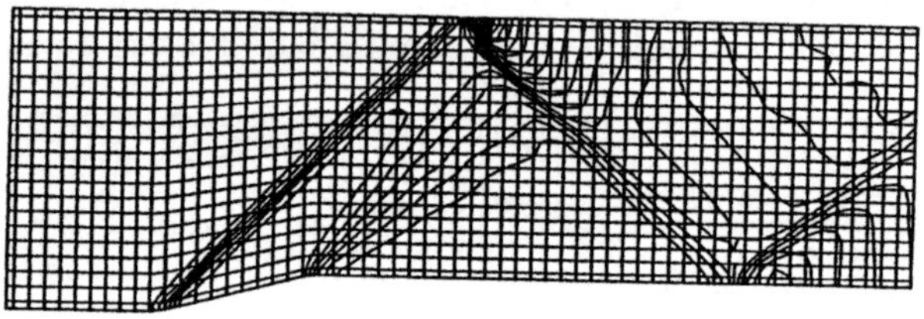

Fig. 4b. *Isobars with classical formulation, 15°-ramp problem, M=2.0.*

Next, we considered the $M = 2.0$ flow through a channel with a 15° compression ramp followed by a 15° expansion corner. In this case, the ramp shock is slightly weakened by the corner expansion fan and results in a Mach reflection from the upper wall. Computed isobars for a 61 by 21 grid for the two schemes are given in Figs. 4a and 4b. The Mach number distribution along the upper wall and the pressure distribution along a line located at 52.5% of the channel height are shown in Figs. 5a and 5b respectively. These figures again confirm our findings that the rotated scheme accurately reproduces the wave-like features of these flows with a higher level of accuracy than corresponding classical upwind schemes (as noted by the resolution of the expansion wave in Fig. 5b). Although the shock angle is steeper (close to 45°) for this case, the rotated formulation still produces a shock whose thickness is reduced by about 50% of the classical result.

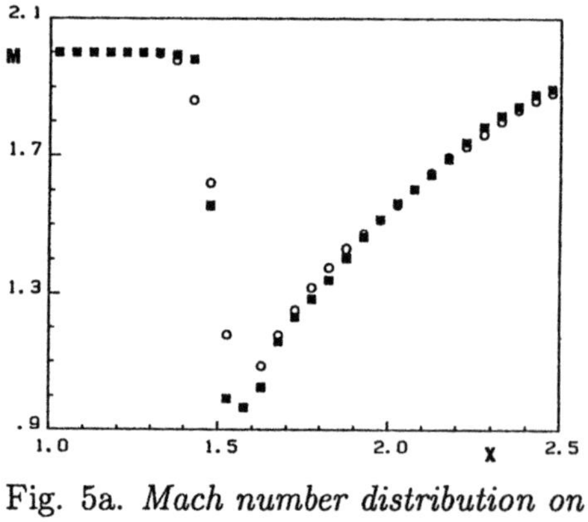

Fig. 5a. *Mach number distribution on upper wall, 15°-ramp problem, M=2.0.*

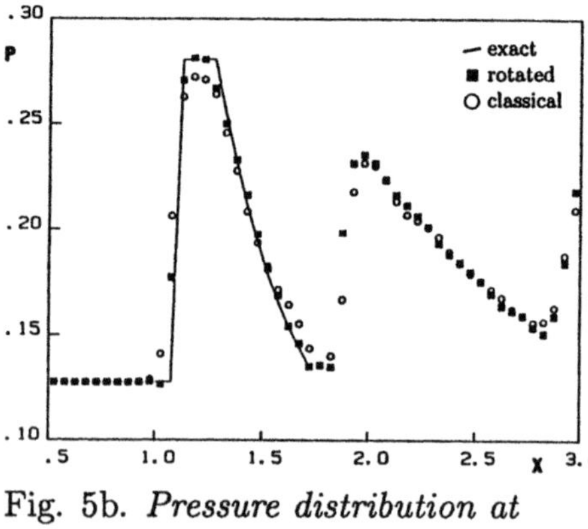

Fig. 5b. *Pressure distribution at 52.5% channel height.*

The final case presented is a $M = 1.4$ flow through a channel with a 4% circular-arc bump on the lower wall with a height-to-chord ratio of 1. The computed Mach contours for the rotated and classical schemes on a grid of 61 by 32 (with 32 of the 61 on the bump surface) are presented in Figs. 6a and 6b, respectively. The main features of the flow, including the λ-shock on the upper wall and the interaction of the reflected shock with the trailing-edge shock, are evident. We have affirmed grid convergence for this case with calculations of 48 and 64 cells in the normal direction and along the surface of the bump. In this case the shock angles are nearly 60° and the advantages of the rotated formulation should be diminished. Nonetheless, the shock width has been reduced by roughly 25–30%. The λ-shock and the subsonic region behind the Mach stem appear to be accurately described. The minimum Mach number computed for the grid with 32 cells on the bump is 0.89. For the grids of 48 and 64 cells on the surface of the bump, this value is reduced to 0.84 and 0.81, respectively.

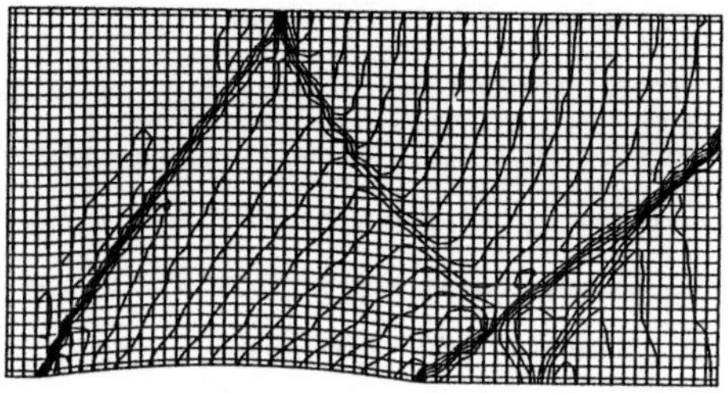

Fig. 6a. *Mach contours with rotated formulation, 4% circular-arc bump in channel, M=1.4.*

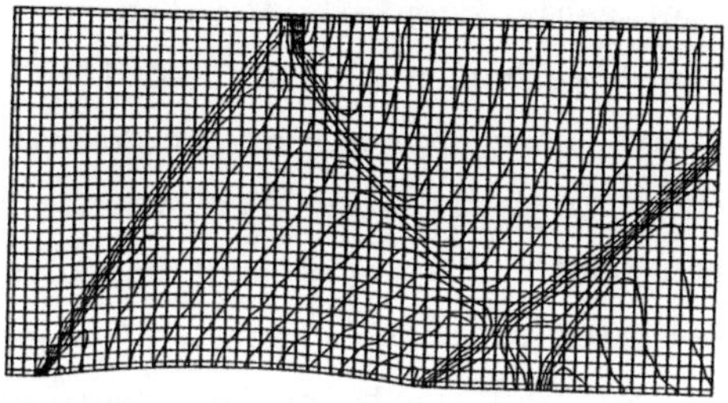

Fig. 6b. *Mach contours with classical formulation;* $\Delta M = 0.05$.

ACKNOWLEDGMENT

The research of the first author was supported by the Italian agency *MURST*. The second author acknowledges the assistance of the National Science Foundation under the U.S.-Italy Cooperative Science Program, grant INT–8814895.

REFERENCES

1. Dadone, A. and Grossman, B., "A Rotated Upwind Scheme for the Euler Equations," AIAA Paper 91–0635, Jan. 1991, (to appear *AIAA J.*, **30**, No. 10, 1992.).
2. Levy, D. W., Powell, K. G. and Van Leer, B., "An Implementation of a Grid-Independent Upwind Scheme for the Euler Equations", AIAA 89–1931–CP, June 1989, (submitted to *J. Comput. Physics*).
3. Dadone, A.,"A Physical-Numerical Treatment of Impermeable Boundaries in Compressible Flow Problems", Proceedings of the Fourth International Symposium on Computational Fluid Dynamics, Davis CA, Sept. 1991, pp. 258–263.

ON A METHOD TO CONSTRUCT GODUNOV-TYPE SCHEMES

Valery V. Dedesh

TsAGI, Central Aerohydrodynamics Institute, 140160. Zhukovsky-3, Moscow Region, Russia, fax: (095)271.00.19

INTRODUCTION

Shock capturing schemes to compute discontinuous solutions of gas-/hydrodynamics must be monotone. First-order accurate Godunov-type schemes are monotone, but methods of their construction [1,3-6] are sometimes impracticably complex. The Godunov-type schemes, constructed using the proposed method PPV (Polynomially Presented numerical Viscosity matrix [2]) are more convenient (from different points of view), than other schemes. In particular, PPV schemes are extremely cheap, easy-to-program, do not need characteristic transformations to operate and allow to bypass the problem of nonuniqueness of eigenvector basis [6], arising when computing multi-D, turbulent or chemical/thermal nonequilibrium flows.

DEFINITIONS

A numerical solution of gasdynamics conservation laws,

$$\partial w/\partial t+\partial f(w)/\partial x=0 \tag{1}$$

where w is a conservative variables vector and f(w)-the corresponding vector of fluxes, is sought on the uniform grid, whose steps in time τ and space h are chosen to verify the stability condition (CFL) $|a_k|\cdot\lambda\leq\mu\leq1$, $(\lambda=\tau/h)$ for every eigenvalue a_k (real due to hyperbolicity of (1)) of the matrix $A(w)=\partial f(w)/\partial w$. With the following definitions:

---

x_i-abscissa of the i-th grid point

$x_{\frac{1}{2}}=(x_0+x_1)/2$-abscissa of the interface between x_0 and x_1

w_0, w^0-the value of w at the moment $t=0, t=\tau$ at x_0

$\Phi_{\frac{1}{2}}$-the value of Φ at the interface $x_{\frac{1}{2}}$ at the moment t=0

$\Delta_{\frac{1}{2}}\Phi=\Phi_1-\Phi_0$

---

the conservative schemes to numerically compute (1) are

$$w^0 = w_0 - \lambda \cdot (f_{\frac{1}{2}} - f_{-\frac{1}{2}}), \tag{2}$$

where $f_{\frac{1}{2}} = f_{\frac{1}{2}}(w_0, w_1)$, a function of two arguments, is either

$$f_{\frac{1}{2}} = \tfrac{1}{2} \cdot (f_0 + f_1 - Q_{\frac{1}{2}} \cdot \Delta_{\frac{1}{2}} w) \tag{3a}$$

for "FDS" schemes [3] or

$$f_{\frac{1}{2}} = \tfrac{1}{2} \cdot (f_0 + f_1 - Q_1 \cdot w_1 + Q_0 \cdot w_0) \tag{3b}$$

for "FVS" schemes [4,5]. The numerical viscosity matrices **Q** in (3) are introduced to duly account for signal distributions in (1).

THE METHOD

Let the positive matrices **Q** commute with **A**: $\mathbf{A} \cdot \mathbf{Q} = \mathbf{Q} \cdot \mathbf{A}$, then schemes (3) are stable and monotone, if the conditions

$$\lambda^2 \cdot \mathbf{A}^2 \leq \lambda \cdot \mathbf{Q} \leq \mathbf{I} \tag{4a}$$

$$\lambda \cdot |\mathbf{A}| \leq \lambda \cdot \mathbf{Q} \leq \mathbf{I} \tag{4b}$$

respectively are met for every eigenvalue of **Q** and **A**. Here, the eigenvalues of $|\mathbf{A}|$ are <u>moduli</u> of those of **A** and **I** is a unit matrix. Within the method PPV [2] the matrices Q are expressed as polynomes in **A**:

$$Q = q_0 \cdot I + q_1 \cdot A + q_2 \cdot A^2 \tag{5}$$

The coefficients $q_{0,1,2}$ are functions of the three distinct eigenvalues u-c,u,u+c of **A** (u is a gas velocity and c-speed of sound, with $M=u/c$). To construct schemes, optimal from the point of view (4) one should choose the spectrum of **Q** to be close to that of $|\mathbf{A}|$. The Table 1 contains $q_{0,1,2}$ for the (so-called in what follows) "two-wave", "three-wave" and "smooth" models of PPV, for the subsonic case $|u|<c$ (when the three PPV models differ).

For the three-wave PPV model ($\mathbf{Q} \equiv |\mathbf{A}|$), the eigenvalues of **Q** are $|u-c|, |u|, |u+c|$. The spectrum of $\mathbf{Q} \approx |\mathbf{A}|$ is $|u-c|, c, |u+c|$ for the two-wave and $|u-c| \cdot (1-(M+1)^2/2), c \cdot (1+M^2)/2, |u+c| \cdot (1-(M-1)^2/2)$ for the smooth PPV model. The two-wave PPV model is the simplest; the smooth model is nicknamed so for continuous differentiability of all its q's (especially in sonic and stagnation points) and is almost as simple as the two-wave one; the three-wave PPV model is the most exact.

All the PPV models in the supersonic case $|u|>c$ are the same with $q_0 = q_2 = 0, q_1 = \mathrm{sign}(M)$.

FIRST ORDER SCHEMES

Using (5) with $q_{0,1,2}$ of Table 1, one writes (3a) as

$$f_{\frac{1}{2}}=\tfrac{1}{2}\cdot(f_0+f_1-(q_0)_{\frac{1}{2}}\cdot\Delta_{\frac{1}{2}}w-(q_1)_{\frac{1}{2}}\cdot\Delta_{\frac{1}{2}}f-(q_2)_{\frac{1}{2}}\cdot A_{\frac{1}{2}}\cdot\Delta_{\frac{1}{2}}f) \tag{6}$$

We stress here that the matrix $A_{\frac{1}{2}}$ and coefficients $(q_{0,1,2})_{\frac{1}{2}}$, used in (6) to represent $|A|_{\frac{1}{2}}$, are computed using Roe [3] averaging procedure, so that the relation $A_{\frac{1}{2}}\cdot\Delta_{\frac{1}{2}}w=\Delta_{\frac{1}{2}}f$ holds exactly. The schemes (2) with (6), using the two- or three-wave PPV models are capable of resolving exactly the motion of the shock, formed by two constant states w_0 and w_1 and moving with speed D, and in particular of capturing the stationary shock between x_0 and x_1.

For the schemes (3b) one uses (5) and the exact relation $A\cdot w=f$ [4] to write

$$f_{\frac{1}{2}}=\tfrac{1}{2}\cdot(f_0+f_1-(q_0)_1\cdot w_1-(q_1)_1\cdot f_1-(q_2)_1\cdot A_1\cdot f_1+ \tag{7}$$
$$+(q_0)_0\cdot w_0+(q_1)_0\cdot f_0+(q_2)_0\cdot A_0\cdot f_0),$$

the second subscript in expressions like $(q_0)_1$ indicates the node, where this coefficient is computed.

The three-wave model of PPV gives for (3a) the algorithm of Roe [3] and for (3b)-of Steger-Warming [4]. The two-wave model of PPV leads to schemes, not widely used today (see e.g.[1]). The scheme (3a) with the smooth model of PPV is unstable, whereas (3b) with the same model yields a modified scheme of van Leer [5]. The three models of PPV differ in the subsonic case $|u|<c$, for the supersonic case $|u|>c$ they all yield first order accurate upwind schemes. The CFL condition for all of them is $\mu\cong 1$.

SECOND ORDER SCHEMES

To construct second order schemes one introduces the linear distribution of initial data within every computational cell:

$$w(\xi)=w_0+\xi\cdot\nabla_0 w,\quad \xi\in[-\tfrac{1}{2},\tfrac{1}{2}], \tag{8}$$

with the vector $\nabla_0 w=\text{minmod}(\Delta_{\frac{1}{2}}w,\Delta_{-\frac{1}{2}}w)$ [6], which serves to compute $\nabla_0 f=f(w(\tfrac{1}{2}))-f(w(-\tfrac{1}{2}))$. Let

$$q^{\pm}_{0,2}=\pm\tfrac{1}{2}\cdot q_{0,2},\qquad q^{\pm}_1=\tfrac{1}{2}\cdot(1\pm q_1), \tag{9}$$

then the vectors w_L and w_R, playing the role of w_0 et w_1 in (3a,b) are:

$$w_L=w_0+\tfrac{1}{2}\cdot[(1-\lambda\cdot(q^+_0)_0)\cdot\nabla_0 w-\lambda\cdot(q^+_1)_0\cdot\nabla_0 f-\lambda\cdot(q^+_2)_0\cdot A_0\cdot\nabla_0 f] \tag{10}$$
$$w_R=w_1-\tfrac{1}{2}\cdot[(1+\lambda\cdot(q^-_0)_1)\cdot\nabla_1 w-\lambda\cdot(q^-_1)_1\cdot\nabla_1 f-\lambda\cdot(q^-_2)_1\cdot A_1\cdot\nabla_1 f]$$

with $q_{0,1,2}$ and A computed in the corresponding nodal point, whichever is the model of PPV.

Remark: if in (10) $q_0=q_2=0$, $q^{\pm}_1=1$, one arrives at the second-order scheme MUSCL due to van Leer [6].

NUMERICAL RESULTS

To check the proposed algorithms, we have conducted a lot of calculations of real and academic flow problems, for different eigensystems and with very encouraging results. Fig.1 depicts the isobaric curves of the Navier-Stokes flow of an incompressible fluid (with pseudocompressibility $\beta=1$) past the RAE2822 airfoil, computed using the second-order two-wave PPV model FDS scheme and direct Newton method, realized in the TsAGI **A+M** [7] CFD package for large scale computations.

Acknowledgements

The author is greatly indebted to Profs. V.Ya.Neiland, V.V.Rusanov, P.J.Zandbergen, P.E.Babikov and M.Napolitano, for their support and valuable discussions.

REFERENCES

1. A.Harten,P.D.Lax,B.van Leer, SIAM Review,(1983),V.25,P.35-61.
2. V.V.Dedesh, DAN USSR,(1991),V.321,P.36-39.(in Russian)
3. P.L.Roe, J.Comput.Phys.,(1981),V.43,P.357-372.
4. J.L.Steger,R.F.Warming, J.Comput.Phys.,(1981),V.40,P.263-293.
5. B.van Leer, ICASE Report 82-30,Sept.1982,(unpublished).
6. J.-L.Montagné,H.C.Yee,M.Vinokur,AIAA Journal,(1987),V.27, P.1332-1346.
7. I.V.Egorov,O.L.Zaycev,J.Comput.Math.&Math.Phys.,(1991),V.31,P.300-313. (in Russian)

Table 1: The coefficients of subsonic† ($|u|\le c$) expansion of $Q\approx|A|$

coef-fici-ent.	Numerical viscosity matrix (PPV) model:		
	two-wave	three-wave	smooth
q_0	$c\cdot(1-M^2)$	$c\cdot\|M\|\cdot(1-M^2)$	$\frac{1}{2}\cdot c\cdot(1-M^2)^2$
q_1	M	$M\cdot(2\cdot\|M\|-1)$	$\frac{1}{2}\cdot M\cdot(3-M^2)$
q_2	0	$(1-\|M\|)/c$	0

† - for supersonic flows ($|u|>c$) $q_{0,2}=0$, $q_1=\text{sign}(M)$.

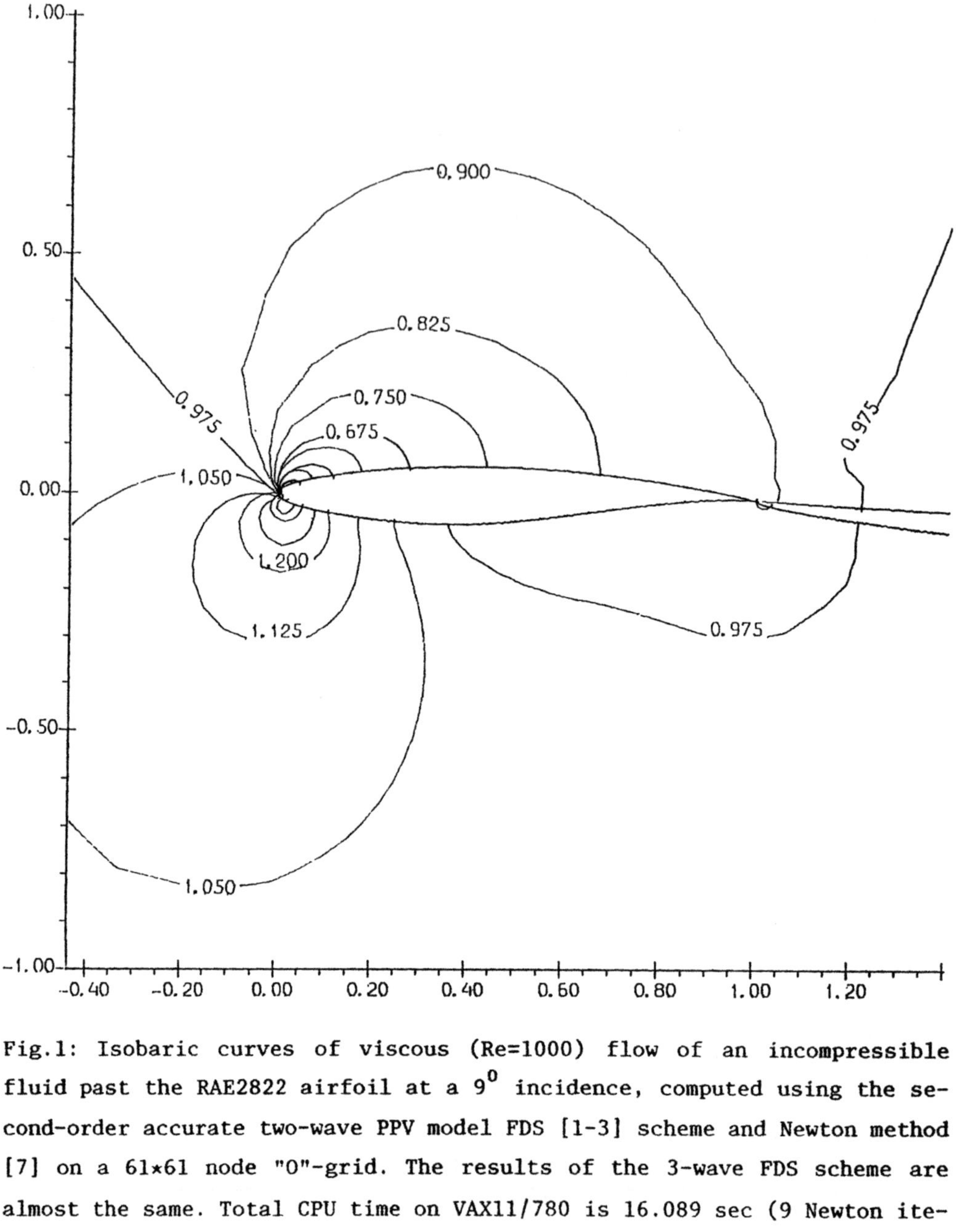

Fig.1: Isobaric curves of viscous (Re=1000) flow of an incompressible fluid past the RAE2822 airfoil at a 9^0 incidence, computed using the second-order accurate two-wave PPV model FDS [1-3] scheme and Newton method [7] on a 61*61 node "O"-grid. The results of the 3-wave FDS scheme are almost the same. Total CPU time on VAX11/780 is 16.089 sec (9 Newton iterations)

3-DIMENSIONAL UPWIND EULER SOLVER USING KINETIC FLUX VECTOR SPLITTING METHOD

S Deshpande[*], S Sekar[§], M Nagarathinam[§], R Krishnamurthy[§], P Sinha[§], P Kulkarni[*]

* Department of Aerospace Engineering, I.I.Sc., Bangalore, 560012
§ Defence Research & Development Laboratory, Hyderabad, 500258
INDIA

1.0 Introduction

Euler codes based on Kinetic Flux Vector Splitting (KFVS) method due to Deshpande[1] and Mandal[2] have been used to solve variety of 2-D problems covering Mach numbers range varying from subsonic to hypersonic speeds([2],[3]). In this paper KFVS method has been used to develop a 3-D code called BHEEMA(Boltzmann Hypersonic Euler Equation Solver for Missile Aerodynamics) to compute inviscid hypersonic flows around reentry configurations.

2.0 Kinetic Flux Vector Splitting Method

The KFVS approach of constructing an upwind scheme for Euler equations due to Deshpande & Mandal [1,2], starts with the Boltzmann equation of Kinetic theory of gases

$$\frac{\partial f}{\partial t} + \vec{v}\cdot\frac{\partial f}{\partial \vec{x}} = J(f,f) \ . \tag{1}$$

which governs the spatio-temporal evolution of the velocity distribution f which is a function of t, $\vec{x}$ and the molecular velocity $\vec{v}$. The structure of the collision term J(f,f) does not matter in the present study as it vanishes in the Euler limit in which case the velocity distribution function f is a Maxwellian given by

$$F = \frac{\rho}{I}\left(\frac{\beta}{\pi}\right)^{3/2}\exp\left[-\beta(\vec{v}-\vec{u})^2 - I/I_o\right] \ , \tag{2}$$

where

$\beta = 1/(2RT)$, ρ = mass density, R = gas constant,

T = temperature, $\vec{v}$ = molecular velocity, $\vec{u}$ = fluid velocity,

γ = ratio of the specific heats

I = internal energy variable corresponding to nontranslational degrees of freedom which enforces the prescribed value of γ for the gas.

$I = (5-3\gamma)RT/(2\gamma-2)$.

Introducing a moment function vector Ψ by

$$\Psi = [\ 1,\ \vec{v}\ , I + v^2/2\ ]^T \tag{3}$$

and the notation

$$\langle\Psi,F\rangle = \int_0^{\infty} dI \iiint_{-\infty}^{+\infty} d\vec{v}\ \Psi\ F \ . \tag{4}$$

$\vec{U} = \langle \psi , F \rangle \equiv [\ \rho, \rho u\ , \rho v\ , \rho w\ , \rho e\ ]$ and $\vec{G} = \langle\psi\ , \vec{v}F\rangle$.

the Euler equations reduce to the compact form

$$< \Psi , \frac{\partial F}{\partial t} + \vec{v}\cdot\frac{\partial F}{\partial \vec{x}} > = 0 \tag{5}$$

2.2 KFVS Finite Volume Scheme

The Euler equations in the integral form are

$$\frac{\partial}{\partial t}\iiint_V \langle\Psi , F\rangle\, d^3x + \iiint_{\partial V} <\psi , \vec{v} \cdot \frac{\partial F}{\partial \vec{x}} > d^3x = 0 \tag{6}$$

where V is the volume of cell with boundary ∂V, and $d^3x=dx_1dx_2dx_3$. Using eq.(4) and the divergence theorem, eq(6) gives

$$\frac{\partial}{\partial t}\left(\iiint_V d^3x \int_0^\infty dI \iiint_{-\infty}^{\infty} d^3v\ \Psi\ F\right) = - \iint_{\partial v} dS \int_0^\infty dI \iiint_{-\infty}^{\infty} d^3v\ \vec{v}\ .\vec{n}\ F . \tag{7}$$

where dS is a surface element on the boundary ∂V and $\vec{n}$ is its outward normal.

The equation (7) can be space-discretised as

$$\frac{\partial}{\partial t}\left(\iiint_V d\vec{x}^3 \int_0^\infty dI \iiint_{-\infty}^{\infty} d\vec{v}^3\ \Psi F\right) = -\sum_{l=1}^{6} \int_0^\infty dI \int_{-\infty}^{\infty} dv_{nl} \int_{-\infty}^{\infty} dv_{t1l} \int_{-\infty}^{\infty} dv_{t2l}\, \Psi_l \vec{v}\circ\vec{n}_l\, |s_l|\, F_{sl} \tag{8}$$

where i,j,k are cell indices, V_{ijk} is the cell volume, S_l is any of six (for hexahedral cell) or five (for a prismatic cell) bounding surfaces of V_{ijk}, $\vec{n}_l$= outward normal. Velocity components normal and tangential to the surface S_l are v_{nl}, v_{t1l}, v_{t2l} and ψ_l, F_{sl} are the moment function vector and the Maxwellian written in terms of v_{nl}, v_{t1l}, v_{t2l}.

The above equation can be written compactly as

$$\left(\frac{\partial U}{\partial t}\right)_{ijk} V_{ijk} = -\sum_{l=1}^{6} |s_l|\ G_{sl} \quad , \tag{9}$$

where the flux normal to S_l is given by

$$G_{sl} = <\Psi_l , v_{nl}F_{sl}> \tag{10}$$

The upwinding property can be enforced by splitting in the following manner

$$v_{nl}\ F_{sl} = v^+_{nl}\ F^+_{sl} + v^-_{nl}\ F^-_{sl} \tag{11}$$

$$\text{where} \quad v^+_{nl} = (v_{nl}+|v_{nl}|)/2 \ ; \quad v^-_{nl} =(v_{nl}-|v_{nl}|) / 2 \tag{12}$$

The Velocity distribution F^+_{sl} and F^-_{sl} can be suitably extrapolated depending on the order of the numerical scheme. In the case of the first order scheme the velocity distributions F^+_{sl} and F^-_{sl} for a typical hexahederal finite volume are given by

$$F^+_{s1} = \left(F_1\right)_{i,j,k} , \quad F^-_{s1} = \left(F_1\right)_{i-1,j,k} , \quad F^+_{s2} = \left(F_2\right)_{i,j,k} ,$$

$$F^-_{s2} = \left(F_2\right)_{i+1,j,k} , \quad F^+_{s3} = \left(F_3\right)_{i,j,k} , \quad F^-_{s3} = \left(F_3\right)_{i,j-1,k}$$

$$F^+_{s4} = \left(F_4\right)_{i,j,k} , \quad F^-_{s4} = \left(F_4\right)_{i,j+1,k} , \quad F^+_{s5} = \left(F_5\right)_{i,j,k} ,$$

$$F^-_{s5} = \left(F_5\right)_{i,j,k-1}, \quad F^+_{s6} = \left(F_6\right)_{i,j,k}, \quad F^-_{s6} = \left(F_6\right)_{i,j,k+1} \tag{13}$$

If we denote the flux going out of a cell as G^+_{sl} and the flux coming into the cell as G^-_{sl}, then eq.(9) can be recast as

$$G_{sl} = \left(G^+_{sl}\right) + \left(G^-_{sl}\right) = \langle\psi, v_{nl} F_{sl}\rangle = \langle\Psi, v^+_{nl} F^+_{sl}\rangle + \langle\Psi, v^-_{nl} F^-_{sl}\rangle \tag{14}$$

Using the above expression for G_{sl} in the eq. (10) gives first order cell centered finite volume KFVS scheme.

2.3 Boundary Condition at Wall Surface

A flow tangency boundary condition has been treated using the Kinetic Characteristic Boundary Condition (KCBC). This is based on the specular reflection model of Kinetic Theory of gases, according to which, the normal component of the velocity of a gas molecule gets reversed after impact on the wall, while the tangential components remain unchanged. The reflected velocity distribution F_R is given in terms of incident distribution F_I by

$$F_R(v_{t1w}, v_{t2w}, v_{nw}, I) = F_I(v_{t1w}, v_{t2w}, -v_{nw}, I) \ . \tag{15}$$

The fluxes for the solid boundary surfaces can now be obtained in a manner similar to the split fluxes for other surfaces. It may be mentioned that implementation of the above boundary condition does not require any dummy or ghost cells.

The stability condition for the first order KFVS finite volume scheme [2] is given by

$$\Delta t \le V_{ijk} \Big/ \left[\sum_{l=1}^{3} \left[u_{nl} + 3\sqrt{RT} \ |S_l| \right] \right] \tag{16}$$

3 Finite Volume Network

The finite volume network is constructed by generating a 3-d grid around a configuration (consisting of the blunt cone-cylinder -flare with a set of cruciform lifting surfaces) either by rotating a two dimensional algebraic grid in the meridian plane in the case of purely axisymmetric regions or by using the stacked grid method in the case of regions with lifting surfaces. The algebraic grid generator is based on a distance function approach and is used for the region up to the lifting surface. The stacked grid method is employed in regions with lifting surface and beyond. The basic grid generator employed in generating a grid in a cross-sectional plane is an elliptic grid generator based on Thomas-Middelcoff[4] procedure containing two grid control functions Φ and Ψ. A typical finite volume is constructed from the corresponding quadrilaterals in two proximate cross-sections by using the ruled surfaces from which the geometrical properties of finite volume are computed. Transparency condition[5] is employed in the region behind the trailing edges of the lifting surfaces in order to avoid the extremely rapid turning of the grid lines.

4 Results and Discussions

The code 'BHEEMA" has been validated against standard shapes such as cone, hemi-sphere, and then applied to the configuration to obtain pressure distribution on the body and the aerodynamic coefficients. The Mach numbers covered are from 4 to 14 with angles of attack 0° and 2°. Fig.1 shows the pressure distribution along the body for Mach 6 at zero angle of attack. Table 1 shows the results obtained. These are compared with the data available from semi empirical theory and wind tunnel tests[6,7]. Limited post processing of data has been done to obtain the flow field around the configuration. Fig.2 shows the iso-pressure contours for the configuration without the lifting surfaces(for Mach no.4 α=2 deg.).

From the comparison, the following conclusion can be drawn.

(i) The predictions by CFD are close both to theory and experiment within about 10 % in most of the cases.

(ii) In the case of axial force coefficient, the deviation is more from the experiment than theory especially in the case of Mach 4. One reason, probably, could be due to the base pressure drag measured is on the higher side, which when subtracted from the over all axial force measured by the balance resulted in lower figure.

(iii) As far as normal force coefficient is concerned it is seen that CFD prediction is very close to experiment for Mach 8 whereas for lower Mach numbers the code seems to over predict by about 15 % .

(iv) As regards pitching moment coefficient and center of pressure the code has predicted close to experiment.

(v) For Mach beyond 8, experimental results are not available for comparison.

(vi) From the above observations it may be concluded that the performance prediction for M > 8 is expected to be better than for Mach number less than 8 (which is already within 10%) in view of the asymptotic behaviour of the performance parameters.

ACKNOWLEDGEMENT

The authors are grateful to Dr. APJ Abdul Kalam, Director, D.R.D.L., Gen. Sundaram Associate Director D.R.D.L. for extending valuable financial and administrative support and to Prof. R. Narasimha, Director, NAL for allowing the use of MAGNUM computer during the course of this work.

REFERENCES

1. Deshpande S.M. "Kinetic theory based new upwind methods for inviscid compressible flows",AIAA Paper 86-0275,1986.

2. Mandal J.C. "Kinetic upwind method for inviscid compressible flows", Ph.D. thesis, IISc, Bangalore, India, 1989.

3. Mathur J S.and Weatherill N, Institute for Numerical Methods in Engg., Report No.CR/666/91 of Dept. of Civil Engg., University College of Swansea, U.K.

4. Thomas P.D. and Middlecoff J.F.,AIAA Journal 18, p 652, 1980

5. Borrel M., Montagne J K.,Diet J., Guillen Ph. and Lorden J., "Upwind scheme for computing supersonic flows around a tactical missile", Rech. Aerosp -n , 1988.

6. Panneerselvam S., Balu G. and Sharma R.K., DRDL Report No.5120.1029.521,1991.

7. Vinaya Kumar, NAL Report No. EA 9107, 1991.

Table 1. Comparison with Experiments

Mach	α	C_A		C_N		C_M		X_{cp}	
No.	Degree	CFD	Expt	CFD	Expt	CFD	Expt	CFD	Expt
4	0	0.189	0.153	0.000	0.000	0.000	0.000	--	--
4	2	0.191	0.157	0.110	0.092	0.314	0.268	2.841	2.913
6	0	0.170	0.137	0.000	0.000	0.000	0.000	--	--
6	2	0.172	0.140	0.090	0.076	0.259	0.215	2.858	2.829
8	0	0.155	0.131	0.000	0.000	0.000	0.000	--	--
8	2	0.157	0.134	0.073	0.075	0.232	0.198	2.924	2.713
10	0	0.151	-	0.000	-	0.000	-	-	-
10	2	0.152	-	0.068	-	0.201	-	2.882	-
12	0	0.150	-	0.000	-	0.000	-	-	-
12	2	0.151	-	0.065	-	0.187	-	2.877	-
14	0	0.149	-	0.000	-	0.000	-	-	-
14	2	0.150	-	0.064	-	0.185	-	2.904	-

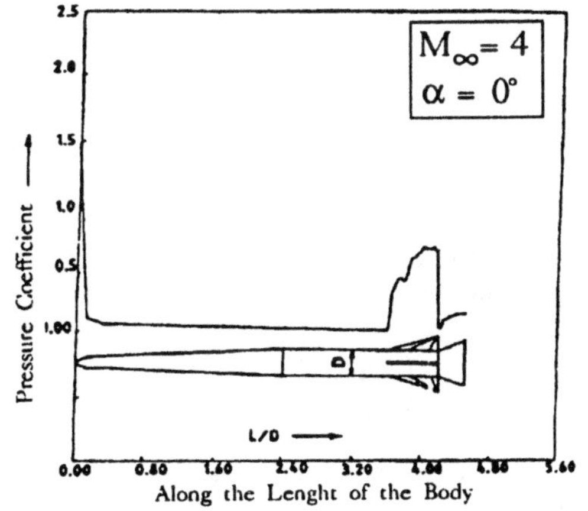

Fig. 1. Pressure distribution along blunt cone-cylinder-flare (with lifting surfaces)

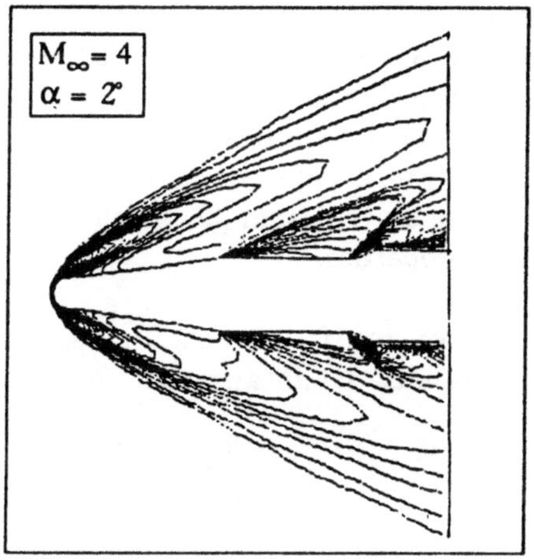

Fig. 2. Iso-pressure contours for the configuration without lifting surfaces

MONOTONE, HIGHER-ORDER ACCURATE, MULTI-DIMENSIONAL UPWINDING

B. Koren and H.T.M. van der Maarel

CWI, P.O. Box 4079, 1009 AB Amsterdam, The Netherlands

1. Introduction

Multi-D upwind discretizations for the steady Euler equations are studied, with the emphasis on both accuracy and solvability. The multi-D upwind schemes to be considered here use neither decoupling of the Euler equations as in [2, 7], nor rotated fluxes as in [1, 5]. The schemes are only based on a rotated interpolation of the left and right cell face states. Per cell face, just as with grid-aligned upwind schemes, only a single numerical flux is computed: the one normal to the cell face. First-order accurate versions of these rotated-interpolation schemes have already been investigated in [3]. In the present paper we make an extension to second-order accuracy for a first-order accurate scheme from [3], which has good solvability properties. We try to maintain these properties.

2. Extension to monotone, second-order accuracy

We consider the linear, scalar, 2-D model equation

$$a\frac{\partial u}{\partial x} + b\frac{\partial u}{\partial y} = 0, \quad 0 \leq \theta \equiv \arctan\left(\frac{b}{a}\right) \leq \frac{\pi}{2}. \tag{1}$$

The extension to second-order accuracy is made for the first-order accurate, four-point compact, rotated-interpolation scheme, with the stencil:

$$\frac{1}{a+b}\begin{bmatrix} \cdot & \cdot & \cdot \\ -a^2 & a^2+ab+b^2 & \cdot \\ -ab & -b^2 & \cdot \end{bmatrix}, \quad 0 \leq \theta \leq \frac{\pi}{2}. \tag{2}$$

Note that the scheme is nine-point compact for the entire θ-range $[0, 2\pi]$. The scheme is differentiable, also at the angles where it switches ($\theta = 0, \frac{\pi}{2}, \pi, \frac{3\pi}{2}$). Further, the scheme's crosswind diffusion is significantly lower than that of the standard, grid-aligned, first-order accurate upwind scheme, whereas it is still positive [3]. If we stick to nine-point schemes which use nearest points only, the natural second-order accurate extension of scheme (2) is the central, rotated-interpolation scheme with stencil:

$$\frac{1}{a+b}\begin{bmatrix} \cdot & \frac{1}{2}b^2 & \frac{1}{2}ab \\ -\frac{1}{2}a^2 & \cdot & \frac{1}{2}a^2 \\ -\frac{1}{2}ab & -\frac{1}{2}b^2 & \cdot \end{bmatrix}, \quad 0 \leq \theta \leq \frac{\pi}{2}. \tag{3}$$

To make scheme (3) monotone, while maintaining second-order accuracy, we first introduce a non-smooth limiter. Next, to even allow successful application of a multigrid-Newton method, we also introduce a smooth limiter. To derive both limiters, consider the following blend of schemes (2) and (3):

$$\frac{1}{a+b}\begin{bmatrix} \cdot & \cdot & \cdot \\ -(1-\omega(r_{\text{hor}}))a^2 & (1-\omega(r_{\text{hor}}))a^2+(1-\omega(r_{\text{skew}}))ab+(1-\omega(r_{\text{ver}}))b^2 & \cdot \\ -(1-\omega(r_{\text{skew}}))ab & -(1-\omega(r_{\text{ver}}))b^2 & \cdot \end{bmatrix}+$$

$$\frac{1}{a+b}\begin{bmatrix} \cdot & \omega(r_{\text{ver}})\frac{1}{2}b^2 & \omega(r_{\text{skew}})\frac{1}{2}ab \\ -\omega(r_{\text{hor}})\frac{1}{2}a^2 & \cdot & \omega(r_{\text{hor}})\frac{1}{2}a^2 \\ -\omega(r_{\text{skew}})\frac{1}{2}ab & -\omega(r_{\text{ver}})\frac{1}{2}b^2 & \cdot \end{bmatrix}, \quad 0\le\omega(r)\le 1, \quad 0\le\theta\le\frac{\pi}{2}, \tag{4a}$$

with $\omega(r)$ the limiter function and r_{hor}, r_{skew} and r_{ver} the following ratios of consecutive solution gradients:

$$\begin{pmatrix} (r_{\text{hor}})_{i,j} \\ (r_{\text{skew}})_{i,j} \\ (r_{\text{ver}})_{i,j} \end{pmatrix} = \begin{pmatrix} (u_{i+1,j}-u_{i,j})/(u_{i,j}-u_{i-1,j}) \\ (u_{i+1,j+1}-u_{i,j})/(u_{i,j}-u_{i-1,j-1}) \\ (u_{i,j+1}-u_{i,j})/(u_{i,j}-u_{i,j-1}) \end{pmatrix}. \tag{4b}$$

Note that in the monotonicity theory of e.g. Sweby [8], a blend is taken of the grid-aligned, first-order accurate upwind scheme and the grid-aligned, second-order accurate, fully one-sided upwind scheme (i.e. the $\kappa=-1$–scheme in terms of Van Leer [4]). In the present paper we follow more the lines of Yee's symmetric TVD approach [9]. However, the novelty is that we also consider a ratio of consecutive solution gradients which is not grid-aligned (r_{skew}). A second difference with the existing, symmetric TVD approach is that we consider ratios of consecutive solution gradients which are defined *per cell center* and *not per cell face.* (The pursuit of compactness requires a cell-center approach, instead of a cell-face approach.) An apparent drawback of this cell-centered way to compute the ratios of consecutive solution gradients, is loss of conservation. However, it can be easily shown that this loss is only $\mathcal{O}(h^2)$.

We proceed by deriving the limiters. With (4b), blended scheme (4a) can be cast into the four-point compact form

$$\frac{1}{a+b}\begin{bmatrix} \cdot & \cdot & \cdot \\ -\xi(r_{\text{hor}})a^2 & \xi(r_{\text{hor}})a^2+\xi(r_{\text{skew}})ab+\xi(r_{\text{ver}})b^2 & \cdot \\ -\xi(r_{\text{skew}})ab & -\xi(r_{\text{ver}})b^2 & \cdot \end{bmatrix}, \quad 0\le\theta\le\frac{\pi}{2}, \tag{5a}$$

where

$$\xi(r)=1+\frac{1}{2}\omega(r)(r-1). \tag{5b}$$

The coefficients in (5a) are then required to satisfy the rules of positivity and finity ($0\le\xi(r)<\infty$). Together with the requirement $\omega(r)\in[0,1]$, this may be combined to

$$\begin{pmatrix} 0 \\ -1 \end{pmatrix} \le \begin{pmatrix} \omega(r) \\ r\omega(r) \end{pmatrix} \le \begin{pmatrix} 1 \\ m \end{pmatrix}, \quad 0\le m<\infty. \tag{6}$$

The accuracy requirements that we impose are

$$\begin{pmatrix} \omega(1) \\ \omega'(1) \end{pmatrix} = \begin{pmatrix} 1 \\ 0 \end{pmatrix}. \tag{7}$$

A proper monotonicity function which satisfies (6) and (7) is e.g.

$$\omega(r) = \max\left[\min(\frac{-1}{r}, 1), \min(1, \frac{m}{r})\right], \quad 1 < m < \infty, \tag{8}$$

where m can be chosen arbitrarily in the indicated range. For all r, limiter (8) simply is the upper bound of the monotonicity domain. The limiter is such that the resulting scheme is centered scheme (3) for all r in the range $[-1, m]$. (By choosing a higher value of m, the limited scheme equals scheme (3) over a wider range of r.) In Fig. 1a we depict the limiter and monotonicity domain, which result for $m = 2$. Because of its non-differentiability, limiter (8) is not suited for a Newton-type solution method.

The second limiter to be presented now *is* differentiable. We assume the form

$$\omega(r) = \frac{1}{\alpha r^2 + \beta r + \gamma}, \tag{9}$$

where $\alpha, \beta, \gamma \in \mathbb{R}$ are unknown constants. By imposing requirements (6) and (7), we get the limiter

$$\omega(r) = \frac{4m(m-1)}{r^2 - 2r + 4m(m-1) + 1}, \quad 1 < m \leq 2, \tag{10}$$

where m can be chosen freely in the indicated range. (For $m > 2$ the limiter no longer satisfies $-1 \leq r\omega(r)$ for all r.) In Fig. 1b we depict a simple example of the present smooth limiter ($4m(m-1) = 1$), together with the corresponding, most tight-fitting monotonicity domain.

3. Numerical results

Although the accurate, non-smooth limiter (8) is not suited for our purposes (a multigrid-Newton method for the Euler equations), to have a reference for smooth limiter (10), we still examine limiter (8)'s performance for the known, rotating cone problem (see e.g. [6]). This problem is governed by a linear, scalar convection equation on a square domain, where the wind field is a given, steady solid-body rotation around the square's center. The exact solution on a uniform, 129×129 finite-volume grid is given in Fig. 2a. Applying an explicit, fourth-order accurate, four-stage Runge-Kutta scheme (with the time step sufficiently small to ensure that the time discretization error is negligible with respect to the space discretization error), we obtain the numerical results given in Figs. 2b – 2d. Non-smooth limiter (8), with $m = 2$, appears to yield monotonicity (Fig. 2c), without reduction of the solution accuracy to that of the rotated, first-order accurate scheme (Fig. 2b). In Fig. 2d we present the solution obtained by smooth limiter (10), with $4m(m-1) = 1$. It appears that this smooth-limiter-solution is only slightly less accurate than the non-smooth-limiter solution given in Fig. 2c.

Next, smooth limiter (10), with $4m(m-1) = 1$, is applied to a steady, 2-D Euler flow with oblique contact discontinuity. The exact solution on a uniform, 32×32 finite-volume grid is given in Fig. 3a; the numerical results are given in Figs. 3b and 3c. Just as for the rotating cone problem, also for this Euler flow problem, the smooth limiter appears to lead to monotonicity (Fig. 3c), without reduction of the solution accuracy to that of the rotated, first-order accurate scheme (Fig. 3b).

It can be concluded that compact, monotone, second-order accurate, rotated-interpolation schemes: (i) are easily implemented, and (ii) may be solved directly by multigrid-Newton iteration.

References

[1] S.F. DAVIS, 'A rotationally biased upwind difference scheme for the Euler equations', *J. Comput. Phys.*, **56**, 65-92 (1984).

[2] CH. HIRSCH, C. LACOR AND H. DECONINCK, 'Convection algorithms based on a diagonalization procedure for the multidimensional Euler equations', *AIAA Paper 87-1163* (1987).

[3] B. KOREN AND P.W. HEMKER, 'Multi-D upwinding and multigridding for steady Euler flow computations', Proceedings of the Ninth GAMM Conference on Numerical Methods in Fluid Mechanics, Lausanne, 1991, *Notes on Numerical Fluid Mechanics* (A. Rizzi and I.L. Ryhming, eds.), Vieweg, Braunschweig (to appear).

[4] B. VAN LEER, 'Upwind-difference methods for aerodynamic problems governed by the Euler equations', Proceedings of the 15th AMS-SIAM Summer Seminar on Applied Mathematics, Scripps Institution of Oceanography, 1983, *Lectures in Applied Mathematics*, **22, Part 2**, 327-336 (B.E. Engquist, S.J. Osher and R.C.J. Somerville, eds.), American Mathematical Society, Providence, Rhode Island (1985).

[5] D.W. LEVY, K.G. POWELL AND B. VAN LEER, 'An implementation of a grid-independent upwind scheme for the Euler equations', *AIAA paper 89-1931* (1989).

[6] S.A. ORSZAG, 'Numerical simulation of incompressible flows within simple boundaries: accuracy', *J. Fluid Mech.*, **49**, 75-112 (1971).

[7] P.L. ROE, 'Discrete models for the numerical analysis of time-dependent multidimensional gas dynamics', *J. Comput. Phys.*, **63**, 458-476 (1986).

[8] P.K. SWEBY, 'High resolution schemes using flux limiters for hyperbolic conservation laws', *SIAM J. Numer. Anal.*, **21**, 995-1011 (1984).

[9] H.C. YEE, 'Construction of explicit and implicit symmetric TVD schemes and their applications', *J. Comput. Phys.*, **68**, 151-179 (1987).

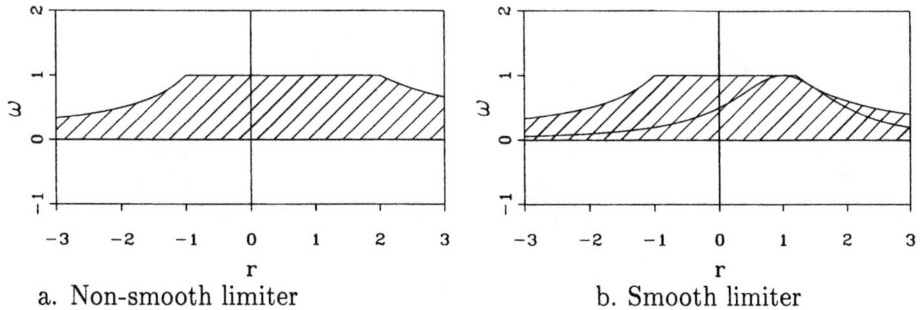

a. Non-smooth limiter $\omega(r) = \max\left[\min(\frac{-1}{r}, 1), \min(1, \frac{2}{r})\right]$.

b. Smooth limiter $\omega(r) = \frac{1}{(r-1)^2+1}$.

Figure 1: Non-smooth limiter, smooth limiter and monotonicity domains.

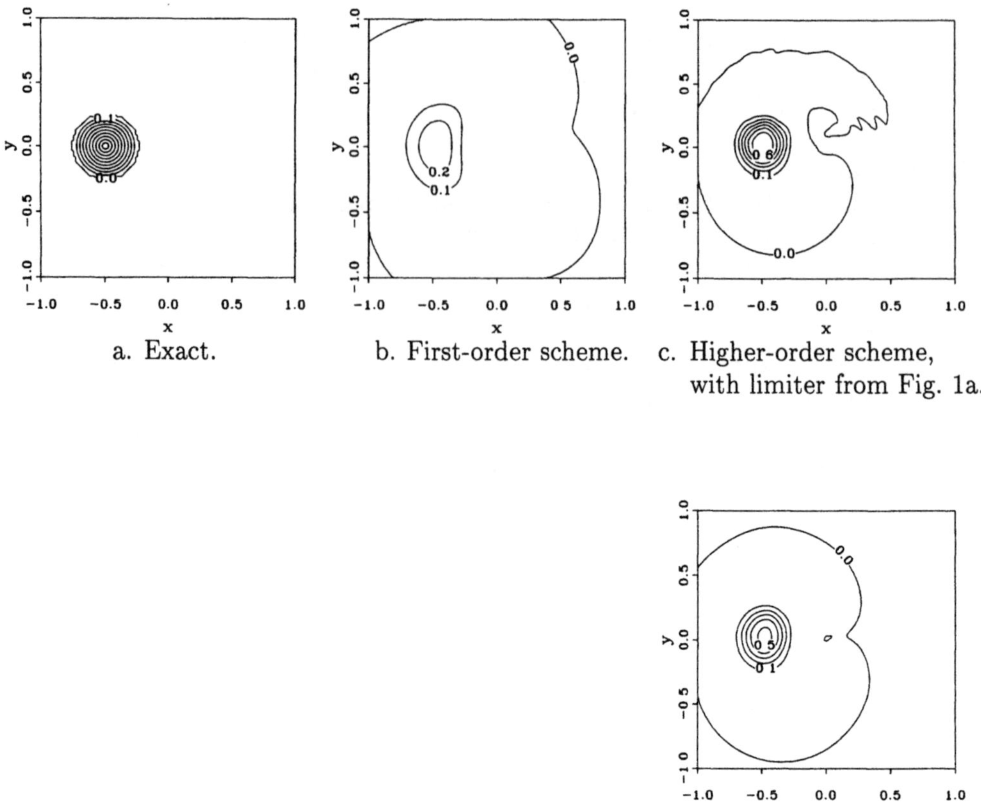

a. Exact.

b. First-order scheme.

c. Higher-order scheme, with limiter from Fig. 1a.

d. Higher-order scheme, with limiter from Fig. 1b.

Figure 2: Solutions rotating cone problem, after a single rotation.

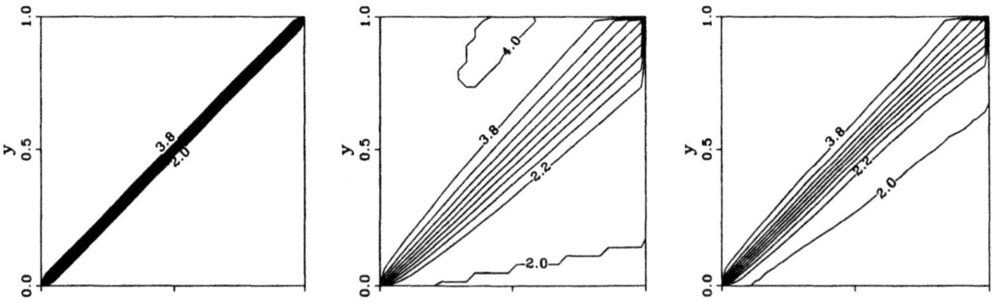

a. Exact.

b. First-order scheme.

c. Higher-order scheme, with limiter from Fig. 1b.

Figure 3: Mach number distributions oblique contact discontinuity.

ON A NEW CLASS OF FLUX SPLITTINGS

M.-S. Liou

MS 5-11, NASA Lewis Research Center, Cleveland, OH 44135

Introduction

In the 1980s, we have witnessed an explosive interest and research in the upwind schemes for its shock capturing capability. Roe [1] gave to this Conference in 1988 an extensive status review of nearly all upwind techniques. A major concern of these upwind schemes is the efficiency/complexity, which becomes even more crucial in dealing with a large set of equations such as those for describing nonequilibrium flows. Recent studies also raised concerns about numerical accuracy of some prominent schemes. Examples are inaccurate viscous-layer solutions by the Van Leer splitting and the so-called "carbuncle shocks" by the Roe splitting. Also, positivity property that is critical in preserving chemical species calculation can not be guaranteed by the Roe or Osher schemes. Clearly, need for improving accuracy, efficiency and robustness for treating a wide variety of problems still remains.

An attempt to address the above issues has been initiated by taking a new approach that is different from the current methods. We propose a remarkably simple (efficient) class of upwind flux splitting schemes and demonstrate in a wide range of problems that its accuracy rivals and in some cases surpasses that of current prominent splitting schemes. The concept is general, allowing rooms for improvement and accomondating ideas from other schemes. Because of avoiding matrix construction, the computation effort for flux evaluation is only linearly proportional to the number of equations (n) solved, rather than $O(n^2)$ as required in some existing schemes. Briefly put, the present scheme features: (1) simplicity and efficiency, (2) accuracy, and (3) positivity-preserving. An early version covering essentially the main ideas is reported in [2]. This paper reports some of the considerable new advances, including a generalization to a class of viable flux formulas, a more in-depth analysis, and further development for unsteady flows.

The Method

To illustrate the concept, let us consider the one-dimensional system of conservation laws for ideal-gas flows:

$$\frac{\partial \mathbf{U}}{\partial t} + \frac{\partial \mathbf{F}}{\partial x} = 0 \qquad (1)$$

where $\mathbf{U}^T = (\rho, \rho u, \rho E)$, the inviscid flux $\mathbf{F}^T = (\rho u, \rho u^2 + p, \rho u H)$, and the specific total energy $E = e + u^2/2 = H - p/\rho$.

The present scheme begins by treating the convective and pressure terms separately. In other words, the convection and acoustic propagation are recognized as two physically distinct (but coupled) processes and thus deserve separate treatment.

Mathematically, we propose to separately deal with the genuinely nonlinear (($u - a, u + a$) pair) and linearly degenerate (u) fields.

We begin by rewriting the flux as a sum of the *convective* and *pressure* fluxes, $\mathbf{F}^{(c)}$ and $\mathbf{P}$ respectively:

$$\mathbf{F} = \mathbf{F}^{(c)} + \begin{pmatrix} 0 \\ p \\ 0 \end{pmatrix} = c\mathbf{\Phi} + \mathbf{P} = \begin{cases} u \begin{pmatrix} \rho \\ \rho u \\ \rho H \end{pmatrix} + \mathbf{P}, & \text{if } c = u \\ M \begin{pmatrix} \rho a \\ \rho u a \\ \rho H a \end{pmatrix} + \mathbf{P}, & \text{if } c = M \end{cases} \tag{2}$$

Depending on whether the convective speed c is either u or M, the vector $\mathbf{\Phi}$, consisting of passive scalar quantities, has the corresponding contents shown in the equations. The pressure flux now contains solely the pressure term. In the finite-volume formulation, the difference among all numerical schemes completely lies in the definition of the numerical flux at the cell interface. The present method proposes a new class of flux definitions. First, the *numerical convective flux* at the interface $L < \frac{1}{2} < R$ is effectively written as:

$$\mathbf{F}^{(c)}_{1/2} = c_{1/2}\mathbf{\Phi}_{L/R}, \tag{3}$$

where $c_{1/2}$ is the interface convective velocity (M or u). Its definition is a major but simple step in the present method. The convectible variable vector $\mathbf{\Phi}$ is then upwinded solely based on the sign of $c_{1/2}$, viz,

$$(\mathbf{\Phi})_{\mathbf{L/R}} = \begin{cases} \mathbf{\Phi}_L, & \text{if } c_{1/2} \geq 0, \\ \mathbf{\Phi}_R, & \text{otherwise.} \end{cases} \tag{4}$$

Note that as $c_{1/2}$ is identically equal to zero, the sign of $c_{1/2}$ is immaterial since the convective flux $\mathbf{F}^{(c)}$ vanishes with $c_{1/2}$. Now the appropriate question to ask at this point is how one expresses the interface convective velocity $c_{1/2}$. Several formulas are appropriate and can be found in [3]. We shall write the interface velocity $c_{1/2}$, as

$$c_{1/2} = c_L^+ + c_R^-, \tag{5}$$

to anticipate that it should allow contributions from the 'L' and 'R' states. To define the 'L' and 'R' components, we use $(u \pm a)$ as basis function for expansion of c, thus allowing interactions of the nonlinear pair. Instead of the convective velocity $u_{1/2}$, Mach number $M_{1/2}$ was formulated in [2]. The split-Mach version introduces extra mixing at the contact discontinuity across which there may be a great difference in sound speed. The difference of using the velocity and Mach-number splittings shows up, e.g., in shock-tube flow, although in most of steady flows, the results are indistinguishable.

One member of the class that has been proven effective employs a second-order polynomial expansion in the subsonic range, which also transitions smoothly at sonic point. Let $c = u$, then

$$u^{\pm} = \begin{cases} (u \pm |u|)/2, & \text{if } |u| \geq a, \\ \pm(u \pm a)^2/4a, & \text{otherwise.} \end{cases} \tag{6}$$

We turn now to the pressure term by writing:

$$p_{1/2} = p_L^+ + p_R^-. \tag{7}$$

Similarly, a whole host of choices are possible for the pressure splitting. A differentiable pair of the '+' and '-' components are given by:

$$p^{\pm} = \begin{cases} p(1 \pm sgn(u))/2, & \text{if } |u| \geq a, \\ p(M \pm 1)^2(2 \mp M)/4, & \text{otherwise.} \end{cases} \tag{8}$$

This completes the definition of the numerical flux **F**.

Putting (3) and (7) together, the interface flux can be recast in the following form:

$$\mathbf{F}_{1/2} = c_{1/2}\,\frac{1}{2}\Big[\mathbf{\Phi}_L + \mathbf{\Phi}_R\Big] - \frac{1}{2}\,|c_{1/2}|\,\Delta_{1/2}\mathbf{\Phi} + \mathbf{P}_{1/2}. \tag{9}$$

where $\Delta_{1/2}\{\bullet\} = \{\bullet\}_R - \{\bullet\}_L$. Here the first term on the RHS is clearly ***not*** a simple average of the 'L' and 'R' fluxes, but rather a weighted average via the convective velocity. The dissipation term has merely a scalar coefficient $|c_{1/2}|$ and requires only $O(n)$ operations for $n-$dimensional vector **F**, in contrast to $O(n^2)$ operations by the Roe or Osher method, as will be seen next. Furthermore, since there is no differentiation (or jacobian matrix) involved in evaluating $\mathbf{F}_{1/2}$, the present method is easily extended to general equation of state and non-equilibrium flows and the cost is only linearly increased with the additional conservation equations considered. Unlike the Roe or Osher scheme, the extension does not yield additional ambiguity such as the definition of averaged or intermediate states. Also, numerical tests strongly suggest entropy-satisfying property by the present method. Last but not the least, the positivity-preserving property can be proved.

Comparison with Other Splittings

In what follows we write the other existing upwind schemes for comparison:

$$\text{Roe}: \quad \mathbf{F}_{1/2} = \frac{1}{2}\Big[(u\mathbf{\Phi})_L + (u\mathbf{\Phi})_R\Big] - \frac{1}{2}\,|\mathbf{A}(\hat{\mathbf{U}})_{1/2}|\,\Delta_{1/2}\mathbf{U} + \frac{1}{2}(\mathbf{P}_L + \mathbf{P}_R), \tag{10a}$$

where $\hat{\mathbf{U}}$ is the Roe-averaged state, **A** the flux jacobian, and $|\mathbf{A}| = \mathbf{A}^+ - \mathbf{A}^-$ in the usual sense. The matrix-vector multiplication is $O(n^2)$.

$$\text{Osher}: \quad \mathbf{F}_{1/2} = \frac{1}{2}\Big[(u\mathbf{\Phi})_L + (u\mathbf{\Phi})_R\Big] - \frac{1}{2}\int_L^R |\mathbf{A}(\mathbf{U})|\,\mathbf{dU} + \frac{1}{2}(\mathbf{P}_L + \mathbf{P}_R), \tag{10b}$$

$$\text{Van Leer/Häner}: \quad \mathbf{F}_{1/2} = \frac{1}{2}\Big[(M\mathbf{\Phi})_L + (M\mathbf{\Phi})_R\Big] - \frac{1}{2}\,\Delta_{1/2}|M|\mathbf{\Phi} + \mathbf{P}_{1/2}, \tag{10c}$$

$$\text{Steger-Warming}: \quad \mathbf{F}_{1/2} = \frac{1}{2}\Big[(u\mathbf{\Phi})_L + (u\mathbf{\Phi})_R\Big] - \frac{1}{2}\,\Delta_{1/2}|\mathbf{A}|\mathbf{U} + \frac{1}{2}(\mathbf{P_L} + \mathbf{P_R}). \tag{10d}$$

In the last expression, the homogeneous property of degree one has been exploited.

The Van Leer/Häner splitting is also seen to require only $O(n)$ operations. The Steger-Warming and Roe splittings differ only in the evaluation of the absolute jacobian, where in the latter the absolute quantity is evaluted using both the 'L' and 'R' states and is taken outside of the difference operator, as also observed in the present method. On the other hand, the so-called flux-vector splittings(FVS) share the similar nature by differencing the absolute quantity. In comparison, the method retains the simplicity and efficiency of the Van Leer scheme, but achieves the high level of accuracy attributed only to the Roe and Osher methods, as borne out by a variety of Euler and Navier-Stokes calculations.

Results and Discussion

We have tested the present method on Euler and Navier-Stokes equations over a wide range of geometries and parameters. We evaluate its effectiveness by considering the accuracy and efficiency. Due to page limitation, we will include only typical results, readers are refered to [3] for more detailed descriptions.

Figure 1 shows the temperature distribution of a conic flow. The present and Roe solutions are essentially identical, while Van Leer splitting produces too thick a boundary layer. For the shock wave/laminar boundary-layer interactions, Fig. 2 displays very good agreement of the present solution with the data at both separation and reattachment points (Note: no C_f data available in the separated region). Figure 3 shows the Mach contours over a circular cylinder, indicating complete symmetry and a well-captured shock. The convergence hitories for the above two flows are given in Figs. 4, along with the effect of the CFL number and spatial accuracy. Noteworthy is that the second-order solution attains slightly faster convergence. Depicted in Fig. 5 is a 3D Navier-Stokes solution of the shuttle orbiter, pressure on the surface and Mach contours on 3 streamwise surfaces. Detailed examination reveals flow separation occuring at the engine pod.

For unsteady tests, Table I summarizes the performance of several schemes over a spectrum of speed regimes, again details are given in [3].

References

(1). P. L. Roe, Lecture Notes in Physics, **323**, 69-78, 1989.
(2). M.-S. Liou and C. J. Steffen, NASA TM 104452, 1991.
(3). M.-S. Liou, NASA TM (in preparation).

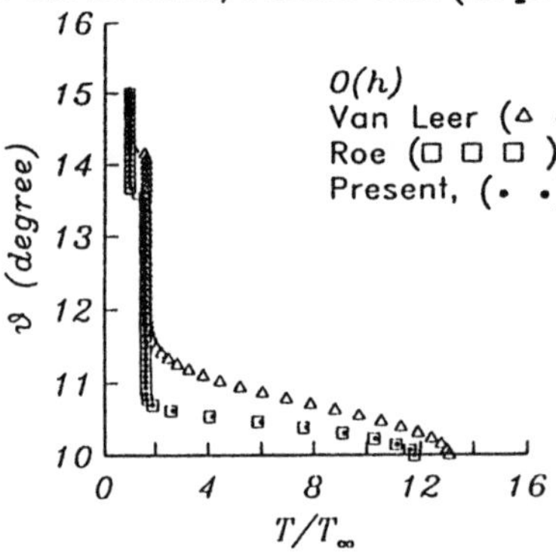

Fig. 1 Conic flow; $M = 7.95$, 65 grid.

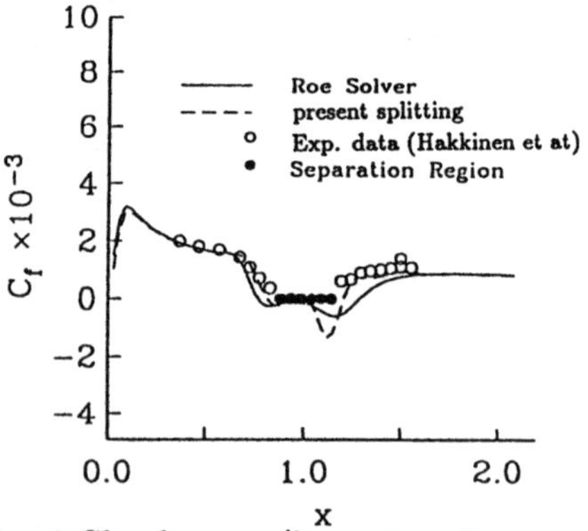

Fig. 2 Shock wave/boundary-layer interaction problem; $M = 2, 75 \times 65$ grid.

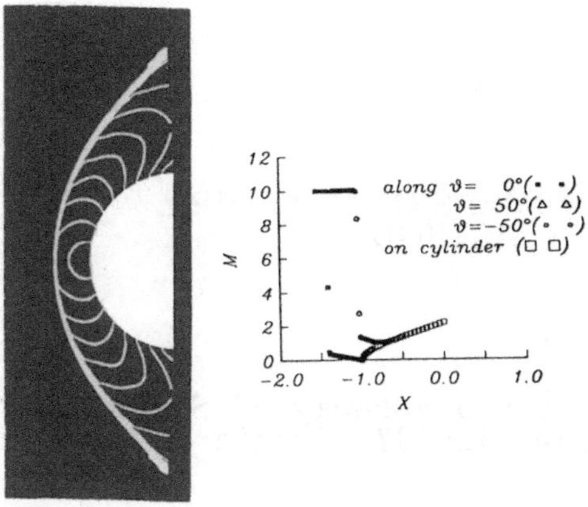

Fig. 3 Euler soultion for circular cylinder; $M = 10, 61 \times 61$ grid.

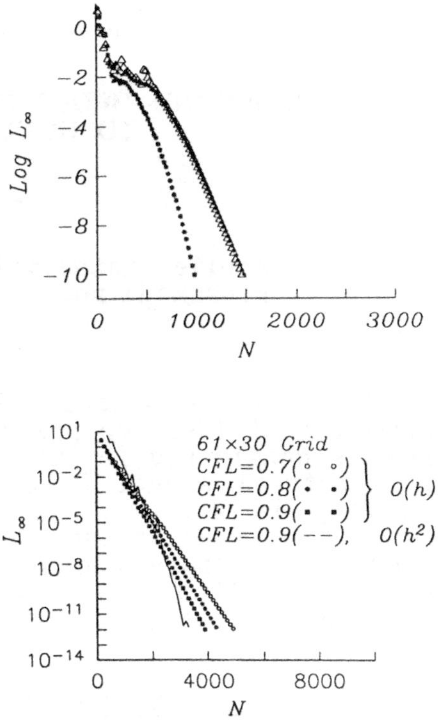

Fig. 4 Convergence history; (a) conic flow, (b) circular cylinder flow.

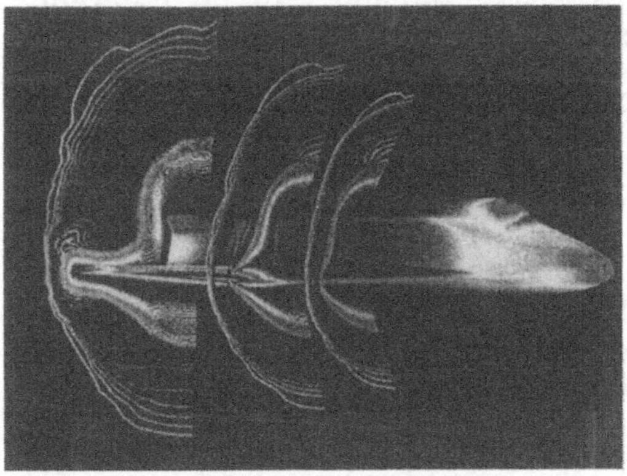

Fig. 5 Laminar solution for shuttle orbiter; $M = 6, Re = 10^4$, and $\alpha = 0°$, $59 \times 59 \times 59$ grid.

Table I Comparison of upwind schemes for shock-tube problems

Problem	Van Leer/Häner	Roe	Osher	Present
Stationary Contact		√	√	√
Slowly Moving Contact		√	√	√
Colliding Flows($M = 0.02$)	√	√	√	√
Colliding Flows($M = 0.20$)	√	√	√	√
Colliding Flows($M = 25.0$)	√	√		√
Receding Flows($M = 0.20$)	√		√	√

(blank indicates failed or unacceptable solution and
√ indicates accurate solution, while degree of accuracy may vary)

A SECOND-ORDER MULTIDIMENSIONAL SEQUEL TO GODUNOV'S METHOD

I. Men'shov

Inst. of Applied Mathematics, Russia Academy of Sciences, Miusskaja sq. 4, Moscow, 125047, Russia

A method of second-order accuracy for integrating the equations of ideal compressible gas flow in multidimensional case is considered in the paper. This method may be regarded as a natural sequel to Godunov's method. The second order accuracy is achieved by taking the distributions of the state quantities inside a computational cell to be linear as regards the space coordinates, rather than uniform as in Godunov's method. Both the case of unsteady equations and the case of steady equations of supersonic flow are considered.

1. Generalized problem of break-up of a single discontinuity.

The base step in Godunov's method is to use the Riemann problem solution (break-up of a single discontinuity) for computing of a numerical flux through the side edges of the computational cell. In the case of uniform distributions the problem is self-similar and has analytical solution. Its complete investigation was given in [1].

If non-uniform distributions are used we arrive at the generalized (nonself-similar) Riemann problem (GRP). In the 1D case it was investigated in [2,3] (for the Lagrangean flow equations) and in [4,5] (for the Euler equations). Further we consider a multidimensional sequel to GRP.

From the mathematical point of view it reduces to the Cauchy problem for the 3D nonsteady Euler equations with the initial data for $t=0$:

(1.1) $\mathbf{f}(\mathbf{r},0)=\mathbf{f}_-(\mathbf{r})$ if $x < 0$ and $\mathbf{f}(\mathbf{r},0)=\mathbf{f}_+(\mathbf{r})$ if $x > 0$

where $\mathbf{r}=(x,y,z)$ = space coordinates, t = time, $\mathbf{f}_\pm$ = are two arbitrary analytical functions ($\mathbf{f}$ denotes the vector of gasdynamic parameters). In a general case constructing the analytical solution to this problem globally in the entire flow domain is practically impossible. However, it may be considered locally near any point ($\mathbf{r}=0$) of initial discontinuity.

To do this we will put the original equations into the x - characteristic form, by introducing the new variables λ, θ

$\lambda = x\, t^{-1}$, $\theta = (x^2 + t^2)^{1/2}$, the specific entropy $s=s(p,\rho)$, the sound velocity $a=a(p,\rho)$ and $c=\rho a$:

(1.2) $cD_{\pm}(u) \pm D_{\pm}(p) = \theta B_{\pm}$,

$D(\mathbf{z}) = \theta \mathbf{B}$, $\mathbf{z} = (v,w,s)^T$,

where (u,v,w) = Cartesian coordinates of velocity, ρ and p = density and pressure, $D_{\pm}$ and D = the operators of differencing along the corresponding x - characteristics, $B_{\pm}$ and $\mathbf{B}$ = the terms which include no derivatives with respect to λ and θ.

Denoting $\mathbf{F} = (u,v,w,p,s)$, we will look for a solution to the system of equations (1.2) in the form

(1.3) $\mathbf{F} = \mathbf{F}_0(\lambda,y,z) + \theta\, \mathbf{F}_1(\lambda,y,z) + O(\theta^2)$

Substituting these expansions into the corresponding equations and considering the zeroth approximation with respect to θ, we find that the functions $\mathbf{F}_0$ may belong among the three classes:

(1.4) $A_0 = \{ \mathbf{F}_0 : \mathbf{F}_0' = 0 \}$

$A_{\pm} = \{ \mathbf{F}_0 : u_0 \pm a_0 = \lambda,\ c_0 u_0' \mp p_0' = 0,\ s_0' = v_0' = w_0' = 0 \}$,

where the prime denotes a derivative with respect to λ. The class A_0 corresponds to a constant (with respect to time) flow, while $A_{\pm}$ to the right (+) and the left (-) rarefaction waves.

The first approximation with respect to θ gives us the system of linear differential equations that determines the functions $\mathbf{F}_1$ in the continuity domain. Coefficients of this system will depend on the functions $\mathbf{F}_0$ (1.4).

By considering the system of equations separately for each of the three classes (1.4), we may prove the following <u>Proposition</u>:

Depending on the class of functions $\mathbf{F}_0$, a general solution to the system of first approximation equations has the forms -

(1.5) I) $\mathbf{F}_0 \subset A_0$

$$I_1^+ = \frac{C_1(u_0 + a_0 - \lambda) + (B_+)_0}{(1+\lambda^2)^{1/2}},\quad I_1^- = \frac{C_2(u_0 - a_0 - \lambda) + (B_-)_0}{(1+\lambda^2)^{1/2}}$$

$$\mathbf{Z}_1 = \frac{\mathbf{C}(u_0 - \lambda) + \mathbf{B}_0}{(1+\lambda^2)^{1/2}},\quad \mathbf{C} = (C_3, C_4, C_5)$$

II) $\mathbf{F}_0 \subset A_{\pm}$

$$I_1^{\mp} = Q_0(C_1 + \Omega),\quad I_1^{\pm} = (0.5 - u_0')I_1^{\mp} + L$$

$$\mathbf{Z}_1 = \frac{c_0}{(1+\lambda^2)^{1/2}}\left[\mathbf{C} \pm \int \frac{\mathbf{B}_0}{a_0 c_0}\, d\lambda\right],\quad \mathbf{C} = (C_3, C_4, C_5)$$

$$Q_0 = \frac{c_0^{3/2}}{(1+\lambda^2)^{1/2}},\ L = -p_0'\left(\frac{\partial a}{\partial s}\right)_0 s_1 - \frac{(\mathbf{B}_\pm)_0}{2(1+\lambda^2)^{1/2}}$$

$$\Omega = -\int \left(\frac{\partial\, ln\, c}{\partial s}\right)_0 u_0' \left(\frac{1+\lambda^2}{c_0}\right)^{1/2} s_1\, d\lambda \pm \int \frac{(B_\mp)_0\, d\lambda}{2 c_0^{3/2} a_0}$$

where $I_1^{\pm} = c_0 u_1 \pm p_1$; C_i $i = 1,\ldots,5$ -are arbitrary constants.

We have thus found in the analytic form the general solution to the first-order approximation equations for both the constant flow and rarefaction wave domains.

It may be shown that the zeroth approximation of GRP (1.1) corresponds to self-similar break-up of a single discontinuity. In this case the entire flow domain is decomposed into a series of subdomains, separated from each other by surfaces of discontinuity of the flow parameters or their derivatives. Each subdomain contains either a constant flow or a rarefaction wave. It is obvious that the spatial variable domains of the functions $\mathbf{F}_1$ coincide with corresponding subdomains given by solving the GRP (1.1) in the zeroth approximation with respect to θ. The solution in these subdomains is known: it is described eitherby relations (1.5) or by (1.6). Hence, one can consider the solution of the GRP in the first approximation to have been constructed, but with the presence of arbitrary constants.

For evaluating these constants we use first the initial data at $\lambda = \pm\infty$ and second the linearized relations at discontinuities, which are reduced to the equality of parameters in the weak discontinuity case or to the Rankine-Hugoniot relations in the shock wave case.

By considering all possible variants of the wave pattern arising in the zeroth approximation (the self-similar case), we may prove that the GRP to the first approximation with respect to θ can be solved completely, i.e. all the constants are determined, together with the discontinuity surfaces. In full detail the proof of this proposition for the 1D case is given in [4]. For the 3D case it is done by analogy with the 1D case.

2. Construction of the generalized Godunov scheme.

Now we return to the system of 3D nonsteady Euler equations:

$$(f_i)_t + div\,(\mathbf{F}_i) = 0, \qquad i=1,\ldots,5$$

where f_i= the gas parameters (density, x–momentum and etc.); $\mathbf{F}_i$= the flux vector. We rewrite it in the integral form for any 3D cell:

$$(2.1)\quad \int f_i dv \Big|_{t=t_0}^{t=t_0+\tau} = -\sum_{k=1}^{6} \int_{t_0}^{t_0+\tau} \int_{\sigma} (\mathbf{F}_i, \mathbf{n})\, d\sigma, \quad i=1,\ldots,5$$

where σ_k defines the cell face, $\mathbf{n}$ is its external normal

t_0 and $t_0 + \tau$ are two time moments.

Relations (2.1) are an exact consequence of the original equations. Using different approximation methods for evaluating the integrals in the right-hand side of (2.1) we obtain different explicit finite-difference schemes, whose approximation properties obviously depend on the precision of the integral computations. We use the solution of the GRP found above to evaluate these integrals in the following manner.

Let the parameter values $\{ \mathbf{f}_0 \}$ be given at the central point of each cell. By using Van Leer's procedure [5] one can obtain the space derivatives of f in the cells. Having linear distributions of the parameters on the left and right of the cell face σ_k, we can consider GRP. Its solution found above has the form (1.3). Denote it by $\mathbf{f}_{rs}$. Substituting this solution into the integrals in the right-hand side of (2.1) we obtain the first order approximation of the integrals:

$$(2.2) \quad \int (\mathbf{F}_i, \mathbf{n}) \, d\sigma = (\mathbf{F}_i(\mathbf{f}_{rs}), \mathbf{n}) \Big|_{\theta = t_0 + \tau/2} |\sigma_k| + O(h^2),$$

where h is a characteristic size of the cell, $|\sigma_k|$ denotes the area of the cell face. Such an approximation of the integrals allows us to obtain an explicit Godunov-type scheme which for smooth solutions and with the time-step limited by a Courant-type inequality has the second-order accuracy both in space and in time.

3. Steady supersonic flow equations.

The approach considered above can be applied to integrating the equations of steady supersonic flow. The role of the GRP is played here by the problem on interaction of two non-constant supersonic flows. This problem can be studied in the first approximation near any point of the interaction surfaces and its analytical solution can be constructed in the explicit form and be used for evaluating the flux. In detail this subject is treated in [6].

4. Preliminary results.

Numerical results given below have been obtained by the classical (1 order approximation) scheme - S1 and by its generalization proposed above (2 order approximation) - S2.

We demonstrate the results of two problems. The first is steady supersonic flow computation in the plane channel. Its geometry is defined by the line $y = 0$ (symmetry axis) and the lines $y = 1$, $y = 0.5x + const$, which conjugate in the smooth manner through a segment of a parabola on the interval $x \in [0, 1.5]$. In the initial cross-section $x = 0$ the uniform flow $(u, v, \rho, p) = (2, 0, 1, 1)$ is given. Figure 1 shows the results of calculations (the density *RO* and the pressure *P*) on a uniform constant grid at cross section $x = 6$ obtained by S1 (80 comput. cells, left) and by S2 (20 comput. cells, right). The solid line corresponds to the scheme S2 (200 comput. cells). The second is computation of a compressible flow around a sphere (M_∞= 2, where $M_\infty$is the free stream Mach number). The results are illustrated in Figures 2, where the Mach number isolines obtained on the uniform grid (32x32 comput. cells) by S1 (left) and by S2 (right) are shown.

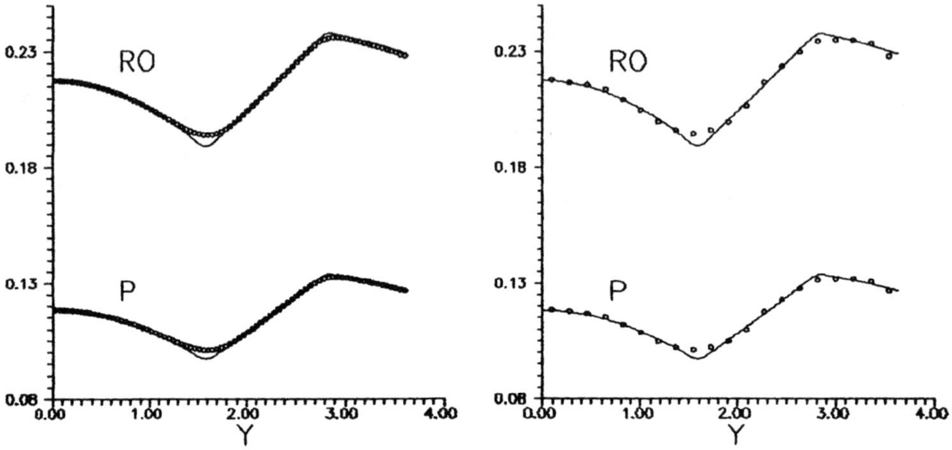

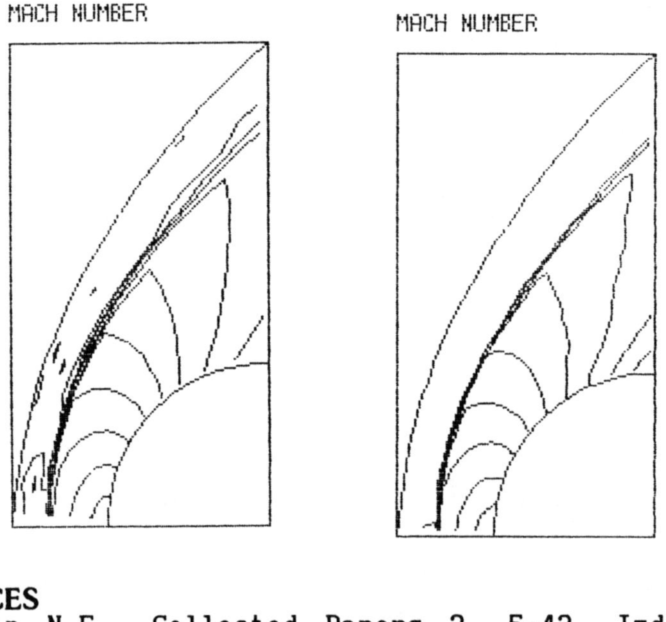

REFERENCES

[1] Kochin N.E., Collected Papers 2, 5-42, Izd.Akad.Nauk SSSR, Moscow,1949 (in Russian).[2] B.van Leer, J.Comp.Phys., 32, 101-136, 1979. [3] Ben-Artzi M., Falkovitz J., J.Comp.Phys. 55, 1-32, 1984. [4] Men'shov I., USSR Comput. Maths.Math.Phys., 30, N5, 54-65, 1990. [5] Men'shov I., Prikl.Matem. i Mech., 1, 86-94, 1991 (in Russian). [6] Men'shov I., Zhur.Vychis.Matem. i Matem.Fis.,32,N2, 311-319 (to be issued in English in USSR Comput.Maths.Math.Phys.).

Preconditioning and Flux Vector Splitting for Compressible Low Mach Number Flow

B. Müller J. Sesterhenn H. Thomann

Institute of Fluid Dynamics, Swiss Federal Institute of Technology, Zurich

1 Introduction

In low Mach number flows, variable density effects can become significant, in particular for low speed flows with heat addition, e.g. in combustion. However, the numerical solution of the Navier–Stokes equations in the low Mach number regime exhibits a stiffness coming from the Euler equations. Since the ratio of the fastest acoustic wave speed and the entropy wave speed is $\frac{u+c}{u} = 1 + \frac{1}{M} \rightarrow \infty \; for \; M \rightarrow 0$, the problem is ill–conditioned for low Mach numbers M. To overcome this problem several methods have been proposed. The wave speeds can be made nearly equal by preconditioning the time derivative of the Euler equations [1],[2]. Thus, the steady state can be reached faster and is not affected by the choice of the preconditioning. The basic idea of flux vector splitting is to identify the "good" non–stiff and the "bad" stiff parts and treat them separately in space and time [3],[4].

We have tested the Roe scheme in a node–centered finite volume formulation either without or with Turkel's or van Leer–Lee–Roe's precondioning for low Mach number quasi 1D nozzle flow. For the same test case, we have investigated a flux vector splitting derived from Abarbanel–Duth–Gottlieb and and a simple convection–pressure splitting.

2 Quasi 1D Euler Equations

To investigate the accuracy of low Mach number simulations we consider the quasi 1D Euler equations for inviscid flow through a Laval nozzle with cross–section $A(x)$:

$$(UA)_t + (FA)_x = Q \tag{1}$$

where $U = (\rho, \rho u, \rho E)^T$ is the vector of the conservative variables. $F = (\rho u, \rho u^2 + p, \rho u H)^T$ is the flux vector, and $Q = (0, p\frac{\partial A}{\partial x}, 0)^T$ is the source term due to varying cross sections. Perfect gas is assumed with $\gamma = 1.4$.

As initial conditions, we choose stagnant flow equal to reservoir conditions U_o, except for the outlet pressure, which is set to the ambient pressure p_e and kept fixed. Since we are interested in the steady state solution, the inlet boundary conditions are: total enthalpy H and entropy fixed at reservoir conditions.

3 Preconditioning

The condition number κ of the Jacobian matrix $\frac{\partial F}{\partial U}$ is given by the ratio $\frac{u+c}{u}$. To eliminate the ill–conditioning for $M \to 0$ the time derivative in (1) should be multiplied by a preconditioning matrix P, such that $\kappa(P^{-1}\frac{\partial F}{\partial U})$ becomes close to one. Then, the hyperbolic system

$$PU_t + F_x = 0 \tag{2}$$

will have wave speeds, which are nearly equal, and the stiffness of the 1D Euler equations for low Mach numbers is removed.

In the simplest case of Turkel's preconditioning for low Mach number flow [1], the acoustic waves are slowed down by multiplying the pressure time derivative in the non–conservative energy equation by $\frac{c^2}{\beta^2}$, from which P can be derived. Choosing $\beta^2 = 2u^2$, the condition number of $P^{-1}\frac{\partial F}{\partial U}$ is $2 + O(M^2)$. In our investigation, the local evaluation of the preconditioning parameter $\beta^2 = 2u^2$ led to disaster, whereas the global evaluation with $\beta^2 = 2u_{max}^2$ worked well.

In van Leer–Lee–Roe's preconditioning [2], each wave is moved with its characteristic speed. Thus, the optimal local condition number of one is attained.

4 Flux Vector Splitting of Abarbanel et al.

The idea of flux vector splitting is to identify the non–stiff and stiff parts of the flux vector and treat them separately:

$$F = (F - S) + S \tag{3}$$

The stiff part S is chosen such that the condition number of $\frac{\partial(F-S)}{\partial U}$ is close to one. $\kappa(\frac{\partial S}{\partial U})$ will be large for low Mach number flow.

Formulating Abarbanel–Duth–Gottlieb's non-conservative splitting based on entropy variables in conservative form we obtained a flux vector splitting (3) [4] with

$$S = \frac{{c_o}^2}{\gamma}(0, \rho, \rho u)^T \tag{4}$$

where c_o is the reference speed of sound. The splitting can be obtained by expressing the pressure p by $c^2\rho/\gamma$ and splitting c^2 into $(c^2 - {c_o}^2) + {c_o}^2$.

In a first approach, we solve the quasi 1D Euler equations using time–splitting. We discretize the two time split equations using a node–centered finite volume approach. For the first equation we choose the explicit Euler scheme. The numerical fluxes at the cell interfaces are evaluated by the Roe scheme using the eigenvalues and eigenvectors of the Abarbanel matrices, *not* those of the split–Jacobians. Because of the special structure of the second equation, it can be easily solved implicitly by integrating its second and third components in this order. More than that, the integration can be done this way without the splitting as a semi–implicit scheme. The stiff part is second-order central differenced.

5 Convection–Pressure Splitting

For the convection–pressure splitting the stiff part in (3) reads:

$$S = (0, p, 0)^T \tag{5}$$

It has been used by Rubin to solve the Reduced Navier–Stokes equations [5]. Since the Jacobian matrices of the non–stiff and stiff parts have positive and non–positive eigenvalues, respectively, for $u > 0$, convection and pressure are discretized upwind and downwind, respectively. Using time–splitting, we obtained for the 1D Euler equations what we termed 'p–downwind' scheme (a source term would be treated in the second step):

$$\left.\begin{array}{ll} \tilde{U}_i & = U_i^n - \frac{\Delta t}{\Delta x}\left[(F-S)_i^n - (F-S)_{i-1}^n\right] \\ U_i^{n+1} & = \tilde{U}_i^n - \frac{\Delta t}{\Delta x}\left[\tilde{S}_{i+1} - \tilde{S}_i\right] \end{array}\right\} \tag{6}$$

A linear stability analysis for $M = 0.1$ indicates that the p–downwind scheme has a stability limit of $CFL \leq 0.965$, whereas the explicit Euler scheme applied to the convection–pressure split equations requires $CFL \leq 0.1$ [4]. To understand the reason for the higher stability of the p–downwind scheme, we linearize the intermediate pressure $\tilde{p}$ for isentropic flow (omitting superscript n): $\tilde{p}_i \cong p_i + \frac{\partial p_i}{\partial \rho}\Delta\tilde{\rho}_i$. Inserting the residual of the continuity equation, we obtain $\tilde{p}_i \cong p_i - c_i^2\frac{\Delta t}{\Delta x}(\rho u_i - \rho u_{i-1})$. Inserting this relation in the second time split equation (6) the comparison with the unsplit scheme shows that the time splitting error introduces a second–order numerical dissipation in the momentum equation.

When solving the convection–pressure split equations by the implicit Euler scheme, the explicit part is discretized second–order in space using flux extrapolation. Since the implicit part is treated first–order in space, underrelaxation is employed. For fast convergence to the steady state, it is necessary to use implicit characteristic boundary conditions and to make sure that the physical boundary conditions are not violated by linearization.

6 Results

The following figures were calculated for a Laval nozzle with $A(x) = (7 - 3cos((x - \frac{1}{2})2\pi)/40$ for $0 \leq x \leq \frac{1}{2}$ and $A(x) = (5 - cos((x - \frac{1}{2})2\pi)/40$ for $\frac{1}{2} \leq x \leq 1$ using 100 equidistant points. $p_e/p_o = 0.99$ leads to $M_{min} \approx 0.07$ and $M_{max} \approx 0.18$. The source term Q in (1) is discretized pointwise. Either the explicit or implicit Euler scheme is employed in time. The explicit Roe scheme preconditioned by [1] or [2] and the semi–implicit flux vector splitting derived from [3] converge only slightly faster than the explicit Roe scheme using characteristic boundary conditions. With fixed boundary conditions, characteristic time stepping [2] is only about 2 times more efficient than local time stepping.

The first–order flux vector splittings are about equally accurate and more accurate than the first–order Roe scheme (Fig. 1). The error in pressure is measured by $\| (p - p_{exact})/(p_o - p_{exact}) \|_2$. The wiggles of the error at the boundaries when using (4) indicate typical problems of time splitting with the intermediate boundary conditions. The second–order p–downwind scheme is more accurate than the second–order Roe scheme (Fig. 2). However, the second–order λ - scheme of Moretti is even more accurate [6]. Since Roe's scheme and the one derived from [3] evaluate the fluxes in a similar way, they produce qualitatively the same error in total enthalpy, whereas the p–downwind scheme yields the exact constant value much more accurately (Fig. 3). The mass flux error is smallest for the flux vector splittings and largest for Roe.

For the explicit first–order schemes, about 10'000 time steps are required to reach the steady state, and twice as many for the second–order schemes, because the CFL number has to be halved. Using the implicit scheme, steady state is reached in about 50 time steps (Fig. 4). The CFL number is ramped from 100 to 2000 in the first 20 time steps.

7 Conclusions

We tested five explicit methods to solve the quasi 1D Euler equations for compressible low Mach number flow through a Laval nozzle. All of them converge similarly slowly to the steady state, although at least the preconditioning methods and the semi–implicit flux vector splitting were expected to perform better. The reason is probably due to the use of reflecting boundary conditions. For low Mach number flow, the error in pressure of the flux vector splitting methods is lower than that of Roe's scheme.

Although about five times more costly per time step, the implicit Euler scheme with underrelaxation and proper implicit boundary treatment is about 100 times more efficient than the explicit Euler scheme, when applied to convection–pressure splitting or Roe's scheme even without preconditioning. Because of its accuracy and simplicity, we plan to test the convection–pressure splitting in 2D.

8 References

1. E. Turkel: Preconditioned Methods for Solving the Incompressible and Low Speed Compressible Navier-Stokes Equationes. JCP, Vol. 72, 1987, pp. 277-298.

2. B.Van Leer, T.Lee, P. Roe: Characteristic Time-Stepping or Local Preconditioning of the Euler Equations. AIAA Paper 91-1552-CP, 1991.

3. S.Abarbanel, P.Duth, D.Gottlieb: Splitting Methods for low Mach Number Euler and Navier Stokes Equations. Computers & Fluids, Vol. 17, 1989, pp. 1-12.

4. J.Sesterhenn, B.Müller, H. Thomann: Flux-Vector Splitting for Compressible Low Mach Number Flow. 4th Intern. Conf. on Hyperbolic Problems, Taormina, 1992.

5. S.G.Rubin: RNS/EULER Pressure Relaxation and Flux Vector Splitting. Computers & Fluids, Vol. 16, 1988, pp. 485-490.

6. N.Botta, J.Sesterhenn: Deficiencies in the Computation of Nozzle Flow. Research Report No.92-05, Seminar für Angewandte Mathematik, ETH Zurich, 1992.

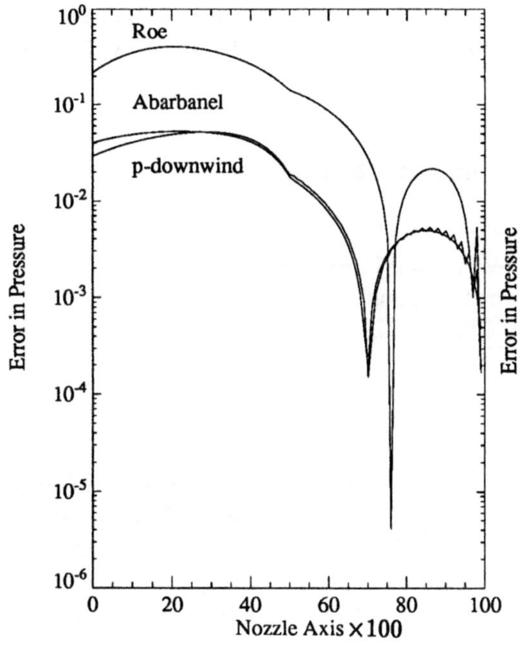

Fig. 1: Error in pressure of
1st order schemes

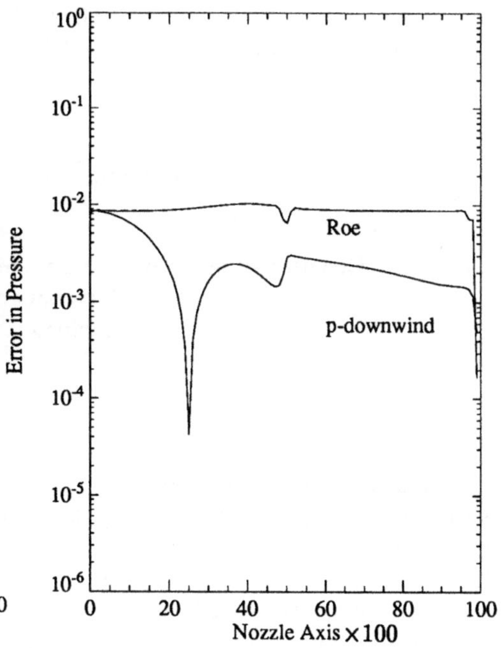

Fig. 2: Error in pressure of
2nd order schemes

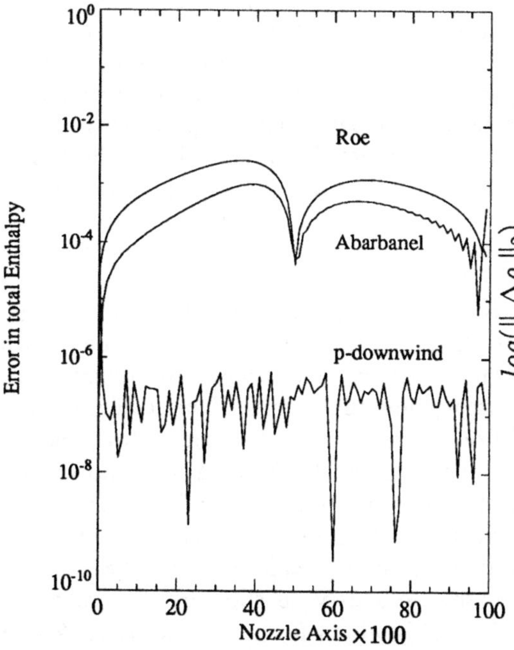

Fig. 3: Error in total enthalpy of
1st order schemes

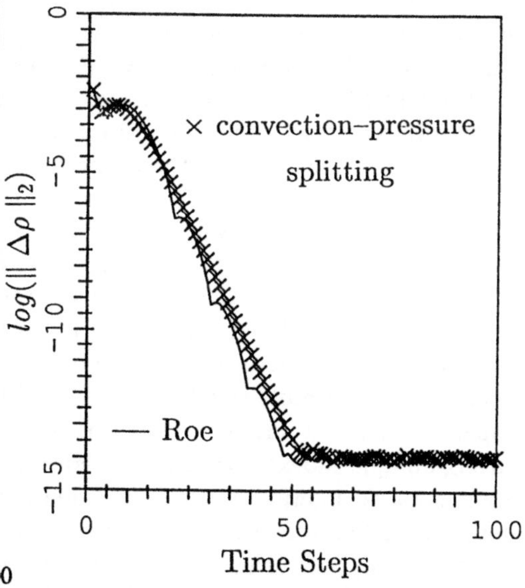

Fig. 4: Convergence history of
implicit 2nd order schemes

A THREE-DIMENSIONAL PRESSURE FLUX-SPLIT RNS FORMULATION: APPLICATION TO SUBSONIC/SUPERSONIC FLOW IN INLETS AND DUCTS

H.S. Pordal, S.G. Rubin and P.K. Khosla

Department of Aerospace Engineering and Engineering Mechanics
University of Cincinnati
Cincinnati, Ohio 45221

Introduction

In a series of papers, see e.g., [1,2], the present authors have developed a primitive variable formulation for the computation of large Reynolds number (Re) flows with viscous-inviscid interaction. It has been shown that for a significant class of such flows the dominant flow physics can be accurately represented by an asymptotic approximation, viz., a reduced Navier-Stokes (RNS) system plus a deferred corrector (DC) that includes all higher-order diffusion terms, to the full Navier-Stokes (NS) equations. This formulation clarifies the role of both surface and farfield boundary conditions and is applicable for flows with moderate regions of axial and secondary flow reversal, for sharp capturing of strong shock waves and contact discontinuities, and for both steady and transient behavior.

The lowest-order RNS approximation consists of the full Euler equations plus all boundary layer diffusion terms. All acoustic influences contained in the NS equations are retained, and, therefore, the RNS system allows for upstream or elliptic influence. A pressure-based form of flux-vector splitting is considered. This allows for direct solver, factored or relaxation techniques. For the latter a global relaxation procedure for the pressure, and for velocities in reverse flow regions, results. This form of flux-splitting is quite different from more conventional characteristic-based methods [2]. Specifically, the convective and acoustic fluxes are treated 'independently'. Therefore, the appropriate domain of dependence, or the upstream influence, is automatically represented by the discrete form of the acoustic and convective gradients.

In the present study, this procedure has been further developed for three-dimensional configurations. For the RNS solutions of reference [3], full cross flow diffusion has been retained in the axial and azimuthal momentum equations, and, central differencing was specified for the discretization of the azimuthal convective terms. This technique has been found to be quite suitable for external flow geometries, for low speed flow, and for supersonic flow, wherein the crossflow Mach number is small. For high speed flows with strong oblique shocks, that are associated with supersonic crossflow, and for generalized application to internal and external geometries, the RNS discretization has now been updated, so that the secondary flow momentum equations retain only the lowest-order crossflow (η,ζ) diffusion terms, that are required to satisfy no slip surface conditions. The higher-order terms are included in the DC. This allows for mid-point discretization that more appropriately represents the secondary flow physics. As in two-dimensions, explicit artificial viscosity is not required, in any coordinate direction, with pressure-based flux-splitting. The inherent numerical viscosity is sufficient to capture strong shocks, and this can be minimized on fine meshes.

Discretization

The present investigation is concerned with the extension of the pressure based flux-split, RNS + DC, formulation to more complex applications and three-dimensional flows. The objective is to develop a discrete system that is founded on (1) pressure-based flux-splitting for the axial (ξ) gradients, (2) grid line or cell centering, in one or both of the secondary coordinate (η,ζ) directions, i.e., a trapezoidal or 'box' concept for equations that do not introduce diffusion terms in the lowest order RNS system, (3) central differences for directions in which diffusion does appear in the lowest order RNS system, and (4) in regions where secondary flow shocks occur, the equations are grid point centered in η or ζ or both directions, depending on the shock orientation, and pressure-based form of flux-vector splitting is applied in that direction. This allows for accurate shock capturing in (ξ,η) and/or (ξ,ζ) planes. For shockless regions, the discretization pattern is shown in fig. 1.

As with the two-dimensional system [4,5], the trapezoidal or box-like discretization is such that surface boundary conditions are required only for the velocities and temperature; the surface pressure is determined by the solver with this discrete prescription. Also, for positive outflow velocity, an outflow boundary condition is required only for the pressure or pressure gradient, as appropriate. For external flow, far field conditions are specified only for the axial and azimuthal velocities, pressure and temperature. The normal component of velocity is determined from the discrete continuity equation. This allows for efficient application of a segmented domain decomposition multi-grid technique, that conserves all flow properties at cell interfaces [7].

Solution Procedure

A global relaxation procedure that involves the computation of all flow variables on successive (in ξ) crossplanes is applied. The pressure, and velocities in reversed flow regions, is updated in subsequent steps. For purely supersonic flows, the solution is obtained in a single marching step. When the negative flux contributions to the pressure gradient are omitted, a PNS methodology is recovered for attached flows. For reverse flows, the negative fluxes contribute in recirculation regions.

The algebraic crossplane equations are solved by a sparse-matrix direct solver. The choice of solution algorithm is dictated by considerations of robustness and consistency. In two dimensions, the direct solver has allowed for the solution of complex flow problems, where iterative techniques fail [4,5]. For example, the computation of transient flow in a supersonic inlet leads to an accurate description of unstart and restart phenomena. High speed moving shocks and high frequency transient shock-boundary layer interaction leads to a complex flow pattern that could be captured only with the direct solver. In addition, for two-dimensional steady problems, the direct solver, which introduces downstream boundary conditions immediately, is competitive, in terms of computational times, with relaxation methods. A simple domain decomposition strategy that allows for the use of the direct solver in small blocks around shocks and recirculation regions, and applies line relaxation elsewhere, has also been very effective.

Straightforward application of a direct solver to three-dimensions, even if only in a planar relaxation mode, can be computationally expensive. A more efficient application strategy has been developed. The uniformity of the coefficients in the equations at neighboring crossplanes allows for a reduction in the number of LU decompositions that are required for the direct solver inversion. During the marching process, the LU decomposition is frozen for several successive crossplanes, with the residuals iterated to a prescribed tolerance. This process only requires inexpensive repeated use of back substitution. When the number of iterations required to zero out the planar residual increases above a prescribed value, LU decomposition is again necessary [3]. When this technique degrades, e.g. in large regions of axial flow reversal, alternating direction planar relaxation can be applied to improve convergence.

Results

The three-dimensional RNS + DC, pressure-based flux split, direct/relaxation flow solver has been applied to compute flow fields with free-stream Mach numbers varying from low subsonic (incompressible) to transonic to supersonic.

(i) Incompressible flow in a three dimensional channel (height h) with a back step height=0.5h, at a laminar Reynolds number Re_h of 133, is investigated. Figure 2 shows the velocity vectors in the XY plane at various Z locations. Three dimensional effects are significant only in the vicinity of the side walls, the central portion of the flow field is predominantly two dimensional in nature. The velocity vectors and streamlines at the mid Z plane are depicted in fig. 3. A comparison of the separation length with earlier two dimensional calculations [6,7] show reasonable agreement. For larger Re_h, three-dimensionality plays a more significant role [7].

(ii) Transonic flow in a converging-diverging, three-dimensional nozzle with a square cross-section is evaluated. The density contours (shown in fig. 4) at several XY planar locations depict a sharp three-dimensional 'normal' shock pattern that lies very close to the inlet throat. The location is consistent with that predicted by quasi-one-dimensional theory.

(iii) Three-dimensional turbulent flow, at unit Reynolds number 1.1×10^6 per inch, along a symmetric corner formed by two intersecting wedges at a Mach number of three is considered. A transition location obtained from the experimental data [8] is included. Corner flows are characterized by strong three-dimensional inviscid-viscous interactions. The oblique shock generated by the one surface interacts with the boundary layer of the other surface and a secondary 'normal' shock forms. This generates a strong shock-boundary layer interaction. To capture the three-dimensional intersecting shock pattern, pressure-based flux-splitting as described previously is specified in all three coordinate directions. Density contours at various axial locations are shown in fig. 5. These compare well with the experimental data [8]. Detailed density contours at X=1. are depicted in fig. 6; once again comparison with the experimental data is good. The shock location is reasonably well predicted. The surface pressure distribution is compared with the experimental data in fig. 7. Although the shock location is reasonably well predicted, the surface pressure at the corner is somewhat lower than the data for the coarse grid in the boundary layer near the leading edge.

(iv) Preliminary results of computations for the pressure distribution on the centerbody of a NASA P2 inlet are shown in fig. 8. These are in good agreement with the data of reference [9], even though the tunnel conditions are highly non-uniform and it is difficult to obtain an exact set of inflow and upper boundary values. Comparisons on the cowl surface are not in good agreement. This problem has been observed by others in earlier studies of similar geometries. The Mach number distribution downstream of the throat is shown in fig. 9. The captured shocks are quite good and the qualitative behavior is in agreement with the data.

References

1. Rubin, S.G. and Khosla, P.K., "A Review of Reduced Navier-Stokes Computations for Compressible Viscous Flows", J. Comput. Syst. in Eng., 1, pp. 549-562, 1990.
2. Rubin, S.G. and Tannehill, J., "Parabolized/Reduced Navier-Stokes Computational Techniques", Ann. Rev. in Fluid Mechanics, 24, pp. 117-44, 1992.
3. Himansu, A., Khosla, P.K. and Rubin, S.G., "Three-Dimensional Recirculating Flows", AIAA Paper No. 89-0552, 1992.
4. Pordal, H.S., Khosla, P.K. and Rubin, S.G., "A Flux Split Solution Procedure for Unsteady Flow Calculations", (to appear in J. of Fluids Engineering), 1992.
5. Pordal, H.S., Khosla, P.K. and Rubin, S.G., "Unsteady Flow In Inlets", AIAA J., 30, 3, pp. 711-17, 1992.
6. Thompson, M.C. and Ferziger, J.H., "An Adaptive MG Technique for the Incompressible NS Equations", J. Comp. Phys., No. 85, pp. 94-121, 1989.
7. Rubin, S.G. and Srinivasan, K., "Adaptive MG Domain Decomposition Solutions for Viscous Interacting Flows", Proceedings of Fifth Symposium on Numerical and Physical Aspects of Aerodynamic Flows, 1992.
8. West, J.E. and Korkegi, R.H., "Supersonic Interaction in the Corner of Intersecting Wedges at High Reynolds Numbers", AIAA J., 10, 5, 1972.
9. Gnos, A.V., Watson, E., Seebaugh, W.R. and Sanator, R.J., "Investigation of Flow Fields Within Large-Scale Hypersonic Inlet Models, NASA TNS-7150, 1973.

Acknowledgement

The work was supported by the NASA Lewis Research Center (D.R. Reddy, Technical Monitor), under Grant No. Nag 3-1178.

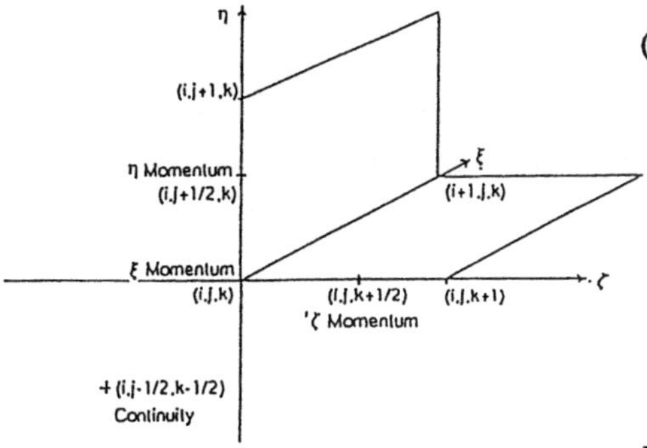

Fig. 1. Discretization pattern

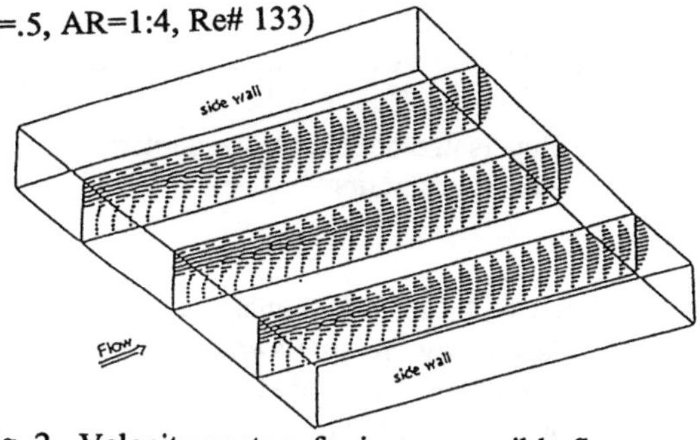

Fig. 2. Velocity vectors for incompressible flow in a back step channel

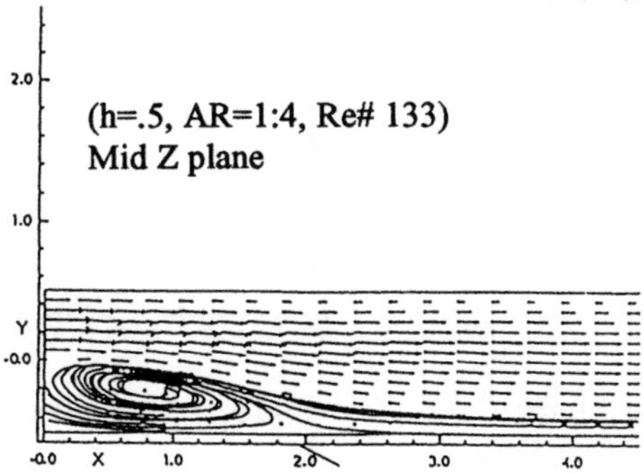

Fig. 3. Streamlines for flow in backstep channel

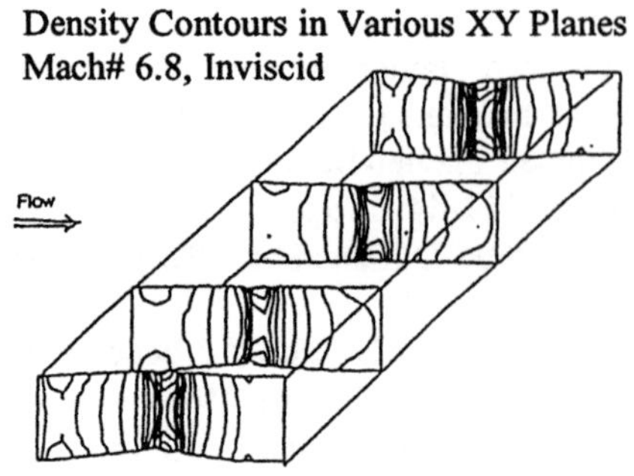

Fig. 4. Transonic flow in a converging-diverging nozzle: density contours

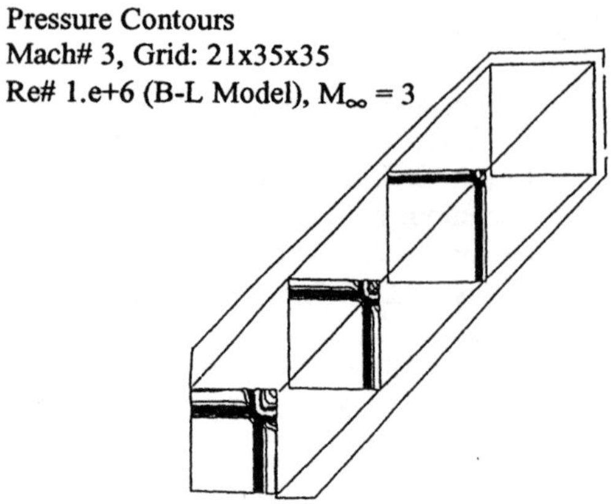

Fig. 5. Supersonic flow along double-wedge corner: pressure contours

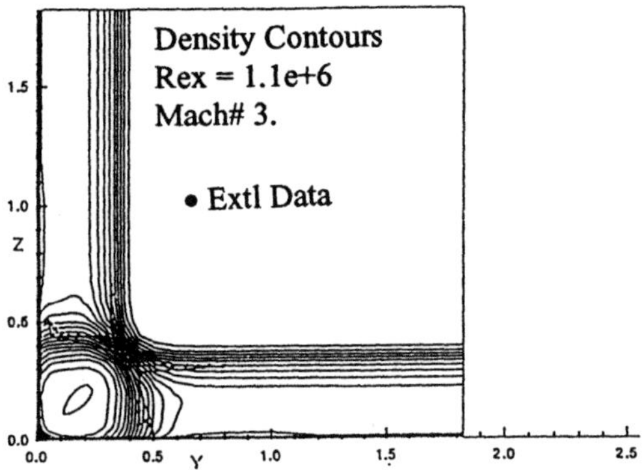

Fig. 6. Supersonic double-wedge corner: density contours

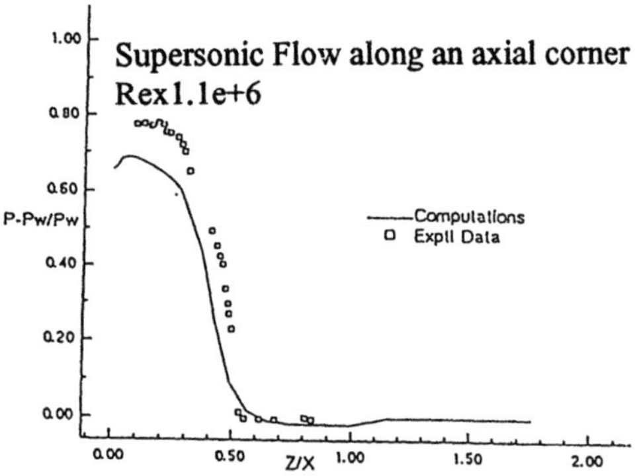

Fig. 7. Supersonic double-wedge corner: pressure distribution in the crossplane

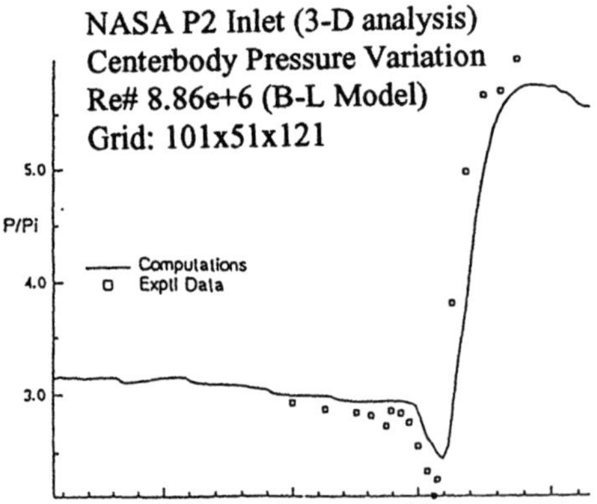

Fig. 8. Supersonic flow along P2 inlet centerbody: pressure distribution

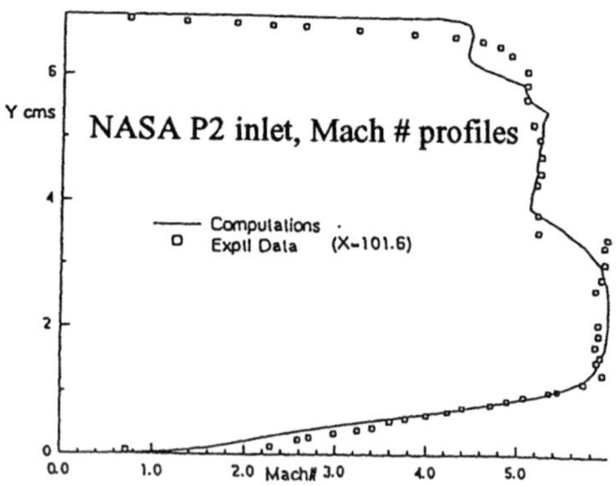

Fig. 9. Mach number profile for P2 inlet

AN IMPROVED WAVE MODEL FOR MULTIDIMENSIONAL UPWINDING OF THE EULER EQUATIONS

P.L. Roe and Lisa Beard

Department of Aerospace Engineering
University of Michigan

INTRODUCTION

An essential element in the development of the truly multidimensional upwind schemes reviewed by Bram van Leer in these proceedings (see also [1]) is the reduction, in each computational cell, of the Euler equations to a set of independent scalar problems. In one dimension the difference of two succesive nodal states $\mathbf{u}_{j+1} - \mathbf{u}_j$ can be projected onto the eigenvectors of the local Jacobian matrix to produce a unique decomposition into simple waves, whose effects can be treated separately in the numerics. In n dimensions, n independent differences along the edges of a simplicial element define a linear variation of the fluid state within that element. However, the decomposition into simple waves is not now unique because such waves can have infinitely many orientations. The contribution made in this paper is to view the problem as one of kinematic analysis. This leads to a particularly natural decomposition.

ANALYSIS

Consider a region of space, filled with fluid, and small enough that all fluid properties can be taken to vary linearly within it. The gradients of the primitive variables can be assembled into a 5×3 matrix, thus;

$$\mathbf{p}_x = \begin{bmatrix} \rho_x & \rho_y & \rho_z \\ u_x & u_y & u_z \\ v_x & v_y & v_z \\ w_x & w_y & w_z \\ p_x & p_y & p_z \end{bmatrix}. \tag{1}$$

A "wave model" for this region will be defined as a set of plane waves, each satisfying the Euler equations, whose superposition will reproduce all the elements of (1).

The central block of nine terms in (1),

$$\mathbf{D} = \begin{bmatrix} u_x & u_y & u_z \\ v_x & v_y & v_z \\ w_x & w_y & w_z \end{bmatrix}. \tag{2}$$

is known as the 'deformation tensor' [2]. A geometric interpretation of this tensor is that a set of fluid particles initially lying on the surface of a sphere with radius dr will, after a short time dt, be found on the surface of an ellipsoid, whose equation is

$$\mathbf{x}^T[\mathbf{I} - (\mathbf{D} + \mathbf{D}^T)dt]\mathbf{x} = (dr)^2. \tag{3}$$

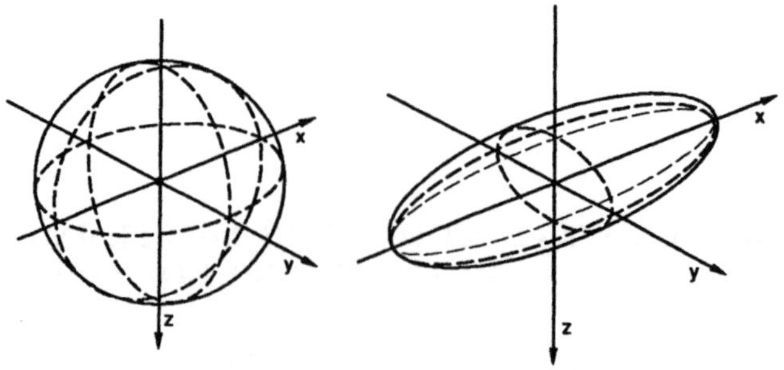

Figure 1: A fluid element before (left) and after (right) an infinitesimal distortion.

This ellipsoid depends only on the symmetric part $\mathbf{S} = \frac{1}{2}(\mathbf{D} + \mathbf{D}^T)$ of $\mathbf{D}$. The anti-symmetric part $\mathbf{A} = \frac{1}{2}(\mathbf{D} - \mathbf{D}^T)$ of $\mathbf{D}$, represents a rotation, and to first order does not affect the shape or the orientation of the ellipsoid. It is usual to conduct kinematic analyses in the (orthogonal) coordinates defined by the principal axes of the ellipsoid. These axes can be found by noting that they are the eigenvectors of the matrix $\mathbf{S}$. In these axes the deformation tensor has a symmetric part that is purely diagonal,i.e.

$$\mathbf{D} = \mathbf{S} + \mathbf{A} = \begin{bmatrix} u_x & 0 & 0 \\ 0 & v_y & 0 \\ 0 & 0 & w_z \end{bmatrix} + \begin{bmatrix} 0 & -\Omega_z & \Omega_y \\ \Omega_z & 0 & -\Omega_x \\ -\Omega_y & \Omega_x & 0 \end{bmatrix}. \tag{4}$$

where the $\{\Omega_i\}$ are components of vorticity,

$$\Omega_x = \frac{1}{2}(w_y - v_z), \quad \Omega_y = \frac{1}{2}(u_z - w_x), \quad \Omega_z = \frac{1}{2}(v_x - u_y).$$

Note that the velocities are now also measured in the principal axis system. The axes can always be labelled to form a right-handed system such that $u_x \geq v_y \geq w_z$.

Each simple wave that might be proposed as an element of the wave model has an associated deformation tensor. We will only include waves having a deformation tensor of the correct form; those having no off-diagonal terms in their symmetric part. For example, any acoustic wave produces a deformation tensor that is purely symmetric, but the only cases without off-diagonal terms are

$$\begin{bmatrix} 1 & 0 & 0 \\ 0 & 0 & 0 \\ 0 & 0 & 0 \end{bmatrix}, \quad \begin{bmatrix} 0 & 0 & 0 \\ 0 & 1 & 0 \\ 0 & 0 & 0 \end{bmatrix}, \quad \begin{bmatrix} 0 & 0 & 0 \\ 0 & 0 & 0 \\ 0 & 0 & 1 \end{bmatrix} \tag{5}$$

which represent waves propagating along one of the principal axes.

The deformation tensor for a shear wave is always trace-free. It can be shown that the only cases with symmetric parts that are diagonal are

$$\begin{gathered} \begin{bmatrix} 1 & -1 & 0 \\ 1 & -1 & 0 \\ 0 & 0 & 0 \end{bmatrix}, \quad \begin{bmatrix} -1 & -1 & 0 \\ 1 & 1 & 0 \\ 0 & 0 & 0 \end{bmatrix}, \quad \begin{bmatrix} 1 & 0 & 1 \\ 0 & 0 & 0 \\ -1 & 0 & -1 \end{bmatrix}, \\ \begin{bmatrix} -1 & 0 & 1 \\ 0 & 0 & 0 \\ -1 & 0 & 1 \end{bmatrix}, \quad \begin{bmatrix} 0 & 0 & 0 \\ 0 & 1 & -1 \\ 0 & 1 & -1 \end{bmatrix}, \quad \begin{bmatrix} 0 & 0 & 0 \\ 0 & -1 & -1 \\ 0 & 1 & 1 \end{bmatrix}. \end{gathered} \tag{6}$$

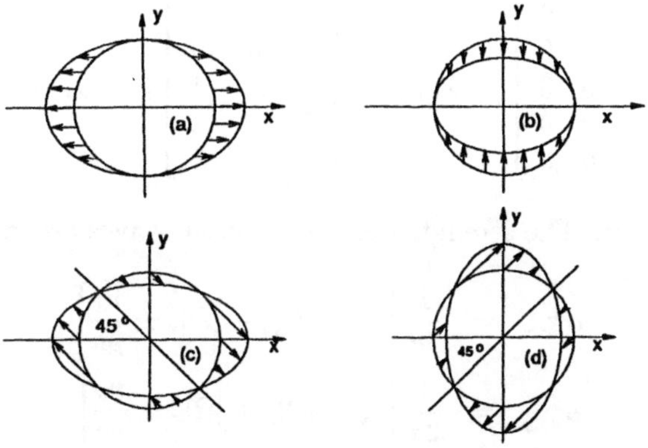

Figure 2: Patterns of deformation due to simple waves in two dimensions

In each of these cases, the planes in which the velocity is constant bisect some pair of coordinate directions.

The situation is most easily visualised in two dimensions. Figure 2 shows the deformation patterns corresponding to the four simple wave tensors that align with the principal axes. In (a) the deformation is due to an acoustic wave travelling in either the positive or negative x-direction, in (b) it is due to an acoustic wave travelling in the y-direction. In (c) and (d) the deformation is due to a shear wave, and we see that two orientations for such a wave are possible. For only one of these is the x-axis the major axis of the ellipse, as our convention assumes. The rule is that for positive vorticity the shear wave lies in the first quadrant, and for negative vorticity (shown here) in the second quadrant.

These considerations select, as candidates for the wave model, four acoustic waves and one shear wave. In three dimensions they select six acoustic waves and three shear waves. In fact, in n dimensions they select $2n$ acoustic waves and $n(n-1)/2$ shear waves. Since linear variation of n velocity components, together with pressure and density, yields $n(n+2)$ pieces of information, of which $n(n-1)/2$ are used to orient the principal axes, this leaves n degrees of freedom unaccounted for. These are precisely the number required to specify a single entropy wave, and the $n-1$ components of its unit normal; these degrees of freedom are simply omitted in considering isentropic flow.

For the full three-dimensional Euler equations, the waves strengths are determined from

$$\begin{bmatrix} \rho_x & \rho_y & \rho_z \\ u_x & -\Omega_z & \Omega_y \\ \Omega_z & v_y & -\Omega_x \\ -\Omega_y & \Omega_x & w_z \\ p_x & p_y & p_z \end{bmatrix} = \sum_{j=x,y,z} \alpha^{\pm}_{ac,j} \mathrm{T}^{\pm}_{ac,j} + \sum_{j=x,y,z} \Omega_j \mathrm{T}_{rot,j} + \mathrm{T}_{en}. \tag{7}$$

Typical examples of the tensors appearing on the right are

$$
\mathrm{T}^{\pm}_{ac,x} = \begin{bmatrix} \pm\rho & 0 & 0 \\ a & 0 & 0 \\ 0 & 0 & 0 \\ 0 & 0 & 0 \\ \pm\rho a^2 & 0 & 0 \end{bmatrix}, \quad \mathrm{T}_{rot,x} = \begin{bmatrix} 0 & 0 & 0 \\ 0 & 0 & 0 \\ 0 & 1 & -1 \\ 0 & 1 & -1 \\ 0 & 0 & 0 \end{bmatrix}, \quad \mathrm{T}_{en} = \begin{bmatrix} s_x & s_y & s_z \\ 0 & 0 & 0 \\ 0 & 0 & 0 \\ 0 & 0 & 0 \\ 0 & 0 & 0 \end{bmatrix},
$$

where $s_j = \rho_j - p_j/a^2$. The strengths of the acoustic waves are given by

$$
\begin{aligned}
\alpha^{\pm}_{ac,x} &= \frac{1}{2a}\left[u_x - |\Omega|_y - |\Omega|_z \pm \frac{p_x}{\rho a}\right], \\
\alpha^{\pm}_{ac,y} &= \frac{1}{2a}\left[v_y - |\Omega|_z + |\Omega|_x \pm \frac{p_y}{\rho a}\right], \\
\alpha^{\pm}_{ac,z} &= \frac{1}{2a}\left[w_z + |\Omega|_x + |\Omega|_y \pm \frac{p_z}{\rho a}\right].
\end{aligned}
$$

This model differs from the author's earlier proposals [3] by incorporating rotational effects more neatly. There, one version left the vorticity as a isotropic effect, providing no clue how to model its convection, whereas another oriented the shear wave with the particle path, which is only valid for steady flow. A subsequent version [4] oriented the shear normal to the pressure gradient, which is no help if there is no pressure gradient. Computational experience to date is that codes based on the present model are much more likely to converge in a satisfactory manner, although other aspects such as boundary conditions are also influential.

IMPLEMENTATION

The model can be applied to the conservative form of the Euler equations by means of a local linearisation within each element. The particular linearisation described in [5] ensures a conservation property that guarantees correct shock capturing. The individual waves are updated in scalar fashion using the 'NN' scheme described in [6]. For scalar problems the order of accuracy of this scheme appears to be about 1.6 [7]. Three-dimensional versions of these schemes are reported in [8].

We have not yet attempted any elaborate calculations using the method, although it performs well on a variety of standard two-dimensional tests, such as the ramp flow shown in Figure 3. Since some of the justification for following this line of research comes from a dissatisfaction with current Navicr-Stokes solvers, the performance of the scheme on isolated shear waves is of particular interest. Figure 4 shows results for the shear wave separating flows at Mach 2.5 and 2.0 on a 30×30 mesh. When the flow direction is aligned with any of the diagonals in the mesh shown, the numerical solution, which is observed to be stable against large perturbations, spans just two mesh intervals, and does not grow away from the inflow boundary. The worst case is when the flow direction roughly bisects a pair of mesh directions. Before leaving the domain the shear spreads over about six intervals, although the results are distinctly better than most grid-aligned upwind schemes.

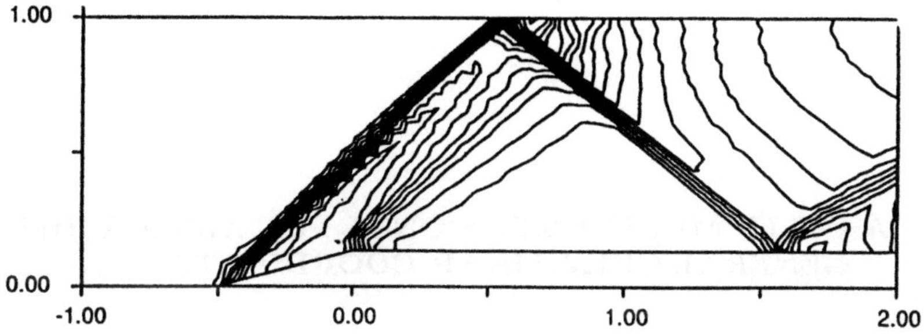

Figure 3: Pressure contours for flow over a ramp in a windtunnel. The calculation uses a 20×60 mesh.

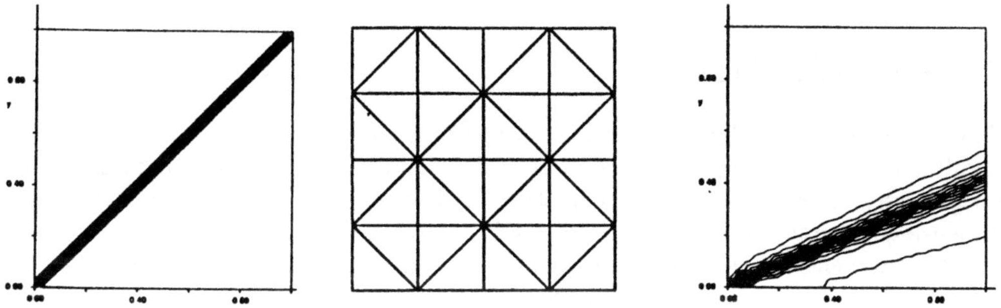

Figure 4: Mach number contours for shear waves at angles of 45° and 22.5^o to the mesh.

FUTURE DEVELOPMENTS

These results suggest some interesting speculation about the practical use of these ideas. Although the directional properties of the grid have been virtually removed from the wave recognition discussed above, they are still present, perhaps inevitably, in the numerical update, especially for linearly degenerate waves such as shear waves and contacts. They can be minimised by choosing a grid having a sufficient number of edges aligned with the streamlines, or perhaps, for unsteady flows, with other wave fronts. This suggests new ways to assess the quality of adaptive grids, particularly unstructured ones, although we have only recently begun to explore this area.

REFERENCES

1. H.Deconinck, Workshop on Algorithms for the 1990s, NASA Langley, 1992, Springer, to appear.
2. R.Aris, *Vectors, Tensors, and the Equations of Fluid Mechanics*, Dover, 1962.
3. P.L.Roe, *J.Comput. Phys.*, **63**, p 458, 1986.
4. P.de Palma, H.Deconinck, R.Struijs, Von Karman Institute TN 172, 1990.
5. P.L.Roe, R.Struijs, H.Deconinck, *J.Comput. Phys.*, to appear.
6. P.L.Roe, ICASE Report 90-75, 1990.
7. J-D Müller, P.L. Roe, Conf. on Modeling and Simulation, Pittsburg, 1992.
8. G.Bourgois et al., 1st European CFD Conference, Brussels, 1992, to appear.

GC-FCT : A FULLY MULTIDIMENSIONAL FCT ALGORITHM FOR GENERAL CURVILINAR COORDINATES

A. Rompteaux J.L. Estivalezes

ONERA-CERT-DERMES 2 Av Belin 31055 Toulouse FRANCE

1 Introduction

Since the seventies, high resolution FCT algorithms have been widely used to simulate highly transient compressible flows [1] as well as instabilities, turbulence [2] and combustion phenomena in cartesian geometries. All those calculations are based on one dimensional fourth order phase accurate FCT algorithm designed by Boris and Book [3]. Multidimensional problems are treated with time-step and directional splitting. A finite element version of FCT algorithm has been developped by Löhner and al. [4] for arbitrary triangular unstructured meshes. In this paper we propose a finite volume application of the multidimensional approach of Zalesak [5] to general curvilinear structured finite volume meshes.

2 Numerical resolution

2.1 Equations of the computed flow

The Euler equations can be written in the conservation form :

$$\frac{\partial U}{\partial t} + \frac{\partial F}{\partial x} + \frac{\partial G}{\partial y} = 0 \qquad U = \begin{cases} \rho \\ \rho u \\ \rho v \\ e \end{cases} \qquad F = \begin{cases} \rho u \\ \rho uu + p \\ \rho vu \\ eu + pu \end{cases} \qquad G = \begin{cases} \rho v \\ \rho uv \\ \rho vv + p \\ ev + pv \end{cases} \tag{1}$$

Here ρ, p, e, u, v denote respectively the density, pressure, total energy, components of the fluid velocities in the direction x et y of a cartesian system.

2.2 FCT concept

High resolution monotonicity preserving schemes must be developped in order to properly simulate strong non linear discontinuities present in flows under consideration. First, let us recall Zalesak generalisation of the one dimensional FCT scheme of Boris and Book, which can be considered as a truly multidimensional high resolution scheme. We consider a set of conservation laws given by a system of Partial Differential Equations like 1. The idea behind FCT is to combine a high order accurate (dispersive - ripples) scheme with a low order (monotonic but spreading) scheme in such a way that in regions where the variable under consideration varies smoothly, the high order scheme is used, whereas in the regions where the variable varies abruptly, the schemes are combined in a conservative way to assure a monotonic and accurate solution. In an other way, FCT limits the amount of the antidiffusion fluxes in such a way that no new over or undershoots are created.

2.3 Finite volume consideration

We consider an arbitrary quadrilateral cell mesh. The variables are stored at the cell centres. The system of equations 1 is integrated over a cell $\Omega_{i,j}$:

$$\int_{\Omega_{i,j}} \frac{\partial Q}{\partial t} d\Omega + \int_{\Omega_{i,j}} div \begin{pmatrix} F \\ G \end{pmatrix} d\Omega = 0 \tag{2}$$

By the Green formula, after spatial discretization, this equation is approximated by :

$$\frac{\partial Q}{\partial t} = -\frac{1}{S_{i,j}} \left[\tilde{F}_{i+\frac{1}{2},j} - \tilde{F}_{i-\frac{1}{2},j} + \tilde{G}_{i,j+\frac{1}{2}} - \tilde{G}_{i,j-\frac{1}{2}} \right] \tag{3}$$

where $S_{i,j}$ is the cell area and $\tilde{F}$ and $\tilde{G}$ are the contravariant fluxes defined by :

$$\tilde{F}_{i+\frac{1}{2},j} = +F_{i+\frac{1}{2},j}\; y_{\eta_{i+\frac{1}{2}}} - G_{i+\frac{1}{2},j}\; x_{\eta_{i+\frac{1}{2}}} \; and \; \tilde{G}_{i,j+\frac{1}{2}} = -F_{i,j+\frac{1}{2}}\; y_{\xi_{j+\frac{1}{2}}} + G_{i,j+\frac{1}{2}}\; x_{\xi_{j+\frac{1}{2}}} \tag{4}$$

2.4 General Curvilinear Flux Corrected Transport

Zalesak [5] proposed a new formulation of the monodimensional algorithm of Boris and Book which generalizes easily to multidimensions. His aim was to deal with problems where time-splitting technics (used to compute multidimensional phenomena with a monodimensional algorithm) may produce unacceptable numerical results. Here [6], we apply this approach to a curvilinear mesh (non orthogonal) via a finite volume method (Löhner [4] uses it in a finite element code).

FCT algorithm can be decomposed in the following steps [5] :

- Compute fluxes by the monotonic scheme (first order) : $\tilde{F}^L_{i+\frac{1}{2},j}$ and $\tilde{G}^L_{i,j+\frac{1}{2}}$.
- Compute fluxes by a second order scheme : $\tilde{F}^H_{i+\frac{1}{2},j}$ and $\tilde{G}^H_{i,j+\frac{1}{2}}$.
- Compute antidiffusive fluxes by difference :

$$AF_{i+\frac{1}{2},j} = \tilde{F}^H_{i+\frac{1}{2},j} - \tilde{F}^L_{i+\frac{1}{2},j} \; and \; AG_{i,j+\frac{1}{2}} = \tilde{G}^H_{i,j+\frac{1}{2}} - \tilde{G}^L_{i,j+\frac{1}{2}}$$

- Compute a transported and diffused solution :

$$Q^{TD}_{i,j} = Q^{o}_{i,j} - \frac{dt}{S_{i,j}} \left[\tilde{F}^L_{i+\frac{1}{2},j} - \tilde{F}^L_{i-\frac{1}{2},j} + \tilde{G}^L_{i,j+\frac{1}{2}} - \tilde{G}^L_{i,j-\frac{1}{2}} \right] \tag{5}$$

- Limit the antidiffusive fluxes :

$$AF^C_{i+\frac{1}{2},j} = C_{i+\frac{1}{2},j} \times AF_{i+\frac{1}{2},j} \; and \; AG^C_{i,j+\frac{1}{2}} = C_{i,j+\frac{1}{2}} \times AG_{i,j+\frac{1}{2}}$$

where $C_{i+\frac{1}{2},j}$ and $C_{i,j+\frac{1}{2}}$ are dimensionless scalar coefficients calculated to avoid apparition of new extrema of ρ.

• Compute the new solution :

$$Q^n_{i,j} = Q^{TD}_{i,j} - \frac{dt}{S_{i,j}}\left[AF^C_{i+\frac{1}{2},j} - AF^C_{i-\frac{1}{2},j} + AG^C_{i,j+\frac{1}{2}} - AG^C_{i,j-\frac{1}{2}}\right] \qquad (6)$$

The limitation procedure of Zalesak (calculated in step 5) is obtained by computation of the worst situation which can happen to ρ in the sixth step of the above algorithm. So we compute :

$$\begin{cases} P^+_{i,j} = \max(0, AF^1_{i-\frac{1}{2},j}) - \min(0, AF^1_{i+\frac{1}{2},j}) + \max(0, AG^1_{i,j-\frac{1}{2}}) - \min(0, AG^1_{i,j+\frac{1}{2}}) \\ P^-_{i,j} = \max(0, AF^1_{i+\frac{1}{2},j}) - \min(0, AF^1_{i-\frac{1}{2},j}) + \max(0, AG^1_{i,j+\frac{1}{2}}) - \min(0, AG^1_{i,j-\frac{1}{2}}) \end{cases} \qquad (7)$$

where $P^+_{i,j}$ (resp. $P^-_{i,j}$) represent the maximum (resp. minimum) antidiffusive fluxes which come in (resp. out) cell (i,j). AF^1 and AG^1 represent the first component of the vectors AF and AG. Then we evaluate the maximum amount we can gain (resp. loss) in the considered cell and the maximum fraction of antidiffusive fluxes we can gain (resp. loss) in cell (i,j) without producing overshoot (resp. undershoot) :

$$\begin{cases} M^+_{i,j} = (\rho^M_{i,j} - \rho^{TD}_{i,j})S_{i,j} \\ M^-_{i,j} = (\rho^{TD}_{i,j} - \rho^m_{i,j})S_{i,j} \end{cases} \qquad \begin{cases} R^+_{i,j} = \min\left(1\ ,\ M^+_{i,j}/P^+_{i,j}\right) \\ R^-_{i,j} = \min\left(1\ ,\ M^-_{i,j}/P^-_{i,j}\right) \end{cases} \qquad (8)$$

To obtain the scalar coefficients C, we finaly test the sign of the considered antidiffusive flux and prevent it from producing an undershoot (in the cell it goes out) or an overshoot (in the cell it goes in) :

$$C_{i+\frac{1}{2},j} = \begin{cases} \min\left(R^+_{i+1,j}, R^-_{i,j}\right) & if\ AF_{i+\frac{1}{2},j} \geq 0 \\ \min\left(R^+_{i,j}, R^-_{i+1,j}\right) & if\ AF_{i+\frac{1}{2},j} < 0 \end{cases} \qquad (9)$$

The choice we made here for the maximum and minimum values that appear in equation (8) is $\rho^M_{i,j} = \max\left(\rho^{TD}_{i-1,j}, \rho^{TD}_{i,j}, \rho^{TD}_{i+1,j}, \rho^{TD}_{i,j-1}, \rho^{TD}_{i,j+1}\right)$ and similar for $\rho^m_{i,j}$, but many others are possible.

2.5 Application to the Euler equations

We still need to express the high and low order fluxes. Here we select a second order centered scheme or a Mac Cormack scheme for the high order. At the moment, tests with high order MUSCL [7] versions of Van Leer Flux Splitting are under progress. For the low order scheme we use a generalisation to multidimensional of the low order scheme of the original Boris and Book algorithm or a first order Van Leer approach. Second order temporal accuracy is implemented by a Runge-Kutta 2 method.

To improve the calculation of the limited antidiffusive fluxes, we first limit them in each direction, and then apply the fully multidimensional limiters to these fluxes (cf Zalesak [5]). As recommended by Löhner [4], it seems important to introduce "system character" in the limiters by combining them for all equations of the system. For the Euler case, we choose (after numerical experiments) to apply to the four equations, the limiter obtained for the density to synchronize antidiffusive fluxes.

A high level of vectorization has been achieved on this code : a performance of 5 μs/timestep/point is obtained on a CRAY XMP116. The repartition of the CPU time during a calculation is described in [6] and illustrates the important part of the limitation process - 40% - compared to the time spent in the "flux" subroutine (evaluation of the high order, low order and antidiffusive fluxes) - 25% - and the "calcul" one (application of the different flux balances) - 25% -. We can also note the high speed of these routines (other 100 Mflops) and the tiny contribution (less than 5%) of the input-output and boundary conditions application.

3 Numerical results

Shock wave diffraction from an obstacle : This test case was proposed by Yee [8] and is studied here to evaluate the capability of the curvilinear scheme to treat with strong shocks. This is a complex flow presenting both supersonic and subsonic zones, steady and unsteady shocks. Our results agree very well with Yee's calculations (with various TVD schemes) on a comparable mesh of 320×96 cells. We present the density contours at different time steps on figure 1. The three-shock system which

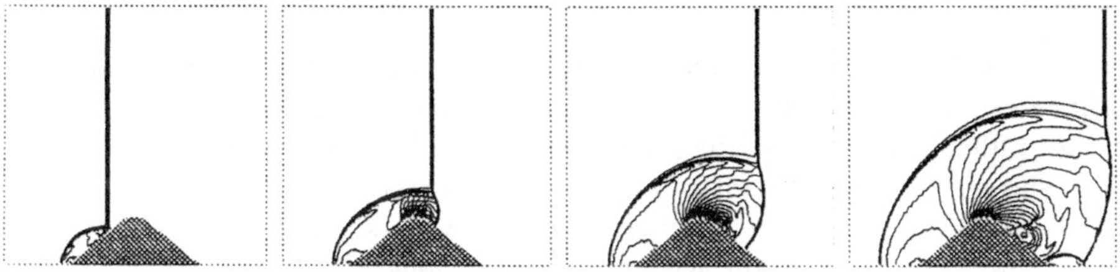

Figure 1: Time evolution of isopycnics

appear when the incident shock impacts the obstacle, is well predicted and a new one begins to form on the last image when the Mach shock reaches the end of the obstacle. One can see a good capture of all the shocks even through curvilinear mesh. This case has also been used [6] to compare the solutions obtained with different choices for low and high order schemes. Only little differences in the quality of the simulation were noticed (if the low order scheme remain monotonic) although the CPU cost increase fastly with the complexity of these schemes.

Shock wave diffraction from a cylinder : This test case uses the same Mach number as Yang [9] for the unsteady shock. It permits also comparisons with experimental schlieren. The time evolution of the density contours is presented on

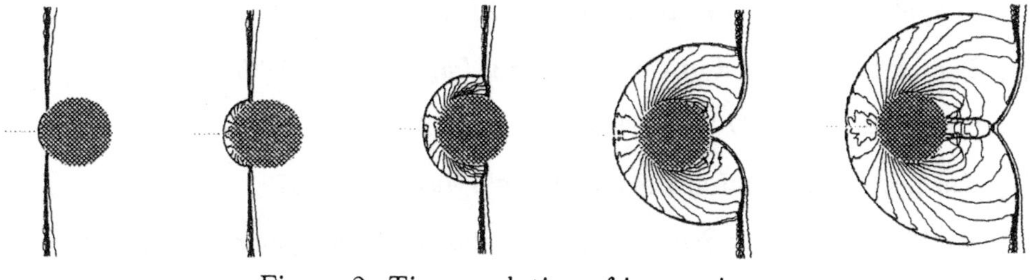

Figure 2: Time evolution of isopycnics

figure 2 and shows a good simulation of all the physical phenomena except the complex viscous effects (last image) which appear in the wake on the experimental results.

Transonic flow over a circular cylinder : This case was proposed for a Workshop [10] in 1986. The time evolution of the pressure contours is presented on figure 3 and is similar to the results of Pandolfi and al.. Although initial and boundary

Figure 3: Time evolution of iso-pressure

conditions are symetric, the flow naturely loses this symetry and leads to a perfectly unsteady periodic solution.

4 Conclusion

An application of FCT algorithm to arbitrary geometries via a finite volume formulation has been presented. It uses the multidimensional approach proposed by Zalesak and the influence of the different schemes to use has been studied. Numerical examples show sharp shock capturing and well predicted physics. A fully vectorized and optimized code has been developped for use with a supercomputer. Points under progress are Navier-Stokes and axisymetric equations.

References

[1] Book D., Boris J., Kuhl A., Oran E., Picone M., and Zalesak S. In *Lecture Notes in Physics*, 7th International conference on numerical methods in fluid dynamics, 1981.

[2] Grinstein F.F., Hussain F., and Oran E.S. *European Journal of Mechanics*, 9:499–525, 1990.

[3] Boris J.P. and Book D.L. *Journal of Computational Physics*, 11:38–69, 1973.

[4] Löhner R., Morgan K., Vahdati M., Boris J.P., and Book D.L. *Communications in applied numerical methods*, 4:717–729, 1988.

[5] Zalesak S.T. *Journal of Computational Physics*, 31:335–362, 1979.

[6] Rompteaux A. PhD thesis, ENSAE, Juin 1992.

[7] Van Leer B. *Journal of Computational Physics*, 32, 1979.

[8] Yee H.C. In *Lecture Series in Computational Fluid Dynamics*, 1989-04.

[9] Yang J.Y., Y. Liu, and H. Lomax. *AIAA Journal*, 25(5):683–689, 1987.

[10] Dervieux A., B. Van Leer, J. Periaux, and A. Rizzi, editors. GAMM, Fried. Vieweg & Sohn, 1986.

DIRECT NUMERICAL SIMULATION OF THE THREE-DIMENSIONAL BREAKDOWN TO TURBULENCE IN COMPRESSIBLE BOUNDARY LAYERS

H. Bestek*, A. Thumm‡, H. Fasel+

* Institut für Aerodynamik und Gasdynamik, University of Stuttgart, Germany
‡ Currently at Behr GmbH, Stuttgart, Germany
\+ AME Department, University of Arizona, Tucson, Arizona, USA

INTRODUCTION

The understanding of transition to turbulence in compressible boundary layers is rather poor. As for incompressible flows, the first stage of the transition process is adequately described by linear stability theory, if the background disturbance level is very small. In detailed surveys Mack [1, 2] reviewed the compressible linear stability theory and discussed its role with respect to transition in compressible boundary layers. For the subsequent nonlinear stages there is no guidance from experiments. Transition experiments are very difficult to perform, and most of the few experiments for controlled disturbances were related to the initial linear regime of disturbance development. For the time being, it is assumed that transition evolves through similar stages as in incompressible flows. Only recently, theoretical results concerning the initial three-dimensional development became available based on compressible secondary instability theory (see [3, 4] for example). But whether these secondary instability mechanisms represent relevant mechanisms at low supersonic Mach numbers, where according to linear stability theory the most unstable disturbances are three-dimensional, is unknown. To answer this question detailed experimental investigations are required preferable in combination with numerical/theoretical studies.

In this paper, we report on numerical investigations of possible routes to turbulence in compressible flat-plate boundary layers at low supersonic Mach numbers. Results from direct spatial simulations for a new type of breakdown to turbulence at M_∞=1.6, initiated by only a pair of oblique disturbance waves, are presented that demonstrate the development of "honeycomb"-like structures (instead of staggered or aligned lambda-vortices arising from secondary instability) which were not observed before in other investigations.

NUMERICAL MODEL

The numerical method is based on a "spatial" model. As shown in Fig. 1, a finite rectangular box is selected to represent a certain region of a boundary-layer flow on a flat plate extending from $x=x_o$ to $x=x_N$ in the streamwise direction. In the normal direction, the integration domain extends from y=0 to $y=y_M$ and typically covers several boundary-layer thicknesses. In the spanwise direction, the flow is assumed to be periodic with the domain extending from z=0 to $z=\lambda_z$, where λ_z is the wavelength in z-direction. 2-D and 3-D disturbance waves with

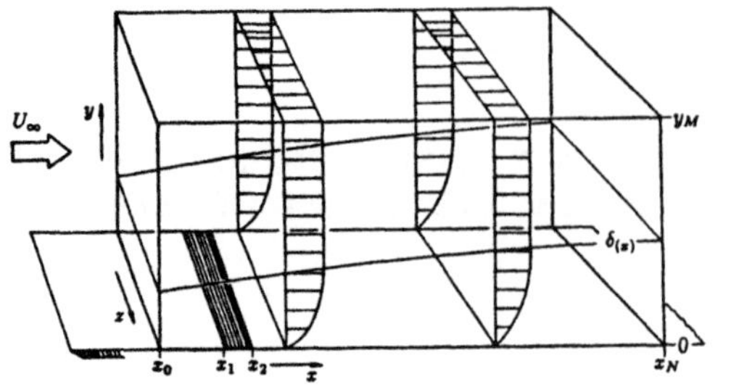

Fig. 1: Integration domain and disturbance strip

prescribed frequency and amplitude are introduced into the domain by periodic blowing and suction through a narrow disturbance strip as discussed in [5]. The streamwise linear and nonlinear evolution of these disturbances are calculated by solving the unsteady 3-D Navier-Stokes equations. Thus the numerical model can be regarded as a numerical simulation of a "controlled" transition experiment in which some sort of a wavemaker is used to generate controlled disturbance waves.

The flowfield is described by the three-dimensional Navier-Stokes equations, the continuity equation, the energy equation, and the thermodynamic equation of state for a compressible perfect gas. The velocity components u, v and w in the x-, y- and z-direction are made dimensionless using the free-stream velocity u_∞. Lengths are non-dimensionalized with respect to a reference length L. Reference values for the temperature T, viscosity μ, conductivity ϑ, and density ρ are their free-stream values (indicated by subscript ∞). The pressure is normalized with $\rho_\infty u_\infty^2$, and the internal energy e with u_∞^2. With these definitions, the global Reynolds number is $Re = u_\infty L \rho_\infty / \mu_\infty$. We assume a calorically perfect gas with the ratio of specific heats $\varkappa = c_p/c_v = 1.4$. The Prandtl number $Pr = \mu_\infty c_p / \vartheta_\infty$ is assumed constant and equal to 0.71, and for the dependence of the viscosity on the temperature $\mu = \mu(T)$ Sutherland's law is used. For the calculations discussed here the free-stream temperature was $T_\infty = 300K$ and the global Reynolds number was 10^5. The calculations were performed on a Cray 2.

For the boundary conditions at the inflow boundary, boundary layer profiles as obtained from the solution of the compressible boundary layer equations, are used. At the wall, all velocity components are zero (except for v at the blowing and suction strip). For the temperature, adiabatic wall conditions are assumed for the base flow, i.e. $\partial T/\partial y = 0$, while for the disturbance temperature $T' = 0$ is assumed (in agreement with linear stability theory). At the outflow boundary, linear wave assumptions are made for the disturbances. At the free-stream boundary, exponential decay of the disturbances is assumed, with boundary conditions derived from compressible linear stability theory [6].

The governing equations are solved in conservative form. The numerical method that we developed is a combination of finite-difference approximations of fourth-order accuracy in the x- and y-directions and spectral approximations (Fourier modes) in the z-direction. The time integration is fully explicit employing a MacCormack-type time stepping, which yields second order accuracy in time direction. The nonlinear terms are computed in physical space. For transforming from spectral to physical space and vice versa highly vectorizable FFT-routines are employed. A detailed discussion of the numerical method is given by Thumm [6].

The numerical method was carefully validated by extensive comparisons with results from compressible linear stability theory [6], and by comparison with nonlinear secondary instability behaviour of incompressible flows for which the code was run at a low Mach number $M_\infty = 0.2$ [7]. Recently, an additional validation of our code has been provided by Masad and Nayfeh [8], who compared their results from (subharmonic) compressible secondary stability theory with our numerical results [9], and found very good agreement.

NUMERICAL RESULTS

In order to investigate the role of secondary instability mechanisms, we have performed a large number of simulations, both for subharmonic and fundamental resonance, at Mach numbers 0.4, 0.8 and 1.6. Results of these calculations were discussed elsewhere [6, 9], and indicated that secondary instability might not be always the relevant mechanism at low supersonic Mach numbers, as fairly high amplitudes of the 2-D wave ($u'_{2D} \approx 5\%$) were found to be necessary to turn on the resonance mechanism. Therefore, we started to search for possible other routes to turbulence in such boundary layers. Based on results from compressible linear stability theory we performed a simulation for $M_\infty=1.6$, where the boundary layer was forced by only a pair of 3-D waves. We observed a surprising development as these disturbances quickly became highly nonlinear.

For this calculation, the integration domain extended in the x- and y-directions from $x_o=0.595$ to $x_N=8.875$, and from $y=0$ to $y_M=0.125$. The step sizes used were $\Delta x=0.00926$, $\Delta y=0.00125$, and $\Delta t=0.000697$. The spectral approximations were truncated at K=15, i.e. 31 spectral modes were used for the discretization in spanwise direction. At the blowing and suction strip, only a pair of 3-D disturbance waves with finite amplitude $u'_{3D} \approx 1\%$, and frequency $F_{1,1}=5.0$ were introduced. The spanwise wave number was $\gamma=2\pi/\lambda_z=10.6$ corresponding to a wave angle of $\psi \approx 45^o$. Here, the 3-D disturbance components are denoted as modes (n, k) where n stands for the frequency as the integer multiples of the fundamental frequency and k denotes the multiples of the spanwise wave number.

Results of this simulation are shown in Figs. 2-5. In Fig. 2, the streamwise development of the maximum value of typical wave components for the u' disturbance, obtained from a Fourier analysis of the twelfth time period, is presented semilogarithmically. In Fig. 2a, the unsteady disturbance components $u'_{1,k}$ (k=1-15), and in Fig. 2b the steady disturbance components $u'_{o,k}$ (k=0-14) are plotted. Both figures show that higher harmonic disturbance components are generated which are strongly amplified and saturate downstream. Most important, for the unsteady disturbance components only odd modes (1, 3) . . . (1, 15) and for the steady disturbance components only even modes (0, 2). . . (0, 14) appear. The underlying mechanism for this kind of nonlinear generation of higher harmonics was found to be completely different from the secondary instability mechanism that is responsible for the fundamental (or subharmonic) breakdown process. From the disturbance input mode (1, 1), modes (0, 2), (0, 0), (2, 0), and (2, 2) are generated by direct nonlinear interactions. From these, mode (0, 2), representing a longitudinal vortex, grows fastest and rapidly reaches a high amplitude. Higher harmonics are then generated by direct nonlinear interactions between the modes (1, 1) and (0, 2) and/or resulting higher modes. More details can be found in [6].

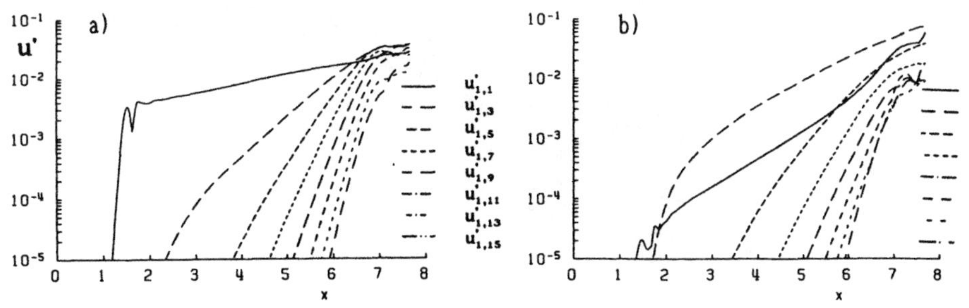

Fig. 2: Amplification curves for the maximum of the u'-disturbance
a) unsteady components $u'_{1,k}$ (k=1-15); b) steady components $u'_{o,k}$ (k=0-14)

It is therefore not surprising that the structures associated with the nonlinear disturbance development are also different from the lambda-shaped vortices observed at the fundamental (or subharmonic) breakdown process. Such structures may be visualized by iso-surfaces of the vorticity components ω_x and ω_z that can be calculated from the velocity components u, v, and w.

In Figs. 3 and 4 iso-surfaces $|\omega_x|$ = const. and ω_z=const. are plotted in three-dimensional perspective view, visualizing instantaneous flow structures in a part of the three-dimensional integration domain. The streamwise section $5.4 \leq x \leq 7.6$, considered here, covers about 3 wave lengths λ_x of the fundamental 3-D disturbance mode (1, 1), while the z-direction covers one wave length λ_z. The iso-surfaces consist of iso-contours $|\omega_x|$ = const., and ω_z = const., respectively, calculated at successive downstream x positions and plotted in perspective representation.

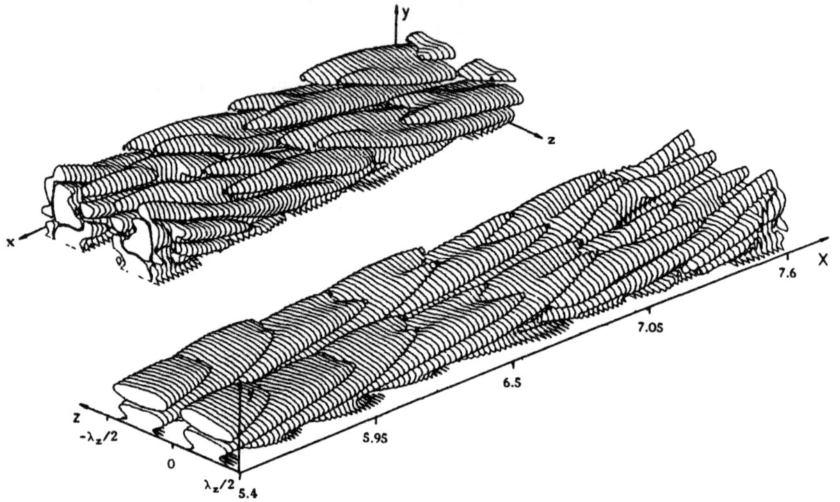

Fig. 3: Perspective view of instantaneous longitudinal vortical structures. Visualization by iso-surfaces $|\omega_x|$=1. Upper left figure: view from upstream.

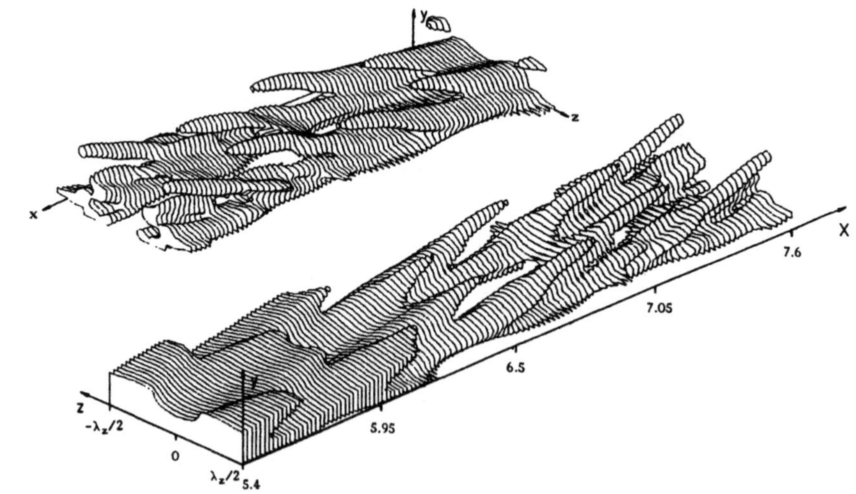

Fig. 4: Perspective view of instantaneous high-shear layers. Visualization by iso-surfaces ω_z=30. Upper left figure: view from upstream.

The iso-surfaces $|\omega_x|$=const. in Fig. 3 visualize vortical structures oriented in streamwise direction. There are no indications for lambda-shaped vortices. In the contrary, the vortical structures are closely spaced in streamwise direction, with one lying on top of the preceeding, and splitting into two tips while evolving downstream. The iso-surfaces ω_z=const. in Fig. 4 enclose the regions of high shear, and it is obvious that two high-shear layers per wave length λ_z in the z-direction appear, organized in the streamwise direction in a staggered pattern.

In Fig. 5, the flow structures are visualized by numerically generated time-lines as they would be generated experimentally by a horizontal (hydrogen bubble) wire placed at x=5, y=0.0125 (critical layer). For the fundamental breakdown, numerically generated time-lines develop into lambda-vortex structures (see [10]). In Fig. 5, however, the structures show a "honeycomb"-like pattern, that is quite different from the patterns of aligned or staggered lambda vortices.

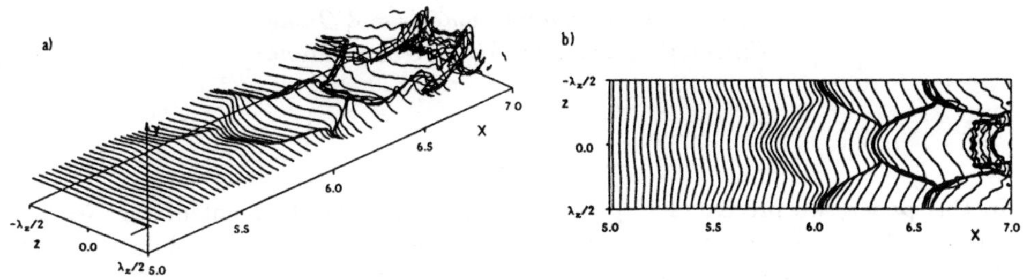

Fig. 5: "Honeycomb"-like pattern of flow structures visualized by time-lines. a) perspective view; b) view from above.

The results demonstrate that realistic numerical simulations of compressible transition phenomena based on the spatial model are already feasible with the current generation of supercomputers. Extreme care has to be placed on the validation of the numerical method. Once validated, such simulations allow for the exploration of physical phenomena that are not previously known from experiments, and enable considerable insight into the relevant physical mechanisms. In addition to experiments and theory, numerical simulations have become a powerful tool in transition research and, therefore, will play an increasingly important role in the future, in particular for high-speed flows.

This research was supported by the German Federal Ministry of Research and Technology (BMFT), Bonn, FRG

REFERENCES

1. Mack, L.M.: AGARD Report No. 709, pp. 3-1 to 3-81 (1984)
2. Mack, L.M.: "Stability of Time-Dependent and Spatially Varying Flows", D.L. Dwoyer, M.Y. Hussaini, eds., Springer, New York, pp. 164-187 (1985)
3. El-Hady, N.M.: AIAA Paper 89-0035 (1989)
4. Masad, J.A., Nayfeh, A.H.: Physics of Fluids A, 2, pp. 1380-1392 (1990)
5. Fasel, H., Rist, U., Konzelmann, U.: AIAA J. 28, pp. 29-37 (1990)
6. Thumm, A.: Dissertation, Universität Stuttgart, Germany (1991)
7. Fasel, H.: Proc. 3rd IUTAM-Symp. "Laminar-Turbulent Transition", Toulouse, France, Sept. 11-15, 1989, D. Arnal, R. Michel, eds., Springer, Berlin (1990)
8. Masad, J.A., Nayfeh, A.H.: submitted for Physics of Fluids A (1991)
9. Thumm, A., Wolz, W., Fasel, H.: Proc. 3rd IUTAM-Symp. "Laminar-Turbulent Transition", Toulouse; 1989, D. Arnal, R. Michel, eds., Springer, Berlin (1990)
10. Rist, U., Fasel, H.: "Boundary Layer Transition and Control Conference", Cambridge, U.K., April 8-12, 1991, The Royal Aeronautical Society, London

THE NONLINEAR GALERKIN METHOD APPLIED TO THE SIMULATION OF TURBULENCE IN A CHANNEL FLOW

T. DUBOIS AND R. TEMAM

The Institute for Applied Mathematics & Scientific Computing
Indiana University, Bloomington, IN 47405
and
Laboratoire d'analyse numérique d'Orsay
Université Paris-Sud, 91405 Orsay, France

1. INTRODUCTION

In this paper, we present new progress made in the development of the Nonlinear Galerkin method : its extension to nonhomogenous flows, namely to the channel flow problem. It is well known that the number of degrees of freedom of a turbulent flow increases as a power of the Reynolds number. But, based on a spectral decomposition, we observed that most of the wavelengths (the small wavelengths) carry only a small part of the total energy. Hence, we have developped news algorithms which take into account the effect of these small eddies in a simplified way. In fact, the Nonlinear Galerkin method is a dynamical adaptative scheme combining space and time discretization.

2. THE NONLINEAR GALERKIN METHOD APPLIED TO THE CHANNEL FLOW PROBLEM

We define, in space dimension 2, a channel $(0, L) \times (-1, +1)$. In this channel, we consider a flow which is driven by a constant pressure gradient and governed by the incompressible Navier-Stokes equations :

$$
\begin{cases} \dfrac{\partial \vec{u}}{\partial t} - \nu \Delta \vec{u} + \vec{\omega} \times \vec{u} + \vec{\nabla} p = 0 \\ div(\vec{u}) = 0 \end{cases} \tag{1}
$$

where $\vec{\omega} = \vec{\nabla} \times \vec{u}$ is the vorticity, $\vec{u} = \vec{u}(\vec{x}, t)$ is the velocity vector and $p = p(\vec{x}, t)$ is the pressure term.

For the boundary conditions, we assume that $\overrightarrow{u}$ is periodic in the x_1-direction (i.e. $\overrightarrow{u}(\overrightarrow{x}+L.e_1,t) = \overrightarrow{u}(\overrightarrow{x},t)$) and that $\overrightarrow{u}$ satisfies Dirichlet conditions in the x_2-direction (i.e. $\overrightarrow{u}(e_2,t) = 0$). Moreover, we require for p the following condition :

$$p(L.\overrightarrow{e_1},t) = p(0,t) + K, \quad K > 0 \text{ given}$$

Using Poisson's equation, we can express p as a function of $\overrightarrow{u}$ and then rewrite (1) in the form :

$$\frac{\partial \overrightarrow{u}}{\partial t} - \nu \Delta \overrightarrow{u} + B(\overrightarrow{u},\overrightarrow{u}) = 0 \tag{2}$$

where $B(\overrightarrow{u},\overrightarrow{u}) = \overrightarrow{\omega} \times \overrightarrow{u} + \overrightarrow{\nabla} p(\overrightarrow{u})$.

It is appropriate, due to the boundary conditions, to expand $\overrightarrow{u}$ in a Fourier series in the x_1-direction and in a Cehbyshev polynomial series in the x_2-direction :

$$\overrightarrow{u}(\overrightarrow{x},t) = \sum_{k\in\mathbb{Z},\ell\in\mathbb{N}} \hat{u}(k,\ell)e^{ikx_1}T_\ell(x_2) \tag{3}$$

where $T_\ell(x_2)$ is the Chebyshev polynomial of order ℓ.

For fixed $N = (n_{x_1}, n_{x_2})$, the classical pseudo-spectral method consists in finding an approximation $\overrightarrow{u_N}$ of $\overrightarrow{u}$:

$$\overrightarrow{u_N}(\overrightarrow{x},t) = \sum_{k=1-\frac{n_{x_1}}{2}}^{\frac{n_{x_1}}{2}} \sum_{\ell=0}^{n_{x_2}} \hat{u}(k,\ell)e^{ikx_1}T_\ell(x_2) \tag{4}$$

that satisfies the following equation :

$$\frac{d\overrightarrow{u_N}}{dt} - \nu \Delta \overrightarrow{u_N} + P_N B_N(\overrightarrow{u_N},\overrightarrow{u_N}) = 0 \tag{5}$$

and the boundary conditions. Here, B_N corresponds to a pseudo-spectral approximation of B. To solve (5), we use an algorithm proposed by Kim, Moin and Moser (1987).

The Nonlinear Galerkin method proceeds as follows. For a given integer $n_i < n_{x_1}$, we define a decomposition of $\overrightarrow{u_N}$, in the x_1-direction, into small and large eddies by :

$$\overrightarrow{u_N} = \overrightarrow{y_{n_i}} + \overrightarrow{z_{n_i}} \tag{6}$$

where :

$$\begin{cases} \overrightarrow{y_{n_i}}(\overrightarrow{x},t) = \displaystyle\sum_{k=1-\frac{n_i}{2}}^{\frac{n_i}{2}} \sum_{\ell=0}^{n_{x_2}} \hat{u}(k,\ell)e^{ikx_1}T_\ell(x_2) \\ \overrightarrow{z_{n_i}}(\overrightarrow{x},t) = \Big(\displaystyle\sum_{k=1-\frac{n_{x_1}}{2}}^{\frac{-n_i}{2}} + \sum_{k=\frac{n_i}{2}}^{\frac{n_{x_1}}{2}} \Big) \sum_{\ell=0}^{n_{x_2}} \hat{u}(k,\ell)e^{ikx_1}T_\ell(x_2) \end{cases} \tag{7}$$

Hence, $\overrightarrow{y_{n_i}}$ represents the large eddies and $\overrightarrow{z_{n_i}}$ represents small eddies of the flow. At this point, we introduce two projectors P_{n_i} and Q_{n_i} defined by $\overrightarrow{y_{n_i}} = P_{n_i}\overrightarrow{u_N}$ and $\overrightarrow{z_{n_i}} = Q_{n_i}\overrightarrow{u_N}$. Applying these projectors to (5), we obtain a coupled system for $\overrightarrow{y_{n_i}}$and $\overrightarrow{z_{n_i}}$:

$$(8) \qquad \begin{cases} \dfrac{d\overrightarrow{y_{n_i}}}{dt} - \nu\Delta\overrightarrow{y_{n_i}} + P_{n_i}B_N(\overrightarrow{y_{n_i}},\overrightarrow{y_{n_i}}) + P_{n_i}\tilde{B}_N(\overrightarrow{y_{n_i}},\overrightarrow{z_{n_i}}) = 0 \\ \dfrac{d\overrightarrow{z_{n_i}}}{dt} - \nu\Delta\overrightarrow{z_{n_i}} + Q_{n_i}B_N(\overrightarrow{y_{n_i}},\overrightarrow{y_{n_i}}) + Q_{n_i}\tilde{B}_N(\overrightarrow{y_{n_i}},\overrightarrow{z_{n_i}}) = 0 \end{cases}$$

where $\tilde{B}_N(\phi,\psi) = B_N(\phi,\psi) + B_N(\psi,\phi) + B_N(\psi,\psi)$.

As it was shown in the periodic case (Foias, Manley and Temam (1987,88)), $\overrightarrow{z_{n_i}}$ (resp. $\frac{d\overrightarrow{z_{n_i}}}{dt}$) is small compared to $\overrightarrow{y_{n_i}}$ (resp. $\frac{d\overrightarrow{y_{n_i}}}{dt}$). Hence, $\overrightarrow{z_{n_i}}$ may be computed less frequently in time : we neglect its evolution during a computed number of iterations. Due to intermettency phenomena, even if $\overrightarrow{z_{n_i}}$ (resp. $\frac{d\overrightarrow{z_{n_i}}}{dt}$) is small, it may have strong bursts and become non negligible during a short transient period (see Figure 1). Hence, the cut-off value n_i can not be fixed uniformly in time. So, we have developped a multilevel adaptative procedure that we describe in the next section.

3. Implementation of the method

We introduce several intermediate levels of discretization n_i smaller than n_{x_1} :

$$n_1 < \ldots\ldots < n_i < n_{i+1} < \ldots\ldots < n_{x_1}$$

where the n_i's are appropriate for the Fast Fourier Transforms.

We consider two real constants θ_1 and θ_2 with $\theta_2 > \theta_1$ and define two levels n_{1min} and n_{1max} as follows :

$$(9) \qquad \frac{|\overrightarrow{z_{n_i}}|_{L^2}}{|\overrightarrow{y_{n_i}}|_{L^2}} < \theta_1 \;\; (\text{resp. } \theta_2) \quad \text{for each} \quad n_i > n_{1max} \;\; (\text{resp. } n_{1min})$$

Assuming that $\overrightarrow{u_N}$ is known at time t_n, we compute these levels according to the procedure (9). According to some criterias, we define a characteristic time T_{cycle}. Now from t_n to $t_n + T_{cycle}$, the cut-off value n_i varies in time as follows :

> n_i decreases from n_{1max} to n_{1min}, grows up from
> n_{1min} to n_{1max} and so on, until $t_n + T_{cycle}$ is reached.

Hence, n_i describes a succession of classical V-cycles. For a given time t_m between t_n and $t_n + T_{cycle}$, we assume that the corresponding level is n_i. Then, the integration of the system (8) is as follows :

$$\begin{cases} \text{the } \overrightarrow{y_{ni}} \text{ equation is integrated by the classical method} \\ \text{while } \overrightarrow{z_{n_i}} \text{ is frozen : } \overrightarrow{z_{n_i}}(t_m) = \overrightarrow{z_{n_i}}(t_{m-1}) \end{cases}$$

When $t_n + T_{cycle}$ is reached, we perform several time steps on the finest grid (level), i.e. we integrate the full system (8). Then, we start again a new cycle as we did at time $t = t_n$. Figure 2 shows the variation in time of n_{1min} and n_{1max}.

4. Numerical results

In this section, we first describe numerical results obtained with the Nonlinear Galerkin methods on this problem. Figure 3 shows the evolution of the L^2_ω norm of the errors made by computation of an exact solution. Computations are performed with the classical and the Nonlinear Galerkin methods : the curves are essentially the same. In this case, the viscosity ν is equal to 1.0 and the time step is $\Delta t = 10^{-3}$. The numerical scheme is globally of order 2 in time. On such computation, we can easily control the accuracy of both methods and test each code. Figure 4 shows the CPU time used by each program : a gain of approximately 25% is obtained with the nonlinear Galerkin method.

Figure 5 and 6 show the contours of the vorticity respectively at time $t = 0$ and $t = 10$. In this case, the solution is not known a priori. The viscosity ν is equal to 5.0×10^{-4} and the time step is $\Delta t = 10^{-3}$. The length of the channel is 2π and $K = \frac{\nu}{L}$. The total number of modes required is $N = (128 \times 160)$.

Concluding remarks

In this paper, we have described numerical tests performed with the Nonlinear Galerkin method on the $2D$ channel flow problem (a nonhomogeneous flow with nonperiodic boundary conditions). These results show that the Nonlinear Galerkin method is well adapted to this problem. As for the periodic case, significant gain in computed time is obtained and the solutions obtained by the two methods are of the same order. Developpements and more physical results on this problem will be reported elsewhere.

Aknowledgement

This work was partially supported by the National Science Foundation under grant NSF-DMS-9024769 and by the Department of Energy under grant DOE-DE-FG02-92ER25120. All the computations presented in this article were done on the NCSA CRAY2 at University of Illinois.

Figure 1 : $G(t) = \frac{|\vec{z}_{n_1}|_{L^2}}{|\vec{y}_{n_1}|_{L^2}}$

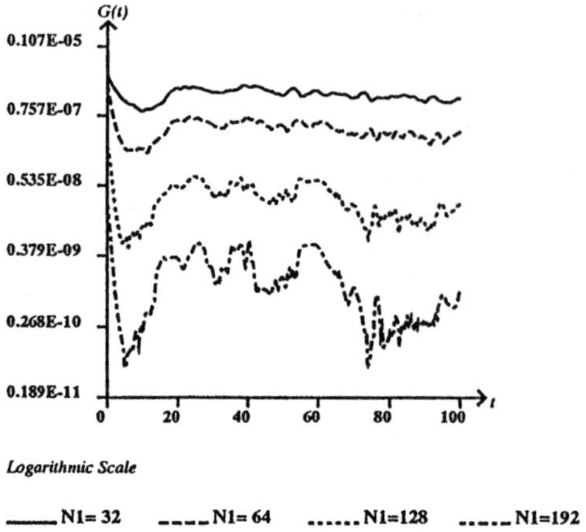

Figure 2 : $G(t) = (n_{1min}, n_{1max})$

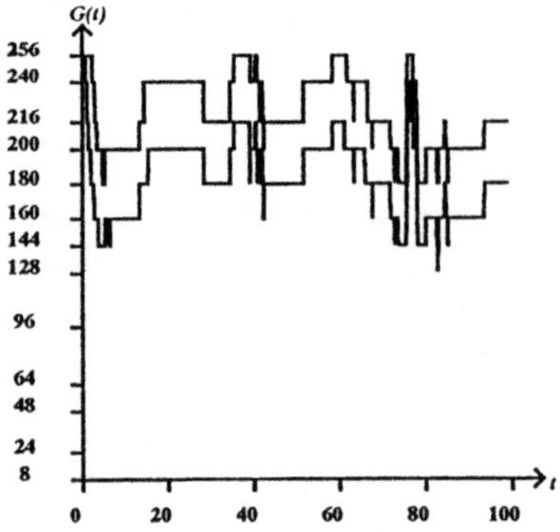

Non linear Galerkin method

Figure 3 : $G(t) = |\vec{u}_{computed} - \vec{u}_{exact}|_{L^2_\omega}$

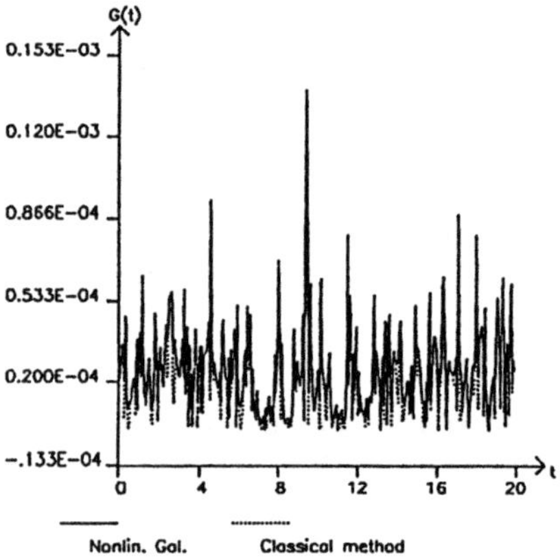

Figure 4 : $G(t) =$ CPU time (second)

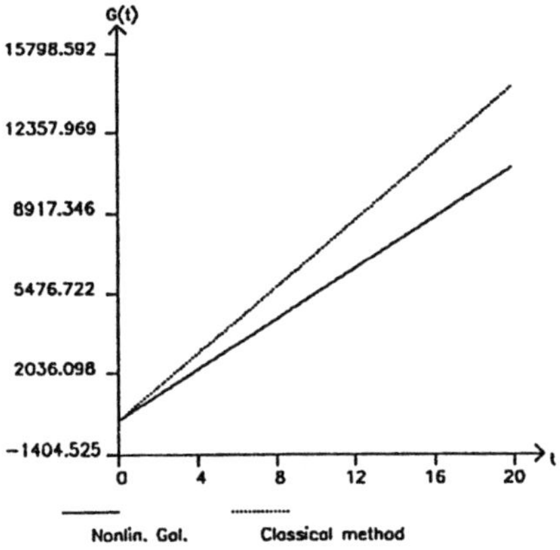

Figure 5

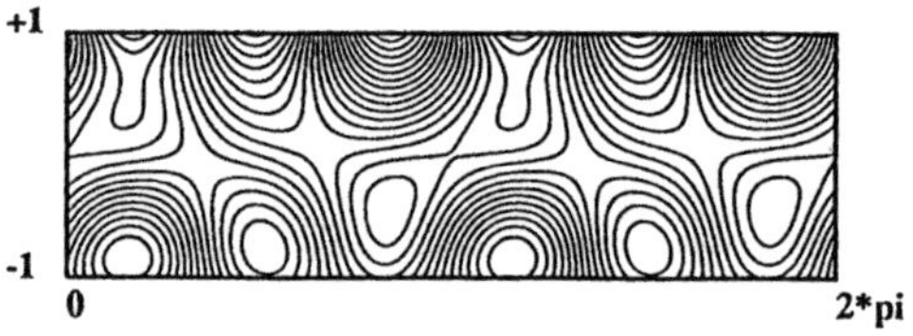

isovorticity lines at t = 0

Figure 6

isovorticity lines at t =10

UNSTEADY AND TRANSITIONAL SEPARATED FLUID FLOWS - DIRECT SIMULATION

V.GUSHCHIN and V.KONSHIN

Institute for Computer Aided Desing Russian Academy of Sciences, 2-nd Brestskaya str., 19/18, Moscow, 123865, Russia, phone: /007/095/2509541, Fax: /007/095/2508928, telex: 411700808 ICAD

Introduction

The most part of phenomena taking place in the unsteady and transitional fluid flows around finite bodies is characterised by the existance of the large scale organized structures in them (laminar and turbulent boundary layers, free shear layers, near wake, jets, etc.). Supposing that the dynamics of such structures is described by hydrodynamic equations without any turbulence models it is possible to use Navier-Stokes equations for the description of it's behaviour in the boundary layer and in the near wake for unsteady and transitional flow regimes around finite body using special tranformation of coordinates. To do this it is necessary to constuct numerical method which possesses by some special properties. The finite difference scheme of method must be second order of accuracy in space variables, to possess by minimal scheme viscousity, to be workable in wide range of Reynolds numbers and to be monotonic.

Numerical method and finite-difference scheme

One of the version of the Splitting on physical factors Method for Incompressible Fluid (SMIF) [1], which has been generalized now for three-dimensional problems [2], non-homogeneous fluids [3] and flows with a free surface [4] is used here for the considered problem.

For the adequate numerical modelling of hydrodynamic flows with large gradients of parameters it is necessary to construct finite-difference scheme with high-order of approximation which has a minimal scheme viscosity and is monotonic and stable over a wide range of Reynolds numbers. The construction of an inhomogeneous (hybrid) finite-difference scheme, which satisfies the requirements pointed above, will now be considered by taking the linear model equation

$$f_t + af_x = 0, \quad a = \text{const} \tag{1}$$

as an example.

Let us introduce a finite-difference approximation of *Eq.*(1)

$$\frac{f_i^{n+1} - f_i^n}{\tau} + a\,\frac{f_{i+1/2} - f_{i-1/2}}{h} = 0 \tag{2}$$

and inverstigate the class of difference schemes which can be written in the from of a two parameter family in the following manner:

$$f_{i+1/2} = \begin{cases} \alpha\, f_{i-1}^n + (1-\alpha-\beta)\, f_i^n + \beta\, f_{i+1}^n, & \text{if } a>0 \\ \alpha\, f_{i+2} + (1-\alpha-\beta)\, f_{i+1}^n + \beta\, f_i^n, & \text{if } a<0 \end{cases} \tag{3}$$

It is known [5] that it is impossible to construct a homogeneous monotonic difference scheme of higher than the first order of approximation for *Eq.*(1). A monotonic scheme of higher order can therefore only be constructed either on the basis of second-order homogeneous schemes using smoothing operators, or on the basis of hybrid schemes using different criteria involving switching over from one scheme to another (depending on the nature of the solution) [6], possibly with the use of smoothing [7].

Here we consider a hybrid finite difference scheme which is based on the combination of modified upwind scheme (MUS) and modified central difference scheme (MCDS).

For the MUS (β=0) the requirement of the minimal scheme

viscosity imposes the following condition: $\alpha = -0.5(1 - C)$, where Courant number $C = \tau\cdot|a|/h$.
For the MCDS (α=0) the same requirement gives: β=0.5(1−C).
Both of them are of the second order of accuracy and have zero scheme viscosity. Both of them are monotonic when

$$\left|f_{i+1}^{n} - f_{i}^{n}\right| = \delta \left|f_{i}^{n} - f_{i-1}^{n}\right|. \tag{4}$$

where $\zeta(C)\le\delta\le\sigma(C)$ and $\zeta(C)=(1-C)/[2(2-C)]$, $\sigma(C)=2(1+C)/C$.
The resulting switch condition may be written as follows:

$$\text{if } (a\cdot f_{x}\cdot f_{xx})_{i+1/2} \begin{cases} \ge 0, \text{ then MUS}, \\ <0, \text{ then MCDS}. \end{cases} \tag{5}$$

In the test case of $Eq.$(1) with the initial data $f(0,x)=1$, if $x\le0$, $f(0,x)=0$, if $x>0$ and $a=1$, $h=0.2$, $C=0.5$, the function f_{i}^{n}, obtained using scheme of the first order of accuracy [5] (curve 1), of the second order of accuracy [8] (curve 2), of the third order of accuracy [9] (curve 3) and using consided here scheme (curve 4) is shown in Fig.1 for t=15 (n=150).

The same approach is used for the approximation of the convective terms in the equations of motion for 2-D and 3-D problems with the central difference for the viscous terms, and may be used for the convective terms in the temperature, energy or density transfer equations both for the compressible and for the incompressible fluid flows.

Some numerical results

At Reynolds numbers 10<Re<40 steady symmetrical separated flows around circular cylinder is stable. At least two regimes exist at Re>40 − unstable symmetrical and stable periodic. The periodic flow which is realized at given fixed Reynolds number doesn't depend on the way of its receiving (influx-outflux of the different intensity, rotation, etc.).

The dependence on time of the total lift coefficient C_y is shown in Fig.2 for Re=55, 100 and 225.

At $Re=100$ only the alternative separation from both sides of the cylinder exists. At moderate Reynolds numbers ($200<Re<500$) the unsteady flow is essentially complicated by the existence of the secondary effects: the instability in separated shear layers during some part of the period, the secondary vortexes on the rear part of cylinder surface, the secondary vortexes near reattachment point.

Our calculations, which are based on the adequate reproduction of large scale vortex structures arising in the unsteady separated flows around finite body (boundary layer, free-shear layer, near wake etc.) allow to simulate transitional fluid flows also. The comparison of numerical results (time-average surface pressure distribution) with the experimental data [10] for $Re=4\cdot10^5$ shown in Fig.3, where also shown experimental data [11] for laminar flows ($Re=10^5$) and experimental data [12] for turbulent flow ($Re=8.6\cdot10^6$).

The crisis of total drag coefficient and sharp rise of the Strouhal number take place and are simulated numerically (without any turbulence models) for the critical Reynolds numbers ($Re \sim 4\cdot10^5$), what is in a good agreement with an experimental data [13].

REFERENCES

1. O.M.Belotserkovskii, V.A.Gushchin and V.V.Schennikov. U.S.S.R. J.Comp.Math.Math.Phys., 15, 190-200 (1975).
2. V.A. Gushchin, U.S.S.R. J.Comp.Math.Math.Phys.,16, 251-256, (1976.)
3. V.A. Gushchin, U.S.S.R. J.Comp.Math.Math.Phys., 21, 190-204 (1981).
4. O.M.Belotserkovskii, V.A.Gushchin and V.N.Konshin, U.S.S.R. J.Comp.Math.Math.Phys., 27,181-191 (1987).
5. S.K.Godunov. Matem.Sbornic, 47, 271-306, (1959).
6. B.Van Leer. J.Comput.Phys., 23, 3 263-276, (1977).
7. J.P.Boris and D.L.Book. J.Comput.Phys.,11, 1,38-69, (1973).
8. R.W.Mac-Cormac. AIAA Paper, 2, 69-354, 1-7, (1969).

9. O.M.Belotserkovskii and A.S.Kholodov. Sixth Internetional Conf. on Numerical Methods in Hydrodynamics. Collection of Papers, 11, Inst.Prik.Matem.Acad.Nauk SSSR, Moscow, 37-47, (1978).
10.N.M.Bychkow, V.V.Larichkin. Inst.Teor.and Appl.Mech. SB AS USSR, 1658, 1-57, (1986).
11.A.Fage, V.M.Falkner. ARC RM, 1369, (1921).
12.A.Roshko. J.Fluid Mech., 10, 345-356, (1961).
13.V.A.Gushchin, V.N.Konshin. Int.J. Numer.Meth.in Engr., 34, 469-472, (1992).

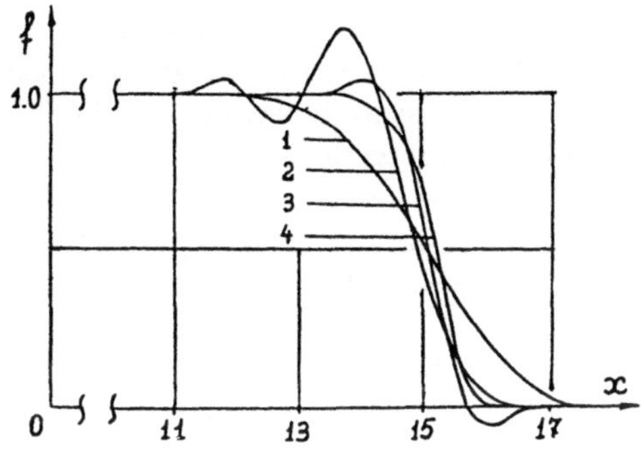

Fig.I. t=I5. I-first order scheme [5]; 2-second order scheme [8], 3-third order scheme [9], 4-present data

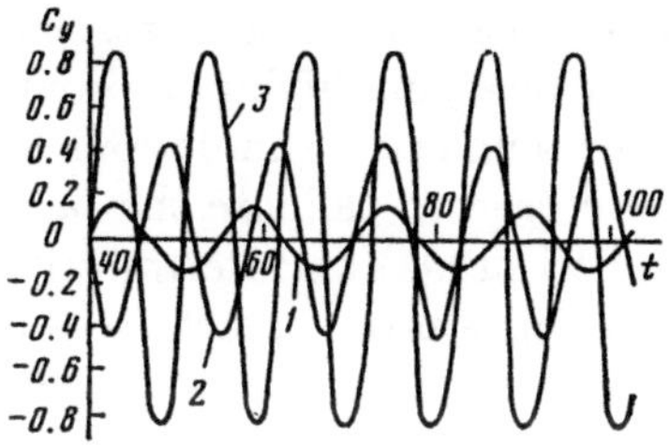

Fig.2. Total lift forth coefficient. I-Re=55, 2-Re=IOO, 3-Re=225

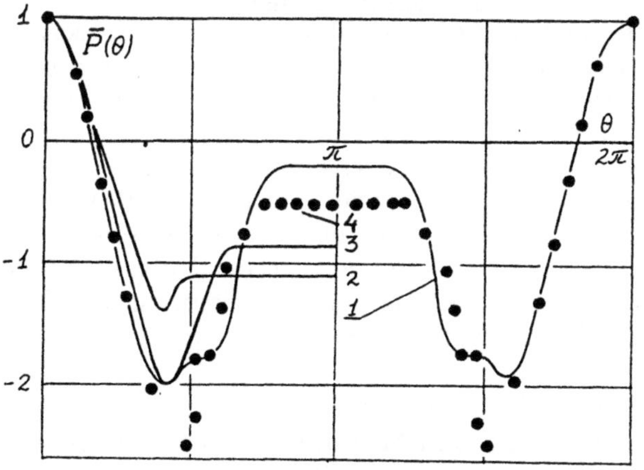

Fig.3. Time-average wall-pressure.
I-present data, Re=400000,
experiments:
2-laminar flow, Re=I00000 [II],
3-turbulent flow, Re=8600000 [I2],
4-transitional, Re=400000 [IO].

SPATIAL SIMULATION OF BOUNDARY-LAYER TRANSITION MECHANISMS

R. D. Joslin, C. L. Streett, and C.-L. Chang*

NASA Langley Research Center, Hampton, Virginia 23665, U.S.A.
*High Technology Corporation, Hampton, Virginia 23666, U.S.A.

Introduction

How disturbances are ingested into a boundary layer, evolve, and lead to transition is of practical engineering importance. To compute this disturbance evolution, direct numerical simulations (DNS) using the full Navier-Stokes equations is the relevant solution procedure; yet, the computational demands for solving this mathematical system are beyond current capabilities for practical engineering design processes. However, in a research environment, solutions to the full system can be used both to study physical mechanisms associated with problems of interest and to test theories which involve simplifying approximations to the full equations. Here, results from the direct numerical simulations of the Navier-Stokes equations in the spatial framework are used to validate the parabolic stability equations (PSE) theory developed by Herbert (1991) and Bertolotti (1991).

For this study, disturbances are ingested into the boundary layer by assuming that linear stability theory can adequately represent the initial disturbance profile. Then amplitudes are somewhat arbitrarily imposed on the profiles. In this manner, both theory and computation can employ the same initial disturbances. First, a single 2-D disturbance is forced, and theoretical results are compared with the computations for the flat plate boundary-layer problem. A pair of oblique waves is, then, forced and similar comparisons are made.

Numerical Methods

The incompressible, unsteady Navier-Stokes equations are solved in disturbance form. Namely, the instantaneous velocities and pressure are decomposed into basic components and disturbance components. The basic component is described by the Blasius similarity profile and the disturbance component is obtained by computation. The governing equations are nondimensionalized with respect to the free-stream velocity, U_∞, the kinematic viscosity, ν, and some length scale at the inflow (say, displacement thickness, δ_o^*). A Reynolds number can then be defined $R_o = U_\infty \delta_o^* / \nu$. The governing equations are subject to homogeneous Dirichlet boundary conditions at the wall and in the far field. At the inflow a disturbance is forced, and at the outflow the buffer-domain technique, introduced by Streett and Macaraeg (1989), is used.

The spatial DNS method is novel in that it obtains high-accuracy gains through using a combination of spectral and high-order compact difference methods. The spatial discretization entails: 4th-order finite differences for the implicit pressure equation and 4th- or 6th-order compact differences for the explicit momentum equations in the streamwise (x) direction; spectral-collocation methods in the wall-normal (y) direction; and full Fourier or Fourier cosine/sine series in the spanwise (z) direction. For time-marching, a time-splitting procedure is used. In the first step of the time-splitting procedure, the pressure is neglected from the momentum equations and intermediate velocities are obtained. The pressure field is next obtained by solving Helmholtz equations. Then, the full time-step velocites are computed with the updated pressure field.

To obtain the pressure field, solutions of 2-D Poisson equations and 3-D Helmholtz equations for each time-step are required, where the pressure is determined in Fourier coefficient space. In order to solve the equations efficiently, a fast elliptic solver is required. For this purpose, the tensor-product, or eigenvector decomposition approach is employed. Danabasoglu, Biringen, and Streett (1990) used the eigenvector decomposition method on a non-staggered grid for the 2-D channel problem. Joslin, Streett, and Chang (1992) extended the approach for the boundary-layer problem. Using the influence matrix method, solutions may be obtained for all Fourier coefficients on a non-staggered grid, or the influence-matrix approach may be used on the 2-D coefficient component, and a direct solver can be used on the 3-D components on a staggered grid. Both approaches have been implemented and tested.

For the time-splitting procedure, implicit Crank-Nicolson differencing is used for normal diffusion terms and an explicit 3-stage Runge-Kutta (RK) method, introduced by Williamson (1980), is used for all remaining terms. This time-splitting procedure consists of three intermediate RK stages, each stage of the following form:

$$\frac{\underline{u}^* - \underline{u}^m}{h_t^m} = C_1^m H^m(\underline{u}) + \frac{C_2^m}{R_o} D^2(\underline{u}^* + \underline{u}^m) \tag{1}$$

$$\nabla^2 \Phi^{m+1} = \frac{1}{h_t^m}(\nabla \cdot \underline{u}^*) \tag{2}$$

$$\frac{\underline{u}^{m+1} - \underline{u}^*}{h_t^m} = -\nabla \Phi^{m+1} \tag{3}$$

where

$$H^m(\underline{u}) = L(\underline{u})^m + C_3^m H^{m-1}(\underline{u}),$$

$$L(\underline{u}) = (\underline{U} \cdot \nabla)\underline{u} + (\underline{u} \cdot \nabla)\underline{U} + (\underline{u} \cdot \nabla)\underline{u} - \frac{1}{R_o}\nabla_{xz}^2 \underline{u}$$

Here, $\underline{u}^*$ are disturbance velocities at the intermediate RK stages, $\underline{u}^m$ are velocities at previous RK stages ($m = 1, 2$ or 3), $\nabla_{xz}^2 = \partial^2/\partial x^2 + \partial^2/\partial z^2$, h_t^m is the time-step size, $C_{1,2,3}^m$ are constants given by Williamson (1980), D^2 is the derivative matrix, and Φ is a pressure-like quantity. Equation (2) is subject to homogeneous Neumann boundary conditions only, so the slip-velocity corrections described by Streett and Hussaini (1986) are used to maintain continuity. The above system (1-3) is solved three times corresponding to the three stages of the Runge-Kutta method. The result after the third stage is the full time-step velocity field.

Results

For the present paper, PSE theory results are compared with the spatial DNS results for 2-D Tollmien-Schlichting (TS) wave propagation and oblique-wave breakdown. The PSE results are obtained using the compressible code of Chang, et al. (1991) in the incompressible limit, $M_\infty \simeq 0$.

2D T-S Wave Propagation. A 2-D Tollmien-Schlichting disturbance with amplitude $A^o = 0.25\%$ is introduced into the boundary layer by a forcing at the computational inflow. Calculations are made with an inflow Reynolds number $R_o = 688.315$

and frequency $F = 86$. The spatial DNS computations are performed on a grid of 2041 uniformly spaced streamwise nodes and 81 wall-normal collocation points. The outflow boundary is located $442\delta^*$ from the inflow boundary, and the far-field, or free-stream, boundary is located $75\delta^*$ from the wall. For the time-marching scheme, the disturbance period is divided into 320 time steps.

In figure 1, maximum streamwise amplitudes for the mean-flow distortion (u_o), fundamental wave (u_1), and first harmonic (u_2) with downstream distance from DNS are compared with the PSE results. Both the fundamental and first harmonic are in good quantitative agreement throughout the linear and weakly nonlinear regions, while some discrepancy occurs with the mean-flow distortion quantity in the nonlinear region where saturation is approached. Further, disturbance profiles at a streamwise location of $R = 1519$ are compared. The TS component and harmonics are in good quantitative agreement even in regions of high gradients. A small discrepancy in the results is revealed in the mean-flow distortion quantity, as drawn out in the wall-normal component (v_o) shown in figure 2. This discrepancy is probably due to the homogenous Neumann boundary conditions imposed in the far-field normal component of the mean-flow distortion equation for PSE theory. This leads to a nonzero normal mean-flow component in the far-field. The DNS approach enforced homogeneous Dirichlet far-field conditions resulting in exponentially decaying disturbances. Also shown in figure 2 is good quantitative agreement for the fundamental wave (v_1) amplitude and profile as predicted by PSE theory and compared to the DNS results.

Oblique-Wave Breakdown. Due to its non-linear nature, no adequate formal theory is available to explain the breakdown process; however, similar mechanisms have been studied by Hall and Smith (1991) using asymptotic methods in the large Reynolds number limit. To quantify the mechanisms of interest in the finite Reynolds number range, one must presently resort to methods such as DNS, or possibly PSE theory, to study the wave interactions. This alternative route to, or mechanism of, transition was previously studied by Schmid and Henningson (1992), Chang and Malik (1992), and Joslin, Chang, and Streett (1992).

Here, an oblique-wave pair is forced at the inflow with amplitudes $A^o = 1.0\%$. The inflow Reynolds number is $R_o = 900$, the frequency is $\omega = 0.0774$, and the spanwise wavenumbers are $\beta = \pm 0.2$. Computations are performed on a grid of 901 uniformly spaced streamwise nodes, 61 wall-normal collocation points, and 10 symmetric-spanwise nodes. In the streamwise direction, the outflow boundary is located $465\delta^*$ from the inflow boundary; the far-field, or free-stream, boundary is located $75\delta^*$ from the wall; and the spanwise-symmetric boundaries are one-half spanwise wavelength, $\lambda_z/2 = \pi/\beta$, apart. For the time-marching scheme, the disturbance period is again divided into 320 time steps.

The computed primary disturbance (1,1), mean-flow distortion (0,0), and the dominant higher-order mode, which is the vortex mode (0,2), are shown with downstream growth in figure 3. The self-interaction of the waves feeds energy into the vortex mode and mean-flow distortion. The interaction of the wave pair with the vortex leads to a rapid cascade of energy to other low-wavenumber modes which have growth-rate characteristics similar to the vortex mode. The vortex and other harmonics (not shown) rapidly overtake the introduced waves (1,1) and breakdown occurs. This is made apparent by a rise in the skin-friction (c_f) curve, as shown in figure 4. Although both the DNS and PSE results near the end of the domain are inadequately grid-resolved, both indicate a similar skin-friction curve rise. In addition, a comparison of the disturbance profiles shows similar good agreement between the PSE and DNS results.

Conclusions

In the present paper, a spatial direct numerical simulation approach has been used to test the validity of parabolized stability equations theory in computing two- and three-dimensional boundary-layer instability mechanisms.

PSE theory has accurately predicted the disturbance development for 2-D wave propagation through saturation. Also, it was shown that oblique-wave pairs self-interact to excite a dominant streamwise vortex structure that can lead to breakdown. The PSE results are in good agreement with the full computations up to the skin-friction rise. A difference in the far-field boundary condition treatment between the DNS and PSE methods was identified as likely the reason for differing mean-flow distortion quantities. How this difference in the mean-flow distortion affects the global prediction capability of PSE theory is yet unknown. In this paper, it has been shown, that for the present flat plate problem, PSE theory results quantitatively agree with the spatial DNS results; however, expanding the range of problems beyond that for which PSE theory has been tested must be made with caution due to the assumptions underlying the theory.

Acknowledgments

The authors extend their thanks to Mr. Gokhan Danabasoglu at the Department of Aerospace Engineering Sciences, University of Colorado at Boulder, for the use of his channel simulation code.

References

Bertolotti, F.P., Ph.D. Thesis, The Ohio State University, (1991).

Chang, C.-L. and Malik, M.R., *Transition and Turbulence*, (Springer-Verlag: New York, 1992).

Chang, C.-L., Malik, M.R., Erlebacher, G., and Hussaini, M.Y., *AIAA Paper No. 91-1636*, (1991).

Danabasoglu, G., Biringen, S., and Streett, C.L., *AIAA Paper No. 90-1530*, (1990).

Hall, P. and Smith, F.T., *J. Fluid Mech.*, **227**, (1991), pp. 641-666.

Herbert, Th., *AIAA Paper No. 91-0737*, (1991).

Joslin, R.D., Streett, C.L., and Chang, C.-L., NASA TP-3205, (1992).

Joslin, R.D., Streett, C.L., and Chang, C.-L., *Transition and Turbulence*, (Springer-Verlag: New York, 1992).

Schmid, P.J. and Henningson, D.S., *Transition and Turbulence*, (Springer-Verlag: New York, 1992).

Streett, C.L. and Hussaini, M.Y., *ICASE Report No. 86-59*, (1986).

Streett, C.L., and Macaraeg, M.G., *Appl. Numer. Math.*, vol. 6, (1989), pp. 123-140.

Williamson, J.H., *J. Comput. Phys.*, vol. 35, no. 1, (1980), pp. 48-56.

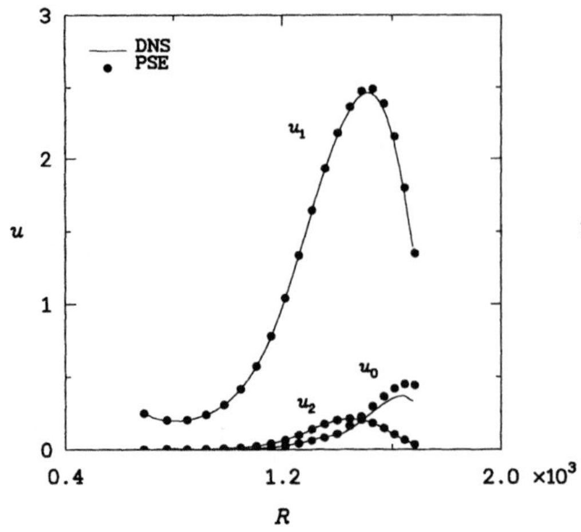

Figure 1. Disturbance Amplitude growth with downstream distance, arising from a two-dimensional TS inflow wave with amplitude $A^o = 0.25\%$.

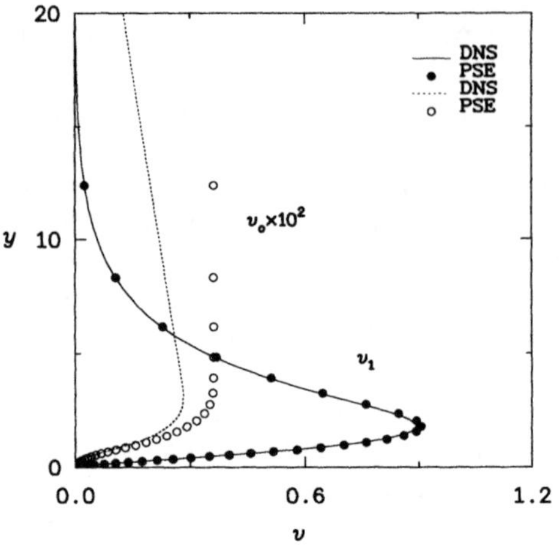

Figure 2. Normal (v) mean-flow distortion and fundamental wave profiles with normal distance from the wall for the local Reynolds number $R = 1519$.

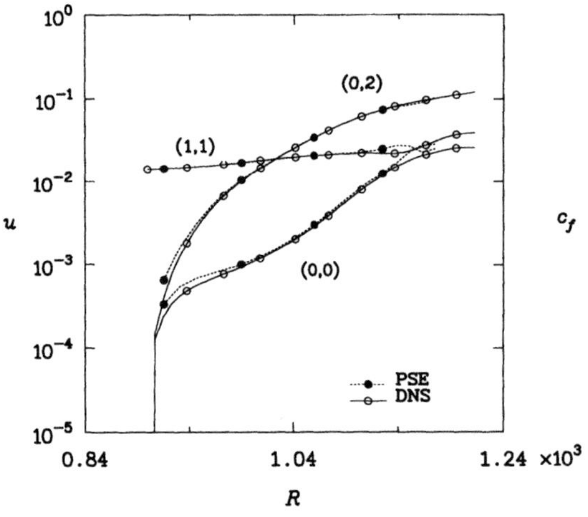

Figure 3. Disturbance Amplitude growth with downstream distance, arising from a pair of oblique-waves, each with amplitude $A^o = 1.0\%$.

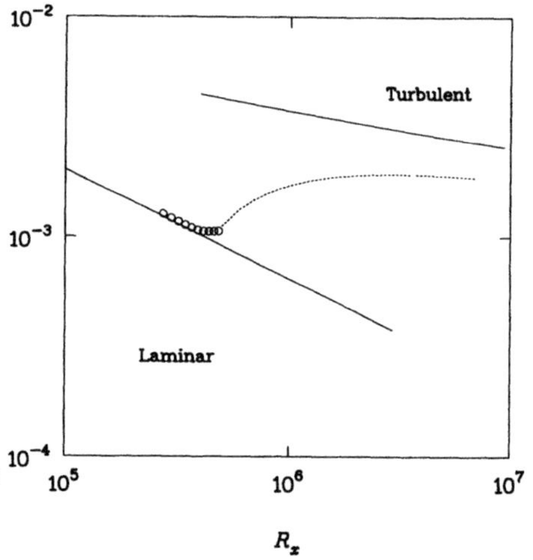

Figure 4. Typical skin-friction curves for laminar, transitional, and turbulent flows with downstream location. The symbols represent computed values.

VORTICAL STRUCTURES AND TURBULENT PHENOMENA IN A PISTON-ENGINE MODEL

J.J.Kloeker†, E.Krause†, K.Kuwahara‡

†Aerodynamisches Institut, RWTH Aachen, Templergraben 55,
D-5100 Aachen, Germany
‡Institute of Space and Astronautical Science, 3-1-1 Yoshinodai,
Sagamihara-shi, Kanagawa, 229, Japan

Introduction

For the development of a new piston-engine a specific power should be obtained with a high efficiency and a minimum of pollution. This requires a deep insight into the combustion and flow processes.

Since the reaction processes are essentially influenced by flow parameters, a numerical calculation of combustion is only possible, if the vortical and turbulent structures of the flow at the end of the compression stroke are known. This requires an exact prediction of the flow. To attain a reliable solution the numerical results should be compared with cycle-resolved experimental measurements of an engine-model. Recent cycle-resolved experimental data of the flow inside the cylinder of a rectangular piston-engine model using the Mach-Zehnder interferometry show large vortical structures dominating the flow during the intake and the beginning of the compression stroke. Before reaching the top-center (TC) the vortical structures decay into turbulence [1], [2].

Physical Problem

The turbulence introduced during the intake stroke with the incoming jet is damped out at the beginning of the compression stroke. Near the TC the three-dimensional perturbations reach a sufficient level to generate smaller vortical structures leading to a complex flow field with turbulent phenomena. The numerical simulation of the flow requires the resolution of the smallest length-scale, i.e. the Kolmogorov-length l_k. Henshaw *et al.* [3] found an estimation of the length-scale for incompressible flows $\lambda_{\min} \equiv (\nu / \overline{|D\vec{u}|_\infty})^{\frac{1}{2}} \leq l_k$. The quantity $\overline{|D\vec{u}|_\infty}$ means the maximum instantaneous velocity gradient in space and time. In the case of the piston-engine model the resolution of this length requires 10^{18} grid points, whereas only $5 \cdot 10^5$ grid points are realized. Due to the relatively coarse grid, preventing the generation of a realistic energy cascade to the smallest length-scales, the flow cannot become fully turbulent. There is a cut-off inside the initial-range changing the behaviour of the numerical solution compared with a physically real flow field.

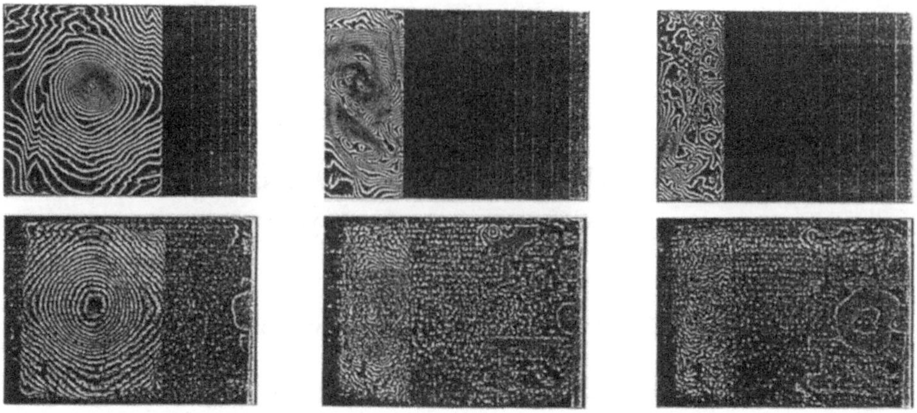

Figure 1: Comparison of the density contours inside a rectangular piston-engine model during compression stroke at 270° CA, 330° CA and 360° CA (from left) (above: integrated values of a three-dimensional simulation, below: Mach-Zehnder interferometry)

Method of Analysis

To evaluate the numerical solution, experiments are performed for the in-cylinder flow of an engine model with a rectangular piston, which is used for flow visualization experiments at the AIA. The Mach-Zehnder interferometry requires a plane flow to obtain density contours. Using a high-speed camera with 6000 frames/sec the cycle-resolved flow field has been shown by Jeschke [2]. An additional temperature measurement at a specific point inside the flow permits the calculation of absolute density contours.

For the numerical simulation the Navier-Stokes equations are integrated for three-dimensional, compressible flows to simulate the vortical structures and the transition to turbulence during the intake and compression stroke of a rectangular piston-engine model. An extended, explicit Godunov-type method, second-order accurate in time and space is employed. Present turbulence models are not able to handle compressible influences in a piston engine and experimental data are insufficient or not available. Therefore neither a turbulence model nor artificial perturbations are used. For details see [5]. For this method, which is fully vectorized, no further artificial damping terms are necessary.

The three-dimensional grid contains $100 \times 100 \times 50$ cells for the case with the flat piston and $120 \times 100 \times 50$ cells for the step-piston. The results of the complete flow field are shown by cross-sections of the density contours. For comparison with experimental data the density values are integrated in the light-direction of the Mach-Zehnder interferometer. This leads to a smoothing of the contours, and some three-dimensional structures disappear.

Results and Discussion

The three-dimensional simulation agrees very well with the interferometry data. As an example the integrated density contours of the compression stroke are shown in Figure 1 with the plane piston at 270° CA, 300° CA and 360° CA(TC) in comparison with experimental

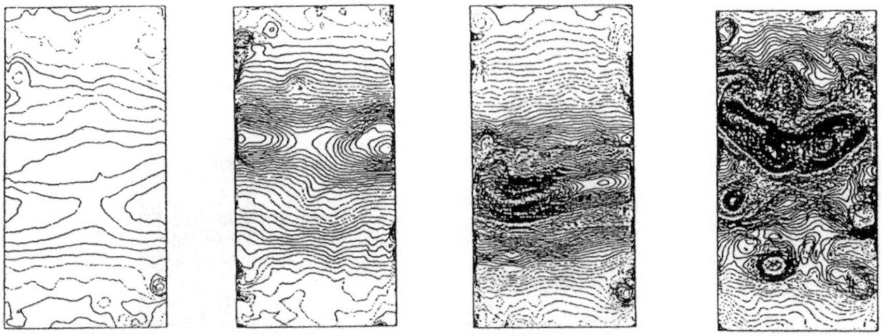

Figure 2: Cross-sections of the density inside a rectangular piston-engine model for the compression stroke at $210° CA$, $270° CA$, $300° CA$ and $360° CA$ (from left) for the y-z-plane at $N_x = 50$

density contours obtained by the Mach-Zehnder interferometry. The integration is carried out in the direction of the light of the interferometer. In addition to satisfying agreement of the vortical structures, transition is predicted in the same chronological order as seen in the experiment.

While the main vortex structure can clearly be recognized at $270° CA$, after the transition into a turbulent flow the influence of the numerical dissipation becomes evident. Since the smallest numerical length-scale is considerably larger compared to the smallest physical length-scale of the energy-spectrum, the spatial fluctuations of the density contours obtained in the computations are less pronounced than those observed in the experiments.

Figure 2 shows several cross-sections for the compression stroke with the view in direction of the piston movement providing an insight to the vortical structures. While at $210° CA$ the main vortical structure is nearly two-dimensional, with progressive compression the main vortex becomes three-dimensional and, finally, decays at $330° CA$. This breakdown of the main vortical structure is called the transition process. First smaller three-dimensional vortices can already be seen at $300° CA$ due to the vortex stretching. The presence of small numerical truncation errors due to the discretization initiates the vortical generation process through vortex stretching. The small perturbations due to the errors are amplified by the non-linear stretching leading to three-dimensional structures with a continuous wave-number spectrum. The transition process is shown in Figure 3 for cross-sections with the view in light-direction. After the transition the flow becomes turbulent due to the decay of the vortices. At TC a turbulent structure can be seen with a minimum vortex size of 3–8 grid cells. Due to the numerical dissipation no smaller vortices can appear. This leads to a relatively large Taylor-length of the size of several grid cells ($l_T = O(\triangle x)$), and in the expansion stroke all vortical structures are diminishing.

The computed flow depends on numerical truncation errors leading to dissipation and dispersion of the solution and influencing the statistical turbulence data of the vortical structures. Although initially the flow is two-dimensional, three-dimensional vortical structures are obtained. The turbulence statistics determined are in qualitative agreement with experimental data of [1]. The data are averaged over 50 time-steps, i.e. $0.5° - 1° CA$ corresponding to the periods of the LDA-measurements in [1]. During the intake stroke as well as the compression stroke the same turbulence intensities are found as in experiments. Since the coherent structures contain the major part of the turbulence energy, the larger vortical structures are investigated without modelling the flow on the subgrid level [4], [5]. In contrast to the LES-model using assumptions for the subgrid-scales, which are still rarely available for compressible flow problems, the influence of the numerical dissipation and

Figure 3: Cross-sections of the density inside a rectangular piston-engine model for the compression stroke at $300° CA$, $330° CA$ and $360° CA$ (from left) for the x-y-plane at $N_z = 25$

their behaviour concerning the numerical cut-off wave-number of the energy-spectrum is investigated.

The energy-spectra of the spacial correlations across the cylinder show very steep gradients of k^{-2} to k^{-4} due to the numerical dissipation. In connection with the large Taylor-length this means a high turbulent energy dissipation, since the cut-off wave number is located inside the inertial range. The trend of the correlation spectra corresponds to the vortex sizes. Only at the expansion stroke the gradients reach a $k^{-\frac{5}{3}}$ behaviour after the dissipation of all larger vortical structures. This means that fully turbulent flows always require a very fine resolution $\Delta x < O(l_k)$. If the solution should also be free of numerical dispersion, i.e. the phase resolution is correct, a resolution $\Delta x << O(l_k^{\frac{3}{2}})$ is required [6]. This leads to a grid with 10^{24} cells for the examined engine-model.

Furthermore essential influences of the inlet and wall boundary conditions and of the initial state of the flow are observed, see also [7]. In the literature simulations always start with a flow at rest. Calculating several strokes demonstrates the importance of improved initial conditions as well as optimal perturbations for the vortex generation. Figure 4 shows several intake strokes for a three-dimensional simulation with a step-piston in comparison with experimental data. The first intake stroke has been performed with an initially quiet flow. After an immediate exhaust stroke the second intake stroke has been simulated resulting in an improved comparison with the experimental density contours. This demonstrates the need of cycle-resolved experimental data as well as of a high computer capacity to simulate several strokes of in-cylinder flows.

Conclusion

The in-cylinder flow of an engine-model with several kinds of pistons is numerically simulated and compared with experimental density contours. The three-dimensional simulation shows an excellent agreement with the cycle-resolved experimental data. The three-dimensional vortical structures are initialized by numerical truncation errors at the boundaries leading to perturbations due to vortex stretching. Though there is a general influence of the boundary shapes, the transition process is predicted in the correct chronological order. Due to the high numerical dissipation with a Taylor-length in the magnitude of the grid cells the gradients of the correlation spectra are to steep. The inertial range appears as late as an equilibrium between the energy transfer of the cascade of the diminishing vortices and the numerical dissipation is reached. But the compression stroke is not influenced by the intake stroke, which can be improved by the simulation of several strokes getting optimized initial conditions.

Figure 4: Comparison of the density contours inside a rectangular piston-engine model with a step-piston for several intake strokes at 90°*CA* (left: first stroke, center: second stroke, right: experiment)

Acknowledgements

The authors gratefully acknowledge the Institute of Computational Fluid Dynamics in Tokyo for making available their computation facilities. The work is partly supported by the German Research Association (DFG).

References

[1] SFB 224: (1990), *Proceeding of the Kolloquium des SFB 224 "Motorische Verbrennung" der RWTH Aachen*, Sept. 24-25, 1990

[2] Jeschke M.: (1992), 'Zyklusaufgel"oste Dichtefelder in einem geschleppten Modellmotor mit quaderf"ormigem Innenraum', *Thesis at the RWTH Aachen*

[3] Henshaw W.D., Kreiss H.O., Reyna L.G.: (1988), 'On the Smallest Scales for the Incompressible Navier-Stokes Equations', *NASA-CR 181628, ICASE Rep.No. 88-8*

[4] Kloeker J.J., Krause E.: (1992), 'Numerical Simulation of Vortical and Coherent Structures in Compressible Jet Flows', *Proc. of the 4th European Turbulence Conf.*, Delft, June 30 - July 4, 1992, (to appear)

[5] Kloeker J.J.: (1992), 'Numerische Simulation einer dreidimensionalen, kompressiblen, reibungsbehafteten Str"omung im Zylinder eines Modellmotors', *Thesis at the RWTH Aachen*

[6] Orszag S.A.: (1971), 'On the Resolution Requirements of Finite-Difference Schemes', *Stud. Appl. Math.*Vol. **50**, pp 395-397

[7] H"anel D., Breuer M., Kloeker J.J., Meinke M.: (1992), 'Computation of Unsteady Vortical Flows', *Proc. 4th Int. Symp. CFD, Davis, Calif., Sept. 9-12, 1991, in: Computer & Fluids* (to appear)

Prediction of Turbulent Lee-Side Vortex Flow

A. Kumar and C.A.J. Fletcher
Department of Mechanical Engineering
University of Sydney, NSW 2006, Australia

1 Introduction

The formation and subsequent breakdown of the lee-side vortices from the leading edges of highly swept wings is an important aerodynamic phenomenon. They have been a topic of investigation since the 1950's when the research and design of delta wing aircraft was initiated. In recent years, with the development of high performance/maneuverable aircraft, there has been intense interest in this area. With the development of high speed supercomputers, over the last decade CFD methods have employed Euler and Navier-Stokes equations to predict progressively more details of the vortex flow behaviour. While computations based on Euler equations [1,2] show that the primary vortex from sharp leading edge wings is well captured, secondary separation is captured in computations based on laminar Navier-Stokes equations [3,4]. Complex vortex interaction on a cranked delta wing is also predicted closely [5].

As the Reynolds number encountered in the real flow is very high and the flow is dominated by a separated shear layer from the leading edge that rolls up to form the primary vortex, inclusion of turbulence effects by appropriate modelling is required for accurate prediction of the flow. Earlier studies have employed the Baldwin-Lomax turbulence model [6,7] and the Johnson-King turbulence model [8] in the study of transonic turbulent vortex flow. It is expected that due to their greater versatility, the two-equations model of turbulence should be applicable to a wider class of complex flows. In [9] a $k-\epsilon$ model of turbulence is reported to give a better prediction of flow over the M-6 wing. In the present study, $k-\epsilon$ and ASM turbulence models are used to compute the incompressible turbulent vortex flow over a delta wing. Application of these turbulence models to lee-side vortex flow prediction appears to be new.

2 Method

The Reynolds-averaged Navier-Stokes equations for steady flow are

$$\frac{\partial}{\partial x_i}(\rho U_i) = 0 \tag{1}$$

$$\frac{\partial}{\partial x_i}(\rho U_i U_j + \delta_{ij} p - \tau_{ij} + \rho\overline{u_i u_j}) = 0 \quad (2)$$

where U_i is the mean velocity, u_i is the fluctuating velocity, ρ is the density, p is the pressure, τ_{ij} are the viscous stresses and $-\rho\overline{u_i u_j}$ are the Reynolds stresses. The Reynolds stresses in (2) are calculated both by a $k-\epsilon$ turbulence model and an ASM turbulence model. Transport equations for the turbulent kinetic energy, k, and the dissipation, ϵ, are

$$\frac{\partial}{\partial x_i}\left(\rho U_i k - \rho C_s (k/\epsilon)\overline{u_i u_j}\frac{\partial k}{\partial x_j}\right) = P_k - \rho\epsilon \quad (3)$$

$$\frac{\partial}{\partial x_i}\left(\rho U_i \epsilon - \rho C_\epsilon (k/\epsilon)\overline{u_i u_j}\frac{\partial \epsilon}{\partial x_j}\right) = (\epsilon/k)(C_{\epsilon 1} P_k - C_{\epsilon 2}\rho\epsilon) \quad (4)$$

The Reynolds stresses are obtained from the ASM approximation

$$(\overline{u_i u_j}/k)(P_k - \epsilon) = P_{ij} + \Phi_{ij} - (2/3)\epsilon\delta_{ij} \quad (5)$$

where P_{ij} ($\equiv -\rho\overline{u_i u_k}\partial U_j/\partial x_k - \rho\overline{u_j u_k}\partial U_i/\partial x_k$) and P_k ($\equiv \frac{1}{2}P_{kk}$) are the production terms for particular Reynolds stresses and k respectively. The term Φ_{ij} represents the pressure-strain behaviour. Detailed expressions and alternative uses of the present formulation are provided by Armfield *et al* [10] and Cho & Fletcher [11]. The use of (5) implies that $(\mathcal{C}+\mathcal{D})\overline{u_i u_j} = (\overline{u_i u_j}/k)(\mathcal{C}+\mathcal{D})k$, where $\mathcal{C}$ and $\mathcal{D}$ are convective and diffusive operators. In the present formulation a two-layer wall function is used to provide the solution and the boundary condition close to the solid surfaces, with the first grid point at $y^+ \approx 50 \sim 100$ from the surface.

A finite volume spatial discretisation of the governing equations is used. Area vectors are used to provide the transformation metrics between physical space and computational space [12]. The control surface values F_n are interpolated in terms of the cell-centered values. On a uniform physical grid the following expression is typical of the interpolation for the convective part of F,

$$F^c_{i-\frac{1}{2}} = F^c_i + \phi_{i-\frac{1}{2}}\{0.5\Delta F_i - (q/3)(\Delta F_i - \Delta F_{i-1})\} \quad (6)$$

where $\phi_{i-\frac{1}{2}}$ is a Roe-type limiter and q allows additional control over dispersion behaviour. The choice $\phi_{i-\frac{1}{2}} = 1.0$ and $q = 0.5$ provide a third order accurate determination of $\partial F/\partial x$ on a uniform grid. The descretised equations are solved sequentially using a strongly implicit procedure at each iteration to obtain velocity components, pressure and turbulence solution. The velocity and pressure solutions are coupled via a SIMPLEC procedure used in conjunction with a velocity potential formulation. The satisfaction of the continuity equation is based on an auxiliary potential method where

$$\underline{q}^c = \alpha\nabla\phi, \quad \rho^c = -\delta\phi/(a^*)^2 \ and \ \ p^c = -\beta\phi \quad (7)$$

and α, β and δ are under-relaxation factors. As part of each iteration the continuity equation is satisfied by the solution of the following transport equation for ϕ,

$$\delta\frac{\partial}{\partial x_i}\left\{\frac{u_i^*\phi}{(a^*)^2}\right\} - \alpha\frac{\partial}{\partial x_i}\left\{\rho^*\frac{\partial\phi}{\partial x_i}\right\} = \nabla\cdot(\rho\underline{q}^*) - \gamma\frac{\partial^4\phi}{\partial\xi^4} \quad (8)$$

The present algorithm is applied to study flow in a wing-body junction by Cho *et al* [13].

3 Results and Discussion

Flow around the 65° round leading edge delta wing of the International Vortex Flow Experiment [2,3] is considered. The wing has a 15% taper and an aspect ratio of 1.38. The cross-section of the wing is a symmetric modified NACA 64A005 section having a maximum thickness of 5% at 0.4c. A C-O mesh consisting of $96 \times 24 \times 24$ cells is generated using a method based on transfinite interpolation.

Figs. 1-2 show the cross-flow velocity vectors and pressure contours of the $k-\epsilon$ solution at angle of incidence of 20° and Reynolds number of 2.369×10^6. These plots and all subsequent plots are at $x/c = 0.9$ plane. Flow separation and the formation of the lee-side vortices is seen from the cross-flow velocity vectors, and the characteristic low pressure in the vortex core regions is seen from the pressure plot. The two turbulence models produce similar mean-flow behaviour. The centre of the primary vortex is at $\eta = 0.724$ and $\zeta = 0.144$, which compares well with the data compiled by Lowson [14]. The present computations do not show secondary separation, which could be due to grid coarseness or, possibly the use of a wall function with the turbulence model.

Figs. 3-4 show the turbulent kinetic energy, non-dimensionalised by square of free-stream velocity, obtained by the $k-\epsilon$ and the ASM turbulence models. The maximum value of turbulence kinetic energy occurs near the minimum pressure point in the $k-\epsilon$ solution. In the ASM solution the maximum k occurs in the boundary layer at the wing surface, and a local maximum of k occurs above the wing. While the level of k is the same on the wing surface in the two solutions, the value of k in the vortex region is higher from the $k-\epsilon$ solution. From the ASM solution the lower value of k and, consequently, Reynolds stresses give rise to a lower production rate. Figs. 5-7 show the transverse plane Reynolds stress components $\overline{vv}$, $\overline{ww}\, and \overline{vw}$. The major contribution to the production of turbulence kinetic energy comes from $\overline{ww}$. Fig. 8 shows the plot of $\overline{ww}/k$. It is seen that gradients of $\overline{u_i u_j}/k$ along the streamline are generally small, except near the wing tip region. This implies that $\mathcal{C}(\overline{u_i u_j}/k)$ is small, in general. In addition the contributions from $\mathcal{D}(\overline{u_i u_j}/k)$ are also small in the region where individual Reynolds stresses are large, except in the vicinity of the wing tip. The ASM approximation is, therefore, generally valid. Further investgation of the ASM solution, particularly the lower level of k predicted in the vortex region, is continuing.

Acknowledgements The authors are grateful to the Australian Research Council (ARC) for their support to the project. The authors are also thankful to Australian Supercomputing Technology for generous computing time provided on the VP2200.

References

[1] A. Rizzi, 'Euler Solutions of Transonic Vortex Flows around the Dillner Wing', *J. Aircraft*, **22**, 325-328, 1985

[2] A. Kumar & A. Das, 'Numerical Solutions of Flow Fields Around Delta Wings Using Euler Equations', *Proc. International Vortex Flow Experiment on Euler Code Validation*, Stockholm, pp. 175-186, 1986

[3] B. Mueller & A. Rizzi, 'Navier-Stokes Simulation of Laminar Flow Over the 65° Round Leading-Edge Delta Wing at Mach = 0.85 and $\alpha = 10^o$', *Proc. International Vortex Flow Experiment on Euler Code Validation*, Stockholm, pp. 269-280, 1986

[4] A. Kumar, 'Flow Calculation Over a a Delta Wing Using the Thin-Layer Navier-Stokes Equations', *Proc. 3rd Intl. Symp. on CFD*, Nagoya, pp. 371-376, 1989

[5] C.-H. Hsu & C.H. Liu, 'Simulation of Leading-Edge Vortex Flows', *Theo. Comp. Fluid Dynamics*, **1**, 379-390, 1990

[6] A. Hilgenstock, 'Validation of Transonic Turbulent Flows Past Delta Wing Configurations', Aero. J., 219-230, 1991

[7] J. Vadyak & D.M. Schuster, 'Navier-Stokes Simulation of Burst Vortex Flowfields for Fighter Aircraft at High Incidence', *J. Aircraft*, 28:638-645, 1991

[8] U. Kaynak, E. Tu, M. Dindar & R. Barlas, 'Nonequilibrium Turbulence Modelling Effects on Transonic Vortex Flows About Delta Wings', AGARD-FDP Vortex Flow Aerodynamics, 1990

[9] Y. Takakura, S. Ogawa & T. Ishioguro, 'Turbulence Models for 3D Transonic Viscous Flows II', *Proc. 3rd Intl. Symp. on CFD*, Nagoya, 680-685, 1989

[10] S.W. Armfield, N.-H. Cho & C.A.J. Fletcher, 'Prediction of Turbulence Quantities for Swirling Flow in Conical Diffusers', *AIAA J.*, 28:453-460, 1990

[11] N.-H. Cho & C.A.J. Fletcher, 'Computation of Turbulent Conical Diffuser Flows Using a Non-Orthogonal Grid System', *Comp. & Fluids*, 19:347-361, 1991

[12] C.A.J. Fletcher, *Computational Techniques for Fluid Dynamics*, 2nd Edn., Vol. 2, p. 53, Springer Verlag, 1991

[13] N.H.Cho, C.A.J. Fletcher & K.Srinivas, 'Efficient Computation of Wing Body Flows', *Lecture Notes in Physics 371, Proc. XII ICNMFD*, pp. 167-171, Springer Verlag, 1990

[14] M.V.Lowson, 'Visualisation Measurements of Vortex Flows', *J. Aircraft*, 28:320-327, 1991

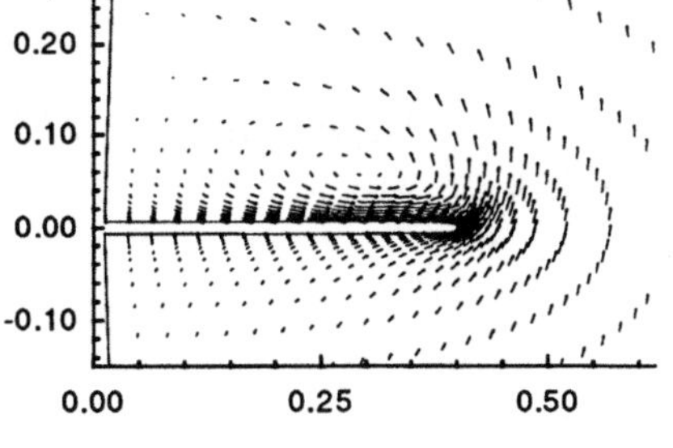

Fig. 1. Cross flow velocity vectors ($k - \epsilon$ model)

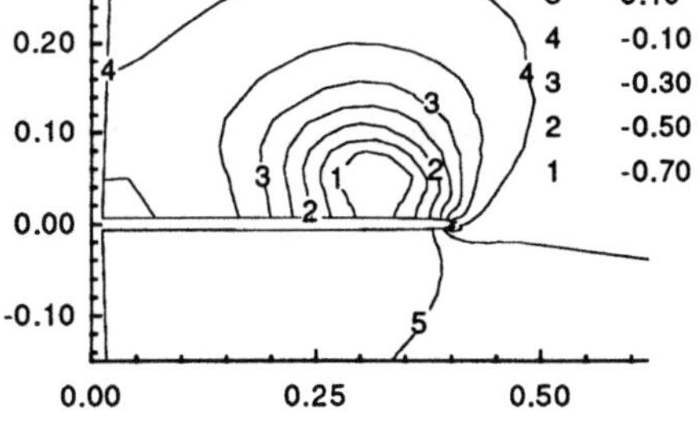

Fig. 2. C_p contours ($k - \epsilon$ model)

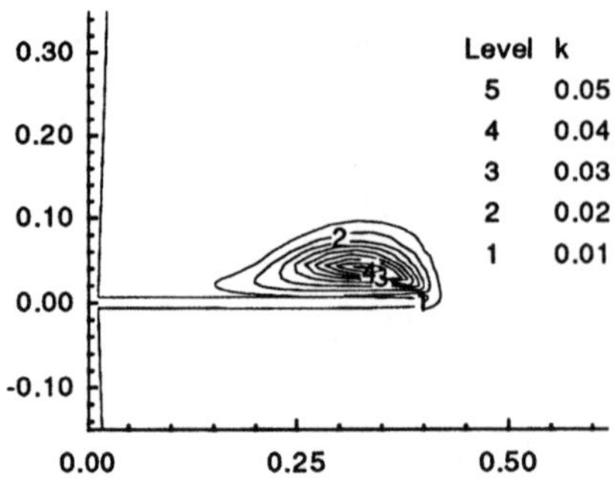

Fig. 3. k contours ($k - \epsilon$ model)

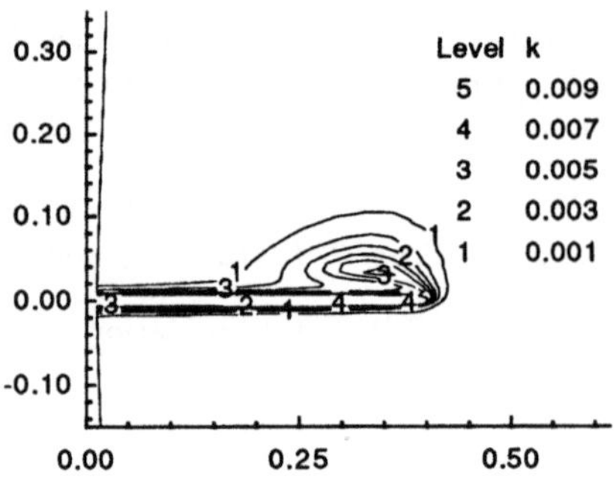

Fig. 4. k contours (ASM model)

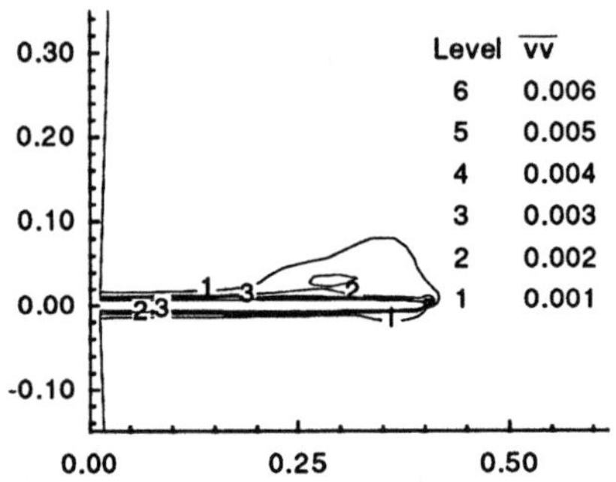

Fig. 5. $\overline{vv}$ contours (ASM model)

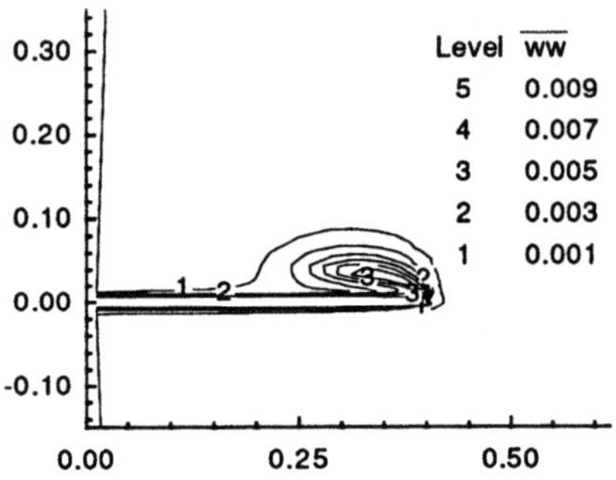

Fig. 6. $\overline{ww}$ contours (ASM model)

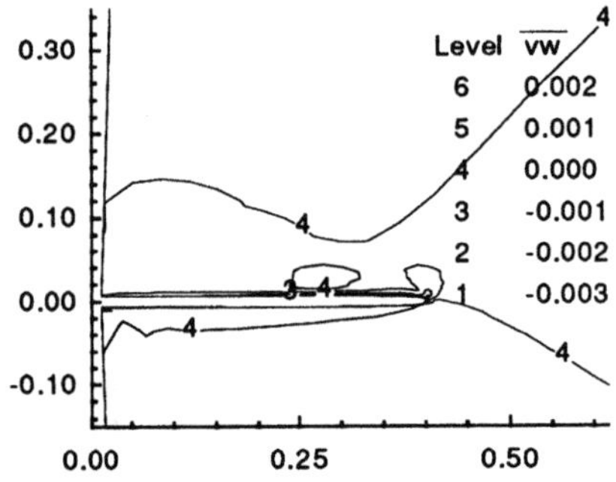

Fig. 7. $\overline{vw}$ contours (ASM model)

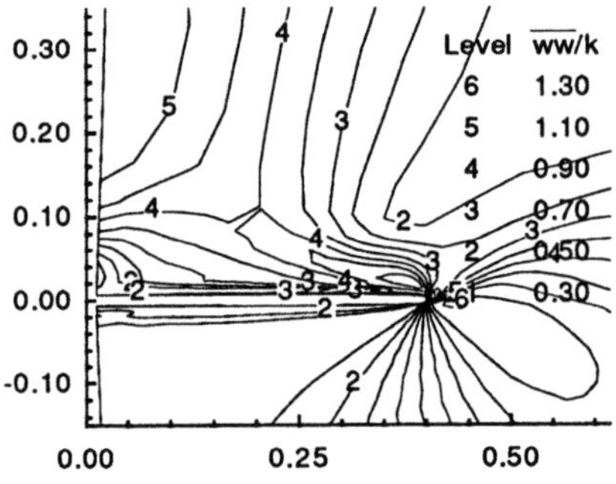

Fig. 8. $\overline{ww}/k$ contours (ASM model)

MODELLING SHOCK/TURBULENT-BOUNDARY-LAYER INTERACTION WITH SECOND-MOMENT CLOSURE WITHIN A PRESSURE-VELOCITY STRATEGY

F-S Lien and M A Leschziner

Department of Mechanical Engineering
University of Manchester Institute of Science and Technology
Manchester, UK

1. INTRODUCTION

Although major parts of transonic flow over aerofoils are practically inviscid, the sheared regions adjacent to the aerofoil's surface can play an important role in determining its aerodynamic performance due to the rapid thickening of the boundary layers induced by the shock. This is particularly so when the shock is sufficiently strong to provoke boundary-layer separation, in which case the details of the separated (post-)interaction region largely determine the position of the shock and materially affect the lift and drag coefficients. The simultaneous presence of shear, severe adverse pressure gradient and recirculation-related curvature close to a solid boundary lead to a complex, highly anisotropic turbulence structure. Extensive experience with such conditions in incompressible flows suggest that sophisticated modelling practices based on second-moment closure are often necessary for a satisfactory prediction of recirculation. In high-speed aerodynamics, this modelling framework is in its infancy, however, but there is, here too, a growing awareness of its potential.

The computational implementation of any turbulence-transport closure can, in principle, be undertaken within a number of alternative solution strategies. The large majority of approaches solve the conservation laws within a density-velocity-based, time-marching framework. This route is well suited to high-speed flow, but difficulties are encountered at low Mach numbers when significant portions of the flow become essentially incompressible. For incompressible flows, in contrast, a pressure-velocity framework is the established route. However, the extension of this method from incompressible to transonic conditions, although previously undertaken, is far from established in multidimensional space. A consistent and general extension is clearly desirable, for it would allow a safe application of the method to flows at all speeds.

This paper focuses on the formulation of a pressure-velocity scheme applicable to both compressible and incompressible conditions, and its performance when applied, in conjunction with second-moment turbulence closure, to shock-induced separation. Particular issues highlighted are: (i) the adaptation of the incompressible algorithm to transonic conditions; (ii) a comparison of predictive performance of variants of the k-ε model with a Reynolds-stress-transport closure for shock-induced separation.

2. COMPUTATIONAL APPROACH

2.1 Numerical Framework

The present solution algorithm sprang from a cell-centred non-orthogonal, fully collocated finite-volume method developed by the authors for incompressible 3D flow [6]. Attention is focused here on the route taken in adapting this procedure to transonic conditions. Before embarking on this topic, however, it is appropriate to summarise the essential elements of the incompressible methodology, which all carry over to the compressible environment.

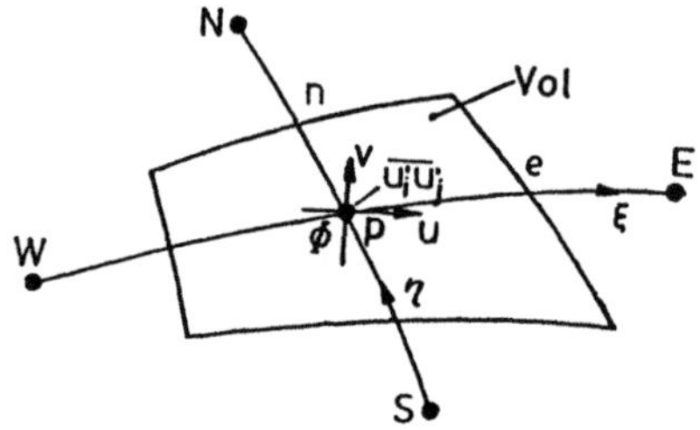

Fig. 1: Finite volume and variable storage

The algorithm employs a fully collocated storage arrangement for all transported properties, Fig. 1, including the Reynolds-stress components. Discretisation involves the integration of the conservation laws over the volume and application of the Gauss Divergence Theorem. Advective volume-face fluxes are approximated using either Leonard's quadratic scheme QUICK [1] or van Leer's MUSCL scheme [2].

At all speeds, mass continuity is enforced by solving a pressure-correction equation which, as part of the iterative sequence, steers the pressure towards a state at which all mass residuals in the cells are negligibly small. In conjunction with a fully collocated approach, this method is known to provoke chequerboard oscillations, reflecting velocity-pressure decoupling. To avoid this, the widely used method of Rhie and Chow [3] is used to interpolate for the cell-face velocities from nodal values. The interpolation can be demonstrated to introduces fourth-order pressure diffusion which removes the chequerboard mode. Particular difficulties arise in relation to the Reynolds stresses. Their collocated storage results in stresses being decoupled from associated 'driving' strains. This must be counteracted by elaborate interpolation practices not dissimilar, in principle, to those applied to pressure and velocity. These practices are given in detail in ref [6].

The starting point of the description of the compressible variant is the mass-conservation equation, written below for any transformed ξ,η domain:

$$(J\rho)_t + (\rho U)_\xi + (\rho V)_\eta = 0 \tag{1}$$

where J is the Jacobian of the transformation and U,V are the contravariant velocities. Within a pressure-correction algorithm, momentum-flux corrections are related to pressure corrections via truncated (i.e. approximated) variants of the momentum equations at the discretised level. Hence, focusing on the eastern and northern faces of the cell in Fig. 1, the requisite truncated relations are:

$$(\rho U)'_e = \overline{DU^\xi}_e(p'_P - p'_E)^{u\xi} y_\eta - \overline{DV^\xi}_e(p'_P - p'_E)^{v\xi} x_\eta \tag{2}$$

$$(\rho V)'_n = \overline{DV^\eta}_n(p'_P - p'_N)^{v\eta} x_\xi - \overline{DU^\eta}_n(p'_P - p'_N)^{u\eta} y_\xi \tag{3}$$

Density and pressure perturbations may be linked via a linearised variant of the total enthalpy-pressure relation,

$$\rho' = \frac{\gamma p'}{(\gamma - 1)(H_o - \frac{u^2+v^2}{2})} \tag{4}$$

To satisfy mass conservation within any iterative step, the contravariant face velocities must be corrected as follows,

$$\rho U \longleftarrow \rho U + (\rho U)' \qquad \rho V \longleftarrow \rho V + (\rho V)' \tag{5}$$

Combining equations (1)-(5) yields the pressure-correction equation,

$$A_P p'_P = \sum_{m=E,W,N,S} A_m p'_m + R_m \tag{6}$$

where R_m is the mass-residual resulting from the difference of contravariant velocities. In essence, the above equation is solved in conjunction with those of momentum fluxes within an overall iterative sequence.

The above sequence is appropriate to subsonic flow in which the conservation equations are elliptic. In supersonic and transonic conditions, the hyperbolic character of the conservation laws must be accounted for in order to retain iterative stability and capture shocks. To this end, an proposal by Hafez et al [4] has been adapted. This proposal - applied by Hafez et al to the solution of the full potential equation - is based on the definition of the *retarded density*, $\tilde{\rho}$. In one-dimensional conditions, $\tilde{\rho}$ is:

$$\tilde{\rho} \equiv \rho - \bar{\mu}\vec{\rho}_x \Delta x \tag{7}$$

in which $\bar{\mu}$ is a 'weighting factor', to be defined later, controlling the degree of retardation. With relation (7), an upwind-biased gradient of the convective flux of any intensive property ϕ may be expressed implicitly as:

$$(\frac{\rho u \phi}{\tilde{\rho}})_x = (u\phi)_x + \underline{[\bar{\mu}(\frac{u\phi}{\rho})\vec{\rho}_x]_x} + HOT \tag{8}$$

where the RHS is an expanded (i.e. explicit) form of the LHS with HOT denoting higher-order terms. The underlined term in equation (8) represents a dissipative mechanism equivalent to upwind-biasing. In the present scheme, upwind-biasing is applied to all convected properties. In particular, the contravariant *convecting* velocities multiplying flux variables are modified via:

$$U \longleftarrow \frac{\rho U}{\tilde{\rho}}, \quad V \longleftarrow \frac{\rho V}{\tilde{\rho}}, \tag{9}$$

To account for directional influences in the upwind-biasing process in two-dimensional conditions, the retarded density is evaluated from:

$$\tilde{\rho} = \rho - \bar{\mu}\rho_s \Delta s = \rho - \bar{\mu}[\cos(\beta)\rho_x \Delta x + \sin(\beta)\rho_y \Delta y] \tag{10}$$

with Fig. 2 giving a graphical interpretation of the above relation. The weighting factor μ takes the form of a Mach-number-dependent *monitor function* which sensitises density retardation to the Mach number via:

$$\bar{\mu} = \max\{0, \kappa[1 - (\frac{M_{ref}}{M})^2]\} \qquad (11)$$

hence ensuring that no numerical diffusion is introduced for $M < M_{ref}$, where M_{ref} is close to 1, while κ was varied between 1 for approximately normal shocks down to 0.6 for highly oblique shocks. A much more detailed account of the rationale behind and the implementation of density or pressure retardation may be found in ref [6], where retardation is linked the characteristics of the velocity potential equation.

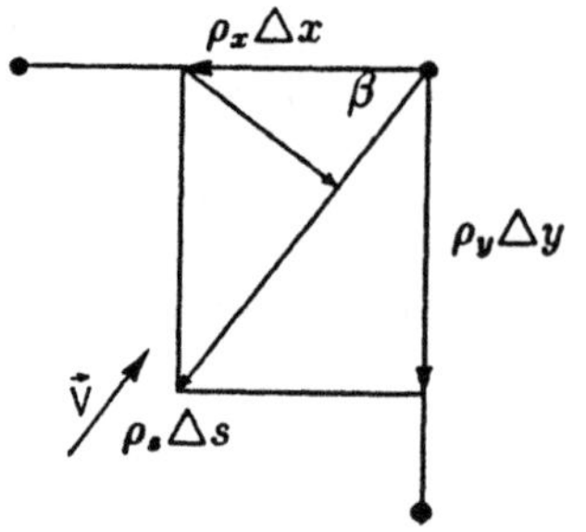

Fig. 2: Schematic of density biasing

2.2 Turbulence Models

Three turbulence models feature in comparisons to be presented below. Two are eddy-viscosity models based on the solution of equations for the turbulence energy k and its dissipation ε. The third is the Reynolds-stress-transport closure of Gibson and Launder [7]. One of the k-ε variants is the "standard" high-Reynolds-number model of Jones & Launder [8], and this operates in conjunction with log-law-based wall laws. The other is a low-Reynolds-number variant [6], formulated so as to conform with length-scale constraints implied by Wolfshtein's one-equation model [9].

3. APPLICATION

In a preliminary validation stage, the present scheme was applied to inviscid transonic and supersonic flows [6] over bumps, the latter involving shock-shock interaction. Calculations were then made for a range of shock-affected boundary layers over plane channel bumps for which detailed LDA measurements were performed by Delery [10]. Of these, the most challenging is the strong-interaction case 'C' in which the turbulent boundary layer separates due to a strong shock at M=1.4. The grid used for high-Re model variants contained 110x50 nodes, while 110x80 nodes were used for the low-Re variant. Grid-independence tests were conducted with MUSCL and QUICK to ascertain that the chosen densities were adequate.

Fig. 3 gives a general view of the predicted shock/boundary-layer interaction region. All models predict similar Mach-contour fields, but the second-moment closure (RSTM) gives rise to the strongest interaction, the most pronounced λ-shock structure and the largest separation zone. Separation is reflected by a characteristic bump pressure plateau following the strong oblique shock, as shown in Fig. 4. Evidently, the k-ε variants fail to capture separation properly, the role of modelling the near-wall being marginal. The RSTM, on the other hand, predicts a far more sensitive response to the shock, resulting in a more extensive recirculation zone. Clearly, however, the rate of post-shock recovery is too slow. This reflects a combination of three causes: first, turbulence is attenuated by the stabilising influence of flow curvature along the shear layer bordering the recirculation zone; second, the oblique shock is so strong that the very weak normal leg of the λ-shock is hardly resolved, the consequence being an insufficient pressure recovery at x/H=3; third, the wall-reflection fragment of the pressure-strain term in the Reynolds-stress model has an excessively attenuating influence on the wall-normal turbulence intensity and hence on the shear stress. This latter weakness has been observed in other flows. More generally, the excessive sensitivity predicted by variants of second-moment closure related to that of Gibson & Launder has been observed, albeit to a lesser extent, in other studies based on entirely different numerical strategies (e.g. Dimitriadis and Leschziner [11]).

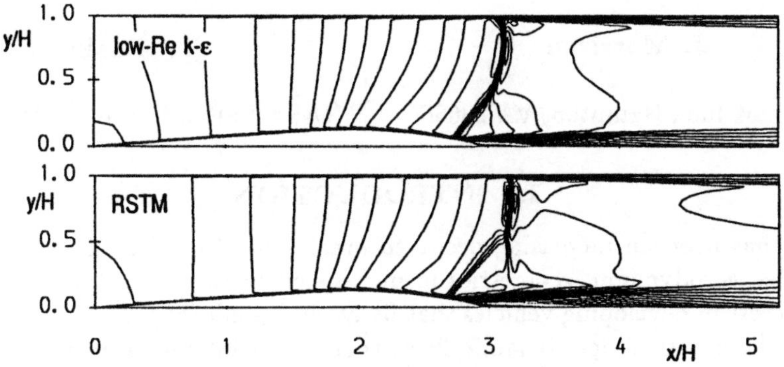

Fig. 3: Predicted Mach-contour fields

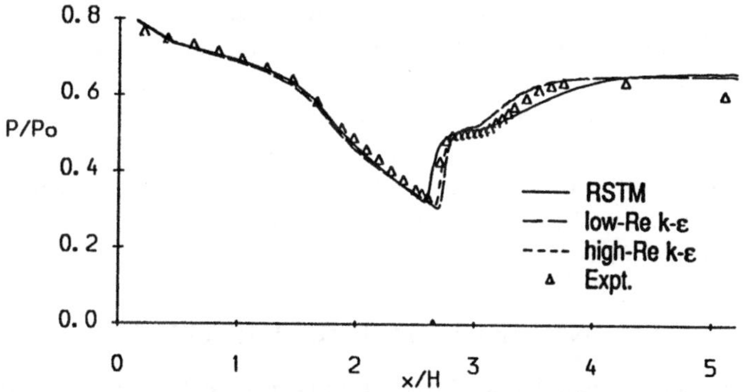

Fig. 4: Bump-wall pressure distributions

REFERENCES

1. Leonard, B.P. (1979), Comp. Meths. Appl. Mech. Eng., 19, p.59.
2. van Leer, B. (1979), J. Comp. Phys., 32, p.101.
3. Rhie, C.M. and Chow, W.L. (1983), J. AIAA, 21, p.1525.
4. Hafez, M., South, J. and Murman, E. (1979), J. AIAA Journal, 17, p.838.
5. Wornom, S.F. and Hafez M.M. (1986), Computers and Fluids, 14, p.131.
6. Lien, F.S. (1992),"Computational Modelling of 2D and 3D Flow Incompressible and Compressible Flow with a Non-Orthogonal Collocated Finite-Volume Method", Ph.D Thesis, University of Manchester.
7. Gibson, M.M. and Launder, B.E. (1978), JFM, 86, p.491.
8. Launder, B.E. and Spalding, D.B. (1974), Comp. Meth. Appl. Mech. Eng., 3, p.269.
9. Wolfshtein, M.W. (1969), J. Heat Mass Transfer, 12, p.301.
10. Delery, J. (1983), J. AIAA, 21, p.180.
11. Dimitriadis, K.P. and Leschziner, M.A. (1990), in Lecture Notes in Physics, 371, Springer Verlag, p.111.

SOLUTION OF COMPRESSIBLE, TURBULENT TRANSPORT EQUATIONS USING A FLUX-DIFFERENCE SPLIT SCHEME

J. Morrison **T. Gatski**

A S & M Inc., Hampton, VA 23666 NASA LaRC, Hampton, VA 23665

I. INTRODUCTION

There has been an increasing desire to predict the turbulent flow behavior over complicated aerodynamic geometries in high-speed flows due to the recent international interest in developing vehicles that fly at supersonic/hypersonic speeds. Unfortunately, the practical aerodynamic flows that are of interest are three-dimensional with strong vortical regions and generally far from equilibrium. Computing these flows requires the use of Reynolds stress transport models. While these models contain improved capabilities for the stress anisotropies and the associated strain histories, their computational stiffness has lessened their appeal to the general user community. The purpose of the present work is to develop a versatile means of implementing Reynolds-stress transport models into aerodynamic codes for general application to high-speed compressible flows. The development of such a numerical capability will allow for the unambiguous comparative testing and evaluation of closure models using the same numerical procedure as well as the same physical and numerical boundary conditions on the same computational grid.

II. NUMERICAL FORMULATION

The Favre-averaged mean Navier-Stokes and turbulent transport equations can be written in vector form in an arbitrary coordinate system (ξ, η, ζ) as [1]:

$$\frac{\partial \hat{Q}}{\partial t} + \frac{\partial(\hat{F} - \hat{F}_v)}{\partial \xi} + \frac{\partial(\hat{G} - \hat{G}_v)}{\partial \eta} + \frac{\partial(\hat{H} - \hat{H}_v)}{\partial \zeta} = S, \tag{1}$$

where $\hat{Q}$ is the vector of dependent variables

$$\hat{Q} = \frac{Q}{J} = J^{-1} \{\rho,\ \rho u,\ \rho v,\ \rho w,\ \rho E,\ \rho \tau_{ij},\ \rho \epsilon\}^T, \tag{1a}$$

τ_{ij} represents the six independent components of the Reynolds stress tensors, F, G, H are the inviscid (convective) fluxes, F_v, G_v, H_v are the viscous (diffusive) fluxes, and S represents the source terms due to production, destruction, and redistribution.

An equation of state is required to complete the system of equations. The perfect gas equation of state is used in this study:

$$p = (\gamma - 1)[\rho E - \frac{1}{2}\rho(u^2 + v^2 + w^2) - \rho k] \tag{2}$$

The presence of the turbulent kinetic energy term, ρk, in the equation of state arises from the Favre-averaging and creates a strong coupling between the mean equations and the normal Reynolds-stress components for the Reynolds-stress models.

II.1 Discretization

The semi-discrete, finite-volume form of Eq. (1) is written as:

$$\begin{aligned}(\partial\hat{Q}/\partial t)_{ijk} + [(\hat{F}-\hat{F}_v)\nabla\xi/J]_{i+1/2,j,k} - [(\hat{F}-\hat{F}_v)\nabla\xi/J]_{i-1/2,j,k} \\ +[(\hat{G}-\hat{G}_v)\nabla\eta/J]_{i,j+1/2,k} - [(\hat{G}-\hat{G}_v)\nabla\eta/J]_{i,j-1/2,k} \\ +[(\hat{H}-\hat{H}_v)\nabla\zeta/J]_{i,j,k+1/2} - [(\hat{H}-\hat{H}_v)\nabla\zeta/J]_{i,j,k-1/2} \\ -(S/J)_{ijk} &= 0\end{aligned} \tag{3}$$

where the fluxes are defined at the interfaces of the computational cell bounding the cell-average value, Q_{ijk}.

Second-order spatial accuracy for the inviscid terms is attained by using the MUSCL scheme of van Leer [2]. The variables interpolated are ρ, u_i, p, τ_{ij}, and ϵ. The min-mod limiter [3] is used to avoid spurious oscillations in the neighborhood of a discontinuity. Other limiters are available and will be investigated in future work to improve monotonicity and convergence behavior.

The remaining terms to be discretized are the diffusive fluxes and the source terms. Consistent with the elliptic nature of the diffusive fluxes, a finite-volume representation of a second-order accurate central-difference operator [3, 4] is employed. Derivatives required in the diffusive flux evaluation at the cell interface are approximated with Gauss's divergence theorem by integrating around an auxiliary cell centered at the interface. Flow variables required at this interface are obtained from arithmetic averaging of neighboring cell averages. Derivatives required for the source terms are also calculated using Gauss's divergence theorem by integrating around the computational cell.

To accomodate geometrically complex configurations, we implement a multi-block procedure which requires C^0 grid continuity. Time integration is performed using an implicit spatially factored approximate factorization scheme.

II.2 Roe Flux-Difference Splitting

The interface flux for the finite volume formulation is calculated in each of the three coordinate directions as the solution of a locally one-dimensional Riemann problem normal to the cell interface using Roe's flux-difference splitting [5].

The Roe-averaged variables are

$$\hat{\rho} = \sqrt{\rho_L \rho_R} \tag{4a}$$

$$\hat{\phi} = \frac{\sqrt{\rho_L}\phi_L + \sqrt{\rho_R}\phi_R}{\sqrt{\rho_L} + \sqrt{\rho_R}} \tag{4b}$$

where

$$\phi = \{H,\, u,\, v,\, w,\, \tau_{ij},\, k,\, \epsilon\}^T \tag{4c}$$

The interface flux can be written as the average of the interface flux calculated from the left state crossing negative running waves and the right state crossing positive running waves:

$$\overline{F}_{i+1/2} = \frac{1}{2}[\overline{F}_R + \overline{F}_L - \left|\widehat{\overline{A}}\right| (\hat{Q}_R - \hat{Q}_L)] \tag{5}$$

This may be written in a more computationally efficient manner as

$$\overline{F}_{i+1/2} = \frac{1}{2}[\overline{F}_R + \overline{F}_L - \sum |\Delta F|] \tag{6}$$

where the flux differences can be derived as:

$$|\Delta F_{\overline{U}}| = \frac{\left|\hat{\Lambda}_{\overline{U}}\right|}{J} \left[(\Delta\rho - \Delta p/\hat{a}^2)\left\{1,\, \hat{u},\, \hat{v},\, \hat{w},\, \hat{q}^2/2 + \hat{k},\, \widehat{\tau_{ij}},\, \hat{\epsilon}\right\}^T \right.$$

$$\left. + \hat{\rho}\left\{0,\, \Delta u - \hat{n}_x\Delta\overline{U},\, \Delta v - \hat{n}_y\Delta\overline{U},\, \Delta w - \hat{n}_z\Delta\overline{U},\, \hat{u}\Delta u + \hat{v}\Delta v + \hat{w}\Delta w - \hat{\overline{U}}\Delta\overline{U} + \Delta k,\, \Delta\tau_{ij},\, \Delta\epsilon\right\}^T\right] \tag{6a}$$

$$\left|\Delta F_{\overline{U}\pm a}\right| = \frac{\left|\hat{\Lambda}_{\overline{U}\pm a}\right|}{J} \frac{\Delta p \pm \hat{\rho}\hat{a}\Delta\overline{U}}{2\hat{a}^2} \left\{1,\, \hat{u} \pm \hat{n}_x\hat{a},\, \hat{v} \pm \hat{n}_y\hat{a},\, \hat{w} \pm \hat{n}_z\hat{a},\, \hat{H} \pm \hat{\overline{U}}\hat{a},\, \widehat{\tau_{ij}},\, \hat{\epsilon}\right\}^T \tag{6b, c}$$

with $\Delta(\cdot) = (\cdot)_R - (\cdot)_L$.

III. RESULTS

An initial test case using Shima's [6] Reynolds-stress model for two-dimensional flow over a 10° compression ramp is presented. The ramp flow is Mach 3 with a mean flow Reynolds number of 10×10^6 and adiabatic wall conditions. The free-stream temperature, T_∞, is set at $300°K$.

The wall boundary conditions are specified as zero velocity ($u_i = 0$) and zero Reynolds stresses ($\tau_{ij} = 0$). At the wall, the pressure is extrapolated from the interior solution with a zeroth-order extrapolation, and the solenoidal dissipation rate is equated to the second derivative of the turbulent kinetic energy (a consequence of evaluating the turbulent kinetic energy equation at the wall). The domain is discretized using a 101 by 51 mesh, with high grid clustering in the near-wall region. The mean velocity and turbulent Reynolds stresses (kinetic energy) that are shown are all normalized by the free-stream mean density and free-stream mean velocity.

The mean velocity profiles are shown in Fig. 1. These profiles are the Favre-averaged Cartesian mean velocities $\tilde{u}_i = (\tilde{u}, \tilde{v}, 0)$ with the domain set between $x = 0.0$ and $x = 1.0$. The start of the ramp is located at $x = 0.5$. The four stations show one set of profiles upstream of the ramp ($x = 0.396$), a second at the start of the ramp ($x = 0.499$), a third slightly downstream of the ramp ($x = 0.598$), and a fourth further downstream along the ramp ($x = 0.808$). The flow is near separation at the start of the ramp, as seen in the $x = 0.499$ plot.

The streamwise variation of the turbulent kinetic energy, obtained from the trace of the normal stress components of the Reynolds stress tensor, is shown in Fig. 2. The downstream development of the turbulent kinetic energy is clearly altered by the presence of the ramp. The ramp causes an increase in the turbulent energy away from the wall at the expense of the peak energy level near the wall ($x = 0.499$). Further downstream, the profile recovers and tends toward its flat-plate distribution.

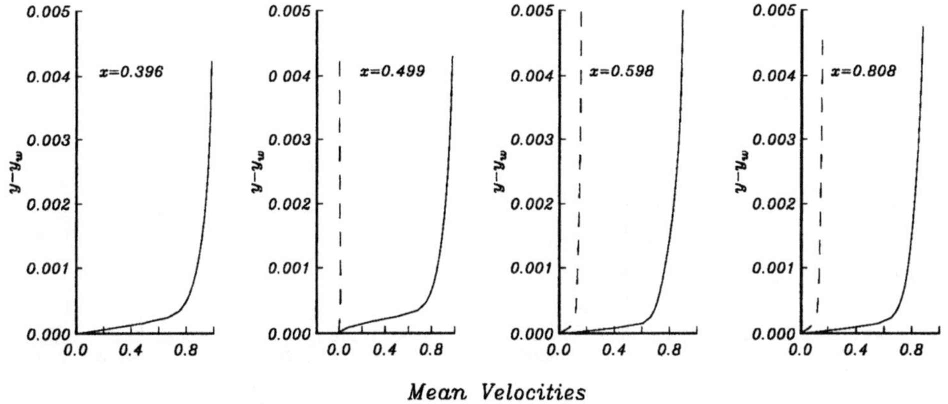

Figure 1: Streamwise variation of mean velocity profiles for 10° compression ramp: $\tilde{u}/U_\infty$, ——; $\tilde{v}/U_\infty$, — —.

The only component of the Reynolds shear stress to survive in this two-dimensional flow is the τ_{xy} component. Figure 3 shows its variation with downstream distance. The shear stress profile is similarly affected by the ramp start as the kinetic energy. In addition, downstream of the ramp start, along the ramp, the Reynolds shear stress goes positive in the near-wall region. While some regions of positive shear stress may be expected near the point of separation, the wide streamwise extent of the positive shear stress values is questionable. This effect is probably due to an imbalance of terms in the near-wall model and/or the grid resolution in this region. Shima's Reynolds-stress model was developed for incompressible flows and may prove to be inadequate for compressible, high Mach number flow fields with a simple variable density extension.

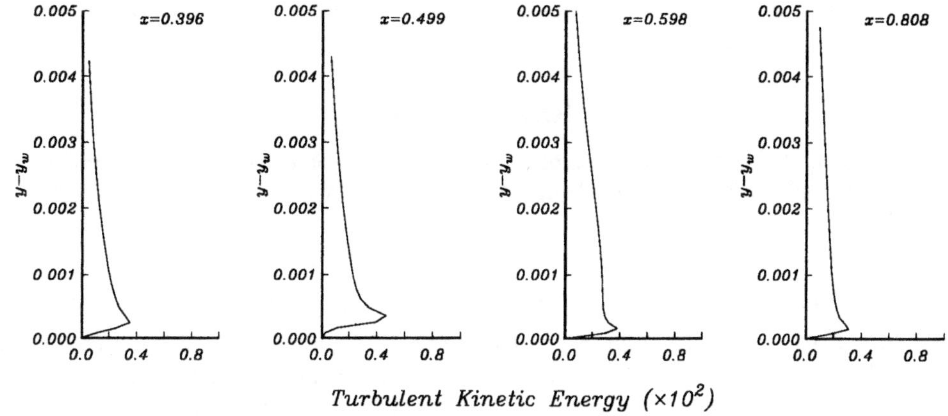

Figure 2: Variation of turbulent kinetic energy profiles for 10° compression ramp.

IV. CONCLUDING REMARKS

The theoretical development of turbulence closure models has become widespread, and it is becoming increasingly apparent that the wide variety of models that have

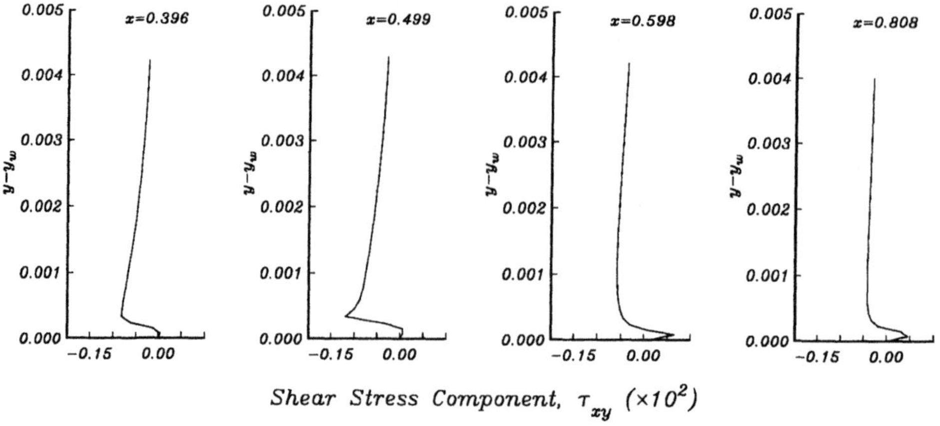

Figure 3: Reynolds shear stress profiles for 10° compression ramp.

been proposed need to be carefully evaluated. In order to do this in a precise, unambiguous fashion, it is necessary to test the various models using the same numerical algorithm with the same physical and numerical boundary conditions on the same computational grid. The present numerical algorithm has been developed for the purpose of solving compressible turbulent flows using phenomenological turbulence models. The procedure outlined has been for Reynolds-stress transport equations which, when coupled with the mean conservation equations, leads to the solution of 12 partial differential equations to determine the compressible flow field. This capability will allow for the application of such models to the solution of complex compressible flow fields.

REFERENCES

[1] Morrison, J.H., "A Compressible Navier-Stokes Solver with Two-Equation and Reynolds Stress Turbulence Closure Models," *NASA Contractor Report 4440*, May 1992.

[2] van Leer, B., "Towards the Ultimate Conservative Difference Schemes V. A Second Order Sequel to Godunov's Method", *J. Comp. Physics*, Vol. 32, pp. 101-136, 1979.

[3] Chakravarthy, S. R., Szema, K-Y., Goldberg, U. C. and Gorski, J. J., "Application of a New Class of High Accuracy TVD Schemes to the Navier-Stokes Equations," *AIAA 23th Aerospace Sciences Meeting*, Paper No. 85-0165, January 14-17, Reno, Nevada, 1985.

[4] Swanson, R. C., and Turkel, E., "A Multistage Time-Stepping Scheme for the Navier-Stokes Equations," *AIAA 24th Aerospace Sciences Meeting*, Paper No. 85-0035, January 14-17, Reno, Nevada, 1985.

[5] Roe, P. L., "Approximate Riemann Solvers, Parameter Vectors, and Difference Schemes", *Journal of Computational Physics*, Vol. 43, pp. 357-372, 1981.

[6] Shima, N., "A Reynolds-Stress Model for Near-Wall and Low-Reynolds-Number Regions," *Transactions of the ASME*, Vol. 110, pp. 38-44, 1988.

DIRECT NUMERICAL SIMULATION OF LAMINAR BREAKDOWN IN HIGH-SPEED, AXISYMMETRIC BOUNDARY LAYERS

C. D. Pruett* and T. A. Zang†

*A.S.&M., Inc., Hampton, VA 23666 USA
†NASA Langley Research Center, Hampton, VA 23665 USA

1 Introduction

In the past decade, direct numerical simulation (DNS) of transition to turbulence has attained considerable maturity for incompressible flows, now complementing theoretical analyses and physical experiments; however, the physics of compressible flow have rendered DNS of transition at high speeds a formidable task, subject to many numerical obstacles [1]. Efficient semi-implicit algorithms, extremely successful for incompressible flows, are not readily adapted to the compressible Navier-Stokes equations (CNSE). Consequently, nearly all simulations of compressible flow are fully explicit and are subject to both advection and viscous constraints on the time step. At very high speeds, the dominant (second) instability modes are of high frequency and slow growth, so that simulations must be highly resolved in time for many disturbance periods. Moreover, the second-mode eigenfunction has a double structure, which necessitates high resolution both at the wall and near the critical layer (Fig. 1). Finally, dissipative methods typically used in aerodynamic applications for shock capturing are inappropriate for DNS. A conservative estimate is that DNS for high-speed flow requires an order of magnitude more computer time than that of a comparable simulation of incompressible flow.

Although it is generally accepted that spatial DNS is the correct computational analog of forced transition experiments, spatial simulation of the complete laminar-turbulent transition process remains beyond the capability of existing supercomputers. For convective instabilities, however, temporal DNS (TDNS) has attained notable success (e.g.,[2]) for incompressible flow. Here, due partly to algorithm improvements, which address many of the issues above, and partly to a generous grant of supercomputer time on the National Aerodynamic Simulator (NAS), we have been able to conduct, for the first time, long-term temporal simulations of laminar breakdown for the high-speed axisymmetric boundary layers along a hollow cylinder in a Mach 4.5 flow and a sharp cone in a Mach 8.0 flow. From initial states perturbed by "second-mode" primary and subharmonic secondary (H-type) disturbances, these well-resolved temporal calculations proceed well into the laminar breakdown regime, characterized by saturation of the primary and secondary instability waves, explosive growth of higher harmonics, rapid increase in the wall skin friction, and rapid decrease in the shape factor.

2 Numerics

In TDNS one examines a small "window" within the flow field in which disturbances are assumed to be spatially periodic and to evolve in time. Mathematically, TDNS is characterized as an Initial-Boundary-Value-Problem (IBVP). Details of the numerical method, nondimensionalization, and boundary and initial conditions can be found in [3,4], summarized below.

Coordinate System and Governing Equations. In the body-fitted coordinate system (x, θ, z) shown in Fig. 2, the compressible Navier-Stokes equations assume the conservative formulation given in [3,4], appropriate for either flat-plate, hollow-cylinder, or sharp-cone geometries. For convenience, we cast the energy equation in terms of pressure.

Initial and Boundary Conditions. Parameter studies based on temporal linear stability theory (LST) [5] and on secondary instability theory (SIT) [6,7] identify subharmonic secondary instability triggered by a second-mode primary instability as a likely route to transition at very high speeds. Accordingly, the initial condition is formulated by the superposition of the mean flow (assumed to be parallel), an axisymmetric second-mode primary disturbance (Fig. 1), and a subharmonic secondary disturbance ensemble of four wave components. The mean flow is obtained from the spectrally accurate boundary-layer code of [8]. Following [9], small forcing terms are subtracted from the governing equations to exactly cancel the initial steady-state residual due to the parallel-flow assumption. Regarding boundary conditions, periodicity is imposed in the streamwise and azimuthal directions. At the wall, no-slip conditions are imposed on velocities, temperature is assumed to be constant at its adiabatic-wall value, no condition is imposed on density, and pressure is derived from the equation of state. At the far-field boundary, disturbances are assumed to vanish.

Spatial and Temporal Discretization. For the long-duration simulation of instability waves, it is essential that dissipation and dispersion errors be kept extremely small. For this purpose, spectral methods and high-order central-difference schemes are well suited. In the periodic streamwise and azimuthal directions, we exploit Fourier spectral-collocation methods [10]. In the wall-normal direction, an option exists to compute derivatives either by Chebyshev spectral collocation or by the 6th-order compact-difference method of [11] with boundary stencils modified according to [12]. Due to numerical stability considerations, the latter method is favored for long-duration simulations. A finely tuned pair of analytic mappings in z concentrate grid points at the wall and near the critical layer [9] where sharp gradients exist. With adequate resolution and appropriate mapping parameters, disturbance growth rates extracted from the TDNS are accurate to 5–6 significant digits, relative to LST and SIT. In the azimuthal direction, the option exists to invoke symmetry, whereby the computational effort is reduced by a factor of two. Present results were obtained with symmetry enforced. No de-aliasing procedure is implemented in the present calculations. Rather, the computations are kept "well-resolved" at all times, with grid refinements as necessary, based on criteria involving the decay of the Fourier (Fig. 3) and Chebyshev spectral coefficients. Temporal advancement is fully explicit, by means of a 3rd-order, low-storage Runge-Kutta method [13], for which the time step is maintained automatically at its maximum allowable size, based on numerical stability analyses of model advection and diffusion problems.

3 Results

Due to space limitations, we present results of the Mach 4.5 simulation only. The geometry is a hollow cylinder (ϕ=0 in Fig. 2) with a sharp leading edge, the axisymmetric analog of the flat plate. The curvature (i.e., the ratio between boundary-layer displacement thickness, δ^*, and radius) is 0.1. Reynolds number based on edge conditions and δ^* is 10,053. Constant values of 1.4 and 0.7 are assumed for the ratio of specific heats and the Prandtl number, respectively. Viscosity and thermal conductivity vary according to Sutherland's law. Edge temperature is 110°R. The wavenumbers (α, β) (based on δ^*) defining the primary $(\alpha, 0)$ and secondary $\{(\alpha/2, \pm\beta), (3\alpha/2, \pm\beta)\}$ disturbances are (2.53,2.10). Initial RMS amplitudes (relative to freestream temperature) of the primary and secondary disturbances are 8.5% and 1.7%, respectively.

The simulation was initiated with a relatively coarse resolution of 12×6×96. Final resolution was 96×48×144, for which the computation consumed about 12 CPU hours/period on a Cray Y-MP, and beyond which it was deemed impractical to refine further. At its termination at about 59 periods of disturbance oscillation, the simulation had advanced well into the breakdown stage of laminar-turbulent transition, as indicated by saturation of the fundamental (1,0) and secondary modes {(1/2,1),(3/2,1)} (Fig. 4), explosive growth of higher harmonics (not shown), rapid increase in the skin friction coefficient, C_f (Fig. 5a), and rapid decrease in the shape factor, H (Fig. 5b).

Flow visualization reveals that the numerical results qualitatively replicate two previously unexplained phenomena observed in high-speed transition experiments: the appearance of so-called "rope-like waves" [14] and the "precursor transition" effect, in which a region of transitional flow originates near the critical layer and spreads toward the body at a relatively constant shallow angle (Fig. 6) [15]. "Schlieren" images, numerically generated from the DNS data, clearly show structures of remarkably "rope-like" appearance at the onset of laminar breakdown (Fig. 7). Additional flow-field analyses, including video sequences, reveal the rope-like waves to be artifices of secondary instability which arise from the two-dimensional projection of staggered "lambda"-vortices. Moreover, Reynolds stresses spatially reconstructed from the temporal data, via a "Gaster" transformation (Figs. 8), show the precursor transition effect to be a highly nonlinear phenomena unexplained even qualitatively by linear theories (Figs. 8a,b). Whereas SIT accounts for a Reynolds stress peak near the critical layer (Fig. 8b), only the fully nonlinear simulation predicts the migration of the Reynolds stress peak toward the wall (Fig. 8c), as observed in experiments (e.g., [15]).

The dominant role played by the (0,2) mode in Fig. 4, suggests an interesting numerical experiment: how does suppression of streamwise vorticity affect transition? To obtain Fig. 9, we have repeated the simulation with the same initial conditions, but with the energy in the (0,2) mode artificially suppressed by zeroing its contents in Fourier space at each Runge-Kutta stage. Figure 9 compares the energy contents of the fundamental and principal subharmonic modes of this "adulterated numerical simulation" (ANS) with the DNS. For the ANS, the primary and secondary disturbances saturate at energy levels at least an order of magnitude lower than when the (0,2) mode is present. Moreover, after 80 periods, there is less than a 5% change in either H or C_f, indicating that transition has not yet commenced in the ANS. The numerical results suggest that streamwise vorticity is essential to laminar breakdown.

References.

[1] L. Kleiser and T. A. Zang, *Annu. Rev. Fluid Mech.* **23**, 495 (1991)

[2] T. A. Zang and M. Y. Hussaini, in *Nonlinear Wave Interactions in Fluids*, eds. R. W. Miksad, T. R. Akylas, and T. Herbert, AMD-87, ASME, New York, 131 (1987)

[3] C. D. Pruett and T. A. Zang, *AIAA Paper No. 92–0742* (1992)

[4] C. D. Pruett and T. A. Zang, *Theoret. Comput. Fluid Dynamics*, **3**(6) (1992, to appear)

[5] L. M. Mack, in *Special Course on Stability and Transition of Laminar Flow*, ed. R. Michel, AGARD Report No. 709, 3–1 (1984)

[6] L. L. Ng and G. Erlebacher, *Phys. Fluids A* **4**(4), 710 (1992)

[7] C. D. Pruett, L. L. Ng, and G. Erlebacher, *Phys. Fluids A*, **3**(12), 2910 (1991)

[8] C. D. Pruett and C. L. Streett, *Int. J. Numer. Meth. Fluids* **13**, 713 (1991)

[9] G. Erlebacher and M. Y. Hussaini, *Phys. Fluids A*, **2**(1), 94 (1990)

[10] C. Canuto, M. Y. Hussaini, A. Quarteroni, and T. A. Zang, *Spectral Methods in Fluid Dynamics*, Springer-Verlag, New York, (1988)

[11] S. K. Lele, *AIAA Paper No. 89–0374* (1989)

[12] M. H. Carpenter, D. Gottlieb, and S. Abarbanel, *ICASE Report No. 91–71* (1991)

[13] J. H. Williamson, *J. Comput. Phys.* **35**, 48 (1980)

[14] A. Demetriades, *J. Appl. Mech.* **99**(1), 7 (1977)

[15] M. C. Fischer and L. M. Weinstein, *AIAA J.* **10**(7), 957 (1972)

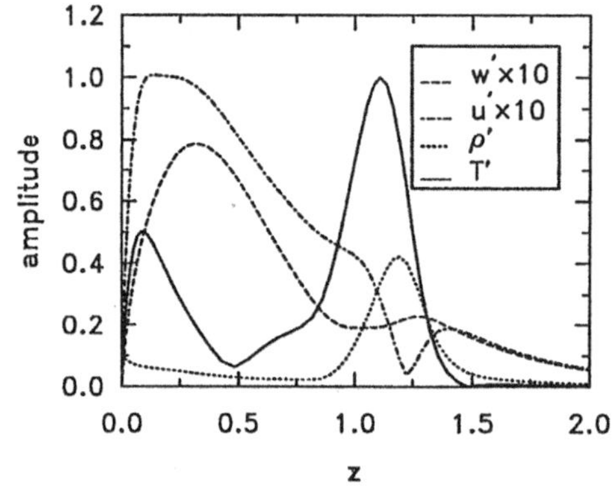

Fig. 1 Second-mode eigenfunction: cylinder, Mach 4.5

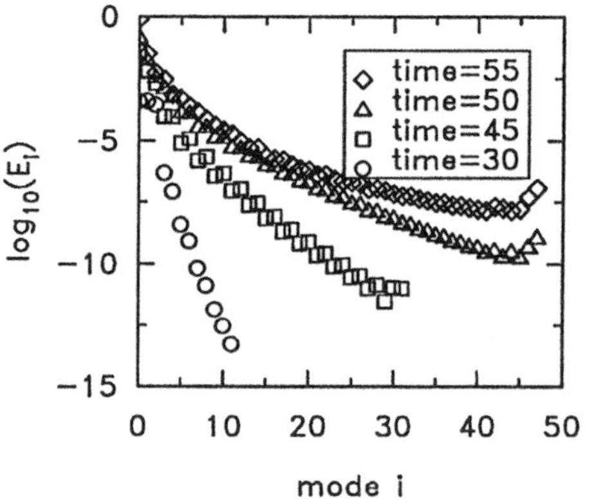

Fig. 3 Streamwise resolution spectra

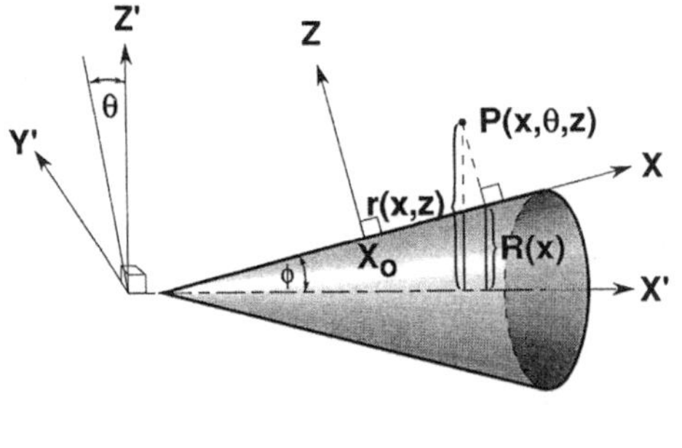

Fig. 2 Body-fitted coordinate system

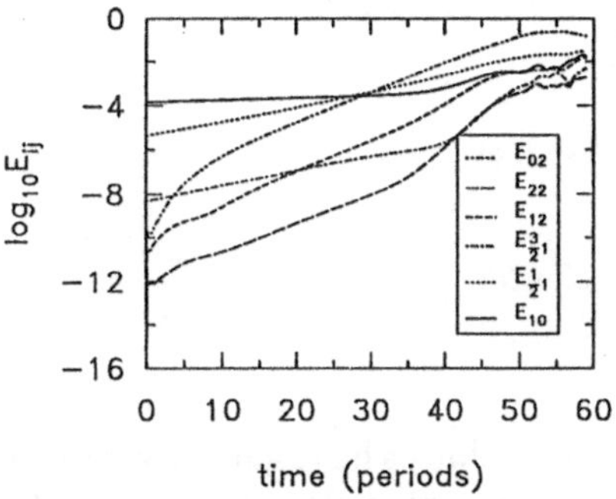

Fig. 4 Evolution of the energy content in selected Fourier harmonics

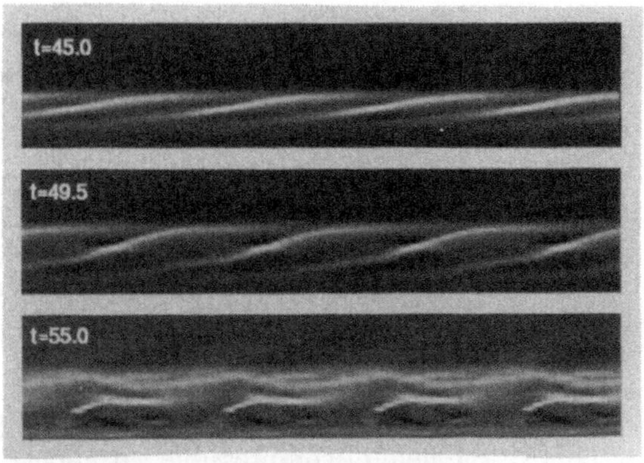

Fig. 7 Numerical "schlieren" images of "rope-like" structures at selected times

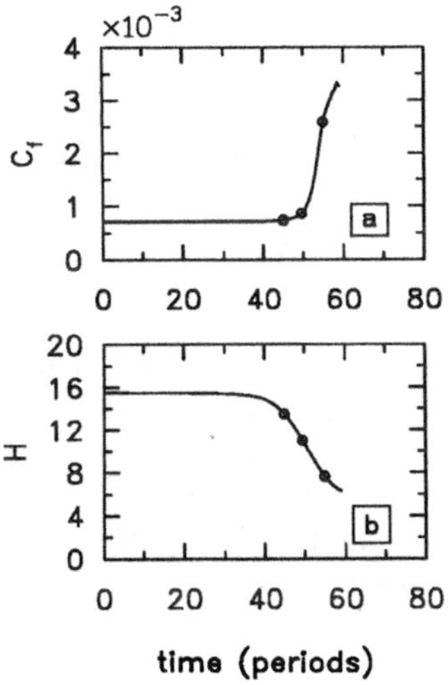

Fig. 5 Temporal evolution of averaged a) wall skin-friction coefficient, and b) shape factor

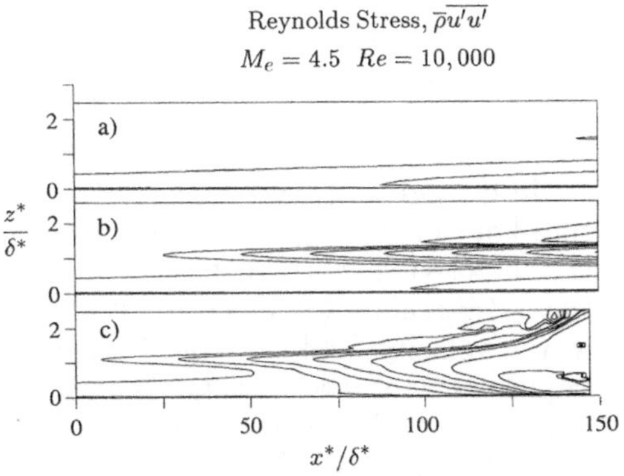

Fig. 8 Spatially-reconstructed Reynolds stress iso-levels, in equal logarithmic increments a) DNS, 2D primary only; b) SIT; c) DNS, 3D

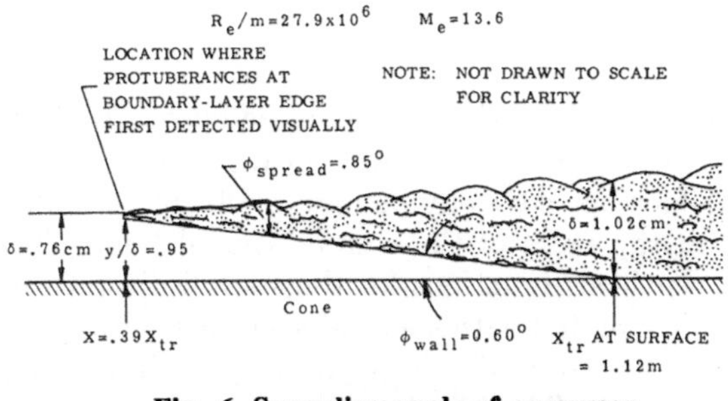

Fig. 6 Spreading angle of precursor transition: Fischer and Weinstein, 1972 [15], Mach 14, helium

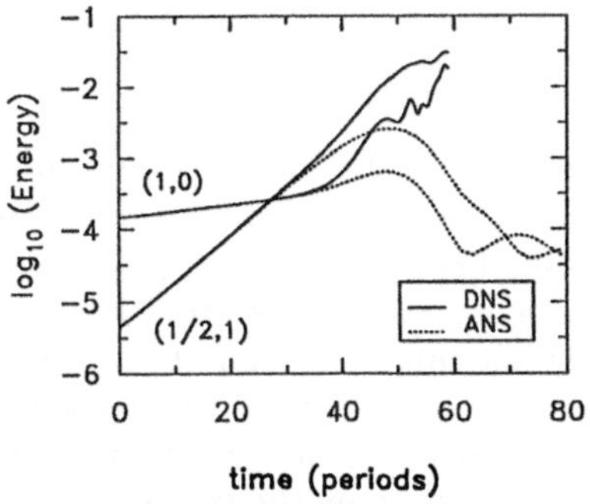

Fig. 9 Energy content of fundamental and subharmonic modes

DIRECT NUMERICAL SIMULATION OF COMPRESSIBLE TURBULENCE IN A HOMOGENEOUS SHEAR FLOW

S. Sarkar and G. Erlebacher

ICASE
NASA Langley Research Center
Hampton, VA 23665

1.Introduction

The need for advances in the technolgy of high-speed vehicles has spurred vigorous activity in the field of compressible turbulence. In the present paper, we investigate compressibility effects in a shear flow through a direct numerical simulation (DNS). Both spectral and higher-order finite difference schemes are used and their relative merits are discussed in the paper.

The particular flow that we consider is homogeneous shear flow. Homogeneous shear flow refers to the problem of spatially homogeneous turbulence sustained by a parallel mean velocity field $\overline{\mathbf{u}} = (Sx_2, 0, 0)$ with a constant shear rate S (cf. Fig. 1). Although, it is a simplification of the inhomogeneous free shear layers encountered in practice, it is an attractive model problem because the crucial mechanisms of sustenance of turbulent fluctuations by a mean velocity gradient, and the energy cascade down to the small scales of motion are both present in this flow. DNS has been used to study homogeneous shear flow by Rogallo (1981), and Rogers and Moin (1987) for the incompressible case. Recently, Blaisdell, Mansour and Reynolds (1991) have performed DNS for the compressible case.

In compressible turbulence, density fluctuations and waves propagating with the speed of sound are generated by the solenoidal velocity field which then interact with the underlying, vortical turbulence. The acoustic velocity component can exhibit wave-steepening and a tendency to form shocks. Furthermore, discretization errors in the velocity field can lead to the computed density field violating the positivity requirement on the density. Thus, DNS of compressible flows has more stringent requirements than corresponding incompressible flows.

2. Numerical Method

The compressible Navier-Stokes equations are numerically solved using a spectral collocation technique. Adopting the approach of Rogallo (1981), explicit dependence on the mean flow $\bar{u}_1$ is removed by writing the equations in x_i^*, a frame of reference moving with the mean flow. The relation between x_i^* and the lab frame x_i is

$$x_1^* = x_1 - Stx_2 \quad , \quad x_2^* = x_1 \quad , \quad x_3^* = x_3$$

Here S denotes the constant shear rate $\bar{u}_{1,2}$. Since the governing equations do not have any explicit dependence on the spatial coordinates x_i^*, and because the homogeneous shear flow problem, by definition, does not have any boundary effects, periodic boundary conditions are allowable on all the faces of the computational box. The periodic boundary conditions allow Fourier basis functions for the spatial discretization and consequently the derivatives are evaluated with spectral accuracy. Since the nonlinear terms in the governing equations are computed in physical space (a feature of collocation methods), aliasing errors occur. The aliasing error can be removed by appropriate truncation of the higher wave number modes. Since truncation as a tool for dealiasing is unavailable for non-periodic domains, we perform numerical experiments to determine the necessity of dealiasing in turbulence simulations

Recently, higher-order compact finite difference schemes have been used for the simulation of turbulent flows. Finite-difference schemes are attractive because (1) they are better suited for complex geometries, (2) they are more amenable to parallelization due to less communication requirements than FFT's, and (3) they are cheaper for a fixed number of points. However, these finite difference schemes, though higher-order than those generally used in CFD, lead to a less accurate estimate of derivatives relative to a spectral method. Thus, it is of interest to determine whether the higher discretization error incurred with the compact scheme outweighs advantages such as less computational cost. We compare a sixth-order compact scheme proposed by Lele (1989) with the spectral collocation method in this paper.

The time advancement scheme for a DNS must meet criteria regarding accuracy, stabilility and storage. A third-order, low storage, Runge-Kutta scheme is used here. Numerical experiments have shown that a CFL number of 0.5 gives acceptable accuracy for velocity and thermodynamic spectra.

3. Spectral vs. a higher-order finite difference scheme

In this section, we compare results of the spectral collocation nethod with those of a sixth-order compact scheme proposed by Lele (1989). The 6th order compact method evaluates the first derivative(f_i') and second derivative (f_i'') at node i by solving the tridiagonal implicit system

$$\begin{aligned} f_i' + \frac{1}{3}(f_{i-1}' + f_{i+1}') &= \frac{14}{9}\left(\frac{f_{i+1} - f_{i-1}}{2\Delta x}\right) + \frac{1}{9}\left(\frac{f_{i+2} - f_{i-2}}{4\Delta x}\right) \\ f_i'' + \frac{2}{11}(f_{i-1}'' + f_{i+1}'') &= \frac{12}{11}\left(\frac{f_{i+1} - f_{i-1}}{\Delta x^2}\right) + \frac{3}{11}\left(\frac{f_{i+2} - f_{i-2}}{4\Delta x^2}\right) \end{aligned} \tag{1}$$

The stencil for evaluation of derivatives is relatively compact, however there is the added expense of (tridiagonal) matrix inversion. The compact scheme, Eq. (1), requires about 35 CPUs/iteration while the Fourier collocation scheme requires about 50 CPUs/iteration. However, with respect to computing the first derivative of a given field , the highest 'resolvable' wave number on a N^3 grid is approximately $N/4$ for the compact scheme, compared to $N/2$ for a spectral scheme.

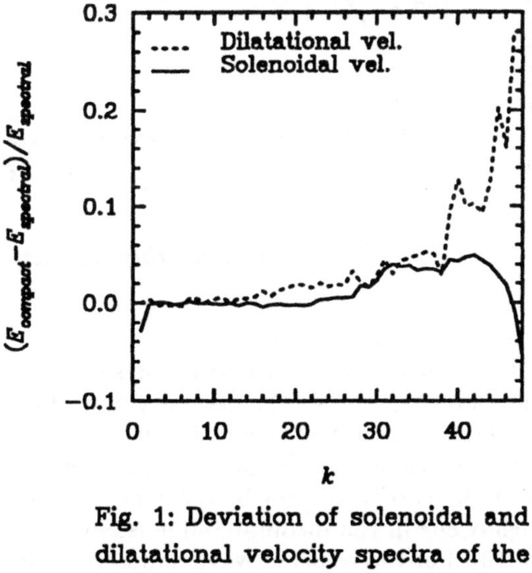

Fig. 1: Deviation of solenoidal and dilatational velocity spectra of the compact case from the spectral one.

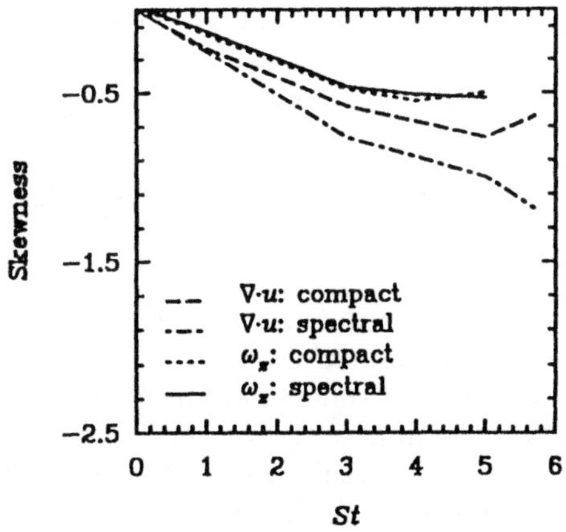

Fig. 2: Evolution of skewness.

Figs. 1-4 compare spectral and sixth-order compact results for some statistical moments and pointwise variables for a 96^3 simulation. Fig. 1 shows the deviations of the compressible (dilatational) spectrum $E^C(k)$ and the incompressible (solenoidal) spectrum $E^I(k)$ from corresponding results of the spectral method. The comparison is at a non-dimensional time of $St = 5$ when the turbulent Reynolds number $Re_\lambda = 30$ and the turbulent Mach number $M_t = 0.38$. Though the error is negligible for $k < 24$, it is high for the larger wave numbers. Thus, single-point statistics dominated by large scales such as kinetic energy and Reynolds stresses are represented well by the compact scheme. For example, the turbulent kinetic energy K deviates by only 0.05% from the spectral result. The enstrophy $\overline{\omega_i'\omega_i'}$ and the turbulent dissipation rate ϵ deviates by about 5%. It is interesting that the much larger discrepancy in the small scale statistics does not significantly contaminate a large scale quantity such as K. Higher-oder moments of velocity gradients are more sensitive to the accuracy of the discretization scheme. Figs. 2-3 compare the skewness and flatness of dilatation $d' = \nabla \cdot \mathbf{u}$ and spanwise vorticity ω_z obtained by the two numerical schemes. The skewness and flatness of ω_z do not differ much between the two schemes. However the compact scheme gives significantly lower values for the dilatation skewness and flatness. This is related to the irrotational velocity being more skewed and intermittent than the rotational field. Fig. 4 shows the evolution of maximum vorticity. This is a pointwise quantity in contrast to the volume-averaged statistics considered in the previous figures. The compact scheme gives values which are somewhat lower than the spectral scheme.

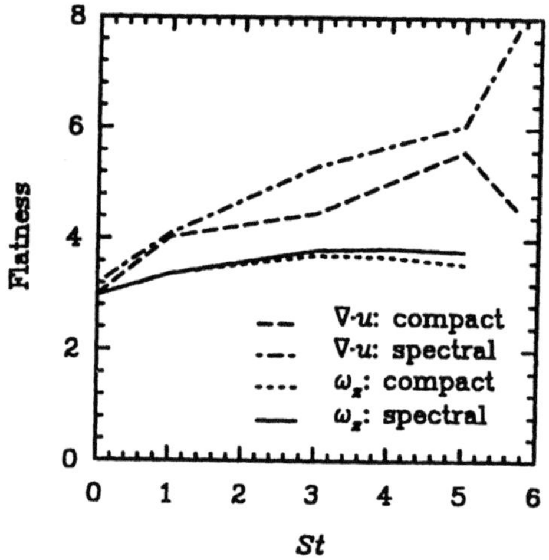

Fig. 3: Evolution of flatness.

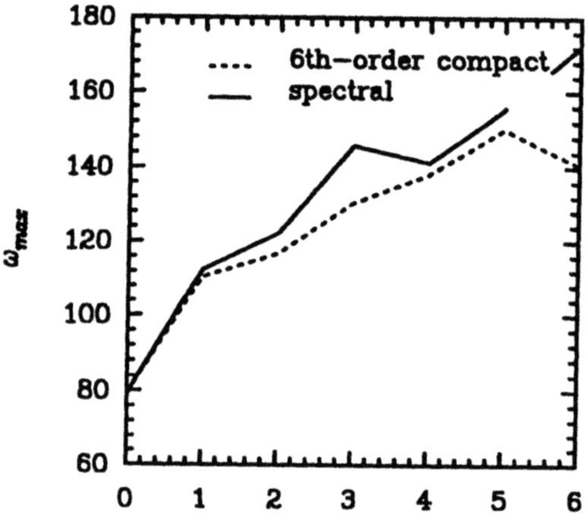

Fig. 4: Evolution of maximum vorticity.

4. An evaluation of truncation as a tool for dealiasing

The evaluation of non-linear terms in the governing equations in physical space - a feature of collocation methods - leads to aliasing errors. The necessity of dealiasing for retaining accuracy (cf. Canuto et al.(1988)) is a controversial issue. There are theoretical results which show that the aliasing error is of the same order as the truncation error. In this section we consider the effect of dealiasing by spectral truncation. For quadratic products such as $\mathbf{uu}$ which occur in the incompressible Navier-Stokes equations, retaining only the lowest two-third of the resolved wave numbers leads to dealiased results. For dealiasing cubic products such as $\rho\mathbf{uu}$ only one-half of the resolved wave numbers showld be retained; however, because of relatively small density fluctuations in the simulations we use the two-third rule.

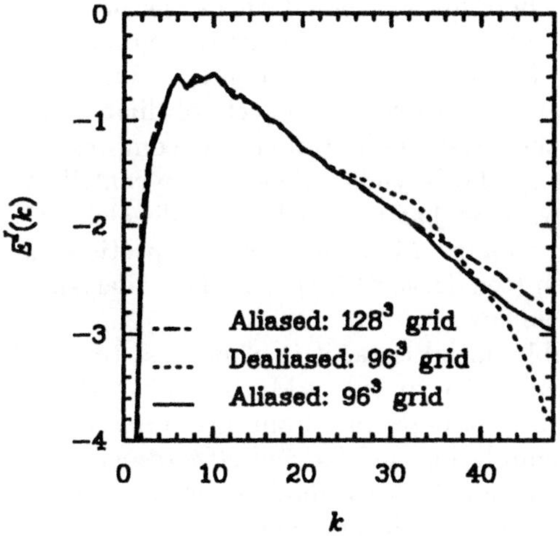

Fig. 5: Effect of dealiasing on the spectrum of the solenodial velocity. Elapsed time is *St*=5.

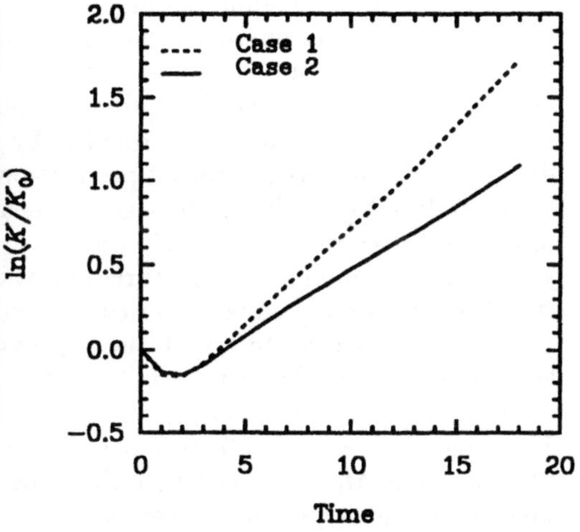

Fig. 6. Compressibility effect on the growth of turbulent kinetic energy.

Fig. 5 contrasts the solenoidal velocity spectrum $E^I(k)$ from a dealiased run to an aliased run; both runs were performed on a 96^3 grid. The difference in the range of interest $0 < k < 32$ is small and restricted to a small band of high wave numbers. The spectrum in the dealiased 96^3 case does not decay in the highest wave numbers, $25 < k < 32$, as fast as the non-dealiased case. This suggests that the truncation of the highest one-third of the spectrum for the purpose of dealiasing (wave numbers $32 < k < 48$ in this case) may lead to inaccuracies in turbulence simulations due to the removal of a physically required sink for the energy cascade down the spectrum. A 128^3 simulation with the same initial data was performed to check this possibility. The high-resolution calculation (dash -dotted line in Fig. 5) is indistinguishable from the 96^3 aliased run confirming that, at least for this particular flow case, spectral truncation (for dealiasing) may deteriorate the accuracy of the computation.

Second-order moments such as K and ϵ show slight differences between the aliased and dealiased cases. However, higher-order moments, especially for the irrotational component are significantly underpredicted by the procedure of dealiasing through truncation.

5. Compressibility effects in homogeneous shear flow

Detailed results regarding the influence of compressibility in homogeneous shear simulations are avilable in Sarkar, Erlebacher and Hussaini (1991, 1992). A brief resume of the results is given here. A primary conclusion of the study is that the growth rate of the turbulent kinetic energy decreases with increasing compressibility - a phenomenon which is similar to the reduction of turbulent velocity intensities observed in experiments on supersonic free shear layers. Fig. 6 shows the time evolution of turbulent kinetic energy K for two cases. At time $t = 0$, both cases have identical solenoidal velocity fields, zero density fluctuations, and identical pressure and temperature fluctuations. However, Case 2 has half the mean speed of sound of Case 1. The initial M_t is 0.2 and 0.4 for Cases 1 and 2, respectively. It is clear from Fig. 2 that Case 2 has a reduced growth rate compared to Case 1. We note that the reduction of kinetic energy growth rate with increased compressibility is a consistent trend in all our simulations. The K transport equation was considered in order to understand this compressibility effect. There are two new terms - the pressure-dilatation and compressible dissipation - in the K equation. The pressure-dilatation transfers energy from the velocity field to the pressure field in homogeneous shear flow, while the compressible dissipation represents dissipative effects of dilatational motion. It seems that sinks provided by the pressure-dilatation and compressible dissipation, along with reduced production by the Reynolds shear stress contribute to the reduced growth rate of kinetic energy. DNS has been helpful in developing a model for the compressible dissipation based on a low-Mach number asymptotic analysis, and for modeling the pressure-dilatation (see Sarkar(1992)) based on analysis of the Poisson pressure equation in compressible flow.

Since both cases start with incompressible initial data, it is of interest to know how 'compressible' does the turbulence become? The turbulent Mach number M_t and the normalized rms density $\rho_{rms}/\overline{\rho}$ are measures of departure from incompressibility. The largest M_t and $\rho_{rms}/\overline{\rho}$ obtained in the simulations are 0.6 and 20% respectively. We note that $M_t = 0.6$ is larger than the turbulent Mach numbers encountered in supersonic and hypersonic shear layers and boundary layers. For purposes of theoretical analysis, and analysis of the numerical data, it is convenient to separate the stochastic variables into compressible and incompressible components. For homogeneous flow, the Helmholtz decomposition gives a unique partition of the velocity field into incompressible (solenoidal) and compressible (irrotational) components by $\mathbf{u} = \mathbf{u}^I + \mathbf{u}^C$, where $\nabla \cdot \mathbf{u}^I = 0$ and $\nabla \times \mathbf{u}^C = 0$. Similarly, it is possible to decompose the pressure field into $p = p^I + p^C$ where fluctuating density effects are contained in the compressible pressure p^C. DNS shows that although $\mathbf{u}^C$ has rms values which are typically less than 15% of $\mathbf{u}^I$, the rms p^C is as large as p^I. One of the important characteristics of the velocity field is the extent and topology of supersonic regions. Flow visualization of the database has shown that supersonic regions occur in small discrete patches which are elongated in the streamwise direction. It has been theorized in the past that, at high enough turbulence Reynolds number, some of these supersonic regions may be associated with eddy shocklets.

References

Blaisdell, G.A., Mansour, N.N, and Reynolds, W.C., *Stanford University Report No. TF-50* (1991).

Canuto, C., Hussaini, M.Y., Quarteroni, A., and Zang, T.A., *Spectral Methods in Fluid Dynamics*, (1988), Springer-Verlag.

Lele, S. K., submitted to *J. Comp. Physics* (1989).

Rogallo, R.S., *NASA TM 81315* (1981).

Rogers, M.M., and Moin, P., *J. Fluid Mech.*, **176** (1987) 33-66.

Sarkar, S., Erlebacher, G., and Hussaini, M.Y., *Theor. Comput. Fluid Dyn.*, **2** (1991) 291-305.

Sarkar, S., Erlebacher, G., and Hussaini, M.Y., *Turbulent Shear Flows 8*, Munich, Germany, (1992), Springer Verlag, to appear.

S. Sarkar, *Phys. Fluids A*, to appear,(1991).

NUMERICAL INVESTIGATION OF TURBULENT FLOW IN BOUNDARY LAYER AT DIFFERENT REYNOLDS NUMBERS

I. Simakin, S. Grubin

INTECO srl, Via Mola Vecchia 2A, 03100 Frosinone, ITALY

Incompressible viscous fluid flow in a boundary layer on a flat plate with a zero pressure gradient is considered. It is assumed that the undisturbed flow does not depend on the streamwise x variable, i.e. local-parallel approximation used. At the initial time moment three-dimensional disturbances (Tollmien-Schlichting waves with small amplitutde) is imposed on the laminar flow. These disturbances seem to be periodic in x, y - directions and exponentially decaying in the external flow in the normal z-direction. The time evolution of the flow from laminar up to developed turbulent regime was computed by means of direct integration of the problem for 3D Navier-Stokes equations. To integrate this problem in time the following semi-implicit scheme is used:

$$(\overset{(S)}{\mathbf{V}}{}^{k+1} - \mathbf{V}^{k})/\tau = \mathbf{F}\,[\overset{(S-1)}{\mathbf{V}}{}^{k+1/2}] - \overset{(S)}{\mathbf{L}} - \nabla \overset{(S)}{P} + (1/R)\,\overset{(S)}{\mathbf{D}} \qquad (1)$$

$$\nabla \overset{(S)}{\mathbf{V}}{}^{k+1} = 0 \qquad (2)$$

where $\mathbf{V} = (u, v, w)$ is the velocity disturbance (main flow $(U_B(z), 0, 0)$ is produced as a result of the action of the "force" $\mathbf{f} = (-(1/R)U''_B(z), 0, 0,)$, $U_B(z)$ is the Blasius velocity profile),

$\overset{(S)}{\mathbf{V}}{}^{k+1/2} = (\overset{(S)}{\mathbf{V}}{}^{k+1} + \mathbf{V}^{k})/2$, $\overset{(0)}{\mathbf{V}}{}^{k+1/2} = 2\mathbf{V}^{k} - \mathbf{V}^{k-1/2}$, $S \geq 1$ is number of iterations,

$P = p + (1/2)\mathbf{V}^2$, p is the pressure; $\mathbf{F}[\mathbf{V}] = [\mathbf{V}, \omega]$, $\omega = [\nabla, \mathbf{V}]$ is the vortex

Let $\overset{(S)}{\mathbf{V}} = \overset{(S)}{\mathbf{V}}{}^{k+1/2}$ then: $\overset{(S)}{\mathbf{L}} = U_B\, \partial \overset{(S)}{\mathbf{V}}/\partial x - \mathbf{i}\, U_B'(\overset{(S)}{w} - 2\overset{(S-1)}{w})$

$$\overset{(S)}{\mathbf{D}} = (\partial^2/\partial x^2 + \partial^2/\partial y^2)\,[\overset{(S)}{\mathbf{V}} - \mathbf{k}(w - \overset{(S-1)}{w})] + \partial^2/\partial z^2\,(\mathbf{i}\,\overset{(S-1)}{\tilde{u}} + \mathbf{j}\,\overset{(S-1)}{\tilde{v}} + \mathbf{k}\,\overset{(S)}{w})$$

$\overset{(0)}{\tilde{u}} = (\hat{u} + u^k)/2$, $\hat{u}$ is found from the following equation:

$$\hat{u} = u^k + (\tau/2R)\,\partial^2/\partial z^2\,(\hat{u} + u^k), \qquad \overset{(S-1)}{\tilde{u}} \equiv \overset{(1)}{u}{}^{k+1/2} \quad \text{for} \quad S \geq 2;$$

$\overset{(0)}{\tilde{v}}$, $\overset{(S-1)}{\tilde{v}}$ are calculated in the same way.

The following boundary conditions are stated for: $\hat{u}, \hat{v}, \overset{(S)}{V}{}^{k+1}, \overset{(S)}{P}$

$$\begin{pmatrix} V \\ P \end{pmatrix}(x,y,z,t) = \begin{pmatrix} V \\ P \end{pmatrix}(x+X,y,z,t) = \begin{pmatrix} V \\ P \end{pmatrix}(x,y+Y,z,t)$$

$$V(\mathbf{x},t)\big|_{z=0} = 0, \qquad (3)$$

$$V(\mathbf{x},t)\big|_{z\to\infty} \to 0(\exp),$$

The problem (1) - (3) is written in dimensionless variables that are related with dimensional ones in the following way: $\mathbf{x}_* = \mathbf{x}\delta,\ t_* = t\delta/U_\infty,\ \mathbf{V}_* = \mathbf{V}U_\infty,\ p_* = pU_\infty^2$, where: $\delta = (\upsilon\ell/U_\infty)^{1/2}$, ℓ is the distance from the leading edge to the integration domain, U_∞ is the flow velocity at $z \to \infty$, υ is the kinematic viscosity coefficient, the Reynolds number is: $R = U_\infty\delta/\upsilon = (R_x)^{1/2}$. The following criterion is used to exit from iteration process:

$$<(\mathbf{F}[\overset{(S-1)}{V}{}^{k+1/2}], \overset{(S)}{V}{}^{k+1/2})>_x < \varepsilon \quad \text{where} \quad <.>_x = <.>_{xyz} = \frac{1}{XY}\int_0^X\int_0^Y\int_0^\infty .\ dxdydz$$

Schemes (1), (2) at S≥2 have the second order of accuracy on τ. To approximate (1)-(3) in streamwise and spanwise directions x, y the spectral method is used. In normal the direction the collocation method with transformation of variable $z = c_1\bar{z} + c_2\bar{z}^k$, $k \geq 2$ is used (the line for $\bar{z}$ is dropped in the following consideration). The approximate solution is sought in the form:

$$V^k(\mathbf{x}) = \sum_{m=-M}^{M}\sum_{n=-N}^{N}\sum_{q=1}^{Q}\hat{V}^k_{mnq}\cdot E_{mn}(x,y)\cdot H_q(z),$$

$$P^k(x) = \sum_{m=-M}^{M}\sum_{n=-N}^{N}\left[\hat{p}^k_{mno}\, e^{-z/2} + \sum_{q=1}^{Q}\hat{P}^k_{mnq}\cdot H_q(z)\right]\cdot E_{mn}(x,y), \qquad (4)$$

where: $E_{mn}(x,y) = \exp(i\alpha_m x + i\beta_n y)$, $\alpha_m = \alpha_o m$, $\beta_n = \beta_o n$, $\alpha_o = 2\pi/X$, $\beta_o = 2\pi/Y$, $H_q(z) = ze^{-z/2}L_{q-1}(z)$, $L_q(z)$ = Laguerre polynomials. It is seen from (4) that the approximate solution satisfies the boundary conditions (3).

By means of the spectral analysis [1, 2] the investigation of the stability and accuracy in the linear approximation was performed. The spectral characteristics of the linearized differential problem (the eigenvalue problem for the Orr-Sommerfeld equation) were calculated. The same was done for the linearized scheme (1)-(4), i.e $\mathbf{F} \equiv 0$ (the scheme is called (1)-(4) as SCMB01). The comparison of the two spectra shows the accuracy and stability of the computational algorithm. A suitable time step and the necessary accuracy of space resolution are also estimated by this algorithm.

The following scheme (SCMB02) was investigated. The convective terms were approximated in time using the Adams-Bashford scheme and for the rest of Navier-Stokes equations the Crank-Nicolson scheme was used; the space approximation was the same as for SCMB01. It should be noted that the schemes, which are analogues of the SCMB02, were used in a number of works where the initial stage of transition to turbulence in the boundary layer was investigated.

In this study the time evolution of the infinitesimal perturbations was considered and the solution of the linearized problems was presented as the Tollmien-Schlichting wave:

$$\mathbf{V}(\mathbf{x},t) = \mathbf{V}(z)\exp[i\alpha(x - Ct) + i\beta y] \quad (\text{for } \alpha = 0 \; \mathbf{V}(\mathbf{x},t) = \mathbf{V}(z)\exp[i\beta(y - Ct)])$$

where α and β are the real values, and C is the complex value.

In Table 1 for the supercritical Reynolds number R = 580 and various wave numbers $\alpha_m = \alpha_0 m$, $\beta_n = \beta_0 n$ the senior (in the imaginary part) eigenvalues C of the Orr-Sommerfeld equation are shown (C corresponds to the exact values with the rounding accuracy). The eigenvalues $\bar{C}$ which were obtained for the two schemes with the number of collocation points Q time step τ and the normal transformation parameters c_1, c_2, k ($z = c_1\bar{z} + c_2\bar{z}^k$) are also shown.

For scheme SCMB01, the values $\overset{(S)}{\bar{C}}$ are shown for the first and third iteration, $\overset{(S)}{\bar{C}}$ for S≥4 are practically the same as for S = 3. The comparison of $\bar{C}$ with the exact values shows, that at moderate numbers of points Q and comparatively large time step τ the scheme SCMB01 is stable and has the high accuracy at transitional Reynolds numbers in a wide region of wave numbers.

At the same parameters Q and τ the scheme SCMB02 has a better accuracy (in comparison with SCMB01 at S =1) in the characteristic part of the spectrum where the physical instability takes place. It is due to the second order accuracy on τ for this scheme. The significant failure of the SCMB02 scheme is the instability at large wave numbers (see Table 1).

Table 1. R=580, α_0=0.179, $\beta_0=\alpha_0/2$; for schemes: Q=32, τ=0.1, c_1=0.1, c_2=0.008, k=2

(m, n)	**C for Orr-Sommerfeld problem**	**$\overset{(s)}{\bar{C}}$ for scheme SCMB01**		**$\bar{C}$ for scheme SCMB02**
		s = 1	**s = 3**	
(1, 0)	0.3641229 + i0.0079597	0.3641041 + i0.0080793	0.3641613 + i0.0079372	0.3641598 + i0.0079475
(0, 1)	0.0000000 - i0.0001543	0.0000000 - i0.0001667	0.0000000 - i0.0001667	0.0000000 - i0.0001667
(1, 1)	0.3766307 + i0.0019978	0.3766258 + i0.0021141	0.3766925 + i0.0019680	0.3766892 + i0.0019778
(2, 0)	1.0000000 - i0.0006172	0.9998932 - i0.0006193	0.9998932 - i0.0006197	1.0005344 - i0.0006090
(2, 1)	1.0000000 - i0.0006558	0.9998932 - i0.0006579	0.9998932 - i0.0006583	1.0005344 - i0.0006476
(2, 2)	1.0000000 - i0.0007716	0.9998932 - i0.0007735	0.9998932 - i0.0007740	1.0005344 - i0.0007634
(10, 0)	1.0000000 - i0.0030862	0.9973387 - i0.0030135	0.9973427 - i0.0030622	1.0136324 - i0.0016327
(0, 10)	0.0000000 - i0.0015431	0.0000000 - i0.0015441	0.0000000 - i0.0015441	0.0000000 - i0.0015441
(10, 10)	1.0000000 - i0.0038578	0.9973378 - i0.0037667	0.9973428 - i0.0038276	1.0136359 - i0.0024232
(30, 0)	1.0000000 - i0.0092586	0.9766679 - i0.0074687	0.9769634 - i0.0086362	1.1345299 + i0.0591940
(0, 30)	0.0000000 - i0.0046293	0.0000000 - i0.0046296	0.0000000 - i0.0046296	0.0000000 - i0.0046296
(30, 30)	1.0000000 - i0.0115733	0.9766021 - i0.0093339	0.9769664 - i0.0107952	1.1350534 + i0.0564223

Two runs were performed at $R = 580$ with the different intervals of periodicity:

run 1 - $X = Y = 2\pi/0.0895$

run 2 - $X = Y = 2\pi/0.03$

The initial energy of perturbations was sufficiently small in both runs, $E_0 = 10^{-3}$. A moderate number of basic functions were used in the solution representation: in x, y, z directions *63 x 32 x 63* for run 1, and *127 x 64 x 63* for run 2 (this number of basic functions was used during the last stage of the runs, i.e. for simulation of developed turbulent regime). The following transformation was used $z = c_1\bar{z} + c_2\bar{z}^k$, $c_1 = 0.12$, $c_2 = 0.003$, $k = 2$; the time step was $\tau = 0.1$. The CPU time per step for run 1 was 15 sec. with a total time of 20 hours on the CONVEX 201 and for run 2 was 51 sec. per step with a total time of 55 hours.

At the initial stage of these runs, there is a weak nonlinear interaction and a slow rise of skin friction and pulsation energy may be observed. When the energy of pulsations reaches the critical value of about $E \sim 10^{-2}$ the transition to developed turbulent flow occurs which is accompanied by a sharp increase of friction and energy of pulsations (especially on the small scales).

Fig. 1 illustrates the two-point correlations for streamwise component of velocity in the x-direction $R_{uu}(x)$, and for normal one in the y-direction $R_{ww}(y)$ at $z/\delta_t = 0.3$, for runs 1 and 2; δ_t turbulent boundary layer thickness. The time averaging for these and other characteristics was performed at time interval $\Delta t \geq 10^2$. It may be observed that correlations are considered fall off to small values. The analogous result was obtained also for $R_{uu}(y)$, $R_{ww}(x)$, $R_{vv}(x)$, $R_{vv}(y)$ at different points of z. Consequently, sufficiently large intervals of periodicity were chosen.

The mean values of skin friction, pulsation energy, profile of velocity, Reynolds stress and turbulence intensities, etc., in run 1 for developed turbulent flow, slightly differ from those obtained in run 2 (see Fig. 2). Moreover, these characteristics have a good conformity with experimental data (see Figs. 2, 3) and computation results [3]. The mean flow non-uniformity in streamwise direction were taken into account in [3], and a great number of basic functions were used in the solution representation (in x, y, z directions *170 x 128 x 42*). It should be noted that turbulent regime at $R = 580$ ($R_\theta = 876$) was obtained in a natural way, i.e. the transition from laminar to developed turbulent flow was simulated.

In computations at Reynolds numbers $R = 1000, 1400$ the same number of basic functions was used as in run 2. Fig. 4 shows the experimental values of the local friction coefficient C_f' and the shape factor H for different Reynolds numbers R_θ [4]. Also, results of our computations and [3] are presented there. The value of C_f' obtained for $R_\theta = 2550$ is lower than the experimental value. Note that generally, mean characteristics of turbulent flows (velocity profile, second-order moments etc.) obtained at $R_\theta = 1640, 2550$ are in satisfactory agreement with experimental data (see Fig. 3).

In conclusion, it has been shown that within the framework of the model considered for the Navier-Stokes equations using the above method, it is possible to simulate turbulent flow in a boundary layer on flat plate in a wide region of Reynolds numbers.

1. Rozhdestvensky B. L., Simakin I. N. - Chisl. Metody Mekh. Sploshnoy Sredy, Novosibirsk, 1982, v. 13, pp.89-128.
2. Simakin I.N. - Keldysh Inst. Appl. Maths. Preprint No. 49, 1983.
3. Spalart P. R. - J. Fluid Mech., 1988, v. 187, pp.61-98.
4. Purtell L. P., Klebanoff P. S., Buckley F. T. - Phys. Fluids, 1981, v. 24, pp 802-811.
5. Kim H. T., Kline S. J., Reynolds W. C. - Report MD-20, 1968, Stanford Univ., CA.

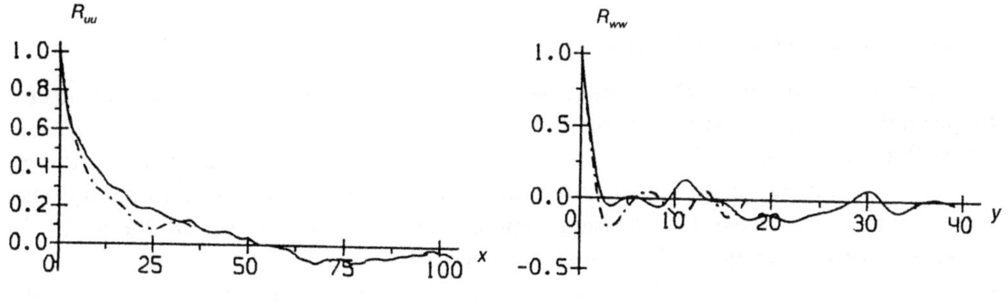

Fig. 1. Two-point correlations in streamwise *(x)* and spanwise *(y)* directions at $z/\delta_t = 0.3$; —•— for run 1, —— for run 2; δ_t - turbulent boundary layer thickness.

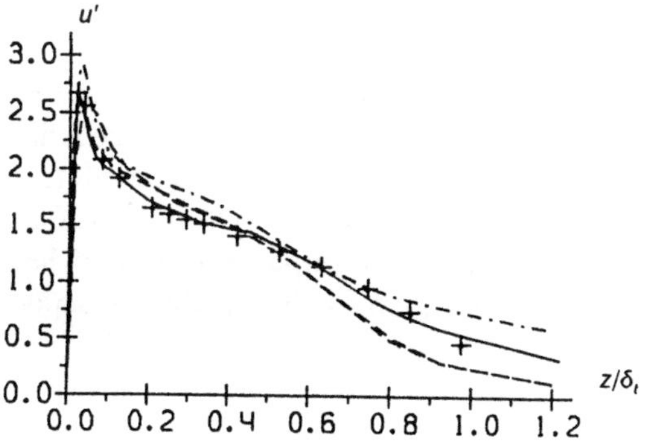

Fig. 2. Streamwise component of r.m.s. velocity fluctuations. Calculations: —•— for run 1, —— for run 2, $R_\theta = 876$; — — —, $R_\theta = 670, 1410$, [3]. Experiment: +, $R_\theta = 1340$, [4].

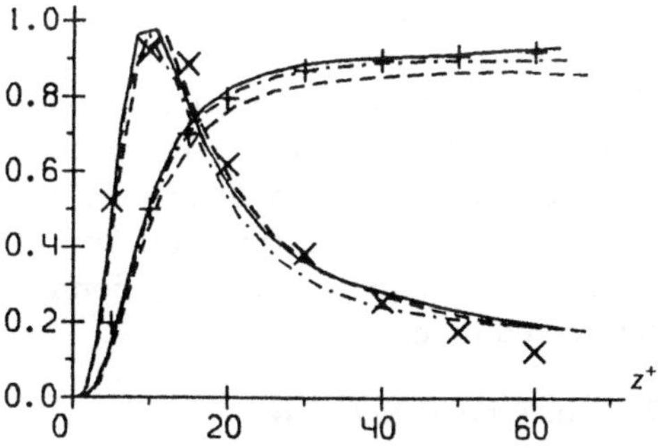

Fig. 3. Reynolds stress - 1, production - 2 (x4) near the wall. Calculations: — — —, $R_\theta = 876$, —•— $R_\theta = 1640$, —— $R_\theta = 2550$. Experiment: +, (1); x, (2); [5].

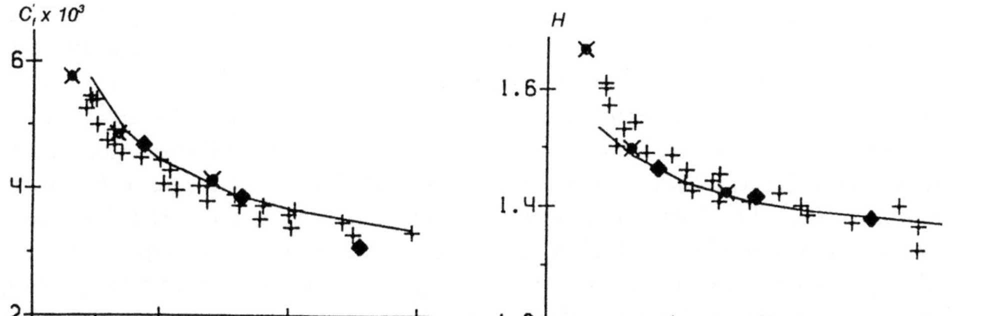

Fig. 4. Local friction coefficient C_f' and shape factor *H*. Calculations: ♦, our results; ✱, [3]. Results of different experiments from [4], —— ; +.

NEAR-WALL TURBULENCE MODELS AND NUMERICAL SOLUTION OF THE REYNOLDS-AVERAGED NAVIER-STOKES EQUATIONS USING UNSTRUCTURED GRIDS

L. Stolcis* and L.J. Johnston

Department of Mechanical Engineering, UMIST
PO Box 88, Manchester M60 1QD, England

1. Motivation and Objectives

The long term aim of the present work is the development of a numerical method to predict the flow around complex three-dimensional aerodynamic configurations. These flows are generally turbulent in nature and involve complex physical phenomena such as interacting shear layers, streamline curvature, shock wave/boundary layer interactions and extensive separation regions. The Reynolds-averaged Navier-Stokes equations for compressible flow must be solved in order to deal with these flows in a satisfactory manner. Further, the turbulent shear layers are not necessarily thin, which has implications for the choice of the turbulence model used to close the set of governing mean-flow equations. Fully-representative aerodynamic configurations also have a geometric complexity which makes the generation of a suitable computational grid a non-trivial task. However, unstructured computational grids offer a convenient means of overcoming these problems and are used in the present work. A number of issues related to integration of the turbulence model into the unstructured flow solver are currently being investigated using a two-dimensional method. This method has been successfully applied to the computation of multi-element high-lift aerofoils, both at low-speed and transonic flow conditions [1], [2]. The important issue of how to treat the molecular-viscosity dominated near-wall region is the particular aspect considered here.

2. Numerical Scheme

A cell-centred, finite-volume spatial discretisation of the governing mean-flow and turbulence-transport equations is used, together with an explicit multi-stage time-stepping scheme to reach the steady-state solution. Numerical dissipative terms are added explicitly to the discretised mean-flow equations, in order to suppress odd-even decoupling and to enable clean capturing of shock waves. The magnitude of these additional terms is controlled within boundary layer and wake regions so as to minimize their influence on the viscous flow development. A high-Reynolds number formulation of the two-equation k-ϵ turbulence model has been implemented to facilitate the computation of complex turbulent flowfields. This model requires the solution of two additional modelled transport equations, for the turbulent kinetic energy k and its dissipation rate ϵ. The basic numerical algorithm adopted for the mean-flow equations needs to be modified for application to the turbulent transport equations, due to stability problems associated with the strongly non-linear source terms appearing in the k and ϵ equations. These problems have been overcome temporarily by using a first-order upwind spatial discretisation of the convective flux terms in the k and

*Present address: CRS4, Via Nazario Sauro 10, 09123 Cagliari (Italy)

ϵ equations [1]. Extensive experience with the present method indicates that accurate solutions can be achieved provided that a second-order scheme is used for the mean-flow equations. The main point of concern with the flow solver is the use of wall function boundary conditions for the turbulence-transport equations, since with this approach the molecular-viscosity dominated near-wall region is not resolved.

3. Near-Wall Formulation

For turbulent flow developing over a smooth flat plate in zero pressure gradient, experiments indicate that the velocity distribution across the fully-turbulent, near-wall region follows the semi-logarithmic law-of-the-wall

$$u_T/u_\tau = ln(y^+E)/\kappa\ , \qquad u_\tau = \sqrt{\tau_{wall}/\rho_{wall}}\ , \qquad y^+ = y_n u_\tau \rho/\mu \tag{1}$$

u_T is the mean-velocity component parallel to the wall, u_τ is the friction velocity, τ_{wall} is the wall shear stress and y_n is the distance from the wall. $\kappa = 0.41$ and E is a roughness parameter, equal to 9 for smooth walls.

3.1 wall function approach

In this approach the turbulence-transport equations are solved only in the fully-turbulent region, by setting the first near- wall computational cell to be in the range $30 < y^+ < 300$. The equations themselves are not solved in these near-wall cells, and it is assumed that there is local equilibrium between production and dissipation of turbulent kinetic energy. If the dissipation length scale is assumed to be directly proportional to y_n, then the near-wall values of k and ϵ become

$$k = u_\tau^2/\sqrt{c_\mu}\ , \quad \epsilon = u_\tau^3/\kappa y_n \tag{2}$$

where $c_\mu = 0.09$ and u_τ is obtained from the law-of-the-wall. The mean-flow equations are solved in the near-wall cells together with the no-slip boundary condition, employing the wall shear stress obtained from the friction velocity u_τ. The wall function approach is ideally-suited to implementation in an unstructured flow solver since information beyond the first computational cell adjacent to the surface is not required.

3.2 low-Reynolds number formulation

The k and ϵ equations can be modified to enable their integration through the viscous sublayer to the wall, by introducing near-wall damping functions in the modelled source terms and the turbulent viscosity relation. Additional terms are also generally required to ensure the correct behaviour of k and ϵ in the molecular-viscosity dominated region. The resulting low-Reynolds number turbulence model removes the uncertainties associated with the wall function approach. However, such formulations require very dense computational grids near the wall, to enable resolution of the high near-wall gradients of turbulence quantities. The source terms in the k and ϵ equations can become very large in this same region, increasing the stiffness of the discretised turbulence-transport equations and thereby reducing the stability properties of the numerical method. In order to circumvent these problems, it is possible to solve the k and ϵ equations only in the high-Reynolds number regions, and use a low-Reynolds number one-equation model in the near-wall region. The attraction of a one-equation near-wall formulation is that such a model requires a less-dense computational grid in the viscous sublayer region, increasing the computational efficiency and convergence properties of the numerical method. The near-wall part of the one-equation turbulence model of Mitcheltree et al [3] has been employed in the present work. In regions adjacent to the surface, where $y^+ < 50$, the mean-flow equations and the turbulent kinetic energy transport equation are solved, whilst the turbulent length scales are prescribed via algebraic relations. The turbulent viscosity coefficient μ_t and the dissipation rate of k in the near-wall region are given by

$$\mu_t = c_\mu \rho k^{1/2} L_\mu \, , \qquad \epsilon = k^{3/2}/L_\epsilon \tag{3}$$

with L_μ and L_ϵ being two algebraic length scales

$$L_\mu = c_1 y_n[1 - exp(-R_t/A_\mu)] \, , \qquad L_\epsilon = c_1 y_n[1 - exp(-R_t/2c_1)] \tag{4}$$

which involve a turbulent Reynolds number R_t

$$R_t = \rho k^{1/2} y_n/\mu \tag{5}$$

The damping functions in the two length scales are introduced to mimic the correct behaviour as the wall is approached, where the flow development is dominated by the molecular viscosity. The model contains two constants $A_\mu = 70$ and $c_1 = \kappa c_\mu^{-3/4}$. The high-Reynolds number $k - \epsilon$ model is employed when $y^+ > 50$.

4. Results and Conclusions

A full description of the present numerical method together with an extensive evaluation is given by Stolcis [4]. Figures 1 and 2 show results comparing the two near-wall formulations for Cases 9 and 10 of the well-known RAE 2822 transonic aerofoil data-set. The surface pressure distributions show some small differences in upper surface pressure levels, and the one-equation near-wall formulation results in a more downstream shock wave location. The principal area of difference between the two formulations is in the upper surface skin friction distributions. The wall function approach produces too high levels of skin friction upstream of the shock wave, and fails to predict the experimentally-observed shock-induced separation for Case 10. The mean-velocity profiles predicted by the two near-wall formulations are in good agreement with each other, and agree reasonably well with experiment, apart from in the immediate vicinity of the shock wave. It can be concluded that the wall function approach gives acceptable results, apart from a reluctance to produce shock-induced separation, and so merits further investigation.

5. References

[1] Stolcis, L. and Johnston, L.J. "Computation of the viscous flow around multi-element aerofoils using unstructured grids", 9th GAMM Conference on Numerical Methods in Fluid Mechanics, EPFL, Lausanne, Switzerland, September 25-27, 1991

[2] Stolcis, L. and Johnston, L.J. "Application of an unstructured Navier-Stokes solver to multi-element airfoils operating at transonic maneuver conditions", AIAA Paper 92-2638, 10th AIAA Applied Aerodynamics Conf., Palo Alto, CA, June 22-24, 1992

[3] Mitcheltree, R., Salas, M. and Hassan, H. "A one-equation turbulence model for transonic airfoil flows", AIAA Paper 89-0557, 1989

[4] Stolcis, L. "Computation of the turbulent flow development around single- and multi-element aerofoils", PhD thesis, University of Manchester, 1992

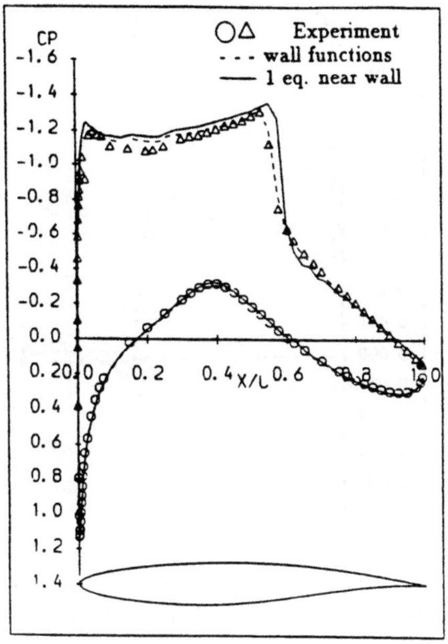

a) surface pressures

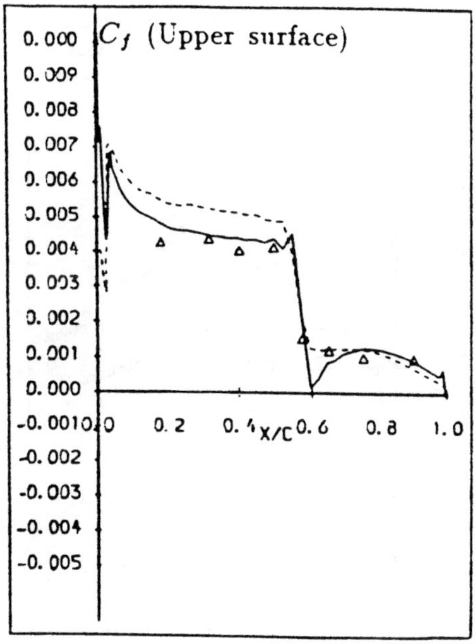

b) upper surface skin friction

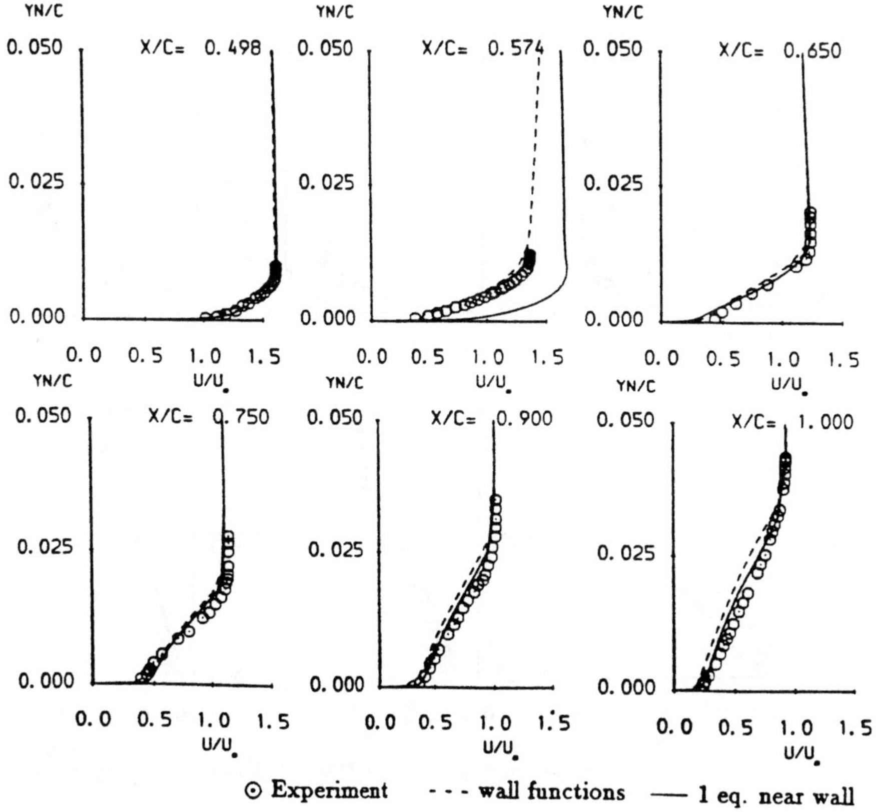

c) upper surface mean-velocity profiles

Figure 1: Results for RAE 2822 - Case 9

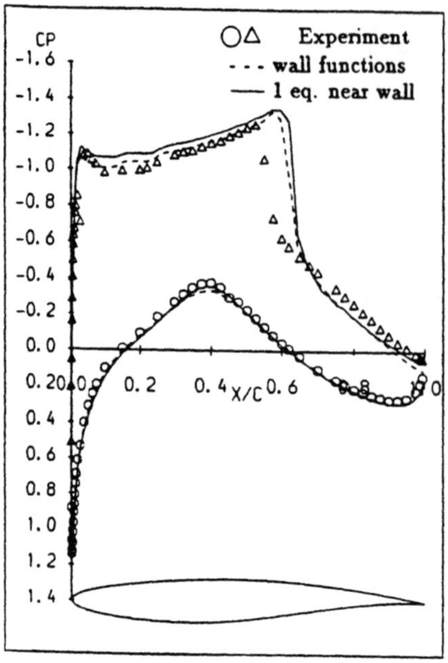

a) surface pressures

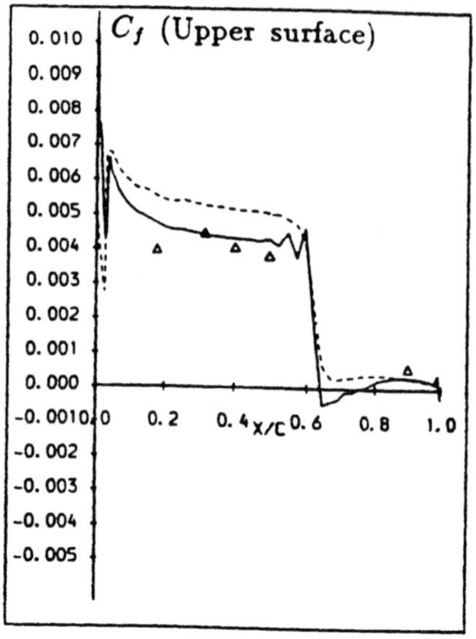

b) upper surface skin friction

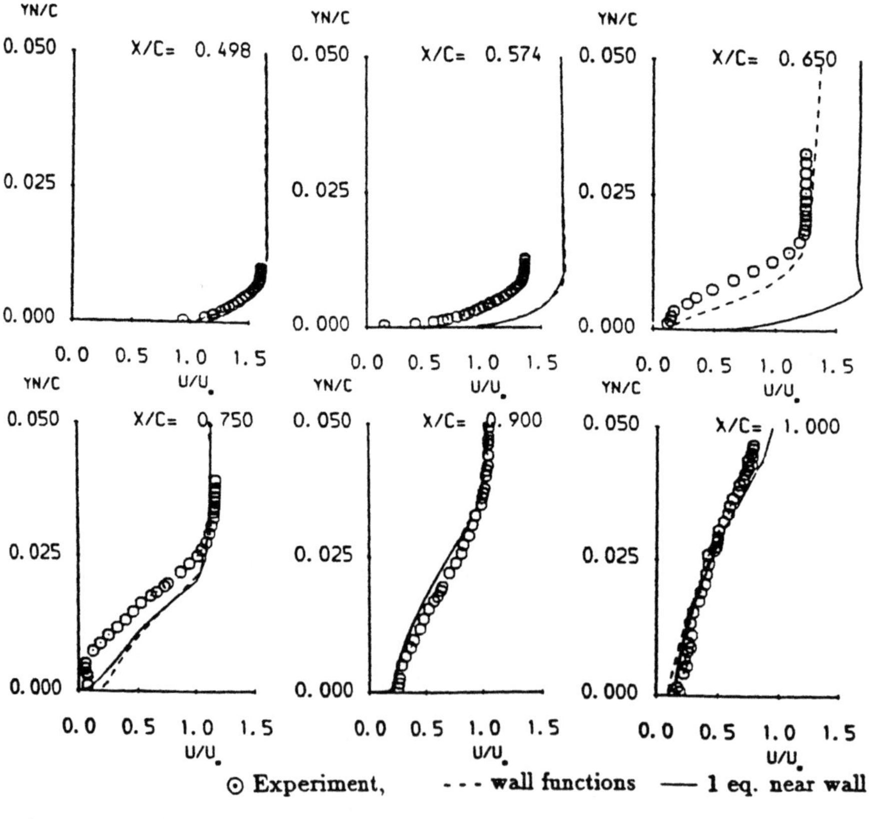

c) upper surface mean-velocity profiles

Figure 2: Results for RAE 2822 - Case 10

SIMULATION OF INSTABILITIES IN A BOUNDARY LAYER WITH A ROUGHNESS ELEMENT

G. Danabasoglu, S. Biringen and C. L. Streett

University of Colorado, Boulder, CO, USA
NASA Langley Research Center, Hampton, VA, USA

1. Introduction

The investigation of boundary layer instability in the presence of an isolated roughness element, and therefore increased receptivity of the boundary layer is a challenging computational problem. Individual roughness elements in a distributed roughness environment actually form quasi two-dimensional effects at their leading and trailing edges (Morkovin, 1990). Therefore, the case of the isolated, two-dimensional roughness element, which is the focus of the present study, has relevance to the problem of distributed roughness.

Because of strong upstream influences, the proper simulation of this problem requires the use of the spatial approach. Several numerical studies, using the time-dependent, full Navier-Stokes equations succesfully simulated transition over smooth flat plates and in channels, assuming the temporal rather than the more realistic spatial evolution. Recently, due to increased capabilities of supercomputers, the spatial numerical simulations became feasible. Among the latter, Danabasoglu, Biringen & Streett (1991) developed a method to study the spatial evolution of instabilities and instability control in channel flows. In this work, we modified the numerical procedure of Danabasoglu et. al (1991) for boundary layer flows with a two-dimensional isolated roughness element to study the spatial development of instability waves, focusing on the effects of roughness size and roughness location. Recently, the experiments of Boiko, Dovgal, Kozlov & Shcherbakov (1990) showed that inviscid frequencies can exhibit greater amplification rates compared to the unstable Tollmien-Schlichting (TS) frequencies; therefore, we also investigated the effects of inflow frequencies.

2. Solution Procedure

The present study considers the two-dimensional, time-dependent, incompressible, full Navier-Stokes equations discretized on a non-staggered, stretched grid by a mixed finite difference/Chebyshev collocation matrix method. The governing equations are integrated by a time-splitting procedure which employs the implicit Crank-Nicolson scheme for the normal direction diffusion terms. All of the other terms are treated explicitly using a third-order, compact Runge-Kutta method (Streett & Hussaini, 1986). For the solution of the Poisson/Laplace equations, the capacitance matrix and eigenvalue decomposition methods are utilized. The inflow perturbations are obtained from the spatial modes of the Orr-Sommerfeld equation, and at the outflow, the buffer domain technique is applied. The details of the numerical scheme are given in Danabasoglu et. al (1991).

One of the difficulties associated with the roughness element simulations is the singularity at the corner points of the element. The singularity stems from the fact that the homogeneous Neumann boundary condition cannot be defined at the corner points. Although there are numerous methods in literature to alleviate this problem, we chose to apply the multidomain method given by Macaraeg & Streett (1986) along the normal direction. In this procedure, the corner points are assigned to the upper domain, thus eliminating the singularity by defining the normal direction implicitly. In addition, because each subdomain contains a separate set of Chebyshev polynomials, which can vary in number, as many collocation points as necessary can be applied independently in each subdomain. Consequently, expected sharp gradients in the vicinity of the roughness element can be resolved conveniently.

The boundary layer simulations naturally require mesh clustering next to the wall in addition to the regular Chebyshev stretching. The application of the multidomain method further necessitates the use of mapping functions along the normal direction, because the subdomains may not coincide with the Chebyshev interval.

In all the cases, once the base flow is obtained using the total velocity equations, the perturbation equations are applied for the instability computations.

3. Results and Discussions

The simulations discussed in this section particularly consider the effects of the roughness size and roughness location; for this purpose, four different configurations are investigated. In each simulation, small amplitude perturbations are imposed as the inflow conditions and the growth rates of the instability waves, especially in the amplification region in the separation zone behind the roughness element, are computed and compared with the parallel linear theory.

For the first two simulations in which two different roughness heights, $s = 0.0982\delta^*$ (small) and $s = 0.245\delta^*$ (medium), are considered, the elements are placed upstream of branch I (stable region) of the neutral stability curve of the Blasius flow; the local Reynolds number at the roughness location is kept at $Re_{\delta^*} = 550$. Here, $Re_{\delta^*} = u^*_\infty \delta^* / \nu^*$ is the Reynolds number, and u^*_∞, δ^* and ν^* are the free stream velocity, boundary layer local displacement thickness and kinematic viscosity, respectively. The small roughness element simulations conducted at $Re_{\delta^*_0} = 450$ and $\omega_R = 0.063$ reveal that the spatial development of the instabilities is not influenced by the presence of the roughness element when compared with the smooth plate computations obtained using the same parameters. Here, δ^*_0 is the boundary layer displacement thickness used as the length scale, and ω_R is the inflow perturbation frequency.

The streamwise distribution of $(u'_{rms})_{max}$ for the medium roughness element case obtained with $Re_{\delta^*_0} = 500$ and $\omega_R = 0.07$ is presented in Fig. 1 (curve A), where $(u'_{rms})_{max}$ denotes the root-mean-square values of the streamwise perturbation velocity, and the maximum is obtained along the normal direction. The figure also includes the distribution obtained from a simulation for the smooth plate with the same parameters (curve B). According to Fig. 1, we observe that the growth of the disturbances is slightly higher in the presence of the roughness element. However, no significant deviation from the smooth plate boundary layer distributions is observed throughout the computational domain.

In the presence of a roughness element, because the parallel linear stability analysis based on the most inflectional streamwise velocity profile indicated the existence of high frequency inviscid instability at high Reynolds numbers, for the subsequent simulation, we consider the medium roughness element and increase the disturbance frequency to $\omega_R = 0.16$, keeping the other parameters the same. Note that this value of ω_R is stable according to the linear theory. The $(u'_{rms})_{max}$ distribution given in Fig. 1 (curve C) shows the decay of the instability waves in agreement with the linear theory. This behaviour is due to the insignificant deviation of the base velocity profiles from the Blasius distributions.

Next, we investigate a case with a large roughness element ($s = 1.01\delta^*$), placed at $Re_{\delta^*} = 712.8$ which is in the unstable region of the smooth plate stability curve, and set $Re_{\delta^*_0} = 625.2$ and $\omega_R = 0.0928$. Because ω_R is in the TS instability range, this case is referred to as the TS frequency case. The streamwise variation of $(u'_{rms})_{max}$ (curve A in Fig. 2) shows significant amplification in the separation zone at this frequency and the instabilities attain finite amplitudes. In the linear region located in the separation zone ($77.42 \le x \le 91.32$) we measure the spatial growth rate of the disturbances as $\alpha_I \simeq -0.12$ which is significantly higher than the amplification rate for the Blasius flow ($\alpha_I = -0.11643 \times 10^{-2}$).

In order to compare the growth rates of high (inviscid) frequencies with TS frequencies, we consider a higher frequency disturbance field, $\omega_R = 0.1856$, which is in the inviscid shear (mixing) layer instability range. Figure 3 presents the contour plots of u' (streamwise perturbation velocity) and ω'_z (spanwise perturbation vorticity) obtained after 26 TS periods. These figures show that the disturbances are amplified in the separation zone, and they attain finite amplitudes quite rapidly in a very short distance. The $(u'_{rms})_{max}$ distribution is also presented in Fig. 2 (curve B) for a comparison with the TS frequency case, revealing a higher growth rate ($\alpha_I \simeq -0.19$) than the TS frequency in the linear region in agreement with Boiko et. al (1990). The figure illustrates that the amplitude of the inviscid frequency case is about 6 times lower than the TS frequency case immediately upstream of the roughness element. In the next simulation, the $(u'_{rms})_{max}$ amplitude is matched for both frequencies upstream of the element by adjusting the inflow perturbation amplitude. It is observed that even though the measured growth rates are the same for both inviscid frequency simulations, the disturbances reach higher amplitudes for the latter case (curve C in Fig. 2) confirming that for the same amplitude upstream of the roughness element, the shear layer frequency attains higher finite amplitudes in the separation zone compared to the TS frequency. Next, we investigate the evolution of instabilities created by a lower frequency, $\omega_R = 0.015$, which is stable according to the smooth plate linear stability theory. The $(u'_{rms})_{max}$ distribution (curve D in Fig. 2) reveals the amplitude increase of the instability waves downstream of the separation zone which is associated with the wavelength conversion observed at this frequency.

Finally, we place the medium roughness element in the unstable region of the smooth plate stability curve, at $Re_{\delta^*} = 712.8$, and investigate the effects of the roughness location and size. For this purpose, we use $\omega_R = 0.0928$ and $\omega_R = 0.1856$. The $(u'_{rms})_{max}$ distributions for both cases, including the distributions from simulations with the same parameters for the smooth plate, are given in Fig. 4. The figure reveals no significant deviations from the smooth plate cases supporting the idea that significantly larger elements are necessary to instigate higher spatial growth rates.

4. Conclusions

The computational procedure which involves the multidomain method was successfully used to study the spatial development of instability waves in the presence of a two-dimensional, isolated roughness element.

The simulations disclosed that for the cases considered here, if the element is placed upstream of branch I of the neutral stability curve, the disturbances created by inviscid frequencies decay due to the small deviations of the base velocity profiles from the Blasius solution. Also, the study with the medium roughness element located in the unstable region of the stability curve revealed no significant variations from the flat plate distributions.

The investigations with the large roughness element, placed in the unstable region of the stability curve, led to significant amplification in the separation zone for both TS and inviscid frequencies. However, the inviscid frequencies which are in the shear layer instability region disclosed higher amplification rates.

Acknowledgements

This work was supported by NASA Langley Research Center under Grant NAG-1-1161 and by ONR Grant No. ONR00014-91-J-1086.

References

BOIKO, A. V., DOVGAL, A. V., KOZLOV, V. V. & SHCHERBAKOV, V. A. 1990 'Flow Instability in the Laminar Boundary Layer Separation Zone Created by a Small Roughness Element,' *Fluid Dynamics* **25**, 12.

DANABASOGLU, G., BIRINGEN, S. & STREETT, C. L. 1991 'Spatial Simulation of Instability Control by Periodic Suction Blowing,' *Phys. Fluids A* **3**, 2138.

MACARAEG, M. G. & STREETT, C. L. 1986 'Improvements in Spectral Collocation Discretization Through A Multiple Domain Technique,' *Appl. Num. Math.* **2**, 95.

MORKOVIN, M. V. 1990 'On Roughness-Induced Transition: Facts, Views, and Speculations,' in *Instability and Transition*, Vol. I, edited by M. Y. Hussaini and R. G. Voigt, pp. 281-295, Springer-Verlag, New York.

STREETT, C. L. & HUSSAINI, M. Y. 1986 'Finite Length Effects in Taylor-Couette Flow,' *ICASE Report No:* 86-59.

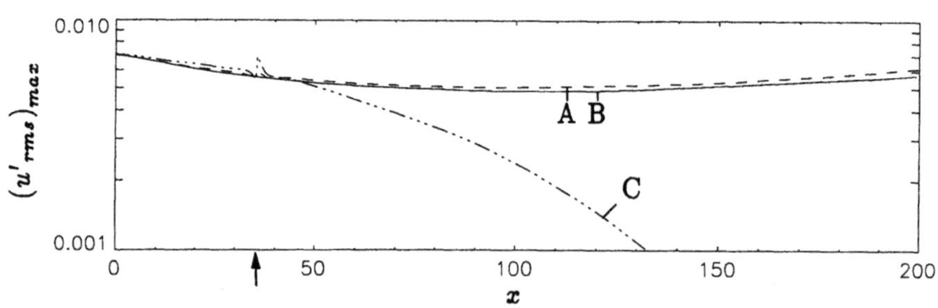

Figure 1. Streamwise distributions of $(u'_{rms})_{max}$: with roughness, $\omega_R = 0.07$ (A), without roughness, $\omega_R = 0.07$ (B), and with roughness, $\omega_R = 0.16$ (C). Medium roughness element is located at $x = 35.41$ ($Re_{\delta^*} = 550$). The starting location of the roughness element is indicated by an arrow in the figures.

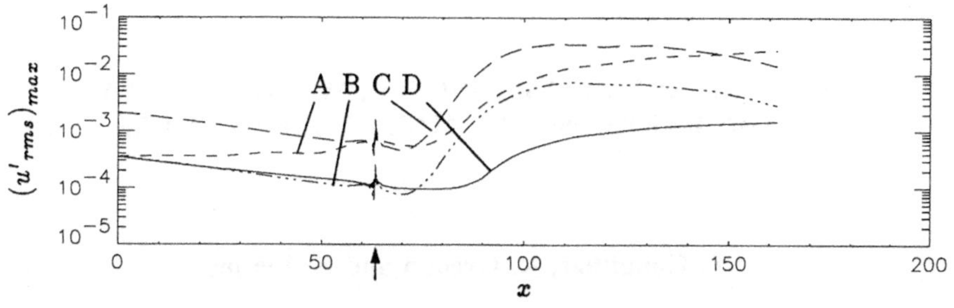

Figure 2. **Streamwise distributions of $(u'_{rms})_{max}$: $\omega_R = 0.0928$, $A_{2d} = 0.0005$ (A); $\omega_R = 0.1856$, $A_{2d} = 0.0005$ (B); $\omega_R = 0.1856$, $A_{2d} = 0.003$ (C); and $\omega_R = 0.015$, $A_{2d} = 0.0005$ (D). Roughness element is located at $x = 63.34$. Large roughness element. A_{2d} is the amplitude of the inflow perturbations.**

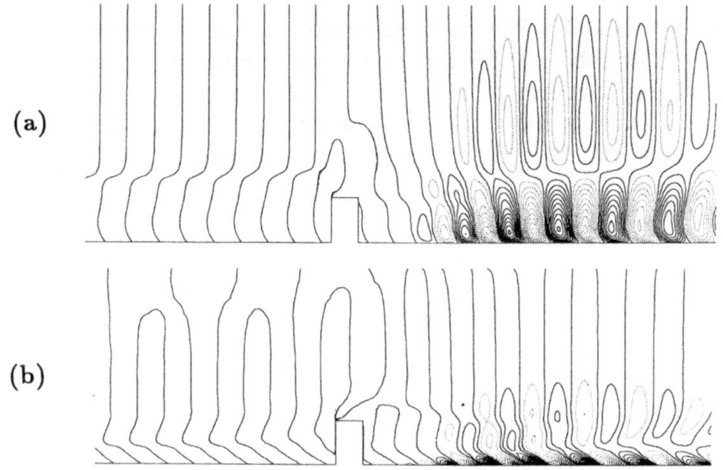

Figure 3. **Contour plots of (a) u' (contour intervals= 0.001) and (b) ω'_z (contour intervals= 0.005). Normal direction is stretched by a factor of 10. Flow direction is from left to right. Large roughness element instability with $\omega_R = 0.1856$.**

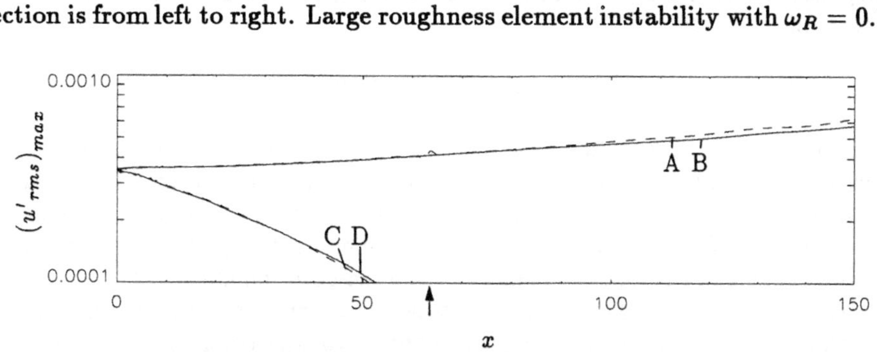

Figure 4. **Streamwise distributions of $(u'_{rms})_{max}$: with roughness, $\omega_R = 0.0928$ (A), without roughness, $\omega_R = 0.0928$ (B), with roughness, $\omega_R = 0.1856$ (C), and without roughness, $\omega_R = 0.1856$ (D). Medium roughness element is located at $x = 63.34$ ($Re_{\delta^*} = 712.8$).**

TRANSONIC AND SUPERSONIC FLOW CALCULATIONS AROUND AIRCRAFTS USING A MULTIDOMAIN EULER CODE.

V. Couaillier, R. Grenon and N. Liamis

ONERA BP 72 - 92322 Chatillon (France)

INTRODUCTION

In this paper we present a 3-D Euler numerical method based on the explicit scheme proposed by Ni [1] combined with the implicit stage proposed by Lerat et al. [2]. This method has been implemented in a 3-D code with multidomain approach and treatment of the boundary conditions based on the characteristic relations. In this paper are given the main features of the method, and results are presented for the solution of transonic and supersonic flows around a supersonic transport aircraft with a five domain decomposition.

EXPLICIT STAGE

We briefly recall the discretization that we use [3]. We consider the system of Euler equations written in a conservative and compact form :

$$\frac{\partial f}{\partial t} + div\, F = 0 \qquad (1)$$

where f denotes the conservative variables and F the fluxes. We use the following one-step explicit Lax-Wendroff scheme to obtain a predicted value at time t^{n+1} for each point of the mesh :

$$f^{n+1,E}{}_{ijk} = f^{n}{}_{ijk} - (\Delta t\; div\, F^{n})_{ijk} + \left[\frac{\Delta t^2}{2}\, div\, (A^{n}\, div\, F)\right]_{ijk} \qquad (2)$$

where Δt denotes the time step and A denotes the jacobian matrix of the fluxes.
The space-discretization is carried out in two stages, following the idea of Ni [1].
i) The first order term $div\, F$, at a mesh point M, is obtained from an arithmetic mean of the values of $div\, F$ at the cell centers surrounding the point M. Each of these values is calculated by a contour integral around the corresponding cell.
ii) The second order term $div\, (A\; div\, F)$ is calculated at point M by an approximation of a contour integral around the staggered control volume containing the point M, by using the "distribution formulas" introduced by Ni [1]. The use of this formulas allows to get values at the boundary points without introducing any extrapolation formula.

ARTIFICIAL VISCOSITY

Due to the non dissipative property of the scheme in the sense of Kreiss, a fourth order linear dissipation D_4 is added. A second order non linear dissipation D_2 is also added in order to capture the flow discontinuities correctly. This treatment is analogous to that proposed by Jameson et al. [4] with a boundary numerical treatment introduced by Eriksson [5], and its implementation is done as follows (see [3] for more details) :

$$f^{n+1,ED}{}_{ijk} = f^{n+1,E}{}_{ijk} + \left(\frac{\Delta t}{V}\right)_{ijk} [\, D_2(\varepsilon_2, f^n) + D_4(\varepsilon_4, f^n)\,]_{ijk} \tag{3}$$

V being the volume of the staggered cell. The coefficients ε_2 and ε_4 depend on the local geometry and on a sensor evaluating second differences of the local aerodynamic field. Because it is important to detect contact discontinuities, a combination of pressure and velocity differences is used.

IMPLICIT STAGE

The implicit stage [2] preserves the space centered approach, the conservative property, the second order accuracy and the dissipative (or non-dissipative) aspect of the explicit stage. This implicit stage consists in solving the following factorised operator, with the following alternating direction form (ADI technique) :

$$[\, \Pi_{l=i,j,k}\, IS_l\,]\, R_{ijk} = R^*{}_{ijk} \tag{4}$$

where R^* and the operator IS_l are defined by :

$$R^*{}_{ijk} = f^{n+1,ED}{}_{ijk} - f^n{}_{ijk} \tag{5}$$

$$IS_l \;=\; [\, 1 + \frac{\beta_l}{2}\left(\frac{\Delta t}{V}\right)_{ijk} \delta_l \left(\frac{\Delta t}{V}\rho^2(A\,.\,n_l)\,\delta_l\right)] \tag{6}$$

The space operator δ is defined by $\delta_i\,\phi = \phi_{i+1/2} - \phi_{i-1/2}$, and $\rho\,(A\,.\,n_l)$ is the spectral radius of the matrix $A^n\,.\,n_l$, n_l being the surface vector of the staggered cell in the l direction.

The parameters β_l are chosen in order to ensure the stability. From a 3-D linear stability analysis of a transport equation (scalar case) and of the Euler system of equations, assuming a cartesian and uniform grid, one can determinate their values ensuring the unconditionnal stability of the method [6].

The new value $f^{n+1,I}$ is obtained from the implicit stage as follows:

$$f^{n+1,I}{}_{ijk} = f^n{}_{ijk} + R_{ijk} \tag{7}$$

We use fictitious points to apply the numerical boundary condition (of Neumann type or of Dirichlet type) on the residual R in the implicit stage.

PHYSICAL BOUNDARY CONDITIONS AND SUBDOMAIN COUPLING

The subdomain coupling is based on the use of an arithmetic mean between the values of each domain (different before the coupling), these values being defined at mesh points or being interpolated. It is important to remark that, for subdomains with coincident points, by using "distribution formulas" and arithmetic mean treatment at matching boundary points, one obtain the same space discretisation for inner and matching points. The numerical boundary conditions and the subdomain coupling are applied before the implicit stage, to maintain the conservative aspect of the scheme, and after the implicit stage.

The treatment of the boundary conditions is based on the use of the characteristic relations as proposed by Viviand et al. [7]. For each point of the mesh, we get the final value of f at time t^{n+1} by solving the following system of equations :

$$B \,.[f^{n+1}{}_{ijk} - f^{n+1,J}{}_{ijk}\,] = 0 \qquad (8)$$

where B denotes the matrix the column of which are the left eigenvectors of the jacobian matrix associated with the direction normal to the boundary. Each relation corresponding to information coming outside from the computationnal domain is discarded and replaced by boundary conditions.

FLOW AROUND A SUPERSONIC TRANSPORT AIRCRAFT

We present here calculations performed with this method for subsonic, transonic and supersonic flows around a supersonic transport aircraft. The computational domain is restricted to the half-aircraft configuration and is divided in five sub-domains as shown in figures 1 and 2. The five different domains contain respectively (along i,j and k directions) 161x10x21 points, 161x9x13 points, 161x64x25 points, 161x11x25 points and 161x73x25 points, which makes a total of 648347 points. The mesh on the aircraft surface is presented on figure 3. This mesh exhibits two singularities before and behind the aircraft body.

Each calculation is initialised by a uniform flow corresponding to the upstream conditions. The three calculations have been performed with a CFL number equal to 4, and the convergence is reached when three order decrease of the residual (corresponding to a stagnation of the different global value levels) is obtained. Figure 4 (resp. figure 5 and figure 6) present the isobaric lines on the upper aircraft surface for the transonic flow (resp. for the supersonic flow and for the subsonic flow). For the subsonic case streamlines starting from the leading edge are plotted in figure 7.

The upstream conditions for the transonic flow are : $M_\infty = 0.95$ and $\alpha_\infty = 4°$. The convergence is obtained after 2000 iterations. One can observe that the numerical solution of the flow, and in particular the shock wave, is not disturbed across the boundaries between subdomains.

The upstream conditions for the supersonic flow are : $M_\infty = 2$ and $\alpha_\infty = 4°$. The convergence is obtained after 750 iterations. As expected for this supersonic case, one see the trace on the fuselage of the trailing edge upper surface shock.

The upstream conditions for the subsonic flow are : $M_\infty = 0.25$ and $\alpha_\infty = 16°$. The convergence is obtained after 3000 iterations. This flow is strongly vortical, as it is observed from the aspect of the streamlines. One can notice the low pressure region on the upper surface of the wing, corresponding to the trace of the vortex.

REFERENCES

[1] Ni R.H. , A Multiple-Grid Scheme for Solving the Euler Equations, AIAA J., vol. 20, n° 11, 1982.

[2] Lerat A., Sidès J. and Daru V., An Implicit Finite-Volume Method for Solving the Euler Equations, Lecture Notes in Physics, 170, pp. 343-349, 1982.

[3] Couaillier V., Multigrid Methods for Solving Euler and Navier-Stokes Equations in two and three Dimensions, Proceedings of the eigth GAMM-Conference on Numerical Methods in Fluid Dynamics, Notes on Numerical Fluid Mechanics, Vol. 29. Edited by Pieter Wesseling, Vieweg, Braunschweig, 1990.

[4] Jameson A. and Schmidt W., Some Recent Developments in Numerical Methods for Transonic Flows, Computer Methods in Applied Mechanics and Engineering 51, pp. 467-493, North-Holland, 1985.

[5] Eriksson L.E., Boundary Conditions for Artificial Dissipation Operators, FFA TN 1984-53, 1984.

[6] Liamis N., Méthodes implicites de résolution des équations d' Euler pour des écoulements tridimensionneles instationnaires dans des turbomachines, Thèse de Doctorat, Université Paris VI, to be published.

[7] Viviand H. and Veuillot J.P., Méthodes pseudo-instationnaires pour le calcul d'écoulements transsoniques, Publication ONERA n° 1978-4, 1978.

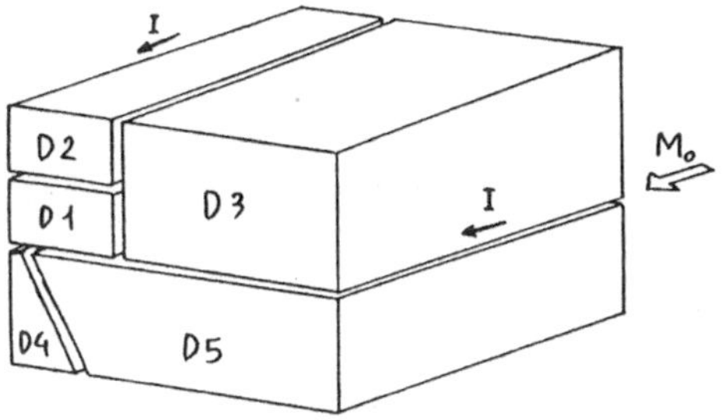

Figure 1 : Domain Decomposition

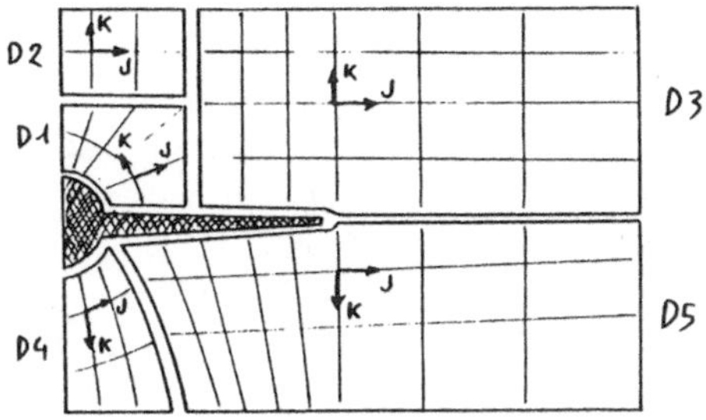

Figure 2 : Mesh Decomposition

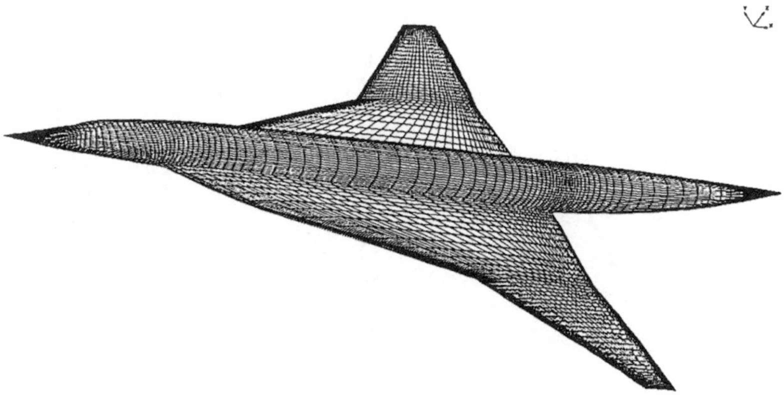

Figure 3 : Mesh on the body

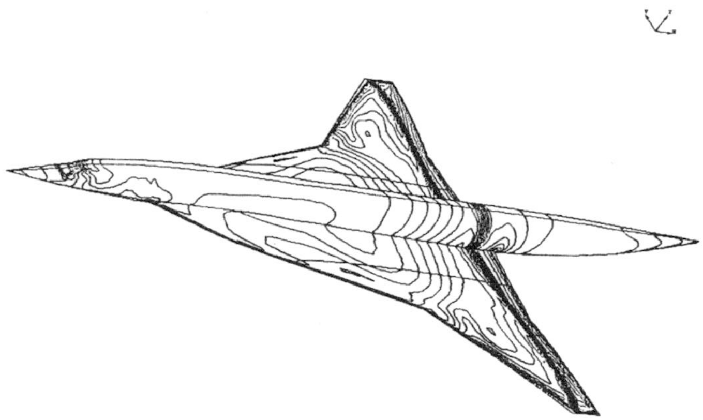

Figure 4 : $M_\infty = 0.95$ $\alpha = 4°$ - Isobaric lines

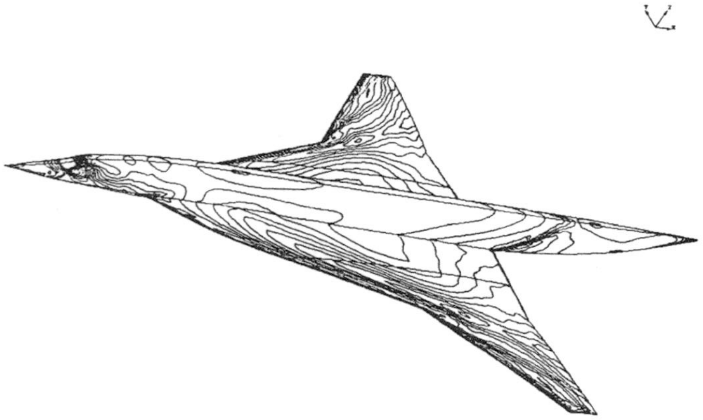

Figure 5 : $M_\infty = 2.0$ $\alpha = 4°$ - Isobaric lines

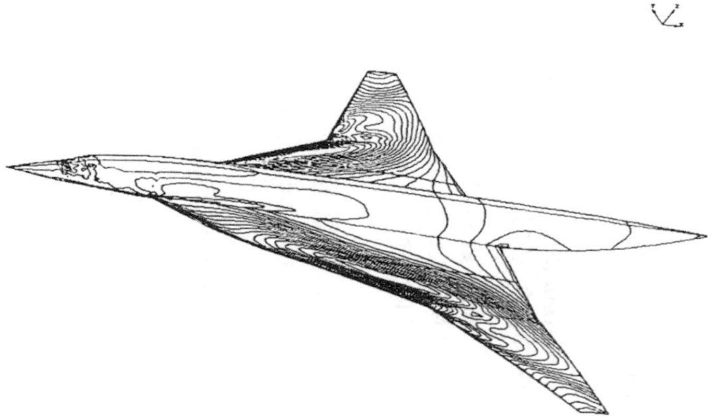

Figure 6 : $M_\infty = 0.25$ $\alpha = 16°$ - Isobaric lines

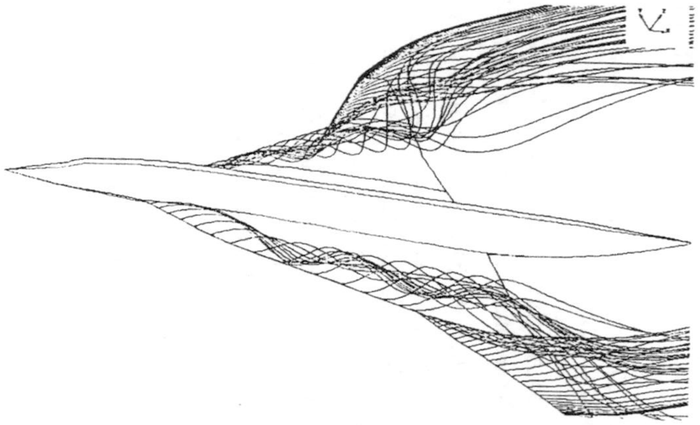

Figure 7 : $M_\infty = 0.25$ $\alpha = 16°$ - Sreamlines

SIMULATION OF 2D EXTERNAL INCOMPRESSIBLE VISCOUS FLOWS BY MEANS OF A DOMAIN DECOMPOSITION METHOD

J.-L. Guermond, S. Huberson, W.-Z. Shen

LIMSI-CNRS, BP n° 133, 91403 ORSAY Cedex.

1. INTRODUCTION – Visualization of incompressible viscous flows shows that vorticity concentrates in wakes and advection dominates viscous diffusion as the Reynolds number increases. These conditions are favourable for simulating such flows by means of particle methods. Furthermore since this class of methods is grid-free, it is suitable for tackling problems with moving boundaries. However, particle methods are inaccurate as viscous effects are of the same order as that of advection. Hence, in boundary layers, methods which are adapted to parabolic problems are needed (*eg.* finite differences, finite elements, *etc.*). The remarks above led us to develop a domain decomposition method that combines advantages of both approaches (*cf.* [3] [5]).

2. FORMULATION OF THE PROBLEM – Consider p moving solids $(S_i)_{i=1,\dots,p}$ in a Galilean frame of reference $(O, \boldsymbol{i}, \boldsymbol{j})$ of $\mathbb{R}^2$. Let $\boldsymbol{k} = \boldsymbol{i} \times \boldsymbol{j}$, define $(\boldsymbol{O}_i)_{i=1,\dots,p}$ origins of reference for each solid, and let $\boldsymbol{v}_i$ (resp. Ω_i) be the velocity of $\boldsymbol{O}_i$ (resp. the angular velocity of S_i). The solids are immersed in an incompressible Newtonian fluid which is at rest at infinity. The fluid domain, denoted by $\mathcal{D}$, is decomposed into $p+1$ open subdomains so that $\mathcal{D} = \mathcal{D}_0 \cup_{i=1,\dots,p} \overline{\mathcal{D}_i}$, where the subdomains $\mathcal{D}_i$ are homeomorphic to a ring. It is hereafter assumed that the domain decomposition has been done so that convective effects are dominant in $\mathcal{D}_0$. Let B_i (resp. Γ_i) be the interface between $\mathcal{D}_i$ and S_i (resp. $\mathcal{D}_0$), and $\boldsymbol{n}_i$ be the outward normal to the boundary of $\mathcal{D}_i$ for $i = 0, \dots, p$. In $\mathcal{D}_0$ the Navier–Stokes equations are formulated in terms of velocity and vorticity $(\boldsymbol{u}_0, \omega_0)$ and are approximated by means of a particle method, whereas in each subdomain $\mathcal{D}_i$ they are formulated in terms of stream function and vorticity (ψ_i, ω_i) and are approximated by means of finite differences. Let $T > 0$ and $N \in \mathbb{N}$, approximations of $(\boldsymbol{u}_0, \omega_0)$ and (ψ_i, ω_i) are sought in parallel in the time interval (t_k, t_{k+1}) where $\delta t = T/N$ and $t_k = k\delta t$ for $0 \leq k \leq N$

3. SOLUTION IN $\mathcal{D}_0$ – Let $\boldsymbol{v}_\infty(t)$ the mean velocity of the p solids. The fluid motion is studied in a frame of reference which moves with velocity $\boldsymbol{v}_\infty(t)$. In $\mathcal{D}_0$ the advection–diffusion equation of ω_0 is approximated by:

$$\partial\omega_0^{k+1}/\partial t + \nabla.(\omega_0^{k+1}\boldsymbol{u}_0^k) = \nu\nabla^2\omega_0^{k+1}, \tag{3.1}$$

$$\boldsymbol{u}_0^k = -\boldsymbol{v}_\infty + \int_{\mathcal{D}} \omega^k \nabla G \times \boldsymbol{k} \mathrm{d}v + \sum_{j=1}^{p} \int_{B_j} [(\boldsymbol{n}_j \times \boldsymbol{v}_{ej}) \times \nabla G + (\boldsymbol{n}_j.\boldsymbol{v}_{ej})\nabla G] \mathrm{d}l, \tag{3.2}$$

G is the Green function of the Laplace operator in $\mathbb{R}^2$, and $\boldsymbol{v}_{ej} = \boldsymbol{v}_j + \Omega_j \times (\boldsymbol{y} - \boldsymbol{O}_j)$.

Wellposedness of the problem requires that some transmission condition through Γ_i is imposed on ω_i. Such a condition is obtained by taking into account the fact

Figure 1: Streaklines about two cylinders, $g^* = 2.5$.

that in the vicinity of Γ_i viscous diffusion is dominated by advection. Hence, (3.1) can locally be considered as a hyperbolic equation whose right hand side, $\nu\nabla^2\omega_0^{k+1}$, can be explicited and considered as a source term. For this kind of problem, Dirichlet conditions are imposed on the subset of Γ_i where the flow enters $\mathcal{D}_0$ (*eg.* see [2], [4]):

$$j = 1, \ldots, p, \qquad \omega_0^{k+1}(\boldsymbol{x}) = \omega_j^k(\boldsymbol{x}), \text{ if } \boldsymbol{u}_0^k(\boldsymbol{x}).\boldsymbol{n}_0(\boldsymbol{x}) < 0 \tag{3.3}$$

Problem (3.1) (3.2) as presented above is approximated be means of a particle method that take into account Dirichlet data (3.3) (see [3] for details on this technique).

4. SOLUTION IN $\mathcal{D}_i$ – For each subdomain $\mathcal{D}_i$, the fluid motion is studied in a non-inertial frame of reference that is linked to S_i, and the Navier–Stokes equations are formulated in terms of stream function of the relative velocity and the vorticity of the absolute velocity. Hence, the PDE's to be solved are:

$$\partial\omega_i^{k+1}/\partial t + \nabla.(\omega_i^{k+1}\nabla \times (\psi_i^{k+1}\boldsymbol{k})) = \nu\nabla^2\omega_i^{k+1} \tag{4.1}$$

$$\nabla^2\psi_i^{k+1} = 2\Omega_i - \omega_i^{k+1} \tag{4.2}$$

The system above is complemented by the following boundary conditions on B_i:

$$\psi_i^{k+1} = \psi_{B_i}^{k+1}, \quad \frac{\partial\psi_i^{k+1}}{\partial n} = 0, \text{ and } \quad \int_{B_i} \frac{\partial\omega_i^{k+1}}{\partial n_i} \mathrm{d}l = \frac{\dot{\Omega}_i}{\nu}\int_{B_i} [\boldsymbol{k} \times (\boldsymbol{y} - \boldsymbol{O}_i)].\mathrm{d}\boldsymbol{l} \tag{4.3}$$

Furthermore, transmission conditions need to be enforced so that continuity of the physical variables along with their flux is ensured across Γ_i. The conditions in question depend on the local nature of the PDE's to which ψ_i and ω_i are solutions. Since ψ_i is solution to an elliptic problem (4.2) whatever the flow nature, a transmission condition on ψ_i must be enforced everywhere on Γ_i. Such a condition is provided by a Green identity based on (4.2) and (4.3):

$$\psi_i^{k+1} = -\psi_{ei} - \int_D \omega^k G \mathrm{d}v + \sum_{j=1}^{p} \int_{B_j} [(\psi_{ej} + \psi_{B_j}^k)\frac{\partial G}{\partial n_j} - G\frac{\partial\psi_{ej}}{\partial n_j}]\mathrm{d}l, \tag{4.4}$$

$$\text{where,} \qquad \psi_{ej}(\boldsymbol{x}) = \boldsymbol{v}_j.[(\boldsymbol{x} - \boldsymbol{O}_j) \times \boldsymbol{k}] - \Omega_j|\boldsymbol{x} - \boldsymbol{O}_j|^2/2. \tag{4.5}$$

Note that (4.4) is global, *i.e.* it transmits the whole spectrum of information to each subdomain at once, whereas classical Dirichlet-Neumann coupling conditions (*eg.* see

Figure 2: Streaklines about two cylinders, $g^* = 1$.

[2] [4]) poorly transmit low frequencies. By using the same arguments as that of §3, transmission of information on ω_i is achieved by:

$$\omega_i^{k+1}(\boldsymbol{x}) = \omega_0^k(\boldsymbol{x}), \text{ if } \boldsymbol{u}_i^k(\boldsymbol{x}).\boldsymbol{n}_i(\boldsymbol{x}) < 0. \tag{4.6}$$

As far as information transfert is concerned, this condition is sufficient. Nevertheless, since (4.1) is approximated by means of a centered finite differences scheme, a boundary condition for ω_i^{k+1} on Γ_i is required. Since the flow regime is almost hyperbolic, the piece of information that is missing on the subset of Γ_i where the flow goes out is obtained by doing an approximate Lagrangian integration of (4.1):

$$\omega_i^{k+1}(\boldsymbol{x}) = \omega_i^k(\boldsymbol{x} - \boldsymbol{u}_i^k\delta t) + \nu\delta t\nabla^2\omega_i^k(\boldsymbol{x} - \boldsymbol{u}_i^k\delta t), \text{ if } \boldsymbol{u}_i^k(\boldsymbol{x}).\boldsymbol{n}_i(\boldsymbol{x}) \geq 0. \tag{4.7}$$

The (ψ_i, ω_i) problem as formulated above is linearized and solved by means of a finite differences method that has been developed in [1].

5. NUMERICAL EXAMPLES – The present method has been coded and tested; comparisons with experimental data have shown reasonable agreement (see [3] for details on tests). In the three examples shown below we try to emphasize the versatility of the present approach and give some flavor of its possibilities.

The first example concerns the flow about two interacting cylinders. In figures 1 and 2 are shown the streaklines about two impulsively started cylinders at times $t = 80$ (fig. 1) and $t = 70$ (fig. 2). The Reynolds number based on the diameter of the cylinders is equal to 110. In each case the finite differences domains are composed of rings the width of which are set to one cylinder radius. The dimensionless gap g^* between the cylinders (i.e. gap/diameter) is equal to 2.5 and 1 in case 1 and case 2 respectively. It is clearly shown here that the stable flow regime consists of two out of phase Karman streets. The phase between the two vortex sheddings depends on the dimensionless gap g^*. A stability analysis of the wake interactions by means of the present method is under way. Since the finite differences subdomains are disconnected, it is possible, at the same numerical cost, to let the cylinders oscillate.

In figures 3, 4, and 5 we present numerical simulations of the flow about tandem airfoils. The leading airfoil oscillates in pitch and the rear one is fixed. This configuration may be viewed as a model for the rotor/stator interaction in turbines and rotating machines. The fluid domain is decomposed into three subdomains as shown

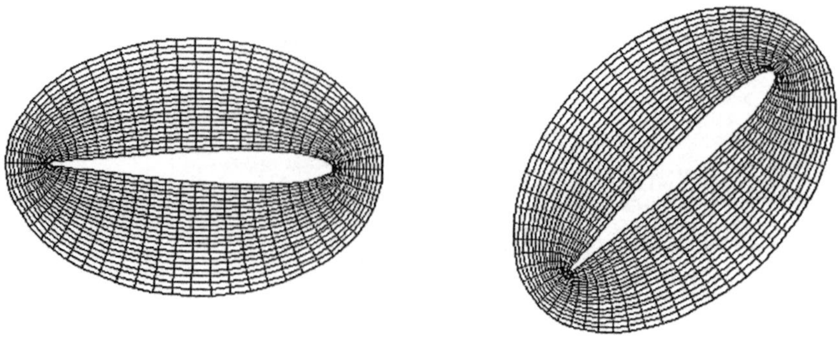

Figure 3: Definition of the finite differences subdomains for the tandem airfoils.

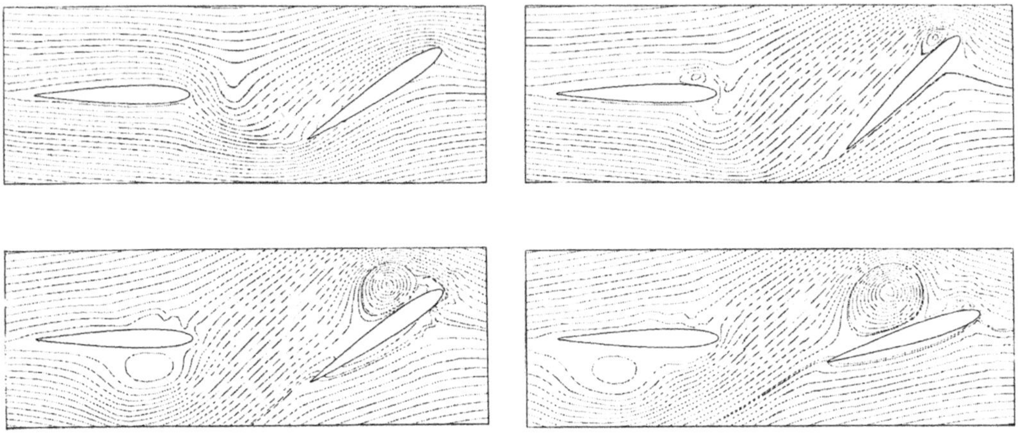

Figure 4: Streamline patterns about impulsively started tandem airfoils.

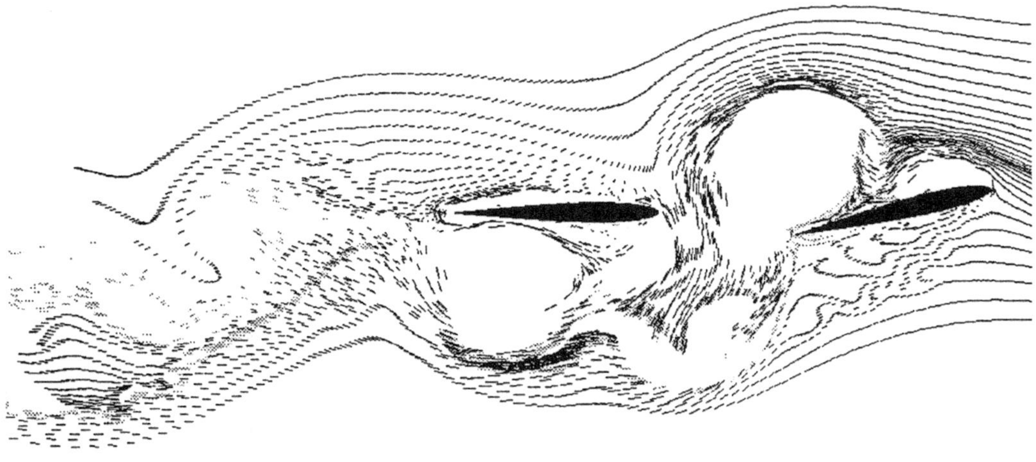

Figure 5: Instantaneous velocity field about the tamdem airfoil at time $t = 10$. Notice the large eddies shed by the leading airfoil.

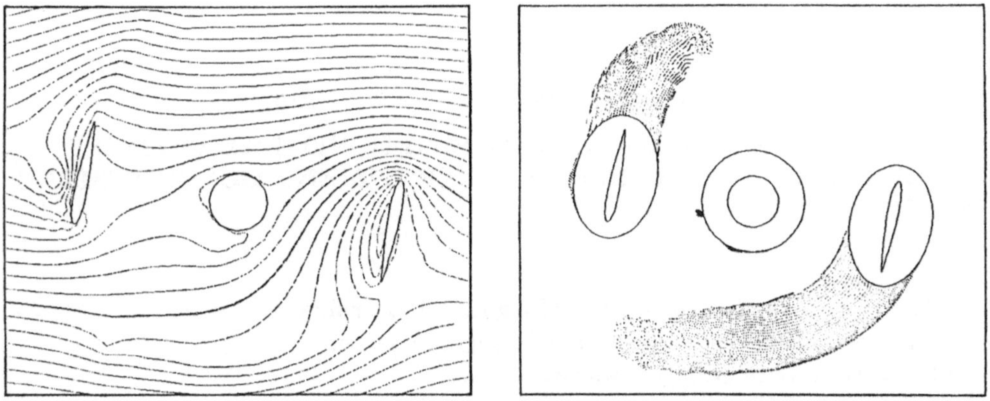

Figure 6: Shed particles and streamline patterns about a Darrieus wind turbine at $t = 2$.

in figure 3. Each airfoil is embedded in a small finite differences subdomain. Since each subdomain moves with the airfoil it embeds, no regridding is required as time evolves.

In figure 4 we present streamline patterns about the impulsively started tandem airfoils at times $t = 1.5$, 2.5, 3.5 and 4.5. Shown here is the interaction between the rear airfoil and the starting vortex that has been shed by the leading airfoil. The Reynolds number $v_\infty C/2\nu$ is set to 3000, the reduced frequency of the oscillating airfoil $fC/2v_\infty$ is equal to 0.2 and $\alpha_{\max} = 45^o$.

In figure 5 is shown the instantaneous velocity field at time $t = 10$. One may verify in figures 4 and 5 that the velocity field is smooth across the interfaces of the subdomains. Note that the tandem airfoil problem or other problems of this kind would be difficult to treat by means of classical global approaches, since for these class of methods the flow domain would have to be either regridded or deformed at each time step.

The third examples concerns the simulation of a Darrieus wind turbine. In figure 6 is shown shed particles and streamline patterns about a Darrieus-like wind turbine at $t = 2.2$ after an impulsive start. There are four subdomains. The wind turbine rotates in the anti-clockwise direction and the fluid moves from right to left with velocity v_∞. denote by R be the windmill radius, $\Omega R/v_\infty = 2.16$ and $v_\infty 2r/\nu = 3000$.

References

[1] O. Daube, T. P. Loc, *Journal de Mécanique*, 17, 5, 1978, p. 651-678.

[2] P. Le Tallec, A direct introduction of some domain decomposition techniques for advection diffusion problems, *Ecole d'Eté d'Analyse Numérique CEA-EDF-INRIA*, Bréau-sans-Nappe, 1991.

[3] J.-L. Guermond, S. Huberson, W. Z. Shen, Simulation des écoulements visqueux externes bidimensionnels par une méthode de décomposition de domaines, *rapport interne LIMSI 91-9*, Août 1991.

[4] A. Quarteroni, Domain decomposition methods for numerical solution of partial differential equations, *University of Minnesota Supercomputer Institute research report UMSI 90/246*, 1990.

[5] Wen-Zhong Shen, Calcul d'écoulements tourbillonnaires visqueux incompressibles par une méthode de couplage, thèse de doctorat, Univ. Paris XI, (in press).

DOMAIN DECOMPOSITION COMBINED WITH ADAPTIVE REMESHING FOR PROBLEMS OF TRANSIENT COMPRESSIBLE FLOW

O. Hassan[1], E.J. Probert[2], K. Morgan[3], J. Peraire[4]

[1] CDR, Innovation Centre, University College, Swansea

[2] EBMS, University College, Swansea

[3] Department of Civil Engineering, University College, Swansea

[4] Department of Aeronautics, Imperial College, London

Introduction

In recent years there has been an upsurge of interest in the application of unstructured grid methods to the solution of problems involving high speed compressible flows. An attractive feature of unstructured grid methods is that they can be easily combined with adaptive mesh procedures so that the solution quality may be enhanced in an 'optimal' manner. In this paper an adaptive finite element method for the solution of transient compressible flow problems in two dimensions is described.

Mesh Generation and Adaptation

The computational domain is represented by an unstructured assembly of linear triangular elements. The approach advocated for the generation of such a mesh is a generalization of the advancing front technique with the distinctive feature that elements, i.e. triangles, and points are generated simultaneously[1]. This enables the generation of elements of variable size and stretching. The mesh adaptation is performed automatically during the computation[2,3] and is achieved by local regeneration of the grid, using an error estimation procedure. Error indicators are used to mark those regions where a finer/coarser mesh is required and those regions are then remeshed, to achieve the desired mesh spacing, using the general triangular mesh generator.

Solution Algorithm and Domain Decomposition

The two dimensional equations governing compressible inviscid flow are considered in the conservation form for the cases of both plane and axisymmetric formulations. A Taylor Galerkin explicit time stepping scheme, which has been described in detail previously[4], is employed for the solution algorithm. This is a relatively inefficient method for the solution of problems of transient compressible flow on meshes exhibiting large variations in elements sizes, because of the stability restrictions placed on the allowable timestep size. However the efficiency of the scheme can be improved by combining it with a domain decomposition method[3,5]. Instead of using a single timestep throughout the computational domain, the objective now is to group the

elements according to the maximum allowable timestep size and to advance the solution independently within each group. Time accuracy of the procedure is maintained by appropriate interchange of information across the boundaries of the groups. The implementation of the method proceeds algorithmically as follows:
1. before advancing the solution, compute the maximum eigenvalue λ_{max} of the system at each node.
2. compute the allowable stability timestep $\Delta t_e = h_e c_s / \lambda_{max}$ for each element where c_s is the safety factor and h_e is a representative element length. In the present implementation the element length is taken to be the minimum element height.
3. determine the minimum timestep Δt_{min} allowed on the mesh and sort the elements into m regions according to the ratio $\Delta t_e / \Delta t_{min}$ e.g.

region	**all elements e for which**
1	$\Delta t_{min} \leq \Delta t_e < 2\Delta t_{min}$
2	$2\Delta t_{min} \leq \Delta t_e < 4\Delta t_{min}$
3	$4\Delta t_{min} \leq \Delta t_e < 8\Delta t_{min}$

4. for each region, m, find the boundary nodes of the region. Overlap the regions by adding two layers of elements to each region m from the regions $p > m$.
5. advance the solution one global timestep.
The number of regions allowed should not exceed some specified maximum. However, this will not necessarily be the optimum number of regions. The domain decomposition technique can be expected to prove most useful when used in conjunction with adaptive refinement techniques, where large variations in the element sizes over the computational domain will certainly be encountered.

Numerical Examples

The performance which may be achieved by the adaptive remeshing approach for transient flows is illustrated by considering three numerical examples.

The first consists of a NACA 0012 airfoil pitching harmonically about the quarter chord at a reduced frequency based on semichord of k = 0.0814. The free stream Mach number is $M_\infty = 0.599$ and the angle of attack is $\alpha_0 = 4.86^o$. The calculations were performed for three cycles of motion to obtain a periodic solution. Results were obtained for the airfoil pitching with an amplitude of $\alpha_1 = 2.44^o$. The calculated results, shown in Figure 1, compare favourably with the experimental data [6].

The second example consists of an axisymmetric simulation of the blast wave produced by an exploding vessel. A sketch of the experimental apparatus for this problem is shown in Figure 2. The problem involves the simulation of the flow field around a vessel whose lid is initially at rest but which then moves upwards under the influence of the initial high pressure region inside the vessel. Initially the air inside the vessel was at a pressure of 40 times atmospheric, while the air in the chamber was at atmospheric pressure. The results of the calculation are displayed in Figure 2. During the initial phase of the transient the predictions are seen to be in good agreement with the experiment[7], but the magnitude of the peak at two transducers is underpredicted by the numerical method.

The final problem considers the rupture of a circular pipe which is passed through a concentric cylindrical chamber. A sketch of the experimental apparatus for this problem is shown in Figure 3. Initially the air inside the pipe is at a density of 110 kg/m^3 (based on a pressure of 1420 psi). At this pressure the pipe fails and the two halves of the pipe then move apart. The variation of the velocity of each pipe half was computed using the difference in pressure between the pipe and the external chamber. The computed pressures, shown in Figure 3, compare favourably with experiment[8].

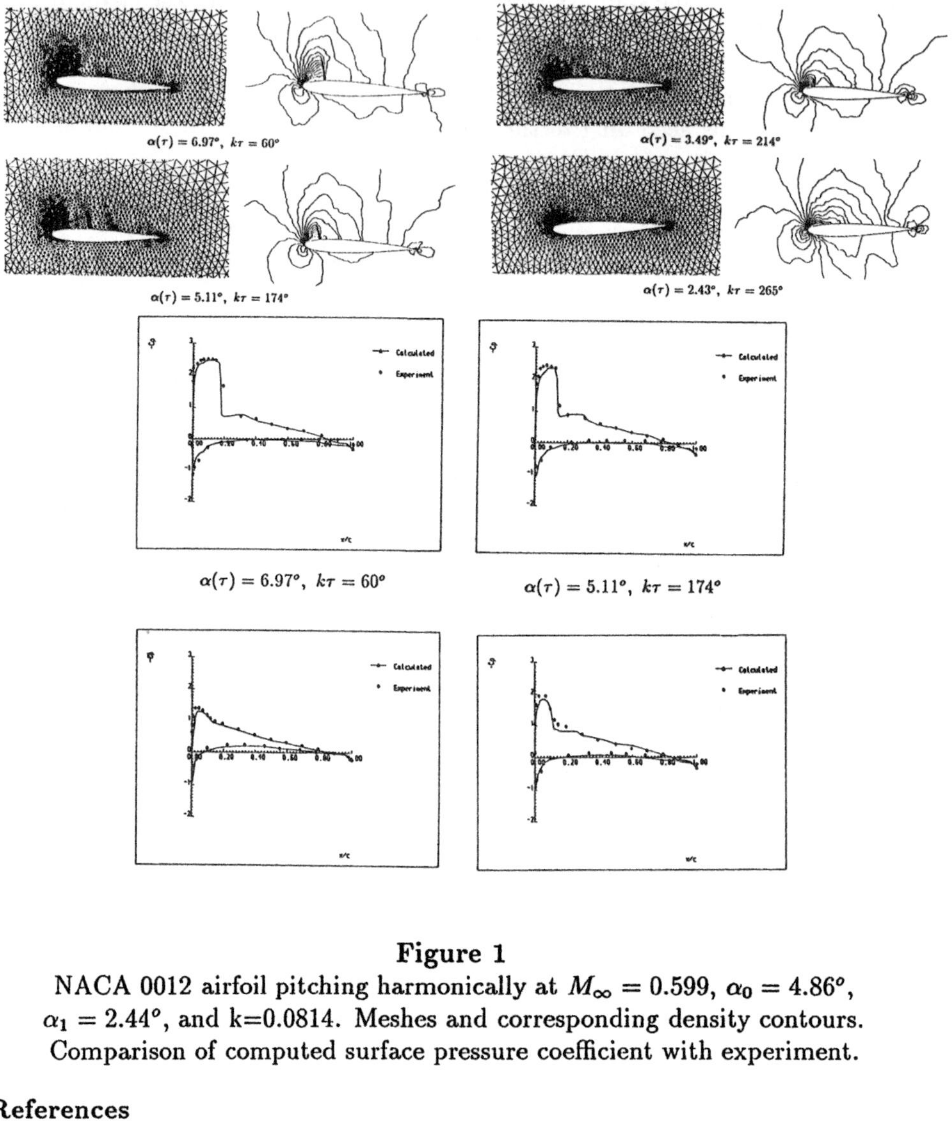

Figure 1

NACA 0012 airfoil pitching harmonically at $M_\infty = 0.599$, $\alpha_0 = 4.86°$, $\alpha_1 = 2.44°$, and k=0.0814. Meshes and corresponding density contours. Comparison of computed surface pressure coefficient with experiment.

References

[1] J. Peraire, M. Vahdati, K. Morgan and O.C. Zienkiewicz, 'An adaptive mesh regeneration procedure for compressible flows', J. Comp. Phys. 72, 449-466, 1987.

[2] E.J. Probert, O. Hassan, J. Peraire and K. Morgan, 'An adaptive finite element method for transient compressible flows with moving boundaries', Int. J. Num. Meth. Engng., 32, 751-765, 1991.

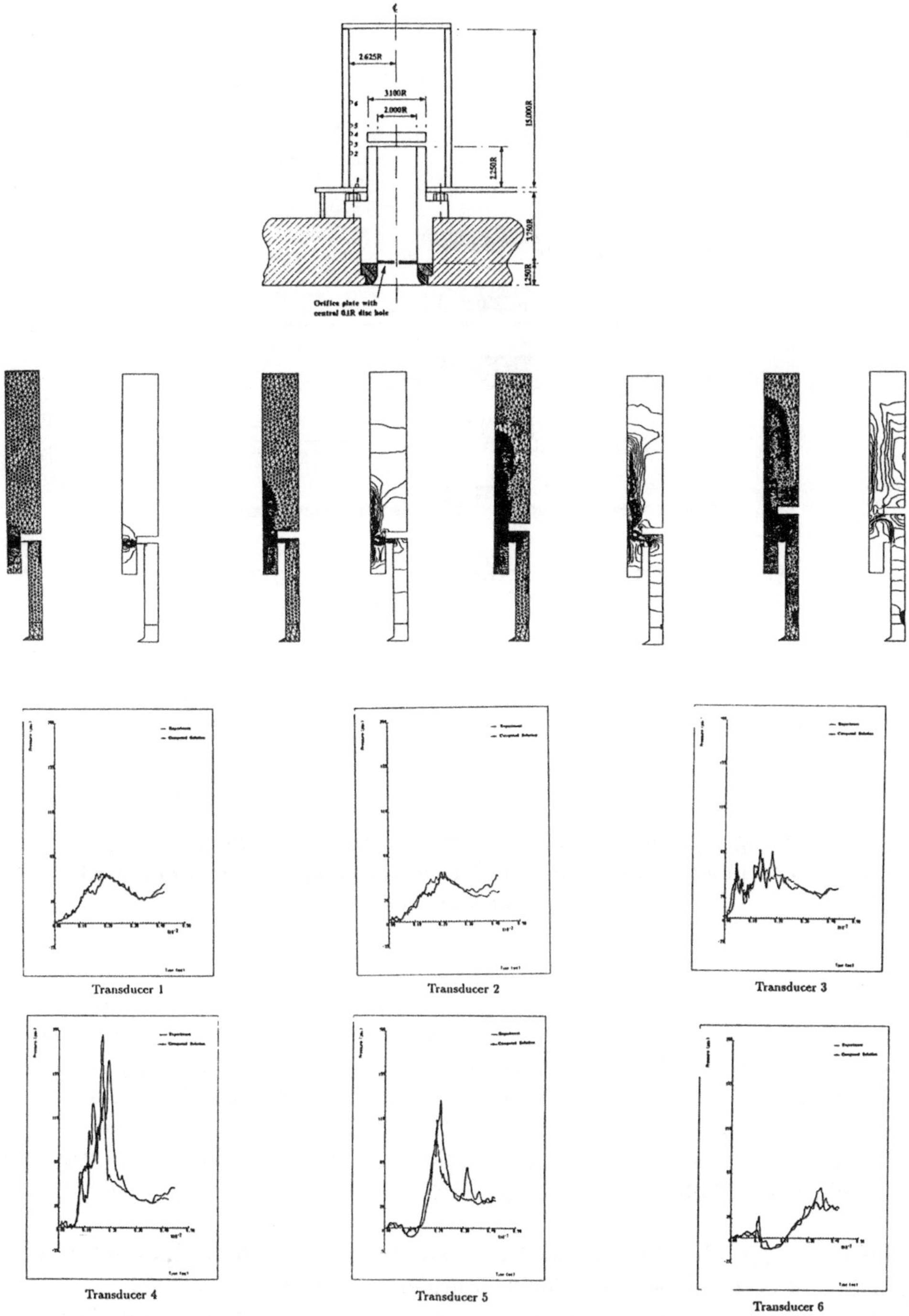

Figure 2
Exploding pressure vessel simulation. Experimental apparatus. Sequence of meshes and Mach number contours during the computation. Computed pressures at transducers.

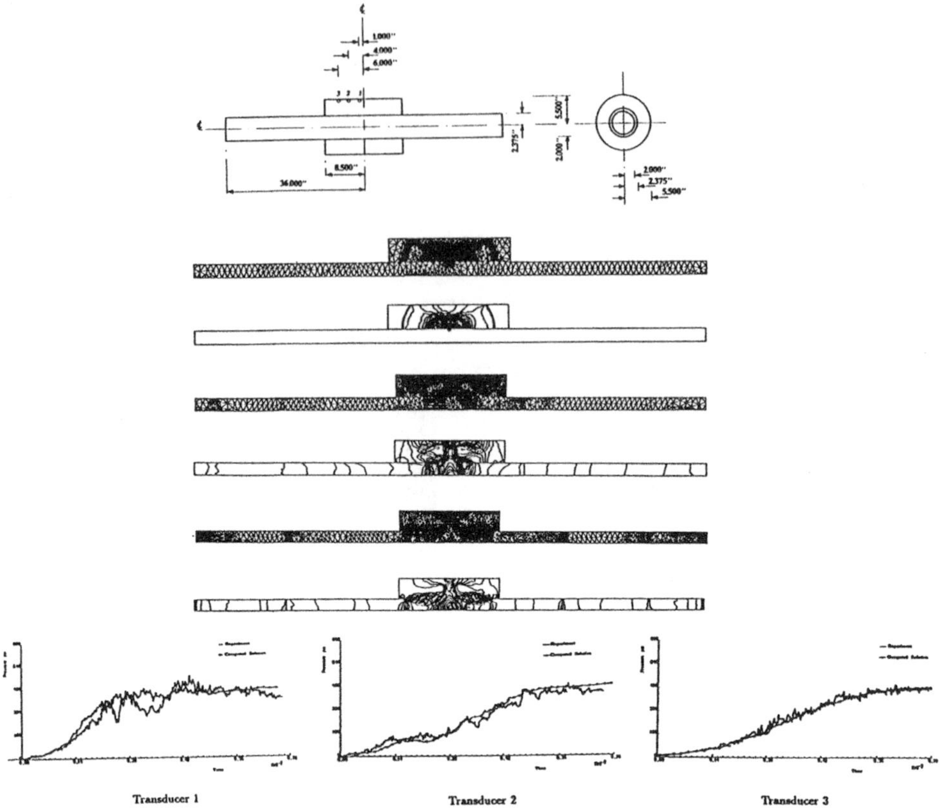

Figure 3
Rupturing pipe simulation. Experimental apparatus. Sequence of meshes and Mach number contours during the computation. Computed pressures at transducers.

[3] O. Hassan, E.J. Probert, K. Morgan and J. Peraire, 'Adaptive finite element methods for transient compressible flow problems', chapter to appear in Adaptive Meshing and Error Estimates, Computational Mechanics Publications, 1992.
[4] J. Peraire, L. Formaggia, J. Peiro, K. Morgan and O.C. Zienkiewicz, 'Finite element Euler computations in 3D', AIAA Paper 88-0032, 1988.
[5] R. Löhner, K. Morgan and O.C. Zienkiewicz, 'The use of domain splitting with an explicit hyperbolic solver', Comp. Meth. Appl. Mech. Engng. 45, 313-329, 1984.
[6] R.H. Landon, 'NACA 0012 oscillating and transient pitching, Data set 3', Compendium of unsteady aerodynamic measurment, AGARD-R-702, 1982.
[7] M.R. Baum, 'Pressure vessel rupture within a chamber: the pressure history on the chamber wall', Experimental and Fluid Science, 1992 (to appear).
[8] A.G. Hutton, Private communication, 1991.

A NUMERICAL PROCEDURE TO SOLVE VISCOUS FLOW AROUND ARBITRARILY MOVING BODIES

W.Jia and Y.Nakamura

Dept. of Aeronautical Engineering, Nagoya University, Nagoya, 464-01, Japan

1. INTRODUCTION

A numerical treatment of the flow around arbitrarily moving bodies is of great importance. As long as a single body is considered, we can adopt a moving grid with the body. However, this approach can hardly be applied to the general case where several bodies move arbitrarily. This paper presents a numerical procedure to solve the general case by introducing an unsteady multi-domain technique.

2. GOVERNING EQUATIONS AND COORDINATES

The incompressible N-S equations are treated by use of three kinds of coordinates, i.e., (x_i, t): the static Cartesian coordinates; (ξ_i, τ): the generalized coordinates moving with the body surface; $(\tilde{x}_i, \tau)$: the Cartesian coordinates moving with the body as long as a rigid body motion is considered. The relationship between the three coordinates is represented as

$$(x_i, t) \Longrightarrow (\tilde{x}_i, \tau) \Longrightarrow (\xi_i, \tau) \tag{1}$$

$$t = \tau \quad , \quad x_i = x_i\,(\xi_k, \tau) \quad , \quad \xi_i = \xi_i\,(\tilde{x}_k) \tag{2}$$

The N-S equations are written in the unsteady generalized coordinate as follows:

$$\frac{\partial}{\partial \xi_i}(J^{-1}U_i) = 0 \tag{3}$$

$$\frac{\partial}{\partial \tau}\left(J^{-1}u_k\right) + \frac{\partial}{\partial \xi_i}\left[J^{-1}\left(U_i\, u_k + \frac{\partial \xi_i}{\partial x_k}p\right)\right] = \frac{1}{Re}\frac{\partial}{\partial \xi_i}\left(J^{-1}g_{ij}\frac{\partial u_k}{\partial \xi_j}\right) \tag{4}$$

$$U_i = \frac{\partial \xi_i}{\partial t} + u_j\,\frac{\partial \xi_i}{\partial x_j} \quad , \quad J = \frac{\partial(\xi_1, \xi_2)}{\partial(x_1, x_2)} \quad , \quad g_{ij} = \frac{\partial \xi_i}{\partial x_k}\frac{\partial \xi_j}{\partial x_k} \tag{5}$$

and for a rigid body motion:

$$U_i = \frac{\partial \xi_i}{\partial \tilde{x}_k}\left(\frac{\partial \tilde{x}_k}{\partial t} + u_j\frac{\partial \tilde{x}_k}{\partial x_j}\right) \ , \ J = \frac{\partial(\xi_1, \xi_2)}{\partial(\tilde{x}_1, \tilde{x}_2)} \ , \ g_{ij} = \frac{\partial \xi_i}{\partial \tilde{x}_k}\frac{\partial \xi_j}{\partial \tilde{x}_k} \tag{6}$$

where u_k is the velocity component in the static Cartesian coordinates and U_i the contravariant velocity component in the unsteady generalized coordinates. J is the Jacobian and g_{ij} are the transformation metric tensor.

The Boundary Conditions (BC) on the body surface are given as follows:

$$u_i = u_i(t) \ , \ U_i = 0 \tag{7}$$

$$\frac{\partial p}{\partial \eta} = \xi_y(J^{-1}u)_\tau - \xi_x(J^{-1}v)_\tau \tag{8}$$

where ξ, η denote the tangential and normal directions to the surface.

3. NUMERICAL SCHEME

The governing equations are spatially discretized by the finite volume method in the unsteady generalized coordinates, where the convective terms are approximated by the Generalized QUICK method[1], [2] and the others by the second order approximation. The time integration is performed by a two step method[3], where the Adams-Bashforth method is applied to the convective terms and the Crank-Nicolson method to the viscous terms. The pressure Poisson equation with Neumann BC is solved between the two sub-steps at every time step. To obtain a fully converged solution for this problem, a new method is applied[3]. Let the original Poisson equation with Neumann BC be

$$\nabla^2 p = D \quad , \quad \left.\frac{\partial p}{\partial n}\right|_{BC} = S \tag{9}$$

which is divided into a Poisson equation with Dirichlet BC and a Laplace equation with Neumann BC.

$$\nabla^2 p_1 = D \quad , \quad p_1|_{BC} = 0 \tag{10}$$

$$\nabla^2 p_2 = 0 \quad , \quad \left.\frac{\partial p_2}{\partial n}\right|_{BC} = S - \left.\frac{\partial p_1}{\partial n}\right|_{BC} \tag{11}$$

$$p = p_1 + p_2 \tag{12}$$

These equations are solved through three steps. The first step solves the Poisson equation under Dirichlet BC by the checkerboard SOR method. The second step calculates the boundary values of the Laplace equation with Neumann BC by the boundary element method. The third step solves the Laplace equation with known boundary values again by the checkerboard SOR method. This method produces a fully converged solution with a fairly rapid convergence rate than the SOR method applied to the original Poisson equation with Neumann BC.

4. MULTI-DOMAIN TECHNIQUE

The static main grid is employed to cover the whole computational region, where the sub-grid system is attached to each moving body. The sub-grids overlap the main-grid and, possibly, each other. The governing equations are separately solved on each grid. To make the procedure global, a mask function δ is introduced instead of dividing the main grid into several sub-regions.

$$u_{\text{new}} = (1-\delta)u + \delta\, u_{\text{replace}} \quad , \quad \delta = \begin{cases} 0 & \text{no replacement} \\ 1 & \text{replaced} \end{cases} \tag{13}$$

The computation is therefore performed for all interior grid points including the connecting boundary.

The data transfer between each grid is performed by interpolating values at the boundaries (Fig. 1) at every time step. To make the interpolation robust and insensitive to the mesh quality, the triangular linear function is employed.

$$u = \phi_1 u_1 + \phi_2 u_2 + \phi_3 u_3 \tag{14}$$

where ϕ_1, ϕ_2, ϕ_3 are the aera coordinates of the point • in the triangle 1-2-3.

The three vertices of a triangle are automatically searched so as to satisfy the following condition:

$$|\phi_1| + |\phi_2| + |\phi_3| = 1 \tag{15}$$

To save the CPU time, the searching area is localized, where use is made of the information on the locations of interpolated points in the previous time step.

4. RESULTS

The procedure is fully vectorized. First, it is applied to solve the flow around an oscillating cylinder at $Re = 314$, the center of which moves as $x_0 = -0.5\cos 2\pi t$, $y_0 = 0$. For this case, there exists a pressure distribution even at initial due to body acceleration (Fig.2). Figure 3 shows velocity vectors during one cycle. A comparison of pressure contours between multi-domain and single-domain solutions is made in Fig.4 which verifies the accuracy. Figure 5 shows another application to the flow around two closely separated cylinders oscillating in the opposite directions as $x_1 = -0.5\cos 2\pi t$, $y_1 = -0.8$, and $x_2 = 0.5\cos 2\pi t$, $y_2 = 0.8$ at $Re = 314$. This case is an example of complicated overlapping. Good agreement of physical quantities is obtained for each grid in the overlapped region. The time variations of aerodynamic coefficients are plotted in Fig.6 and compared with the potential flow solution. For one cylinder case, the results of multi-domain and single-domain solutions fall on the same line. It is seen that the phase of the viscous flow solution is slightly delayed and the amplitude is greater than the potential flow solution. Finally, an application to a three body problem ($Re = 1000$) is shown in Fig.7, which proves the versatility of the procedure.

5. CONCLUDING REMARKS

A general procedure of treating incompressible viscous flow around several bodies with each arbitrary motion is developed and verified by complicated overlapping problems.

REFERENCES

1. B.P.Leonard: A Stable and Accurate Convective Modeling Procedure Based on Quadratic Upstream Interpolation, Computer Methods in Applied Mech. and Eng., 19, 1979, pp. 59-98.
2. Y.Nakamura and Y.Takemoto: Solution of Incompressible Flows Using a Generalized Quick Method, Proc. 1st ISCFD, Tokyo, edited by K.Oshima, 2, 1986, pp. 285-296.
3. W.Jia, Y.Nakamura and M.Yasuhara: An Efficient Pressure Poisson Solver for Incompressible Flows, Proc. 4th ISCFD, Davis, edited by H.A.Dwyer, 1, 1991, pp. 539-544.

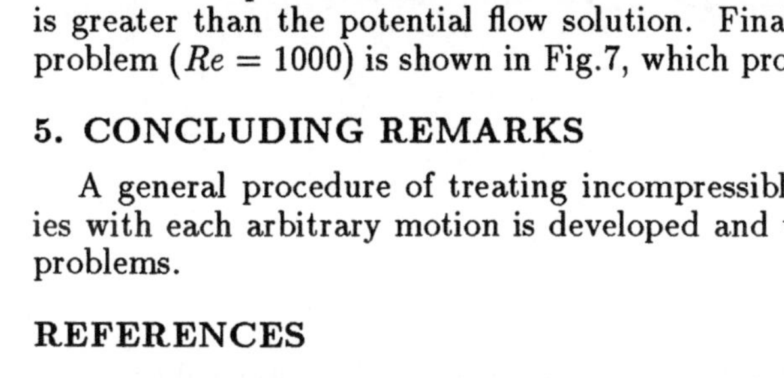

Fig. 1 (a) Interpolation and (b) point searching

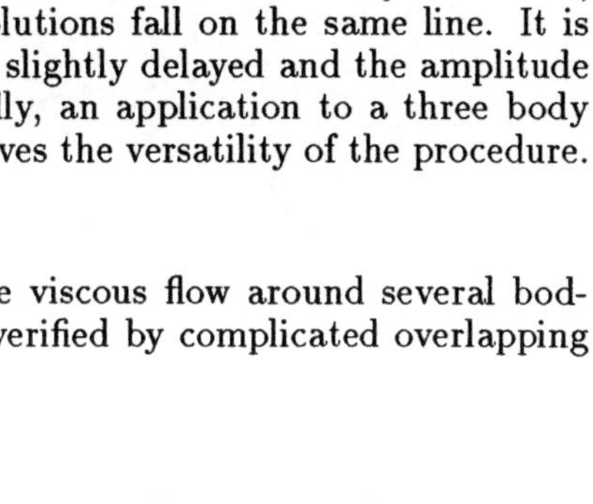

Fig. 2 Pressure contours around an oscillating cylinder at initial

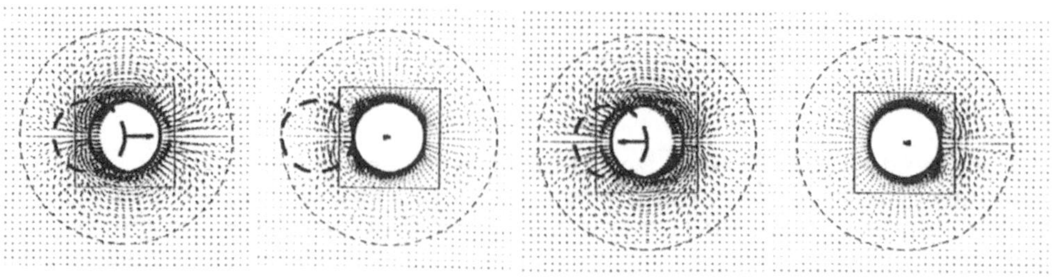

(a) (b) (c) (d)

Fig. 3 Velocity vectors around an oscillating cylinder at $Re = 314$
(a) $\omega t = \pi/2$, (b) $\omega t = \pi$, (c) $\omega t = 3\pi/2$, (d) $\omega t = 2\pi$

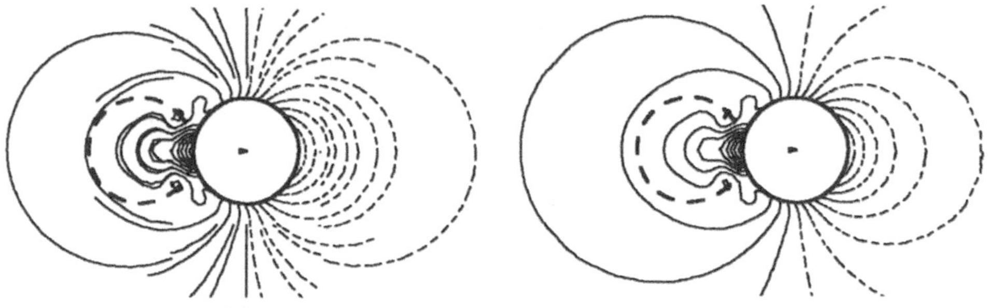

(a) (b)

Fig. 4 Pressure contours around an oscillating cylinder at $Re = 314$ and $\omega t = \pi$
(a) Multi-domain solution and (b) single-domain solution

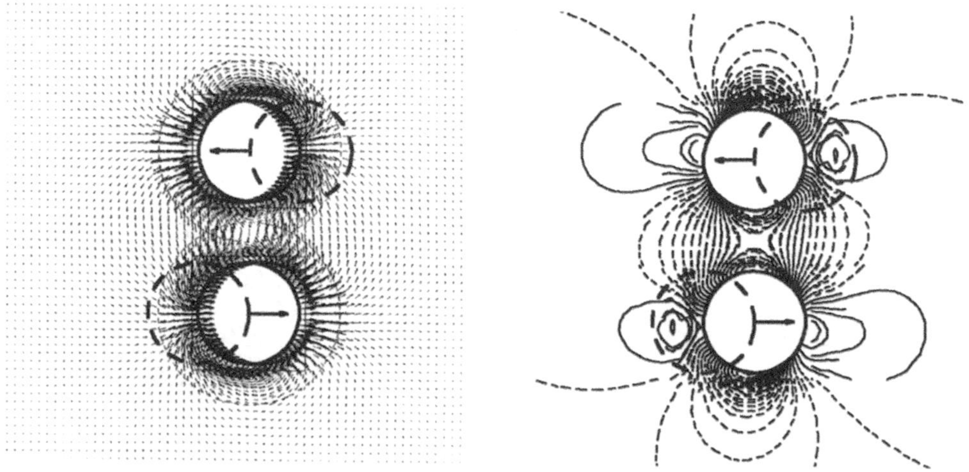

(a) (b)

Fig. 5 Computational results of two closely separated cylinders oscillating in the opposite directions at $Re = 314$ and $\omega t = \pi/2$ (a) Velocity vectors and (b) pressure contours

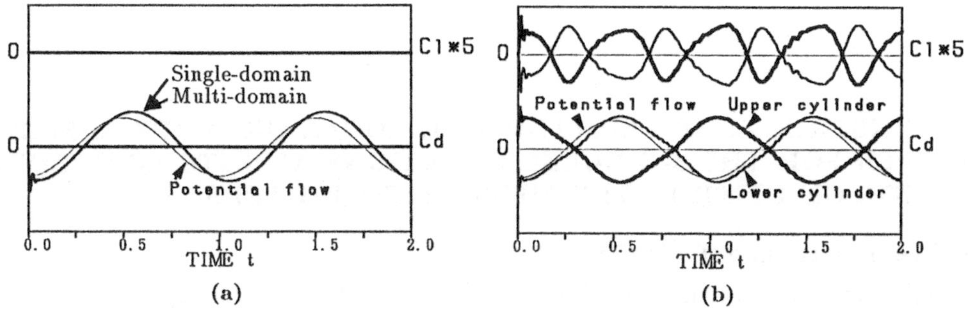

Fig. 6 Time variations of aerodynamic coefficients
(a) One cylinder case and (b) two cylinder case

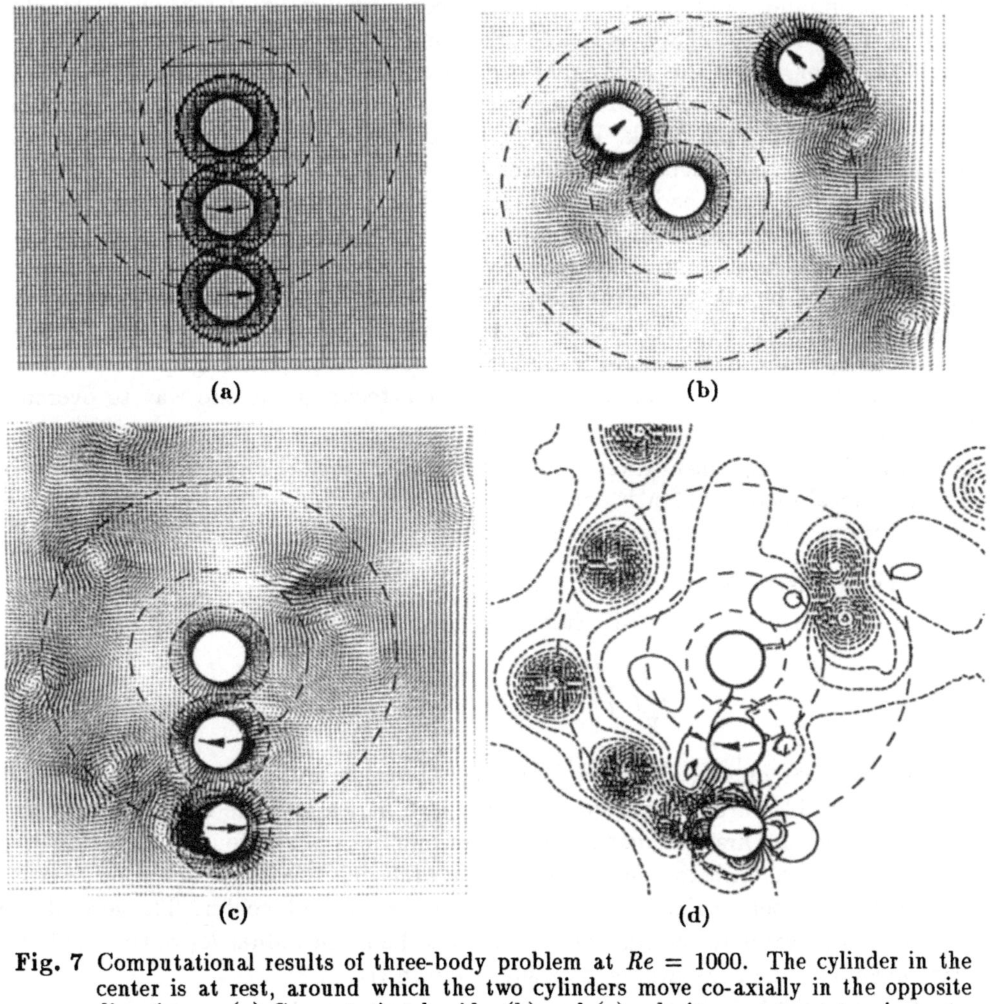

Fig. 7 Computational results of three-body problem at $Re = 1000$. The cylinder in the center is at rest, around which the two cylinders move co-axially in the opposite directions. (a) Computational grids, (b) and (c) velocity vectors at two instants, (d) pressure contours after one cycle.

AN UNSTRUCTURED MULTI-BLOCK LOCAL GRID REFINEMENT SOLVER FOR THE 3-D EULER EQUATIONS AND ITS IMPLEMENTATION ON DISTRIBUTED MEMORY COMPUTERS

P. Weinerfelt† and T. Schönfeld‡

† Department of Mathematics, Linköping University, S-581 83 Linköping, Sweden

‡ C.E.R.F.A.C.S., 42, Av. Gustave Coriolis, F-31057 Toulouse, France

Multi-block Techniques for Structured and Unstructured Grids

Most computers today are based on several processors and therefore the development of techniques to parallelize codes has become important. A tool widely used for parallelizing a CFD program is to split the computational domain into sub-domains (or blocks), for which the flow equations are solved independently and in parallel. Commonly, the multi-block technique is employed on structured grids resulting in a regular data set in each block. However, if the data structure is altered by facilities such as local mesh refinement the number of nodes per block can differ quite substantially resulting in a bad load balancing on parallel machines. In our approach to local grid refinement an unstructured data environment is introduced which permits the creation of blocks, not necessarily related to the original ones. These domains can be obtained by applying a splitting technique which decomposes an unstructured mesh into a number of blocks equal to the number of processors available. The use of the unstructured data system results in a decrease of computing speed due to the indirect addressing. However, the parallelization technique is one way to overcome this problem.

In this work the implementation of an unstructured three-dimensional flow solver on a parallel local memory MIMD computer is presented. As a target machine the Intel iPSC/2 Hyper Cube has been chosen.

Numerical Discretization of the 3-D Euler Equations

For the discretization of the 3-D Euler equations, that write in integral form as

$$\iiint_{\mathcal{V}} \frac{\partial}{\partial t}\vec{W}\, d\mathcal{V} = -\iint_{\partial\mathcal{V}} \overline{\overline{F}} \cdot \vec{n}\, dS$$

with $\vec{W} = (\rho, \rho u, \rho v, \rho w, \rho E)^T$, we choose a cell-vertex finite volume approach. The primitive variables are associated with the mesh nodes and the mean flux across a face is obtained by first calculating the fluxes functions at the four vertices and then averaging these quantities. Summing up in a first step the fluxes across the six cell faces yields the mean value of the rate of change at the cell center. The second step is to relate this value to the cell vertex. For this the flux-residual $\vec{\mathcal{R}}_k$ of the eight cells that surround node i are averaged as suggested in [1].

Since central difference schemes suffer from an inherent dispersive behavior and in order to rule out non-physical solutions, a supplementary, so-called artificial dissipation, term must be added to the convective terms. The well-known model consisting of blended second and fourth differences is modified into an unstructured formulation and total smoothing in operator notation takes the form: $D_i(W) = D_i^{(2)} - D_i^{(4)}$, with the undivided Laplacian $D^{(2)}$ and the bi-harmonic operator $D^{(4)}$. The unstructured formulation is well-suited for the application to hanging nodes at the interface between coarse and fine cells. In the absence of (at least) one neighbor the summation is done over a number of differences with the surrounding points (less than six).

The inviscid boundary condition of zero flux transport through solid walls is imposed by first setting the normal flux components to zero and secondly by forcing velocity to be tangential to the wall. At the far field the method of Riemann invariants serves to ensure non-reflecting boundary conditions. Periodicity across the interior block boundaries is obtained by means of auxiliary cells. A second layer of dummy cells becomes necessary to accommodate the treatment of the smoothing terms.

To advance solution in time an explicit Runge-Kutta three-stage scheme is employed in connection with the local time stepping convergence acceleration technique.

The Adaptive Local Grid Refinement Method

The solution quality of flow calculations is strongly related to the number of mesh points of which an increase means an increased resolution of flow features; a benefit, however, which is achieved at the expense of higher computational costs. In general dominating salient flow features like shocks or vortices cover only a relatively small part of a typical flow field, whereas large regions are of low activity.

The method of local grid refinement is one of the basic ways to improve flow resolution locally. The underlying idea is to fix the global grid and to add nodes exclusively to domains of physical interest. Local grid refinement is of particular importance for three-dimensional problems for which limiting the number of grid points is crucial [2].

Our approach to local refinement is to apply an unstructured data system to regular grids of hexahedra cells. With this technique the rigidity of the ordered (i, j, k)-indices is broken and replaced by a set of pointers. The method provides a fair amount of flexibility in refining the grid on several levels and permits the solution adaptive refinement of arbitrarily shaped regions. It is exactly the unstructured environment that, in combination with the local grid refinement method, gave motivation to implement the code on a local memory computer. Here we restrict ourselves to a brief survey of the method used, for a detailed description we refer to [3].

The refined zones are obtained by a simple sub-division of each coarse grid cell into eight smaller hexahedra. Each cell is divided independently from its neighbors. The topological similarity of coarse and fine cells permits the application of the flow algorithm to the entire flow field except to nodes at the refinement interfaces. For the update of these so-called hanging points we modify the residual distribution coefficients. The inviscid terms are evaluated by performing a special line integration at interface cells which includes the hanging nodes.

Since the precise location of flow features is unknown at the time of mesh generation, the concept of flow adaption is indispensable. Feedback about the solution during calculation is obtained in the form of some error indicator. Typically, the error is great in regions of large gradients in the flowfield which lead to the exploitation of flow quantity gradients to drive adaption.

The Unstructured Domain Splitting

A structured single block mesh can easily be split into a set of equal sized sub-blocks, which provides a good load balancing on a multi-processor computer. However, if facilities such as local mesh refinement are applied the number of nodes per block can differ quite substantially resulting in a bad load balancing. It hence is preferable to use an unstructured blocking of the computational domain. The recent development of unstructured flow algorithms together with the introduction of massively parallel computers has lead to an increasing investigation of such partitioning techniques. We will here mention four different algorithms:

- recursive coordinate bisection (RCB)
- recursive spectral bisection (RSB)
- recursive graph bisection (RGB)
- methods of wave front type

The first three approaches have been investigated in [4]. The last (more heuristic) method has recently been applied to the 2-D Euler equations discretized on an unstructured grid [5]. In the present work the program of [6], which uses the RSB algorithm, has been employed to split a 3-D mesh of hexahedral elements. The domain is decomposed by using a specific eigenvector of the Laplacian matrix $L(G)$ associated with the grid graph. If the graph G is connected then λ_2, the second largest eigenvalue, is negative and its magnitude is a measure of the connectivity of the graph. Associating the components of the corresponding eigenvector $\vec{x}_2$ with the grid points we obtain a weighting of the vertices. Sorting these weights in increasing order and decomposing them into two halves provides a partitioning of the domain into two sub-domains, a procedure that can be repeated recursively. The most demanding part of the RSB algorithm is the computation of the $\vec{x}_2$. This is done using a special version of the Lanczos algorithm. The domains created by RSB are often well-balanced and compact which is not the case for the alternative algorithms mentioned above [4].

Implementation on the iPSC/2 Hyper Cube

The unstructured multiblock approach yields automatically an algorithm which is well-suited for parallelization on local memory MIMD computers. We have implemented the flow solver on an iPSC/2 Hyper Cube with 32 nodes. Each block is mapped on a node processor doing the local computations independently. After every time step, exchange of information between the blocks is done by means of message passing between the processors. Tasks such as loading the node programs and checking of the convergence criteria are handled by the host processor.

Results

The test configuration chosen for the implementation of the 3-D Euler solver on the Hyper Cube is the transonic flow past an isolated delta wing at moderate incidence. Local grid refinement is used to improve the resolution of the leading edge vortices. This case has been used throughout for all the Hyper Cube performance tests. It is pointed out that the memory limit (4 Mbyte per processor) represents a severe restriction on the ability to run realistic 3-D tests. A considerable amount of work has been spent on code modifications to enable the use of grids as large as possible.

The first test in Fig.1 shows the timing obtained from a splitting of a global grid with $41 \times 13 \times 15$ points into maximal 32 blocks. As can been seen from the performance graph a speed-up is obtained that is at first linear, but is then followed by a flattening of the curve. The reason for this behavior is that the internal communication overhead becomes more significant as the number of blocks increases.

In a further test, which is more important for practical applications, the result of scaling the problem is studied (Fig.2). The number of grid points per processor is kept constant ($17 \times 7 \times 11$), which corresponds to the largest possible mesh fitting onto one processor. A nearly linear speed-up was obtained for each doubling of the number of processors and hence refining the global grid.

Next, the load balancing of block based local grid refinement, the main issue of this work, has been studied. As expected, maintaining the original blocks results in a bad load balancing. It can be seen that those blocks containing refined domains spend nearly 90% of their execution time on computations (Fig.3), while the non-refined blocks use more than half of the run time on communication and idle time (Fig.4).

A remedy to this problem is the automatic domain splitting approach described above. Before the application to a grid refinement case we tested the RSB-algorithm on a non-refined grid. A total of 13530 points is partitioned into 32 sub-domains of either 422 or 423 points. The well-balanced distribution of nodes on the processors can be seen in Fig.5 and 6. Finally, a locally refined grid with a total of 16709 points (3179 additional nodes embedded into the $33 \times 10 \times 41$ global grid) is partitioned into 32 blocks of either 522 (27 blocks) or 523 (5 blocks) nodes each. Note again the small deviation of only one grid point. The corresponding timing charts in Fig.7 and 8 confirm the good balancing achieved. All processors spend similar amounts of time on the flow calculations and on message passing which limits the idle time.

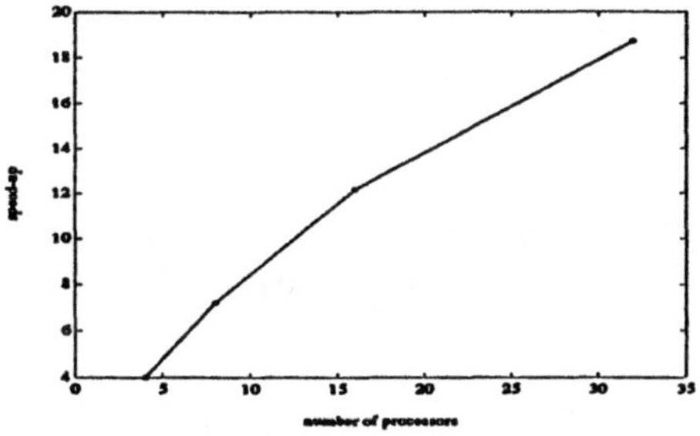

Fig.1: Speed-up for splitting a fixed global mesh

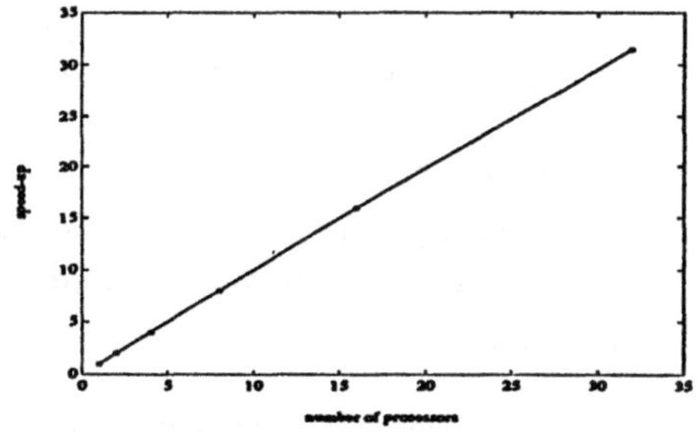

Fig.2: Speed-up for scaling by constant number of nodes per processor

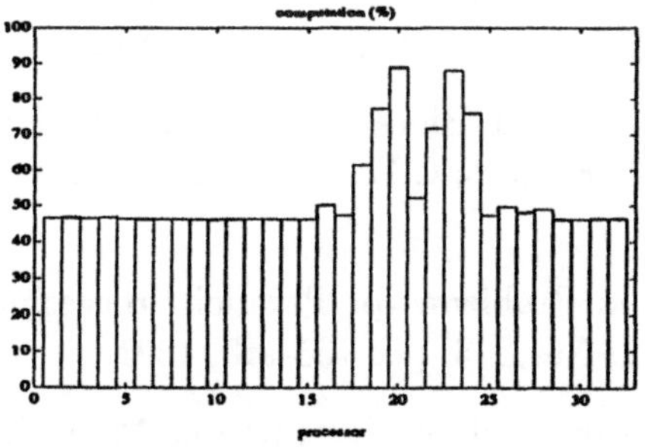

Fig.3: Computation time as a percentage of total run time for each processor (structured blocking with refinement)

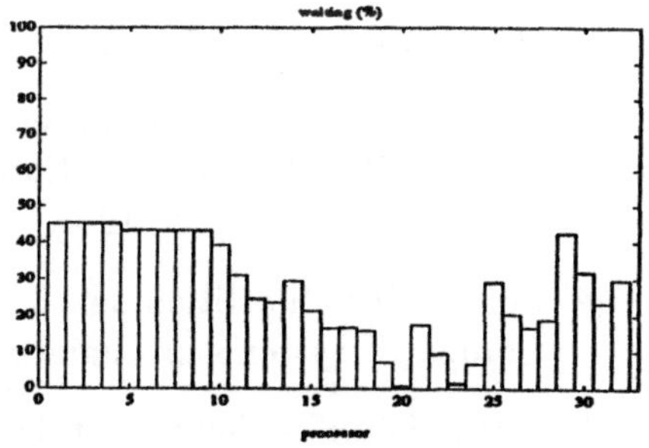

Fig.4: Idle time as a percentage of total run time for each processor (structured blocking with refinement)

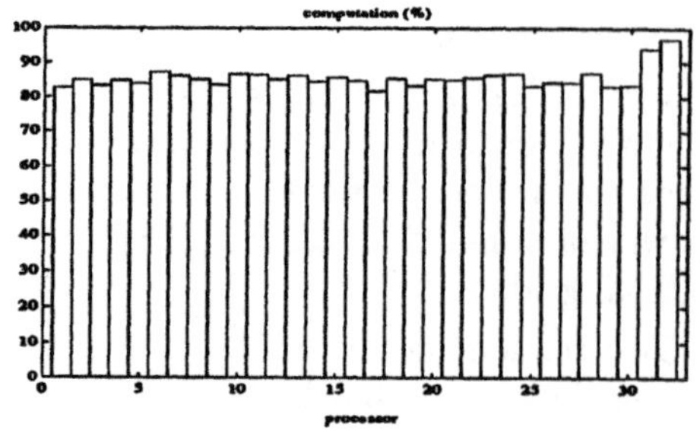

Fig.5: Computation time as a percentage of total run time for each processor (unstructured blocking without refinement)

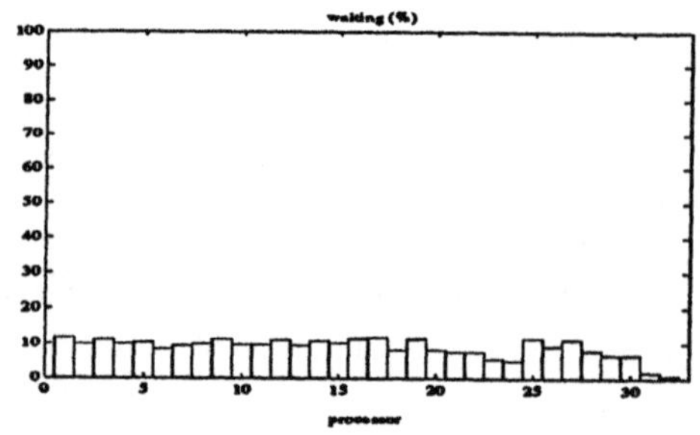

Fig.6: Idle time as a percentage of total run time for each processor (unstructured blocking without refinement)

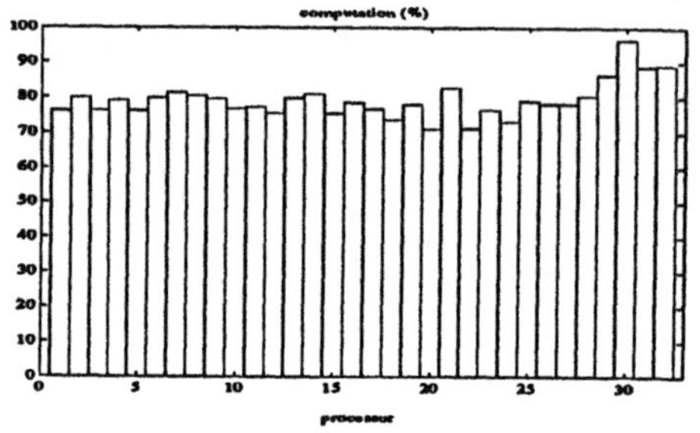

Fig.7: Computation time as a percentage of total run time for each processor (unstructured blocking with refinement)

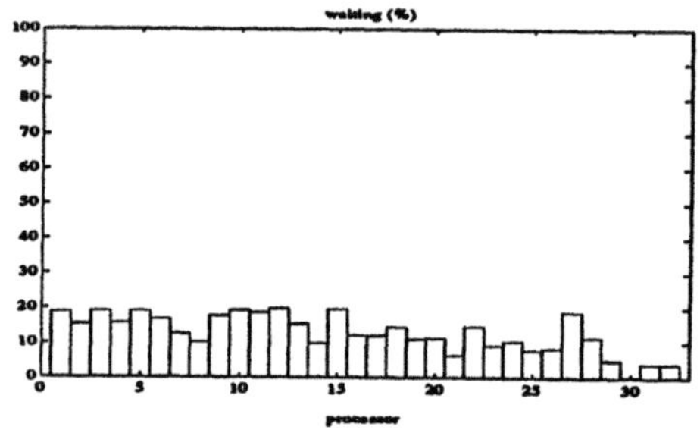

Fig.8: Idle time as a percentage of total run time for each processor (unstructured blocking with refinement)

Acknowledgements

We acknowledge the use of the automatic partition program written by Alex Pothen (Pennsylvania State University) and Horst Simon (NASA Ames Research Center).

References

[1] Hall, M.G.: "Cell-Vertex Multigrid Schemes for Solution of the Euler Equations", Proceedings of the *Conference on Numerical Methods for Fluid Dynamics*, Reading/U.K., 1985.

[2] Baker, T.J., Jameson, A., Vermeland, R.E.: "Three-Dimensional Euler Solutions with Grid Embedding", AIAA Paper 85-0121, Reno, January 1985.

[3] Schönfeld, T.: "Local Mesh Enrichment for a Block Structured 3-D Euler Solver", BRITE/EURAM EuroMesh 1042-Project, C.E.R.F.A.C.S.-Report PR/RF/92/11, March 1992.

[4] Simon, H.: "Partitioning of Unstructured Problems for Parallel Processing", Computer Systems in Engineering, Vol. 2, No. 2/3, pp. 135-148, 1991.

[5] Weinerfelt, P.: "Solution of the 2-D Euler Equations on Unstructured Multiblock Grids Using Parallel Computers", Techn. Rep., Dept. of Math., Linköping Univ.

[6] Pothen, A., Simon, H., Liou, K.-P.: "Partitioning Sparse Matrices with Eigenvectors of Graphs", Siam J. Matrix Anal. Ap., Vol.11, No.3, pp.430-452, 1990.

A SECOND-ORDER TVD METHOD FOR THE SOLUTION OF THE 3D EULER AND NAVIER-STOKES EQUATIONS ON ADAPTIVELY REFINED MESHES

M. J. Aftosmis
U.S. Air Force Wright Laboratory, Wright-Patterson Air Force Base, OH 45433-6553

1. OVERVIEW

The accurate simulation of flows with features spanning many length scales presents a major challenge to Computational Fluid Dynamics. This issue becomes especially critical in 3D flows over complex configurations, or even flows over simple configurations at extreme flow conditions, where multiple interacting 3D features must be correctly represented. The last decade has seen the emergence of two basic approaches to this fundamental difficulty. High resolution upwind schemes attempt to represent shocks and contact discontinuities with as few cells as possible, while unstructured, adaptive meshes automatically refine the local mesh dimension in an attempt to resolve the scale of the local flow physics.

The present work combines these two techniques by developing an adaptive, high resolution, upwind algorithm for the simulation of inviscid and viscous flows. This work extends the two dimensional inviscid procedure of Ref. [1] to the simulation of the three dimensional Navier-Stokes equations. The method is node based and written within a finite volume framework which uses hexahedral cells to discretize the second order Upwind TVD algorithm of Harten and Yee[2]. An unstructured implementation of this operator permits grid adaptation by division of mesh cells in one or more directions. The viscous terms in the Navier-Stokes equations are modeled with central differencing using a newly formulated unstructured discretization based upon the 2D work of Kallinderis and Baron[3].

Preliminary results are presented for a variety of test cases. These examples focus primarily on topologically simple problems intended to investigate the accuracy of the inviscid modeling, viscous modeling, and the ability of the adaptation method to find, resolve and faithfully represent features in complex 3D flows with multiple interacting inviscid and viscous structures. A built-in central difference option offers the ability to compare upwind and central difference results on identical adapted meshes.

2. NUMERICAL METHOD

The numerical method relies upon Harten and Yee's Upwind TVD algorithm to integrate the inviscid fluxes while the viscous terms use central difference modeling. This spatial discretization is advanced in time through a modified Runge-Kutta procedure, and both the inviscid and viscous integration proceeds on a mesh of auxiliary cells formed by connecting the centroids of the physical mesh cells. Figure 1 shows the construction of these auxiliary cells in a 3D mesh. The state vector of conserved quantities is stored at nodal locations.

2.1 INVISCID DISCRETIZATION

The upwind discretization uses a Roe type approximate Reimann solver and the flux function is based upon the "modified flux" approach of Harten[2]. Like many upwind methods, the flux function takes a (*Central difference + Dissipation*) form where the "dissipative flux" modifies the central flux in accordance with local wave propagation to recover an upwind stencil at each node. This dissipative flux also contains a flux function which includes the limiter and has the additional role of bringing the entire discretization to second order accuracy away from local extrema. The method is applied in multi-dimensions by essentially a superposition of the 1D upwind operators in each spatial direction.

The fact that the method relies on a superposition of 1D operators and uses this (*Central + Dissipation*) form suggests a natural formulation on unstructured hexahedral control volumes. This choice is supported by the 2D inviscid results contained in Ref. [1]. Here the unstructured algorithm entirely reproduced the shock capturing and slip surface representation of the structured formulation by Kroll and Rossow[4].

The upwind operator for the inviscid flux terms requires five points in each spatial direction and it requires special attention in the current unstructured environment. Ref. [1] details the construction of the full operator by a sequence of nearest neighbor operations.

2.2 VISCOUS DISCRETIZATION

The Euler scheme[1] was extended to the full Navier-Stokes system by a separate discretizaton of the viscous components of the flux density tensor. The total update to the state vector at any node, $\Delta \mathbf{U}_i$, then becomes the summation of inviscid and viscous contributions.

$$\Delta \mathbf{U}_i = (\Delta \mathbf{U}_i)_{Inviscid} + (\Delta \mathbf{U}_i)_{Viscous} \qquad \{1\}$$

The viscous discretization uses central difference modeling based upon the 2D ideas of Kallinderis and Baron[3] and the 3D work of Radespiel *et al.*[5]. Most importantly, the use of central differencing for these terms allows the scheme to be written extremely compactly using only node-to-cell and cell-to-node communication.

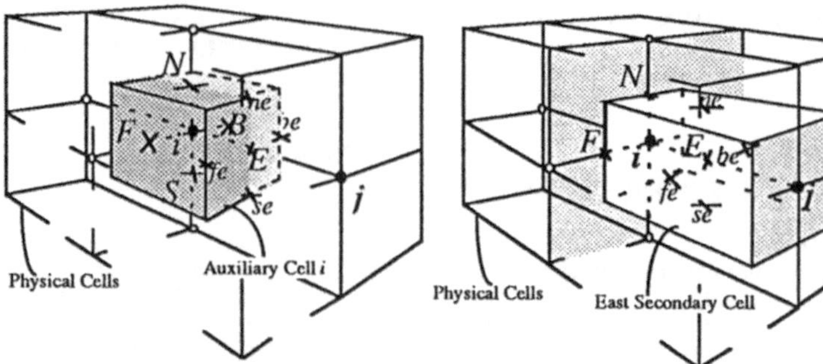

Figure 1. Construction of auxiliary and secondary cells for inviscid and viscous discretization.

This discretization may be illustrated by considering the formation of the model second derivative term u_{xx} at node i. Assuming the control volume V_i (pictured in left of Figure 1) has planar faces and applying the Gauss Divergence Theorem over the surface of this volume, ∂V_i, gives

$$(u_{xx})_i = \frac{1}{V_i}\oint_{\partial V_i} (u_x) n_x dS = \frac{1}{V_i}\left[(u_x)_E S_x^E - (u_x)_W S_x^W + (u_x)_N S_x^N - (u_x)_S S_x^S + (u_x)_F S_x^F - (u_x)_B S_x^B\right] \quad \{2\}$$

The x-component of the unit normal vector, n_x, and the surface vector, $\mathbf{S} = (S_x, S_y, S_z)^T$, are both taken oriented in positive coordinate directions. With the auxiliary cells constructed as described above, each edge incident upon node i will always pierce the face of an auxiliary cell. Thus, Eq.{2} contains one contribution from each edge incident upon node i. This observation becomes increasingly important when integrating more general control volumes.

Equation {2} makes use of the first derivatives, u_x, at the N, S, E, W, F, B faces. A second surface integral, similar to Eq.{2}, provides the derivatives at the centers of these faces through the construction of a set of secondary control volumes surrounding each of the edges incident on node i. Figure 1 (right) illustrates this secondary cell construction around the E face of auxiliary cell i.

$$(u_x)_E = \frac{1}{V_E}\oint_{\partial V_E} u n_x dS = \frac{1}{V_E}\left[u_j S_x^j - u_i S_x^i + u_{ne} S_x^{ne} - u_{se} S_x^{se} + u_{fe} S_x^{fe} - u_{be} S_x^{be}\right] \quad \{3\}$$

V_E is the volume of the secondary cell, and the surface vectors on the faces of this control volume are constructed by averaging those from the faces of the auxiliary cells surrounding nodes i and j (the nodes which define the end points of the edge that pierces the E face of V_i). Similar constructions around the rest of the edges which meet at i form the remaining first derivatives needed to evaluate Eq.{2}. Chosing these secondary control volumes leads to a discretization which is conservative, damps odd-even mode oscillations, is free stream preserving on arbitrary meshes, and second-order accurate on any smooth mesh.

The location of the first derivatives along the edges of the physical cells and the viscous update at the nodes of physical cells permits both of the surface integrals (Eqs.{2} and {3}) to be evaluated by a single sweep through the physical mesh cells. This implementation is an extension of that described in Ref. [6] and relies on the fact that under the assumptions of a Newtonian fluid with the Stokes hypothesis, the viscous fluxes become a simple linear combination of first derivative quantities.

2.3 Adaptation

Grid adaptation occurs through directional division of the cells in the physical mesh. A feature detection algorithm marks cells for division after the solution has progressed sufficiently on a given mesh. Ref. [1] contains details of this detection routine in two dimensions. After first examining the domain with an undivided second difference of pressure to find "shock" cells, this routine re-scans only the remaining cells in the computation to locate "smooth" features. This second scan relies upon an undivided second difference of velocity magnitude to locate regions of rapidly changing shear stress and an undivided first difference of density to locate inviscid features in the flow. The cells tagged in the second scan are evaluated on a statistical basis as outlined in Ref. [6]. The division routine operates recursively in a direction-by-direction manner and can divide the cells directionally when features are almost mesh aligned. Directional division is usually only applied at the finest level of adaptation[6,1]. After the cell division process is complete, the entire mesh is smoothed by a sequential application of Laplacian and curl operators to distribute any stretching and skewing error throughout the mesh. This final step is important since the adaptation process reduces the truncation error in the solution only if the mesh remains smooth as it is refined.

2.4 Interfaces

An essential complication of hexahedral cell division is the appearance of interfaces between different levels of physical cells. Both the inviscid and viscous discretizations treat interfaces conservatively, which leads to a first order stretching error in the solution perpendicular to interface boundaries[1]. The inviscid discretization integrates these nodes directly, while the viscous part uses a specially constructed routine to conservatively integrate nodes along the interface. Numerical experiments in both two and three dimensions have demonstrated that the inviscid treatment is both smooth and robust[1].

3. Preliminary Numerical Results

The first test case is designed to establish the accuracy of the inviscid Upwind TVD discretization. Figure 2 provides an overview of this test case. It is a topologically simple example in which an inviscid 3D double wedge corner flow is simulated at M_∞=2.98 with wedge angles δ_1=δ_2=9.49°. The final mesh contained ~70,000 nodes which is about 14 times fewer than the roughly 980,000 which would be required to provide the same shock resolution on a globally refined mesh. Directional cell division was permitted at the finest mesh level and the mesh plane inset in Fig. 2c shows that the cells have adapted themselves to match the orientation of wedge and corner shocks. Examining the flow structure in a cross-flow plane at x=0.8 permits a direct comparison of Mach contours with both experiment and other previous numerical computations[7]. Additionally, the undisturbed wedge flow away from of the corner interaction allows a direct comparison with inviscid gasdynamic theory. The agreement with experiment, theory and the previous numerical simulation is very good, and the shock spans no more than two cells.

A subsonic flat plate boundary layer provides an initial assessment of the behavior of the viscous discretization. This quasi-2D flow was computed at two levels of resolution using both TVD and central difference operators for M_∞=0.5 and Re_L=5,000. This simulation used an entropy cut-off in the TVD scheme of 0.1 and a fourth difference dissipation coefficient of $^1/_{128}$ in the central difference calculation. The domain extended $2L$ upstream, along, and above the plate. Figure 3a displays u-velocity profiles of both schemes with five and 13 points in the boundary layer. With only five points in the boundary layer, the Upwind TVD method has already essentially reproduced the Blasius profile. This result contrasts sharply with that from the central difference scheme at this level of resolution. With 13 points in the boundary layer, the two integrations produce essentially identical results. At this higher level of resolution, the central difference result produces almost no evidence of the "viscous overshoot" evident in the five point case.

Figure 3b shows skin friction development along the flat plate. This plot reports results for both the central and TVD discretizations at two levels of resolution and compares these with the Blasius relation C_f=0.664$(Re_x)^{-1/2}$. With 13 points in the boundary layer (Re_L of 5,000), the theoretical skin friction is predicted well by about Re_x = 100, or 2% of the plate length from the leading edge.

The final example examines the performance of the detection algorithm in resolving multiple interacting features in 3D viscous flow. The test case presented in Figure 6 considers Mach 1.2 flow over the 65° cropped delta wing tested in Ref. [8] at 20° incidence angle. The laminar, half-span calculation used a Reynolds number of 480,000 based on the root chord, while the experiment was conducted at Re_C=2.4-5.3x10^6. The wing in the calculation was not truncated at the trailing edge, but instead, was extended downstream to the outflow boundary (located at approx. x=1.1c).

This flow is characterized by a strong steady primary vortex which separates at the sharp leading edge. Underneath this vortex, a cross-flow shock develops which induces separation of the boundary layer, and "locates" the secondary separation. The tertiary separation lies between the the cross-flow shock and sharp leading edge. On the windward side, a bow shock develops to turn the flow around the wing. The combination of interacting viscous and inviscid features makes this a discriminating problem for evaluating the capabilities of the feature detection algorithm for realistic, complex 3D flows.

Figure 4a contains an overall view of the computation The wing is depicted with the mesh on the starboard, and density contours on the port side of the delta wing. These contours show the footprints of the primary vortex, cross-flow shock, secondary and tertiary vortices on the wing surface. The final adapted mesh used 999,000 nodes, placing about 260 points chordwise and 150 points spanwise (at the t.e.) on the upper surface of the wing. Figure 4b shows further details of the flow as it passes through the plane located at the trailing edge through contours of total pressure loss. The figure clearly shows the adaptation pattern responding to capture the shock triggered separation underneath the primary vortex, as well as resolution of the feeding sheet and the entrained tip vortex. This indicates that the feature detector successfully located these weaker features, while still responding to the strong vortical structures and bow shock. Figure 4c contains symmetry plane Mach contours, and a view of the corresponding adapted grid while Figures 4d and 4e compare this calculation with wind tunnel results from Ref. [8]. The computed surface streamlines reproduce the overall surface surface shear pattern from the experimental oil flow visualization including formation of the secondary and tertiary separations. The tertiary vortex in the calculation does not appear to form quite as early or as fully as in the experiment and this is believed to be a Reynolds number effect. Figure 4e is a direct comparison of the Cp distribution at x=0.8c. Starting at the symmetry plane, the line plot agrees with the data underneath the primary vortex, through the cross-flow shock and secondary separation point, and begins to differ only in the location of the tertiary separation. Despite the Reynolds number difference, this comparison is quite reasonable. Nevertheless, it should be re-emphasized that this example focuses on the performance of the detection algorithm and was not conducted as a rigorous attempt to reproduce wind tunnel data.

4. ACKNOWLEDGEMENTS

This work was accomplished within a scientific exchange program between the U.S.A.F and the Deutsche Forschungsanstalt für Luft und Raumfahrt e.V. during the author's stay at DLR Braunschweig. Financial support for this exchange was provided by the SEEP program and is gratefully acknowledged. Computing support was provided by both the DLR and the Phillips Laboratory Supercomputing Center at Kirtland AFB. The author would like to sincerely thank Dr. N. Kroll for his invaluable contributions to this work and the exchange as a whole. Additionally, the many contributions of U. Hermann and Dr. R. Radespiel are gratefully acknowledged.

5. REFERENCES

[1] Aftosmis, M., and Kroll, N., "A Quadrilateral Based Second-Order TVD Method for Unstructured Adaptive Meshes," *AIAA 91-0124*, 1991.

[2] Yee, H.C., Harten, A. "Implicit TVD Schemes for Hyperbolic Conservation Laws in Curvilinear Coordinates," *AIAA Jol.*, Vol 25, pp. 266-274, 1987.

[3] Kallinderis, Y. and Baron, J.R. "Adaptation Methods for a New Navier-Stokes Algorithm," *AIAA 87-1167*, 1987.

[4] Kroll, N., and Rossow, C., "A High Resolution Cell Vertex TVD Scheme for the Solution of the Two and Three Dimensional Euler Equations," *Proceedings of the 12th Conference on Numerical Methods in Fluid Dynamics*, Oxford, UK, 1990.

[5] Radespiel, R., Rossow, C., and Swanson, R. C., "Efficient Cell-Vertex Multigrid Scheme for the Three-Dimensional Navier-Stokes Equations," *AIAA Jol.*, Vol. 28, No. 8. pp. 1464-1472, 1990.

[6] Kallinderis, Y.G.: "Adaptation Methods for Viscous Flows," PhD.thesis Massachusetts Institute of Technology, Dept. of Aeronautics and Astronautics, Cambrige, MA, 1988.

[7] Kutler, P., "Supersonic Flow in the Corner by Two Intersecting Wedges," *AIAA Jol.* Vol.12. pp. 577-578, 1974.

[8] Erickson, G.E., Schreiner, J.A., Rogers, L.W., "On the Structure, Interaction, and Breakdown Characteristics of Slender Wing Vortices at Subsonic, Transonic, and Supersonic Speeds," *AIAA-89-3345*, 1989.

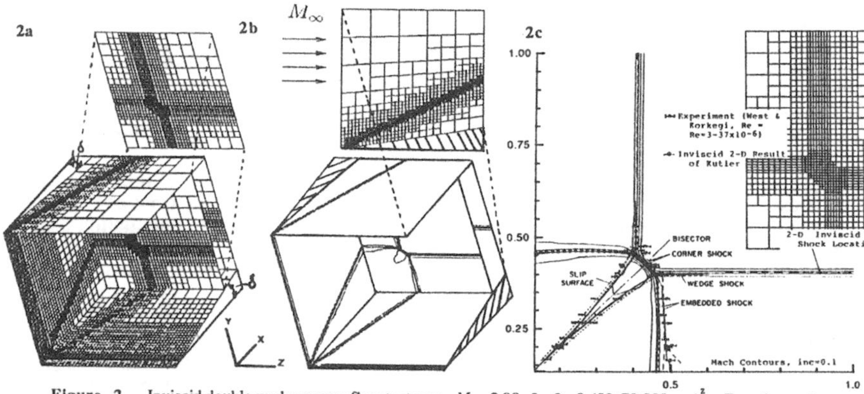

Figure 2. Inviscid double wedge corner flow test case. M_∞=2.98, δ_1=δ_2=9.49°, 70,000 nodes. Experimental results in 2c at plane located at x=0.8 from Ref. [7].

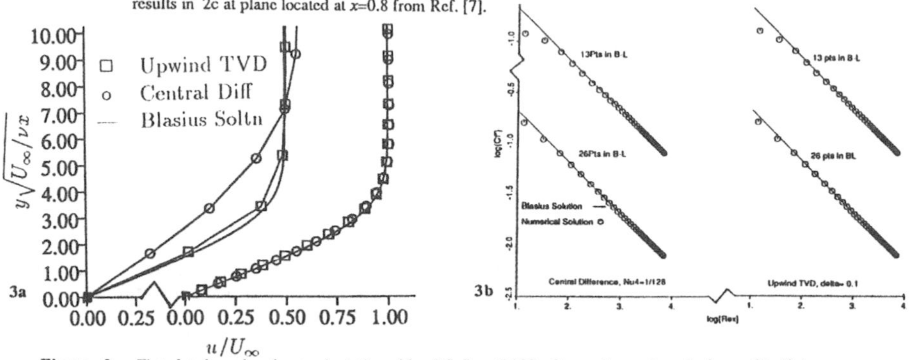

Figure 3. Flat plate boundary layer calculation. M_∞=0.5, Re_L=5,000. Comparison of u-velocity profile (3a) and skin friction (3b) with Blasius solution at different levels of mesh resolution.

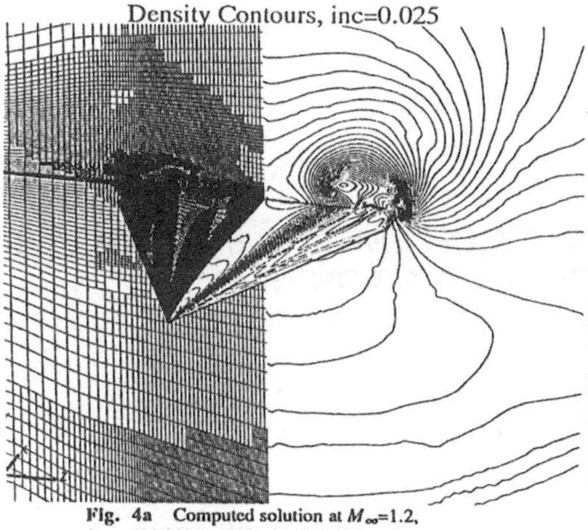

Fig. 4a Computed solution at M_∞=1.2, Re_C=480,000, α=20° with 999,000 nodes.

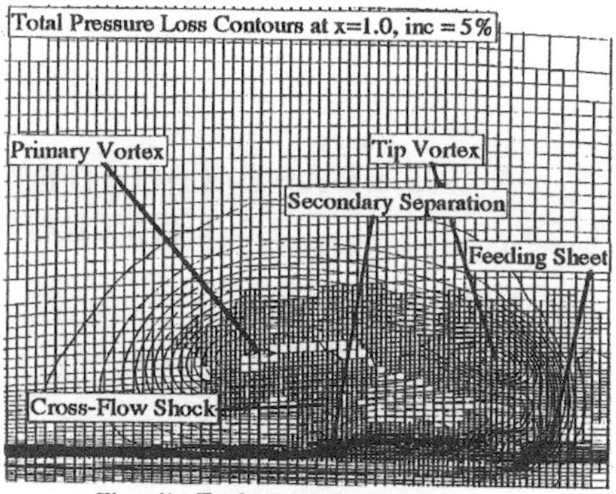

Fig. 4b Total pressure loss contours at t.e. (x=1.0) showing shock induced separation.

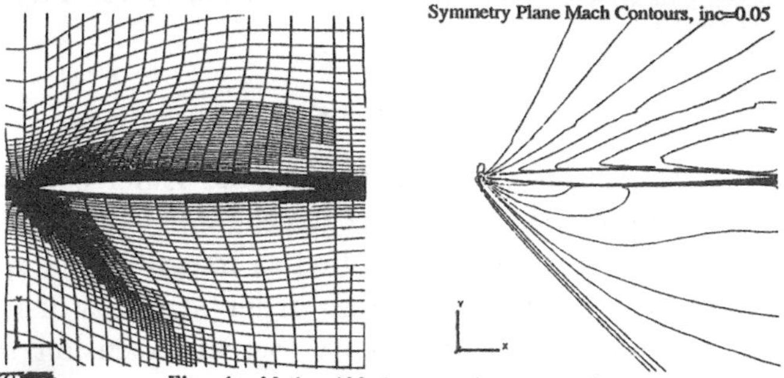

Fig. 4c Mesh and Mach contours in symmetry plane.

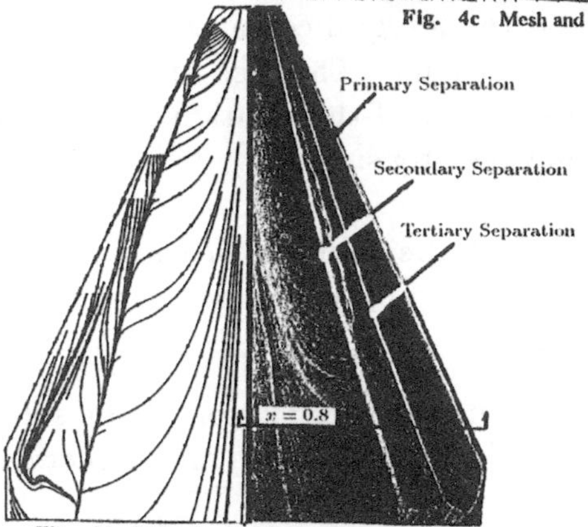

Fig. 4d Right: Results from experimental oil flow visualization [8] (Re_C=2.4-5.3x10^6). Left: Computed surface streamlines from numerical solution (Re_C=4.8x10^5).

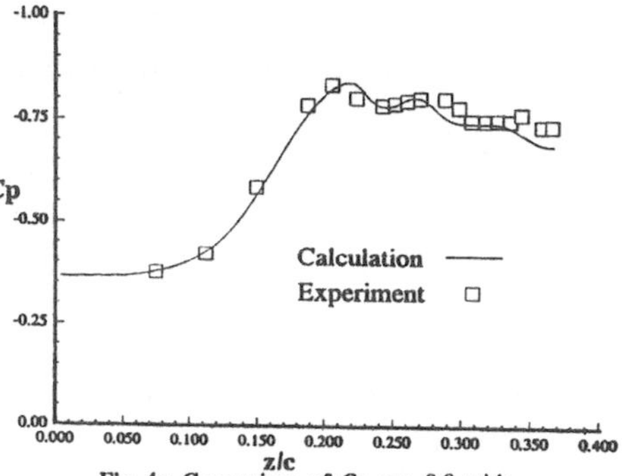

Fig. 4e Comparison of Cp at x=0.8 with experimental results from Ref. [8].

A 3-D LEAST-SQUARES UPWIND EULER SOLVER FOR UNSTRUCTURED MESHES

T. J. Barth

NASA Ames Research Center, Moffett Field, CA, USA

§Introduction

An upwind finite-volume discretization of the Euler equations in integral form on arbitrary unstructured grids is developed. The present scheme uses a variant of Godunov's method [1] applied to irregularly shaped control volumes. The method is extended to high order accuracy using the MUSCL procedure due to van Leer [2,3] and its logical extension to unstructured grids [4]. Recall that in Godunov's method the integral cell averages are the basic unknowns (degrees of freedom) in the scheme. The task at hand is to evaluate the flux integral given these cell averages of the solution. The basic solution process is summarized in the following steps:

Reconstruction. Given integral averages of the solution in each control volume, reconstruct a piecewise polynomial which approximates the behavior of the solution in each control volume.

Flux Quadrature. From the piecewise polynomial description of the solution, approximate the flux integral by an upwind flux function and numerical quadrature.

Evolution. Given a discretization of the flux integral, evolve the system in time using any class of implicit or explicit schemes. (In the present calculations, a four stage Runge-Kutta time stepping scheme is used with space varying time step.) This results in new integral cell averages of the solution. The solution process can then be repeated.

Reconstruction

Perhaps the most challenging task is the reconstruction of piecewise polynomials given cell averaged solution data. The situation is depicted in one space dimension in figures 1a-d.

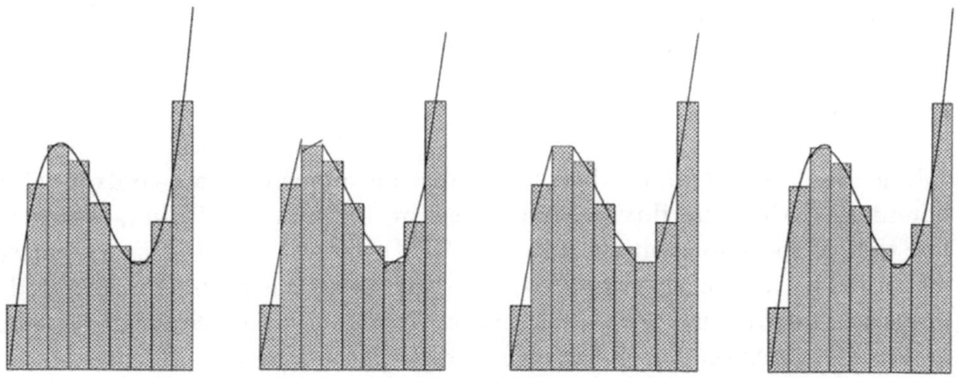

(a) (b) (c) (d)

Figure 1. (a) Cell-averaging of quartic polynomial, (b) piecewise linear reconstruction, (c) piecewise linear reconstruction with monotonicity enforcement, (d) piecewise quadratic reconstruction.

Figure 1b shows the linear reconstruction of a quartic polynomial given integral cell averages. To prevent spurious oscillations from occurring in numerical solutions near discontinuities and steep gradients, monotonicity of the reconstructed data is enforced (see figure 1c). The method can be extended to high order accuracy by encreasing the order of the reconstructed piecewise polynomials as shown in figure 1d. These concepts extend naturally to unstructured grids.

The process of solution reconstruction from integral cell averages is greatly simplified by exploiting the following (well-known) observation: *in the case of piecewise linear reconstruction, we can dispense with the notion of cell averages as unknowns by reinterpreting the unknowns as pointwise values of the solution sampled at the centroid of the control volume.* This simplifies the task of reconstruction considerably.

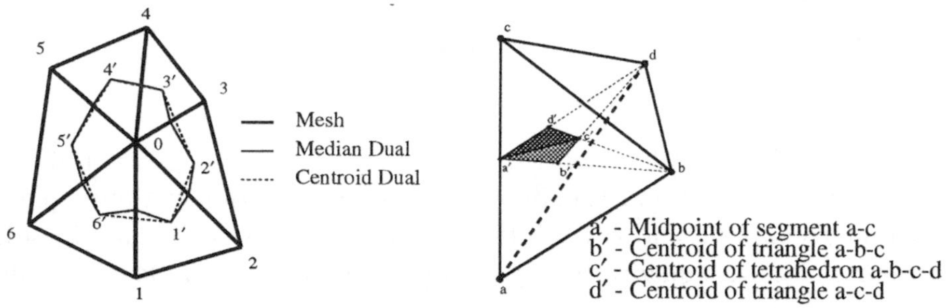

Figure 2. (a) 2-D triangulation with median dual, (b) portion of 3-D median dual for tetrahedral mesh.

In the present formulation, control volumes are formed from a median dual tessellation of the mesh triangulation. The median dual tessellation is illustrated in figures 2a-b for two and three space dimensions. For steady-state computations on dual control volumes, another important observation is that pointwise values of the solution can be placed at vertices of the mesh (which need not be exact centroids of the median dual control volumes) without apparent loss of accuracy. This further simplifies the reconstruction procedure.

Flux Quadrature

Since the reconstructed data varies discontinuously across control volume boundaries, the true flux is supplanted by a "numerical" flux, which is a function of the two solution states. In the present case, Roe's flux function [6] is used because of its simplicity. The flux integral is then approximated by numerical quadrature, see [4] for details. *It should be stressed that the actual choice of numerical flux function plays a diminishing role of importance as the order of reconstruction is increased.* This point is illustrated in figure 3 for schemes utilizing solution reconstruction of increasing order of accuracy together with the scalar equivalent of Roe's flux function.

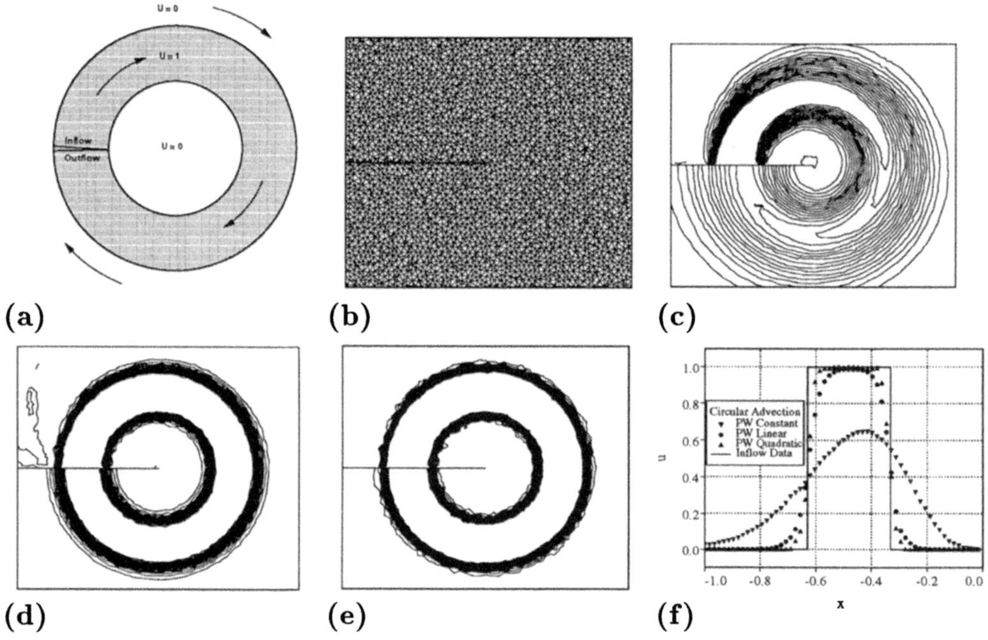

Figure 3. Circular advection ($u_t + \overrightarrow{\lambda} \cdot \nabla u = 0,\ \overrightarrow{\lambda} = (y, -x)^T$) of a linear discontinuity. (a) Schematic of flow, (b) mesh, (c) solution contours, piecewise constant reconstruction, (d) solution contours, piecewise linear, (e) solution contours, piecewise quadratic, (f) solution comparison at inflow and outflow.

§Simplified Least-Squares Reconstruction

For the present three-dimensional algorithm, we consider the simplified linear reconstruction procedure presented in reference [4]. Consider a vertex v_0 and suppose that the solution varies linearly over the support of adjacent neighbors of the mesh with local cyclic index i. The scaled projection of the solution gradient along an edge $e(v_i, v_0)$ can be calculated by $(\nabla u^h)_0 \cdot (\vec{\mathbf{r}}_i - \vec{\mathbf{r}}_0) = u_i - u_0$. A similar equation could be written for all edges incident to

v_0 subject to an arbitrary weighting factor:

$$\begin{bmatrix} w_1\Delta x_1 & w_1\Delta y_1 & w_1\Delta z_1 \\ \vdots & \vdots & \vdots \\ w_n\Delta x_n & w_n\Delta y_n & w_n\Delta z_n \end{bmatrix} \begin{pmatrix} u_x \\ u_y \\ u_z \end{pmatrix} = \begin{pmatrix} w_1(u_1-u_0) \\ \vdots \\ w_n(u_n-u_0) \end{pmatrix} \tag{2}$$

or in symbolic form $\mathcal{L}\,\nabla u = \mathbf{f}$, where $\mathcal{L} = \left[\vec{\mathbf{L}}_1 \quad \vec{\mathbf{L}}_2 \quad \vec{\mathbf{L}}_3\right]$ in three dimensions. Exact calculation of gradients for linear u is guaranteed if any three row vectors $w_i(\vec{\mathbf{r}}_i - \vec{\mathbf{r}}_0)$ span all of three space. This implies linear independence of $\vec{\mathbf{L}}_1$, $\vec{\mathbf{L}}_2$, and $\vec{\mathbf{L}}_3$. The system can then be solved via a Gram-Schmidt process, i.e.,

$$\begin{bmatrix} \vec{\mathbf{V}}_1 \\ \vec{\mathbf{V}}_2 \\ \vec{\mathbf{V}}_3 \end{bmatrix} \left[\vec{\mathbf{L}}_1 \quad \vec{\mathbf{L}}_2 \quad \vec{\mathbf{L}}_3\right] = \begin{bmatrix} 1 & 0 & 0 \\ 0 & 1 & 0 \\ 0 & 0 & 1 \end{bmatrix} \tag{3}$$

The row vectors $\vec{\mathbf{V}}_i$ are given by $\vec{\mathbf{V}}_i = \frac{\vec{\mathbf{U}}_i}{(\vec{\mathbf{L}}_i \cdot \vec{\mathbf{U}}_i)}$ where

$$\vec{\mathbf{U}}_i = (l_{kk}l_{jj} - l_{jk}l_{jk})\vec{\mathbf{L}}_i - (l_{kk}l_{ij} - l_{jk}l_{ik})\vec{\mathbf{L}}_j - (l_{jj}l_{ik} - l_{jk}l_{ij})\vec{\mathbf{L}}_k$$

with $l_{ij} = (\vec{\mathbf{L}}_i \cdot \vec{\mathbf{L}}_j)$ for all cyclic permutations of i, j, k

Note that reconstruction of N independent variables in $\Re^d$ implies $\binom{d+1}{2} + d\,N$ inner product sums. Since only $d\,N$ of these sums involves the solution variables themselves, the remaining sums could be precalculated and stored in computer memory. Using an edge data structure, the calculation of inner product sums can be calculated for *arbitrary* combinations of polyhedral cells. In all cases linear functions are reconstructed exactly. This formulation provides freedom in the choice of weighting coefficients, w_i. These weighting coefficients can be a function of the geometry and/or solution. Classical approximations in one dimension can be recovered by choosing geometrical weights of the form $w_i = 1/|\vec{\mathbf{r}}_i - \vec{\mathbf{r}}_0|^t$ for values of $t = 0, 1, 2$. Numerical computations shown in the next section were performed with $t = 1$.

§Sample Numerical Calculations

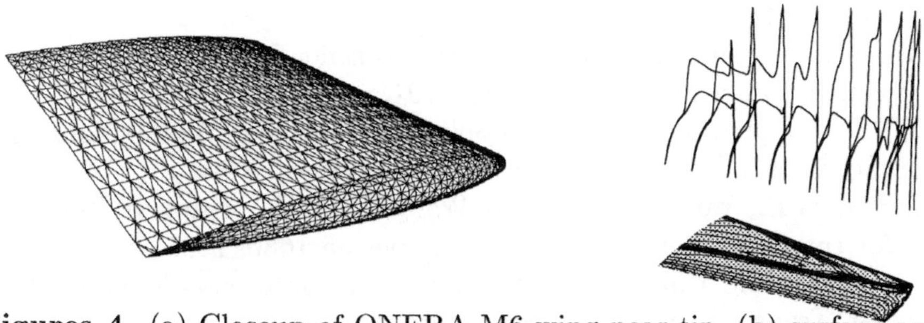

Figures 4. (a) Closeup of ONERA M6 wing near tip, (b) surface pressures and C_p distributions at several span stations.

Three-dimensional transonic flow computations over the ONERA M6 wing geometry have been computed using the method described above. Numerical calculations with $M_\infty = .84$ and $\alpha = 3.06°$ were carried out using the upwind code with least-squares gradient and Galerkin gradient [4] calculations. Structured mesh results by Thomas, van Leer, and Walters [7] using a 161x41x31 mesh and experimental data [8] at a Reynolds number of 11.7 million are used for comparison. Figure 4b shows surface pressure contours on the wing surface and C_p profiles at several span stations. Pressure contours clearly show the lambda type shock pattern on the wing surface. Figures 5a-b compare pressure coefficient distributions at two span stations on the wing, y/b=.44 and .65. The comparison with existing codes and experiment is very reasonable. Computation times are about 1.25 hour on a CRAY Y-MP.

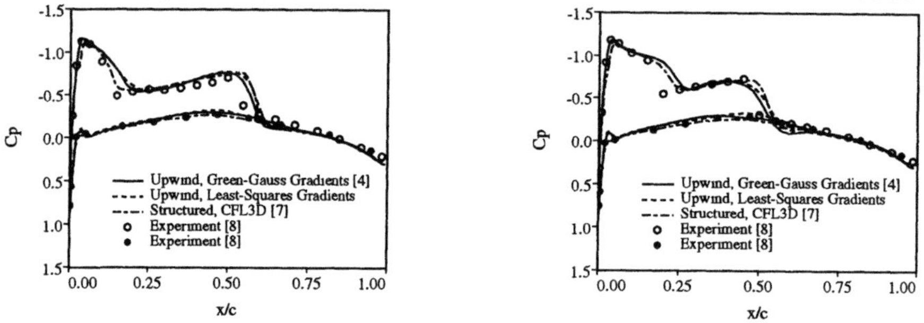

Figure 5. Pressure coefficient distributions at span stations, (a) y/b=.44, (b) y/b=.65.

1. Godunov, S. K., "A Finite Difference Method for the Numerical Computation of Discontinuous Solutions of the Equations of Fluid Dynamics", Mat. Sb., Vol. 47, 1959.

2. Van Leer, B., "Towards the Ultimate Conservative Difference Scheme. IV. A New Approach to Numerical Convection", JCP, Vol. 23, 1977, pp. 276-299.

3. Van Leer, B., "Upwind-Difference Methods for Aerodynamic Problems Governed by the Euler Equations," Lecture Notes in Applied Mathematics, Vol. 22, 1985.

4. Barth, T. J., and Frederickson, P. O., "Higher Order Solution of the Euler Eqns on Unstructured Grids Using Quadratic Reconstruction", AIAA-90-0013, Jan. 8-11, 1990.

5. Barth, T. J.,"A Three-Dimensional Upwind Euler Solver for Unstructured Meshes," AIAA Paper 91-0721, January 1991.

6. Roe, P.L.,"Approximate Riemann Solvers, Parameter Vectors, and Difference Schemes", J. Comput. Phys., Vol 43, 1981.

7. Thomas, J.L., van Leer, B., and Walters, R.W., "Implicit Flux-Split Schemes for the Euler Equations," AIAA paper 85-1680, 1985.

8. Schmitt, V., and Charpin, F., "Pressure Distributions on the ONERA M6-Wing at Transonic Mach Numbers," in "Experimental Data Base for Computer Program Assessment," AGARD AR-138, 1979.

DISCONTINUOUS FINITE ELEMENT EULER SOLUTIONS ON UNSTRUCTURED ADAPTIVE GRIDS

F. Bassi[1], S. Rebay[1], M. Savini[2]

[1]Dip. di Energetica, Politecnico di Milano, Piazza Leonardo da Vinci 32, 20133 Milano, Italy
[2]CNPM–CNR, Via Francesco Baracca 69, 20068 Peschiera Borromeo (MI), Italy

Introduction

In recent years numerical solutions of Euler equations in problems with discontinuities have been greatly improved by using total variation stable schemes, mainly developed in the finite difference framework. Various TVD (Total Variation Diminishing) and ENO (Essentially Non-Oscillatory) schemes have proven to be accurate and reliable for the application to a large class of problems involving strong discontinuities. These methods, based on one-dimensional algorithms, are extended to multidimensional problems by directional splitting, thus implying the use of structured grids. However, the extension of these higher order schemes to the solution of Euler equations on unstructured grids, is not immediate: related work can be found in Barth [1], in Harten and Chakravarthy [2].

More recently, Cockburn *et al.* [3,4] introduced a discontinuous finite element method in which accuracy is obtained by means of a high-order polynomial approximation inside the elements and a physically sound upwind treatment of the discontinuous solution at elements interfaces is employed. This method is perfectly suited for numerical discretizations on unstructured grids since it does not need wide stencils to improve the accuracy. Unfortunately, as usual in the high-order finite difference schemes, some kind of limiting is needed at discontinuities to enforce total variation stability. The limiting procedure advised by Cockburn *et al.* [3,5] for triangular meshes and that by Bey and Oden [6] for quadrilateral meshes have a strong theoretical background and have proved effective in several test cases. However they cause the numerical scheme to loose the strict local nature offered by the unlimited discontinuous Galerkin approach.

The aim of this work is to investigate by numerical experiments a limiting procedure that relaxes the time evolution of the degrees of freedom that express the deviation from the mean value of the solution inside each element. This rather crude approach has been found to yield very accurate results in a number of one-dimensional unsteady problems and has highlighted the fact that this method should need a very weak limiting. We have also experimented this relaxation technique for two-dimensional steady state applications in order to find if this approach can be useful to drive the solution toward the steady state.

In this paper we present computational results for several test cases, including a shock tube problem, a prototype of a shock-turbulence interaction, the flow in quasi-one-dimensional diverging nozzle, a double Mach reflection, a regular shock reflection.

Discontinuous Galerkin formulation

Consider the Euler equations written in strong conservation form

$$\frac{\partial u}{\partial t} + \nabla \cdot \mathbf{F}(\mathbf{u}) = 0. \tag{1}$$

By multiplying by a test function v, integrating over the domain Ω and performing an integration by parts we obtain the weak statement of the problem

$$\int_\Omega v \frac{\partial u}{\partial t} d\Omega + \int_\sigma v\mathbf{F}(u) \cdot \mathbf{n} d\sigma - \int_\Omega \nabla v \cdot \mathbf{F}(u) d\Omega = 0, \tag{2}$$

where σ denotes the boundary of Ω.

Let's denote by Ω_h a subdivision of the domain Ω into a collection of nonoverlapping elements (segments in 1D and triangles in 2D). A discrete analogue of eq. (2) is obtained by considering functions u_h and v_h which are polynomials P^k of degree less or equal to k inside each element, i.e.,

$$u_h = \sum_i U_i(t)\phi_i(\mathbf{x}), \qquad v_h = \sum_i V_i\phi_i(\mathbf{x}), \tag{3}$$

where U_i and V_i denote the degrees of freedom of the numerical solution and of the test function respectively, and ϕ_i denote the shape functions.

By substituting these expression for u and v in equation (2), the semidiscrete equations for a generic element E can be written as

$$\frac{d}{dt}\int_E v_h u_h d\Omega + \int_e v_h \mathbf{F}(u_h) \cdot \mathbf{n} d\sigma - \int_E \nabla v_h \cdot \mathbf{F}(u_h) d\Omega = 0, \tag{4}$$

where e denotes the boundary of element E.

Due to the discontinuous function approximation, flux terms are not uniquely defined at element interfaces. It is at this stage that the successful nonoscillatory methodology of finite difference (volume) schemes is introduced in the present method. The flux function $\mathbf{F}(u)$ appearing in the second term of eq. (4) is in fact replaced by a consistent numerical flux function $H(u_l, u_r)$ which depends on the two interface states u_l and u_r; among the several numerical flux functions proposed in the literature, in the present work we have used the exact Godunov flux for both the one-dimensional and the two-dimensional problems.

By replacing the flux function $\mathbf{F}$ with the numerical flux H and by evaluating the integrals with appropriate Gauss quadrature formulas, we finally obtain from eq. (4) the system of ordinary differential evolution equations for the element degrees of freedom U_i

$$M\frac{dU_i}{dt} + R_i(U) = 0, \tag{5}$$

where M denotes the element mass matrix and R the vector of residuals obtained by assembling elemental contributions given by eq. (4).

The time integration of this system is accomplished by means of the Runge-Kutta time stepping method with coefficients advised by Shu [7].

Limiting Procedure

We regard the numerical solution in each element as the sum of the mean component and the deviation from the mean, i.e.,

$$u_h(\mathbf{x}, t) = \bar{u}_h(t) + \delta_h(\mathbf{x}, t). \tag{6}$$

Starting from an arbitrary set of shape functions (for example the cardinal functions we have actually used), we express the solution inside each element by using a new set of shape functions which allow to introduce as new degrees of freedom the solution mean and the linear and quadratic components of the deviation from the mean.

The evolution equations for these new degrees of freedom read as

$$\frac{d\bar{U}}{dt} + \bar{R}(U) = 0, \qquad \frac{d\Delta_i}{dt} + \alpha_i R_{\Delta_i}(U) = 0, \qquad 0 \le \alpha_i \le 1. \tag{7}$$

where $\bar{U}$ denotes the degree of freedom corresponding to the solution mean and Δ_i denote the remaining degrees of freedom. The coefficients α_i appearing in eq. (7) have been introduced to relax the time evolution of the deviation components.

Also when dealing with steady state problems we found useful to relax the time evolution of the deviation components during the transient phase of the computations. Note that the steady solution is independent from the adopted relaxation coefficient.

Numerical Results

The method has been applied to one- and two-dimensional problems both unsteady and steady. The considered one-dimensional applications are a shock tube problem, a simulation of a shock turbulence interaction and a steady quasi-one-dimensional flow in a divergent nozzle. The initial conditions for the unsteady problems and the geometry of the divergent nozzle are as follows:

$$(\rho_l, u_l, p_l) = (1, 0, 1)\,, \qquad (\rho_r, u_r, p_r) = (0.125, 0, 0.1)\,. \tag{8}$$

$$\begin{aligned} (\rho_l, u_l, p_l) &= (3.857143, 2.629369, 10.333333)\,, & x < -4, \\ (\rho_r, u_r, p_r) &= (1 + 0.2\sin(5x), 0, 1)\,, & x \ge -4. \end{aligned} \tag{9}$$

$$s(x) = 1.398 + 0.347\tanh(0.8x - 4), \;\; 0 \le x \le 10. \tag{10}$$

The prescribed inlet Mach number and the exit-to-total pressure ratio for the divergent nozzle computation are, respectively, $M = 1.26$ and $p_e/p_t = 0.746$.

In these one-dimensional computations we used quadratic elements and fixed the CFL number to 0.2 according to the linear stability analysis of Cockburn and Shu [3]; the relaxation parameter α has been set to 2/3 for the quadratic component of the deviation, and to 1 (no relaxation) for the remaining components: this value of the coefficient α has been found by numerical testing and represent the limiting case to obtain a stable algorithm with a very limited amount of numerical dissipation. It is surprising that even such a crude choice for the coefficient α yields good results.

The results displayed in Figures 1–3 show that for this class of problems the adopted discontinuous finite element approach allows to obtain highly resolved discontinuities with only very slight over- and under-shoots; in particular, it is worth noting that the density distribution of the shock-turbulence interaction problem is very accurately described even with a mesh of only 100 elements.

Next we present the results obtained for the following standard two-dimensional test problems: the double Mach reflection and the regular shock reflection. For these problems we used P^1 elements, $CFL = 0.2$ and $\alpha = 1/2$. In order to avoid any smoothing of the results, we have plotted the solution inside each element retaining the discontinuities at the element interfaces. In the double Mach reflection test case a Mach 10 shock that impinges on a 30 deg ramp. The computational mesh consisting of 8196 elements and has been constructed by means of the grid generation method described in Rebay [9]. The isodensity lines show that the results are comparable to the best ones presented by Woodward and Colella [8] for the same mesh density. The regular shock reflection example involves an oblique shock reflecting from the lower boundary of a rectangular domain. To enhance the shock resolution, this test case has been computed on the adaptively refined mesh.

Conclusions

A discontinuous finite element method has been employed to compute unsteady and steady shocked flows. The one-dimensional test cases show that it is possible to obtain highly resolved and very accurate solutions by using time evolution limiting without the need of sophysticated forms of spatial limiters that often degrade too much the favourable behaviour displayed by the solutions obtained using the unlimited scheme. The time relaxation limiting procedure has been used also for unsteady and steady two-dimensional problems: in these cases we do not obtain as good results as in the one-dimensional problems. This behaviour is probably due to the use of one-dimensional Riemann solvers to model wave propagation at element interfaces that do not correctly describe the two-dimensional situation. It is likely that some form of spatial limiting has to be employed to control the solution as long as this one-dimensional treatement is employed.

In any case, the favourable behaviour displayed by the proposed scheme with constant relaxation parameters α_i, allows to presume that the eventual form of spatial limiting, possibly necessary in more difficult problems, could be very weak and local.

In the future we plan to carry out a more complete analysis to understand more deeply the time relaxation limiting approach and to investigate other forms of limiting which can be used successfully in conjunction with the discontinuous Galerkin method.

REFERENCES

[1] Barth, T. J. and Frederickson, P. O. (1990). "Higher Order Solution of the Euler Equations on Unstructured Grids Using Quadratic Reconstruction". *AIAA-90-0013*.

[2] Harten, A., Chakravarthy, S. R. (1991). "Multi-Dimensional ENO Schemes for General Geometries". *ICASE Report No. 91-76*.

[3] Cockburn, B. and Shu, C.-W. (1989). "TVB Runge-Kutta Local Projection Discontinuous Galerkin Finite Element Method for Conservation Laws II: General Framework". *Math. Comp.*, *52*, 411–435.

[4] Cockburn, B. and Shu, C.-W. (1989). "TVB Runge-Kutta Local Projection Discontinuous Galerkin Finite Element Method for Conservation Laws III: One Dimensional Systems". *J. Comput. Phys.*, *84*, 90–113.

[5] Cockburn, B., Hou, S. and Shu, C.-W. (1990). "The Runge-Kutta Local Projection Discontinuous Galerkin Finite Element Method for Conservation Laws IV: The Multidimensional Case". *Math. Comp.*, *54*, 545–581.

[6] Bey, K. S. and Oden, J. T. (1991). "A Runge-Kutta Discontinuous Finite Element Method for High Speed Flows". *AIAA paper no. 91-1575-CP*, 541–555.

[7] Shu, C.-W. and Osher, S. (1988). "Efficient Implementation of Essentially Non-Oscillatory Shock-Capturing Schemes". *J. Comput. Phys.*, *77*, 439–471.

[8] Woodward, P. and Colella, P. (1984). "The Numerical Simulation of Two-Dimensional Fluid Flow with Strong Shocks". *J. Comput. Phys.*, *54*, 115–173.

[9] Rebay, S. "Efficient Unstructured Mesh Generation by means of Delaunay Triangulation and Bowyer-Watson Algorithm". to appear in *J. Comput. Phys.*.

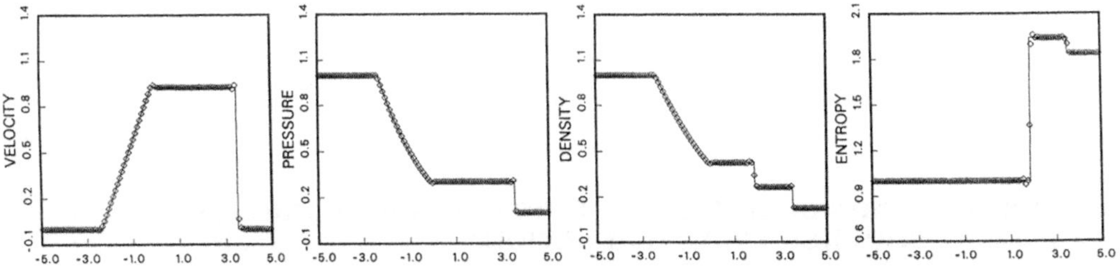

Figure 1: Numerical solution of Sod problem with 100 elements.

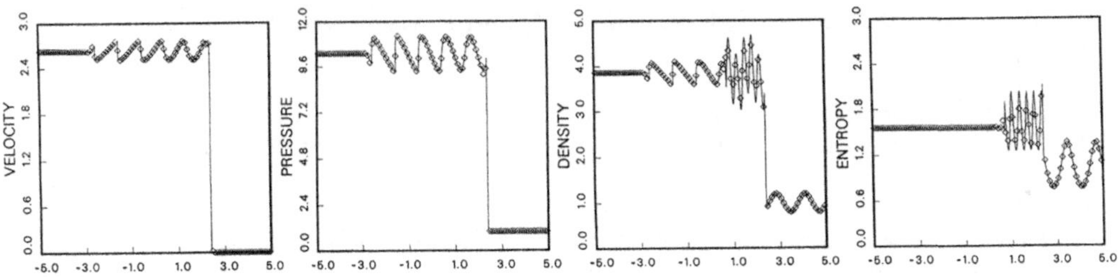

Figure 2: Numerical solution of the shock-turbulence interaction problem with 100 elements.

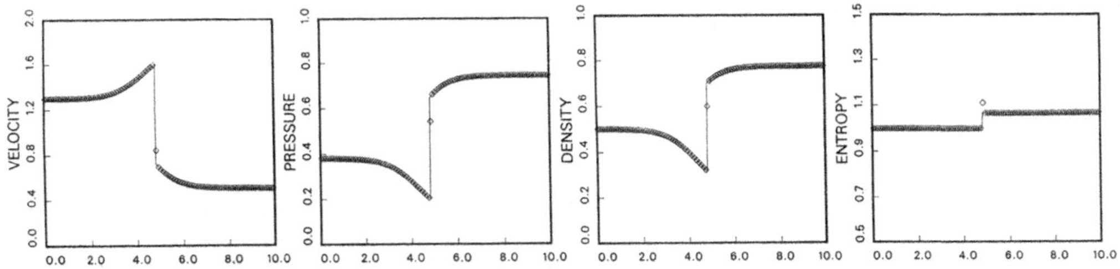

Figure 3: Numerical solution of divergent nozzle flow with 100 elements.

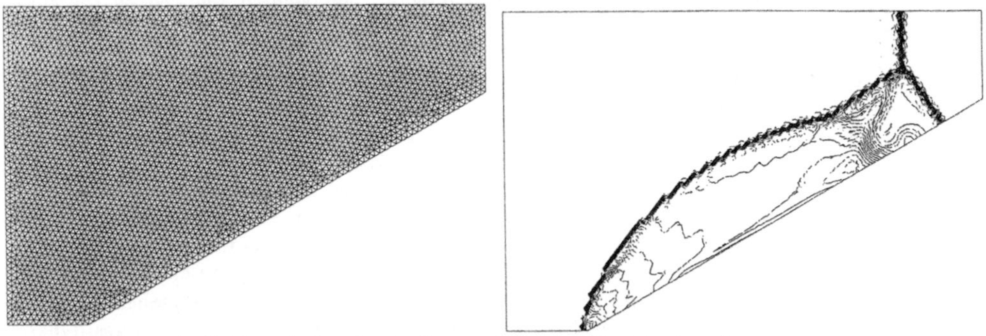

Figure 4: Grid and density isolines of the double Mach reflection problem.

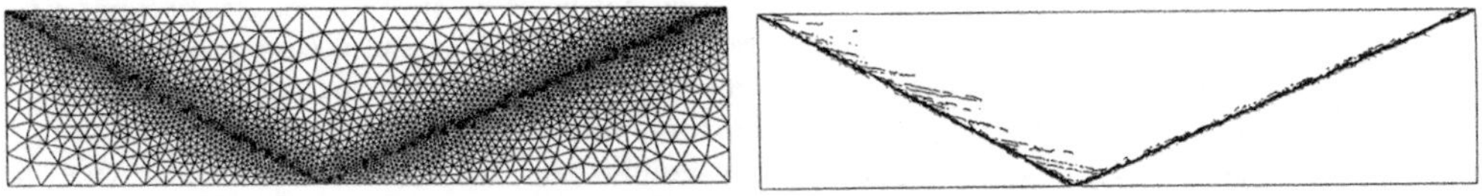

Figure 5: Grid and Mach isolines of the regular reflection problem.

HIGH-RESOLUTION SCHEMES AND UNSTRUCTURED GRIDS IN TRANSIENT SHOCKED FLOW SIMULATION

A.A.Fursenko[†], D.M.Sharov[‡], E.V.Timofeev[‡], and P.A.Voinovich[†]

[†]Advanced Technology Center, P.O.Box 160, St.Petersburg 198103, Russia
[‡]A.F.Ioffe Physical-Technical Institute, Russian Academy of Sciences, St.Petersburg 194021

Introduction

A new high-accurate, efficient and flexible approach combining the advantages of TVD finite-volume schemes and adaptive unstructured grids in numerical simulation of strongly transient shocked gas flows is proposed. Inviscid, non-heat-conducting gas model is considered. The mass, momentum and energy conservation laws written in an integral form underlay the mathematical model. For two dimensions the governing equations are:

$$\int\limits_{\mathrm{Vol}} U dV \Big|_{t_0}^{t} + \int\limits_{t_0}^{t} dt \int\limits_{\sigma} (F_x \nu_x ds + F_y \nu_y ds) = 0;$$

$$U = \begin{Bmatrix} \rho \\ \rho u_x \\ \rho u_y \\ \rho e \end{Bmatrix}, \qquad F_x = \begin{Bmatrix} \rho u_x \\ \rho u_x^2 + p \\ \rho u_x u_y \\ (\rho e + p) u_x \end{Bmatrix}, \qquad F_y = \begin{Bmatrix} \rho u_y \\ \rho u_y u_x \\ \rho u_y^2 + p \\ (\rho e + p) u_y \end{Bmatrix};$$

where the conventional notations are used. Vol represents a gas volume bounded by the closed surface σ with the outward normal $\vec{\nu} = (\nu_x, \nu_y)$.

The shock-capturing scheme

A considerable progress achieved in shocked gas flow simulation during the recent decade is related to the rapid development of high-resolution schemes realizing new ideas and principles to provide high-order accuracy and nonoscillatory solution: TVD, TVB and ENO schemes. Today one can count more than ten different schemes of the above class proposed by various authors. Most of the schemes in the originals differ from each other both in designing mode and in formulation. In spite of the distinctions, an attempt has been made to range these schemes according to a certain system and select the best ones for a given class of gasdynamic problems on the base of the comparative analysis using a special set of test problems. The classification system and comprehensive testing are based on consecutive consideration of 1-d linear transport equation, 1-d and 2-d gas dynamic test problems [1,2].

As a result of the above testing a second-order Godunov-type scheme, proposed by Rodionov [3] for steady-state supersonic flows and modified for transient flows by the authors, has shown better properties in view of the compromise between computational efficiency and accuracy. It does better also for unstructured grids with arbitrary shaped control volumes. The scheme represents a two-step predictor-corrector method providing second order temporal and spatial accuracy for smooth one-dimensional solutions. The predictor step does not employ the Riemann solver which provides its high efficiency:

a. Predictor. For each control volume related to nodal point i we obtain predictor solution $\tilde{U}_i$:

$$\tilde{U}_i \mathrm{Vol}_i = U_i^n \mathrm{Vol}_i - \Delta t \sum_j \left\{ F_x\left(v_{ij}^{\mathrm{in}}\right) \nu_x \Delta S_{ij} + F_y\left(v_{ij}^{\mathrm{in}}\right) \nu_y \Delta S_{ij} \right\},$$

where subscript j denotes grid points surrounding node i; superscript n – time level; Vol_i – value of control volume i; ΔS_{ij} – area of volumes i and j interface; $\vec{\nu} = (\nu_x, \nu_y)$ – unit vector of the outward normal to the control volume i surface; v_{ij}^{in} – primitive gas dynamic variables inside of the control volume surface; and

$$v_{ij}^{\text{in}} = v_i + 0.5\Delta v_{ij}^{\text{in}};$$

$$\Delta v_{ij}^{\text{in}} = \text{limiter}\left\{(v_i - v_{i-1}), (v_j - v_i), (v_j - v_{i-1})\right\};$$

$$\Delta v_{ij}^{\text{out}} = \text{limiter}\left\{(v_j - v_i), (v_{j+1} - v_j), (v_{j+1} - v_i)\right\};$$

$$\text{limiter}(a, b, c) = 0.5[\text{sign}(a) + \text{sign}(b)] \min(2|a|, 2|b|, .5|c|)$$

$$\Delta v_{ij}^{\text{in}} = -\Delta v_{ji}^{\text{out}};$$

where subscripts "$i-1$" and "$j+1$" correspond to the points which are placed on the straight line (i, j), so that $\overrightarrow{(i-1, i)} = \overrightarrow{(i, j)} = \overrightarrow{(j, j+1)}$.

We remark here that in case of structured grids, points $i-1$, i, j, $j+1$ usually belong to the same grid line, when for unstructured grids we have to take an additional assumption about gas dynamic values in points $i-1$ and $j+1$ because there is no grid lines in this case.

It should be noted that even though the predictor step is not conservative because "inside" and "outside" fluxes through the control volume surface differ from one another, the results of the predictor step are used only to compute the fluxes for the corrector step which provides the conservative property of the scheme in general.

b. Corrector. For each volume i we have for time level $(n+1)$:

$$U_i^{n+1}\text{Vol}_i = U_i^n\text{Vol}_i - \Delta t \sum_j \left\{F_x(w_{ij})\,\nu_x \Delta S_{ij} + F_y(w_{ij})\,\nu_y \Delta S_{ij}\right\},$$

where w_{ij} – primitive gas dynamic variables obtained from the Riemann problem solution for $\tilde{v}_{ij}^{\text{in}}$ and $\tilde{v}_{ij}^{\text{out}}$;

$$\tilde{v}_{ij}^{\text{in}} = .5\left(\tilde{v}_i + v_i + \Delta v_{ij}^{\text{in}}\right); \qquad \tilde{v}_{ij}^{\text{out}} = .5\left(\tilde{v}_j + v_j - \Delta v_{ij}^{\text{out}}\right);$$

$\tilde{v}_i$, $\tilde{v}_j$ are the results of the predictor step; and $\tilde{v}_{ij}^{\text{in}} = \tilde{v}_{ji}^{\text{out}}$.

The timestep Δt is chosen so that the Courant stability criterion is satisfied [4].

Unstructured finite-volume method

The following reasons stimulate CFD researches to give up traditional structured grids: 1. structured grids are hardly being fitted to complicated geometries; 2. it is hard to construct a structured grid of a desirable quality at points of geometry singularities (e.g. symmetry center or axis); 3. structured grids cannot be adapted to the solution containing numerous or diversely shaped discontinuities, especially in transient case. The advantages of the unstructured grids consist in natural grid conforming to complicated geometries and in possible adaptation of the grid to the solution peculiarities.

We have used unstructured grids composed of triangular area elements in two dimensions, which provides more isotropic spatial discretization as compared to traditional rectangular grids. Usually, unstructured grids are used together with the finite element methods (for instance, FEM FCT [5]), however, the most accurate shock capturing technique for the above problems is provided by the finite-volume TVD approach [6]. The finite-volume discretization for TVD scheme is analogous to the method described in [6].

Let us consider two-dimensional gas flow in a cartesian coordinate system. We establish a triangular grid on the computational domain. The vertices of the grid triangles correspond to the nodes in which dependent gas dynamic variables are given. The control volumes are constructed around each node as shown in Figure 1. The control volume side ∂S_{ij} placed between nodes C_i and C_j is formed by straight lines connecting the centers of triangles $C_iC_jC_k$ and $C_iC_jC_l$ with the middle of segment C_iC_j. The scheme calculating gas dynamic fluxes through the volume side ∂S_{ij} requires gas dynamic variables in nodes C_i and C_j and also in points C_{i-1} and C_{j+1} which do not coincide with mesh nodes. Values v_{i-1} and v_{j+1} are obtained using interpolation/extrapolation procedure based on the triangles $C_iC_mC_n$ and $C_jC_pC_q$ respectively.

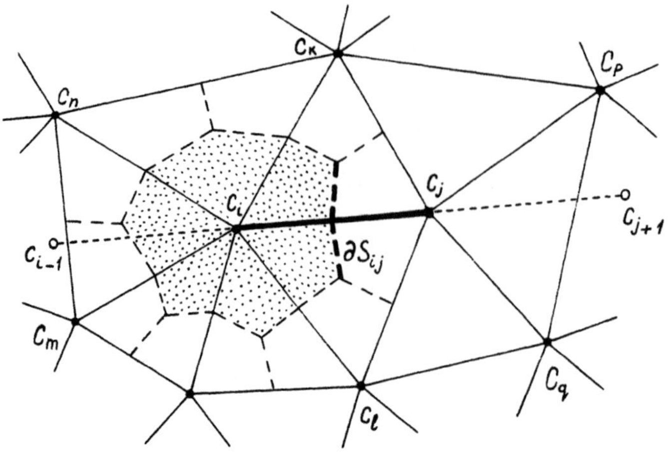

Figure 1. Control volume

The local grid adaptation

The unstructured grid automatically adapts to the solution using refinement/coarsening technique similar to [5], which saves CPU time and computer memory consumption by more than an order of magnitude as compared to the uniform refinement. Unlike [5], the original data structure underlaying the developed computer code is based on mesh nodes assembling, which fits better to the control-volume method and allows to reach excellent efficiency. The data structure needs 7 words per mesh node to storage cross-references for fixed grids and 14 for adaptive grids. No memory to storage "parent" and "son" cells is required. The refinement/coarsening procedure called at each time step takes about 1% of CPU time consumed by the scheme.

Figure 2 shows the six allowable modes for the triangular cells' division and three for their confluence [5].

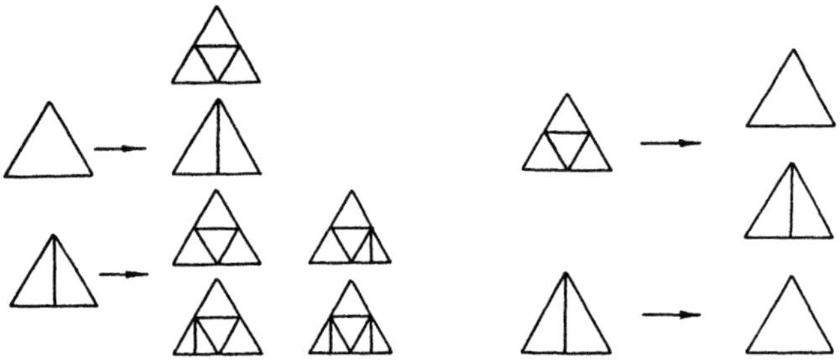

Figure 2. Allowable refinement/coarsening cases

Following reference [5], we utilize here the refinement/coarsening criterion:

$$E_i = \max_j(E_{ij}),$$

where j denotes the neighbors of the node i (see Figure 1), and

$$E_{ij} = \frac{|W_j - 2W_i + W_{i-1}|}{|W_j - W_i| + |W_i - W_{i-1}| + c_n}$$

$$c_n = \epsilon\left(|W_j| + 2\,|W_i| + |W_{i-1}|\right)$$

According to E_i values all the nodes are marked. If $E_i > T_r$, all the triangles of mutual vertex C_i should be fractionized (refinement). If $E_i < T_c$, the node C_i might be removed (coarsening).

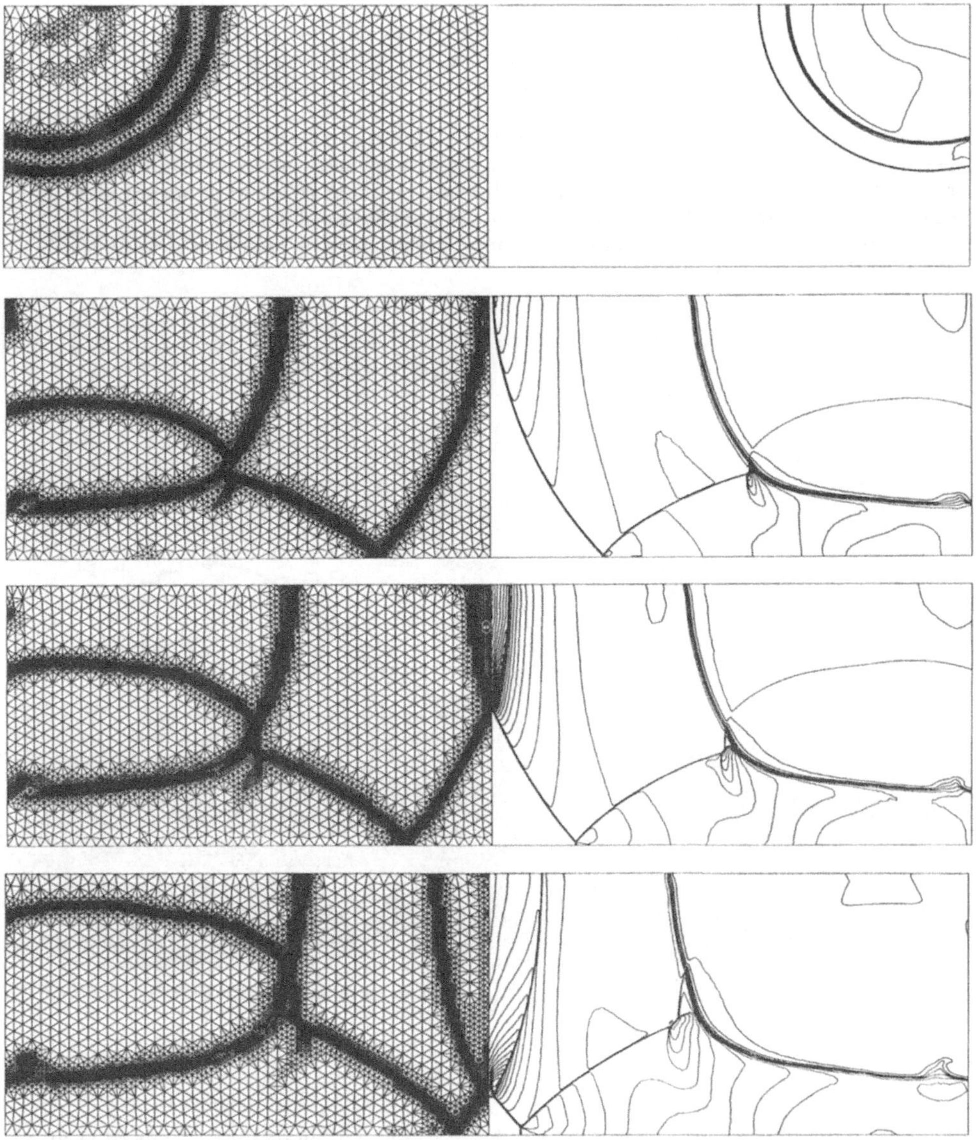

Figure 3. The time history of a toroidal shock wave interaction with the symmetry axis and wall (density contour lines and the respective grids)

The typical values of the refinement and coarsening thresholds and filter coefficient are: $T_r = 0.02$, $T_c = 0.007$, $\epsilon = 0.05$. As a key variable W we have used gas density. The refinement algorithm is described in detail in [4].

Numerical examples

Several gasdynamic problems have been solved using the developed method and code: shock wave diffraction over a 90 degree corner [7,8], initially plane shock wave propagation through plane channel bends [2], toroidal shock collapse near a wall, shock wave diffraction over a circular cylinder and some others. For lack of space, consider here only the last two problems.

The time history of toroidal shock collapse near a wall is shown on Figure 3. The toroidal shock is formed by instant gas heating in a volume shaped like an anchor-ring. Approaching the symmetry axis the shock becomes stronger and reaching the axis reflects, which results in extremely high pressure and temperature. The incident shock reflects on the symmetry axis and wall irregularly, forming Mach disc and cylindrically shaped Mach stem respectively. The simultaneous interaction of the converging Mach stem with the axis and the Mach disc reflection from the wall increase gas pressure and temperature much higher than by the first collapse. As to numerical simulation, we remark here that a high resolution is required because the shock collisions of interest occur in very small regions as compared to the spatial scale of the computational domain.

Figure 4 shows two instant density contour level plots and the respective grids for shock wave $M_s = 3$ diffraction over a circular cylinder.

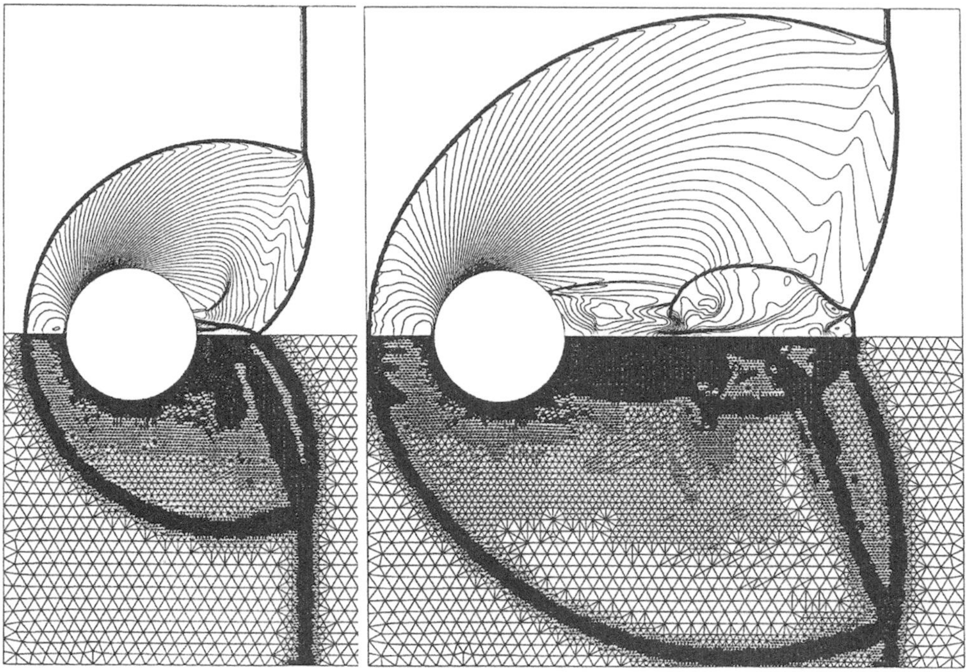

Figure 4. Shock wave $M_s = 3$ diffraction over a circular cylinder (density contour lines and the respective grids)

References

1. Iliin S.A., Timofeev E.V., et al, Preprints A.F.Ioffe Phys. Tech. Institute, USSR Academy of Scienses N1383, N1550 Leningrad (1991). (in russian).
2. Fursenko A.A., Sharov D.M., Timofeev E.V., and Voinovich P.A., *Computers & Fluids*, **21**, p.p.377-396 (1992).
3. Rodionov A.V., *Zh. Vychisl.Mat.Mat.Fiz.*, **27**, p.p.1853-1860 (1987).
4. Voinovich P.A., and Sharov D.M., Preprints A.F.Ioffe Phys. Tech. Institute, USSR Academy of Scienses N1534, N1547 Leningrad (1991). (in russian).
5. Lohner R., AIAA Paper N 87-0555 (1987).
6. Fezoui F., and Stoufflet B., *J. Comp. Phys.*, **84**, p.p.174-206 (1989).
7. Fursenko A.A., Oshima K., Sharov D.M., Timofeev E.V., and Voinovich P.A., Collection of Technical Papers of the 4th Int. Symp. on CFD, **1**, p.p.371-376 (1991).
8. Takayama K., and Inoue O., Posters presented at 18th ISSW. *Shock Waves*, **1**, p.p.301-312 (1991).

VALIDATION OF AN UNSTRUCTURED EULER CODE FOR THE SIMULATION OF TRANSONIC FLOWS PAST A COMPLETE AIRCRAFT

E. Hettena, L. Formaggia, A. Antonellini, V. Selmin

Alenia Aeronautica, Divisione Velivoli Difesa,
Corso Marche 41, 10146 Torino, Italy

Introduction

The exponential growth of computer speed and storage capacity as well as algorithm sophistication has allowed application of advanced numerical methods to practical design problems in many engineering disciplines such as fluid dynamics and aerodynamics. More and more use is being made of numerical methods in analyzing complex flowfields than has been possible in the past. Intelligent use of such capacity can be very helpful in eliminating the poorer designs and in allowing promising design configurations to be developed with less reliance on extensive wind-tunnel testing. In addition, CFD offers the opportunity to obtain detailed flowfield information; some of which is either difficult to measure in a wind-tunnel or is compromised by wall effects, provided that results are accurate as well as cost effective. The problem of code validation is thus of crucial importance.

In this paper, an exercise performed in Alenia Aeronautica in order to validate an Euler code for the computation of transonic flows past a new generation of fighter aircraft, is reported. Through this exercise, we want, above all, to assess the reliability of the unstructured grid generator and solver which were both developed in house.

Mesh generation

An optimal mesh generator should be able to handle complex aeronautical configurations with the minimum effort required by the user. This is a very difficult task to achieve, and it is very likely that a full automation of the gridding process of any realistic 3-D configuration will never be achieved in practice. However, it is felt that the use of unstructured grids can provide the flexibility for a highly automated mesh generation process. The algorithm adopted here is a variation of the front advancing technique, in the form which has been originally developed by J. Peraire et al., and which is described in some detail in [1,2]. The major characteristics of this method are:

- It allows to vary, almost arbitrarily, the mesh density over the computational model.
- It allows to generate anisotropic meshes, i.e. meshes where the elements may be "stretched" along a particular direction.
- It generate nodes and elements at the same time.
- It follows a "geometrical hierarchy" which is suited for interfacing with existing CAD system.

Adaptivity may provide a better solution accuracy without an excessive increase of the number of grid points. Unstructured grids lend themselves to a variety of mesh adaptation methodologies such as mesh movement, mesh refinement and remeshing. The generation process may be subdivided into two stages:

1 Boundary discretization

2 Domain triangulation.

In stage (1) the computational domain boundary is completely covered with a surface mesh formed by triangular facets. In stage (2), a mesh formed by thetrahedral elements is generated, starting from the boundary surface and advancing inside the domain. The *generation front* is here defined as the dynamic data structure which, at any stage of the generation process, will contain the connectivity of the triangular faces which are currently available to form the base for a new element. Everytime a new element is generated, the front is updated by appropriately inserting the tetrahedron faces, so that at the end of the generation process the front is empty.

Euler solver

The code UES3D developed by Alenia Aeronautica D.V.D. solves the 3-D Euler equations on unstructured grids and is able of engine intake/outlet massflow simulation [3]. It is based on an explicit time marching scheme employing a node-centered finite-volume-based central difference spatial discretization.

Spatial approximation is obtained from the integral equations of mass, momentum and energy in a polygonal control volume C_i which is built surrounding each vertex S_i of a finite element type grid, which consists of tetrahedrons. In particular, the flux balance is evaluated by splitting the control volume boundary into panels, so that each panel is shared by a pair of cells C_i and C_j. Thus, the flux balance reduces to the sum of the contributions due to each panel over which we assume that the fluxes are constant. A second-order spatial approximation is obtained by computing the fluxes at the interface as the average value. Non linear second-order and linear fourth-order damping terms are added for stability and shock-capturing properties. Time integration is performed explicitly using a multistage algorithm. Convergence is accelerated with the aid of local time stepping and residual smoothing.

Numerical experiments

An exercise performed in order to validate the previous tools for the computation of transonic flows past a very complex aircraft configuration is described here. Figure 1 illustrates the model inside the wind tunnel, which emphasizes the geometry complexity. This includes foreplanes, tippods, stubpylon+launcher+missile, flap jackfairings, ... The experimental model, manufactured in 1:13 scale, is instrumented with 800 pressure taps on port wing upper surface, starboard wing lower surface fuselage and fin-rudder. The model is provided with a 5 components main balance (C_{n}, C_{m}, C_{y}, C_{yaw}, C_{roll}). The aim of the tests was to provide pressure data from the wings, fin and fuselage and to integrate these pressures to obtain aerodynamic loads and load distributions. They have been successively used for code validation. Our interest, here, will be mainly on pressure data measured on 7 wing sections as indicated in Fig. 2.

The surface definition consists of 154 surfaces provided by a CAD system. Due to the numerous amount of surfaces, one week has been necessary to generate the surface mesh which is shown in Fig. 3. However, the 3-D mesh has been completely generated in one night of Workstation. The grid obtained was made of 141339 nodes forming 763566 elements with 17450 nodes on the airplane skin.

Fig. 1: View of the experimental model

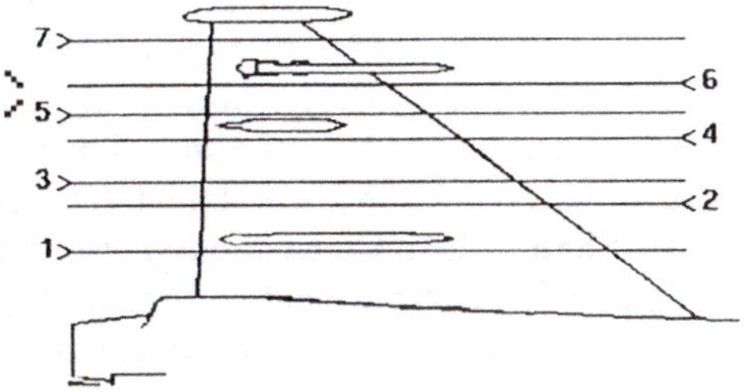

Fig. 2: Location of the wing pressure taps section

Fig. 3: Surface mesh

The Euler code is quite fully vectorized. In particular, a coloring algorithm has been employed in order to avoid data dependencies. Time integration is usually performed using a four-step multistage algorithm where the artificial viscosity terms are evaluated twice. The scheme uses approximately 45 microseconds of CPU time per node and per cycle on a single processor CRAY Y-MP 2E computer. Residual averaging technique requires thirty percents of this time. Steady state solution (decrease of four order of magnitude for the residual) of a typical transonic flow past a complete configuration using about 150000 nodes can then be obtained in less than one hour of CPU time. The memory required is about 70 words per node. A modification of the geometrical definition and the related solution can be obtained in less than a day. Consequently, this numerical strategy can be profitably used in an industrial environment. The Iso-C_p lines on the aircraft surface, obtained at $M_\infty = 0.8$ with a 4° angle of attack and a 2° yaw angle, are given in Fig. 4. In this computation an engine simulation has been made by imposing the mass flow ratio.

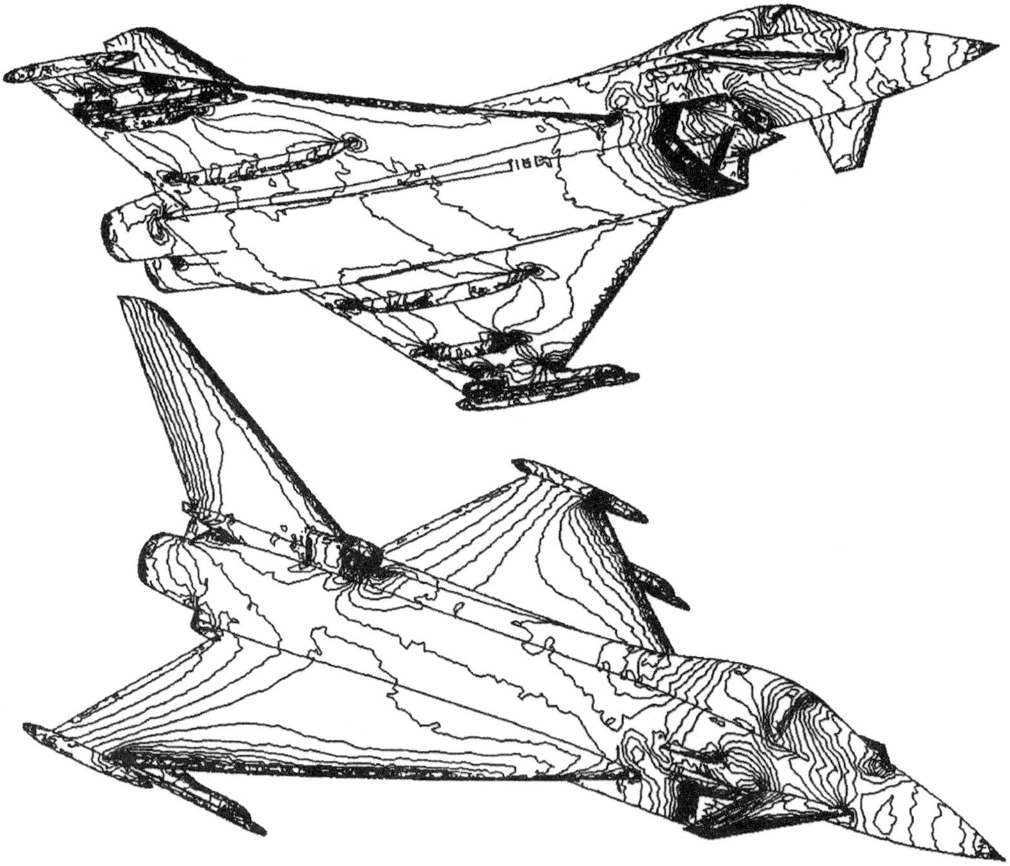

Fig. 4: Iso-C_p contours

Figure 5 finally shows a comparison between measured and computed pressure coefficients at 6 different starboard wing sections. The correlation between experimental and numerical data is quite satisfactory. A good accordance has been found on the windward side of the wing. Some discrepancies appear on the leeward side, especially on the outboard sections due to stronger viscous effects which are not taken into account in our numerical model.

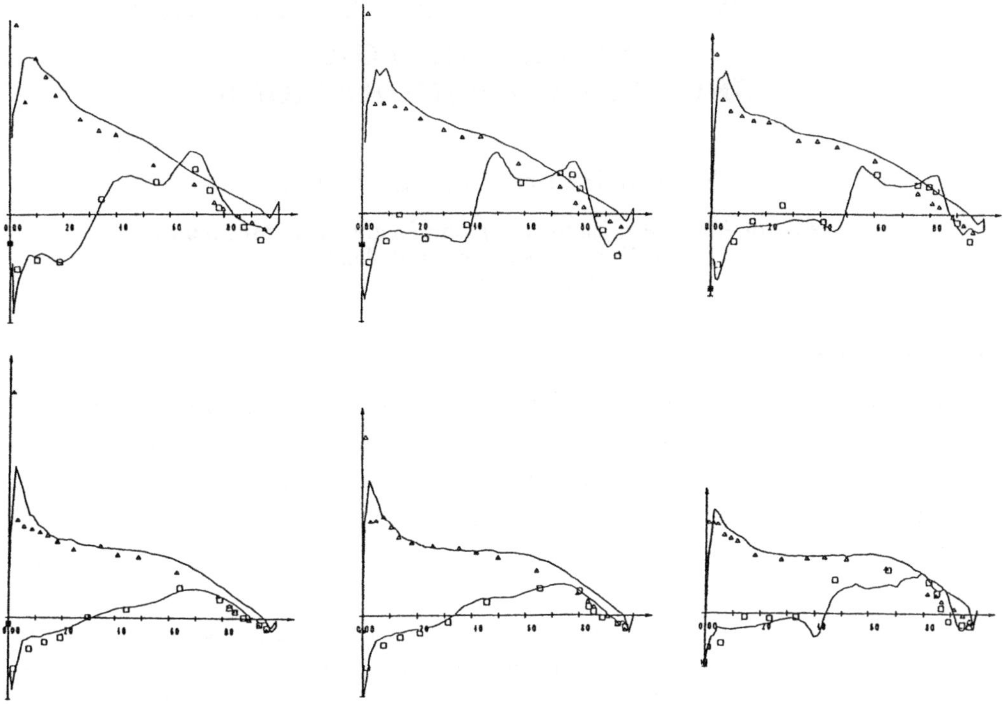

Fig. 5: Numerical (lines) and experimental (symbols) C_p values for 6 wing sections

Conclusion

A numerical procedure to solve inviscid compressible flows about complex geometries has been described. Numerical solutions of transonic flows around a complete fighter aircraft has been compared with available experimental data. The results are quite satisfactory except in regions where viscous effects are important, which is a known limitation of the numerical model. Overall time for grid generation and solution are sufficiently low to use these tools for production.

References

[1] Peraire J., Morgan K., Peiró J.: "Unstructured Finite Element Mesh Generation and Adaptive Procedures in CFD", in Application of Mesh Generation to complex 3-D configurations AGARD CP-464 (1990)

[2] Formaggia L.: "An Unstructured Mesh Generation Algorithm for Three Dimensional Aeronautical Configurations", in Numerical Grid Generation in Computational Fluid Dynamics and Related Fields, A.S. Arcilla et al. Eds., Elsevier Science Publishers B.V. (North-Holland) pp. 249-260 (1991)

[3] Selmin L., Hettena E., Formaggia L.: "An Unstructured Node Centered Scheme for the Simulation of 3-D Inviscid Flows", in Proceedings of the First European Computational Fluid Dynamic conference (1992).

SEMI-UNSTRUCTURED PRISMATIC GRID GENERATION FOR AIRCRAFT CONFIGURATIONS

Y. Kallinderis* , **S. Ward†**

Dept. of Aerospace Engineering and Engineering Mechanics
The University of Texas at Austin
Austin, TX 78712

INTRODUCTION

Simulation of flows around three-dimensional bodies is a major issue in computational fluid mechanics. Geometrical and flow-field complexity combine to make 3-D computations a pacing item. Generation of a body-conforming grid has proven to be a difficult task [1]. Generation of a structured grid may be significantly dependent on geometry and user proficiency. A radical alternative to structured meshes is to use tetrahedra to fill the computational domain. Tetrahedral grids provide flexibility in 3-D grid generation since they can cover complicated topologies easier compared to the hexahedral meshes [1, 6]. However, regions of high gradients, such as a boundary layers, require very thin tetrahedra, which is very difficult to generate.

Prismatic cells, however, are a suitable choice as their semi-unstructured nature allows for very thin cells in the direction normal to the surface while maintaining geometric flexibility in the lateral direction. The prismatic grid region consists of triangular faces that cover the body surfaces, with quadrilateral faces extending in the direction normal to the surface. The topology of a prismatic grid is shown in Figure 1. Employment of a directionally structured grid has some important advantages over fully unstructured grids. First, an implicit numerical scheme that would alleviate stiffness of computations, due to a larger allowable time-step compared to corresponding explicit schemes, is possible in the "structured" direction. Second, implementation of algebraic turbulence models on semi-unstructured grids is much simpler when compared to a fully unstructured grid case [2]. Third, in comparison to tetrahedral grids with comparable resolution, the prismatic grid requires significantly less memory, which is essential to three-dimensional Navier-Stokes computations. The structure of the prismatic grid in one of the directions can be exploited in order to reduce storage to the amount required for two-dimensional Navier-Stokes solvers with triangles [4]. Up to the present, very

*Assistant Professor

†Graduate Research Assistant

few prismatic grid generators have been under development [3, 5]. The present prisms generator is a coupled algebraic / elliptic method.

GENERATION OF PRISMATIC GRID

An unstructured triangular grid is employed as the starting surface to generate a prismatic grid. This grid, covering the body surface, is marched away from the body in distinct steps, resulting in the generation of structured prismatic layers in the marching direction (Figure 1). The process can be visualized as a gradual inflation of the body's volume. There are two main stages in the algebraic grid generation process. In the first, the destination of the marching surface is determined by employing a new technique based on *voxels*; in the second, the nodes are positioned on that surface by determining the *marching* vectors corresponding to the nodes of the previous surface. The marching scheme reduces the curvature of the previous marching surface at each step while ensuring smooth grid spacing.

Direct control of grid orthogonality and spacing is a main advantage of the algebraic method. Algebraic grid generators may yield a grid that is non-smooth and that may overlap. A grid generator must produce a grid with no overlapping faces, which may occur especially in concave regions. In both stages of the present algebraic method elliptic-type steps are employed in the form of Laplacian smoothing.

Smooth Voxel Generation

A voxel is a three-dimensional element used to approximate a point in space. It may have any shape which will entirely fill a domain when voxels are placed adjacent to each other, thus conserving the volume. In the present work, parallelepiped voxels are used. The voxel representation of an object comprises all voxels partially occupying any part of the object. Figure 2 shows the voxel representation of the surface of an F-16A aircraft.

In general, the voxel representation is externally bounded by quadrilateral faces which do not intersect the object. It is this aspect of voxel representations which allows them to be used to generate a conformal surface. The voxel face structure is smoothed in a manner which prevents voxel nodes from moving toward the marching surface. Laplacian-type of smoothing is applied a number of times (typically 10-20). The resulting smooth surface will be termed the *target* surface. Nodes forming the triangular faces of the prismatic grid need to be placed on this surface. Figure 3 shows the corresponding smoothed voxel representation of the same aircraft surface as in Fig. 2. It should be emphasized that the resulting *target* surface has reduced surface curvature compared to the marching surface. Concave regions are "filled-in", while convex areas are smoother.

Node Normals and Marching Vectors

One main issue of the present method is determination of the vectors along which each node of the triangular surface of the previous layer of prisms will march (*marching* vectors). The initial step is calculation of the vectors that are normal to the surface at the nodes. Such normals are frequently difficult to define, since the grid marching surface may comprise discrete faces of discontinuous slope. Furthermore, nodes that lie on concave surfaces must march away from the concavity. Also, as the node marches, it must not intersect the current marching surface prior to intersecting the target surface.

This situation is avoided by enforcing a *visibility condition* [3], which constrains the marching vectors such that the new node position is visible from all faces surrounding the node. This region is the *visibility region.* It has the shape of a polyhedral cone extending outward from the node as shown in Fig. 4. To simplify the constraints, a *visibility cone* with a circular cross-section and centered on the normal vector is constructed at the node. This cone lies completely within the visibility region. The normal vectors are then defined using a method described in [3].

Advancement and Smoothing of Marching Surface

The nodes on the previous surface are connected to the points of intersection of the *target* surface and the marching vectors. Marching to a point on the target surface may result in a reduction in face area caused by "convergence" of the marching grid lines. This will yield a non-uniform mesh; eventually, overlapping may occur. This situation may arise during growth of concave regions due to decrease in the surface area available for node placement. The nodes on the target surface are redistributed by applying a Laplacian-type operator to the marching directions so that the surface elements change their areas and curvature smoothly.

Distribution along the Marching Lines

Flexibility in specifying grid-spacing along the marching lines is crucial for accuracy of Navier-Stokes computations. The marching process generates a relatively small number of prisms occupying a relatively large volume and is therefore an undesirable grid for Navier-Stokes computations. This is remedied by generating new prisms with user-specified dimensions *within* the original prisms.

The prisms generated by marching the surface form a skeleton mesh of valid prisms within which any number of new cells may be generated. A scheme is employed which distributes new nodes along each one of the marching lines emanating from points on the body surface. In other words, the marching directions are maintained, but the marching distances are modified. This is accomplished by performing a cubic-spline fit to each of the marching lines using the prism node locations for the spline knots. New nodes are then distributed along the splined lines. The distribution is such that the new node positions satisfy certain grid spacing requirements.

AIRCRAFT APPLICATION

An F-16A aircraft geometry was chosen as a case for the developed grid generator, since the complexity and singularities of the surface are a severe test for the method. The main features of the configuration are the forebody, canopy, leading-edge strake, wing, shelf regions, and inlet. The surface triangulation consists of 2408 triangular faces , and of 1259 points. Generation of a prismatic grid around this geometry is quite complex, and especially so at the junctions between the different aircraft components. Two parts of the generated prismatic grid require examination in terms of quality. The first is the grid formed by the triangular faces of the prisms (unstructured part), while the second is the grid formed by the quadrilateral faces (structured part).

Two views of the initial and grown surfaces are shown in Figures 5 and 6. The effect of the marching process is similar to inflating of the original body volume. It is observed that the distribution of points on the grown surface is quite smooth. The highly singular regions at the aircraft nose, the wing leading and trailing edges, the wing tip, the canopy, as well as the inlet, have been smoothed-out on the grown surface. Furthermore, the grid spacing on the grown triangular surface is smoother compared to the initial triangulated body surface. The singular concave regions at the junctions between the wing and the fuselage, as well as between the engine inlet and the body have been "filled-in", and the grid is more uniform over those regions compared to the initial grid on the body. A view of the structured part of the prismatic grid is shown in Fig. 7. The grid spacing on the quadrilateral surface is quite uniform. Furthermore, the marching lines emanating from the aircraft surface are quite smooth, including the ones that correspond to singular points.

The number of prismatic cells is 96320 and the number of prismatic grid nodes is 50360. The required computing time for the prismatic grid was 1742 seconds on a Sun station of 2 mflops of speed.

Main advantages of the developed grid generator are its simplicity, low computation cost, its direct control of grid orthogonality, spacing and smoothness, as well as its generality for treatment of complex 3-D geometries.

Acknowledgement

This work was supported by AFOSR Grant 91-0022, and monitored by Dr. L. Sakell. This support is gratefully acknowledged.

References

[1] T. J. Baker, *Transonic Symposium, NASA Langley Research Center, Hampton, Virginia, April 1988.*

[2] Y. Kallinderis, *Journal of the American Institute of Aeronautics and Astronautics*, Vol. 30, No. 3, pp. 631-639, March 1992.

[3] Kallinderis, Y., Ward, S., 10th Applied Aerodynamics Conference, Paper 92-2721, Palo Alto, Ca, June 1992.

[4] Y. Kallinderis, 10th Applied Aerodynamics Conference, Paper 92-2697, Palo Alto, Ca, June 1992.

[5] K. Nakahashi, AIAA paper 91-0103 , 1991.

[6] J. Peraire, J. Peiro, L. Formaggia, K. Morgan and O.C. Zienkiewicz, AIAA Paper 88-0032, January 1988

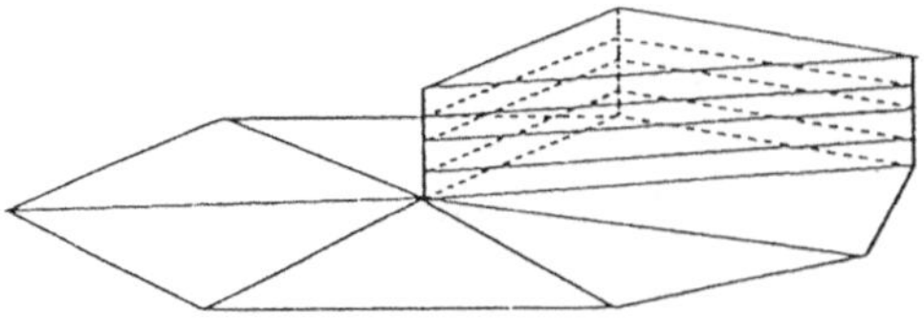

Figure 1: Prismatic Grid Topology

Figure 2: Voxel Representation of F-16A

Figure 3: Smoothed Voxel Representation of F-16A

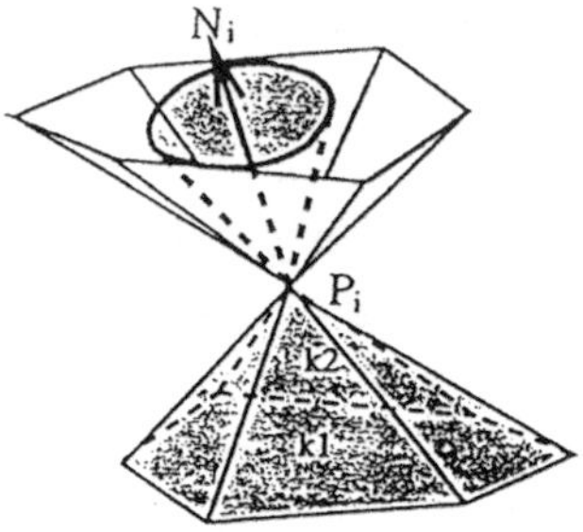

Figure 4: Visibility region

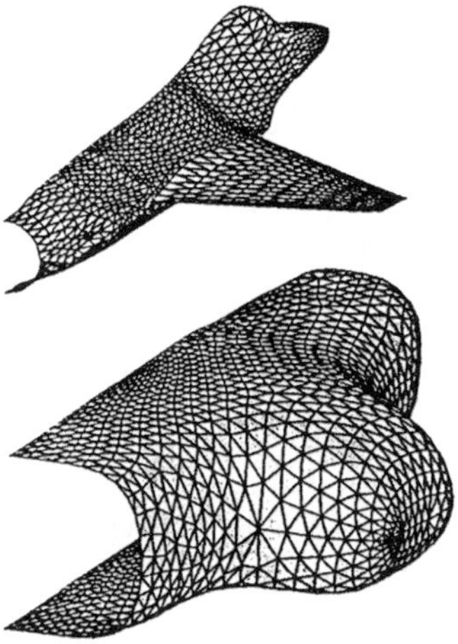

Figure 6: Aftview of Initial and Outermost Prism Surface

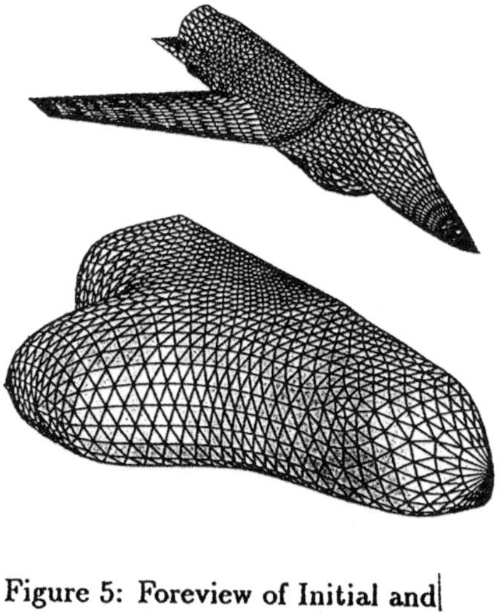

Figure 5: Foreview of Initial and Outermost Prism Surface

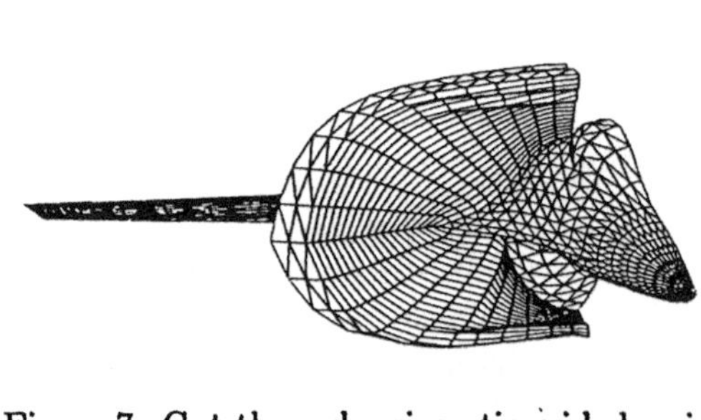

Figure 7: Cut through prismatic grid showing filling of concavities near the strake/fuselage junction

BLOCK FULL MULTIGRID ADAPTIVE SCHEME FOR THE COMPRESSIBLE EULER EQUATIONS

A.KANARACHOS and N.PANTELELIS

P.O.Box 64078, 15710, National Technical University of Athens, GREECE

1. INTRODUCTION

The multigrid methods are well established as fast and robust solvers but further development is essential in order to accomplish the solution of complex equations in a few work units. Although multigrid methods were proposed initially as grid adaptation procedures only few adaptive multigrid schemes such as the MLAT [1], the FAC [2] and other similar schemes [3] have been developed and applied only to elliptic type equations. On the other hand, adaptive schemes for hyperbolic systems of equations have not been developed in a systematic way taking advantage of the favourable multigrid environment [4] although some attempts have been made [5]. At the present study a dynamically adaptive scheme is introduced including a solution error prediction [8] and a composite grid structure of rectangular blocks entirely based on a welltried multigrid method [7]. For the integration and relaxation of the Euler equations an unfactored implicit upwind finite volume scheme is employed. Using the proposed method stable and very accurate results are obtained in a few number of work units.

2. FINITE VOLUME DISCRETIZATION

The basic equations are the conservative time-marching Euler equations in two dimensions and for the solution of the hyperbolic system a cell-centered finite volume method with a Newton linearization scheme for the evolution in time is used, allowing high CFL numbers (100-200):

$$\frac{\Delta U}{\Delta t} + (A^n \Delta U)_\xi + (B^n \Delta U)_\eta = -(E^n_\xi + F^n_n) = -\,\mathrm{Res}\,(U^n) \qquad (2.1)$$

where E and F are the flux vectors normal to the ξ and η faces respectively, while A and B are the jacobians of the fluxes E and F. Upwind differencing of the flux vectors is used to reach diagonal dominant system of equations, hence a symmetric collective point Gauss- Seidel relaxation scheme is implemented efficiently. For the flux calculations a linear locally one- dimensional Riemann solver (Godunov- type) is employed thus, the homogeneous property of the Euler fluxes [6] is guaranteed. The mean values of the conservative variables at both sides of the faces are used as flow variables for the Riemann solver. For the calculation of the fluxes E and F the conservative variables are extrapolated up to third order (MUSCL- type) in computational space at both sides of each face (using two volumes on each side of the corresponding face) depending on the sign of the eigenvalues while sensors are used to guarantee monotonic behavior of the solution.

3. THE BLOCK ADAPTIVE MULTIGRID METHOD

The Block Adaptive Multigrid (BAM) method is based on three main parts: the fast nonlinear multigrid solver, the truncation error prediction and the composite grid solver, while an efficient solution strategy is essential in order to obtain a fast and robust dynamically adaptive method.

With respect to the multigrid acceleration the FAS formulation with the "alternate point of view" [7] is implemented, considering the finer grid levels as devices to increase the spacial accuracy of the solution and the coarser grid levels as devices to accelerate the solution process. The formulation of the problem is independent of the grid level (coarse or fine) and the type of grid (local or global), adding to the RHS of eq.(2.1) the fine-to-coarse defect correction (τ). Thus, the common formulation of the problem for the n-th grid level, when m denotes the corresponding finest grid level and m+1,..,N the next coarser levels, is given by:

$$L^n \cdot \Delta U_n = -Res^n(U_n) + \tau_n^{n+1} \tag{3.1}$$

considering that:

$$\tau_n^{n+1} = \Sigma_n^{n+1} L_n \cdot \Delta U_n - L_{n+1} \cdot (I_n^{n+1} \Delta U_n) \qquad \text{and} \qquad \tau^m = 0 \tag{3.2}$$

As the multigrid coarsening is cellwise (a next coarse volume is formed by four finer ones) the restriction operator (I) for the physical variables is the weighted volume average while the restriction (Σ) for the generalized residuals Res and τ is the summation of the residuals of the corresponding fine volumes, considering that the flux integrals of the inner common faces are canceled so flux conservation is maintained. For the coarse-to-fine direction only the prolongation of the ΔU variables is required as no relaxation work is done, so cellwise bilinear interpolation is sufficient.

In order to determine the erroneous regions of the computational domain where refinement is required, a reliable error estimation should be employed. Based on the Richardson extrapolation an estimation of the truncation error is formulated as $Q(h) \cdot u = c\, h^p$ where Q is the differential operator, u is the physical solution, h is the mesh size, p is the local order of accuracy and c is an unknown factor which is considered as grid independent. Guided by the physical interpretation of the truncation error, the difference of the residuals between the fine and the corresponding coarse volumes can provide a reliable local estimation of the solution error and moreover a prediction of the error for the next finer grid level. Due to the Newton linearization currently used, the Res(U) differential operator is more suitable than the $L \cdot \Delta U$ operator for the truncation error estimation as the former is insensitive to the relaxation errors maintaining the accuracy of the solution. So the truncation error estimation ($T = Q(h) \cdot u$) for the current grid and the prediction for the next finer grid are given respectively by:

$$T_n^{n+1} = \Sigma_n^{n+1} Res_n(U_n) - Res_{n+1}(I_n^{n+1} U_n) \qquad \text{and} \qquad T_{n-1}^n = 2^{-p}\, T_n^{n+1} \tag{3.3}$$

This error predictor requires only an additional quarter of the currently integration work, it is valid for nearly convergent solution ($Res_n(U_n) \neq 0$) and moreover converges fast to its steady state value. In order to specify the order of accuracy (p) the sensing functions that are used to control monotonicity near discontinuities by the integration process are used also to calculate the reduction of the order of accuracy at these regions.

In order to realize the adaptive solution for complex problems several additional issues have to be settled. A composite grid structure is employed which enables the solution of a non uniform grid using a union of uniform subgrids. When the subgrids are restricted to rectangles (blocks) in the computational plane [4] significant simplifications to the data structure and the interface treatment are obtained. To suppress the errors not consistent with the solution method at the artificial boundaries due to the existence of "hanging" nodes, extra artificial boundary conditions are required, namely flux conservation, a compatible variable extrapolation scheme and a special relaxation process. According to the multigrid coarsening a compatible refinement process is chosen hence, two fine volume faces form a coarse one. Thus, the flux conservation at the artificial boundaries is achieved explicitly, calculating first the fluxes at the fine subgrid faces, while the fluxes of the adjacent faces of the coarse subgrid are the summation of the corresponding fine grid fluxes. In order to calculate the fine grid fluxes at the artificial boundaries the same integration subroutine is used but an accurate extrapolation towards the adjacent coarse volumes is required as shown in fig.1 for the one dimensional example. Thus, from the coarse volumes 1,2 and the fine ones 3,4 in fig.1 the two virtually equivalent fine volumes 5,6 are calculated using the same MUSCL- type extrapolation scheme which is used by the integration process, considering that the order of accuracy and the compatibility with the initial integration scheme are maintained. Finally, for the flux vector splitting and the relaxation process the interface problems are handled by the multigrid cycle. As shown in fig.1 only the fine block (block 1) is relaxed at the first grid level, while at the next multigrid level the coarse block (block 2) is relaxed together with the restricted fine one preserving the block structure throughout the multigrid cycle using a "horizontal" communication scheme among blocks.

With respect to the dynamically adaptive multigrid strategy a modified Full Multigrid scheme is applied. That is: starting from a global coarse grid with an acceptable solution accuracy after convergence (or after a fixed amount of work) the truncation error estimation is applied and where the error estimate is above a threshold the corresponding volumes are flagged and grouped into rectangular blocks. Then, the computational domain is decomposed to the appropriate blocks where only those which contain the flagged volumes are refined to the next grid level and the solution procedure continues until the entire computational domain is below the truncation error level and the solution has totally converged. This strategy has the benefit of a continuous procedure taking advantage of the most accurate available solution it converges faster towards the best solution, hence no wasted operations occur.

4. RESULTS

In order to validate the efficiency of the proposed method two test cases are investigated: the first case is a NACA-0012 airfoil for Mach 0.85 and angle 1.0 degree while the second case is the RAE-2822 airfoil for 0.73 Mach and 2.79 degrees. A Work Unit (W.U.) is defined by the CPU-time required for a global finest grid relaxation sweep in lexicographic order. Results for the multigrid implementation to global fixed grids [7] as well as for the truncation error estimation [8] can be found for inviscid and viscous flows.

For both test cases an automatic adaptation process as described previously is applied starting with two global multigrid levels with 64x14 volumes at the finest grid level. The user supplies only the number of the additional grid levels and the truncation error threshold, which for the present test cases were kept the same: two additional refinement levels and a three-fold reduction for the initial truncation error levels. It is important that throughout the solution process the multigrid convergence rate was maintained and the overhead for the interface manipulations was only 2 % for a nine block structure with respect to an equivalent global grid. Thus, the required work for the solution of an adaptive grid is proportional to the number of volumes with the same convergence rate as the global multigrid solution, i.e. the acceleration of the BAM method with respect to the global multigrid equals the reduction of the number of volumes.

For the first case the convergence of the drag coefficient (solid lines) and the truncation error estimation (dashed lines) with respect to the number of volumes are given in fig.2 where a great similarity between the error and the accuracy is shown. In fig.3 the superiority of the BAM method with respect to the single grid and the global multigrid schemes is shown, where a 14-fold acceleration was achieved with respect to the single grid case using 3.2 times less volumes for practically the same accuracy (0.5% discrepancy of the Cd). The final composite grid structure and the corresponding mach contours are shown in fig.4 where the quality of the solution is exactly the same as for the global solution even where shock crosses the artificial boundaries.

Similar efficiency was achieved for the second test case where the finest refinement regions spread only at the upper region of the airfoil (fig.6). Using the same truncation error threshold the total number of volumes is less than the previous case, so a 17 times acceleration was achieved with respect to the single grid and a four-fold reduction of the volumes for almost the same accuracy (Cl discrepancy is 0.2 %). The convergence histories of the error reduction and the lift coefficient are shown in fig.5 while in fig.6 the final composite grid together with the isomach contours are given.

5. CONCLUSIONS

The great advantages of the Block Adaptive Multigrid (BAM) method were exhibited. The incorporation of numerous efficient schemes into the BAM method is very promising as the ultimate target of solving complex problems in just a few work units is feasible, maintaining the robustness, the simplicity and the accuracy of the single grid code. Although many issues still have to be settled, the extension to the viscous three dimensional problems as well as the implementation of the BAM method to parallel machines are expected to be straight forward and very promising. On the other hand, in order to improve the adaptation capabilities and the efficiency of the BAM method a moving grid point scheme should also be considered as the grid alignment towards certain flow features is essential in some problems in combination with the present grid refinement method.

ACKNOWLEDGEMENTS

This work was partially supported by CEC BRITE/EURAM Project Aero 0018. The authors wish to thank MBB GmbH for providing the EuFlex code.

REFERENCES

[1] **A.Brandt,** Multi-level Adaptive solutions to boundary-value problems, Math. Comp., 31, pp. 333-390

[2] **S.McCormick,** Multilevel Adaptive Methods for Partial Differential Equations (SIAM, Philadelphia, 1989).

[3] **W.Joppich,** Proceedings of the 3rd European Conference in Multigrid methods, eds W.Hackbusch and U.Trottenberg, Int.Ser.Num.Math, 98 (Birkhaeuser Verlag, Basel, 1991) pp. 241-252

[4] **M.J.Berger and A.Jameson,** Automatic adaptive grid Refinement for the Euler Equations, AIAA J., **23** (1985), pp. 561

[5] **J.F.Dannenhoffer,III,** Adaptive methods for partial differential equations, eds J.Flaherty et al, (SIAM, Philadelphia, 1989) pp.68-82.

[6] **A.Eberle,** Characteristic flux averaging approach to the solution of the Euler equations, VKI Lecture series 1987

[7] **A.Kanarachos and N.Pantelelis,** An alternative multigrid method for the Euler and Navier Stokes equations, J.Comp. Phys. (Submitted)

[8] **A.Kanarachos, N.Pantelelis and I.P.Vournas,** Multigrid methods for the acceleration and the adaptation of the transonic flow problems (Vieweg Verlag, Braunschweig, 1992) to be published.

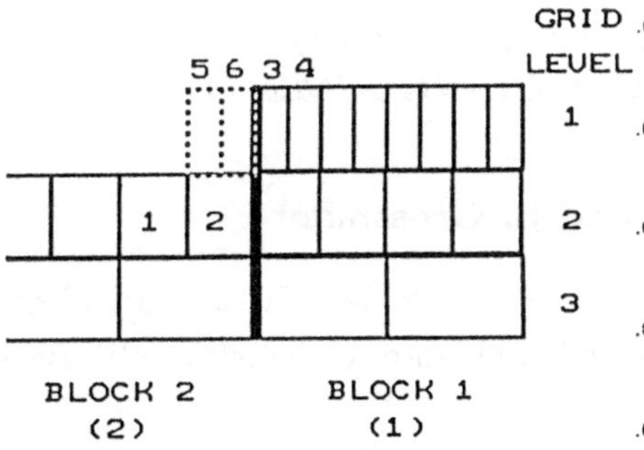

FIG. 1 One dimensional example of the handling of the composite grid structure.

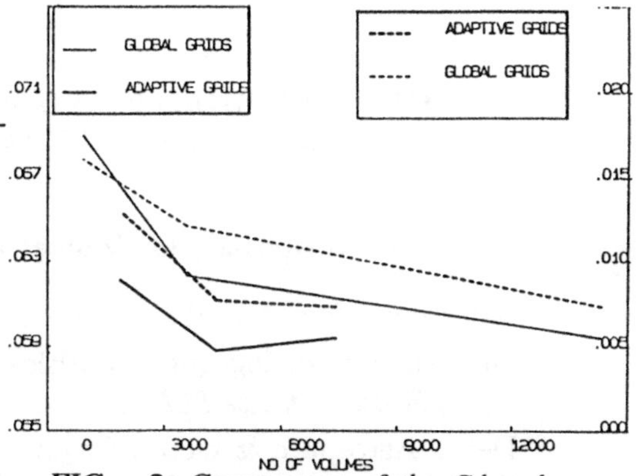

FIG. 2 Convergence of the Cd and the max truncation error with respect to the number of volumes (CASE 1).

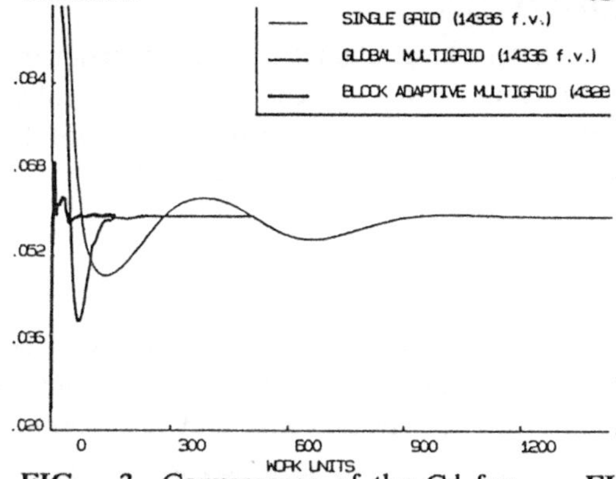

FIG. 3 Convergence of the Cd for single grid, multigrid and the BAM methods (CASE 1).

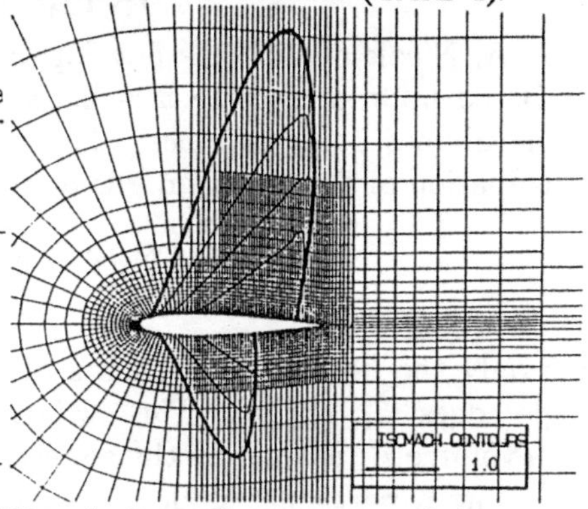

FIG. 4 Isomach contours and the composite grid structure (CASE 1).

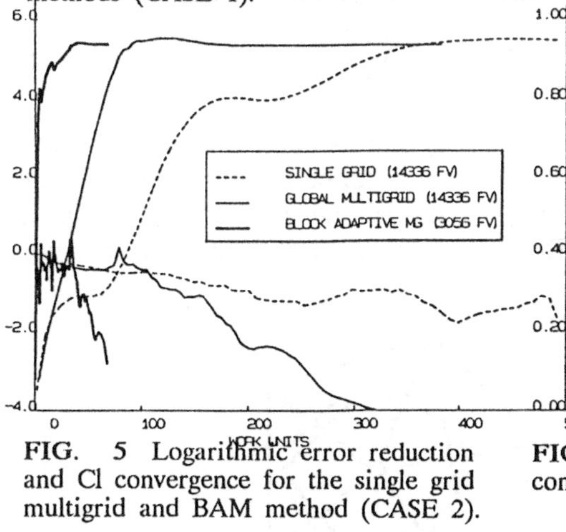

FIG. 5 Logarithmic error reduction and Cl convergence for the single grid multigrid and BAM method (CASE 2).

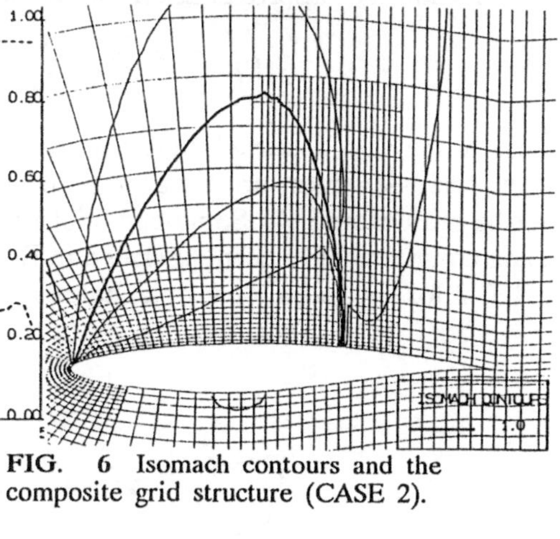

FIG. 6 Isomach contours and the composite grid structure (CASE 2).

INCOMPRESSIBLE NAVIER-STOKES SOLUTIONS ON UNSTRUCTURED GRIDS USING A CO-VOLUME TECHNIQUE

A. Lippolis*, G. Vacca† and B. Grossman‡

* Istituto di Macchine ed Energetica, Politecnico di Bari, 70125 Bari ITALY

† Dipartimento di Ingegneria e Fisica dell'Ambiente, Università della Basilicata, 85100 Potenza ITALY

‡ Dept. Aerospace & Ocean Engineering, Virginia Polytechnic Institute & State University, Blacksburg, VA 24061 USA

INTRODUCTION

R. A. Nicolaides[1–3] has developed an interesting algorithm for unstructured grids based upon a discrete *div-curl* formulation on a dual set of complementary control volumes or *co-volumes.* The main features of the approach includes the approximation of $\nabla \cdot \vec{V}$ within a triangular mesh τ, see Fig. 1a, such that

$$\iint_{\tau} \nabla \cdot \vec{V}\, dS = \oint_{\partial\tau} \vec{V} \cdot \hat{n}\, dl \approx \sum_{i} q_i h_i\,, \tag{1}$$

where only the normal components of velocity $q = \vec{V} \cdot \hat{n}$ and the lengths of the triangle edges, h_i, appear. The *curl* of $\vec{V}$ is approximated on the complementary control volume τ', formed by the connection of all the circumcenters of the triangles surrounding a particular node, as shown in Fig. 1b. For a Delaunay triangulation, the resulting polygon co-volume is the Vornoi diagram for the nodes of the triangulation. An approximation for the vorticity may be found from

$$\iint_{\tau'} \vec{\omega} \cdot \hat{n}\, dS = \oint_{\partial\tau'} \vec{V} \cdot \hat{t}\, dl \approx \sum_{j} q_j l_j\,, \tag{2}$$

where it may be observed that the same velocity components which are normal to the triangle edges, are tangential to the sides of the co-volume. The directed lengths are denoted l_j. Some candidate approaches to the solution of the Navier-Stokes equations have been described in Ref. 1 and calculations have been presented in Ref. 3 for the radial flow in a channel.

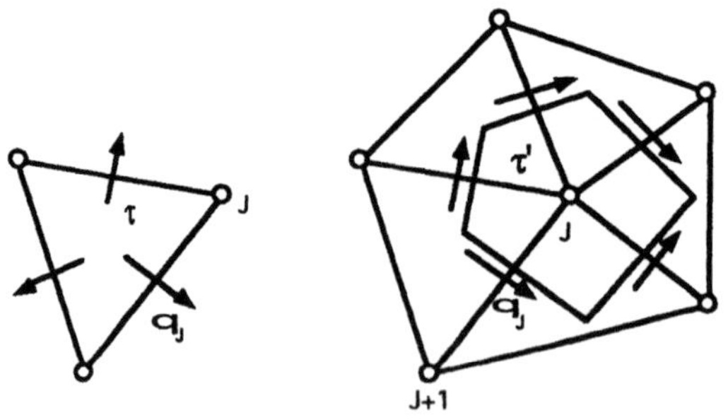

Fig. 1 Triangular and Complementary Control Volumes

FORMULATION

We have developed new procedures to apply this algorithm to the Navier-Stokes equations within a finite-volume formulation on an unstructured mesh of triangles and their associated co-volumes. The vorticity-velocity form of the governing equations is used in integral conservation-law form, with the vorticity transport equation,

$$\frac{d}{dt}\iint_{\tau'} \omega\, dS + \oint_{\partial\tau'} \left(\omega\vec{V} - \nu\nabla\omega\right)\cdot\hat{n}\, dl = 0\,, \tag{3}$$

the continuity equation,

$$\oint_{\partial\tau} \vec{V}\cdot\hat{n}\, dl = 0\,, \tag{4}$$

and the vorticity definition

$$\iint_{\tau'} \omega\, dS = \oint_{\partial\tau'} \vec{V}\cdot\hat{t}\, dl\,. \tag{5}$$

The discretization of the (3)–(5) may be performed in a straightforward manner using (1) and (2), with the exception of the non-linear convection terms. One can note that the discretization of $\oint_{\partial\tau'} \omega\vec{V}\cdot\hat{n}\, dl$, will involve the average of ω at neighboring nodes times the velocity normal to each edge of the co-volume. However, we have only defined the velocities normal to the triangle sides, which are tangential to the co-volume edges. A reconstruction of the velocity field from the normal velocity components in the triangular domain is performed by assuming that the vector velocity is constant in each triangle. The vector velocity may be then determined from any two cell-edge normal velocities. The tangential velocities along the side of each triangle are then easily constructed. The value of the normal velocity to the co-volume edge is then taken as a weighted average of the tangential velocity over the edges of the neighboring triangles. We have verified this form of convective differencing will have the added benefit that it enforces $div\,\vec{V} = 0$ on the co-volume. (This procedure was independently verified recently in Ref. 3). The system is closed by specifying the normal and tangential velocities on the boundaries. The boundary vorticity is computed from Eq. (3) for the partial co-volume centered on the boundary node as illustrated in Fig. 2. For cases where the domain is multiply connected, such for the flow about an airfoil, one additional boundary condition is required. We specify the normal velocity at a single cell-edge near the outer boundary of the domain.

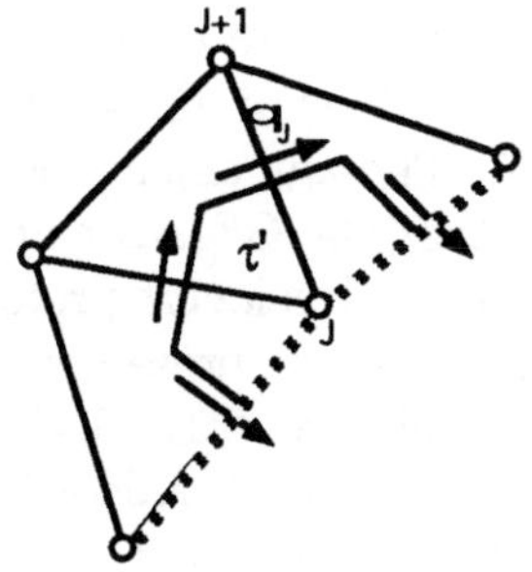

Fig. 2 Complementary Control Volume Near Boundary

This procedure has been incorporated with a fully implicit time integration. A steady-state solution is developed using an infinite time step and a direct method to solve the resulting sparse matrix system. Due to the quasi-linearization of the non-linear convective term in Eq. (3), the solution will be iterative. For the two-dimensional applications considered here we have implemented a stream function, ψ defined at the triangle vertices which reduces the memory requirements for the direct solver. The cell-edge velocity, q is simply the difference in ψ at the two vertices divided by the edge length h. The co-volume averaged vorticity will involve ψ at the cell center and ψ at neighboring vertices. The resulting matrix is assembled into banded form of reduced matrix sparseness by opportunely reordering the ψ_j variables. A direct method based on a generalized Thomas algorithm is used for the matrix solution.

Once this system is solved for the vorticity and velocity components, the pressure may be evaluated from a discretized form of

$$\iint_{T'} \left[\nabla \left(\frac{p}{\rho} + \frac{\vec{V} \cdot \vec{V}}{2} \right) - \vec{V} \times \vec{\omega} \right] \cdot \hat{n} \, dS = 0 \, . \tag{6}$$

Eq. (6) gives an implicit relationship for the total pressure involving only the triangle-edge normal velocity components and the node-centered values of vorticity.

RESULTS AND DISCUSSION

Several test cases have been computed to assess the accuracy, efficiency and reliability of this approach. First we considered the flow in a driven cavity at Reynolds number of 100 using a triangular grid formed by bisecting uniform rectangular grids of 16×16 and 48×48 to form 512 and 4,608 triangles, respectively. The computed vorticity distribution along the moving wall is presented in Fig. 3 and is in good agreement with the results of Napolitano and Walters[4]and those of Ghia[5] *et al.* Next a $R_e = 10,000$ driven cavity was computed on grids formed by bisecting 64×64 and 128×128 rectangular grid which has been hyperbolically stretched. The moving-wall vorticity distribution shown in Fig. 4 agrees well with the results of Ref. 5 which used a uniform 256×256 grid. (*N.B.*, the $R_e = 10,000$ case is used as numerical test case. Physical steady-state solutions to the unsteady Navier-Stokes equations may not exist for the driven cavity at this Reynolds number).

Calculations of the incompressible Navier-Stokes flow over an NACA 0012 airfoil at $R_e = 500$ and 2000 have been performed on an unstructured grids formed by Delaunay triangulation[6] containing 4096 and 9216 nodes. The 9,216 node grid is shown in Fig. 5. The complete grid extended 12 chord lengths in each direction. Surface pressure distributions are presented in Fig. 6 for the $R_e = 500$ case and in Fig. 7 for the $R_e = 2000$ case and are compared with the results of the code CFL3D[7]. We note that the calculations at $R_e = 500$ appear to be mesh-converged and agree with those of Ref. 7. The agreement for the $R_e = 2000$ is not as good and is probably due to the fact that the CFL3D calculations had a finer mesh near the airfoil surface. Interpolated axial velocity profiles for the $R_e = 2000$ case at locations of $x/c = 0.2$ and $x/c = 0.8$ are presented in Figs. 8 and 9, respectively. Although the agreement is relatively good, the lack of resolution of the boundary layer at the $x/c = 0.2$ location for the 9216 node grid is evident.

These preliminary results are encouraging and establish the feasibility of the co-volume technique for two dimensions. Further studies are required to evaluate the efficiency of the procedure. The real test of the method will come for three-dimensional applications.

ACKNOWLEDGMENTS

The research of the first two authors was partially supported by the *Consiglio Nazionale delle Richerche.* The third author acknowledges the assistance of the National Science Foundation under the U.S.-Italy Cooperative Science Program, grant INT–8814895. The authors would like to thank Dr. D. Mavriplis of ICASE for providing the airfoil grid-generation program and Dr. C. Rumsey of NASA Langley for providing the Navier-Stokes results from CFL3D.

REFERENCES

1. Nicolaides, R. A., "Flow Discretization by Complementary Volume Techniques", Proceedings of the 9th AIAA Computational Fluid Dynamics Meeting, AIAA-89-1978-CP, Buffalo NY, June 1989, pp. 464–469.
2. Nicolaides, R. A., "Direct Discretization of Planar *div-curl* Problems", ICASE Report No. 89–76, Oct. 1989.
3. Nicolaides, R. A., "Covolume Algorithms", Proceedings of the 4th International Symposium on Computational Fluid Dynamics, Davis CA, Sept. 1991, pp. 861–866.
4. Napolitano, M. and Walters, R. W., "An Incremental Block-Line-Gauss-Seidel Method for the Navier-Stokes Equations", *AIAA Journal,* **24**, No. 5, May 1986, pp. 770–776.
5. Ghia, U., Ghia, K. N. and Shin, C. T., "High-Re Solutions for Incompressible Flow Using the Navier-Stokes Equations and a Multigrid Method", *J. Comput. Phys.,* **48**, 1982, pp. 387–411.
6. Mavriplis, D. J., "Adaptive Mesh Generation for Viscous Flows Using Delaunay Triangulation", ICASE Report No. 88–47, Aug. 1988.
7. Thomas, J. L. and Walters, R. W., "Upwind Relaxation Algorithms for the Navier- Stokes Equations", Proc. AIAA 7th CFD Conf., AIAA–85–1501–CP, July 1985.

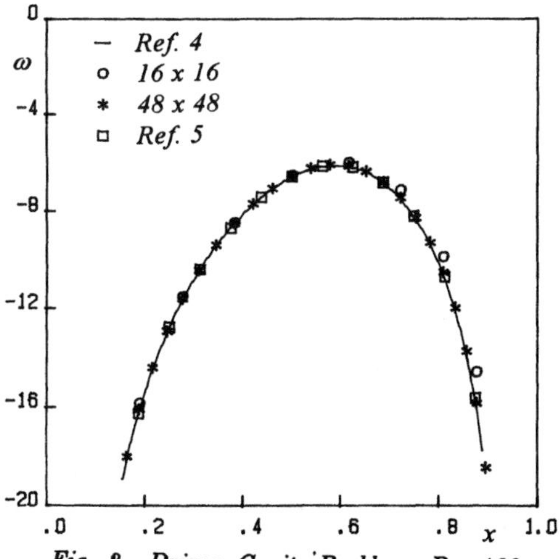

Fig. 3 - Driven Cavity Problem Re=100
Wall vorticity distribution

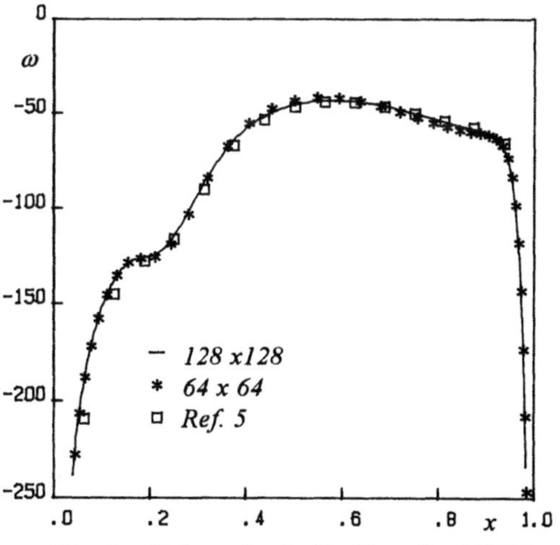

Fig. 4 - Driven Cavity Problem Re=10000
Wall vorticity distribution

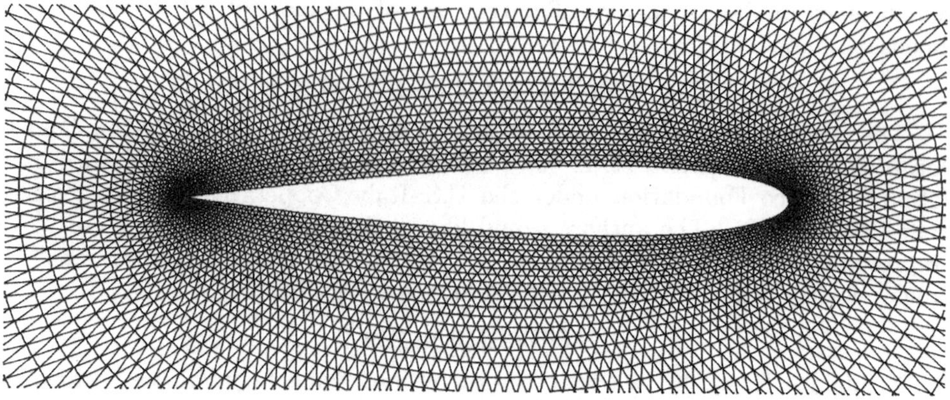

Fig. 5 - Delaunay triangulated 9216-node grid around an NACA 0012 Airfoil

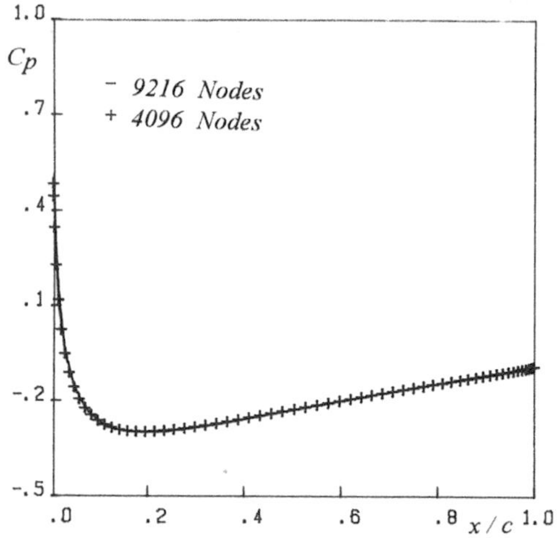

Fig. 6 - NACA 0012 Airfoil, Re=500, $\alpha = 0$
Surface Pressure Distribution

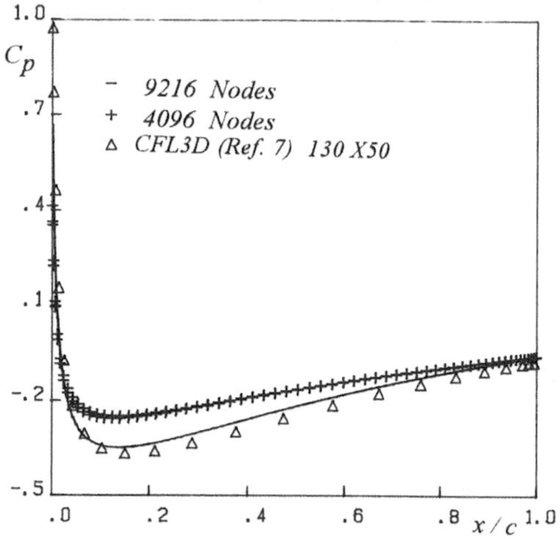

Fig. 7 - NACA 0012 Airfoil, Re=2000, $\alpha = 0$
Surface Pressure Distribution

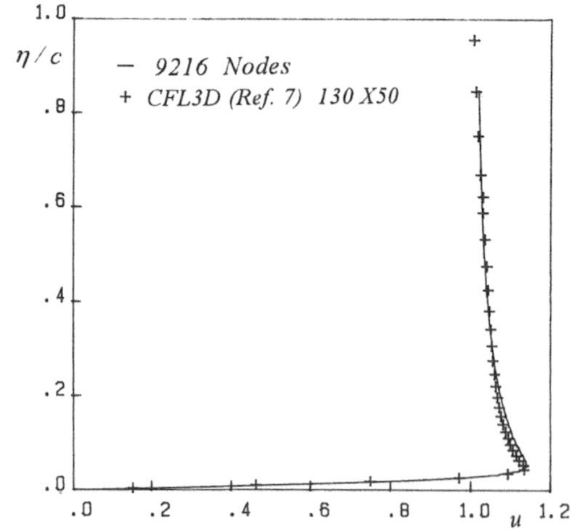

Fig. 8 - NACA 0012 Airfoil, Re=2000, $\alpha = 0$
Velocity normal to surface at x/c=.2

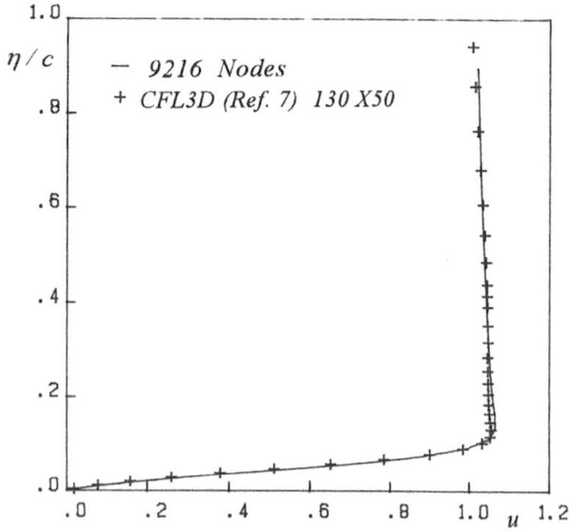

Fig. 9 - NACA 0012 Airfoil, Re=2000, $\alpha = 0$
Velocity normal to surface at x/c=.8

DELAUNAY BASED TRIANGULATIONS FOR THE NAVIER-STOKES EQUATIONS WITH MINIMUM USER INPUT

J.-D. Müller[1,2], P.L. Roe[1], H. Deconinck[2]

[1] Department of Aerospace Engineering, The University of Michigan, USA,
[2] CFD-Group, Von Karman Institute for Fluid Dynamics, Sint-Genesius-Rode, Belgium.

1. Introduction

Over the last decade solutions of the Euler equations on unstructured meshes have become a standard procedure. The slightly increased time and storage requirements of the unstructured solvers are outweighed by the drastic reduction in man and machine power for the generation of unstructured grids as they can be generated in a mostly automatic fashion starting with only the boundary nodes as user input. Still, in order to have control over the distribution of the vertices in the mesh, additional user input in the form of a user generated background grid [1] or a node cloud derived from structured grids [2] is required by the currently known algorithms. This becomes even more important for grids for Navier-Stokes calculations where the mesh structure has to resolve very thin shear layers.

The method presented here is based on a philosophy of truly minimal user input. In the case of isotropic meshes for the Euler equations minimal user input consists of the set of boundary nodes only. In the case of stretched meshes for Navier-Stokes calculations, the thickness of the viscous layer and the desired maximum aspect ratio are also needed to make the problem well-posed. However, simple qualitative information about the boundary surfaces will be used to strongly improve grid quality.

Our method is a Steiner triangulation, i.e. starting from an initial triangulation, nodes are inserted incrementally in a frontal manner until a final grid with an appropriate node distribution is achieved. The concept is based on the Delaunay triangulation [3] and its construction principle to connect closest nodes is exploited for the automation of the grid generation process. Moreover, Delaunay grids have many interesting properties, including the fact that a Delaunay triangulation maximizes the minimum angle of all possible triangulations with a given set of vertices and is in this sense the smoothest triangulation. An extensive documentation on Delaunay methods can be found in [4].

2. Background Mesh

The Delaunay triangulation of all boundary nodes is computed as an initial triangulation to begin the node generation process. This triangulation provides at no extra cost a suitable background mesh to interpolate a local value of desired distance between nodes at any point. It will be assumed here that this desirable distance is a piecewise linear function of position, interpolated between the nodal values of a triangle in the background grid. The spacing h at that node is computed as the average distance to its two neighboring nodes on the boundary.

As this initial mesh is a Delaunay triangulation, only closest nodes have been connected. In other terms, the circumcircle of each triangle does not contain any other vertex of the grid. We can exploit this property to find a second set of nodes at a user specified distance, say δ, to define the viscous region. In this region the interpolated stretching σ decays from the user defined maximum aspect ratio at the solid boundaries to an isotropic value used in the inviscid domain. If the radius of the circumcircle of a triangle with a solid boundary face is larger than δ, a node normal to the face at distance δ can be introduced without any interference with other regions. The stretching at this node will be the isotropic value. Two layers interfere if the radius is smaller than δ and the triangle in question connects different solid boundaries. In that case the stretching will decay only to an appropriate fraction of the stretching at the solid boundaries.

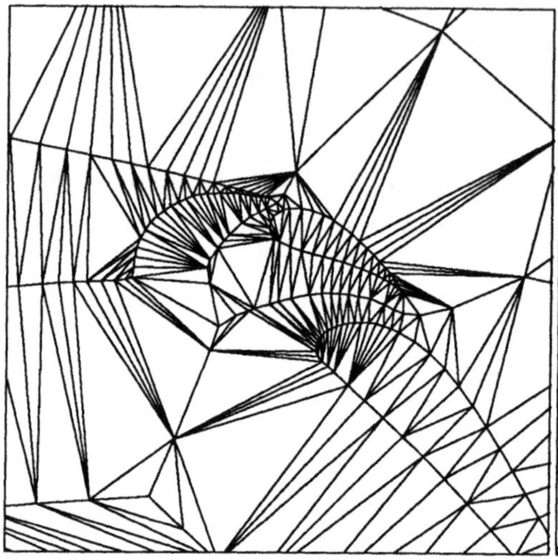

Figure 1: Background grid with viscous regions.

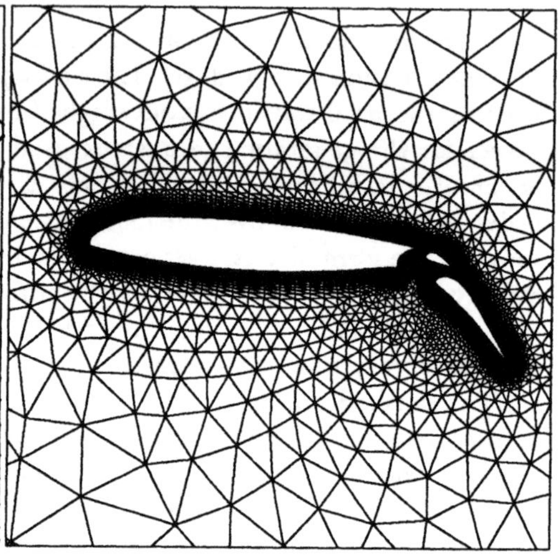

Figure 2: Close-up of the three element aerofoil with viscous region.

The spacing values assigned to these nodes can be interpolated on the initial triangulation. We take the spacing to be the smaller value of interpolated spacing h and the viscous scale. The viscous scale ν is the length of the longest solid boundary face divided by the local stretching value σ as will be seen in section 3. Note that by omitting the viscous modifications we obtain a background grid for meshes for Euler calculations.

3. Internal node generation

We employ frontal techniques to construct nodes from frontal edges to refine the existing triangulation. The process of generating and connecting interior nodes for our initial triangulation of the boundaries can be distinguished into three different parts: building wedges with the viscous scale ν in the viscous regions along solid boundaries, building triangles with the viscous scale ν in the remaining viscous regions, building triangles with the inviscid scale h in the rest of the domain. While building triangles in the viscous and inviscid regions is essentially the same process with different scales, the building of wedges is a fundamentally different procedure.

3.1 Building wedges

Shear layers expose high gradients normal to the layer while the gradients tangential to the layer remain low. Refining isotropically in both directions is not affordable. In the framework of minimum user input, grid stretching can be aligned with attached boundary layers or user-specified wakes. Once the layer separates from the surface, solution adaptive refinement procedures have to be used to modify the grid. Similarly, the grid ought to be less stretched in regions where these 'boundary layer assumptions' don't hold.

An important problem of stretching in triangular meshes is the control of the angles in the cells. If the maximum angle approaches 180°, the truncation error in that cell becomes exceedingly high. More or less the only reliable way to avoid large angles while retaining high aspect ratios is to emulate structured quadrilateral grids in the region where the shear layer is to be expected and choose a dividing diagonal. Isotropic meshing should take place in the regions far from the surfaces or where the layers begin to develop.

It thus seems natural to define a viscous scale ν, the length of the longest solid boundary face divided by the local stretching value. Starting from the string of nodes defining the solid boundary, a string of nodes is created around each previous string with distance ν between the nodes until the viscous regions are filled. The string links the nodes that an average surface normal can be calculated to displace the nodes by ν. This will create cells with large aspect ratios where mesh spacing h is coarse and more isotropic cells where the user decided anyhow to have finer spacing like e.g. at a corner (figure 3). Isotropic meshing with nearly equilateral triangles is more appropriate wherever the user chose such a fine boundary discretization that ν is larger than the length of the frontal edge. In a similar way the strings of nodes are split around corners in the geometry as the flow features around a corner cannot be assumed a priori and isotropic refinement has to be used. The effects of reduced stretching and reverting to isotropicity can be seen in figure 3.

A Delaunay triangulation maximizes the minimum angle in the mesh. On the other hand, the solvers impose constraints on the maximum cell angle rather than the minimum angle. While the differences in Min-Max and Max-Min triangulations are minor in isotropic grids, the differences are fundamental in stretched meshes. Consequently, we have to apply constraints to the triangulation while introducing nodes in the viscous regions. As the possible connections between the different strings of nodes are known, the triangles that properly connect between strings can be flagged and exempted from retriangulation. Hence, the algorithm to build layers of wedges can be cast in the following steps:

```
for each string of nodes
    open a new string of nodes
    for each node in the string
        evaluate the local stretching value on the background grid
        build a new node at distance ν to the node normal to the string
        check whether that node lies within the proper viscous region
        check whether that node is properly spaced with the nodes in the grid
        if properly spaced and located
            append the new node to the new string
            introduce the node into the constrained triangulation
```

```
            protect the newly formed wedges between strings
        fi
    rof
    close the new string of nodes
rof
```

The actual implementation is slightly more complicated to properly evaluate the normal at the ends of the strings and to allow for an interruption of the string in case one of the checks fails or a corner is detected.

3.2 Building triangles in the viscous regions

Once the viscous layers have been filled with wedges, the remaining areas in the viscous regions around corners and edges and the rest of the computational domain are filled with isotropic triangles. The only difference between viscous and inviscid regions is the evaluation of the local length scale. In order to match the short sides of the wedges, the length scale for triangles in the viscous region has to be ν.

3.3 Building triangles in the inviscid region

In contrast to Advancing Front techniques, tracking of the front is not required for the generation of isotropic triangles as we provide over a closed Delaunay triangulation at any stage of the process. A frontal edge is defined as a face shared between a well shaped cell with all three sides of similar length and a badly shaped cell with an obtuse or acute angle opposite to the shared edge. For each frontal edge a new node is constructed on the perpendicular bisector into the badly shaped cell such that the distances between the new node and the two vertices forming the edge approximate h, the inviscid length scale. A further check is required to make sure that the new node is sufficiently distant from the remaining nodes in the grid and from the other new nodes. Nodes that exhibit bad spacing are either merged with other nodes or discarded. With these new nodes in place, the Delaunay algorithm is re-run and the process can be repeated until all bad triangles have vanished. More details about the isotropic node generation process and examples can be found in [5,6].

The computational cost of the isotropic node generation process can be shown to be $O(N \log N)$, hence the algorithm is asymptotically optimal. For an Euler grid around the three element aerofoil configuration in figures 1 to 4, 2047 nodes were created in 27 seconds on a DEC 5000. The viscous generation runs roughly four times as fast, although the implementation is not yet optimal. Further improvements in efficiency could be achieved with the use of an efficient data structure.

4. Examples

Figures 1 to 4 give the triangulation around a three element aerofoil. Figure 1 gives a detail of the background mesh. The thickness of the viscous layer is .1 chord lenghts. Note how the border of the viscous region separates the different sections. Figure 2 gives a view of the entire aerofoil. The wedges smoothly match the isotropic triangles at the outer edge of the viscous region. The maximum aspect ratio of 10 is reached on the surface at the middle of the main aerofoil. Figure 3 gives the same detail as figure 1 but for the computational grid. Also the isotropic triangles constructed in the viscous regions match the wedges. Figure 4 gives a close up of the trailing edge of the main flap where the the different length scales are smoothly blended.

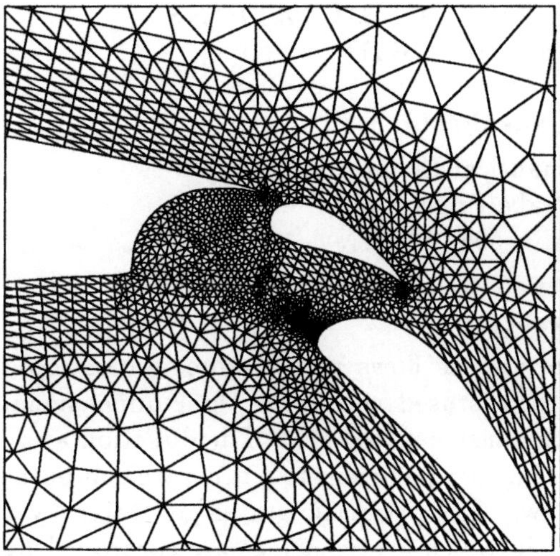

Figure 3: Detail in the region between the three elements.

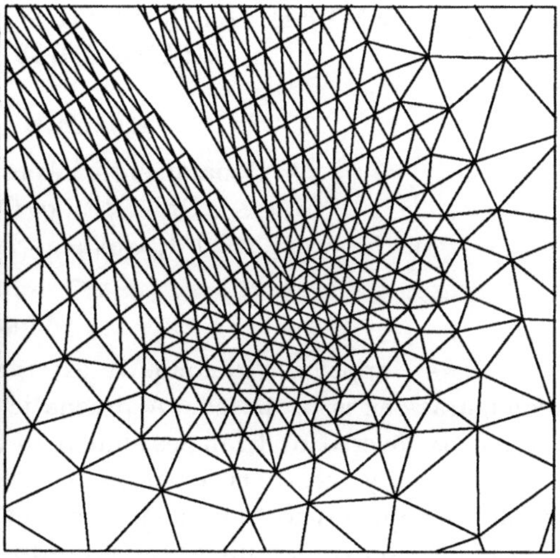

Figure 4: Detail at the trailing edge of the main flap.

5. Conclusions

A fully automatic grid generation algorithm has been developed that can generate highly stretched unstructured meshes for Navier-Stokes calculations. Regions where stretching takes place are defined on a background grid that is automatically derived from a triangulation of the boundary nodes. The use of a regular pattern of wedges in the viscous regions eliminates virtually all obtuse angles that deteriorate the solution quality. With a proper boundary discretisation the algorithm reverts smoothly to isotropic grid generation when complex flow features have to be resolved in regions of high boundary curvature. The meshes produced are smooth in the stretched and the isotropic region as well as in the areas where the transition takes place. The method uses only minimum user input and has an asymptotically optimal cost of operations.

6. References

1. J. Peraire, M. Vahdati, K. Morgan and O.C. Zienkiewicz: Adaptive Remeshing for Compressible Flow Computations, Journal of Computational Physics 72, 1987.
2. D. Mavriplis: Adaptive Mesh Generation for Viscous Flows Using Delaunay Triangulation, ICASE Report N0. 88-47, 1988.
3. B. Delaunay: Sur la sphère vide, Bull. Acad. Science USSR VII: Class. Sci.-Mat. Nat. 793-800, 1934.
4. T. Barth: Aspects of Unstructured Grids and Finite Volume Solvers for the Euler and Navier-Stokes Equations, AGARD R-787
5. J.-D. Müller:A Frontal Approach for Node Generation in Delaunay Triangulations, AGARD R-787
6. J.-D. Müller: A Frontal Approach for Internal Node Generation in Delaunay Triangulations, submitted to Int. J. of Num. Meth. in Fluids.

EXTERNAL VISCOUS FLOW COMPUTATIONS USING PRISMATIC GRID

Kazuhiro Nakahashi

Department of Aeronautical Engineering
University of Osaka Prefecture
1-1, Gakuen, Sakai 593, JAPAN

Abstract

A new approach to compute external viscous flows around three-dimensional configurations is proposed. A prismatic grid is used where the three-dimensional surface is covered by triangles similar to the unstructured grid. The direction away from the body surface is structured so as to achieve efficient and accurate computations for high-Reynolds number viscous flows. The prismatic grid is generated by a newly developed marching-type procedure in which grid spacings are controlled by a variational method. The capability of the method is demonstrated by applying it to a viscous flow computation around a complete aircraft configuration.

1. Introduction

Simulations of high Reynolds number viscous flows about a complete aircraft configuration are still far from the practical use in the aerodynamic design. One of the real bottlenecks is the time-consuming procedure to generate an appropriate grid around a complex configuration. In order to circumvent this bottleneck, unstructured grid approaches have been investigated in recent years. However, flow solution schemes on the unstructured grid for the Navier-Stokes equations remain painfully inefficient when compared to the schemes on structured grid. As a result, a new approach that has capabilities of the easy gridding and the efficient computation of the 3-D Navier-Stokes equations is a crucial subject in the current CFD.

The objective of this paper is to propose an approach to compute high-Reynolds number viscous flows around three-dimensional configurations. A prismatic grid[1] as shown in Fig.1 is used as a hybrid approach of the structured grid and the unstructured grid. To use the prismatic grid has several important advantages; it is 1) geometrically flexible for 3-D external flow problems, 2) efficient and accurate for viscous flow computations with the assistance of the structured lines in the grid, 3) easy to implement the Thin-Layer approximation and the Baldwin-Lomax turbulence model, and 4) easy to vectorize and parallel with less computer storage requirement compared to the fully-unstructured approach.

This paper describes the prismatic grid generation method and the solution scheme on the prismatic grid. The capability of the approach is shown by applying it to a viscous flow computation of a wing-fuselage-nacelle configurations.

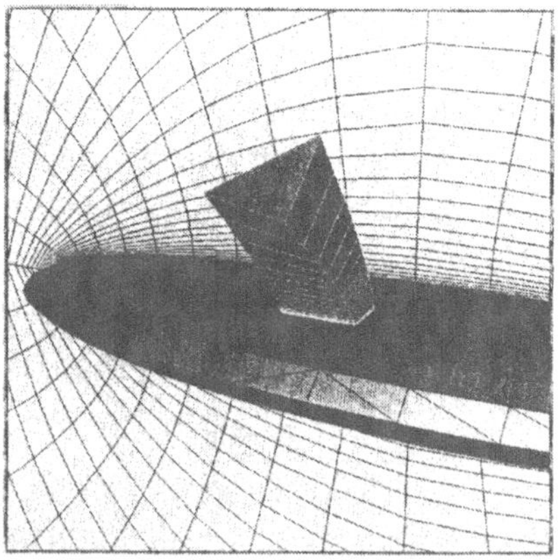

Fig.1 Prismatic grid

2. Prismatic Grid Generation

External flows are assumed so that a marching-type generation procedure[2] is developed to generate the prismatic grid. This is conceptually similar to the hyperbolic grid generation[3] and the Takanashi's method[4] for the structured grid.

As shown in Fig.2, let's define the location of the i-th grid point on the j-th marching front by a position vector $\vec{r}_{i,j}$. The body surface is defined by $\vec{r}_{i,1},\ i = 1, N$. Then we can write,

$$\vec{r}_{i,j} = \vec{r}_{i,j-1} + \Delta r_{i,j} \vec{t}_{i,j} \qquad (1)$$

where $\Delta r_{i,j}$ is the grid spacing in the marching direction, and $\vec{t}_{i,j}$ is a unit vector to the marching direction. Therefore the problem is how to determine the values of $\Delta r_{i,j}$ and $\vec{t}_{i,j}$. In the present procedure, they are separately controlled by using variational approaches so as to generate a prismatic grid around a complex geometry in an automatic manner.

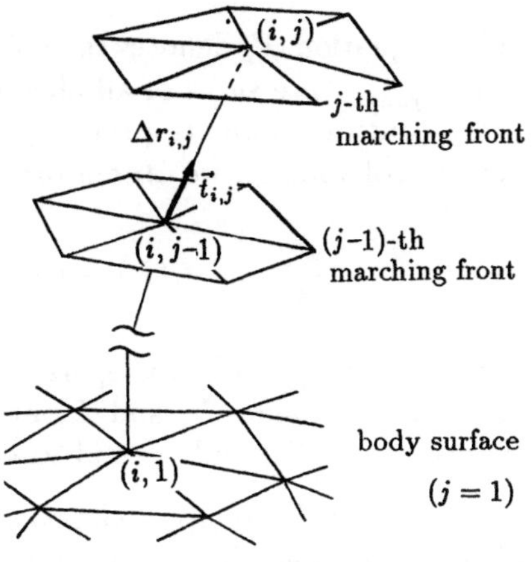

Fig.2 Marching fronts

The spacings in the marching direction, $\Delta r_{i,j}$, is determined by a minimization procedure of the total area of the marching front under an isoperimetric constraint of a constant volume between two marching fronts, $\vec{r}_{i,j-1}$ and $\vec{r}_{i,j}$. This minimization of the marching surface is physically similar to the minimization of the free surface of the liquid by the surface tension. Thus the marching surfaces determined by this concept are expected to be naturally smooth.

The unit vector $\vec{t}_{i,j}$ is determined by two major factors. One is that the surface elements smoothly change their areas and shapes. The other is that the marching lines should be smooth and be close to the normal line to the marching front. An optimization problem is then formulated where a weighted average of two objective functions for the smooth elements and the smooth marching lines are minimized.

3. Flow Solution Method

With an integral form, the Navier-Stokes equations can be written,

$$\frac{\partial}{\partial t} \iiint_V Q dV = - \iint_S F dS + R_e^{-1} \iint_S H dS \quad (2)$$

where $Q = [\rho, \rho v_1, \rho v_2, \rho v_3, e]$ is the vector of conserved dependent variables, and F and H are the convective and viscous flux vectors respectively. The pressure is given by ; $p = (\gamma - 1)[e - \rho v_i v_i]$.

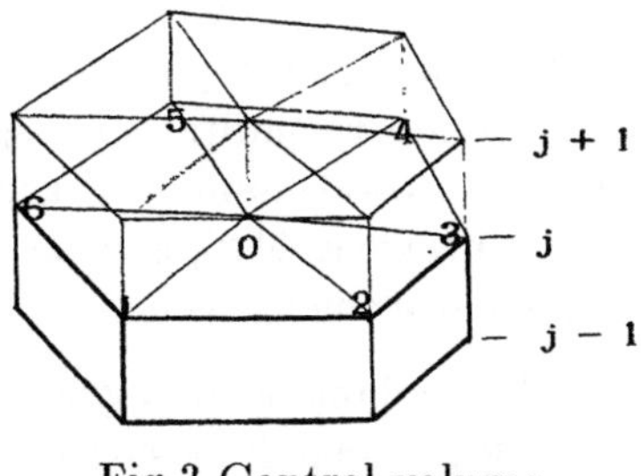

Fig.3 Control volume

The numerical scheme on the prismatic grid employs a finite volume method[1] which is similar to the Jameson's unstructured flow solver[5]. Assuming that flow variables are stored at mesh nodes, we consider a control volume as a union of prisms surrounding each node as shown in Fig.3. By considering the flux balance, we obtain a following discretized form of the equations.

$$\frac{\partial}{\partial t}(\sum_{el} V_e Q) = -(\sum_{face-i} FS_e + \sum_{face-j} FS_e) + R_e^{-1}(\sum_{face-i} HS_e + \sum_{face-j} HS_e) \quad (3)$$

In this equation, V_e denotes the volume of each prismatic element, and $\sum_{el}$, $\sum_{face-i}$ and $\sum_{face-j}$ are sums of all elements, all side faces, and all upper and lower faces of the control volume respectively. Assuming Q in the left-hand side be constant in the control volume, we obtain ordinary differential equations as

$$\frac{dQ}{dt} = -f_i(F) - f_j(F) + R_e^{-1}[f_i(H) + f_j(H)] + D_i + D_j \quad (4)$$

where f_i and f_j are flux operators corresponding to those terms in Eq.(3). On the right-hand side, D_i and D_j are the nonlinear artificial dissipative terms in the directions parallel and normal to the surface, respectively.

The Thin-Layer approximation, which is effective to improve the computational efficiency without decreasing the accuracy, can be easily implemented by just neglecting $f_i(H)$ in Eq.(4) which is the viscous flux parallel to the body surface. The Baldwin-Lomax turbulence model also prefers the structured grid in the direction normal to the surface, thus the present formulation can easily implement the model.

Eq.(4) is integrated in time by using a multi-stage time stepping scheme. In order to accelerate the convergence to steady state, spatially variable time step based on the local CFL condition is used. The implicit residual averaging[6)] is also applied to the equation. The residual, $R(Q)$, at a mesh point is replaces by a weighted average of neighboring residuals. This average can be calculated implicitly by solving tridiagonal equations along the structured lines of the prismatic grid .

4. Results

The method was applied to a wing-fuselage-nacelle configuration of the Boeing 747-200. Fig.4 shows the surface triangular grid on the body. The inlets and outlets of the engine nacelles are covered by dummy surfaces. In the flow field computations, inflow and outflow boundary conditions are specified on these surfaces. Total number of grid points on the entire surface is 4426, and the total number of triangular elements is 8684.

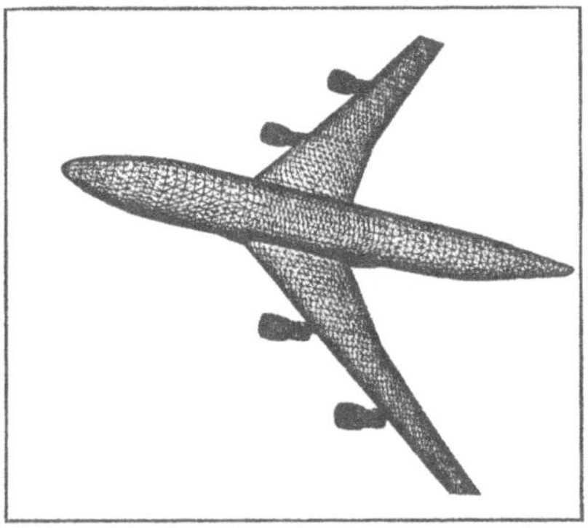

Fig.4 Surface grid

The volume grid is generated by the marching-type method described above. The selected marching surfaces of the volume grid are shown in Fig.5. The index L in the figures is the number of the marching steps counted from the surface which is $L = 1$. Because of the viscous grid, the grid on the tenth front ($L = 10$) is very similar to the surface grid. However, smoothed surfaces can be observed near the junction corners of the wing-fuselage and the strut-wing. Since each marching surface is determined by a condition which is physically similar to the minimization of the liquid surface by the surface tension, the concave sharp corners are smoothed at first. The marching front becomes a semi-sphere at the final stage. The outer boundary which is $L = 61$ for this case is located about twenty times root-chord length from the body surface. The total number of grid points is therefore $4426 \times 61 = 269986$.

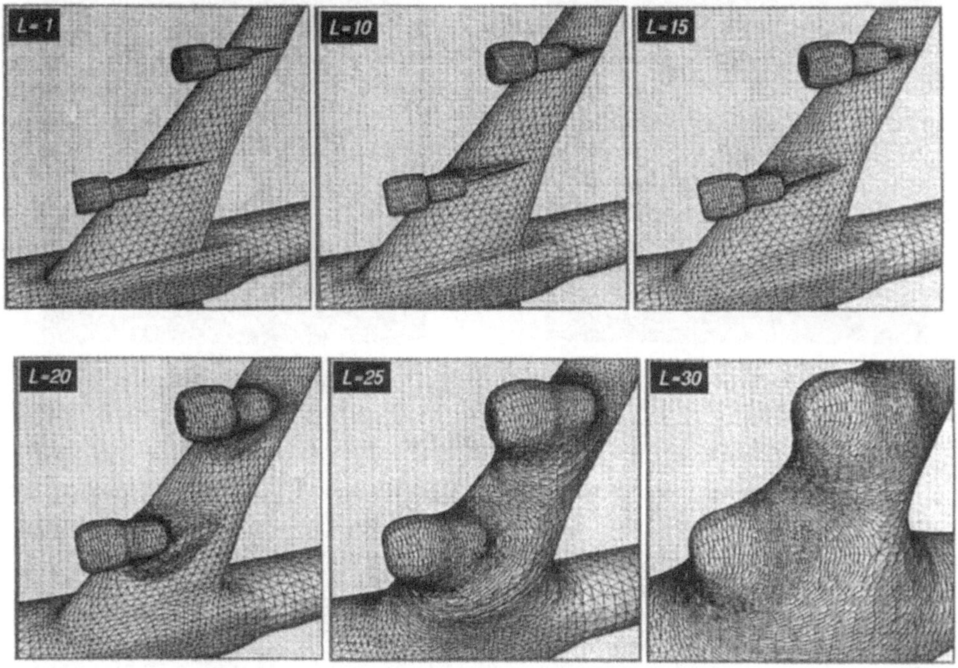

Fig.5 Selected grid surfaces of the prismatic grid

Fig.6 shows the volume grid on its symmetrical plane. On this plane, the prismatic grid becomes a structured O-grid. Fig.7 shows how the prismatic cell structure is constructed in the severely concave region. A prismatic column departs from the inboard strut surface and moves between the inboard and outboard nacelles as shown in the figure. This prismatic cell structure continues to the outer boundary.

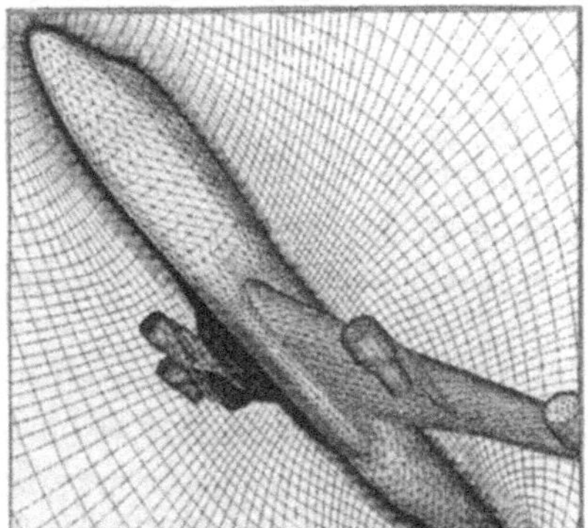

Fig.6 Prismatic grid on symmetrical plane

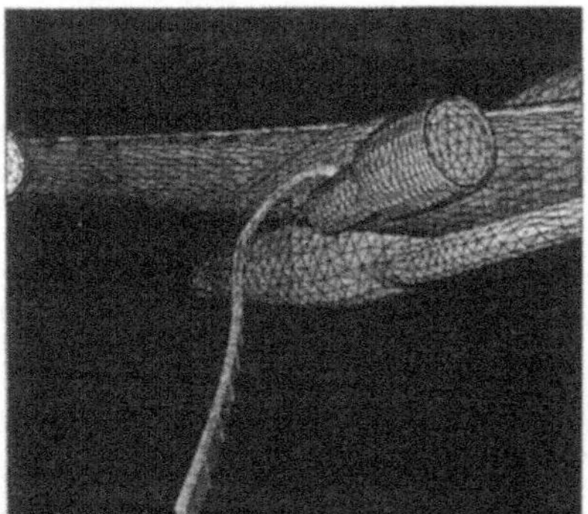

Fig.7 Prismatic cell structure

Fig.8 is a computed surface pressure. The Baldwin-Lomax turbulence model with the Thin-Layer approximation was used. The computed result captures the essential feature of the flow field but more grid points on the body surface will be required for the accurate prediction of the aerodynamic properties.

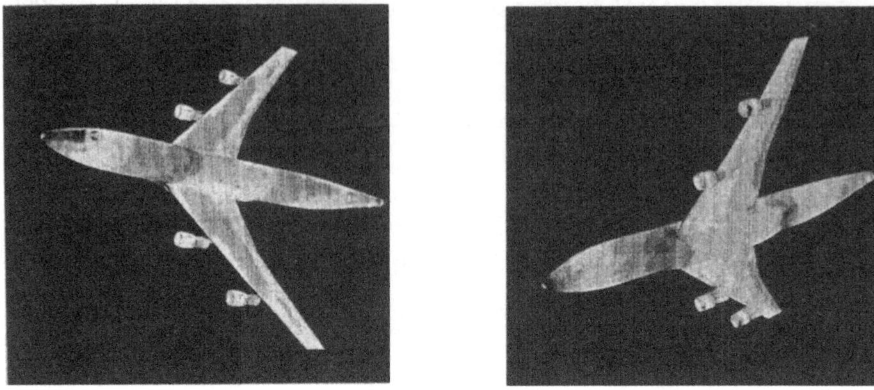

Fig.8 Surface pressure distributions; $M_\infty = 0.8, \alpha = 2°, R_e = 5.0 \times 10^6$

5. Summary

A prismatic grid approach to compute viscous flow fields around complex configurations has been developed. As shown in the result, the present approach is a practical compromise between structured grid and unstructured grid approaches for external viscous flow problems.

Acknowledgments

The major part of this study was done when the author had been visiting NASA Ames Research Center as an NRC Research Associate. The author would like to thank H. Lomax, M. Inouye, and T. Pulliam for their kind help during his stay at NASA Ames. The author also gives his thanks to H. Yoshihara for providing the geometry data of Boeing 747-200 configuration.

References

[1] Nakahashi, K., "A Finite-Element Method on Prismatic Elements for the Three-Dimensional Navier-Stokes Equations", Lecture Notes in Physics, No.323, Springer-Verlag, pp.434-438, 1988.

[2] Nakahashi, K., "Marching Grid Generation for External Viscous Flow Problems", Trans. Japan Soc. Aero. Space Sci., Vol.35, No.108, 1992.

[3] Steger, J.L. and Chaussee, D.S., "Generation of Body Fitted Coordinates Using Hyperbolic Partial Differential Equations", SIAM Journal Sci., Stat. Comput., Vol.1, No.4, pp.431-437, 1980.

[4] Takanashi, S., "A Simple Algorithm for Structured-Grid Generation with Application to Efficient Navier-Stokes Computation", National Aerospace Laboratory TR-1073T, 1990.

[5] Jameson, A. and Baker, T.J., "Improvements to the Aircraft Euler Method", AIAA Paper 87-0452, 1987.

[6] Jameson, A. and Baker, T.J., "Solution of the Euler Equations for Complex Configurations" AIAA Paper 83-1929, 1983.

UNSTRUCTURED GRID GENERATION BY ADVANCING FRONT METHOD USING STRUCTURED BACKGROUND GRIDS*

S. Pirzadeh

ViGYAN, Inc., Hampton, Virginia 23666, U.S.A.

I. Introduction

One of the techniques for generating unstructured grids is the advancing-front method [1-3]. In this approach, the local grid characteristics, such as grid spacings, are commonly controlled by information stored at the nodes of a secondary coarse mesh referred to as the 'background' grid. As the front advances into the field, the grid parameters defining the position of a new point are interpolated from the values stored at the nodes of the background grid cell which encloses the point. As an essential part of the advancing front method, background grids introduce some important grid generation issues. Among these are the simplicity, accuracy, efficiency, and flexibility of background grids.

Typically, a background grid consists of unstructured triangular or tetrahedral cells which enclose the entire domain (Fig. 1a). A background grid of this type is usually generated manually. For complex problems, larger and more sophisticated background grids are required to provide controllable grid point distributions. The manual construction of a large background grid is difficult and time consuming. Furthermore, the task of prescribing spacing parameters for a large number of background nodes is both tedious and inaccurate. Often, several attempts are required to produce an acceptable distribution. A conventional background grid is inflexible to subsequent changes and modifications. Should a portion of a background grid be altered or if more nodes are subsequently desired in a section of the domain, a new background grid must be generated to supply additional information. Interpolation from the nodes of an unstructured background grid requires the storage of mesh coordinates and connectivities and a series of search and check operations to locate the background grid cell which encloses the interpolation point. Each operation requires computation of several Jacobians of the background grid cells and shape functions.

A new type of background mesh has been proposed [4,5] which resolves many of the problems associated with the conventional approach. The new method is based on uniform Cartesian meshes and automatically distributes grid parameters in the field by solution to an elliptic partial differential equation. The methodology is described in 2D space in this paper. The extension to 3D is reported in Ref. [5]. Some sample 2D grids are presented here to demonstrate the power of the technique.

II. Structured Background Grids

Since a background grid is used only for interpolation of grid characteristics and need not conform to the boundaries of the configuration, a simple uniform Cartesian grid can also accomplish the objective sufficiently (Fig. 1b). Associated with a Cartesian background grid is an arbitrary number of user prescribed source elements at which grid parameters are defined. Two types of sources are used in this work: nodal and linear elements. The sources may be positioned anywhere in the field, preferably near the surfaces of the geometry, where a well-controlled distribution of grid points is desired, and at the outer boundaries.

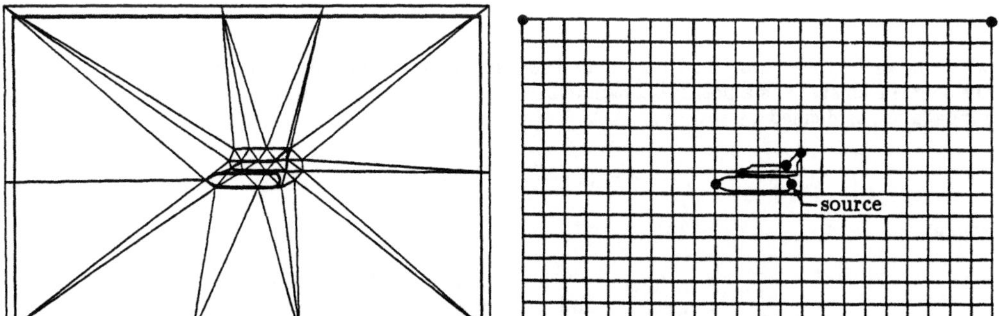

(a) (b)

Fig. 1- Sample background grids. (a) unstructured, (b) structured.

* This work was supported by NASA LaRC, Contract NAS1-18585.

The spatial variation of grid parameters in a field is determined by solving a Poisson equation on the Cartesian grid similar to computing the diffusion of heat from discrete heat sources in a conducting medium. The distribution is modeled by

$$\nabla^2 S = G \tag{1}$$

along with a set of specified boundary conditions. In Eq. (1), S denotes the grid spacing parameter and G is a source term maintaining the spacings in the field. A discretization of Eq. (1) is obtained by assuming that propagation of the spacing parameter to a background grid point p from the source elements and the adjacent grid points is determined by a weighted averaging procedure based on the inverse of distance squared. The outcome is a five-point approximation of Eq. (1) on the Cartesian background grid with the resulting source term

$$G_{i,j} = \sum_{n=1}^{N} \psi_n (S_{i,j} J_n - I_n) \tag{2}$$

where (i, j) represents a background grid node index, N is the total number of source elements in the field, and ψ_n is an intensity factor for the nth source element. The functions I_n and J_n in Eq. (2) are defined as

$$I_n = \begin{cases} S_n/r_n^2 \\ \frac{1}{|\ell_n|}\int_{\ell_n} \frac{f(\ell)}{r(\ell)^2} d\ell \end{cases} \qquad J_n = \begin{cases} 1/r_n^2 & \text{nodal source} \\ \frac{1}{|\ell_n|}\int_{\ell_n} \frac{d\ell}{r(\ell)^2} & \text{linear source} \end{cases} \tag{3}$$

where S_n is the prescribed spacing at the nodal source n with a distance r from the location p, and $f(\ell)$ denotes a linear variation of spacing along the line source n with a length $|\ell_n|$. For linear source elements, parameters are prescribed at the end points of the elements. The implementation of the Gauss-Seidel iterative scheme with successive over-relaxation to the problem yields

$$S_{i,j}^{m+1} = (1-\omega)S_{i,j}^m + \omega\left(S_{i-1,j}^{m+1} + S_{i+1,j}^m + S_{i,j-1}^{m+1} + S_{i,j+1}^m + h^2 \sum_{n=1}^{N} \psi_n I_n\right) \Big/ \left(4 + h^2 \sum_{n=1}^{N} \psi_n J_n\right) \tag{4}$$

where h is the uniform background grid spacing, m is an iteration level and ω is a relaxation parameter ($0<\omega<2$). Equation (4) is solved iteratively on the interior background grid points until convergence when a final smooth distribution of grid spacings is obtained.

The spacing parameter prescribed for a source at a location represents the local dimensional grid size with its effect gradually diffused as the distance from the source is increased. The intensity of a source, ψ_n, controls the extent by which the effect of the element propagates in a field. The higher intensity for a source, the farther the influence of the source can extend into the domain. A directional distribution capability is devised by limiting the source intensities to certain zones and directions. A directional (zonal) intensity for the source element n in relation to the interpolation point p is calculated as

$$\psi_n = a_n \beta + b_n |\phi|^k \tag{5}$$

where a_n and b_n are prescribed parameters specifying the magnitude of the domain of influence of the source element, k is a positive constant and β and ϕ control the zone and direction of the source propagation as defined by

$$\phi = \left(1 - \frac{|\alpha|}{2}\right)\mathbf{v}\cdot\mathbf{u} + \frac{\alpha}{2}|\mathbf{v}\cdot\mathbf{u}| \tag{6}$$

where

$$\mathbf{v} = \mathbf{r}_n/r_n \qquad \text{and} \qquad \beta = \begin{cases} 0 & \text{if } \alpha(\mathbf{v}\cdot\mathbf{u}) < 0 \\ 1 & \text{if } \alpha(\mathbf{v}\cdot\mathbf{u}) \geq 0. \end{cases}$$

A value of 1 for α indicates propagation on one side of the source in the direction of the prescribed unit vector $\mathbf{u}$ whereas -1 specifies propagation on the opposite side. For α equal to 0, propagation would be on both sides of the source.

During the process of advancing front, the grid spacings are interpolated from the values stored at the nodes of the background grid. The interpolation is performed by any conventional method such as bi-linear, inverse distance weighting, etc. Since a background grid is usually very coarse in relation to the geometry of interest, interpolation from only four nodes of a background grid cell may not provide sufficient control of grid point clustering within the cell. For a better point distribution using a coarse background grid, the spacing parameter at a location may be interpolated from the four nodes of the background grid cell which encloses the interpolation point on the front while also taking contributions from the source elements using a weighted averaging procedure. The density (resolution) of a background grid depends on the size of the configuration to be gridded, the desired smoothness of the final grid, and whether interpolation is from both source elements and background grid nodes.

III. Results

Two sample grids are presented in this section to show the capability of the new method. To illustrate the improved grid quality obtained by the new method, two meshes were generated around a NACA 0012 airfoil configuration: one generated with the conventional approach and the other with the new method. The triangular background grid used for the conventional mesh is shown in Fig. 2a having 22 cells and 16 nodes. A corresponding 21 × 21 Cartesian background grid (Fig. 2b) has equal number of source elements to the nodes of the unstructured background grid with the same locations and prescribed spacings. A comparison of the generated grids (Fig. 3) clearly indicates the improvement obtained by the new method. The conventionally generated grid lacks the desired smoothness in distribution, whereas the one generated with the present method exhibits an orderly progression of contours resembling concentric "isotherms" around the configuration.

The capability of the method has been further examined by generating a grid around a complex multi-element airfoil. To complicate the condition even further, the grid has been assumed to be adapted to a hypothetical complex flow field containing a shock wave. The geometry is composed of four airfoil sections set up in a high-lifting arrangement as shown in Fig. 4. A 81×81 Cartesian background mesh for this configuration includes 7 nodal and 2 linear source elements positioned near the airfoil sections (also shown in Fig. 4) and 4 nodal source elements at the outer boundary. All nodal elements have symmetrical intensities except the one at the trailing right-angled corner of the main airfoil which has a one-sided intensity propagating downward and the two elements at the leading edges of the third and fourth airfoil sections which propagate streamwise mostly. The linear element on the flap has a variable spacing and propagates mainly toward the upper surface of the airfoil. Figure 5 shows the generated grid. The grid distribution is smooth and efficiently resolves all the details of the configuration and the flow field. This example demonstrates the high degree of the mesh flexibility and distribution control that can be achieved with the present method. For example, note the miniature grid refinement at the presumed shock location near the nose of the main airfoil in Fig. 5c as compared to the overall size of the domain in Fig. 5a. The refinement has been made by addition of a small linear source element with a weak symmetrical intensity. Also note that the effect of the source element at the trailing corner has been confined only to the lower surface in Fig. 5d without influencing the grid resolution on the upper surface of the thin tail section. A reproduction of a grid such as this using the conventional method would require a complex unstructured background grid with many triangular cells resolving the desired grid clustering around the individual components.

The method has also been extended and applied to 3D problems [5]. Due to page limitation, 3D results are not included in this paper.

IV. Concluding Remarks

A new method for distribution of grid points has been proposed to improve the generation of unstructured grids by the Advancing Front technique. The method is based on Cartesian meshes and solution to an elliptic partial differential equation. Due to simplicity and flexibility of the approach, many of the problems associated with the conventional method are resolved. The construction of new background grids is, for the most part, automatic and requires minimal user interaction. Unlike the conventional method, the new technique distributes spacing parameters among the nodes of a background grid in a systematic manner. The smoothness of a resulting grid is guaranteed by solution of the governing elliptic equation. The new background grids are more flexible to modifications due to the arbitrariness of locations and directional control capability of the source elements which make them suitable for adaptive grid refinement/de-refinement. Interpolation from a uniform Cartesian grid is convenient as no grid connectivity or additional information regarding neighboring cells are required, and no search and check operations are involved for locating interpolation points. In general, the new approach is simple but powerful which relies on a rigorous mathematical basis. The full benefit of the method is realized for generation of 3D grids where the complexity of conventional tetrahedral background grids becomes excessive.

V. References

[1] Löhner, R. and Parikh, P., *Int J Num Meth Flds,* **8** (1988), 1135-1149.
[2] Peraire, J., Morgan, K., and Peiro, J., AGARD CP 464 (1989), 18.1-18.12.
[3] Parikh, P., Pirzadeh, S., and Löhner, R., NASA CR 182090 (1990).
[4] Pirzadeh, S., AIAA Paper, 91-3233 (1991).
[5] Pirzadeh, S., AIAA Paper, 92-0445 (1992).

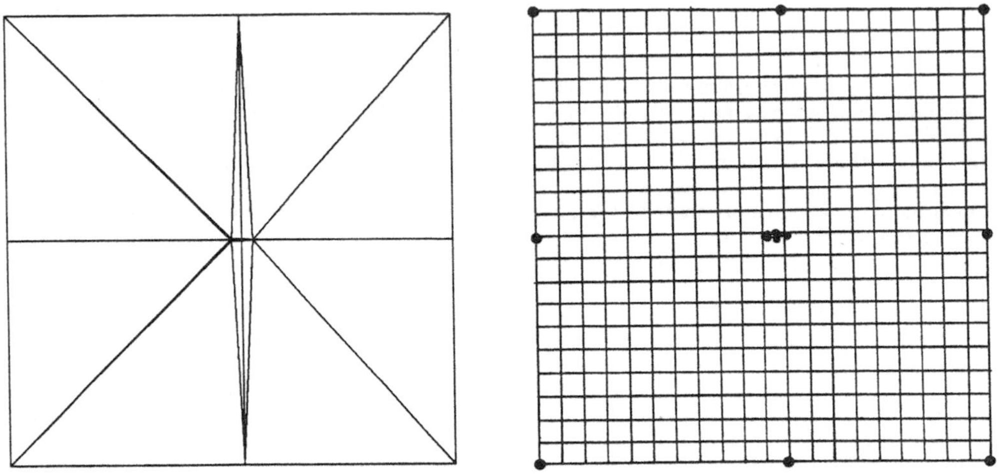

(a) (b)
Figure 2 - Background grids for a NACA 0012 airfoil: (a) unstructured, (b) structured.

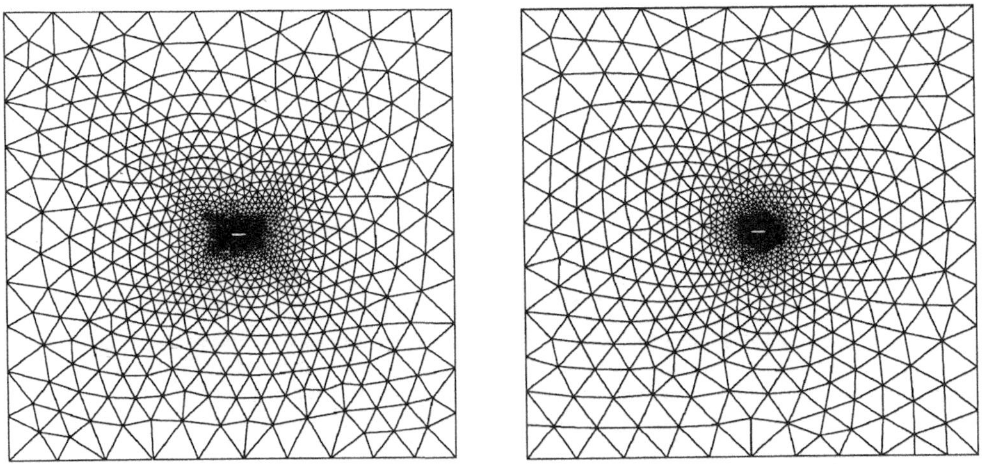

(a) (b)
Figure 3 - Unstructured grids around a NACA 0012 airfoil: (a) old method, (b) new method.

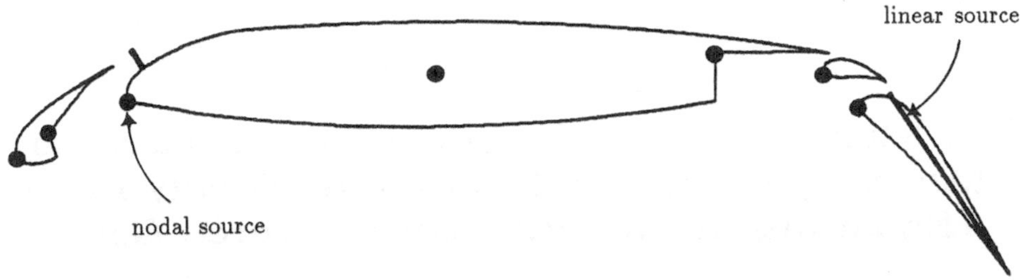

Figure 4 - A 4-element airfoil configuration with prescribed nodal and linear source elements.

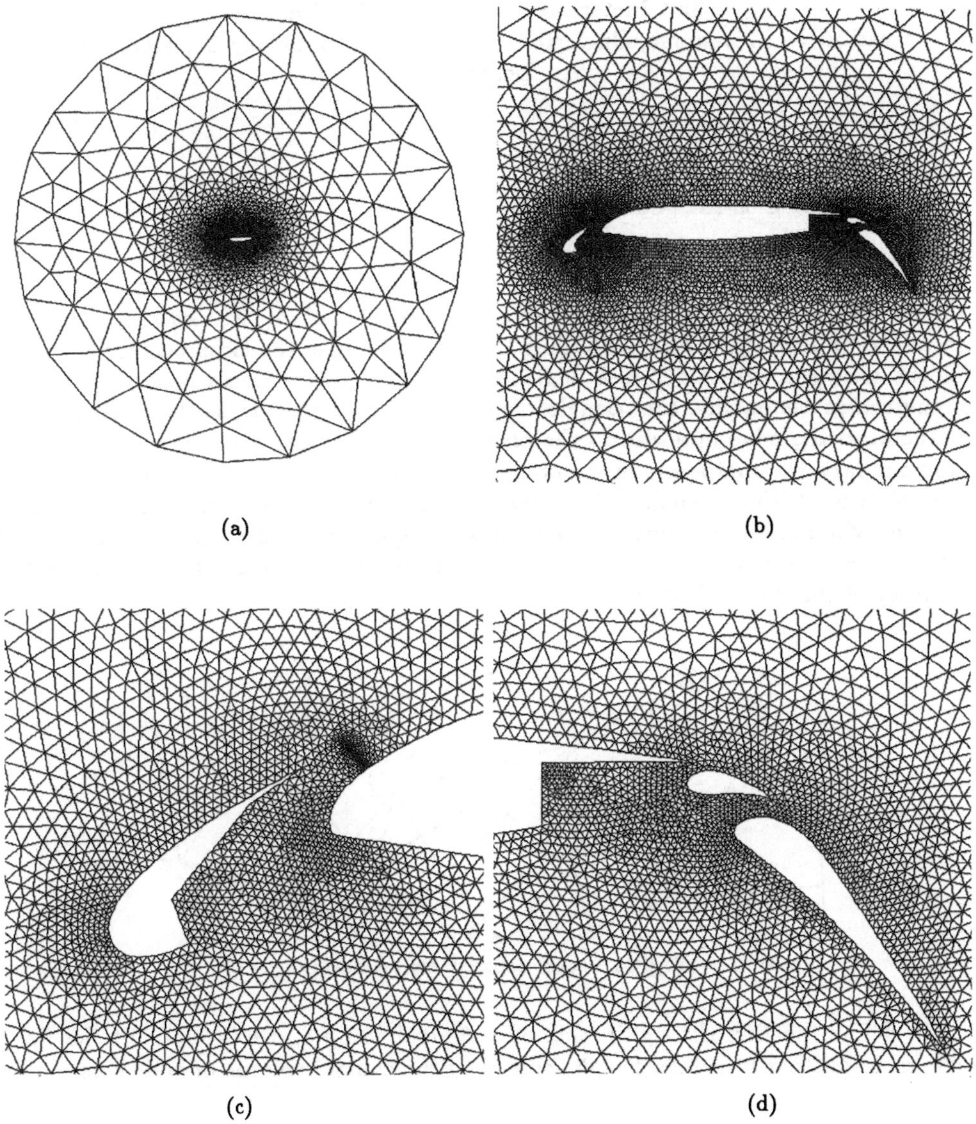

Figure 5 - A triangular grid around a multi-element airfoil configuration.

A Multigrid Finite Volume Method Based on Multistage Jacobi Relaxation for Steady Euler Equations on Adaptive Unstructured Grids

K. Riemslagh [1] , E. Dick [2]

[1]Laboratorium voor Machines en Machinebouw, Universiteit Gent, Sint-Pietersnieuwstraat 41, 9000 Gent,
[2]Institut de Mécanique, Université de Liège, Rue E. Solvay 21, 4000 Liège, Belgium.

1. The Space Discretization

Figure 1 shows part of an unstructured triangular grid. The vertex-centred finite volume method is used. Around every node a controle volume is constructed by connecting *centres* in the cells surrounding the node. We take the circumcentre as *centre* for the triangles with no obtuse angle. For the triangles with an obtuse angle, the circumcentre lies outside the triangle. To correct this, the *centre* is then taken as the midpoint of the edge opposite to the obtuse angle. To close the control volumes on the boundary, the midpoints of the boundary edges are chosen as vertices of the control volumes. With the choice of the *centres* used here, the control volumes coincide with the Voronoi regions, except where obtuse angles occur. The faces of the control volume are as much as possible perpendicular to the cell edges. This improves the accuracy of the flux definition.

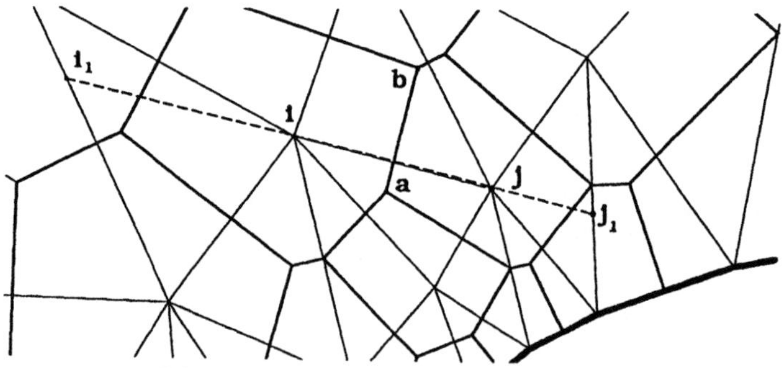

Fig. 1. Vertex-centred discretization.

To define the flux through a side of a control volume, use is made of the flux-difference splitting principle. For the side *ab* of the control volume of node *i* and node

j, Fig. 1, the upwind flux for use in the control volume around node i is defined by

$$F_{i,j} = F_i + A^-_{i,j}\ \Delta U_{i,j}\ \Delta s_{i,j} + S.O. \tag{1}$$

F_i is the flux calculated with the flow variables in nodes i. $\Delta U_{i,j}$ stands for the difference of the vector of conserved variables. $\Delta s_{i,j}$ is the length of the side ab. $A^-_{i,j}$ is the negative part of the discrete Jacobian. To construct the discrete Jacobian, we use here the polynomial flux-difference splitting. This splitting technique was introduced by the second author [1]. Full details of this splitting are given in [2, 3]. The technique is of Roe-type, but is not identical to the splitting formulated by Roe [4]. The precise splitting used is not really relevant for the method we describe here.

The way of writing the flux in form (1) shows the incoming wave components. $S.O.$ denotes the second order correction to the flux. This correction is constructed by the flux-extrapolation technique of Chakravarthy and Osher [5]. For details and examples, illustrating the quality of this second order formulation on structured grids, the reader is referred to [2, 3, 6]. To define the second order correction, points i_1 and j_1 in Fig. 1 are used.

The example to follow is flow around an airfoil. Due to the use of the unstructured grid, we can afford to choose the far-field boundary very far away from the airfoil. In the example, the far-field boundary is a square with sides 100 chord lengths away from the airfoil. As a consequence the upper and lower parts of this boundary can be treated as solid walls while the left and right parts can be treated as inflow and outflow boundaries. This converts the external flow type problem to an internal flow type problem. For inflow and outflow boundaries, the classic extrapolation procedures are used. At solid boundaries, impermiability is imposed by setting the convective part of the flux equal to zero.

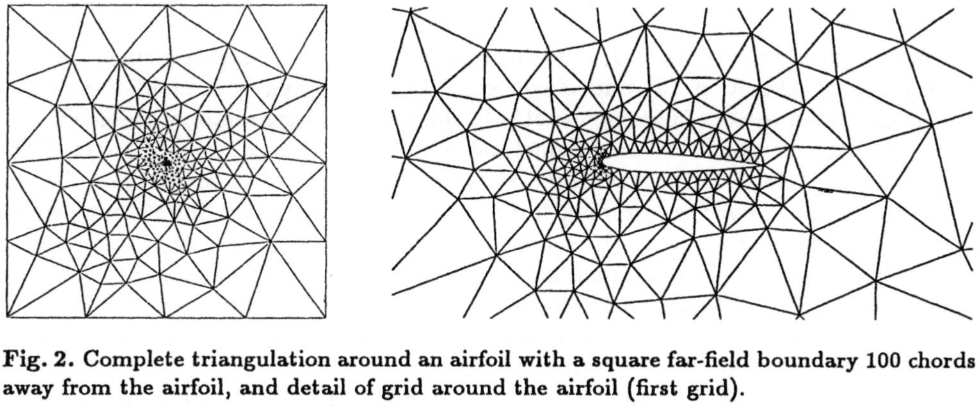

Fig. 2. Complete triangulation around an airfoil with a square far-field boundary 100 chords away from the airfoil, and detail of grid around the airfoil (first grid).

For a point i on a solid boundary, the flux expression (1) refers to a ficticious point j outside the domain. A physical flux expression can be recovered by taking the values of the variables in the ficticious node j equal to the values in the node i. So $\Delta U_{i,j}$ and $S.O.$ vanish. The matrix $A^-_{i,j}$ in (1) is then calculated with the values of the variables in the node i. The impermiability is introduced in the term F_i. As will be discussed in the next section, the matrix $A^-_{i,j}$ at a solid boundary plays an important role in the relaxation method, although it is multiplied with a zero term.

2. The Multigrid Relaxation Method

It is not simple to use a succesive relaxation method on an unstructured grid, since this requires the construction of paths through the grid. A much more natural method is a simultaneous method. A simultaneous method, like the Jacobi relaxation, also has the further advantage of being easily vectorizable and parallelizable. The only drawback is that Jacobi relaxation, at least in its basic form, is not really effective. To repair this, we suggest to bring in multistage into the Jacobi method in the same way as multistage is used for time stepping methods and to use the optimization results with respect to smoothing known for time stepping schemes. This suggestion is not really new. It was first made by Morano et al. [7], but not worked out in detail. A more detailed analysis on the possible multigrid performance was made recently by the authors [8].

We choose here a priori the defect correction multigrid procedure as was used in [2, 3, 6]. This means that the second order part of the flux, $S.O.$ in (1), is updated only on the finest grid and is frozen on all other grids, of course transferred by the usual coarse grid transfer operators.

The Jacobi-relaxation applied to the first order fluxbalance in an interior node reads

$$\sum_j A^-_{i,j}(U^n_j - U^{n+1}_i)\Delta s_{i,j} = 0. \tag{2}$$

The matrices $A^-_{i,j}$ are on the relaxation level n. Using increments $\delta U_i = U^{n+1}_i - U^n_i$, this gives

$$\left(-\sum_j A^-_{i,j}\Delta s_{i,j}\right)\delta U_i + \sum_j A^-_{i,j}(U^n_j - U^n_i)\Delta s_{i,j} = 0. \tag{3}$$

The 4x4 matrix coefficient of δU_i in (3) is non-singular. The difference between (single stage) Jacobi relaxation (3) and single stage time stepping is seen in the matrix coefficient of the vector of increments δU_i. In the time stepping method, the coefficient is a diagonal matrix. In the Jacobi method, the matrix is composed of parts of the flux-Jacobians associated to the different faces of the control volume. The collected parts correspond to waves incoming to the control volume. In the time stepping, the incoming waves contribute to the increment of the flow variables all with the same weight factor. In the Jacobi relaxation the weight factors are proportional to the wave speeds. As a consequence, Jacobi relaxation can be seen as a time stepping in which all incoming wave components are scaled to have the same effective speed. So, as to speak, they all have the same CFL–number.

For a node on a solid boundary, an expression similar to (3) is obtained provided that for a face on the boundary the difference in the first order flux-difference part is introduced as $U^n_i - U^{n+1}_i$, similar to the term $U^n_j - U^{n+1}_i$ which is used for a flux on an interior face. So, in order to avoid a singular matrix coefficient of the vector of increments in (3), this special treatment at boundaries is necessary.

To bring in multistage is now very simple. We use here a three-stage method with the parameters optimized for smoothing, according to Van Leer et al. [9]. The parameters are $\alpha_1 = 0.1481; \alpha_2 = 0.40; \alpha_3 = 1.0; CFL = 1.5$. Equation (3) corresponds to an equation with $CFL = 1$.

3. The Mesh Generation

The technique used here is the well known Delaunay triangulation algorithm formulated by Bowyer [10]. Compared with other grid generation techniques, this algorithm allows a strategy for placing points which is independent of the mechanism to connect the points. The algorithm iteratively generates a grid, by bringing point after point into the grid. Every time, a small portion of the grid is deleted and reconnected to include the new point. This feature of the algorithm leads to easy grid adaption through refinement. The domain is discretized with a mesh generator which is able to handle non-convex geometries. To generate the grid, a description of the domain boundary has to be given. Then points are introduced on the boundaries, based on local curvature and local grid spacing information. With these points a preliminary grid is constructed, conform with the boundary. Then, criteria based on area and aspect ratio of the triangles are used to refine this mesh. This leads to the first grid. Then two more refined grids are constructed using the same basic grid generator by doubling the number of points on the boundary. These three grids are used in a FAS-multigrid method starting on the coarsest grid. The V-cycle is used. The restriction is based on area weighting. As soon as the residual drops below a threshold value, an adaption cycle is started. Using the same procedure as for the other grids, criteria are used to refine the mesh. Instead of being based on mesh properties they are now based on flow properties like pressure and entropy differences. In this way, shock regions, stagnation regions and slip regions are captured accurately. Each newly generated mesh is incorporated into the multigrid procedure.

4. Results

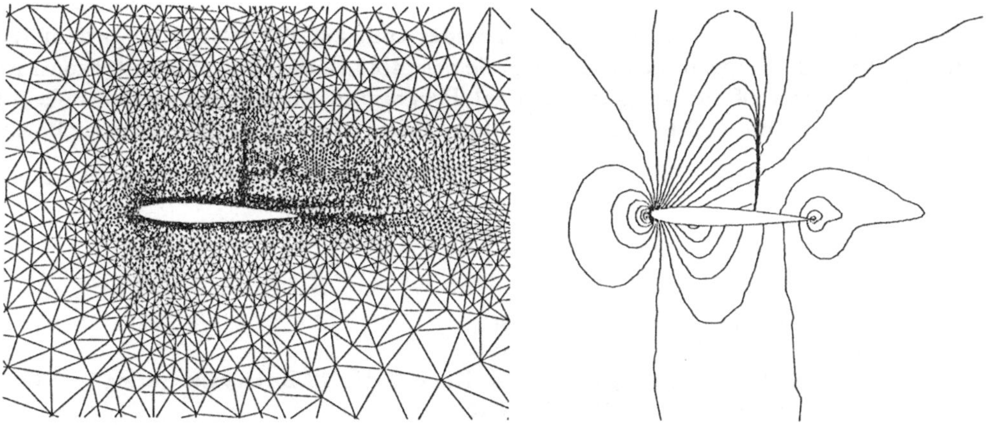

Fig. 3. Part of the final grid and the resulting iso-Mach lines per 0.05.

The foregoing procedure was applied to a flow over a NACA-0012 airfoil at an angle of attack of 1.25 degrees and Mach number 0.80. Three flow adapted grids were generated. The number of points in the three initial grids are : 464, 1420, 2105, and in the flow adapted grids : 3423, 6913, 9089. Figure 3 shows the result.

Figure 4 shows the convergence behaviour. The residual used in the figure is defined as the maximum of the scaled residuals of all equations in all points.

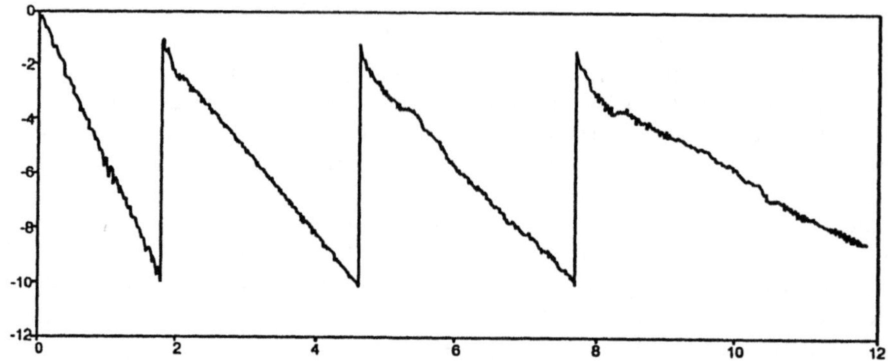

Fig. 4. Convergence behaviour for the second order defect correction scheme (TVD minmod limiter) on the NACA-0012 test case, using three-stage Jacobi. Logarithm of the residual as function of work equivalent to 1000 basic relaxation stages on the finest mesh in the local sequence.

We have showed that the combination of an adaptive unstructured grid generation technique and the multistage Jacobi relaxation method results in an efficient multigrid technique for unstructured grid applications. By the use of suitable refinement criteria shock regions, stagnation regions and slip regions can be well captured.

Acknowledgement

The research reported here was granted under contract IT/SC/13, as part of the Belgian National Programme for Large Scale Scientific Computing and under contract IUAP/17 as part of the Belgian National Programme on Interuniversity Poles of Attraction, both initiated by the Belgian State, Prime Minister's Office, Science Policy Programming.

References

1. E. Dick: J. Comp. Phys. **76** 19-32 1988
2. E. Dick: J. Comp. Phys. **91** 161-173 1990
3. E. Dick: Multigrid Methods III Int. Ser. Num. Math. **98** Birkhauser Verlag 1-20 1991
4. P. L. Roe: J. Comp. Phys. **43** 357-372 1981
5. R. S. Chakravarthy, S. Osher: AIAA paper **85-0363** 1985
6. E. Dick: Int. J. Num. Methods Heat Fluid Flow **1** 51-62 1991
7. E. Morano, M.-H. Lallemand, M.-P. Leclerq, H. Steve, B. Stoufflet, A. Dervieux: GMD-Studien **189** 1991
8. E. Dick, K. Riemslagh: ICFD Conf. Num. Meth. Fluid Dyn. Reading (to appear)
9. B. Van Leer, C. H. Tai, K. G. Powell: AIAA-paper **89-1983** 1989
10. A. Bowyer: Computer Journal **24** 162-166 1981

ADAPTIVITY TECHNIQUES FOR THE COMPUTATION OF TWO-DIMENSIONAL VISCOUS FLOWS USING STRUCTURED MESHES

J. Szmelter[1], A. Evans[2] and N.P. Weatherill[2]

[1] British Aerospace Comercial Aircraft Ltd., Airbus Division, PO BOX 77, Bristol BS99 7AR, U.K.

[2] Institute for Numerical Methods in Engineering, University College of Swansea, Singleton Park, Swansea SA2 8PP, U.K.

1. INTRODUCTION

An initial mesh point distribution within the domain is based on the point distribution on the boundary, and therefore, mesh points will not reflect features which occur in the flowfield, such as, shock waves and boundary layers. Once a flowfield solution has been produced, it becomes possible to adapt the mesh and generate a mesh which reflects the flow conditions more appropriately. Three different adaptive *h*-refinement strategies: mesh point enrichment, mesh movement and their combination are presented and applied to two-dimensional multiblock meshes. The multiblock mesh generation methods are commonly used in aircraft industry to model complex geometries, however, the proposed techniques are readily applicable also to unstructured and hybrid types of meshes [1]

Steady state solutions for inviscid and viscous laminar flows are obtained by the finite volume method, based on the Jameson Runge-Kutta scheme [2,3]. Some applications for turbulent flows using the Baldwin-Lomax turbulence model have also been investigated. Two versions of the finite volume discretization in space, cell vertex and cell centre, are available.

Initial computational grids are obtained using a multiblock mesh generator [4,5]. This approach represents a compromise between a globally structured mesh and a locally connected unstructured mesh. Smooth meshes for complex geometries, with good orthogonality control near the boundaries, can be generated in this way.

2. ADAPTIVITY TECHNIQUES

To simulate any real features of a flow, within the capacity of existing computers, the application of adaptivity is essential. Different adaptivity strategies have been developed. They can be summarized as those based on mesh enrichment, mesh movement, mesh regeneration, *p*-refinement and their combination (see for example [6,7,8,9]). In this paper mesh enrichment, mesh movement and their combination are considered. The mesh enrichment technique, where an existing mesh is locally subdivided into smaller cells, has the ability to accurately model flow features. However, its application poses some difficulties. The first problem lies in the introduction of 'hanging nodes' i.e. nodes which are not fully connected to surrounding points in the mesh. This has been successfully overcome within the solution procedure [10]. Secondly, the mesh enrichment technique becomes computationally expensive with an increasing number of points. In some cases this cost can be reduced by using a derefinement. On the other hand, the mesh movement technique can be performed in a very simple and efficient manner. However, it is well known that the application of such a strategy can result in the adapted mesh having very skewed cells, and special treatments are required to avoid this. Additionally, if the number of points in the initial mesh is too small, the adapted meshes fail to capture flow features effectively. This problem can be particularly severe in viscous boundary layers. Finally, to avoid these problems and to maintain the benefits of the two adaptive strategies, they have been combined.

Mesh enrichment

The automatic procedure for cell subdivision within the quadtree data structure is based on the gradient of some variable. For inviscid flows the pressure has been used, whilst for viscous flows, in the presented numerical examples, the Mach number has been applied. However, any other flow parameter or a measure of local error can be chosen. Other possibilities, such as the error estimation based on the Richardson extrapolation, have also been investigated by the authors.

The choice of the type of subdivision is dictated by the flow feature which is to be detected. Thus, directionaly bisected cells are typically used in the boundary layer and for capturing grid aligned shock waves. To model multidimensional flow features cells are subdivided into four.

Mesh movement

A node movement technique has been devised that is very simple to implement into flow codes, and is applicable to all mesh types. The position of the node 0, $\mathbf{r}_0^{n+1}$, at relaxation level $n+1$, where $\mathbf{r} = (x, y)$, is altered according to

$$\mathbf{r}_0^{n+1} = \mathbf{r}_0^n + \omega \frac{\sum_{i=1}^{M} C_{i0}(\mathbf{r}_i^n - \mathbf{r}_0^n)}{\sum_{i=1}^{M} C_{i0}} \tag{1}$$

where C_{i0} is the adaptive weight function between nodes i and 0, while ω is the relaxation parameter. The summation is performed over all M edges connecting point 0 to i. As applied in the mesh enrichment, the weight function C_{i0} has been taken as a measure of pressure for inviscid flows, and Mach number for viscous flows.

Combined mesh enrichment and mesh movement

Finally, to avoid the disadvantages of the two adaptive methods and to maintain their merits, they have been combined. They are implemented at regular intervals within the flow simulation. The mesh movement should be used to avoid an uneconomical number of points being introduced into the computational mesh, while the mesh enrichment (together with the derefinement) must ensure sufficient mesh point resolution. In addition, if in a cell the mesh movement technique results in an unacceptable measure of skewness, the cell should be subdivided.

3. NUMERICAL EXAMPLES

Fig.(1) shows the Williams two component configuration - initial and embedded meshes. The solution has been obtained for the viscous laminar flow of $M_\infty = 0.15$, $Re = 5000$ and zero incidence. The typical velocity profiles at 1/3 and 2/3 of the chord on the upper surfaces of both aerofoils are given. It can be observed, that there is little change in velocity profiles computed for the main component, obtained on the initial and enriched grid, but a significant improvement can be noticed for the slat.

The viscous laminar flow at Mach number $M_\infty = 0.5$, Reynolds number $Re = 5000$ and zero incidence has been simulated around a NACA0012 aerofoil. A very coarse grid, suitable for the solution of the Euler equations, has been used as a basic grid, which has been adapted using three diferent techniques, mesh enrichment 4 levels, mesh movement and their combination. The resulting computational grids together with the comparison of velocity profiles at 1/3 of the chord, for different levels of embedding and different techniques are provided in Fig.(2)

The subcritical turbulent flow at freestream Mach number $M_\infty = 0.5, Re = 2890000$ and zero incidence has been computed on an embedded mesh. The velocity profiles at 1/3 chord are compared in Fig.(3) with results obtained in experiment [11].

4. CONCLUSIONS

In this paper three different adaptivity techniques have been investigated on the base of structured meshes. All the techniques indicate the significance of using adaptivity for improving computational results. In particular, the technique of combining point enrichment and node movement strategies offers the best compromise. Although, the work presented here used two-dimensional structured meshes, the techniques can be readily applied to hybrid and unstructured meshes. Also, preliminary three-dimensional numerical results have been already obtained by coauthors.

5. REFERENCES

[1] N.P.Weatherill, Generalised mesh and adaptivity techniques for computational fluid dynamics, In The Finite Element Method in the 1990's, O.C. Zienkiewicz Anniversary Volume, (ed. Onate et al.), CIMNE and Springer-Verlag, Barcelona (1991), 398-409

[2] A.Jameson, W.Schmidt and E.Turkel, Numerical solution of the Euler equations by finite volume methods using Runge-Kutta timestepping schemes, AIAA Paper no. 81-1259, (1981)

[3] A.Jameson, J.Baker and N.P.Weatherill, Calculation of transonic flow over a complete aircraft, AIAA Paper no. 86-0103, (1986)

[4] N.P.Weatherill and C.J.Forsey, Grid generation and flow calculations for aircraft geometries, J. of Aircraft, Vol. 22, No. 10, (1985) 855-860

[5] J.F.Thompson, Z.U.A.Warsi and C.Mastin, Numerical grid generation, foundations and applicatons, Pub. North-Holland, 1985.

[6] A.Evans, M.J.Marchant, J.Szmelter and N.P.Weatherill, Adaptivity for compressible flow computations using point embedding on 2-D structured multiblock meshes, Int.J.Num.Meth.Engng, 32, (1991), 896-919

[7] J.F. Dannenhoffer, A comparison of adaptive-grid redistribution and embedding for steady transonic flows, Int.J.Numer.Meth.Engng, 32, (1991) 653-663

[8] J.Peraire, M.Vahdati, K.Morgan and O.C.Zienkiewicz, Adaptive remeshing for compressible flow computations, J.Comp.Phys., 72, (1987) 449-466

[9] L.Demkowicz, J.T.Oden and T.Strouboulis, An adaptive P-version finite element method for transient flow problems with moving boundaries. In: R.H.Gallagher, G.F.Garey, J.T.Oden, O.C.Zienkiewicz (eds.), Finite element in fluids, Vol.6. New York, Wiley (1985)

[10] J.Szmelter, M.J.Marchant, A.Evans and N.P.Weatherill, Solution of the two-dimensional com-

pressible Navier-Stokes equations on embedded structured multiblock meshes, Proceedings of the third International Conference on Numerical Grid Generation, Barcelona (1991), ed. Arcilla A.S., Elsevier Science Publishers

[11] N.P.Weatherill, L.J.Johnston, A.J.Peace and J.A.Shaw ARA Report 70 (1986).

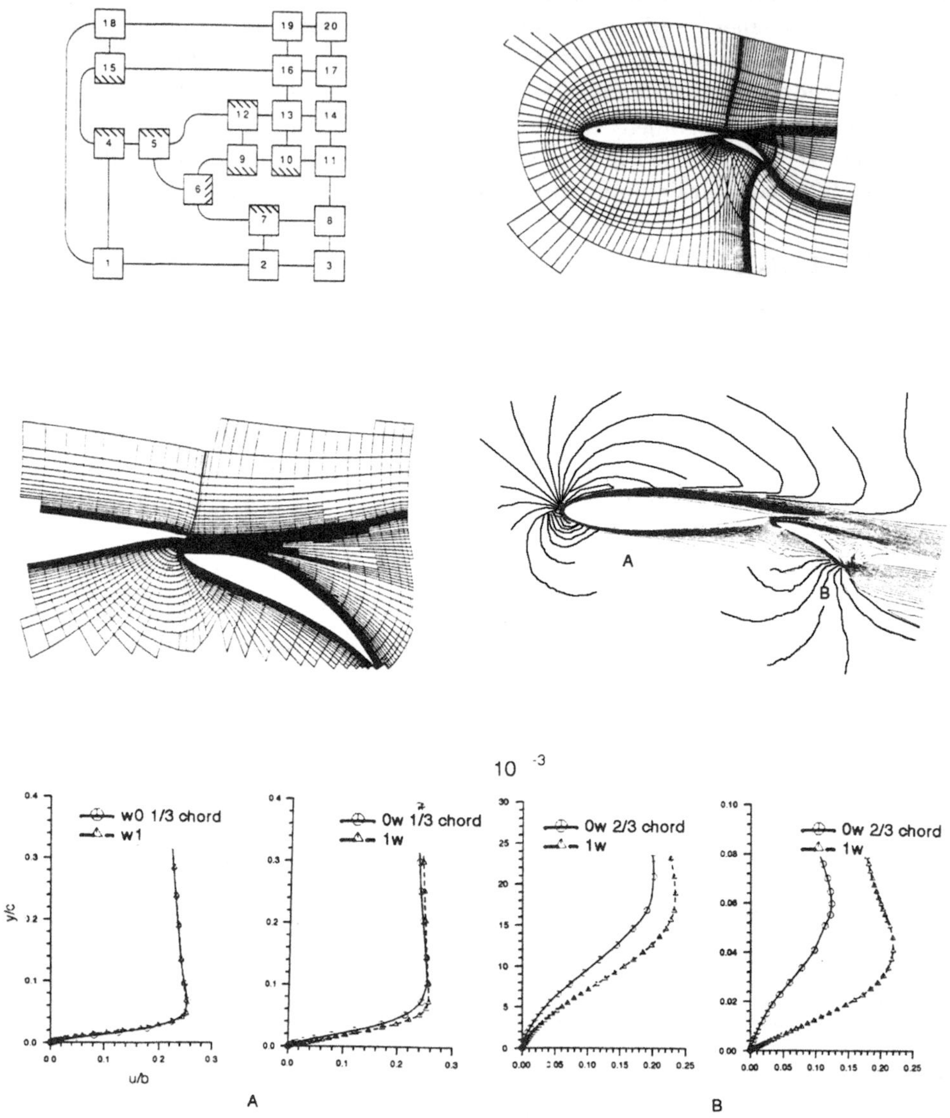

Fig. 1. Embedded Navier-Stokes multiblock mesh, Mach no. contours and velocity profiles. ($M_\infty = 0.15, \alpha = 0°, Re = 5000$).

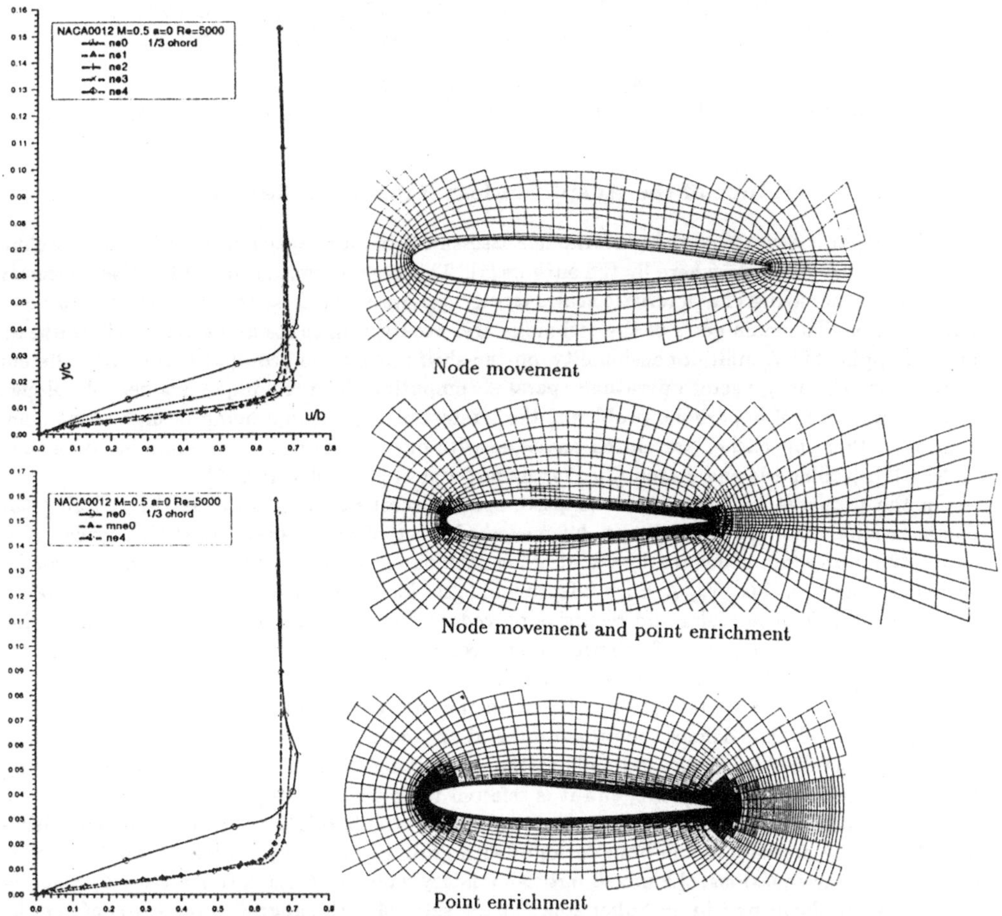

Fig. 2. NACA0012 aerofoil. Adapted meshes and boundary layer velocity profiles 1/3 chord. ($M_\infty = 0.5, \alpha = 0°, Re = 5000$).

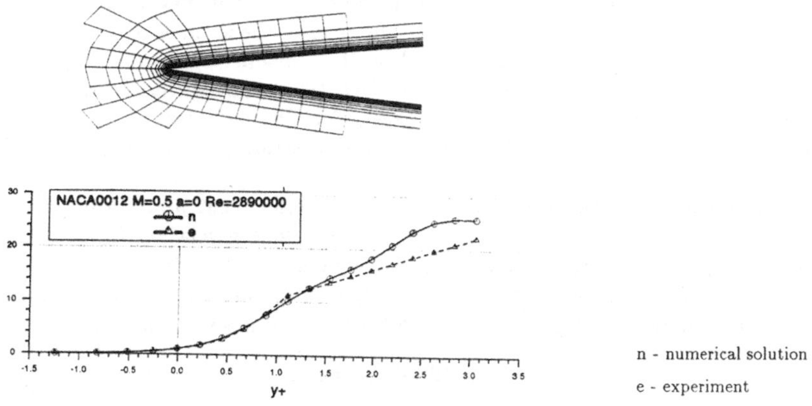

Fig. 3. NACA0012 aerofoil. Embedded mesh and and boundary layer velocity profiles 1/3 chord. ($M_\infty = 0.5, \alpha = 0°, Re = 2890000$).

PERFORMANCE ASSESSMENT OF AN ADAPTIVE MESH REFINEMENT TECHNIQUE FOR DETONATION WAVES

M. Valorani and M. Di Giacinto
Dipartimento di Meccanica e Aeronautica, Università di Roma "La Sapienza,,

— o —

Outline of the Adaptive Mesh Refinement Technique

This paper is devoted to the performance assessment of an adaptive mesh refinement technique (AMR) proposed elsewhere by the authors [1]. This technique is designed to solve efficiently unsteady flows having characteristic scales largely separated, such as those developing in detonation waves. The main goals of the technique are: accuracy in space *and* time, small overhead, general applicability, multidimensionality, off-the-shelf integration (compatibility with different integration schemes), vector operations, parallel computing. The technique has been developed in the framework of hyperbolic problems, the interest of the authors being in high speed combustion. Other approaches presenting analogies with our technique can be found in Berger and Colella [2], Kallinderis and Baron [3], Pervaiz and Baron [4], and Quirk [5].

In the proposed technique, the adaptive space discretization is obtained by tailoring the computational mesh to the solution by means of a set of structured, embedded grids. The embedding procedure is defined by partitioning each parent grid into an even number of equally sized offspring subgrids, N_{part}. Each offspring subgrid has a resolution twice as large as that of its parent grid. This modular and recursive procedure predefines a tree of subgrids of increasing resolution. The total number of subgrids, their geometry, size, location, and resolution level are determined *once and for all at the beginning* of the calculation.

The adaptive time discretization is based on the enforcement of a global CFL condition for all subgrids, which thus defines the proper integration time step of each subgrid. The integration starts from the finer subgrids, and ends at the coarsest ones [1]. The time step corresponding to the root grid is the largest one, and it is referred to as *time stride* [4].

The partitioning of the discrete space domain into subgrids introduces a mixed initial-boundary value problem at the boundaries of each grid. However, by exploiting the hyperbolic properties of the equations, the mixed initial-boundary value problem is converted into a purely initial value problem by adding buffer zones at the sides of each subgrid in order to enforce the interface conditions among adjacent subgrids. Each grid can thus be integrated "independently" from the others.

The process of refinement and derefinement which adapt the computational mesh is automatically driven by a criterium defined on the most significant flow variables [6].

The a-priori definition of subgrids drastically reduces the number of operations related to the post-processing of the information provided by the refinement criterium, thereby reducing the overhead and the complexity of the code. In fact, the refinement criterium "simply" activates or deactivates the integration over predefined grids. The modular, recursive space adaptation allows to perform a new spatial adaptation at all time strides, at low cost, thus achieving tighter space-time mesh adaptation.

On the other hand, the recursive property of the refinement procedure fixes an upper limit to the number of refinement levels, l_{max}, according to the relation $l_{max} = \mathrm{int}[(\log_2 N_{root} - \log_2 N_{part})/(\log_2 N_{part} - 1)]$, where N_{root} is the number of cells of the root grid.

The use of structured grids and the constraint of increasing the resolution by consecutive halvings only, minimize the generation of perturbations due to interpolations.

The test cases

The performance of the AMR technique is analyzed on a classical problem of detonation theory: the initiation of detonation after a shock-wall reflection. This problem was considered by the authors in previous works to validate the reactive Euler solver both on uniform [7] and adapted meshes [6].

A left-running shock, whose shock Mach number is 3.5, travels into an initially quiescent gas mixture, whose molar composition is $2H_2 + O_2 + 3.76N_2$, contained in a shock tube closed at its left end. The pressure, temperature and flow speed used for nondimensionalization are those behind the shock, namely: $p = 0.41\ atm$, $T = 879\ K$ and $V = 591.1\ m/s$. The reference time τ, set as $\tau = 20\ \mu s$, is chosen on the basis of the ignition delay time of the combustible mixture. Therefore the reference length L turns out to be $L = V\tau = 11.8\ mm$. Simulations are carried out on a nondimensional length of 4 ($l = 47.2\ mm$, test # 1), and 1.25 ($l = 14.75\ mm$, test # 2).

TEST # 1	N_{root}	l_{max}	N_{eq}
AMR-N_{part} = 4	256	5	8192
TEST # 2			
UNIF. GRID	512	0	512
AMR-N_{part} = 16	256	1	512
AMR-N_{part} = 8	128	2	512
AMR-N_{part} = 4	64	3	512

Table 1. Parameter selection for the AMR technique

Under these conditions, the thickness of the detonation wave is approximately 0.75 mm wide, whereas the spatial extent over which the simulation is carried out in test # 1 is 3.2 times larger than in test # 2. The ratio ω of the detonation thickness L_c over the domain length L_f measures the relative magnitude of the two main characteristic scales in the phenomenon. It thus provides a measure of the stiffness of the problem, since ω is the inverse of the Damkölher number ($Da = \tau_f/\tau_c = L_f/L_c = 1/\omega$).

Test # 1 requires a very fine resolution in a narrow region which rapidly moves. It thus constitutes a typical example for which the adoption of an AMR strategy becomes highly effective and maybe indispensable. The ratio ω scales as the domain length and thus test # 1 is 3.2 times stiffer than test # 2. The AMR technique can satisfy this demand with $N_{root} = 256$ and $N_{part} = 4$. With 5 levels of refinements allowed, the resolution at the finest level equals that of a uniform grid of $N_{eq} = 8192$ intervals.

Test # 2 is a far less demanding case obtained by halting the calculation before the flow evolution reaches the stage of severest resolution requirement, i.e. before the first hot spot ignition. This test allows an inexpensive comparison of the performance of adapted versus uniform grids. All different parameter combinations used in test # 2 (Table 1) provide, at their finest level the resolution of a uniform grid with $N_{eq} = 512$ intervals.

Performance evaluation

The overall behaviour of the technique is illustrated with the help of the results of test # 1 and test # 2. Results of test # 2 are compared on the basis of accuracy, computational work, CPU time requirements, and storage memory. The *UNIX* subroutine *gprof* provided the detailed time consumption of each segment of the code.

Overall behaviour. A multivariable refinement criterium [6] adapts the resolution to waves of different physical nature, such as the gasdynamic compressions and expansions, or the chemical waves which characterize the dynamics in test # 1 and test # 2. The ability to adapt the mesh to the flow features is especially highlighted by the results of test # 1 (fig. 1a and 1b). Fig. 1a demonstrates how close can the mesh at the finest level of resolution follow the main flow features, although the recursive rule of partitioning does not explicitly incorporates any global pattern recognition strategy [2].

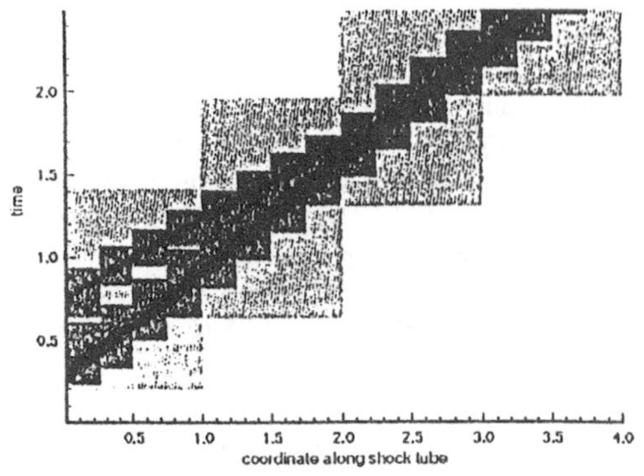

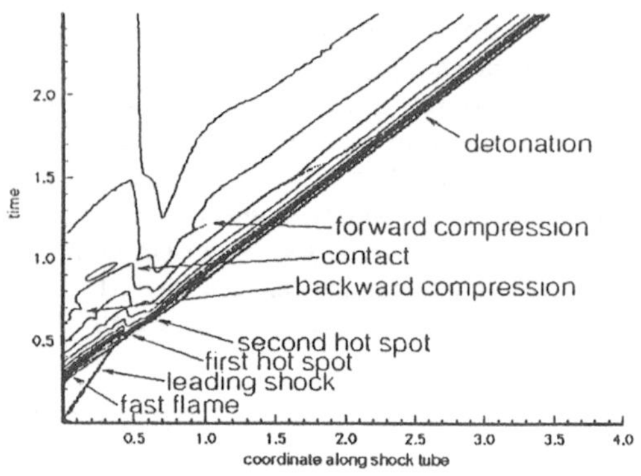

Fig.1a. Contour lines of refinement levels. Fig.1b. Contour lines of temperature.

Results of test # 2 show that the AMR performance can be adjusted to the nature of the flow by a proper choice of the pair (N_{root}, N_{part}), which controls the recursive refinement: a small value of partitions ($N_{part} = 4$) with a relatively coarse root grid ($N_{root} = 64$) provides many levels of resolution to deal with flows with only a few strong singularities; a larger value ($N_{part} = 16$) with a finer root grid ($N_{root} = 256$) provides fewer levels of resolution to deal with flows with a more distributed need of resolution. In this way an optimal choice can be achieved.

Accuracy. A typical snapshot of the structure of the overdriven detonation trailing the leading shock is shown in fig. 2. In this figure, results obtained with the combinations of N_{root} and N_{part} defined in Table 1 for test # 2 are compared with that of a uniform grid. The comparison points out that the states ahead and behind the reaction zone are essentially the same for all combinations of meshes, but the path connecting them shows a weak dependence on the mesh history. The departure from the reference solution - assumed as that provided by the uniform grid with 512 intervals - enlarges as the number of refinement levels increases. This confirms that the reduction of computational work allowed by a large number of resolution levels is counterbalanced by a slight loss of accuracy.

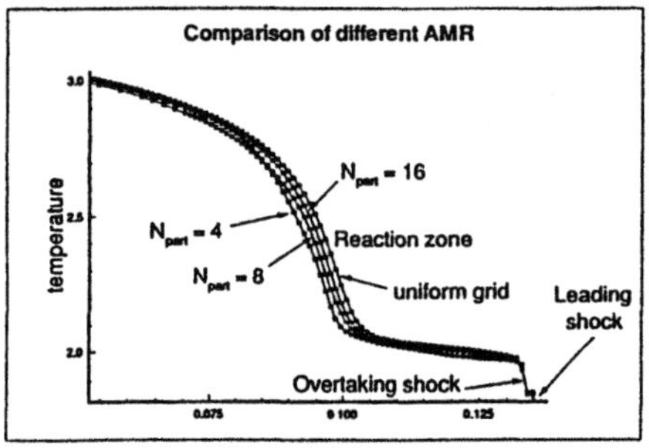

Fig. 2. Comparison among adapted and uniform meshes at $t = 0.165 \times 50\mu s = 8.25\ \mu s$.

Work and CPU time. The total CPU time (CPU_{tot}) absorbed by the calculation is provided by the *gprof* routine.

For a uniform grid the total work can be defined as $W_{unif} = N_c \times N_t \times N_l$, where N_c is the number of cells (or nodes) of the mesh, N_t is the number of integration time step, and N_l is the number of levels of the scheme (two for a predictor corrector scheme). For an adapted mesh, the total work should include the operations related to the AMR routines (activation and deactivation of subgrids, data transfer among parent and offspring grids, etc.). However, it is conventionally defined as $W_{AMR} = N_l \sum_{k=1}^{n_{ts}} [\sum_{i=1}^{n_{sg}} N_{c,i} \times N_{t,i}]_k$, where n_{ts} is the total number of time strides, n_{sg} is the number of active subgrids in a time stride, $N_{c,i}$ the number of cells of each subgrid, $N_{t,i}$ the number of substeps in a time stride for each subgrid.

The largest total work and CPU time saving (95.89 % and 95.73 % respectively) are obtained in test # 1 by setting $N_{root} = 256$ and $N_{part} = 4$ (Table 2). Because of the high degree of stiffness ($\omega = 62.5$) of test # 1, the total CPU time drops from 266.9 hours to 11.4, i.e. a factor of about 23 times smaller. The order of magnitude of this saving is well worth the complexities added by the AMR technique and the slight reduction of accuracy ascertained in the previous paragraph.

The savings are smaller for the less stiff case ($\omega = 19.6$, test # 2). However, the aim of this test was mainly to demonstrate how the saving increase depends on the number of refinement levels (Table 2).

TEST # 1 $\omega = 62.5$	N_{root}	l_{max}	n_{ts}	Work [work u.]	Work % saved	CPU_{tot} (*) [h]	CPU_{tot} % saved
UNIF. GRID	8192	0	9600	157,286,400	0	266.9	0
AMR-$N_{part} = 4$	256	5	300	6,471,082	95.89	11.4	95.73

TEST # 2 $\omega = 19.6$	N_{root}	l_{max}	n_{ts}	Work [work u.]	Work % saved	CPU_{tot} (**) [s]	CPU_{tot} % saved
UNIF. GRID	512	0	320	328,320	0	3,898.4	0
AMR-$N_{part} = 16$	256	1	160	109,312	66.71	1,306.7	66.49
AMR-$N_{part} = 8$	128	2	80	53,682	83.65	654.8	83.20
AMR-$N_{part} = 4$	64	3	40	43,539	86.74	530.9	86.38

Compilation performed with: (*) optimization on, or (**) optimization off.

Table 2. Work and CPU time savings.

The *gprof* routine allows to find the CPU time absorbed by each segment of the code, and thus, to split the quota absorbed by the basic Euler solver (CPU_{ES}) and that by the AMR routines (CPU_{AMR}). Results of such analysis are reported in Table 3. The percent CPU time is referred to the corresponding total CPU reported in Table 2, i.e. $\%CPU = 100(CPU_{(.)} - CPU_{tot})/CPU_{tot}$. The specific CPU time is referred to the corresponding total work reported in Table 2, i.e. Specific $CPU = CPU_{(.)}/W_{(.)}$ [ms/unit of work]. The profiling analysis shows that the AMR overhead for all cases of test # 2 does not rise above 3.17%. This confirms that not only the overall CPU savings are significant but that most of the time is spent by the technique in the integration of the field equations as it happens in a standard uniform grid.

Reactive Euler Solver	CPU_{ES} [s]	$\% CPU_{ES}$	Spec. CPU_{ES} [ms/units of work]
UNIF. GRID	3824.00	98.09	11.65
AMR-$N_{part} = 16$	1264.49	96.76	11.57
AMR-$N_{part} = 8$	633.42	96.73	11.80
AMR-$N_{part} = 4$	513.15	96.66	11.78

Overhead for AMR	CPU_{AMR} [s]	$\% CPU_{AMR}$	Spec. CPU_{AMR} [ms/units of work]
AMR-$N_{part} = 16$	41.53	3.17	0.3799
AMR-$N_{part} = 8$	19.79	3.06	0.3686
AMR-$N_{part} = 4$	16.1	3.03	0.3697

Table 3. Specific CPU time for reactive Euler solver and AMR overhead.

Storage. Table 4 reports the maximum number of active nodes (*active nodes*) and the maximum number of total nodes belonging to the subgrid tree (*tot nodes*) encountered in the whole calculation.

Once more, the largest savings are obtained with the pair $(N_{root}, N_{part}) = (256, 4)$ in test # 1. The savings increase with l_{max} (test # 2 in Table 4). The storage memory may be reduced up to a factor of 4 with respect to that of a uniform grid (test # 1).

The ratio *active/used* nodes decreases as the number of refinement levels increses. When l_{max} increases, it tends to a typical value of 60 % (Table 4).

The number of subgrids created by the automatic procedure increases with l_{max} (Table 4) as well. Although the number of subgrid increases with l_{max} (Table 4), the global overhead requirement decreases, because of the smaller work (Table 2).

Conclusions and Future work

The AMR indeed provides significant savings in CPU time (up to a factor of 25) and storage memory (up to a factor of 4) with only minor loss of accuracy with respect to an equivalent uniform grid calculation. These figures are likely to become even more advantageous when the technique will be extended to two dimensional flows. The overhead CPU time by AMR remains typically below 3% of the total. In general, the best performance is obtained for combination of (N_{root}, N_{part}) providing many refinement levels.

Future work will be devoted to: (i) the extension to two dimensional problems and shock capturing schemes; (ii) the parallelization of the integration procedure.

TEST # 1	N_{root}	l_{max}	tot nodes	saving %	active nodes	saving %	active/tot	max # of grids	average # of grids
UNIF. GRID	8192	0	8192	0	8192	0	1.000	1	1
AMR-$N_{part} = 4$	256	5	2050	75	1199	85.4	0.584	73	46

TEST # 2	N_{root}	l_{max}	tot nodes	saving %	active nodes	saving %	active/tot	max # of grids	average # of grids
UNIF. GRID	512	0	512	0	512	0	1.000	1	1
AMR-$N_{part} = 16$	256	1	409	20.1	396	23	0.968	5	2
AMR-$N_{part} = 8$	128	2	372	27.4	302	41.1	0.811	13	6
AMR-$N_{part} = 4$	64	3	360	29.7	237	53.7	0.658	19	9

Table 4. Storage savings.

References

[1] Valorani, M., Di Giacinto, M., "Adaptive Mesh Refinements for Unsteady Nonequilibrium High Speed Flows", *Ninth GAMM Conf. on Num. Meth. in Fluid Mechanics*, Lausanne, Switzerland (Sept. 1991).

[2] Berger, M.J., Colella, P., "Local Adaptive Mesh Refinement for Shock Hydrodynamics", *J. Comput. Phys.*, vol. 82 (1989) pp. 64-84.

[3] Kallinderis, J.G., Baron, J.R., "Adaptation Methods for Viscous Flows", Comp. Meth. in Viscous Aerodynamics, pp. 163-195 (Elsevier 1990).

[4] Pervaiz, M.M., Baron, J.R., "Spatiotemporal Adaptation Algorithm for Two-Dimensional Reacting Flows", *AIAA J.*, vol. 27, n. 10 (1989).

[5] Quirk, J.J., "An Adaptive Grid Algorithm for Computational Shock Hydrodynamics", Ph.D. Thesis, Cranfield Inst. of Tech., College of Aeronautics (1990-91).

[6] Valorani, M., Di Giacinto, M., "Numerical Simulation of High Speed Reactive Flows with Adaptive Mesh Refinements", *Fourth Int. Conf. on Hyperbolic Problems*, Taormina, Italy (April 1992).

[7] M. Valorani, "Unsteady Detonation Waves: Numerical Simulation with a Quasi-Linear Formulation", *Third Int. Conf. on Hyperbolic Problems*, Uppsala, Sweden (June 1990).

AN EVALUATION OF TURBULENT HEAT TRANSFER PREDICTIONS

D. Chaussee

NASA Ames Research Center Moffett Field, CA 94035 USA

INTRODUCTION

The intent of this work has been to investigate computational aerothermodynamic codes that are capable of calculating the heat transfer accurately for viscous supersonic/hypersonic flow about realistic configurations. In investigating these codes, a systematic study of the various computational parameters and their affect on the heat transfer as compared to experiment is performed. The tools consist of a PNS code, a time-dependent upwind code (UWIN), and a time-dependent finite volume code (CNSFV). The NASA Ames Parabolized Navier-Stokes (PNS) code as developed by Schiff and Steger[1]was used as the mainline procedure to numerically simulate the viscous supersonic/hypersonic flow over a biconic configuration. The baseline turbulence model that has been used in this project is the Baldwin-Lomax model(BL)[2]. The boundary conditions are the usual viscous no slip at the wall, specified wall temperature, and a characteristic procedure is used for the PNS code to fit the bow shock wave which is the outermost boundary. Since the equations are cast in conservation-law form, all discontinuities within the flow domain are predicted accurately. For the time-dependent calculations, the outer most shock wave is captured as well. The various parameters which are investigated consist of: smoothing coefficients, the wall temperature, the radial grid spacing at the surface of the vehicle, two different ways of calculating the heat transfer rate (Qdot), and changing the C_{cp} coefficient in the BL-turbulence model.

COMPUTATIONAL TECHNIQUES

The PNS code[1] uses the Beam-Warming implicit algorithm to solve the parabolized approximation to the Navier-Stokes equations. It is first- or second-order accurate in the marching direction (X) and second-order accurate in the spatial directions. Stability is enhanced by adding dissipation and smoothing to the equations. In addition, a parameter (ε_A) is appended to the diagonal of the marching Jacobian to further improve the stability, especially in the areas of near separation.

The UWIN code[3] uses a TVD-Osher upwind scheme to march in time. This scheme is implemented within an iterative implicit algorithm that permits the solution of the fully implicit, nonlinear form of the finite-difference equations. The solution is advanced in time using a Newton iterative technique. The method is second-order accurate in space and first-order accurate in time.

The CNSFV code[4] employs the Pulliam-Chaussee diagonal algorithm to march in time. The code is multi-zonal and the Navier-Stokes equations are cast in a finite-volume time-marching form. The method is second-order in space and first-order in time.

RESULTS

The case chosen to make the critical evaluations consists of a hypersonic Mach number, $M_\infty = 7.95$ flow, past a $10.5°/7°$ bicone at an angle-of-attack $\alpha = 0°$. The test Reynolds Number is 3.7×10^6 per foot. The model had trips placed near the nose in order to obtain turbulent flow. However, only the trip of .06 consistently gave turbulent flow. The other trips, .013, .03, gave transitional flow over the fore cone. The wall temperature (T_{wall}) of 560° R is nominal, however in some cases the T_{wall} was varied as a function of X based on experimental data. In addition, the constant, C_{cp}, in the BL turbulence model was changed from 1.6 to 2.08 to allow for a better fit of the data. In most cases, the computational grid consisted of 45 points in the meridional direction and 30 points in the radial direction.

The boundary-layer properties are integrated to form integral thickness calculations for shear stress, skin friction coefficient, heating rate, heat-transfer coefficient, and the Stanton number. These values are based on local wall streamline conditions and are referenced to free-stream conditions and the adiabatic wall temperature, T_{total}. The normal variation of the velocity and temperature gradients are calculated using the quadratic fits of local temperature and velocities. The radial derivatives are corrected for the true normal gradient providing an accurate representation of the derivatives. This is the basis for the second order heat transfer calculations. The first order heat transfer calculations use a simple backward difference for the gradients.

The X-variation of ε_A, a stabilizing parameter in the PNS code, is presented in Figure 1. From this figure, it is concluded that for this case this parameter does not have a significant effect on the calculation of heat transfer. It has been found by the author that a value of 0.1 is usually the minimum value that can be used for computations, in the general case. In the figure, a value of 0 corresponds to the experiment. These results have all of the other changeable parameters at their optimum values.

Figure 2 shows the variation of the Stanton number when the radial spacing at the surface of the body is refined. The finer the spacing at the surface, the better comparison between the computation and the best experimental results. Also, it is observed that as the radial spacing asymptotically approaches zero, the Stanton Number evaluations by the various codes are converging to a common curve.

In Figs. 3 and 4, the variation of the Stanton number is presented based on the accuracy of the numerical approximations made in determining the Stanton number. The solid curve is the baseline optimum result based on the first order evaluation of the derivatives that are used in determining the Stanton number. In both figures, the results are improved, as compared to experiment, by using a second-order approximation in the evaluation of the Stanton number. It should be noted that the largest variation occurs on the fore cone. By using the higher approximation, a coarser spacing at the body surface can be used to obtain the same result as the fine spacing solution. This will allow for fewer grid points and a more efficient computational procedure.

Figure 5 presents the comparison between the best experiment and the axial distribution of the error band that is created by the variation in the numerical parameters. In Figure 6, the best numerical result is compared with the scatter of the various wind tunnel tests that were conducted. It is observed that either discipline can make the others result look bad by how the data is presented. It should be understood by both sides (computation and experiment) that a variation in results can occur just by changing various parameters.

In Figure 7, a heat transfer and Stanton Number comparison between all of the CFD code results and the experimental results are presented. The experimental results are deemed to be the best of the many experimental results available. In all cases the heat transfer parameters are calculated using first order derivatives. The PNS and the UWIN results are for a radial spacing, DS = 0.0005 while the CNSFV result is for a DS=0.001. This corresponds to a $Y^+ = 0.5$ and $Y^+ = 1$, respectively. The Stanton Number comparisons yield a reasonable comparison between all of the computational results and the experimental results. The better comparisons are of course for the finer radial grid. The trend of the heat transfer comparisons is consistent with the radial spacing.

SUMMARY

All three codes showed good comparison with heat transfer as long as the first radial spacing was relatively fine ($Y^+ < 0.5$). This was improved further by employing a second order approximation for the radial derivatives that are used in the prediction of the heat transfer. These conclusions are predicated on the rather simple circular cross-section of the conical body.

REFERENCES

1. Schiff,L.B., Steger,J.L., **Numerical Simulation of Steady Supersonic Viscous Flows**, NASA Technical Paper 1749, 1981

2. Baldwin,B.S. and Lomax,H.L., **Thin Layer Approximation and Algebraic Model for Separated Turbulent Flows**, AIAA Paper 78-257, 1978

3. Edwards,T., Chaussee,D., Lawrence,S., and Rizk,Y., **Comparisons of Four CFD Codes as Applied to a Hypersonic All-Body Vehicle**, AIAA Paper 87-2642, 1987

4. Chaussee,D.S. and Klopfer,G.H., **The Numerical Study of 3-D Flow Past Control Surfaces**, AIAA Paper 92-4650, 1992

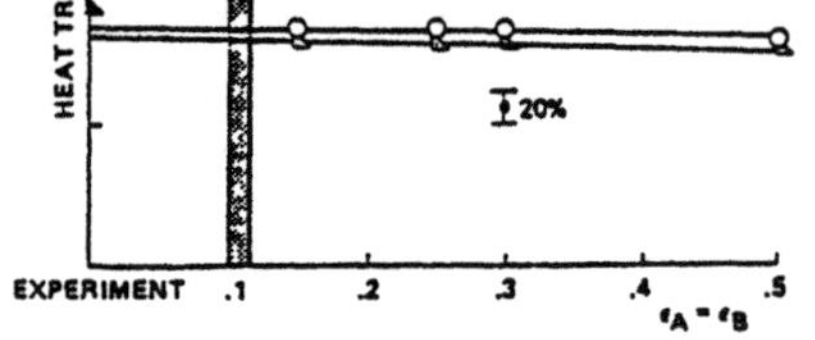

Figure 1. Heat Transfer Variation at Different X/L Stations for Varying ε_A.

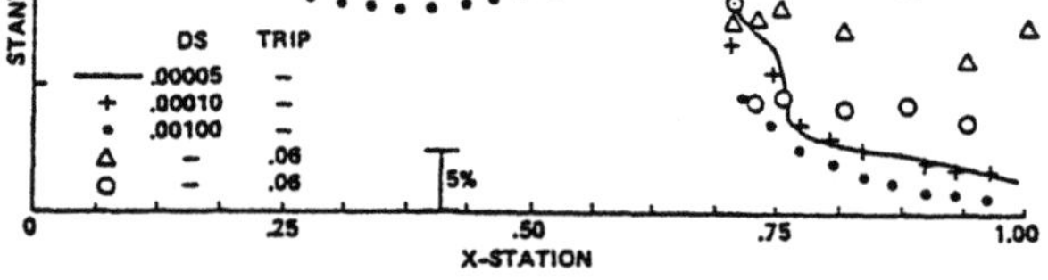

Figure 2. Computational Stanton Number Comparisons with Experiment.

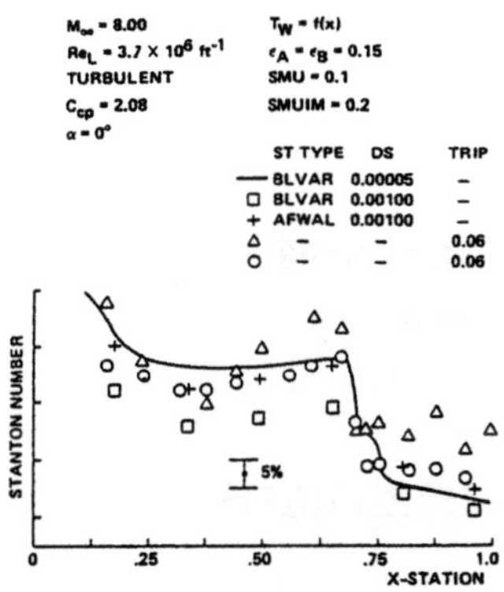

Figure 3. Computational Stanton Number Comparisons with Experiment.

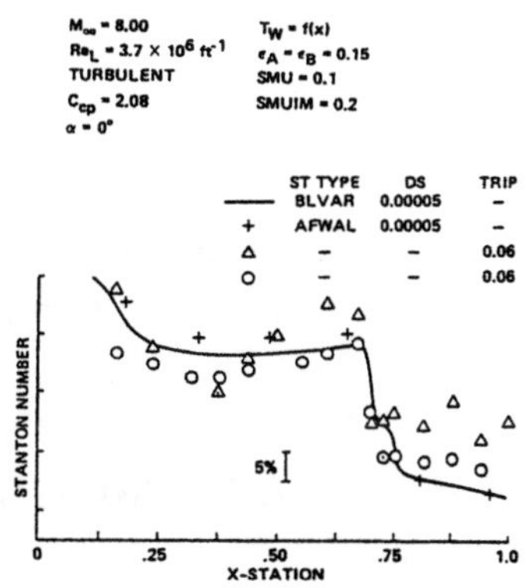

Figure 4. Computational Stanton Number Comparisons with Experiment.

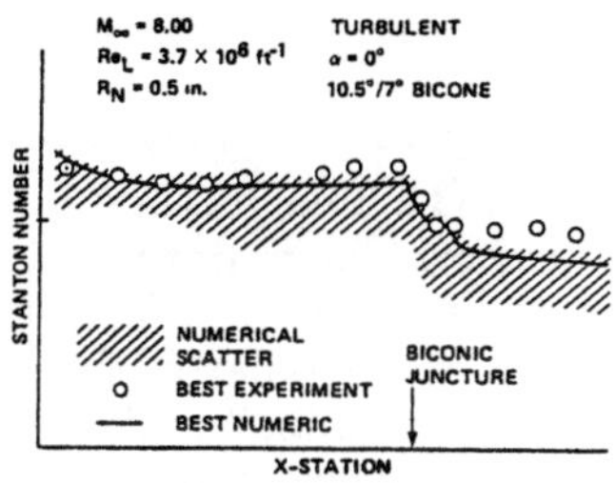

Figure 5. Comparison Between the Best Experiment and the Numerical 'Scatter'.

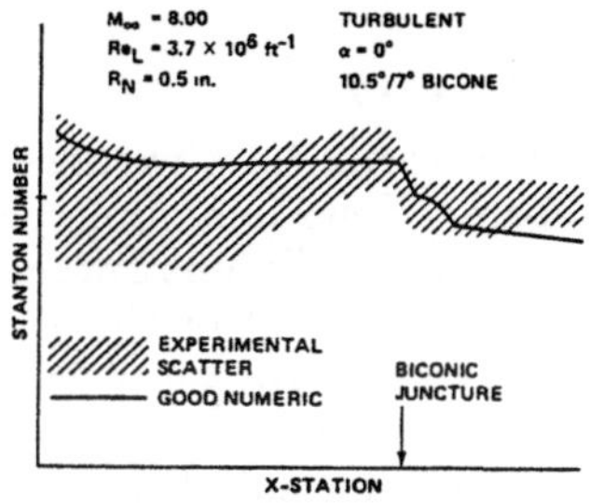

Figure 6. Comparison Between the Best Computation and the Experimental 'Scatter'.

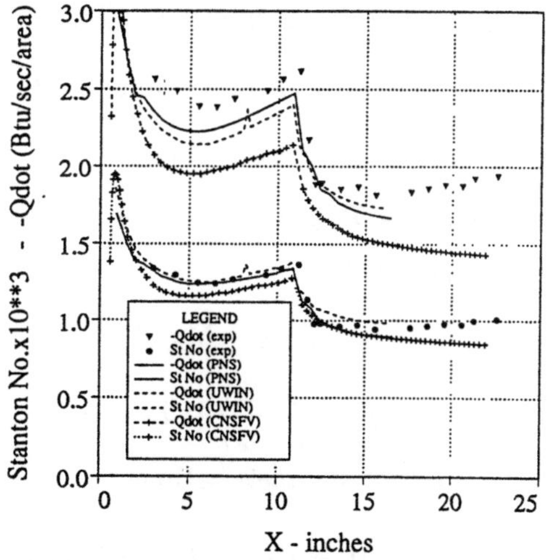

Figure 7. Computational Stanton Number and Heat Transfer Comparisons with Experiment.

Kinetic flux-splitting schemes for hypersonic flows

J.-P. CROISILLE[1] **and P. VILLEDIEU**[2]

[1] Université Paris-Sud, 91405 - ORSAY Cedex - France
[2] CERT/DERI/GAN, BP 4025, 31055 - TOULOUSE Cedex - France

1. Introduction

In this contribution, we present the first results of the extension to hypersonic nonequilibrium flows of a numerical method initiated in **[1]**. This method is of finite volume type on totally unstructured autoadaptive meshes. The major idea is to use both the kinetic origin of the Euler equations and their symmetric form (via the entropic variables). We get a decomposition of the flux of the polar of the entropy, into a convex and a concave part, that corresponds to "incoming" and "outgoing" particles. The differentiation of this decomposition with respect to the entropic variables yields a flux-splitting scheme, formally of order 1.

2. Kinetic flux splitting

We consider the bidimensional Euler equations

$$\partial_t w + \partial_1 f^1(w) + \partial_2 f^2(w) = \sigma(w) \tag{1}$$

with $w = (\rho, \rho\underline{u}, \rho e)$, f^1, f^2 the fluxes in the x^1 and x^2 directions, and $\sigma(w)$ the source term. If $S(w)$ is a Lax entropy for (1), $\varphi(w) = \nabla_w S(w)$ the corresponding entropic variables, a symmetrization function of (1) is

$$\Sigma^*(\varphi, n) = S^*(\varphi)\,(n_t + u^1 n_1 + u^2 n_2) \tag{2}$$

and the flux of the Euler equations is

$$w n_t + f^1 n_1 + f^2 n_2 = \nabla\varphi \Sigma^*(\varphi, n)\,. \tag{3}$$

The underlying pressure law is $p(\varphi) = \dfrac{S^*(\varphi)}{(-\varphi_e)}$ (where $\varphi = (\varphi_\rho, \varphi_u, \varphi_e)$).

We address the special class of symmetrizations

$$\Sigma^*(\varphi, n) = \int_{\substack{v\in\mathbb{R}^2\\ \eta>0}} (n_t + v \cdot n_x) H^*(\varphi \cdot K(v, \eta))\, dv d\eta \tag{4}$$

where $H^*(x)$ is a strictly convex function and $K(v, \eta)$ a collisional vector, typically of the form $K(v, \eta) = (1, v, \frac{1}{2}|v|^2 + g(\eta))$, where $g(\eta)$ is an increasing positive function.

By splitting $\Sigma^*(\varphi, n)$ into incoming and outgoing particles with respect to the parametric vector $\underset{\sim}{n}$ we get the following decomposition

$$\text{(5)}\quad \Sigma^*(\varphi,n) = \int_{n_t+v\cdot n_x>0} (n_t + v\cdot n_x) H^*(\varphi\cdot K(v,\eta))\,dv\,d\eta + \int_{n_t+v\cdot n_x<0} (n_t + v\cdot n_x) H^*(\varphi\cdot K(v,\eta))\,dv\,d\eta\,.$$

The corresponding decomposition of the flux appears as the gradient of the above splitting (summation holds from $i = 0$ to $i = 2$)

$$\text{(6)}\qquad f^i n_i = \underbrace{\nabla_\varphi \Sigma^{*,+}(\varphi,n)}_{(f^i n_i)^+} + \underbrace{\nabla_\varphi \Sigma^{*,-}(\varphi,n)}_{(f^i n_i)^-}$$

In this context, the Euler equations are written directly in symmetric and in kinetic form, and the flux-splitting is a gradient.

3. Physical models

- **The case $H^*(x) = \exp x$.**

We point out that the most classical models emanating from Statistical Mechanics enter in this framework by taking $H^*(x) = \exp x$. The fonction $\Sigma^*(\varphi, \underset{\sim}{n})$ plays in this case a similar role to the one played by the repartition function in Statistical Mechanics. The underlying velocity repartition function is of Maxwellian type. We describe now some examples.

a. Perfect (non polytropic) gas. For gases with pressure law $p = r\rho T$, and internal specific energy $\varepsilon = \varepsilon(T)$ one can show that there exists a unique function $g(\eta)$ such that the function $\Sigma^*(\varphi, n)$ symmetrizes the Euler equations, [2]. However the knowledge of $g(\eta)$ is not necessary for the decomposition formula.

In fact one has (in the case $n_t = 0$)

$$\text{(7)}\quad \begin{aligned} &\Sigma^{*,+}(\varphi,\underset{\sim}{n}) = S^*(\varphi)\Big[\underset{\sim}{u}\cdot\underset{\sim}{n} G_0(Z) + (2rT)^{\frac{1}{2}} G_1(Z)\Big]\,, Z(\varphi) = -\frac{\underset{\sim}{u}\cdot\underset{\sim}{n}}{(2rT)^{\frac{1}{2}}} \\ &G_0(z) = 1/\sqrt{\pi}\int_z^{+\infty} e^{-t^2}dt\ ;\ G_1(z) = 1/\sqrt{\pi}\int_z^{+\infty} te^{-t^2}dt \end{aligned}$$

Differentiation of (7) with respect to φ gives

$$\text{(8)}\qquad \nabla_\varphi \Sigma^{*,+}(\varphi,\underset{\sim}{n}) = G_0(Z) f^j n_j + (2rT)^{\frac{1}{2}} G_1(Z)(w + \widetilde{w})$$

where $\widetilde{w} = (0,0,0,p/2)$.

We point out the validity and the simplicity of formula (8) regardingless to the particular energy law that is considered. (8) generalizes the formula of Pullin [4] and of Deshpande [3].

b. Perfect mixtures of gases.

The symmetrization of a perfect mixture is obtained by summation of the symmetrization of each species in the mixture,[2]. This procedure is here the counterpart in this framework of the notion of perfect mixture in Statistical Mechanics, where the total pressure is the sum of the partial pressures (Dalton law). We emphasize that the split flux for the mixture is just the sum of the partial split fluxes. This gives a very natural counterpart of the physics at the numerical level.

c. Vibrationaly excited diatomic gases [2].

The classical statistical model of the harmonic oscillator can be recovered by taking a collisional vector of the form

$$K(v,\eta,m) = (1, v, \tfrac{1}{2}|v|^2 + g(\eta)), \beta m), \ \eta > 0, \ m \in \mathbb{N}. \tag{9}$$

The entropic variable is $\varphi = (\varphi_\rho, \varphi_u, \varphi_e, \varphi_{\varepsilon_{\text{vib}}})$. The resulting conservative variables are

$$\begin{aligned} w &= (\rho, \rho u, \rho e, \rho\varepsilon_{\text{vib}}) \ \text{ with } \ e = \tfrac{1}{2}|u|^2 + \varepsilon(T), \\ \varepsilon_{\text{vib}}(T_{\text{vib}}) &= \frac{rT^*}{e^{\frac{T^*}{T_{\text{vib}}}} - 1} \ \text{ with } \ \beta = rT^* \ ; T_{\text{vib}} = -\frac{1}{r\varphi_{\varepsilon_{\text{vib}}}}. \end{aligned} \tag{10}$$

The flux is (1-D case)

$$f = (\rho u, \rho u^2 + \rho, (\rho e + p)u, \rho\varepsilon_{\text{vib}} u) \ \text{ with } \ p = r\rho T. \tag{11}$$

The flux-splitting formula is the same as (8) with $\tilde{w} = (0, 0, p/2, 0)$.

- **The case $H^*(x) = x_+^\alpha$, $\alpha > 1$.**

This choice of the function $H^*(x)$ is described in **[2]**. It enables velocity repartition functions with finite supports. It allows flux-splitting formulae for perfect gases different from that with $H^*(x) = \exp x$ but involving different special functions. So the latter are prefered, by now, for numerical purposes.

4. Numerical scheme

The numerical scheme being developped is of finite volume type. It reads

$$\int_{\partial q_h} \left[\nabla_\varphi \Sigma^{*,-}(\varphi_h^{\text{out}}, \underline{n}) + \nabla_\varphi \Sigma^{*,+}(\varphi_h^{\text{in}}, \underline{n})\right] = 0 \tag{12}$$

q_h is an element of the space-time grid, that can be totally unstructured, φ_h^{out} and φ_h^{in} are the piecewise constant outer and inner traces of φ_h on ∂q_h. $\underline{n}$ is the normal to ∂q_h giving orientation. If $q_h = \omega_h \times [t^k, t^{k+1}]$ and p indexes the edges of $\partial\omega_h$, (12) becomes

$$w_h^{k+1} = w_h^k - \frac{\Delta t}{\text{area}(\omega_h)} \sum_p \left[\nabla_\varphi \Sigma^{*,-}(\varphi_h^{\text{out},k+1}, \underline{n}_p) + \nabla_\varphi \Sigma^{*,+}(\varphi_h^{\text{in},k+1}, \underline{n}_p)\right]. \tag{13}$$

On a cartesian grid, (13) appears as a first order, implicit in time, flux-splitting scheme. A nice property of the scheme (13) is the consistency with the entropy inequality.

We can identify the numerical production of entropy on an edge A of the mesh by

$$\int_A [S^i n_i] = \int_{A\cap\partial q_h^1} \bar{\sigma}(\varphi_h) + \int_{A\cap\partial q_h^2} \bar{\sigma}(\varphi_h) \tag{14}$$

where

$$\bar{\sigma}(\varphi_h) = \Sigma^{*,-}(\varphi_h^{\text{in},k+1}, \underline{n}) - \Sigma^{*,-}(\varphi_h^{\text{out},k+1}, \underline{n}) - \nabla_\varphi \Sigma^{*,-}(\varphi_h^{\text{out},k+1}, \underline{n})(\varphi_h^{\text{in},k+1} - \varphi_h^{\text{out},k+1}).$$

The right-hand-side of (14) is negative by the concavity of $\Sigma^{*,-}$.

In fact the explicit version of (13) still preserves this property under a cfl-like condition. Moreover the positivity of the density and of the pressure are proven to be preserved under conditions of the same type, **[5]**.

5. Numerical results

We present results corresponding to an hypersonic chemically reactive flow at incidence 30° over a double-ellipse. The conditions at infinity are $M_\infty = 9.15$, $T_\infty = 260°K$, $\rho_\infty = 1,344 10^{-3} kgm^{-3}$. 5 species $(N_2, 0_2, N0, N, 0)$ are taken in account. The chemical source term is the Arrhenius one with the Gardiner model (3 reactions). The time integration is implicit for the source, explicit for the hydrodynamic.

Moreover a fully automatized refining procedure of the mesh, that can be totally unstructured has been developped. The refining criterion, that is used, is the **local numerical entropy production** (§ 4).

The mesh refining procedure works very well on supersonic or hypersonic flows. Moreover the extension to other reactive models including vibrational nonequilibrium is in progress, based on § 3.c.

Acknowledgments : We would like to thank P.A. Mazet for many helpful discussions. We thank also P. Klotz for her very efficient implementation of the refining procedure and her graphic package.

References

[1] F. BOURDEL, P. DELORME, P.A. MAZET, *Convexity in hyperbolic problems. Application to a discontinuous Galerkin method for the resolution of the multidimensional Euler equations*, Notes on Num. Fluid Mech., vol. 24, Vieweg, 1988.

[2] J.P. CROISILLE, P. DELORME, *Kinetic symmetrizations and pressure laws for the Euler equations*, Prépublications Université d'Orsay n° 91-61. To appear in Physica-D.

[3] S.M. DESHPANDE, *On the Maxwellian distribution, symmetric form and entropy conservation for the Euler equations*, NASA TP 2583, 1986.

[4] D.I. PULLIN, *Direct simulation methods for compressible inviscid ideal-gas flow*, Journal of Comp. Phys. **34**, (1980), 231–244.

[5] P. VILLEDIEU, Thesis, to appear.

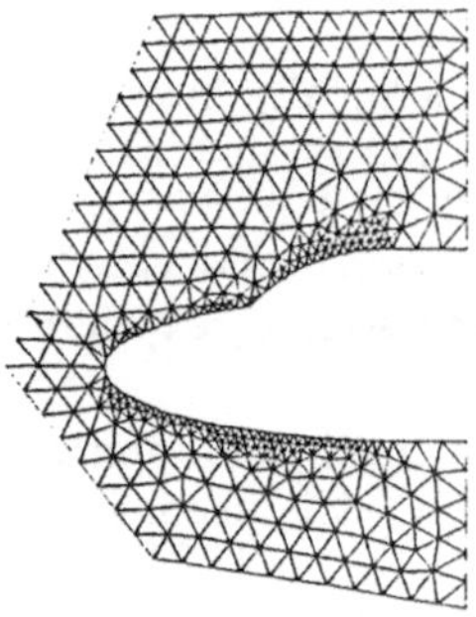

Fig1: Coarse mesh
410 vertices, 1094edges, 685 elements

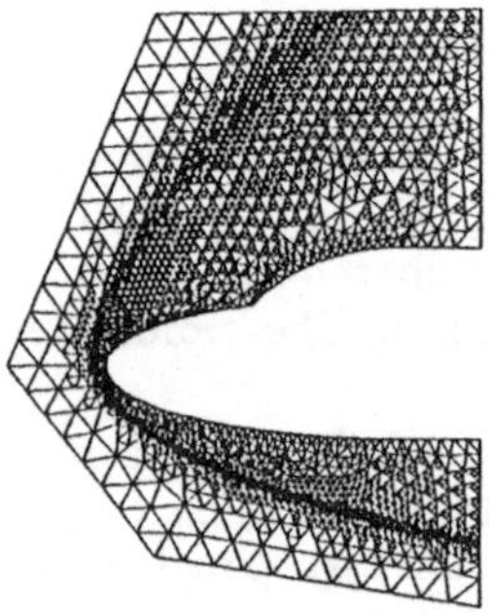

Fig 2: Mesh (after 4 refinement steps)
6012 vertices, 14616 edges, 8605 elements
$M_\infty = 9.15$, reactive case

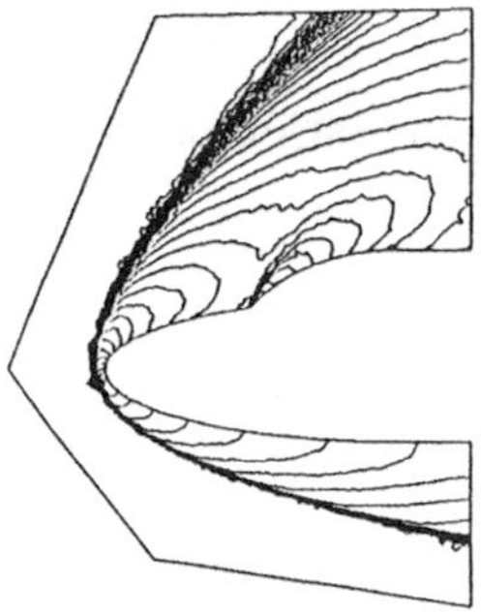

Fig 3: Iso-Mach lines
$7.89.10^{-2} < M < 9.15$
inert case

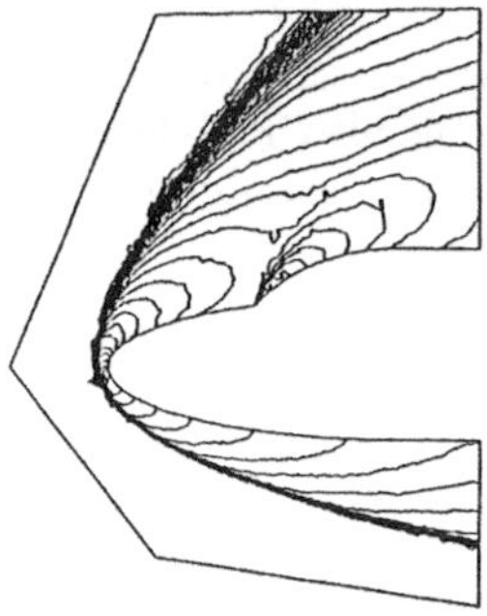

Fig 4: Iso-Mach lines
$7.38.10^{-2} < M < 9.15$
reactive case

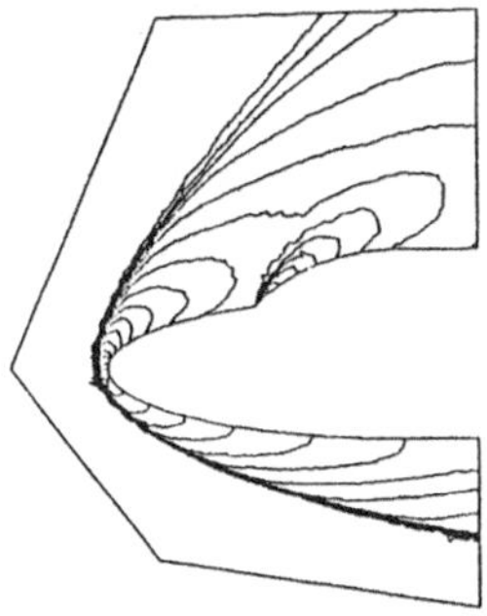

Fig 5: Iso-Temperature lines
$260^0K < T < 3590^0K$
inert case

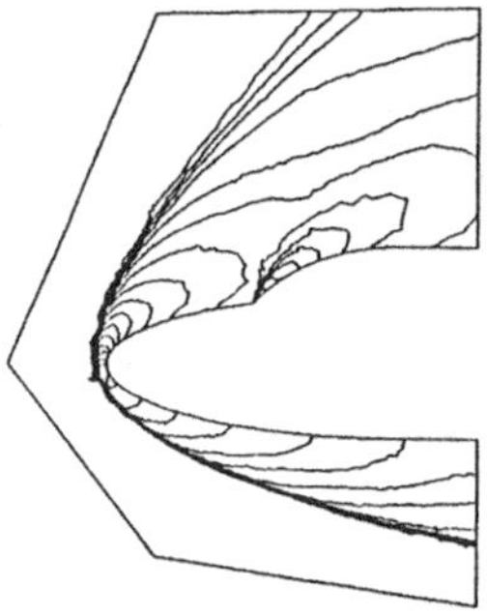

Fig 6: Iso-Temperature lines
$260^0K < T < 3550^0K$
reactive case

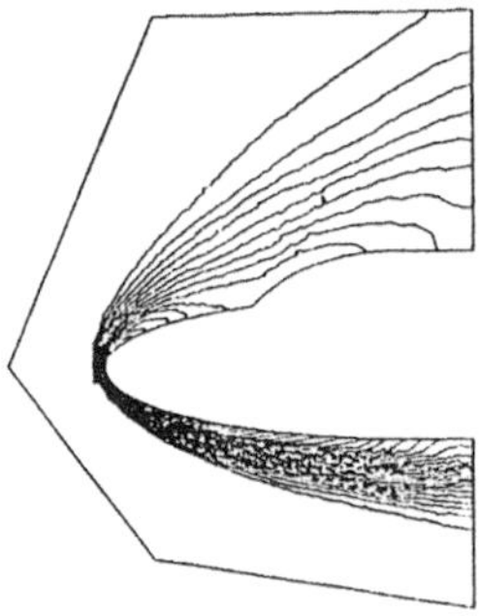

Fig 7: Iso-y^{NO} lines
$9.94.10^{-5} < y^{NO} < 5.87.10^{-3}$

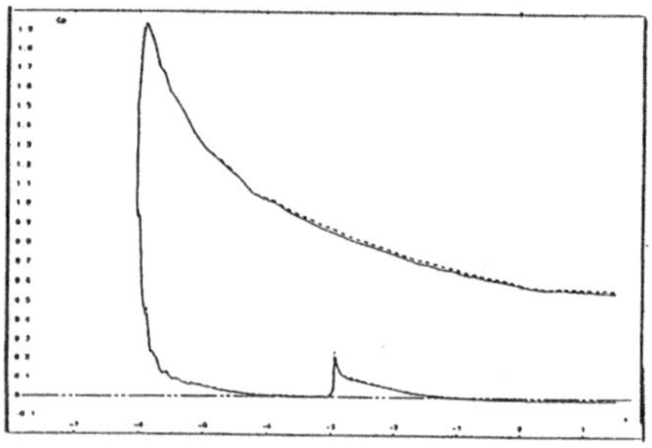

Fig 8: Pressure coefficient on the wall
—:inert case
- -:reactive case

NUMERICAL SIMULATION OF SUPERSONIC MIXING LAYERS USING A FOURTH-ORDER ACCURATE SHOCK CAPTURING SCHEME

Hisaaki Daiguji, Satoru Yamamoto and Kōichi Ishizaka

Department of Aeronautics and Space Engineering
Tohoku University, Sendai 980, Japan

1. INTRODUCTION

The purposes of this paper are first to simulate supersonic mixing layers including shocks by using a fourth-order high-resolution scheme developed by the authors[1], and to preliminarily understand the physical phenomena for the development of hypersonic air-breathing engines.

With regard to the numerical approaches for simulating mixing layers, a number of numerical methods have been already reported. But, most of them are for rather low speed mixing layers and can not simulate any flows having shock waves. For simulating high speed compressible mixing layers, we must fundamentally use something like a shock capturing scheme. And in this case, some investigations using TVD scheme have also been reported[2]-[4]. But, they use the difference scheme with second-order accuracy in space. Therefore, the resolvability of slip surface is still poor, though reasonable shock patterns are obtained. On the other hand, the present method uses a new shock capturing scheme having fourth-order accuracy in space and second-order accuracy in time.

The present scheme is based on the third-order accurate MUSCL upwind-scheme, and is modified by compact additional terms with a parameter ϕ for improving the accuracy. The scheme is usuallly fourth-order, but becomes fifth-order for $\phi = 1/5$. And a fourth-order compact MUSCL TVD scheme is attained for $\phi = 1/3$. This scheme can be easily applied to any existing numerical solvers based on the third-order upwind scheme[5], can capture the slip surface clearly, and is simpler than so-called ENO scheme[6].

2. FUNDAMENTAL EQUATIONS

In this study, we use the three-dimensional compressible Navier-Stokes equations

$$\partial q/\partial t + L(q) \equiv \frac{\partial q}{\partial t} + \frac{\partial F_j}{\partial x_j} + \frac{1}{Re} S = 0 \qquad (1)$$

The equations are solved by an implicit time-marching finite-difference scheme with the delta-form approximate-factorization, the diagonalization and the upwinding. To simulate unsteady flows accurately, the sceond-order Crank-Nicholson method and the Newton iteration are applied to this implicit scheme[7].

3. HIGH-RESOLUTION SCEMES

The fourth(fifth)-order compact MUSCL TVD scheme can be written as

$$F_{i+1/2} = F^{+}(q^{L}_{i+1/2}) + F^{-}(q^{R}_{i+1/2}) \tag{2}$$

where

$$\begin{aligned} q^{L} &= q_i + \frac{1}{6}\text{minmod}[\tilde{D}q_{i-1/2}, b\tilde{D}q_{i+1/2}] + \frac{1}{3}\text{minmod}[\tilde{D}q_{i+1/2}, b\tilde{D}q_{i-1/2}] \\ q^{R} &= q_{i+1} - \frac{1}{6}\text{minmod}[\tilde{D}q_{i+3/2}, b\tilde{D}q_{i+1/2}] - \frac{1}{3}\text{minmod}[\tilde{D}q_{i+1/2}, b\tilde{D}q_{i+3/2}] \end{aligned} \tag{3}$$

$$\begin{aligned} \tilde{D}q_{i-1/2} &= Dq_{i-1/2} - \frac{1-\phi}{4}\bar{D}^3 q_{i-1/2} \\ \tilde{D}q_{i+1/2} &= Dq_{i+1/2} - \frac{1+\phi}{8}\bar{D}^3 q_{i+1/2} \\ \tilde{D}q_{i+3/2} &= Dq_{i+3/2} - \frac{1-\phi}{4}\bar{D}^3 q_{i+3/2} \end{aligned} \tag{4}$$

$$Dq_{i+1/2} = q_{i+1} - q_i \quad , \quad \bar{D}^3 q_{i+1/2} = \bar{D}q_{i-1/2} - 2\bar{D}q_{i+1/2} + \bar{D}q_{i+3/2}$$

$$\begin{aligned} \bar{D}q_{i-1/2} &= \text{minmod}(Dq_{i-1/2}, b_1 Dq_{i+1/2}, b_1 Dq_{i+3/2}) \\ \bar{D}q_{i+1/2} &= \text{minmod}(Dq_{i+1/2}, b_1 Dq_{i+3/2}, b_1 Dq_{i-3/2}) \\ \bar{D}q_{i+3/2} &= \text{minmod}(Dq_{i+3/2}, b_1 Dq_{i-1/2}, b_1 Dq_{i+1/2}) \end{aligned}$$

Equation (3) is made by taking a linear combination of the fourth-order upstream-difference and the fourth-order central-difference. $-1 \le \phi \le 1$. $\phi = -1$ and 1 correspond to the upstream- and central-differences, respectively. If $\phi = 1/5$, then Eq.(3) becomes fifth-order. By taking $\phi = 1/3$, Eq.(4) can be written the same form as

$$\tilde{D}q_{k+1/2} = Dq_{k+1/2} - \frac{1}{6}\bar{D}^3 q_{k+1/2} \quad , \quad (k = -1, 0, 1) \tag{5}$$

Therefore, the existing code using the third-order MUSCL TVD scheme can be easily improved to the fourth-order by using $\tilde{D}q$ of Eq.(5)

4. NUMERICAL RESULTS

In this paper, we perform direct numerical simulations of the time-developing subsonic and supersonic mixing layers (Figure 1). The initial mean velocities $\bar{u}$, $\bar{v}$, $\bar{w}$ are

$$\bar{u} = \frac{1}{2}\tanh(y) \quad , \quad \bar{v} = 0. \quad , \quad \bar{w} = 0. \tag{6}$$

and the perturbations used here are

$$\begin{aligned} u' &= \varepsilon_1\{\phi_r'(y)\cos(\alpha x) - \phi_i'(y)\sin(\alpha x)\} \\ &+ \varepsilon_2\{\phi_r'(y)\cos(0.5\alpha x) - \phi_i'(y)\sin(0.5\alpha x)\} \\ v' &= \varepsilon_1\{\phi_r(y)\alpha\sin(\alpha x) + \phi_i(y)\alpha\cos(\alpha x)\} \\ &+ \varepsilon_2\{\phi_r(y)0.5\alpha\sin(0.5\alpha x) + \phi_i(y)0.5\alpha\cos(0.5\alpha x)\} \\ &+ \varepsilon_3\{\phi_r(y)2\alpha\sin(2\alpha z) + \phi_i(y)2\alpha\cos(2\alpha z)\} \\ w' &= \varepsilon_3\{\phi_r'(y)\cos(2\alpha z) - \phi_i'(y)\sin(2\alpha z)\} \end{aligned} \tag{7}$$

where α is wave number($= 0.4446$). $\phi_r(y)$ and $\phi_i(y)$ are the most unstable eigenfunctions for this wave number[8]. The amplitudes of oscillations ε_1, ε_2 and ε_3 are 0.219, 0.110 and 0.003, respectively. The computational domain is $0 \sim 4\pi/\alpha$ in x direction $-2\pi/\alpha \sim 2\pi/\alpha$ in y direction and $0 \sim 2\pi/\alpha$ in z direction (Figure 1). The relative Mach number M_r is defined as the differnce between two streams.

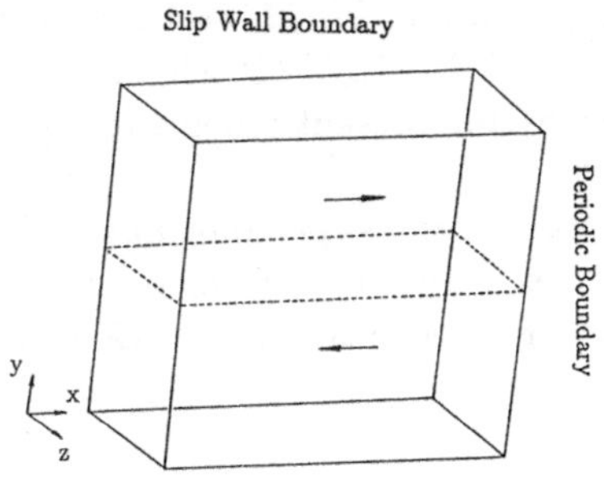

Figure 1 Time-developing mixing layer

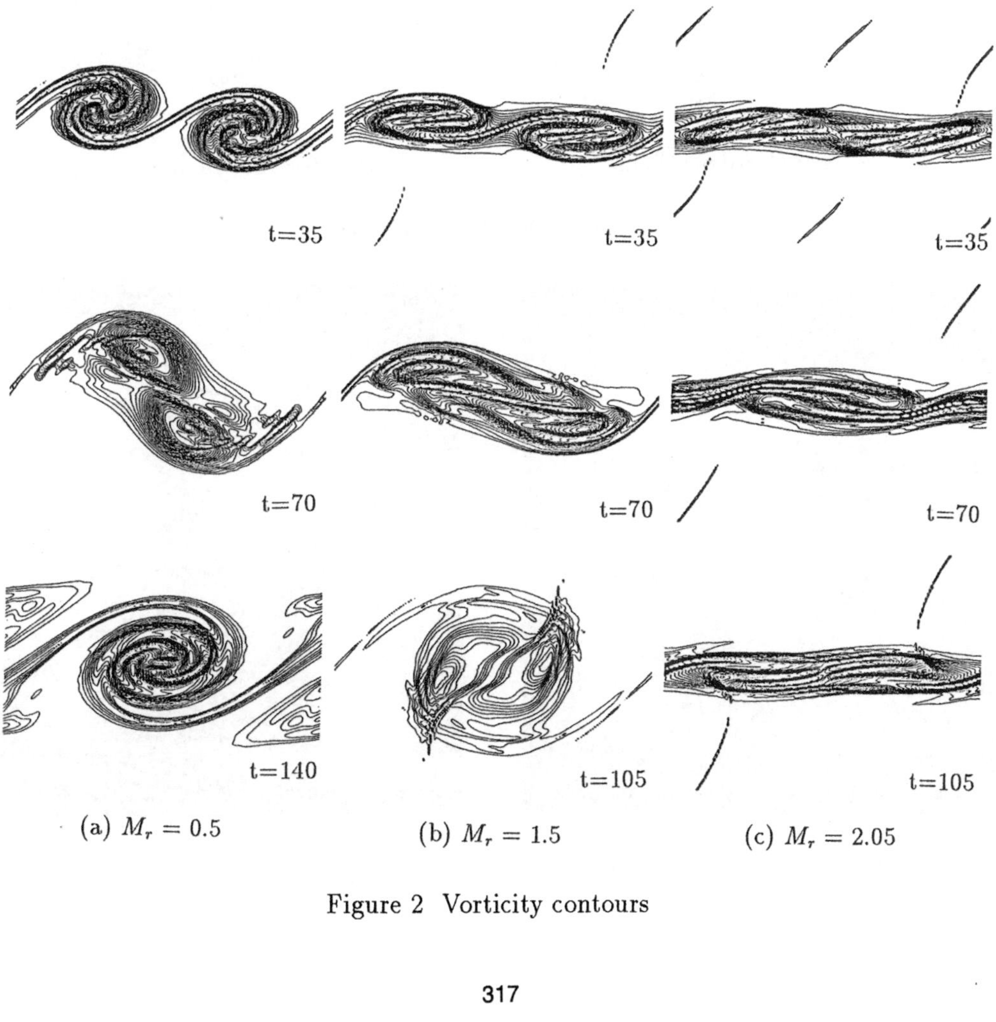

(a) $M_r = 0.5$ (b) $M_r = 1.5$ (c) $M_r = 2.05$

Figure 2 Vorticity contours

We first calculate two-dimensional subsonic and supersonic mixing layers. The computational grid is 181 × 181 uniform grid. Figure 2 shows the calculated vorticity contours at (a) $M_r = 0.5$, (b) $M_r = 1.5$ and (c) $M_r = 2.05$, respectively. $Re = 10000$. For $M_r = 0.5$ and 1.5, the perturbation leads to symmetric roll-up and pairing. These vortex behaviors are similar to the incompressible mixing layers[9]. But for $M_r = 2.05$, kinks appear and the pairing dose not occur clearly. For $M_r = 1.5$ weak nomal shocks appear and for $M_r = 2.05$, oblique shocks are observed. Figure 4 shows a comparison of calculated vorticity contours for $M_r = 1.5$ by the third-order MUSCL TVD scheme[5] and the fourth-order MUSCL TVD scheme. It is found from these figures that resolution of slip surface by the present scheme is superior than the existing third-order scheme.

We next calculate three-dimensional subsonic and supersonic mixing layers. Figure 4 shows the calculated equi-vorticity surfaces of the time-developing subsonic mixing layer and the Mach number distribution on a vertical plane. Considerable three-dimensional effects can be observed.

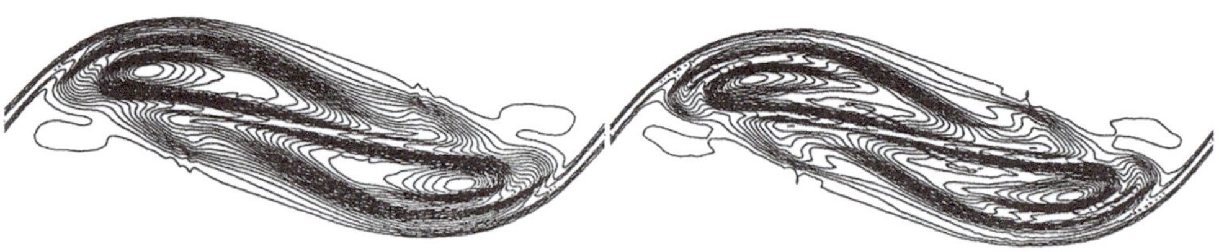

(a) Third-order scheme (b) Present scheme

Figure 3 Vorticty contours

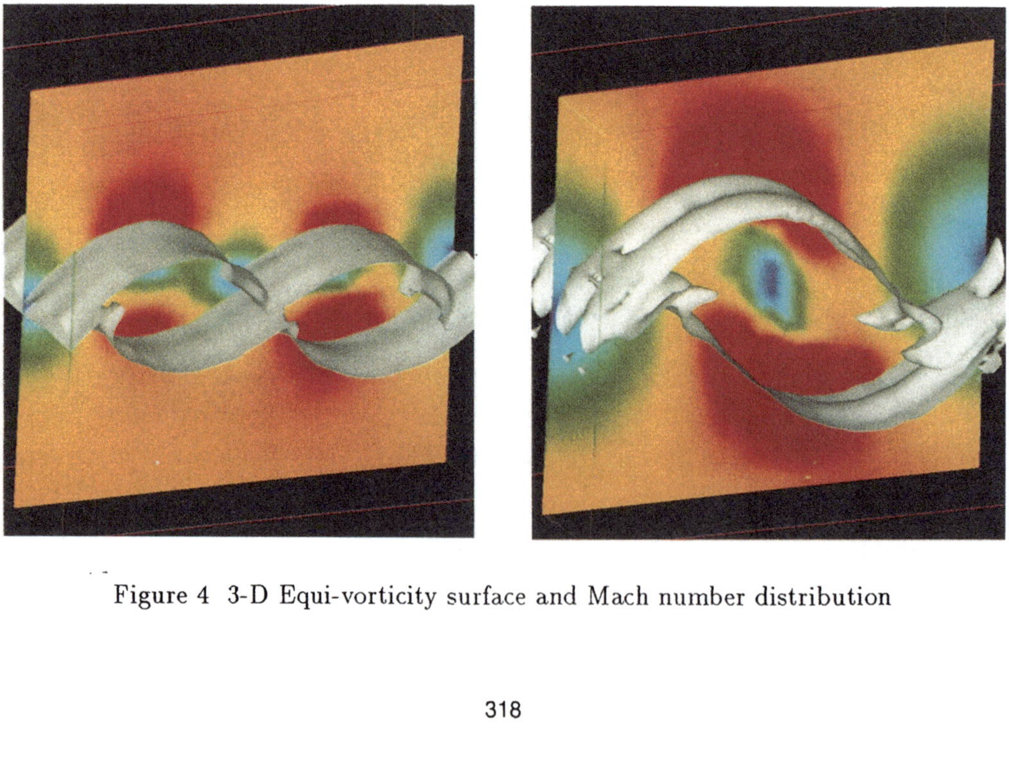

Figure 4 3-D Equi-vorticity surface and Mach number distribution

5. CONCLUSIONS

The direct numerical simulations of the time-developing subsonic and supersonic mixing layers have been presented. We can conclude that the present scheme is considerably excellent than the existing second- and third-order TVD schemes to simulate high speed flows having slip surfaces and shock waves.

References

[1] Yamamoto, S. and Daiguji, H., "Higher-Order Accurate Upwind Schemes for Solving the Compressible Euler and Navier-Stokes Equations," *4th Int. Symp. on Computational Fluid Dynamics-Davis*, **II** (1991), pp.1269-1274.

[2] Soetrisno, M., Riley, J. J. and McMurtry, P., "A Study of Inviscid, Supersonic Mixing Layers Using a Second-Order Total Variational Diminishing Scheme," *AIAA J.*, Vol.27, No.2 (1989), pp.1770-1778.

[3] Klein, R., Munz, C.D. and Schmidt, L., "A Numerical Study of Interfacial Instabilities at High Mach Numbers," *Proc. of the 8th GAMM-Conf. on Numerical Methods in Fluid Mechanics,*, (1989), pp.242-251, Vieweg.

[4] Sandham, N.D. and Yee, H.C., "A Numerical Study of a Class of TVD Schemes for Compressible Mixing Layers," *NASA TM*-102194, (1989).

[5] Yamamoto, S. and Daiguji, H., "Numerical Simulation of Unsteady Turbulent Flow through Transonic and Supersonic Cascades," *Proc. 12th Int. Conf. on Numerical Methods in Fluid Dynamics-Oxford, Lecture Notes in Phys.*, **371** (1990), pp.485-489, Springer-Verlag.

[6] Harten, A., Engquist, B., Osher, S. and Chakravarthy, S. R., "Uniformly High Order Accurate Essentially Non-Oscillatory Scheme III," *J. Comp. Phys.*, Vol.71, No.2 (1987), pp.231-303.

[7] Rai, M. M., "Unsteady Three-Dimensional Navier-Stokes Simulations of Turbine Rotor-Stator Interaction," *AIAA Paper*, 87-2058, (1987).

[8] Michalke, A., " On the Inviscid Instability of the Hyperbolic Tangent Velocity Profile," *J. Fluid Mech.*, Vol.19, (1964), pp.543-556.

[9] Bell, B., J., Glaz, M., H., Solomon, M., J. and Szymczak, G., W., " Application of a Second-Order Projection Method to the Study of Shear Layers," *Proc. 11th Int. Conf. on Numerical Methods in Fluid Dynamics-Williamsburg, Lecture Notes in Phys.*, **323** (1988), pp.142-146, Springer-Verlag.

TRULY TWO-DIMENSIONAL ALGORITHMS FOR RADIATIVE NON-EQUILIBRIUM FLOWS

G. J. Elbert and P. Cinnella

ERC for Computational Field Simulation, Mississippi State, USA

1. OBJECTIVES

This study details the derivation and application of "truly" two-dimensional algorithms for the simulation of radiative heat transfer in flows out of chemical equilibrium. The present approach removes the need for the One-Dimensional Slab Approximation[1], which is commonly used in order to simplify the treatment of the radiative heat transfer term appearing in the governing gasdynamic equations. Moreover, the radiative field is determined over the entire flowfield, instead of being limited to a small region close to the stagnation streamline. The increase in accuracy due to the removal of the 1D Slab approximation is important for the design of reentry configurations, where radiative heat transfer is the predominant form of heat exchange[2].

The radiative calculation is fully coupled with the flow solver[3], although some freezing techniques are implemented in order to save computer time for steady-state problems. Comparisons of the proposed algorithms with the slab theory and the simple case of an emission-dominated gas are shown for test cases in the hypersonic regime.

2. METHODOLOGY

The governing integro-differential equations for a reacting gas in radiative non-equilibrium, neglecting relativistic effects[1], read in generalized coordinates (ξ, η)

$$\frac{\partial}{\partial t}\left(\frac{\boldsymbol{Q}}{J}\right)+\frac{\partial(\tilde{\boldsymbol{F}}-\tilde{\boldsymbol{F}}_v)}{\partial \xi}+\frac{\partial(\tilde{\boldsymbol{G}}-\tilde{\boldsymbol{G}}_v)}{\partial \eta}=\frac{\boldsymbol{W}}{J}, \tag{2.1}$$

where $\boldsymbol{Q}$ is the vector of conserved variables; $\tilde{\boldsymbol{F}}$ and $\tilde{\boldsymbol{G}}$ are the inviscid flux vectors; $\tilde{\boldsymbol{F}}_v$ and $\tilde{\boldsymbol{G}}_v$ are the viscous flux vectors; $\boldsymbol{W}$ is the vector of source terms; and J is the Jacobian of the coordinate transformation[3]. The radiative heat transfer contribution appears explicitly only in the source term vector, in correspondence with the global energy equation, as the negative of the divergence of the radiant heat flux vector, $-\nabla\cdot\boldsymbol{q}^R$. For two-dimensional problems, $\nabla\cdot\boldsymbol{q}^R$ is given by

$$\begin{aligned}\nabla\cdot\boldsymbol{q}^R =& 4\sigma\alpha_P T^4 - 2\int_0^\infty \alpha_\nu\left(\int_0^{2\pi} I_\nu(\infty)e^{-\int_0^{r_\infty}\alpha_\nu dr}d\phi\right)d\nu \\ & -2\int_0^\infty \alpha_\nu\left(\int_0^{2\pi}\int_0^{r_\infty}\alpha_\nu B_\nu\, e^{-\int_0^{w}\alpha_\nu dr}dw d\phi\right)d\nu ,\end{aligned} \tag{2.2}$$

where σ is the Stefan-Boltzmann constant, α_P is the Planck mean absorption coefficient, T is the temperature, α_ν is the absorption coefficient at the frequency ν, I_ν is the specific intensity and B_ν its equilibrium value, ϕ is the angle that describes the direction of propagation of radiation, and dr, dw are line elements along a direction of propagation. In the above, the line integrals are started at point 0, which is the point in the flowfield under consideration, and terminate at point ∞, at a distance r_∞, which is a boundary point. It is apparent from this form of the radiative source term that its value at any point in the flowfield at a given time will depend upon the properties of all the points that are in its "line of sight," that is, all the points lying in the portion of the flowfield that is swept by straight lines (rays) emanating from the point under consideration and terminating either at solid boundaries or in the farfield.

The discretization of the governing equations is performed by means of the finite-volume technique. Flux-split techniques are employed for the inviscid fluxes[4], and central differences are used for the viscous fluxes. The discretized equations are advanced in time using either an LU decomposition or an Approximate Factorization in the computational plane[5].

The numerical evaluation of the radiative source terms, (2.1), involves integration over frequency, over angle, and over length along a direction of propagation. This evaluation should be performed at every time step when time accuracy is of interest. However, for steady-state problems, the radiative source terms can be lagged for a few time steps. The integration over the frequency spectrum can be performed in its simplest form by means of the gray-gas approximation[1]. More realistic approaches include the three-band and the eight-band models[6], whereby the mean absorption coefficient is considered constant over a band, and the integration reduces to a summation over the bands. Presently, preliminary results have been obtained for the one-band (gray-gas) model only, although the multiple band model is being implemented.

The double integration over the geometric parameters (angle and length) is accomplished by superimposing a "radiation grid" on the discretized domain. In particular, for every finite volume in the calculation it is necessary to define the rays that will be considered for the integration over the angle, and the points along each ray for the integration over the length, including the boundary points in the farfield or at some solid wall. Specifically, given a computational cell and a direction of propagation, a ray is started from the cell center and continued until a boundary is reached. After locating the position of the boundary point at the intersection of the ray with the grid boundary, the points along the ray are distributed in accordance with the technique chosen for the line integration. The process is continued until all directions of propagation are exhausted, and is repeated for all volumes in the computational domain.

The integration over the angle (direction of propagation) is performed by partitioning the interval $[0, 2\pi]$ into equally distributed subintervals. Due to the fact that the two limits of integration physically coincide, this procedure is tantamount to using a composite trapezoidal rule (or a two-point open Newton-Cotes formula, for that matter). Several options have been implemented for the integration over the length, including composite trapezoidal and Simpson rules, the use of Richardson's extrapolation, and the implementation of a composite Gauss-Legendre formula.

3. RESULTS

Preliminary results were obtained for a viscous flow at Mach 46.9 over a circular cylinder in air at an altitude of 57.9 km. This problem was studied by Howe and Viegas[7] by means a shock layer solution in equilibrium air using the 1D Slab approximation. It should be noted that their methodology provides a prediction only of the stagnation streamline variables. The present investigation employed 60 volumes in the circumferential direction and 90 in the radial direction. The "radiation grid" at every volume consisted of 36 equally spaced rays and 40 intervals of integration per ray. The composite two-point Gauss-Legendre formula was used for the numerical quadrature along a ray, which resulted in 80 points being employed for the integration. A surface temperature $T_w = 1500\ K$ was imposed as a boundary condition.

Results were obtained for the truly two dimensional radiative heat transfer case, the 1D Slab approximation, the emission-dominated case, and a baseline case with no radiative heat transfer. A Van Leer-type discretization of the inviscid fluxes was employed. The finite-rate chemistry model for air includes five-species and seventeen reactions[5], and is widely used in hypersonic applications.

In a related paper[3], a comparison of CPU time per iteration on a Cray YMP was performed for the reactive case with no radiation, the emission-dominated problem, the 1D Slab, and the full 2D algorithm. Comparing the overheads associated with the inclusion of a radiative model to the baseline time for a reactive calculation, it was possible to conclude that the cost of running an emission-dominated model is negligible. On the other hand, a five to ten percent increase in CPU time is required for the Slab calculation. The full 2D algorithm is more expensive, requiring between a twofold and a fourfold increase in CPU time. This should not be surprising, in light of the fact that the evaluation of the radiative source term involves 36 rays per finite volume instead of two, as in the 1D Slab.

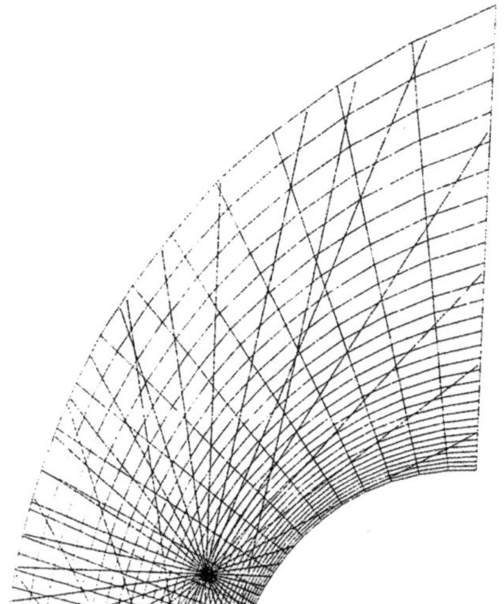

Fig. 1

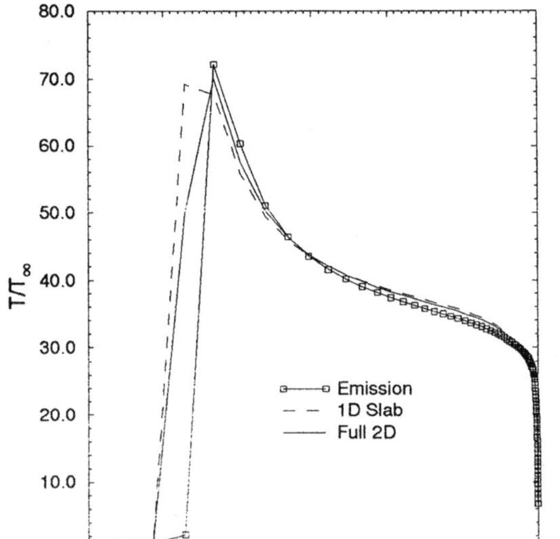

Fig. 2

Fig. 1 shows a symbolic representation of the physical grid, and superimposed on it the "radiative" grid for a single finite volume. It may be noticed that at the symmetry axis the lines of sight are specularly reflected, allowing the calculations to be performed with a quarter of circumference only.

Fig. 2 depicts the temperature along the first row of volumes off of the stagnation streamline. The temperature profiles are affected by the choice of the radiative transfer model. In particular, absorption is re-heating the gas in the shock layer when compared with the emission-dominated case, but the fully two-dimensional algorithm provides the flowfield with some "relieving" effects which ultimately result in lower temperatures when compared with the 1D Slab approximation. The peak temperature is virtually the same for the three cases. The baseline case with no radiation is not presented, because it would be outside of the plotted scale due to the dramatic (and unphysical) increase in temperature levels for a gas that is not allowed to radiate off most of the aerothermodynamic heating associated with a Mach 47 flow.

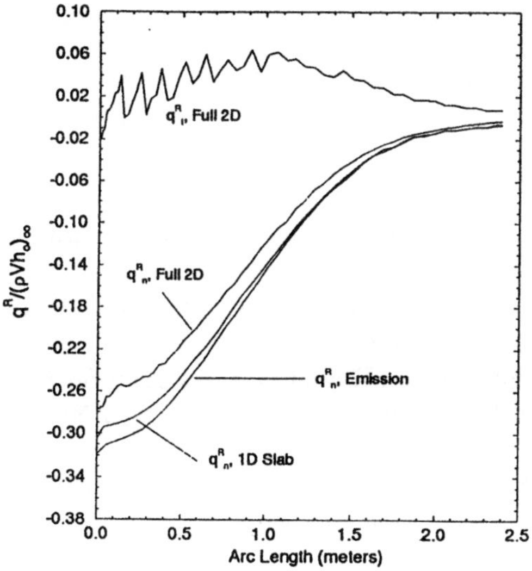

Fig. 3

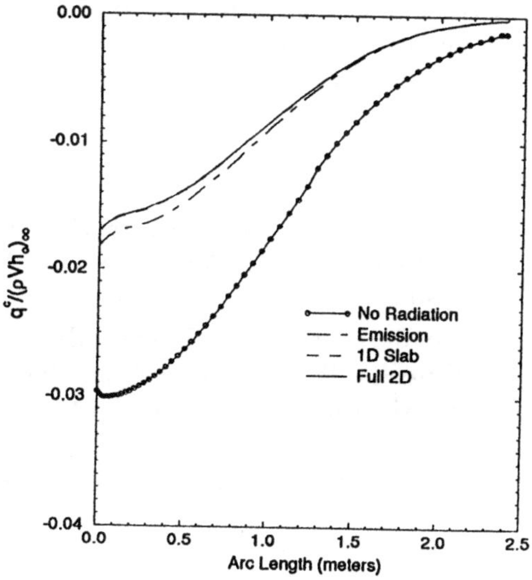

Fig. 4

Fig. 3 shows the component of the radiative heat flux vector normal to the body on the first row of volumes off of the body. Also represented is the tangential component for the two-dimensional case. There is a significant change between the two-dimensional and the one-dimensional predictions. It should be reiterated at this point that both emission-dominated and 1D Slab results are to be considered in the stagnation point region only. Nonetheless, these preliminary results seem to indicate that the approximations made in the 1D Slab theory may produce a significant error. More specifically, the radiative heat flux vector at the stagnation point has a dimensional value of 240 MW/m^2, when evaluated with the 1D Slab theory, and a value of 204 MW/m^2, when calculated by means of the fully two-dimensional algorithm. The difference is about 20%. The convective heat flux vector, which is shown in Fig. 4, ranges in magnitude from one tenth to one twentieth of its radiative counterpart, which confirms the fact that at this extreme Mach number the heat transfer is essentially radiative[6].

The pressure coefficient on the body is in good qualitative agreement with Newtonian theory[2], and is not dramatically affected by radiation modeling. On the other hand, the skin friction coefficient is significantly reduced by the presence of radiative heat transfer when compared with the baseline case with no radiation, although there is no appreciable difference between the predictions of the different radiative models.

4. FUTURE WORK

The present study has found some evidence that the currently accepted One-Dimensional Slab Approximation to the modeling of radiative heat transfer in hypersonic applications can be inaccurate. Moreover, the practical feasibility of truly two-dimensional radiative heat transfer simulations has been demonstrated, at a reasonable computational price when compared with the simplified approach.

Although the technique discussed is fully general and applicable to a wide range of spectrographic models, only the simple gray gas model has been utilized at present, and more refined models are necessary in order to improve the accuracy of the quantitative predictions.

More work is needed in order to extend the present methodology to three space dimensions, and more validation tests are necessary for different geometries.

5. REFERENCES

1. Vincenti, W.C., and Kruger, C.H.Jr., "Introduction to Physical Gas Dynamics", ISBN No. 0-88275-309-6, Krieger, Malabar FL, 1986.
2. Anderson, J.D.Jr., "Hypersonic and High Temperature Gas Dynamics", ISBN No. 0-07-001671-2, McGraw-Hill, New York, 1989.
3. Cinnella, P., and Elbert G.J., "Two-Dimensional Radiative Heat Transfer Calculations for Flows in Themo-Chemical Non-Equilibrium", AIAA Paper No. 92-0121, 1992.
4. Grossman, B., and Cinnella, P., "Flux-Split Algorithms for Flows with Non-equilibrium Chemistry and Vibrational Relaxation", *Journal of Computational Physics*, Vol. 88, No. 1, pp. 131–168, 1990.
5. Walters, R.W., Cinnella, P., Slack, D.C., and Halt, D., "Characteristic-Based Algorithms for Flows in Thermo-Chemical Nonequilibrium", AIAA Paper No. 90-0393, 1990.
6. Anderson, J.D.Jr., "An Engineering Survey of Radiating Shock Layers", *AIAA Journal*, Vol. 7, No. 9, pp. 1665–1675, 1969.
7. Howe, J.T., and Viegas, J.R., "Solution of the Ionized Radiating Shock Layer, Including Reabsorption and Foreign Species Effects, and Stagnation Region Heat Transfer", NASA Technical Report No. R-159, 1963.
8. Moss, J.N., and Simmonds, A.L., "Galileo Probe Forebody Flowfield Predictions", *Progress in Astronautics and Aeronautics*, Vol. 85, pp. 419–445, 1983.

DEVELOPMENT AND VALIDATION OF A HIGH RESOLUTION SHOCK-CAPTURING SCHEME FOR NONEQUILIBRIUM HYPERSONIC FLOWS

S. Jonas, E. Messerschmid, H.-H. Frühauf, O. Knab

Universität Stuttgart
Institut für Raumfahrtsysteme
Germany

1 Introduction

The numerical simulation of high-temperature hypersonic flows calls for an accurate shock resolution as well as for the consideration of nonequilibrium phenomena. Not only surface pressure and temperature, but also the details of relaxation processes are desired in order to understand the nature of these flows and to validate competing thermodynamic gas models. In the work presented here, the equations for air in thermo-chemical nonequilibrium are solved in a fully coupled manner using a high resolution Godunov-type finite volume scheme adapted to the extended system of governing equations. The implicit treatment of fluxes and source terms eliminates the problem of stiffness [1] and yields convergence rates comparable to those of equivalent perfect gas schemes.

2 Governing Equations

Consider the two-dimensional, inviscid flow of dissociating and ionizing air which is taken to be a mixture of eleven species (N_2, O_2, NO, N, O, N_2^+, O_2^+, NO^+, N^+, O^+, e^-). Among those species, 47 different chemical reactions are assumed to take place according to Park [2]. Thermal nonequilibrium is accounted for by means of different temperatures for heavy particle and electron translation (T, T_e) as well as by three temperatures for the vibration of each neutral molecular species (T_{vib,N_2}, T_{vib,O_2}, $T_{vib,NO}$). Accordingly, three conservation equations for vibrational energy and one for electron energy have to be solved. The result is the following system of 16 governing equations:

$$\frac{\partial}{\partial t}\vec{q} + \frac{\partial}{\partial x_i}\vec{f}_i(\vec{q}) = \vec{s}(\vec{q})$$

$$\frac{\partial}{\partial t}\left\{\begin{array}{c} \rho_s \\ \rho u_j \\ \rho e_t \\ \rho_{s'} e_{vib,s'} \\ \rho_e e_e \end{array}\right\} + \frac{\partial}{\partial x_i}\left\{\begin{array}{c} \rho_s u_i \\ \rho u_j u_i + \delta_{j,i}\, p \\ \rho h_t u_i \\ \rho_{s'} e_{vib,s'} u_i \\ (\rho_e e_e + p_e)\, u_i \end{array}\right\} = \left\{\begin{array}{c} M_s \sum_r \nu_{s,r}(R_{f,r} - R_{b,r}) \\ 0 \\ 0 \\ \{Q_{T-V} + Q_{V-V} + Q_{C-V}\}_{s'} + Q_{E-V,s'=1} \\ Q_{T-E} + Q_{C-E} - Q_{E-V,N_2} \end{array}\right\}$$

$$\begin{array}{llll} s = 2, ..., 10 & \text{species index} & r = 1, ..., 47 & \text{reaction index} \\ s' = 1, 2, 3 & N_2, O_2, NO & i, j = 1, 2 & \text{spatial coordinates} \end{array} \qquad (1)$$

The continuity equations for N_2 and e^- are not included in (1). This is because the constant ratio of nitrogen and oxygen nuclei, which is 79/21 for standard air, allows the N_2 density to be expressed as a linear combination of all other heavy particle densities. Furthermore, the assumption of charge neutrality yields the electron density as

$$\rho_e = \sum_{s=ions} \frac{M_e}{M_s}\rho_s \qquad M_s = \text{molecular weight of species s} \qquad (2)$$

The rate $\dot{\omega}_r = (R_{f,r} - R_{b,r})$ of reaction r appearing in the species production term of eq. (1) has been modeled considering the effect of thermal nonequilibrium (coupling vibration-chemistry, [3]). The vibrational energy source term accounts for the exchange of energy with heavy particle translation (Q_{T-V}), with vibration of other molecules (Q_{V-V}) and with electron translation (Q_{E-V}), as well as for the gain and loss of vibrational energy due to chemical reactions (Q_{C-V}):

$$Q_{T-V,s'} = \rho_{s'} \frac{e^*_{vib,s'}(T) - e_{vib,s'}}{\tau_{s'}} \qquad s' = 1, 2, 3 \tag{3}$$

$$Q_{V-V,s'} = \sum_{t \neq s'}^{molecules} \left\{ 0.01\, N_{avo} \sigma_{s't} \sqrt{\frac{8\Re T}{\pi \mu_{s't}}} \rho_{s'} \rho_t \left(\frac{e_{vib,t}}{M_{s'}} \frac{e^*_{vib,s'}}{e^*_{vib,t}} - \frac{e_{vib,s'}}{M_t} \right) \right\} \tag{4}$$

$$Q_{C-V,s'} = M_{s'} \sum_{r = reactions} \begin{cases} - R_{f,r} e_{vib,s'}(T_{vib,s'}) + R_{b,r} e_{vib,s'}(\bar{T}_{vib}) & \text{if } \nu_{s',r} < 0 \\ R_{f,r} e_{vib,s'}(\bar{T}_{vib}) - R_{b,r} e_{vib,s'}(T_{vib,s'}) & \text{if } \nu_{s',r} > 0 \\ 0 & \text{if } \nu_{s',r} = 0 \end{cases} \tag{5}$$

where $e_{vib,s'}(X) = \frac{R_{s'} \Theta_{s'}}{exp(\Theta_{s'}/X) - 1}$ vibrational energy function (harmonic oscillator)

$\bar{T}_{vib} \equiv \frac{\sum_{s'=1,2,3} \rho_{s'}/M_{s'}\, T_{vib,s'}}{\sum_{s'=1,2,3} \rho_{s'}/M_{s'}}$ mean vibrational temperature

The ionized molecules are assigned the mean vibrational temperature defined above. A comprehensive derivation of the electron energy source term can be found in [4].

3 Implicit Solution Method

3.1 Finite Volume Godunov-Type Scheme

The integration of the governing equations (1) over a two-dimensional control volume V_{ij} (Fig. 1) gives

$$\frac{\partial \vec{q}_{ij}}{\partial t} V_{ij} + \sum_{\partial V_{ij}} \vec{\varphi} \Delta s - \vec{s}(\vec{q}_{ij})\, V_{ij} = 0 \tag{6}$$

with $\vec{\varphi} \equiv n_x \vec{f}_x + n_y \vec{f}_y$.

The flux $\vec{\varphi}$ points in the outward normal direction of the respective cell face and is taken to be the solution of a one-dimensional Riemann problem with initial left and right states $\vec{q}^L$ and $\vec{q}^R$. The Riemann solver of Roe [5] is utilized in the present scheme because of its accurate shock resolution which is indispensable if post-shock relaxation phenomena shall be investigated thorougly. In addition, this solver can be adapted to a multiple-temperature model in a straightforward manner:

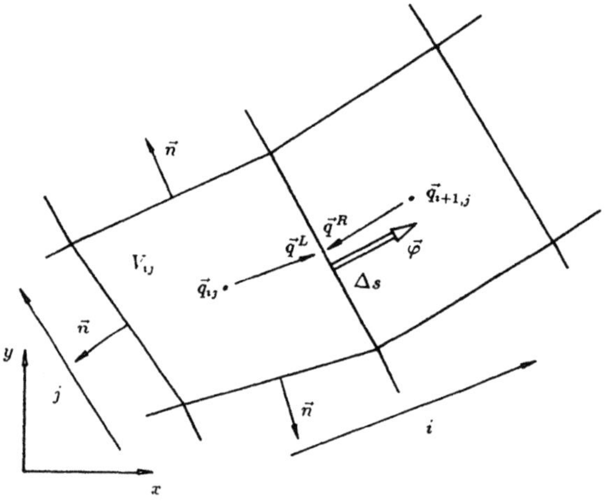

Figure 1: Control volume in a 2D structured mesh

$$\vec{\varphi}_{Roe}(\vec{q}^L, \vec{q}^R) = \frac{1}{2} \left\{ \vec{\varphi}(\vec{q}^R) + \vec{\varphi}(\vec{q}^L) - |\Phi| (\vec{q}^R - \vec{q}^L) \right\} ; \quad |\Phi| = R\, |\Lambda|\, R^{-1}; \tag{7}$$

$$\Lambda = \lceil u_\perp - a,\, u_\perp,\, ...,\, u_\perp,\, u_\perp + a \rfloor; \quad u_\perp = \text{normal velocity} = n_x u + n_y v$$

The normal flux Jacobian $\Phi = \frac{\partial \vec{\varphi}}{\partial \vec{q}}$ is

$$\Phi_{16\times16} = \begin{bmatrix} u_\perp (\delta_{s,t} - \xi_s \mu_t) & \xi_s n_x & \xi_s n_y & 0 & 0 & 0 \\ \beta_2^t n_x - u u_\perp \mu_t & u_\perp + (1-\beta_1)\, u n_x & u n_y - \beta_1 v n_x & \beta_1 \cos\Theta & -\beta_1 n_x & (2/3-\beta_1)\, n_x \\ \beta_2^t n_y - v u_\perp \mu_t & v n_x - \beta_1 u n_y & u_\perp + (1-\beta_1)\, v n_y & \beta_1 n_y & -\beta_1 n_y & (2/3-\beta_1)\, n_y \\ u_\perp (\beta_2^t - h_t \mu_t) & h_t n_x - \beta_1 u_\perp u & h_t n_y - \beta_1 u_\perp v & u_\perp (\beta_1 + 1) & -\beta_1 u_\perp & (2/3-\beta_1)\, u_\perp \\ -\xi_{s'} e_{vib,s'} u_\perp \mu_t & \xi_{s'} e_{vib,s'} n_x & \xi_{s'} e_{vib,s'} n_y & 0 & u_\perp \delta_{s',t'} & 0 \\ -5/3\, \xi_e e_e u_\perp \mu_t & 5/3\, \xi_e e_e n_x & 5/3\, \xi_e e_e n_y & 0 & 0 & 5/3\, u_\perp \end{bmatrix}$$

$s, t = 2, ..., 10 \qquad \beta_2^t = \frac{\partial p(\vec{q})}{\partial \rho_t} \qquad \xi_s = \frac{\rho_s}{\rho} \qquad \delta_{s,t}$ = Kronecker delta

$s', t' = 1, ..., 3 \qquad \beta_1 = \frac{\partial p(\vec{q})}{\partial (\rho e_t)} \qquad \mu_t = \frac{\partial \rho(\vec{q})}{\partial \rho_t}$

This Jacobian is calculated at a Roe-averaged state which is not uniquely defined in the non-perfect gas case. In this work, the formulation of Abgrall [6] is used.
In order to achieve second order accuracy, the left and right states $\vec{q}^L$ and $\vec{q}^R$ are extrapolated linearly from the cell center. The corresponding slopes are evaluated from locally defined characteristic variables. In these variables, slope limiters are applied. Dependent on the kind of limiter used, the result is a TVD or ENO scheme [7].
The steady state solution of (6) is computed with the following implicit integration scheme:

$$\left[I\frac{V_{ij}}{\Delta t} + \frac{\partial}{\partial \vec{q}_{mn}}\left(\sum_{\partial V_{ij}} \vec{\varphi}\,\Delta s\right) - \frac{\partial \vec{s}_{ij}}{\partial \vec{q}_{mn}} V_{ij}\right] \Delta \vec{q}^{\,it}_{mn} = -\left(\sum_{\partial V_{ij}} \vec{\varphi}\,\Delta s - \vec{s}_{ij}\,V_{ij}\right) \equiv -R_{ij}$$
$$\vec{q}^{\,it+1} = \vec{q}^{\,it} + \omega\,\Delta\vec{q}^{\,it} \qquad (8)$$

The parameter ω controlling the update of $\Delta\vec{q}^{\,it}$ is a relaxation parameter ($\omega \leq 1$). The time step Δt shows up on the main diagonal of the implicit operator. In the limit of an infinite time step, solving the unsteady equation with this method corresponds to solving the steady state equation by Newton's method. It has been found that for flows in strong nonequilibrium with source terms of the same magnitude as the inviscid fluxes, an exact evaluation of the source term Jacobian $\partial\vec{s}/\partial\vec{q}$ is crucial for a good performance of the implicit scheme.
Since the Jacobians on the left-hand side are being derived from first order accurate fluxes, a block-pentadiagonal linear system of equations has to be solved in each iteration:

3.2 Implicit Line Relaxation
Consider the pentadiagonal system (8) in a rather formal notation:

$$A^{0,0}_{i,j}\,\Delta\vec{q}^{\,it}_{i,j} + A^{1,0}_{i,j}\,\Delta\vec{q}^{\,it}_{i+1,j} + A^{-1,0}_{i,j}\,\Delta\vec{q}^{\,it}_{i-1,j} + A^{0,1}_{i,j}\,\Delta\vec{q}^{\,it}_{i,j+1} + A^{0,-1}_{i,j}\,\Delta\vec{q}^{\,it}_{i,j-1} = -R_{i,j} \qquad (9)$$

$$\begin{aligned} A^{0,0}_{i,j} &= 16\times 16 \text{ matrix on main diagonal} &&= \text{derivative of } R_{i,j} \text{ with respect to } \vec{q}_{i,j} \\ A^{1,0}_{i,j} &= 16\times 16 \text{ off-diagonal matrix} &&= \text{derivative of } R_{i,j} \text{ with respect to } \vec{q}_{i+1,j} \text{ etc.} \end{aligned}$$

System (9) will be solved for $\Delta\vec{q}^{\,it}$ in an iterative manner. A convenient choice for an initial guess is $\Delta\vec{q}^{\,it,\alpha=0} := \Delta\vec{q}^{\,it-1}$, where α represents an index for the following subiterations:

1. Increment α; for each j, solve the block-tridiagonal system below for $\vec{q}^{\,it,\alpha}$:
$$A^{0,0}_{i,j}\Delta\vec{q}^{\,it,\alpha}_{i,j} + A^{1,0}_{i,j}\Delta\vec{q}^{\,it,\alpha}_{i+1,j} + A^{-1,0}_{i,j}\Delta\vec{q}^{\,it,\alpha}_{i-1,j} = -R_{i,j} - A^{0,1}_{i,j}\Delta\vec{q}^{\,it,\alpha-1}_{i,j+1} - A^{0,-1}_{i,j}\Delta\vec{q}^{\,it,\alpha-1}_{i,j-1} \qquad (10)$$

2. Increment α; for each i, solve the block-tridiagonal system below for $\vec{q}^{\,it,\alpha}$:
$$A^{0,0}_{i,j}\Delta\vec{q}^{\,it,\alpha}_{i,j} + A^{0,1}_{i,j}\Delta\vec{q}^{\,it,\alpha}_{i,j+1} + A^{0,-1}_{i,j}\Delta\vec{q}^{\,it,\alpha}_{i,j-1} = -R_{i,j} - A^{1,0}_{i,j}\Delta\vec{q}^{\,it,\alpha-1}_{i+1,j} - A^{-1,0}_{i,j}\Delta\vec{q}^{\,it,\alpha-1}_{i-1,j} \qquad (11)$$

3. This process is repeated until $\vec{q}^{\,it,\alpha} \approx \vec{q}^{\,it,\alpha-1}$, i.e. until convergence is reached. Thus, system (9) has been solved exactly.

With this solution procedure carried out in each global iteration "*it*", it is possible to reach CFL numbers of 200 for nonequilibrium hypersonic flows containing strong shocks, with about 8 subiterations for the solver described above. Whereas for flows without shocks, there is no upper CFL limit. The line relaxation scheme can be implemented in an extremely efficient way on a vector computer: If the LU decompositions of 16×16 matrices, which must be performed in the block-tridiagonal solver, are unrolled, it is possible to vectorize the solver for system (10) over j and the solver for system (11) over i. Moreover, since the block-tridiagonal operators remain constant in all subiterations, their factorization has to be carried out only once in each global iteration.

4 Validation Results

4.1 Inviscid Flow over a Double Ellipse

The first case for a validation of the presented method is problem 6.2-3 of the Antibes-Workshop [8] which is a $M_\infty = 25$ flow over a double ellipse. Freestream conditions correspond to an altitude of $75\,km$. A rather coarse grid of 74×26 points has been used in order to demonstrate that accuracy is not only a matter of grid refinement.

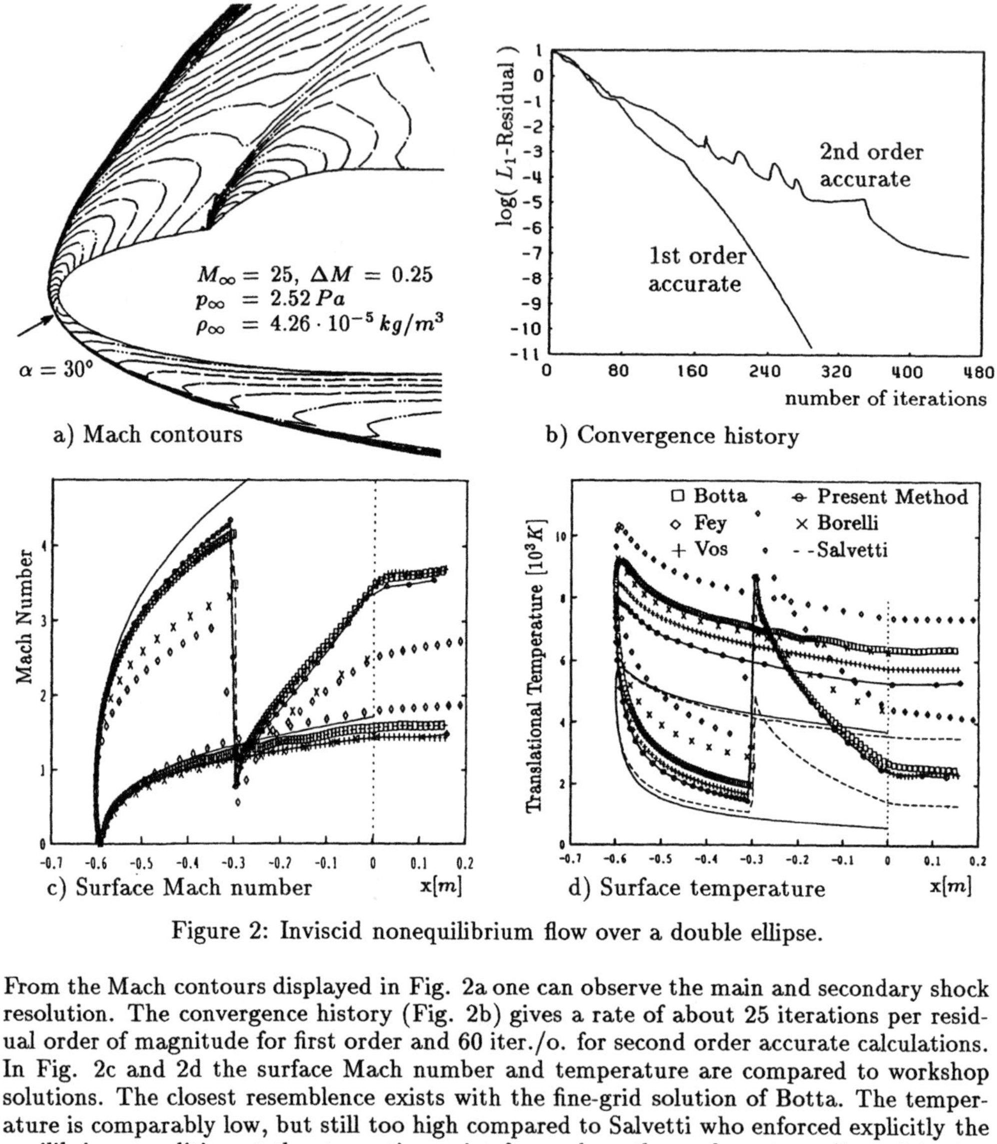

a) Mach contours b) Convergence history

c) Surface Mach number d) Surface temperature

Figure 2: Inviscid nonequilibrium flow over a double ellipse.

From the Mach contours displayed in Fig. 2a one can observe the main and secondary shock resolution. The convergence history (Fig. 2b) gives a rate of about 25 iterations per residual order of magnitude for first order and 60 iter./o. for second order accurate calculations. In Fig. 2c and 2d the surface Mach number and temperature are compared to workshop solutions. The closest resemblence exists with the fine-grid solution of Botta. The temperature is comparably low, but still too high compared to Salvetti who enforced explicitly the equilibrium condition at the stagnation point from where the surface streamline originates.

4.2 High Enthalpy Wind Tunnel Göttingen (HEG)

For the recently constructed HEG shock tunnel, a design calculation of the axisymmetric expansion nozzle has been carried out by Hannemann [9] considering chemical nonequilibrium in 8 different species and a one-temperature formulation (i.e. thermal equilibrium assumption). The total enthalpy at design condition is $44.52\,MJ/kg$, stagnation pressure and temperature are 2023 *bar* and 13780 *K*, respectively. This nozzle flow has been recalculated with the present 11-species, 5-temperature model. Fig. 3a compares the center line

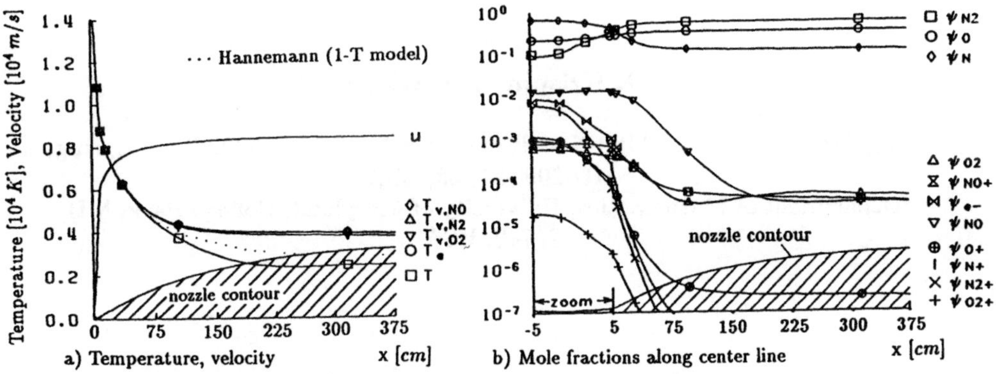

Figure 3: HEG expanding nozzle flow.

results with those obtained by Hannemann. The velocity profiles of both calculations are equal within plotting accuracy. The dotted line indicates the temperature that Hannemann calculated using a one-temperature model. Because of vibrational freezing at $x \approx 50\,cm$, the vibrational temperatures of the presented model do not fall below 3700 *K*, whereas the heavy-particle translational temperature reaches a much lower value. From these results it is obvious that a two-temperature model is the least one needs in order to calculate this flow correctly. The corresponding mole fractions are shown in Fig. 3b where the throat region has been zoomed by a factor of 10 in order to display the recombination processes in that area clearly. The convergence rate for this shock-free flow is 15 iter./o. for the second order scheme.

Acknowledgements
This research was supported by the Konrad Adenauer Foundation and by the Deutsche Forschungsgemeinschaft.

References

[1] S. Jonas, H.-H. Frühauf, O. Knab, Fully Coupled Approach to the Calculation of Nonequilibrium Hypersonic Flows Using a Godunov-Type Method. Proceedings of the First European CFD Conference, Brussels, September 1992.
[2] C. Park, Convergence of Computation of Chemical Reacting Flows, Thermophysical Aspects of Reentry Flows, Progress in Astronautics and Aeronautics, 103, edited by J.N. Moss and C.D. Scott, 1986, AIAA, New York, pp. 478-513.
[3] O. Knab, H.-H. Frühauf, S. Jonas, Multiple Temperature Descriptions of Reaction Rate Constants with Regard to Consistent Chemical-Vibrational Coupling, AIAA 92-2947, AIAA 27th Thermophysics Conference, July 1992, Nashville.
[4] J. H. Lee, Basic Governing Equations for the Flight Regimes of Aeroassisted Orbital Transfer Vehicles, AIAA-84-1729, 19th Thermophysics Conference, 1984.
[5] P. L. Roe, Approximate Riemann Solvers, Parameter Vectors, and Difference Schemes., J. Comp. Phys., 43, 1981, pp. 357-372.
[6] R. Abgrall, Hypersonic Calculations by Riemann Solver Techniques, Computer Physics Communications 65, 1991, pp. 1-7.
[7] A. Harten, Uniformly High Accurate Essentially Non-Oscillatory Schemes III, J. Comp. Phys., 71, 1987, pp. 231-303.
[8] J.-A. Désidéri, R. Glowinski, J. Périaux (Eds.), Hypersonic Flows for Reentry Problems, Vol. II, Springer, 1991.
[9] K. Hannemann, Design of an Axisymmetric, Contoured Nozzle for the HEG, Research Report DLR-FB 90-04, Institut für Theoretische Strömungsmechanik, Göttingen.

Nonequilibrium, Chemically Reacting Flowfields and the Implicit–Explicit Godunov Scheme[1]

J. Krispin [1], H. Glaz [2]

[1]Enig Associates, Inc., 11120 New Hampshire Ave., Suite 500, Silver Spring, MD 20904-2633, USA
[2]Department of Mathematics, University of Maryland, College Park, MD 20742, USA

1 Introduction

A new hybrid implicit–explicit (IE) scheme has been proposed (Collins et al 1990) for systems of hyperbolic conservation laws, as well as their viscous extensions, which has far–reaching potential in many fields. Here, the difference approximation in time is either implicit or explicit, separately for each family of characteristics and for each cell in the finite difference grid, depending on whether the local CFL number for that family is greater than or less than one. This IE strategy is intended for problems with spatially and/or temporally localized stiffness in wave speeds. The method is based on the operator–split second order Godunov scheme (Colella and Glaz 1985). Recently, this methodology has been extended for unsplit differencing (Collins 1992, Collins et al 1992) by basing the differencing strategy on an unsplit second order Godunov scheme (Colella 1990).

The objective of this research effort is the efficient and accurate computation of reactive multiphase boundary layers at both high and low Mach numbers. We believe that the IE strategy, when coupled with mesh refinement approaches and methods for the stable computation of stiff source terms, can be a valuable tool in this field. The purpose of the current study is to further advance and validate this IE approach by investigating the effects of including source terms representing multispecies reacting flow. For the case of a 1D variable area duct converged solutions for both smooth and shocked flows are demonstrated using an extension of the solution methodology of Collins et al 1990. An appropriate extension of the unsplit IE scheme has been developed and implemented for 2D reactive flow with periodic boundary

[1] This work has been supported by the Small Business Innovative Research program of the National Science Foundation, Division of Mathematical Sciences under Grant ISI–9022351; we are also grateful for the support of the San Diego Supercomputer Center in providing computer resources for this project.

conditions; the new scheme is used here for problems analogous to the nonreactive results presented previously (Collins 1992, Collins et al 1992).

2 Numerical Method

The 1D equations of motion for a system of conservation laws with a viscous extension (e.g., compressible flow) and nongeometric source terms (e.g., chemical reaction) in a variable area duct may be written in the form

$$\frac{\partial U}{\partial t} + \frac{\partial}{\partial \Lambda}(A(F^{ad}(U) - b(U, U_x)) + \frac{\partial H}{\partial x} = S(U) \tag{1}$$

where U is the vector of conserved quantities, $F^{ad}(U), H(U)$ are the associated advective and pressure fluxes, respectively, $A(x), \Lambda(x)$ are area and volume coordinates, $b(U, U_x)$ is the viscous flux and $S(U)$ is the vector of source terms. We refer to the references (Glaz et al 1988 and Krispin 1992) for a detailed description of these terms including the reactive air chemistry and the formulation of the viscous flux. The 2D inviscid equations with Cartesian symmetry are

$$\frac{\partial U}{\partial t} + \frac{\partial F}{\partial x} + \frac{\partial G}{\partial y} = S(U)\,. \tag{2}$$

The numerical method for the case $S(U) = 0$, the frozen flow case, is described (for a single atomic specie) in the references (Collins et al 1990 for (1), and Collins 1992, Collins et al 1992 for (2)). For both (1) and (2), the nonlinear system arising from the implicit characteristics is solved without operator splitting of the source terms (which we found to be inefficient as well as inaccurate – incorrect steady states at large CFL – for (1)); for (1), the source terms are discretized by $S(U^n)$, and for (2), by $\frac{1}{2}(S(U^n) + S(U^{n+1}))$. The solution strategy chosen for the discrete nonlinear system arising from (1) is standard linearization for the fluxes. For (2), we use instead a weighted Jacobi iteration procedure (Collins 1992); fifteen iterations per time step are used for the current results.

3 Results and Discussion

Our 1D simulations were made using a standard expanding duct, supersonic inflow boundary conditions and 200 mesh points. The exit pressure was varied for the shocked flows and several cases were studied; by changing various parameters, it is easy to adjust the stiffness of the reaction, the strength of the shock and the degree of smoothing of the viscous terms if present. The initial conditions were set to be linear profiles – which is very far from the steady solution. Some of the results are shown in Fig. 1. The temperature increases to about 3000°K across the standing shock and the different relaxation scales for the three diatomic species can be noted in the figure, and the species concentrations exhibit similar behavior (not shown); the inflow density used in these calculations was $\rho = 0.48582 \times 10^{-5}\mathrm{gm/cm^3}$ and T was near room temperature which leads to a relatively strong shock jump. This parameter combination sets the widely differing relaxation scales. It can be seen that the frozen and nonequilibrium calculations result in slightly different shock locations and postshock profiles – the same exit pressure was used in all three calculations. We have noticed significant variation in the viscous profile with respect to changes in the two parameters viscosity and heat conductivity for this problem. The residual profile (not norm dependent) is typical for problems of this type – a long initial transient as the shock sets up and strong waves interact between the shock and the subsonic outflow followed by a steady decline to machine accuracy; the 'periodic' appearance at this later stage is also expected since the shock profile is changing slightly each time step and there is a characteristic time for acoustic signals to traverse the mesh. The CFL maximum was four (resp., three) for the inviscid (resp., viscous) nonequilibrium calculation presented; there is no limit in practice for the frozen flow case.

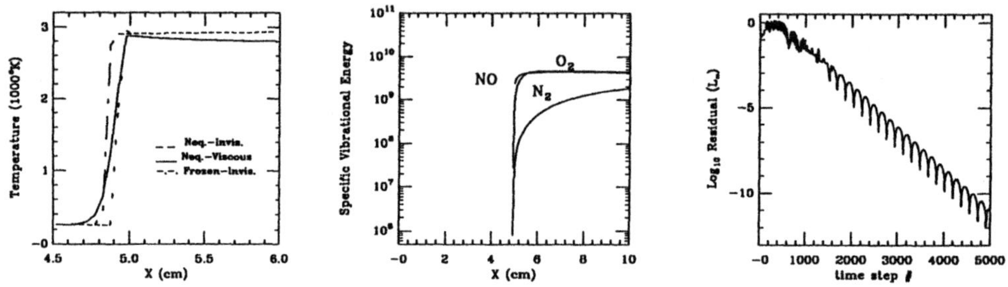

Fig. 1. Chemically reacting shocked flow in an expanding variable area duct. From left to right: temperature vs. distance at steady state compared with the frozen flow solution as well as the viscous flow solution in a small region around the shock location; specific vibrational energies vs. distance; l^∞ residual vs. time step for the total mass equation.

The 2D simulation is inviscid, uses a 64×64 mesh and doubly periodic boundary conditions. The initial velocity field $\mathbf{u} = (u, v)$ is scaled to match the previously studied nondimensional single specie nonreactive calculations (Collins 1992). The initial maximum flow Mach number is about 0.1 in the x–direction and the initial shear jump is set up using a hyperbolic tangent profile for u with just a few interior points. The y–velocity component is given by $v = 0.05 * |u_{\text{max}}| * \sin 2\pi x$ and it is this inviscid perturbation which makes the flow nontrivial and leads to the eventual characteristic rollup. The CFL maximum is determined by insisting that the explicit particle characteristic CFL number be set to 0.9. We note that the vibrational energies are very stiff for this problem at initialization and there is a steep dropoff in maximum temperature which then holds steady. The vorticity field is shown at three times in Fig. 2 where the evolution of the rollup is clear. The mass fraction field evolution shown in Fig. 3 is more intersting because it illustrates the capability of the current more complex scheme to handle intricate flow mixing problems with high resolution detail in the nonequilibrium, reactive case as well as the frozen case (not shown, but the results are similar), even on the coarse mesh used here. For applications, it is perhaps the case that the mixing phenomenology is at least as important as reaction rates, etc. in determining a successful design. The other important point to note concerning these results is the near total lack of oscillatory profiles, strong pressure waves, etc.

which would be inappropriate for the reactive air mechanisms at the chosen flow conditions; thus, the reactive, nonequilibrium scheme is seen to be as stable as the gas dynamics scheme (Collins 1992) in this flow regime. An illustration of this is provided in the plot of div**u** vs iteration count (fifteen iterations per time step) presented in Fig. 4. Although this is not clear from the figure, the divergence always decreases during a time step as the iteration count increases. The overall variation is consistent with an initial transient during which the rollup forms followed by nearly equilibrium mixing at smaller and smaller scales; the magnitude is appropriate for $M = 0.1$ and the expected results of an explicit calculation.

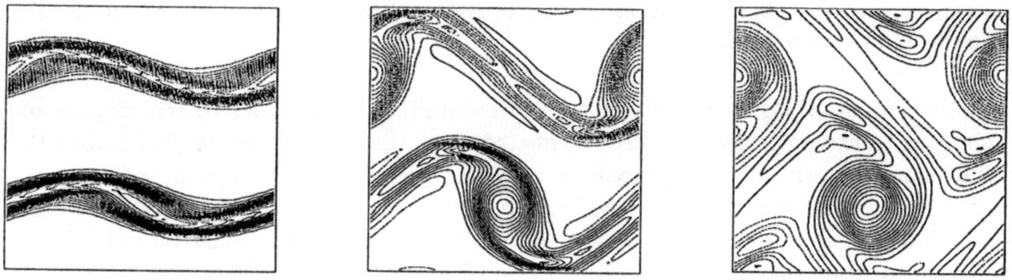

Fig. 2. Vorticity contour plots at $T = 0.8, 1.2, 1.6$ seconds using 30 equally spaced contours; dashed contours represent negative vorticity.

Fig. 3. Contour plots of mass fraction of atomic oxygen at the same times as in Fig. 2; 30 equally spaced contours. The maximum amplitude increases from 5.0×10^{-9} in the left figure to 8.41×10^{-9} in the right figure.

For the 1D calculations we find that the performance criteria depend on the degree of source term stiffness. In particular, a fairly low CFL maximum is required (that is, to obtain converged solutions) during parts of the calculations. Unlike the 1D calculations, it is found that trapezoidal rule differencing for the source terms is

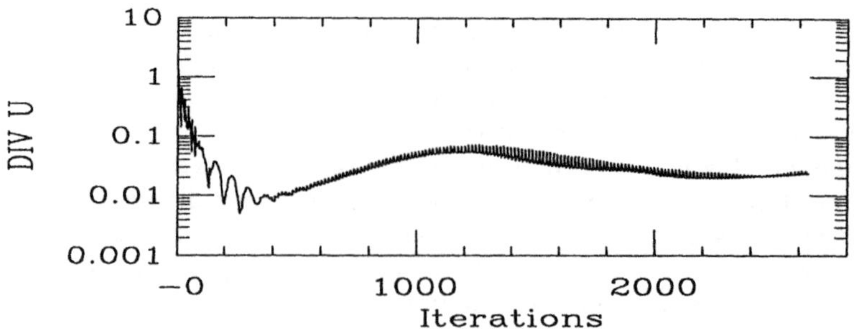

Fig. 4. Plot of max divu over the mesh; divu is computed by differencing the edge values obtained from the charateristic projection operators and Riemann problem solutions (see Colella 1990, Collins 1992).

required for stability for the 2D problem presented. In future work we will explore more complex treatments of stiff source terms (see, e.g., Lafon and Yee 1992, Pember 1992) and, especially, the coupling (analytic and numerical) between wave speed and source term stiffness. Additionally, we will treat two–phase flows and their source terms so that complex shock – boundary layer interactions with ablation can be modelled. An important numerical issue is developing and implementing efficient techniques such as multigrid acceleration for the solution of the large nonlinear systems arising from this method; advances in this area alone may well be sufficient, via obtaining the capability of converging or nearly converging the nonlinear system, for a dramatic improvement in maximum CFL number thereby making the method viable for these more difficult applications.

References

Colella, P. (1990): 'Multidimensional Upwind Methods for Hyperbolic Conservation Laws', J. Comp. Phys. **87**, pp. 171–200

Colella, P., Glaz, H. M. (1985): 'Efficient solution algorithms for the Riemann problem for real gases', J. Comp. Phys. **59**, pp. 264–289

Collins, J. P. (1992): *Implicit–Explicit Godunov Schemes for Unsteady Gas Dynamics*, Ph. D. Dissertation, University of Maryland at College Park, Department of Mathematics

Collins, J. P., Colella, P., Glaz, H. M. (1990): 'An Implicit-Explicit Eulerian Godunov Scheme for Compressible Flow', preprint

Collins, J. P., Colella, P., Glaz, H. M. (1992): 'An Unsplit Implicit–Explicit Godunov Method for Compressible Gas Dynamics', to appear in Proc. First European CFD Conf., ed. Ch. Hirsch (Elsevier)

Glaz, H. M., Colella, P., Collins, J. P., Ferguson, R. E. (1988): 'Nonequilibrium effects in oblique shock-wave reflection', AIAA J. **26**, pp. 698–705

Krispin, J. (1992): 'High–Resolution Calculations in Gas Dynamics', Enig TR 92–2

Lafon, A., Yee, H. C. (1992) 'On the Numerical Treatment of Nonlinear Source Terms in Reaction–Convection Equations', AIAA 92–0419

Pember, R. B. (1992): 'Numerical Methods for Hyperbolic Conservation Laws with Stiff Relaxation II. Higher Order Godunov Methods', UCRL–JC–109097 Pt. 2

AN UPWIND SCHEME FOR OBLIQUE SHOCK/BOUNDARY LAYER INTERACTION IN A SUPERSONIC INLET

S. M. LIANG, R. N. WU and J. J. CHAN

Institute of Aeronautics and Astronautics, National Cheng Kung University, Tainan, Taiwan, R.O.C.

INTRODUCTION

Recent papers[1,2] have indicated that there is a substantial market in development of High-Speed Civil Transport (HSCT), since the HSCT would have greater economic benefit if it would fly supersonically overland or oversea. The design of supersonic inlets for HSCT is one of important research areas. Shock waves in a supersonic inlet can interact with boundary layer either weakly or strongly, depending on the shock strength. A strong interaction can induce a separated flow inside the boundary layer, and has local and global effects on the inflow. The effects can significantly reduce an inlet performance, if the boundary layer is not effectively controlled. Before the use of boundary layer control, it is better to understand the oblique shock/boundary layer interaction.

In the present study, numerical approach is adopted for studying the shock interaction problem. A second-order upwind scheme, originally developed by Coakley[3] and improved by Liang and Chan,[4] is employed, although Total Variation Diminishing (TVD) schemes[5] have been widely used for calculation of the Euler/Navier-Stokes equations. The disadvantage of TVD schemes is that TVD schemes are reduced to first-order accuracy at extreme points so that the solution of skin friction coefficient may not be correctly predicted.

A generic two-dimensional supersonic inlet, as shown in Fig. 1, is constructed for the study of shock/boundary layer interaction. The inlet cowl and centerbody are assumed to be adiabatic. The inlet length is chosen as a characteristic length and is normalized to be unity. Based on the inlet length, the inlet height is set to be 0.2 and the throat height is 0.13. The cowl is a flat plate, beginning at $x = 0.354$. The lower wall (centerbody) starts with a flat plate at the point (0.04, 0) with an inclination angle of 9°, and is followed by a convex wall with radius of curvature $R = 1.137$. The curved surface begins at the point (0.394, 0.056) and ends at the point (0.57, 0.07). Further downstream, a flat plate with an inclination angle of −0.88° is followed and terminates at $x = 0.7$ for saving computation time.

MATHEMATICAL FORMULATION

The equations governing the two-dimensional inlet flow are the compressible Navier-Stokes equations, which can be written in a body-fitted coordinate system (ξ, η) as

$$v\partial_t U + \partial_\xi(F'_\xi + F''_\xi) + \partial_\eta(F'_\eta + F''_\eta) = 0 \qquad (1)$$

where U is the conservative variable, F' and F'' the inviscid and viscous flux vectors, respectively, and v the volume of a control element. More information about Eq. (1) can be found in Ref. 3. Equation (1) associated with the Baldwin-Lomax turbulence model[6] is employed to model the inlet flow and solved numerically.

NUMERICAL METHOD

The implicit method used in the study is based on the Coakley scheme[3] with the following modification in the numerical dissipation:

$$D_{i+1/2} = \overline{S}_{i+1/2}\Delta F'_{i+1/2} + L^{-1}_{i+1/2}(\Lambda^-_{i+1/2}\Delta W_{i+3/2} - \Lambda^+_{i+1/2}\Delta W_{i-1/2})$$

where $\Delta W = L\Delta U$, $\overline{S} = L^{-1}sgn(\Lambda)L$, $\Delta(\ )_{i+1/2} = (\ )_{i+1} - (\ )_i$, and L and Λ consist of the left eigenvectors and eigenvalues of flux Jacobian matrix, respectively. Note that the variable W denotes the characteristic variable and its value is evaluated at the cell interface by Roe's averaging.[7] Although $\overline{S}\Delta F = L^{-1}|\Lambda|\Delta W$, it was found the replacement of $L^{-1}|\Lambda|\Delta W$ in the Coakley scheme by $\overline{S}\Delta F$ was able to resolve a sharp shock wave definition. One- and two-dimensional inviscid flow solutions have verified the merits of the modified numerical dissipation.[4]

RESULTS AND DISCUSSION

All numerical calculations were performed on a Vax-8600 computer. The convergence criterion is set by measuring the l_2 norm of the density errors less than 10^{-5}.

In order to verify the scheme accuracy, the test problem of an incident shock impinging on a laminar boundary layer over a flat plate at freestream Mach number 2, Reynolds number 2.96×10^5, and various incident shock angles, $\theta_s = 31°, 31.5°, 32°$ and 32.6° was considered. The computed wall pressure rise caused by the incident shock is about 1.15, 1.22, 1.31, and 1.41 for $\theta_s = 31°$, 31.5°, 32°, and 32.6°, respectively. For $\theta_s = 31°$ and 31.5°, no flow separation was observed. For $\theta_s = 32.6°$, a pressure plateau was developed, which corresponded to the formation of a reversed flow. In contrast to the experimental results of Hakkinen et al.,[8] the flow started to separate when the pressure ratio of p_2 to p_1 is 1.25, where p_2 and p_1 denote the pressures ahead of and behind the shock reflection, respectively. The result of computed skin friction distribution indicated that the critical value of the pressure rise for separation, p_2/p_1, was about 1.25, which coincided with the experimental value.

Next, consider an inflow with freestream Mach number 2.4 and zero degree angle of attack. The Reynolds number is chosen to be 1×10^6.

Case 1: laminar flows:

Solutions for laminar flows were obtained on five different grids. Basically, the grid spacing in the x direction, Δx, is uniform. For $x < 0.354$, Δx is chosen to be 0.01, and $\Delta x = 0.0035$ for $0.394 < x < 0.7$. For $0.354 < x < 0.394$, the grid spacing Δx is stretched. This is a transition region that connects the coarser grid and the finer grid. The transition grid was found to be able to eliminate the numerical wiggle that occurred on the grid without the transition region. In the y direction, the grid distribution was stretched from the wall. The minimum spacing, Δy_{min}, was selected to be 3×10^{-5} for a 129x55 grid, which was found to be fine enough for resolving the local flow structure in the shock interaction region.

It was found that there are three regions with negative values of the skin friction coefficient. The particle traces in the shock interaction region are shown in Fig. 2a. It is shown that there are two clockwise and one counterclockwise rotating vortices inside the separated region. The largest clockwise vortex, called the primary reversed flow, induced a secondary flow with counterclockwise rotation direction. The secondary flow induced a tertiary flow. A subplot in Fig. 2a sketches the flow structure of the primary reversed flow. Based on the topological rule of skin-friction lines,[9,10] the rule of $(\Sigma_N + \frac{1}{2}\Sigma_{N'}) - (\Sigma_S + \frac{1}{2}\Sigma_{S'}) = 0$ must be satisfied, where Σ stands for the number of nodal or saddle points, the subscripts N, S for the nodal and saddle point, respectively, and the superscript $'$ for the nodal or saddle point on the body surface. In this case, $\Sigma_N = 3$, $\Sigma_{N'} = 0$, $\Sigma_S = 0$, $\Sigma_{S'} = 6$, so this rule is satisfied. Figure 2b shows the corresponding pressure distribution. The adverse pressure gradients are clearly shown, as indicated by an arrow, leading to flow separation. In addition, a saddle point in the pressure distribution, denoted by S, is found. Moreover, several local maxima in the pressure distribution are also shown.

The existence of the critical points in the pressure distribution may reduce the TVD scheme to be first order accuracy, causing a wrong prediction in the pressure and the skin friction. Figures 3a and 3b show the surface pressure and skin friction distributions by using the TVD scheme.[11] Figure 3a indicates that the TVD scheme does not produce a pressure plateau, i.e., the constant pressure region. Thus the

prediction in the structure of local flow separations seems to be not correct, as shown in Fig. 3b.

Case 2: turbulent flows

Figure 4 shows the comparison of local flow structures in the shock interaction region for laminar and turbulent flows. Generally turbulent flows can transport momentum and overcome the adverse pressure gradient better than laminar flows can under the same flow conditions. Thus the size of separation bubble for turbulent flows is smaller than that for laminar flows. Note that the tertiary flow in Fig. 4a is not visible because the velocity field was not plotted at every point and the size of the tertiary flow was relatively too small.

Conclusion

A Navier-Stokes solver using a second-order upwind scheme has been developed to investigate the oblique shock/boundary layer interaction in a supersonic inlet. The upwind scheme was verified to be reasonably accurate and robust. The local flow field in the shock interaction region was investigated in details. An interesting flow structure of multiple separation bubbles for laminar flows was found numerically, which satisfied the topological rule of skin-friction lines.

REFERENCES

1. C. S. Domack, "A Preliminary Investigation of Inlet Unstart Effects on a High-Speed Civil Transport Concept," AIAA Paper 91-3327, 1991.

2. H. L. Wesoky, J. R. Facey and K. P. Shepherd, "Technical Bases for High Speed Civil Transport Environmental Acceptability," AIAA Paper 91-3326, 1991.

3. T. J. Coakley, "Implicit Upwind Methods for the Compressible Navier-Stokes," NASA TM-84364, May 1983.

4. S. M. Liang and J. J. Chan, "An Improved Upwind Scheme for the Euler Equations," *J. Comput. Phys.*, **84**, 461-473, (1989).

5. H. C. Yee and A. Harten, "Implicit TVD Schemes for Hyperbolic Conservation Laws in Curvilinear Coordinates," *AIAA J.*, **25**, 266-274, (1987).

6. B. S. Baldwin and H. Lomax, "Thin Layer Approximation and Algebraic Model for Separated Turbulent Flows," AIAA Paper 78-257, January 1978.

7. P. L. Roe, "Approximate Riemann Solvers, Parameter Vectors, and Difference Schemes," *J. Comput. Phys.*, **43**, 357-372, (1981).

8. R. J. Hakkinen, I. Greber, L. Trilling and S. S. Abarbanel, "The Interaction of an Oblique Shock Wave with a Laminar Boundary Layer," NASA Memo 2-18-59W, 1957.

9. J. C. R. Hunt, C. J. Abell, J. A. Peterka and H. Woo, "Kinematical Studies of the flow around free or Surface-Mounted Obstacles; Applying Topology to Flow Visualization," *J. Fluid Mech.*, **86**, pt. 1, 179-200, (1978).

10. D. J. Peake and M. Tobak, "Three-Dimensional Separation and Reattachment," *AGARD LS-121: High Angle-of-Attack Aerodynamics,* 1-1–1-4, (1982).

11. C. L. Hu, "Numerical Investigation of Thrust-Reversing Nozzles Using an Implicit TVD Scheme," Ph. D. Thesis, Institute of Aeronautics and Astronautics, National Cheng Kung University, Tainan, Taiwan, 1989.

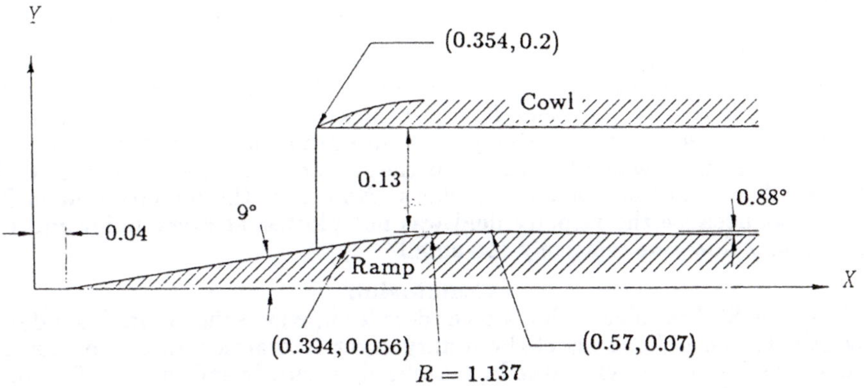

Fig. 1. A sketch of a supersonic inlet.

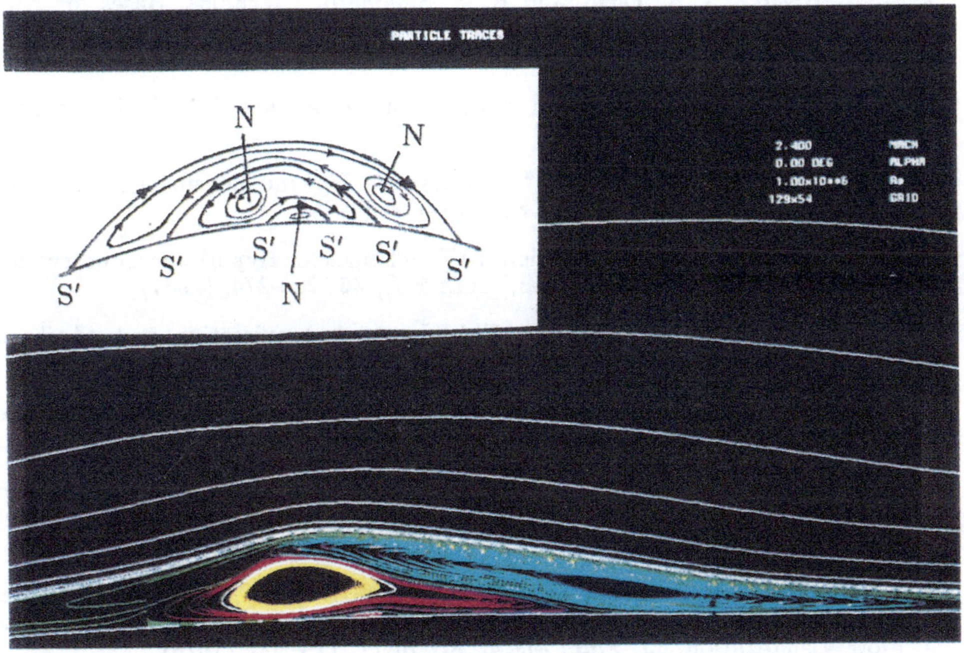

Fig. 2a.

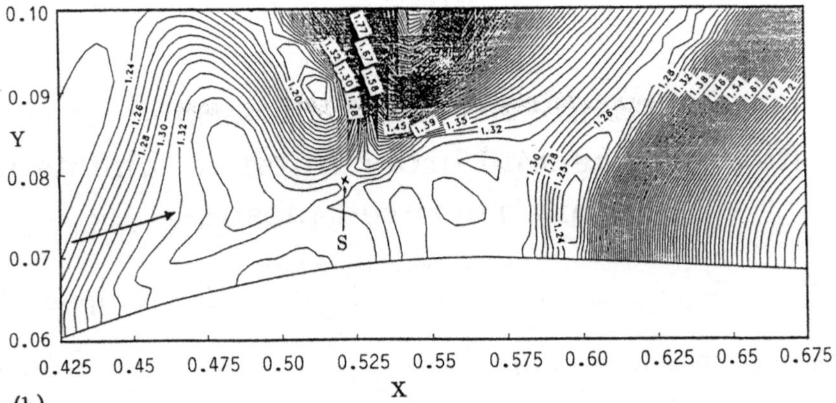

(b)

Fig. 2. (a) particle traces and (b) pressure distribution near the shock/boundary layer interaction region.

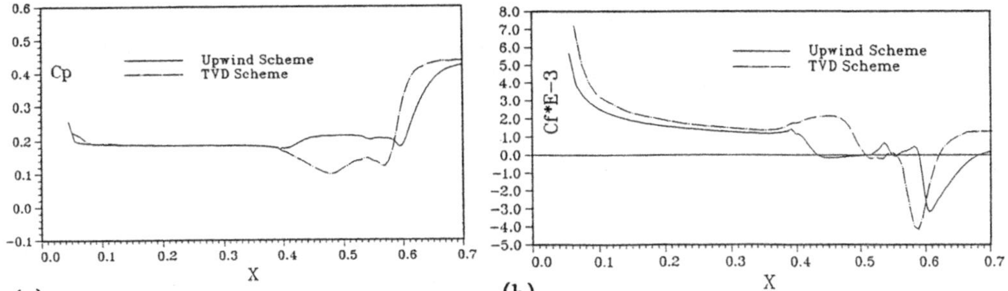

(a) (b)

Fig. 3. Comparison of (a) surface pressure and (b) skin friction coefficient on the centerbody using different schemes.

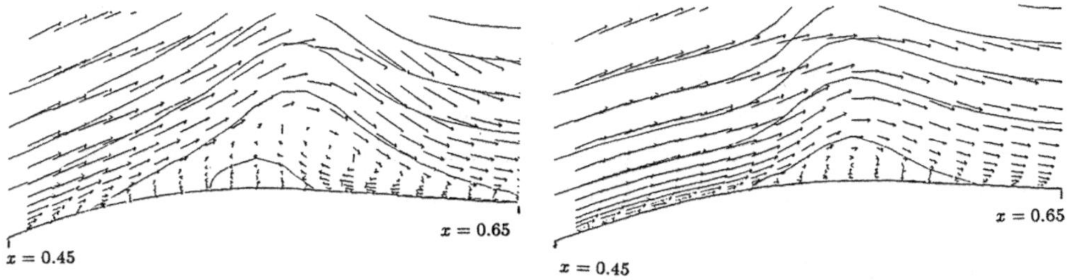

(a) (b)

Fig. 4. Comparison of local flow structures for (a) laminar and (b) turbulent flows.

Calculation of a viscous hypersonic nozzle flow by a second order implicit scheme on unstructured meshes

A. Merlo, R. Abgrall
INRIA Sophia Antipolis, BP 92. 06902 Sophia Antipolis, France

This paper follows a serie by the authors [5, 6] about the numerical simulation in hypersonc wind tunnels. It is divided into three parts : first, we describe briefly our physical model, second, we detail our numerical method then numerical experiments and results are presented.

1 Physical Model

In our model of air, we take into account five species (O, N, NO, O_2 and N_2) and they will correspond to indices 1,2,3.4 and 5. The mixture is made of 21% of Oxygen and 79% of Nitrogen. The diatomic species O_2 and N_2 may be in vibrational non equilibrium. Their vibrational temperature are denoted T_{O_2} and T_{N_2} respectively. The species NO is set to vibrational equilibrium. The pressure p is given by Dalton's law. The internal energy is defined ϵ as the sum of the internal energy of each of the five species; each one is the sum of a rotational-translational part characterized by the temperature T, a vibrational part characterized by a temperature T_{v_i} (for the diatomic species only) and an energy term ($h_i{}^0$) due to possible formation. The vibrational energies are related to T_{v_i} assuming the harmonic oscillator model [3]. The characteristic temperatures θ_i of that model and the enthalpy $h_i{}^0$ are those assumed for the Antibes workshops [5].

1.1 Chemical and Vibrational effects

The chemical model is Park's and the vibrational model is Landau-Teller's. The source terms involved by the chemistry are denoted by $\Omega_i{}^c$ and those by the vibrational effects by $\Omega_i{}^v$. We assume no coupling between dissociation and vibration. The relaxation times are given by [3, 2] : $\tau_\xi{}^{-1} = \sum_{species\ \xi'} X_{\xi'} \tau_{\xi,\xi'}{}^{-1}$ where $\tau_{\xi,\xi'}$ are relaxation times ans $X_{\xi'}$ is the molar concentration of species ξ'.

1.2 Viscous effect

The stress tenser is modelled assuming Stokes' relations. The cinematic viscosity is obtained by Wilke's and Blottner's law. Each species has a diffusion velocity, it is obtained by Fick's law assuming a constant Lewis Number (=1). The vibrational heat fluxes are calculated with Fourier's law which heat conductivity is obtained by Eucken's law. The heat flux is the sum of the vibrational heat fluxes, energy fluxes caused by diffusion of species and a Fourier term that is modelled by the same relationships. The coefficient we have used are those of Candler's thesis [2].

2 Numerical Method

The present numerical method is a direct extension of the implicit scheme developped by Fezoui et al [4] for perfect gases to chemically and vibrationally non equilibrium axisymmetric flows. It uses unstructured meshes and upwind solvers. In the two following paragraphs, we will only present an adaption of that scheme for reacting axisymmetric flows ; a description of the basic method can be found in [4].

2.1 Navier Stokes equation

We integrate the Euler equation adapted to the present case :

$$\frac{\partial W}{\partial t} + \frac{\partial}{\partial x}\left[F(W) + F_v(W)\right] + \frac{\partial}{\partial y}\left[G(W) + G_v(W)\right] = \Omega + H \tag{1}$$

where $W = ([\rho_i]_{i=1,5}, \rho u, \rho v, E, [E_{v_i}]_{i=4,5})^T$ (u and v are the x and y component of the velocity, E is the sum of internal energy and cinetic energy), F and G are the Euler fluxes, F_v and G_v are the viscous fluxes, H is a source term to take into account axisymmetry and $\Omega = ([\Omega_i{}^c]_{i=1,5}, 0, 0, 0, [\Omega_i{}^v]_{i=4,5})$.

2.2 First order scheme

Around each node, one constructs a control volume and one classicaly evaluates a numerical flux. Here, we have chosen an extension of Osher's fluxes detailed in [1]. Lots of test cases have shown that this extension has the same properties as the original Osher's Riemann solver and do not need any entropy fix, even for very strong shocks. It also respects the relative proportion of moles of Oxygen and Nitrogen. The discretisation of the diffusive termes is achieved assuming a P_1 reconstruction of the density, velocity, temperatures and a Galerkin type approximation. To take into account axisymmetry, we start from a conservative 3-D formulation of the scheme on an unstructured mesh obtained from a 2-D mesh of the upper part of the nozzle. If Ox is the axis of symmetry, the 3-D mesh is get from the 2-D one by applying a discrete set of rotation of angle $k \times 2\pi/n$, $0 \leq k < n$ around Ox. Then we particularize the 3-D scheme by taking into account the geometrical symmetries of the flow. This leads to a discretisation of the source term H that is *consistent* with the discretisation of the convectives one and depend on the Riemann solver. The global scheme is exactly conservative.

In order to improve the convergence to the stationary solution an implicit scheme is used following [4]. The Jacobian of the numerical flux is approximated by Steger-Warming's. The chemistry and vibrational effects are *strongly* coupled to the flow terms. At each iteration, the numerical system is solved by point-Gauss Seidel iteration. This yields to a very efficient numerical method, especially for external supersonic flows where the boundary conditions are very easy to take into account.

2.3 Second order scheme

We have used the MUSCL technique as in [4] with some adaptions. One can choose several sets of variables to reach quasi second order of accuracy. The "best" set seems to be the so-called physical variables (velocity, internal energy and total density), mass fractions and specific vibrational energies for O_2 and N_2. If one makes linear interpolation on each mass fraction, one will in some cases change the composition of the mixture since the limitation procedures are higly non linear and since the rule "21%-79%" is expressed through a linear relation between mass fractions The "best" choice here is tho interpolate on Y_2, Y_3 and Y_4 since in the convergent part one have very few diatomic oxygen and in divergent part very few monoatomic nitrogen. The other features of the scheme are classical.

3 Numerical experiment

3.1 Initial Conditions

The the reservoir condition for the nominal case studied here are : Pressure : 1530 Bars, Temperature : 6500 K, composition : obtained assuming equilibrium. The equilibrium model we have used is that obtained from Park's. Vibrational equilibrium is assumed. The wall temperature is 600 K.

We initialize the calculation by using the exact solution of the quasi-1D problem with the reservoir condition stated above and assuming the mixture remains perfect with a ratio of specific heats equal to the ratio of enthalpy and pressure in the reservoir. A first order calculation is done assuming an Eulerian flow. Then this solution is used as an initial guess for the Navier Stokes calculations.

The geometry of the nozzle is made by two cones (convergent 45°, divergent 10°) connected by a circle which radius is 1 cm. That of the throat is 3mm.

3.2 Solution method

The Navier Stokes simulations are done in two stages : we split the mesh into two parts. The first block contains the convergent, the throat and the divergent up to 3 cm after it. The second bock overlaps the first on 1 cm and contains the remaining of the nozzle From the Eulerian intial guess. a first then a second order calculation is made on the first block. We then use the solution on the overlapping region to initialize the second block. The runs are stopped when the L^2 norm of the normalized density residual is less than 10^{-4}. This may seem very large but one should be aware of the difficulty of the problem as it can be seen from the results.

3.3 Mesh

Several meshes have been used. From our experiments, it turns out that the best mesh should have enough points around the throat and that their y-distribution should be as regular as possible. In order to capture reasonably well the boundary layer and because of the very large temperature gradients near the wall, one must have at least 20 points *every where* in the boundary layer which is the case for our mesh. An additional difficulty is that size of the boundary layer varies very much from the convergent to the exit. This mesh is constructed from a 200×60 structured mes (first block : 6000 points, second block 6600 points, overlapping : 600 points).

3.4 Results

Because of lack of space, we present a very limited set of results. A Zoom near the throat of the translational temperature is shown in Figure 1. One clearly see the change of inclination of the temperature lines near the throat. This phenomena is due to the rapid expansion and to the shape of the nozzle, it can bee observed only with a very fine mesh. A plot of the temperatures along is axis is displayed on Figure 2. One clearly see the frozing phenomena as well as the above mentionned pressure wave. A plot of the Mach number (Figure 3) shows that the flow is very little affected by the viscous effects near the axis, the same comparison with all the variables (temperatures, mass fraction) give the same indication. The size of the boundary layer varies very much in the nozzle as can be seen on Figure 4 where $\delta = (y - y_{nozzle})/y_{nozzle}$: the size of the boundary layer varies from less than 1% to 10% of the throat.

4 Conclusion

We present calculations made for a difficult axisymmetric viscous nozzle flow. The results compares well with the results obtained in [7]. This results will be used to assess similarity parameters, such as the binary scaling one. often used to extrapolate experimental results to real flight.

References

[1] R. Abgrall, L. Fezoui & J. Talandier, Extension of Osher's Riemann solver to chemically and vibrationnal nonequilibrium gas flows. Internal Journ. for Num. Methods in Fluid, Vol 14, pp. 935-960, 1992.

[2] G. W. Candler, PhD Thesis, Stanford University, 1988.

[3] W. G. Vincenti, C. H. Kruger, Introduction to physical gas dynamics, R.E. Krieger Publishing Co, Malabar/Florida, 1982

[4] L. Fezoui & B. Stoufflet, A Class of Implicit Upwind Schemes for Euler Simulations with Unstructured Meshes, *Journal of Comp. Physics*, 84(1) pp. 174-206, September 1989.

[5] R. Abgrall, Calculation of an hypersonic nozzle flow, Proceedings of the *Workshop on Hypersonic Flows for Reentry Problems*, Antibes, January 1990, Springer Verlag, January 1992.

[6] R. Abgrall & A. Merlo, Calculation of an Hypersonic nozzle flow, Proccedings of the second *Workshop on Hypersonic Flows for Reentry Problems*, Antibes, 1991, Springer Verlag, to appear

[7] E. Boccacio, D. Zeitoun, M. Imbert & R. Brun. Proccedings of the second *Workshop on Hypersonic Flows for Reentry Problems*, Antibes. 1991, Springer Verlag, to appear

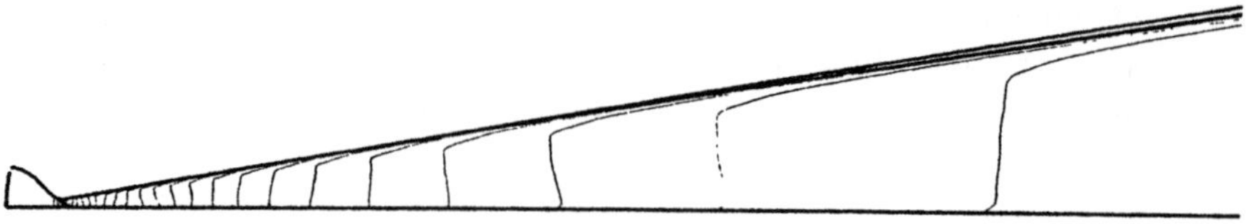

Figure 1 : Isolines of T, max=6468 K, min=296 K, ΔT=200 K

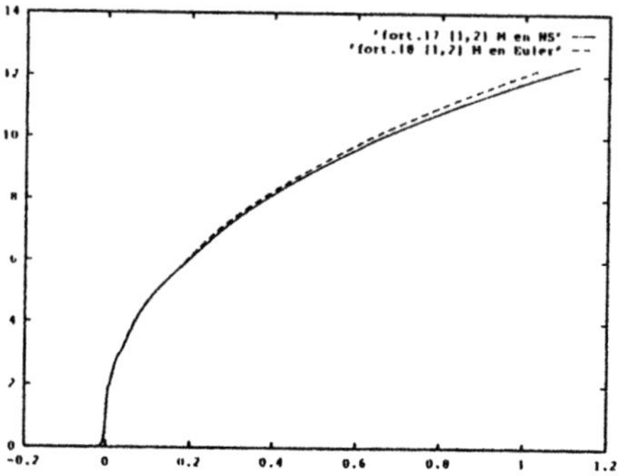

Figure 3 : Plot of the Mach number for Euler (dashed) and Navier Stokes (plain)

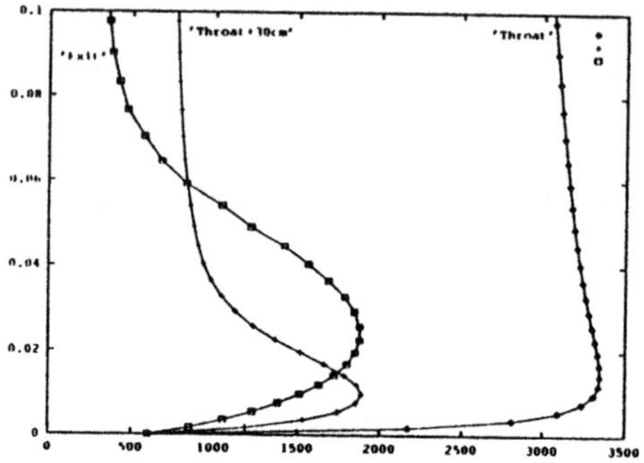

Figure 4 : Plot of the temperature normal to the axis of symmetry

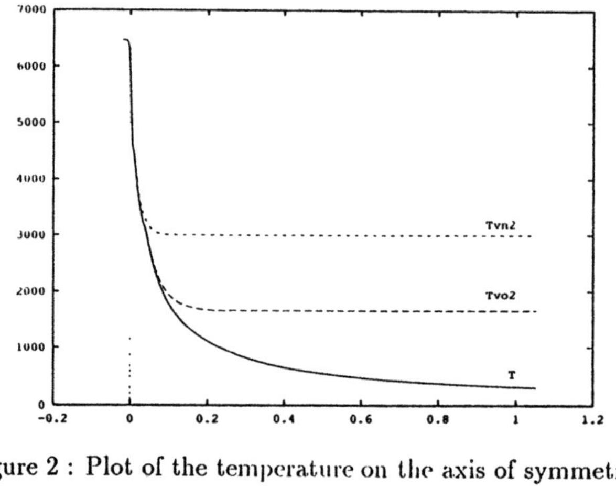

Figure 2 : Plot of the temperature on the axis of symmetry.

SHOCK-BOUNDARY LAYER INTERACTION BY SHOCK FITTING

G. Moretti*, F. Marconi*, M.Onofri+
*Grumman Aerospace Corp., Bethpage, NY, USA
+Università di Roma, "La Sapienza," Roma, Italy

1. Outline of the method

An extension of the λ-scheme with shock fitting [1] to viscous flows can be used efficiently to compute rather complex flow fields. The λ-scheme with shock fitting allows unsteady flows to be analyzed on as coarse a grid as possible. Smearing of shocks can be avoided, thus minimizing the numerical uncertainties associated with shock-boundary layer interactions. Using current second-order accurate shock-capturing codes, either the accuracy is reduced or oscillations occur at shocks. In either case, the overall validity of the computation has not been clearly assessed to date. Adaptive grids or grid clustering may help in resolving geometric singularities, boundary layers, vortex sheets and vortex centers, which spread as functions of the Reynolds number, but these are less efficient in the case of shocks, which are represented better as discontinuities since their thickness is of the order of a few mean free paths.

The technique used in this paper treats convective and diffusive terms in different ways, according to their physical nature. The viscous terms to be added to the right-hand sides of the inviscid equations as source terms are three: two components of the viscous contribution to the momentum equations, produced by the stress tensor, and the local contribution to entropy, produced by dissipation and heat conduction. The viscosity is defined by Sutherland's law. All viscous terms are computed, prior to each level of the λ-scheme, by centered differences. On the rigid walls, both velocity components vanish, no viscous term is computed, temperature follows from either the adiabatic or the isothermal condition, and pressure is computed assuming zero pressure gradient normal to the wall.

The main stream being supersonic, the upstream and downstream boundary conditions are trivial. In the subsonic downstream computational boundary within the boundary layer, zeroth-order extrapolations are used, consistent with the local parabolic nature of the flow.

Orthogonal grids, generated by conformal mapping of the $z(= x + iy)$ physical plane onto a $\zeta(= \xi + i\eta)$ plane are used throughout; they make the coding simpler and the results more accurate. The computational grid is Cartesian. The X-lines are parallel to the image of the wall, located at $Y = 0$. A strong stretching is provided in the Y-direction, to cluster lines near the wall. The stretching is controlled by the Y of the upper boundary of the grid, the number of X-lines and the height, $\Delta\eta$, of the first cell in the mapped plane.

Since (contrary to the case of inviscid λ-schemes) pressure is stored for the calculation of viscous terms, shocks may be detected by looking for inflexions in the pressure distribution. All other shock-fitting features follow the general rules mentioned in [1]. Only shock points located on X-lines are considered.

2. Numerical Results

A large number of calculations were made to test the range of validity of the code. Our first concern was with the response of the viscous subroutine and of the shock-fitting procedure to Reynolds number effects and to wall temperature boundary conditions. Flow over a flat plate was considered first, with free-stream Mach numbers ranging between 2 and 10 and Reynolds numbers between 1,000 and 6,000,000. Grids of different fineness were used, with 60 to 120 intervals along the wall and 40 to 80 intervals in the normal direction; $\Delta\eta$ ranged between .01 and .0001, according to the Reynolds number. The boundary layer is resolved with 10 to 20 grid points. The behavior of the flow near the leading edge confirmed the theoretical results of [2]. From a strictly numerical viewpoint, it is interesting to note that, particularly for low Reynolds numbers, shocks can be fitted in front of the leading edge all the way down; the first shock point, however, tends to move over the wall and gets trapped exactly at the leading edge itself (obviously, we cannot have a steady shock on the wall, where the velocity vanishes). The other shock points remain anchored to the first, and the entire leading edge shock is slightly displaced to the left (Fig. 1); its slope, however, is correct. If fitting of the shock is forbidden for a certain number of horizontal rows above the wall, some coalescence of pressure waves occurs slightly downstream of the leading edge, and the fitted shock proceeds thereafter (Fig. 2). Indeed, the work of [2] clearly shows that the flow at the leading edge starts with a non-continuum region followed by a merged-layer region. In the latter, the thickness of the "shock" and of the boundary layer are of the same scale. Further downstream, shock and boundary layer become distinct and the shock is much thinner than the boundary layer. The merged-layer region is represented by the coalescence of pressure waves above the leading edge in Fig. 2. The strong effect of the free-stream Mach number on the leading edge interaction is evident in Fig. 3, that shows a preliminary attempt to interpret the model of [2] via a numerical analysis. In it, the shock slopes for three values of M_∞ are shown. At station A, the corresponding velocity profiles are plotted. It is clear that, for M_∞=2 and 3 the shocks are already far from the boundary layer, whereas for M_∞=8 the shock is still within the boundary layer and station A may be considered in a region of strong interaction. Convergence of residuals to machine-zero occurs spontaneously in certain cases (Fig. 4). In others, the first shock point may oscillate indefinitely within a cell, because of numerical interactions with the leading edge itself; in this case, if we freeze it when its average displacement is a small fraction of the cell length, say, convergence to machine-zero is reached again quickly. Typical lengths of the runs are 4000 computational steps. At low free-stream Mach numbers (less than 3), comparisons of velocity profiles in the boundary layer with a Blasius profile as modified according to Stewartson [3] are very good (Fig. 5). The wall pressure follows the theoretical trend [4] very closely; see Fig. 6, where results from a shock-capturing calculation also are shown. Adiabatic walls are harder to evaluate, because the wall temperature may become very high, substantially increasing the viscosity. In the context of our current code, it was found that a parabolic approximation for the temperature distribution normal to the wall is a must to obtain a reasonable approximation of the theoretical value (Fig. 7) where, again, results from a shock-capturing calculation also are shown. A final validation test was made, comparing our results with experimental data of Pitot pressures [5] (results omitted here for brevity).

The second series of calculations analyzes the generation of shocks and the possible appearance of a recirculation region on a flat plate followed by a wedge. Here again the same ranges of Mach numbers, Reynolds numbers and $\Delta\eta$ mentioned above were explored. More combinations of grid resolutions in either direction were used, with the number of vertical lines ranging between 60 and 120 and the number of horizontal lines ranging between 40 and 180. Space limitations prevent us from presenting more than one case; we have opted for a 20° wedge in a stream defined by M_∞=3 and R_e=15000; the wall is isothermal, and the wall temperature equals the free-stream temperature. The grid has 60 intervals along the wall and 180 in the normal direction; $\Delta\eta = .0005$. This case has been chosen because three shocks are sufficiently strong for fitting: a leading edge shock, losing strength as it proceeds away from the wall, the separation shock, and the reattachment shock. The two latter shocks coalesce into one within the computed region. Lines of constant X-velocity components are shown in Fig. 8, isobars in Fig. 9. For the sake of comparison, isobars obtained by a shock-capturing code are also shown in Fig. 10. The extension of the recirculation region is seen in detail in Fig. 11, where the value of the X-velocity component at the first grid row above the wall is shown.

A third set of computations was made to analyze the reflection of an oblique shock from a rigid wall. The free-stream Mach number is 2 and the Reynolds number is 296,000. A down-running shock is produced by a 3° deflection. Isobars computed with the shock-fitting code and with a shock-capturing code are shown in Fig. 12 and Fig. 13, respectively. In Fig. 12 the shocks are not marked by × as in previous figures. A comparison of Fig. 12 and Fig. 13 indicates the amount of shock spreading in the shock-capturing results. Such a spreading can significantly distort the computed results.

Acknowledgements

This work was supported in part by the US Air Force Office of Scientific Research, Dr. L. Sakell acting as technical monitor, and partially by the "Agenzia Spazile Italiana".

References

1. G. Moretti, "A Technique for Integrating Two-dimensional Euler Equations," Comp. and Fluids, 15:59-75, 1987.
2. S. Rudman and S.G. Rubin, "Hypersonic Viscous Flow over Slender Bodies with Sharp Leading Edges," AIAA J., 6:1883-1889, 1968.
3. K. Stewartson, *The Theory of Laminar Boundary Layers in Compressible Fluids*, Oxford 1964.
4. M.H. Bertram, "Boundary-Layer Displacement Effects in Air at Mach Numbers of 6.8 and 9.6," NACA TN 4133, 1958.
5. J.M. Kendall, jr., "Experimental Investigation of Leading-Edge Shock-Wave – Boundary-Layer Interaction at Mach 5.8," J.Aero.Sci. 24:47-57, 1957.

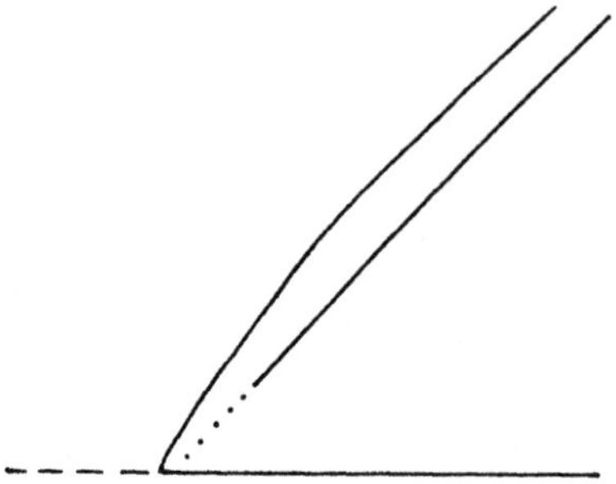

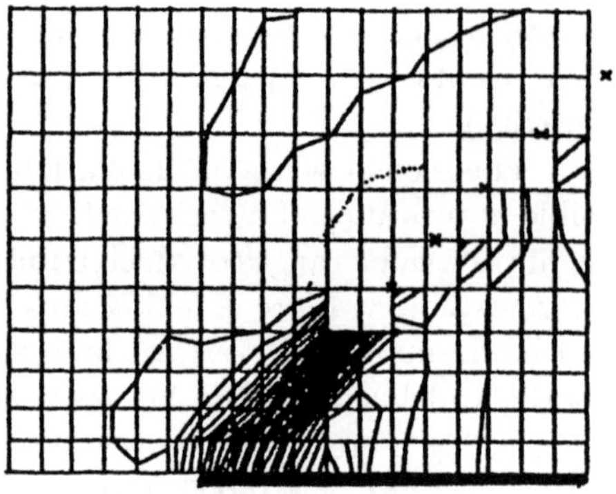

Fig. 1 - Two leading-edge shocks resulting from computations

Fig. 2 - Isobars and shock wave near leading edge of a flat plate

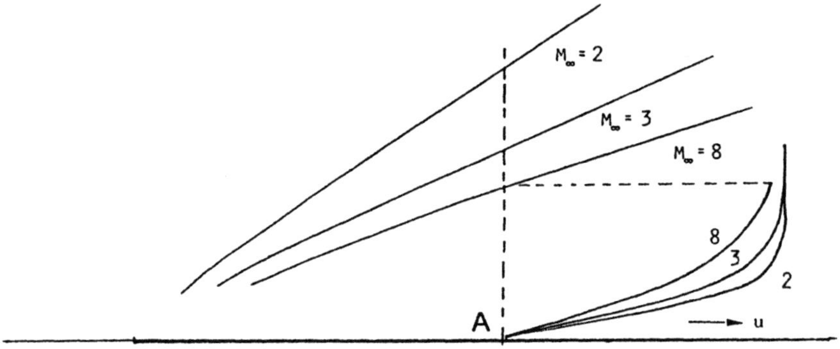

Fig. 3 - Mach number effects on leading-edge shocks

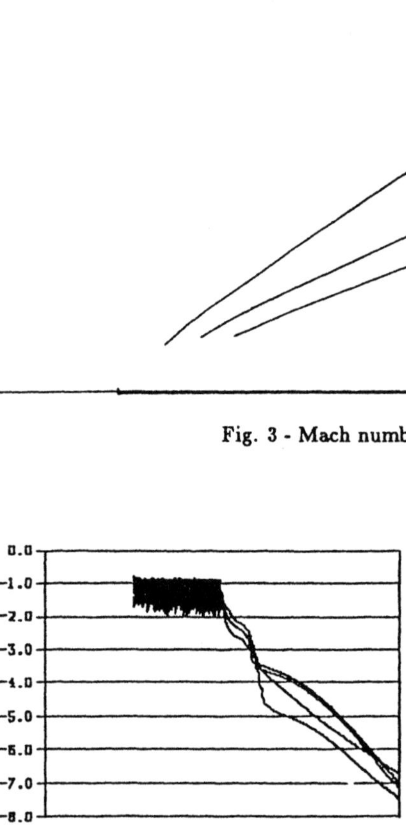

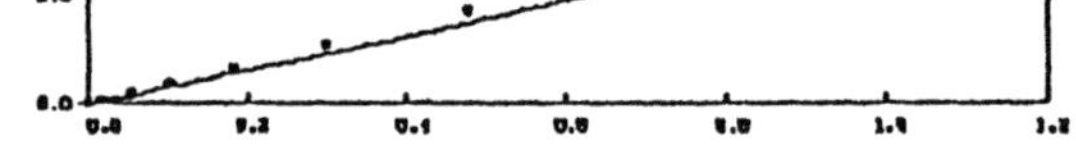

Fig. 4 - Convergence of residuals

Fig. 5 - Velocity profile in the boundary layer

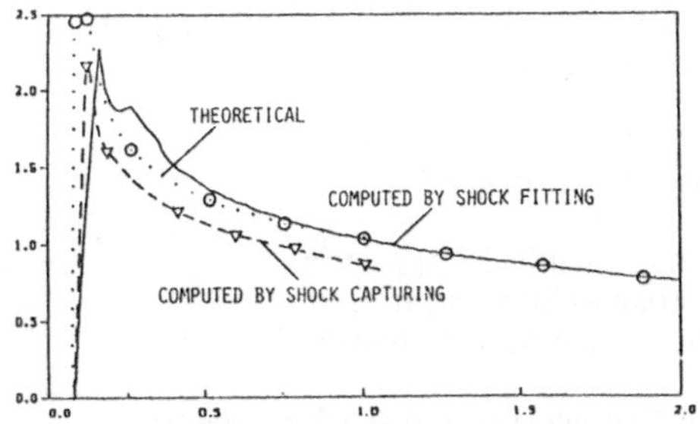

Fig. 6 - Wall pressure on a flat plate

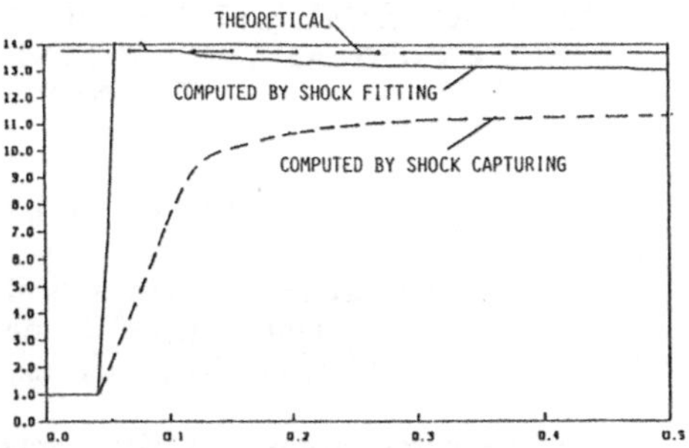

Fig. 7 - Temperature on adiabatic wall

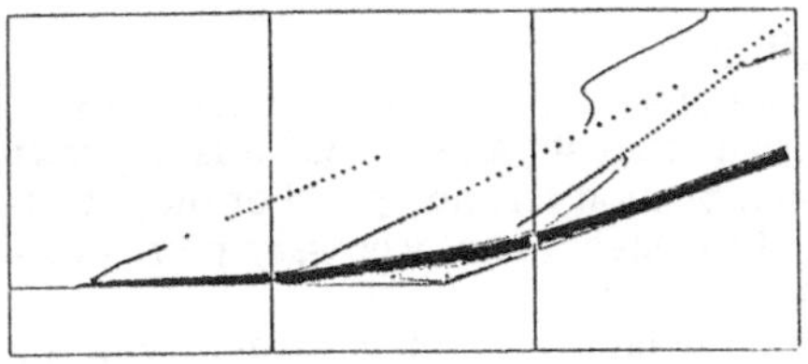

Fig. 8 - Wedge flow - Constant velocity lines

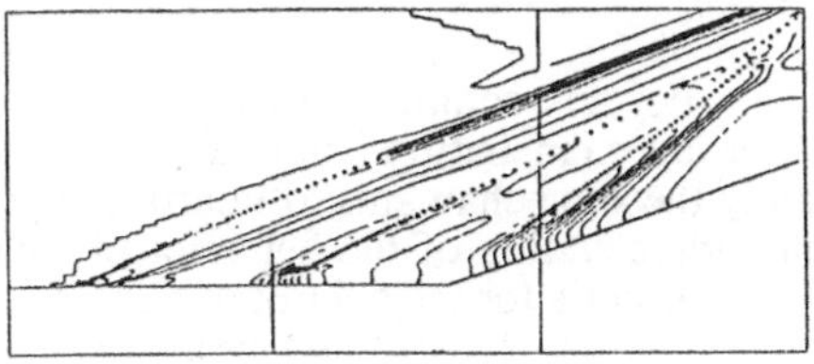

Fig. 9 - Wedge flow - Isobars

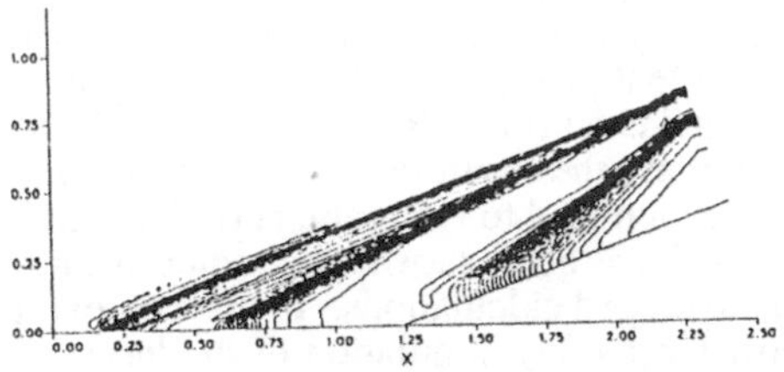

Fig. 10 - Wedge flow - Isobars by a shock-capturing code

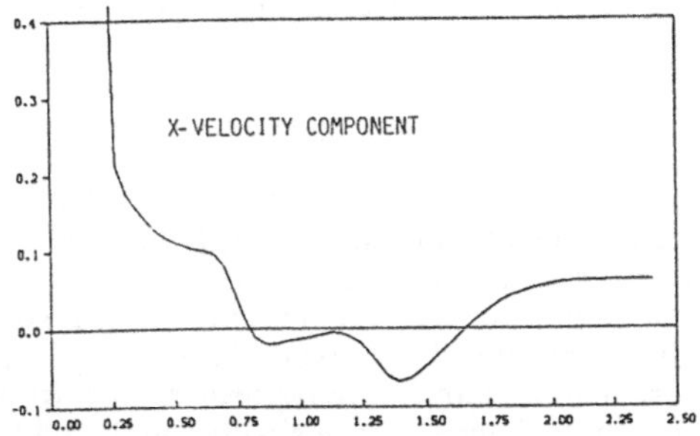

Fig. 11 - Wedge flow - Velocity in the vicinity of the wall

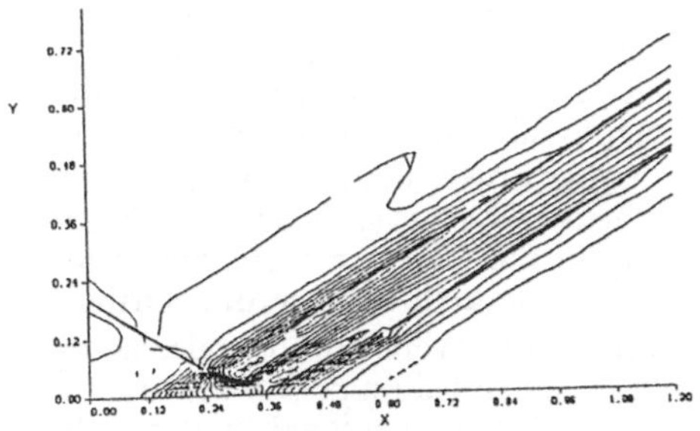

Fig. 12 - Shock reflection from a wall Isobars by a shock-fitting code

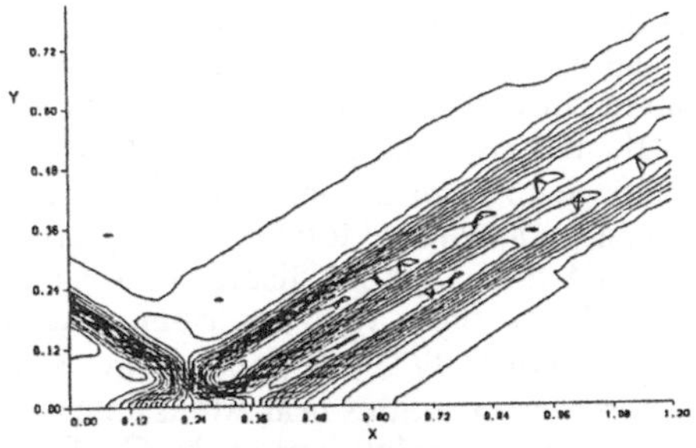

Fig. 13 - Shock reflection from a wall Isobars by a shock-capturing code

NUMERICAL SIMULATION OF SUPERSONIC MIXING LAYER USING A SOLUTION ADAPTIVE FINITE-ELEMENT FLUX-CORRECTED TRANSPORT SHCEME

C. Oh*, E. Loth and R. Löhner*****

Department of Aeronautical and Astronautical Engineering,
University of Illinois at Urbana-Champaign.
* Graduate Research Assistant, **Assistant Professor.

*** Research Professor, CMEE, School of Engineering and Applied Science,
The George Washington University, Washington D.C. 20052.

Introduction

For any scramjet engine, the mixing and combustion process is governed by a free shear layer emanation from the juncture of the air and fuel streams. Thus, it is important to study these mechanisms to improve supersonic combustion technology. For these reasons, an understanding of the behavior of supersonic shear layers is a necessary prerequisite to setting criteria for the control of mixing in supersonic flows.

In this study, spatially evolving supersonic mixing layers were investigated with several sets of initial and boundary conditions applied based on flow characteristics[1]. Finite element techniques were used to take advantage of the triangular grid adaptation capabilities. Experimental results which contain turbulent upstream boundary layers[2], are chosen to test the performance of the simulations.

We may regard a thin shear layer as a discontinuity in the velocity, i.e. a vortex sheet. Such a vortex sheet is unstable since any small disturbance will initially grow exponentially. This instability leads to rolling up of the vortex sheet into a spectrum of vortices which include large scale structures, which drives the turbulent process for shear layers. Thus the large-scale motions are primarily inviscid in nature and exist for all Reynolds number beyond some critical value. The physics of the mixing layer are the result of an essentially inviscid interaction between neighboring vortical structures which lead to entrainment of the unmixed fluids into the shear layer. Large eddy simulations have been shown to predict mean and turbulent features more robustness than Reynolds-averaged calculations. In this context, the unsteady Euler equations were employed to allow large eddy simulation of the large scale structures within supersonic mixing layer in hopes of capturing the mixing processes only as far as the entrainment dynamics.

Numerical Model and Numerical Method

The numerical model used to carry out the simulations consists of the two-dimensional, time-dependent conservation equations of mass density, momentum and energy for the inviscid perfect gas (although artificial viscosity is present). To compute the mixing (as a result of convection), a mixture fraction conservation equation was also included for some of the calculations. The gas dynamic equations for compressible flow of an ideal gas can be discretized to form a conservative scheme[4]. The higher order solution for FEM-FCT is obtained via a two step form of the Taylor-Galerkin scheme, which is second order scheme in space and time. Spatial discretization is performed via the Galerkin weighted residual method.A low-order term contribution is combined with the high-order term contribution through the FEM-FCT[5] formulation to prevent the formation of overshoots or undershoots in the conserved quantities near admissible discontinuities. The global limiter for the first cases was set to be the minimum of the C_{el} for r and re since this conventional limiter yielded monotonic gas flow properties for a minimal contribution of the low order solution in several past studies.[6] However, the code was modified to include the mixture fraction equation with the Euler equation. With the new information of mixing gradients now available, the limiter for the gas dynamic variables was kept the same, but the limiter for mixture fraction equation was applied independently using the C_{el} from mixture fraction.

Grid adaptation was employed to optimize the distribution of grid points by refining areas with high gradients of flow variables and coarsening areas of low gradients of flow variables. This allows a balanced (and efficient) distribution of truncation errors. Two different error indicators were used. For the first cases, density was used as an error indicator since it was found to yield excellent resolution of compressible phenomena such as shock waves. However for the cases which included the mixture fraction equation, the variables of mixture fraction and density were both used since the mixture fraction is a natural indicator of the interface of the two layers - and thus, of entrainment.

Simulation Conditions

Two sets of flow conditions were used. The first set of computations were chosen to examine a large range of convective Mach numbers, i.e. cases A-E. A second set of calculations were performed which duplicate the conditions given for an experimental case of Goebel et al.[7], i.e. case F. This particular case was chosen since both streams were supersonic and the convective Mach number (M_c) was around 0.5, which insured a predominantly two dimensional character of the supersonic mixing layer[8] thus making it a practical case to simulate. Computational domain and boundary conditions for the present calculations are shown in Fig. 1.

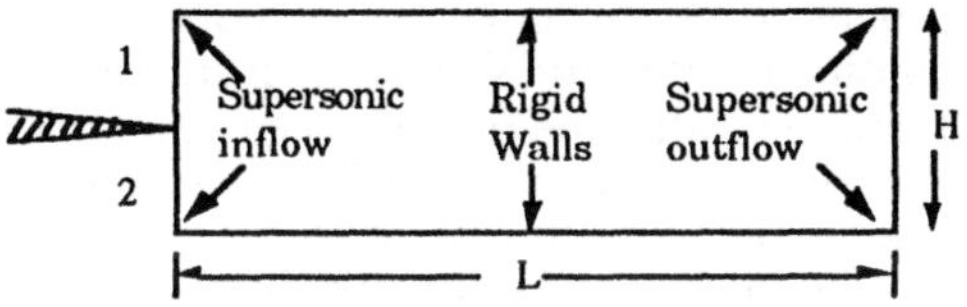

Fig 1. Physical computational domain and boundary conditons

For the error indicator, density was used for cases A-E, whereas density and mixture fraction were used for case F.

For the calculations of set I (cases A-E), a step function profile was used as the initial condition. Three types of initial conditions were applied to the computations for case F to determine the effects on both the mixing layer dynamics and the prediction fidelity with experimental results. These initial conditions included 1) step function, 2) boundary layer based on the paper of Child and Sun.[9] and 3) hyperbolic tangent profiles for all flow variables. All calculations are unforced in the sense that no external frequency is imposed on the flow through boundary or initial conditions.

Results and Discussion

Figs 1 shows mixture fraction Gouraud shaded contours at 12,000 timesteps, which is considered to be a fully-developed state. The classic incompressible eddies were noted along with "flattened" eddies, which appear to be correlated with pairing processes. This is presumably an effect of the compressibility on the large scale structures - and is possibly the major cause of the reduced mixing phenomena. Note, such an mechanism can not be described by linearized analysis.

Figs. 2 is v-velocity Gouraud shaded contours at 12,000 timesteps for a step function. Unstable waves and rotations of vortices and their dynamics yields perturbations of v-velocity, which is on average zero. The extent of these perturbations is proportional to the eddies size and coherency (correlate with Fig.1). Therefore, the source of v-velocity fluctuations, and turbulent intensity of transverse direction (v-rms) mainly originate from large scale structures and their dynamics. It was also noted that coherent structures and strong perturbations appeared to come in low frequency waves, i.e. groups of strong eddies were followed by groups of weaker, less coherent eddies.

Fig. 3 shows the mesh refinement levels for the baseline case at 12,000 timesteps. Each color represents successive refinement levels (blue : 0th level, green : 1st, yellow : 2nd, red : 3rd and purple : 4th refinement level). Mesh refinement was noted to yield high resolution along braids and edges of vortices, as desired. Figure 4 exhibits two line contours of the mixture fraction at 12,000 timesteps, which indicates asymmetric entrainment i.e., the high speed fluid is entrained at a higher rate than that of the lower speed stream. This asymmetric entrainment was not found in the temporally growing mixing layer simulations because periodic boundary conditions prevent this nonlinear phenomena.

Fig. 5 exhibits a vortex pairing process where an upstream vortex has begun to ride the tail of a downstream vortex. The yellow colored region is believed to represent entrained fluid. The mixture fraction contours for the high speed are more diffuse and more convoluted than on the low speed side. Fig. 6 shows velocity vectors colored by the Mach number range. The convective velocity calculated based on Roshko and Papamoschou[2] was subtracted from upper (U_1) and lower (U_2) velocities. The vortices rotate about their cores and show highest speed right over vortex cores and lowest speed under vortex cores.

Fig. 7 shows the variation of the normalized growth rate versus convective Mach number defined by Papamoschou et al.[16] The mixing layer growth rate (db/dx) was normalized by the incompressible visual growth rate $(db/dx)_o$. The cases A-E are used for calculations. It can be clearly seen that the growth rate of the mixing layer decreases with increasing convective Mach number This reduction of growth rate is due to effects of compressibility which correlate well with convective Mach number. The present predictions show this trend as well with reasonable quantitative agreement.

The similarity coordinates used for similarity profile are $h=(y-y_o)/b$ for turbulent analysis. The first 1,500 timesteps after the flow reached a fully-developed state were used to predict mean velocity profiles and the next 7,500 timesteps were used for u-rms, v-rms and Reynolds stress profiles. The step function case (Fig. 8) first diffuses in shape then converges to a similarity hyperbolic tangent profile by 20 cm. The boundary layer case (Fig. 12) shows the effects of the wake up to 15 cm and then converges to a hyperbolic tangent profile. The agreement for the experimental data given at 30-35 cm downstream is quite good two cases. Thus the effect of the initial conditions has only a small affect on the mean profiles. However, turbulence quantities are a more sensitive indicator of flow development.

The u-rms predictions for the step function initial conditions are shown in Figures 9. This result exhibits both a primary peak at the centerline aswell as secondary peaks at values of h near 0.6. The v-rms predictions for the three initial conditions are shown in Figures 10 and 13. Two cases exhibit only a primary peak near the centerline which is consistent with the experimental results. The step function case converges after 20 cm; the boundary case converges monotonically at 20 cm. The Reynolds stress predictions for the three initial conditions are shown in Figures 11and 14. All two cases exhibit primarily negative Reynolds stresses except for the step function case at 20 cm, and to a lesser degree the boundary layer. The only the boundary layer case can be considered to be fully converged. Peak values of two cases are roughly half that of experimental values.

References

1. Pulliam, T.H., NASA-CP-2201, 1982.
2. Papamoschou, D., Roshko, A., J. of Fluid Mech., Vol. 197, 1988, pp. 453-477.
3. Goebel, S.G., Dutton, J.C., AIAA 90-0709.
4. Löhner, R., Morgan, K., Peraire, J. and Vahdati, M., International Journal for Numerical Methods in Fluids, Vol. 7, 1987, pp. 1093-1109.
5. Löhner, R. Morgan, K., and Zienkiewicz, O.C., Computational Methods in Applied Mechanics Engineering, Vol. 51, 1985, pp. 441-465.
6. Loth, E., Baum J., Lohner R., AIAA 90-1655.
7. Goebel, S.G., Dutton, J.C., Krier, H., Renie, J.P., Experiments in Fluids, Vol. 8, 1990, pp. 263-272.
8. Clemens N.T., HTGL Report No. T-274, 1991, Mechanical Engineering Department, Stanford University.
9. Sun, C.C., Childs, M.E., AIAA J. of Aircraft, Vol. 10, 1973, pp. 381-383.

Fig. 1. Mixture fraction Gouraud shaded contours at 12,000 timesteps. The step function is used as a initial condition.

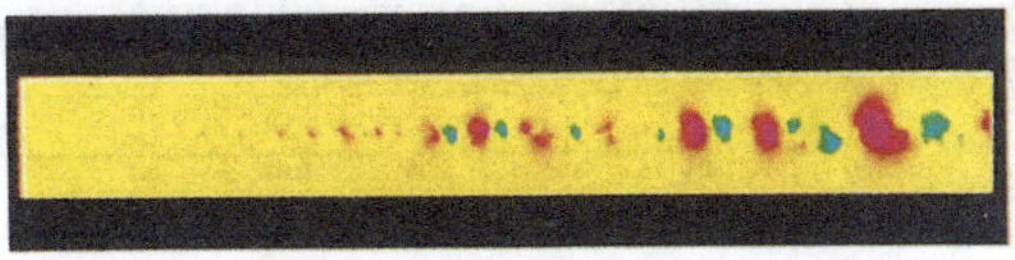

Fig. 2. V-velocity Gouraud shaded contours at 12,000 timesteps. The step function is used as a initial condition.

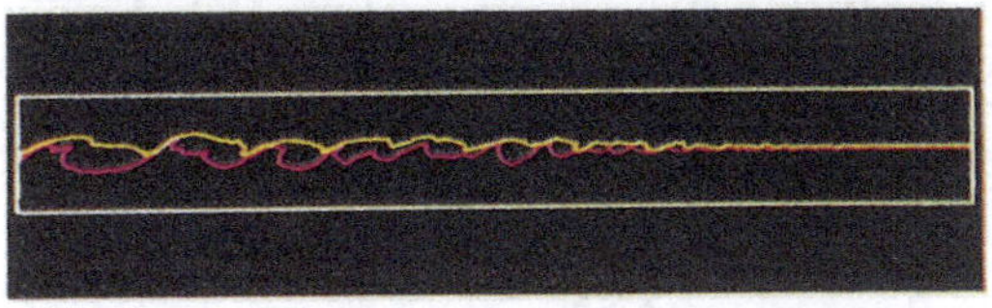

Fig. 3. Two line contour of mixture fraction at 12,000 timesteps. The step function is used as a initial condition.

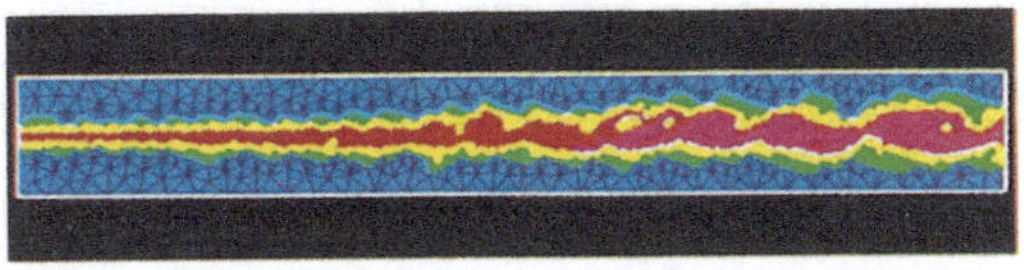

Fig. 4 . The mesh refinement level plot at 12,000 timesteps. The step function is used as a initial condition. (0th level; blue, 1st ; green, 2nd; yellow, 3rd; red, 4th ; purple).

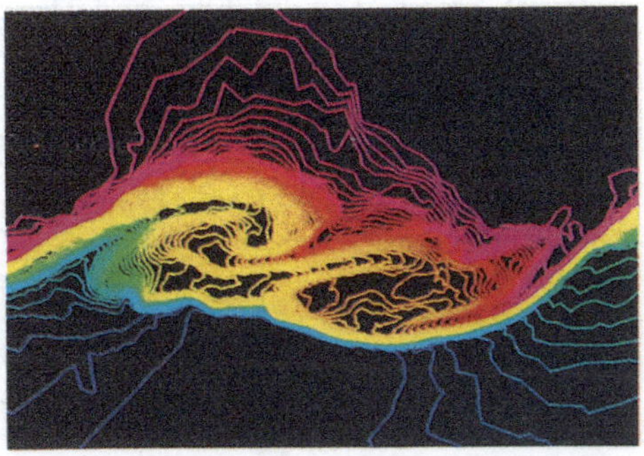

Fig. 5. Mixture fraction line contours at12,000 timesteps. The step function is used as a initial condition. The number of contour lines are 64.

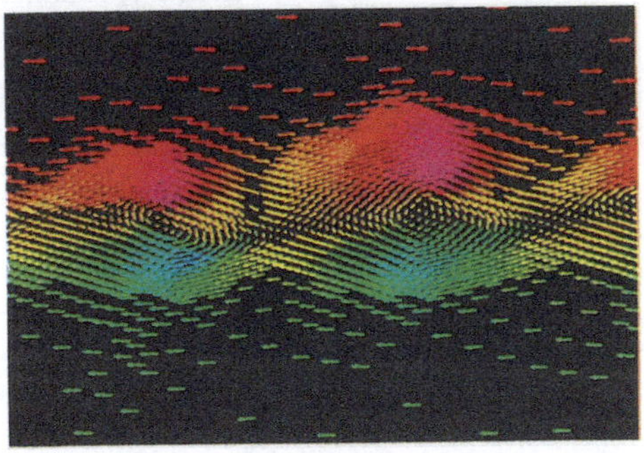

Fig. 6. Velocity vector contour colored by Mach number at 12,000 timesteps. The step function is used as a initial condition.

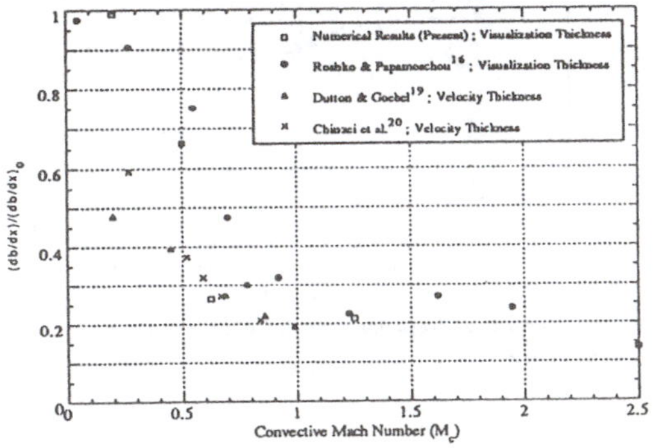

Fig. 7. Normalized Mixing Growth Rate of Cases A-E with Experimental Results

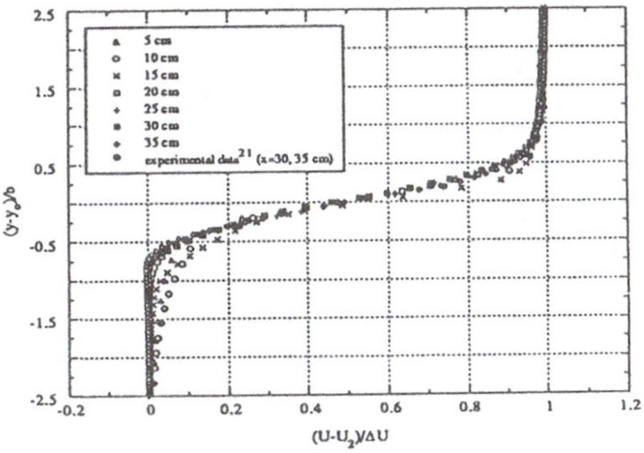

Fig. 8. Normalized Mean Streamwise Volocity of Step Function of Case F.

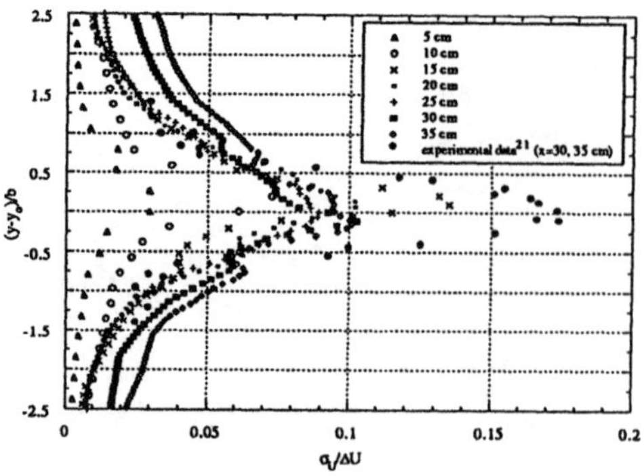

Fig. 9. Normalized Streamwise Turbulence Intensity of Step Function of Case F.

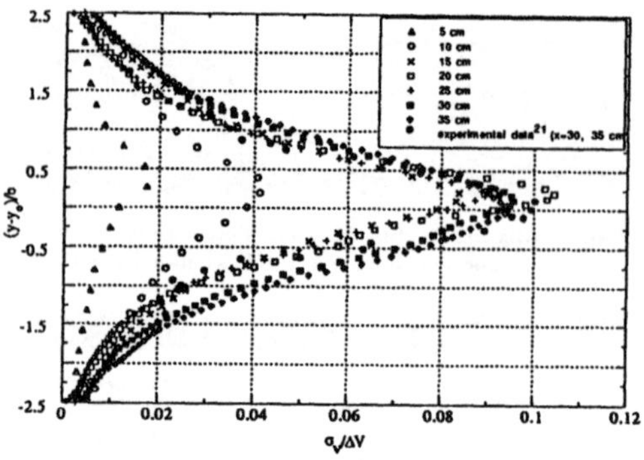

Fig. 10. Transverse Turbulence Intensity of Step Function of Case F.

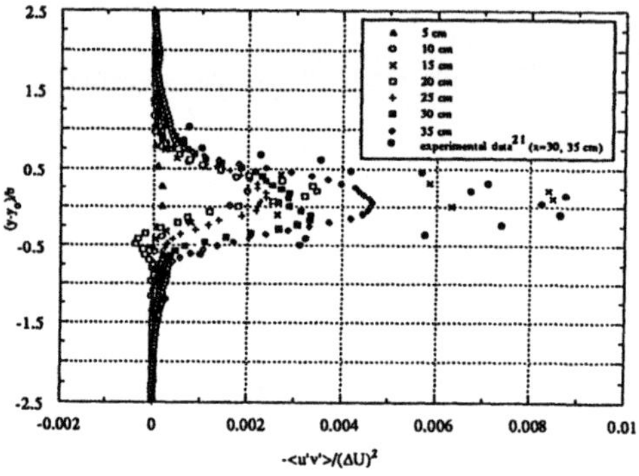

Fig. 11. Normalized Reynolds Stress of Step Function of Case F.

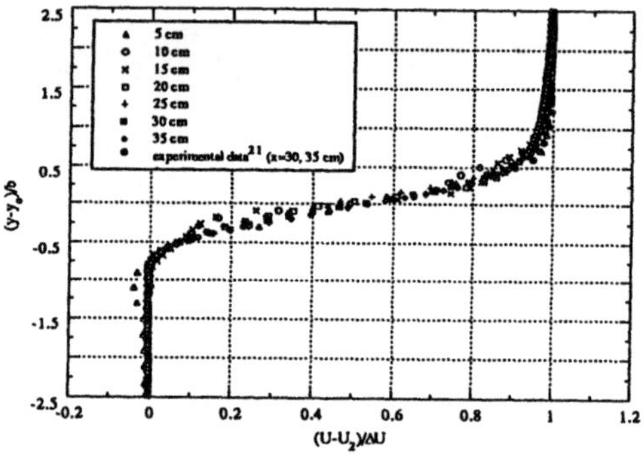

Fig. 12. Normalized Mean Streamwise Velocity of Boundary Layer of Case F.

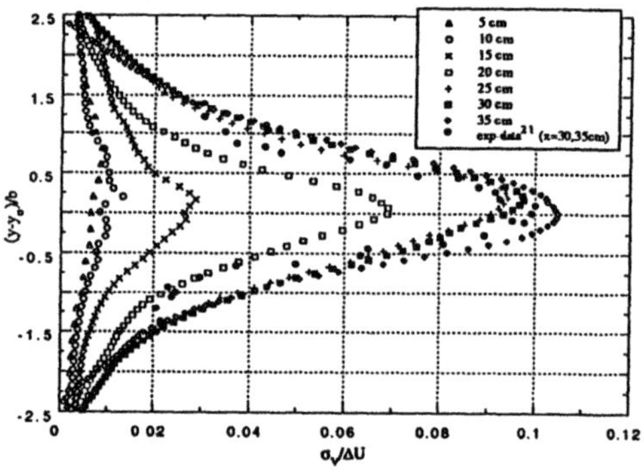

Fig. 13. Transverse Turbulence Intensity of Boundary Layer of Case F.

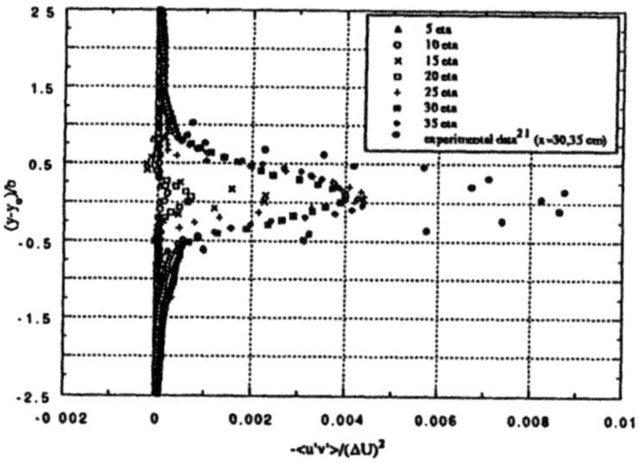

Fig. 14. Normalized Reynolds Stress of Boundary Layer of Case F.

Numerical Solution of Hypersonic Real Gas Flows Using Method of Lines

Nobuyuki Satofuka, Koji Morinishi, Asami Matsumoto and Osamu Ogawa

Kyoto Institute of Technology
Matsugasaki, Sakyo-ku, Kyoto 606 JAPAN

1. Introduction

An explicit method of lines approach[1] has been applied for solving hypersonic real gas flows governed by the Euler and Navier-Stokes equations. The method is based on a finite difference approximation to spatial derivatives and subsequent time integration using the rational Runge-Kutta[2] scheme. Both finite rate and locally equilibrium approaches are adopted in the chemical reaction model for dissociation of the air. Numerical results are presented for the hypersonic flows over a double ellipse/ellipsoid.

2. Governing Equations

The Navier-Stokes equations subject to the two-dimensional general coordinate transformation can be written in dimensionless, conservation-law form as:

$$\frac{\partial \hat{q}}{\partial t}+\frac{\partial \hat{E}}{\partial \xi}+\frac{\partial \hat{F}}{\partial \eta}=\frac{1}{Re}\left(\frac{\partial \hat{R}}{\partial \xi}+\frac{\partial \hat{S}}{\partial \eta}\right), \tag{1}$$

where the conservative quantities $\hat{q}$ and the flux terms $\hat{E}, \hat{F}, \hat{R}$ and $\hat{S}$ are

$$\hat{q}=J^{-1}\begin{bmatrix}\rho\\ \rho u\\ \rho v\\ e\end{bmatrix},\quad \hat{E}=J^{-1}\begin{bmatrix}\rho U\\ \rho u U+\xi_x p\\ \rho v U+\xi_y p\\ (e+p)U\end{bmatrix},\quad \hat{F}=J^{-1}\begin{bmatrix}\rho V\\ \rho u V+\eta_x p\\ \rho v V+\eta_y p\\ (e+p)V\end{bmatrix},$$

$$\hat{R}=J^{-1}\begin{bmatrix}0\\ x_x s_x+x_y t_{xy}\\ x_x t_{xy}+x_y s_y\\ x_x R_4+x_y S_4\end{bmatrix},\quad \hat{S}=J^{-1}\begin{bmatrix}0\\ h_x s_x+h_y t_{xy}\\ h_x t_{xy}+h_y s_y\\ h_x R_4+h_y S_4\end{bmatrix} \tag{2}$$

with the viscous stress,

$$\sigma_x=2\mu\frac{\partial u}{\partial x}-\frac{2}{3}\mu\left(\frac{\partial u}{\partial x}+\frac{\partial v}{\partial y}\right),\quad \tau_{xy}=\mu\left(\frac{\partial u}{\partial y}+\frac{\partial v}{\partial x}\right),\quad \sigma_y=2\mu\frac{\partial v}{\partial y}-\frac{2}{3}\mu\left(\frac{\partial u}{\partial x}+\frac{\partial v}{\partial y}\right),$$
$$R_4=u\,\sigma_x+v\,\tau_{xy}+\frac{\gamma}{\gamma-1}\frac{\kappa}{Pr}\frac{\partial T}{\partial x},\quad S_4=u\,\tau_{xy}+v\,\sigma_y+\frac{\gamma}{\gamma-1}\frac{\kappa}{Pr}\frac{\partial T}{\partial y}. \tag{3}$$

Here ρ, p and T denote the density, pressure and temperature. In the viscous terms, Re is the reference Reynolds number, Pr the Prandtl number, μ the viscosity coefficient and κ the coefficient of thermal conductivity. The equations of state for equilibrium air can be written as:

$$p=(\bar{\gamma}-1)\rho i \tag{4}$$

where $\bar{\gamma}$ is the ratio of specific heats and is a function of ρ and i which is given by Srinivasan et al[3]. For a non-equilibrium model we chose following three reaction model proposed by Park[4].

$$O_2 + N_2 = 2O + N_2$$
$$N_2 + O = NO + N \tag{5}$$
$$NO + O = O_2 + N$$

The so-called contravariant velocities U and V are obtained from the Cartesian velocity components u and v as:

$$U = \xi_x u + \xi_y v\,, \quad V = \eta_x u + \eta_y v \tag{6}$$

Hereafter, the symbol ^, which denotes the vector quantity divided by Jacobian J, may be omitted for simplicity without any confusion, unless a specific note is provided. In the case of $Re\to\infty$, Eqs.(1) reduce to the Euler equations.

3. Numerical Procedure

In our method of lines approach[1], the Navier-Stokes equations (1) are first discretized by the conventional central finite-difference approximation as follows:

$$\frac{d}{dt}q_{i,j} = -(E_{i+1,j} - F_{i-1,j})/(2\Delta\xi) - (F_{i,j+1} - F_{i,j-1})/(2\Delta\eta)$$
$$+ (R_{i+1/2,j} - R_{i-2/1,j})/(Re\Delta\xi) + (S_{i,j+1/2} - S_{i,j-1/2})/(Re\Delta\eta) + D_\xi + D_\eta\,, \tag{7}$$

where subscript i and j denote the grid indexes such as $q_{i,j} = q(i\Delta\xi, j\Delta\eta)$. In order to eliminate spurious oscillation and capture shock wave, the dissipative terms D_ξ and D_η are added in the right hand side of Eq.(7). The dissipative terms used here are based on the second and fourth differences, which was first introduced by Jameson[5]. The term can be written for ξ-direction as:

$$D_\xi = d_{i+1/2,j} - d_{i-1/2,j}\,, \tag{8}$$

with the dissipative flux

$$d_{i+1/2,j} = \left[\varepsilon^{(2)}_{i+1/2,j} \Delta_\xi (Jq)_{i,j} - \varepsilon^{(4)}_{i+1/2,j} \Delta^3_\xi (Jq)_{i-1,j} \right] / (J\Delta t_\xi)_{i+1/2,j}\,. \tag{9}$$

Here Δ_ξ is forward difference operator, Δt_ξ is the certain time step specified later, and $\varepsilon^{(2)}$ and $\varepsilon^{(4)}$ are defined as

$$\varepsilon^{(2)}_{i+1/2,j} = \omega^{(2)} max(\nu_{i+1,j}, \nu_{i,j})\,, \; \varepsilon^{(4)}_{i+1/2,j} = max(0, \omega^{(4)} - \varepsilon^{(2)}_{i+1/2,j}) \tag{10}$$

where $\omega^{(2)}$ and $\omega^{(4)}$ are adaptive coefficients and $\nu_{i,j}$ is obtained from the second difference of pressure.

In order to eliminate the uncertainties of adaptive coefficient and improve the spatial resolution of discontinuities, new dissipative terms are derived from the Yee's TVD scheme[6] as follows:

$$D_\xi = R_{i+1/2,j}\Phi_{i+1/2,j} - R_{i-1/2,j}\Phi_{i-1/2,j}\,,, \tag{11}$$

where R is the matrix whose columns are eigenvectors of the Jacobian $A = \partial E/\partial q$. The elements of the $\Phi_{i+1/2,j}$ denoted by $\phi^m_{i+1/2,j}$ are

$$\phi^m_{i+1/2,j} = \frac{1}{2}\psi(a^m_{i+1/2,j})(g^m_{i,j} + g^m_{i+1,j}) - \psi(a^m_{i+1/2,j} + \gamma^m_{i+1/2,j})\,\alpha^m_{i+1/2,j}\,, \tag{12}$$

with

$$g^m_{i,j} = S\; max\left[0, min(S\alpha^m_{i+1/2,j}, S\alpha^m_{i-1/2,j})\right], \; S = sign(\alpha^m_{i+1/2,j})\,. \tag{13}$$

Here α is the eigenvalues of Jacobian A, and α is given by

$$\alpha_{i+1/2,j} = R^{-1}_{i+1/2,j}\left[(Jq)_{i+1,j} - (Jq)_{i,j}\right]. \tag{14}$$

The detail descriptions of of the function Ψ and γ can be found in the reference (6).

For a time stepping scheme, the rational Runge-Kutta (RRK)scheme of Wambecq[3] is used. The system of ordinary differential equations(7) can be rewritten as:

$$\frac{d\vec{q}}{dt} = \vec{W}(\vec{q}) \tag{15}$$

Then the second order RRK scheme may be written in the following two stage form.

$$\begin{aligned} &\vec{g}_1 = \Delta t\vec{W}(\vec{q}^n) , \ \vec{g}_2 = \Delta t\vec{W}(\vec{q}^n + c_2\vec{g}_1), \\ &\vec{g}_3 = b_1\vec{g}_1 + b_2\vec{g}_2, \\ &\vec{q}^{n+1} = \vec{q}^n + [2\vec{g}_1(\vec{g}_1,\vec{g}_3) - \vec{g}_3(\vec{g}_1,\vec{g}_1)] / (\vec{g}_3,\vec{g}_3), \end{aligned} \tag{16}$$

where superscript n denotes the index of time steps and $(\vec{g}_i,\vec{g}_j)$ denotes the scalar product of two vector $\vec{g}_i$ and $\vec{g}_j$. An efficient second order scheme is given by the coefficients

$$b_1 = 2 , b_2 = -1 , c_2 = 0.5 \tag{17}$$

A constant CFL number is imposed on all grid points, so that the time step Δt is determined locally as

$$\Delta t_{i,j} = C_N \, min\,(\Delta t_\xi , \Delta t_\eta)_{i,j}, \tag{18}$$

with

$$(\Delta t_\xi)_{i,j} = \left(1/(|U| + c\sqrt{\xi_x^2 + \xi_y^2})\right)_{i,j}, (\Delta t_\eta)_{i,j} = \left(1/(|V| + c\sqrt{\eta_x^2 + \eta_y^2})\right)_{i,j}. \tag{19}$$

Here C_N is the CFL number to be given and c is the speed of sound.

In order to accelerate the rate of convergence to a steady state solution, the residual averaging technique[5] is incorporated into the basic scheme. The final stage of the RRK scheme (16) can be written in the following form:

$$q_{i,j}^{n+1} = q_{i,j}^n + r_{i,j}, \tag{20}$$

where r is the residual. In the residual averaging technique,[5] r may be replaced by the implicit average $\bar{r}$ defined as:

$$(1 - \varepsilon_\xi\delta_\xi^2 - \varepsilon_\eta\delta_\eta^2)\bar{r}_{i,j} = r_{i,j}, \tag{21}$$

where δ^2 is the second difference operator. The smoothing parameters ε_ξ and ε_η may be obtained by

$$(\varepsilon_\xi)_{i,j} = max\left\{\frac{1}{4}\left[\left(\frac{\Delta t}{\lambda\,\Delta t_\xi}\right)^2 - 1\right], \varepsilon_{min}\right\}_{i,j}, (\varepsilon_\eta)_{i,j} = max\left\{\frac{1}{4}\left[\left(\frac{\Delta t}{\lambda\Delta t_\eta}\right)^2 - 1\right], \varepsilon_{min}\right\}_{i,j}, \tag{22}$$

where $\lambda = 1.25$ and $\varepsilon_{min} = 0.2$.

4. Numerical Results

Numerical experiments of the present scheme are carried out for hypersonic flows over a double ellipse and a double ellipsoid. At first, numerical solutions of two-dimensional flows over a double ellipse are obtained at free stream Mach numbers of 25 and an angle of attack of 30°. The configuration of the double ellipse is defined by:

$$\begin{aligned} &(\frac{x}{6})^2 + (\frac{y}{1.5})^2 = 1 && x<0, \\ &(\frac{x}{3.5})^2 + (\frac{y}{2.5})^2 = 1 && x<0 , y>0, \\ &x = 2.5 , y = -1.5 && x>0 \end{aligned} \tag{23}$$

Numerical results for inviscid and viscous hypersonic equilibrium flows are shown in *Fig.*1 which displays the iso-Mach number lines for the flow over the double ellipse at M_∞= 25 and $\alpha = 30^\circ$. The free stream conditions are chosen corresponding to the standard atmosphere at 75 *km* (P_∞ =2.52Pa and T_∞ =205.3K). The local Mach numbers are estimated by local velocities and frozen sound speeds. The figure demonstrates the reduction of the stand-off distance of the shock wave compared with the perfect gas flows. *Fig.*2 shows comparison of the computed temperature distribution along the surface for the non-equilibrium gas. The present result for 17 reaction model coincides with those of the other contributers. The reduction of temperature is evident for the case of non-equilibrium gas.

5. Conclusions

The rational Runge-Kutta scheme for the compressible Navier-Stokes equations is applied to the simulation of hypersonic real gas flows for reentry problems. The central difference scheme with the artificial dissipation models gives the converged solutions even at a high Mach number of 25. The facilities of the rational Runge-Kutta scheme for chemically non-equilibrium reactive flows are also demonstrated. The rational Runge-Kutta scheme is confirmed to be reliable even for the analyses of hypersonic reacting flows.

References

[1] Satofuka, N., Morinishi, K., and Nishida ,Y.," Numerical Solution of Two-Dimensional Compreessible Navier-Stokes Equations Using Rational Runge-Kutte Method ", Note on Numerical Fluid Mechanics ,Vol.18 ,Vieweg, pp. 201- 218 (1987)

[2] Wambecq, A., "Rational Runge-Kutta Methods for Solving Systems of Ordinary Differential Equations", Computing 20, pp. 333-342 (1978)

[3] Srinivasan, S., Tannehill, J.C., and Weilmuenster, K.J. , "Simplified curve fits for the thermodynamic properties of equilibrium air", Iowa State University Engineering Research Institute Project 1626 (1986)

[4] Park, C., "On the Convergence of Chemically Reacting Flows", AIAA Paper 85-0247 (1985)

[5] Jameson, A. and Baker, T.J., "Solution of the Euler Equations for Complex Configurations", AIAA Paper 83-1929 (1983)

[6] Yee, H.C., "Upwind and Symmetric Shock-Capturing Schemes", NASA TM 89464 (1987)

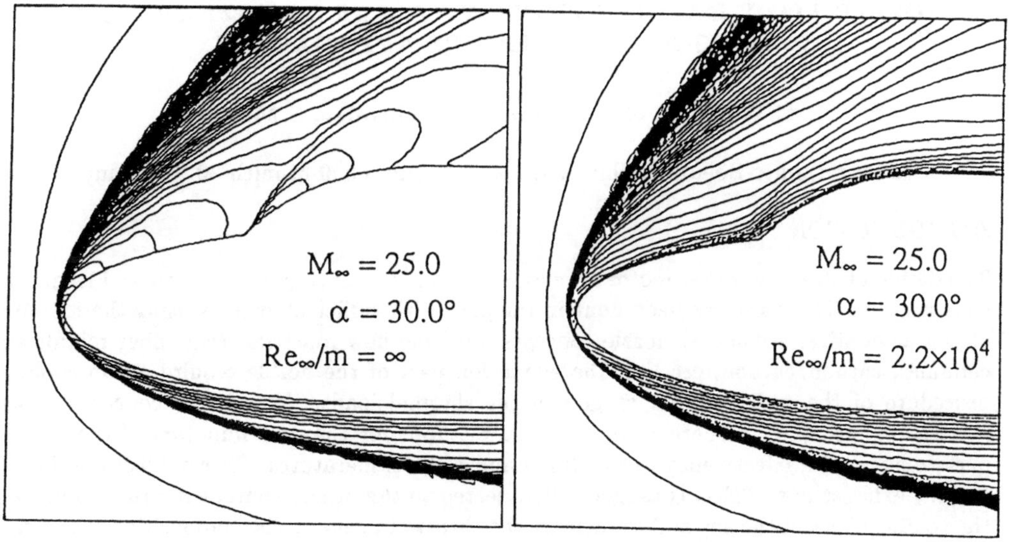

a) inviscid flow

b) viscous flow

*Fig.*1 Iso-value contours for Mach Number, $\Delta M = 0.2$ (equilibrium)

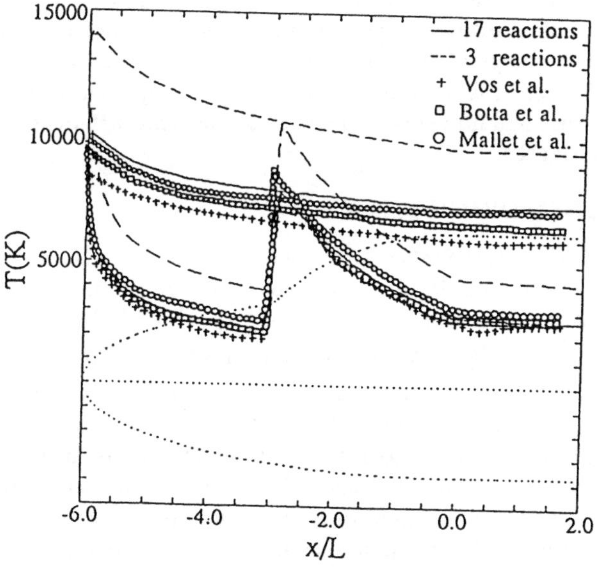

*Fig.*2 Temperature distribution
(non-equilibrium)

NOZZLE FLOW CALCULATIONS WITH GAS INJECTIONS USING INVISCID AND VISCOUS APPROACHES

C.Weiland, G.Hartmann, W.Schröder, S.Menne

Messerschmitt-Bölkow-Blohm AG, Space Group, 8000 Munich 80, Germany

INTRODUCTION

The design of classical rocket motor nozzles is optimized with respect to thrust and length of construction. This task has been done in the past with sufficient success using the method of characteristics. Advanced nozzle concepts applying new materials (e.g. fiber reinforced ceramics, carbon/carbon, etc.) in the expansion part of the nozzle require a film cooling procedure of the wall in order to govern the thermal loads. These new concepts make it necessary to apply modern computational methods for the determination of the nozzle performance, film effectiveness, heat fluxes and wall temperatures. To establish the film a turbine exhaust gas (TEG) is tangentially injected to the nozzle contour in order to utilize the cooling effect of the gas and to increase the thrust. In general the composition of the gas coming from the combustion chamber and the turbine exhaust gas are completely different. Due to the high temperatures of the gases real gas effects have to be taken into account either in thermodynamic equilibrium or non-equilibrium .If different gases interact with each other the numerical description of such flow fields is complicated. In an inviscid approach (solution of the Euler equations) a contact discontinuity occurs whose location is a proiri not known. One way to overcome this problem consists in establishing a fitting procedure. This strategy will be followed here and examples will be discussed in this paper.
However in reality there exist a mixing zone (viscous approach) whose size and extension depends on the turbulent diffusion and the mass diffusion. Since the behaviour of this zone determines the effectiveness of the film cooling it is highly important to predict this zone very carefully. A Navier-Stokes method is extended and adjusted to this kind of flows where the influence on thrust and film effectiveness of laminar and turbulent boundary and mixing layers will be investigated.

THERMODYNAMIC FORMULATION

The Euler [1-3] and Navier-Stokes [4-6] methods are formulated for perfect gas, real gas in thermodynamic equilibrium and non-equilibrium as well. For equilibrium real gas the equation of state in the forms $p(\rho,\varepsilon)$ and $T(\rho,\varepsilon)$ and the derivatives $(\frac{\partial p}{\partial \rho})_\varepsilon, (\frac{\partial p}{\partial \varepsilon})_\rho, (\frac{\partial T}{\partial \rho})_\varepsilon, (\frac{\partial T}{\partial \varepsilon})_\rho$ are needed and implemented in the Euler [1] and Navier-Stokes [5] methods, where p, ρ, ε and T are pressure, density, internal energy and temperature. For air the equation of state can be taken from Mollier-fit routines. For other mixtures of gases, for example hydrogen/oxygen, the equation of state is calculated by the computer code of Gorden & McBride [7] which deals with the equilibrium chemical kinetics of combustion. For gases in chemical non-equilibrium the governing equations have to be completely rearranged, adding the continuity equations for the species concentrations and formulating the relations for the transport mechanisms (viscosity, heat conduction, thermal and mass diffusions). Up to now Park's chemical model of air is used, while the implementation of a suitable hydrogen/oxygen model is underway. The inviscid method is described in [3] and the viscous one in [6].

FLUIDDYNAMIC FORMULATION

Euler Equations

The basic equations governing the inviscid flowfield describe the conservation of mass, momentum and energy. This system is closed by suitable relations between the thermodynamic variables pressure, density, temperature, species concentrations and internal energy. For discretization purposes the basic equations are reformulated quasi-conservatively, which means that a quasi-linear form using conservative variables is employed. The Jacobian matrices are split due to an eigenvalue evaluation (split-matrix method) and are discretized by a finite difference approach. A third-order upwind biased difference formula is used in space. The time integration is carried out by a three step Runge-Kutta scheme which leads to a second-order accurate scheme. Discontinuities like bow shocks and shear layers can be fitted. This makes it necessary to consider time dependent grids which move with the discontinuities during the transient to steady state (for steady freestream and boundary conditions) [1-3].

Navier-Stokes Equations

For perfect and equilibrium real gas applications a finite volume approach of the Navier-stokes equations is used. The inviscid fluxes are discretised by a symmetric TVD scheme with Roe's approximate Riemann solver, while a centered differencing is employed for the viscous fluxes. The final set of algebraic equations is solved by a symmetric point Gauss-Seidel relaxation scheme [5]. The non-equilibrium Navier-Stokes method is based on the Euler scheme of Ref.[3] where the related viscous terms are added [6].

Boundary Conditions

For subsonic and supersonic inflow and outflow boundaries characteristic compatibility relations based on one-dimensional characteristic equations are employed. Usually, near the wall boundary (inviscid flow) the flow is directed towards the wall and consequently all characteristics except one are utilized together with the kinematic wall condition for the determination of the variables at the wall.
As mentioned earlier in case of interaction of various gases with different compositions and thermodynamic states contact surfaces or shear layers arise (inviscid flow) which separate the regions of different gases. The location of this contact surface is a priori not known and is part of the solution. For the calculations (inviscid) discussed in this paper the contact discontinuity is fitted by aligning a coordinate surface to the discontinuity using the suitable one-dimensional characteristic equations on both sides and utilizing the condition that the pressure has to be continuous. This procedure is described in detail in [2].
For viscous flows the usual wall boundary conditions are used. At inflow boundaries suitable boundary layer profiles are prescribed.

COMPARISON OF RESULTS

Euler solutions for nozzle flows with TEG injection are shown in Figs.1 and 2. A hydrogen/oxygen gas mixture for both the main stream ($mix(LOX/H_2) = 7.09$) and the injected stream ($mix(LOX/H_2) = 0.93$) was considered. If the static pressure of the main stream coincides with the one of the injected gas (at the location of injection) the flow is called matched, otherwise unmatched. Fig. 1 shows the matched case and Fig. 2 the unmatched case where the static pressure of the injected gas has twice the value of the combustor gas. In the latter case a pressure wave arises which can clearly be identified, and the location of the shear layer differs from that of the matched case. The influence of chemical non-equilibrium effects compared to equilibrium effects are considered in Figs. 3-6. The Mach number patterns show only small differences but the evaluation of the species concentrations reveals that in the equilibrium case after a modest distance from the throat oxygen and nitrogen are completely recombined and NO does not exist, while in non-equilibrium even at the nozzle exit the gas is still in non-equilibrium (Figs. 5-6). This leads to a considerable reduction in thrust ($F_{equilibrium} = 855[KN]$, $F_{non-equilibrium} = 836[KN]$). Navier-stokes solutions (equilibrium real gas,air) for laminar and turbulent boundary layers are considered in Figs. 7-8. As expected the boundary is thicker in the turbulent case but this affects only a little the Mach number distribution outside the boundary layer and the thrust decreases from $F_{laminar} = 822[KN]$ to $F_{turbulent} = 818[KN]$. Viscous solutions are shown for the nozzle flow with turbine exhaust gases in the Figs.9-13. The total temperature for the combustor gas was $T_o = 3800K$ and for the TEG gas $T_o = 600K$. Perfect gas is assumed and the effect of either laminar or turbulent (Baldwin-Lomax model) flow on the development of the mixing layer is investigated. Figs. 9-10 show Mach number distributions for both cases. In the turbulent case a well developed mixing zone can be identified which touches the nozzle wall and leads to an increase of the adiabatic wall temperature (Fig.13). On the contrary there is only a narrow extension of the laminar mixing zone (Fig.12). The increase of the temperature distribution at the wall beyond the location of the TEG injection is a measure for the film effectiveness. The prediction of this temperature distribution is sensitive to the turbulent model applied, the consideration of mass diffusion and the high temperature effects. These phenomena will be treated in the future.

REFERENCES

[1] *Pfitzner M., Weiland C.:*
AGARD-CP 428, Paper No.22, Bristol (1987).
[2] *Weiland C., Hartmann G., Menne S.:*
AGARD-CP 510, Paper No.32, San Antonio (1991).
[3] *Pfitzner M.:*
AIAA Paper 91-1467, (1991)
[4] *Schröder W., Hartmann G.:*
Computer and Fluids, Vol 21, (1992),pp.109-132
[5] *Schröder W., Hartmann G.:*
To appear in: Notes on Numerical Fluid Mechanics, Vieweg, (1993)
[6] *Menne S., Weiland C., Pfitzner M.:*
AIAA Paper 92-2876, Nashville, (1992)
[7] *Gordon S., McBride B.J.:*
NASA SP- 273, (1976)

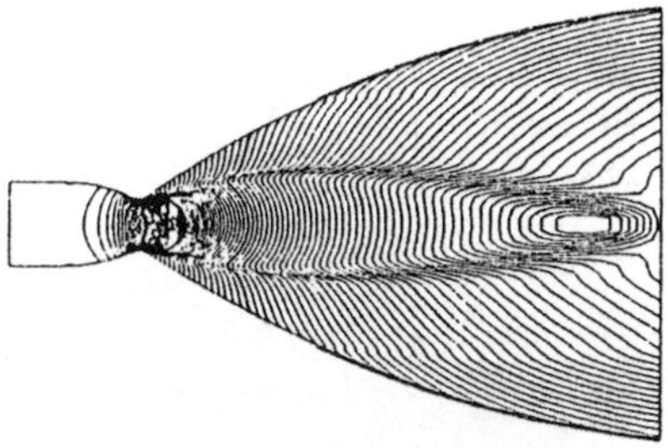

FIG. 1 Euler solution equilibrium real gas; combustion gas ($Mix(LOX/H_2) = 7.09$) ; TEG injection ($Mix(LOX/H_2) = 0.93$); injection matched pressure $1 \cdot 10^5[Pa]$

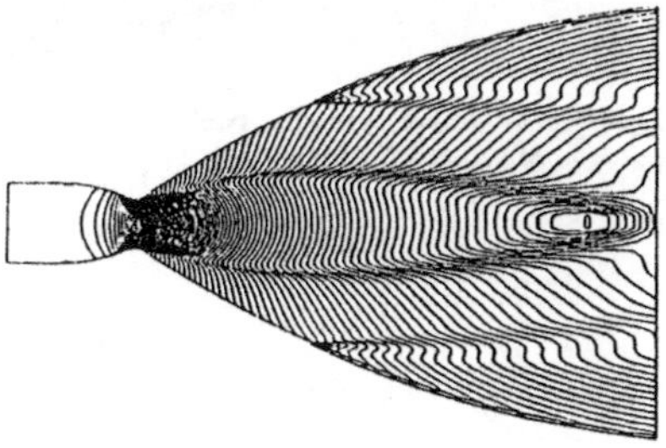

FIG. 2 Euler solution equilibrium real gas; combustion gas ($Mix(LOX/H_2) = 7.09$); TEG injection ($Mix(LOX/H_2) = 0.93$); injection mismatched pressure $2 \cdot 10^5[Pa]$

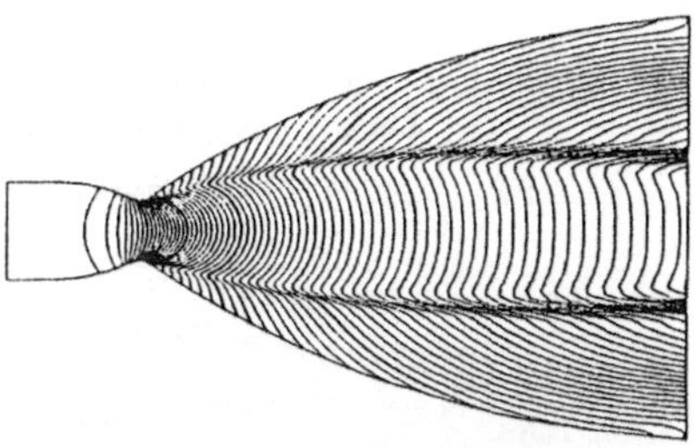

FIG. 3 Euler equilibrium real gas, air flow, Mach number contours $M_{Min} = 0.231$, $M_{Max} = 7.01$, $\Delta M = 0.1$

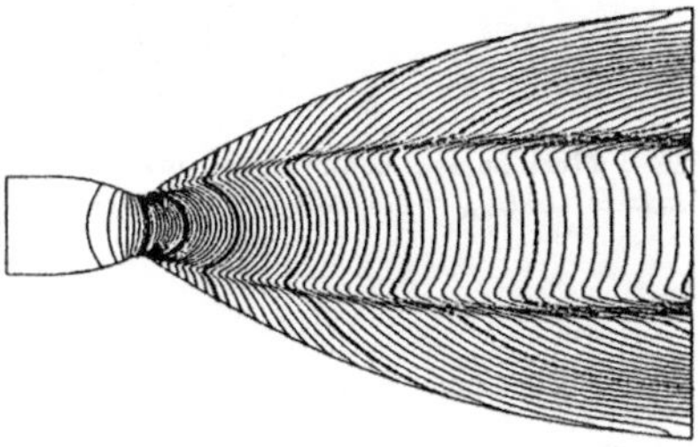

FIG. 4 Euler non-equilibrium real gas, air flow, Mach number contours $M_{Min} = 0.225$, $M_{Max} = 7.06$, $\Delta M = 0.1$

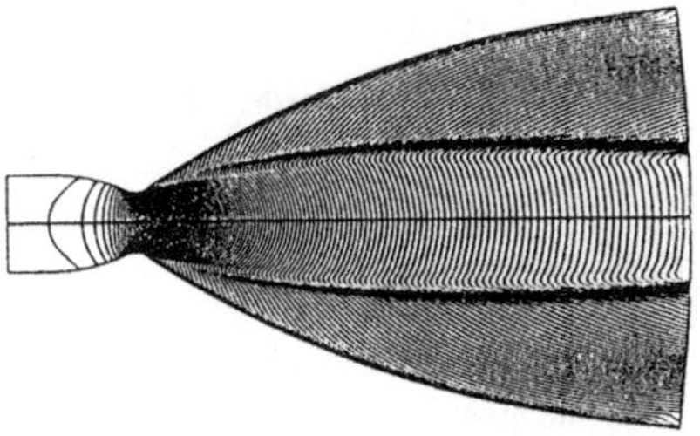

FIG. 7 Navier-Stokes solution; real gas equilibrium; laminar Mach number contours $M_{Min} = 0.$, $M_{Max} = 7.0$, $\Delta M = 0.05$ (120 * 190) cells

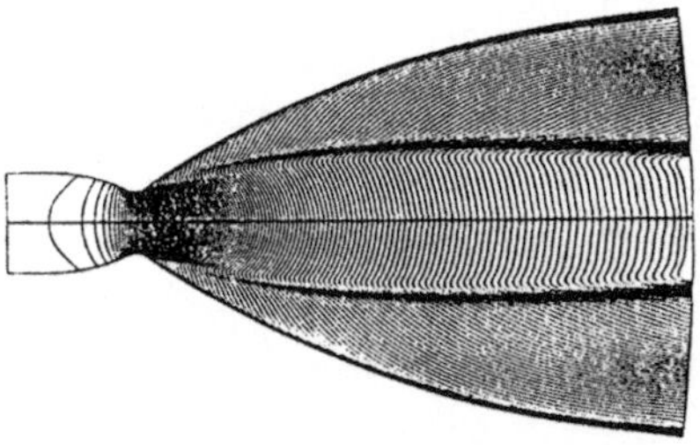

FIG. 8 Navier-Stokes solution; real gas equilibrium; turbulent Mach number contours $M_{Min} = 0.$, $M_{Max} = 7.0$, $\Delta M = 0.05$ (120 * 190) cells

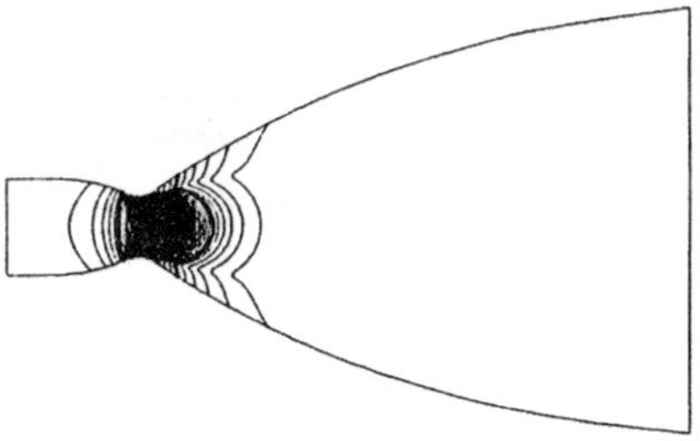

FIG. 5 Euler equilibrium real gas ,air flow, Y_{O_2} concentration $Y_{O_2}^{Min} = 0.192$, $Y_{O_2}^{Max} = 0.233$, $\Delta Y_{O_2} = 0.0005$

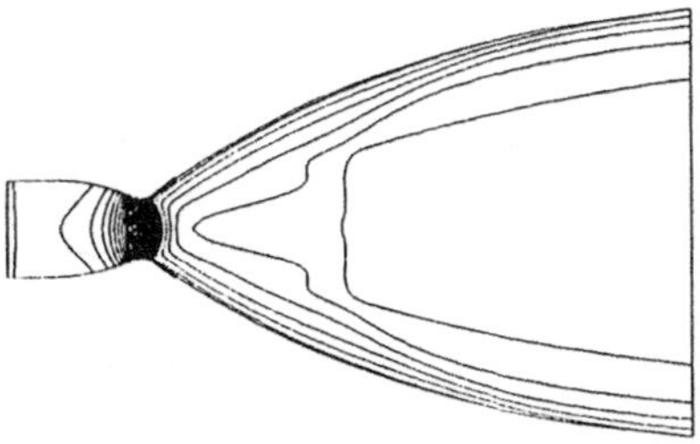

FIG. 6 Euler non-equilibrium real gas ,air flow, Y_{O_2} concentration $Y_{O_2}^{Min} = 0.191$, $Y_{O_2}^{Max} = 0.208$, $\Delta Y_{O_2} = 0.0005$

FIGS. 9 - 14
Nozzle flows with turbine exhaust gas injection (TEG)
perfect gas, $\gamma = 1.17$, adiabatic, Navier-Stokes, $Re = 5.26 * 10^6$

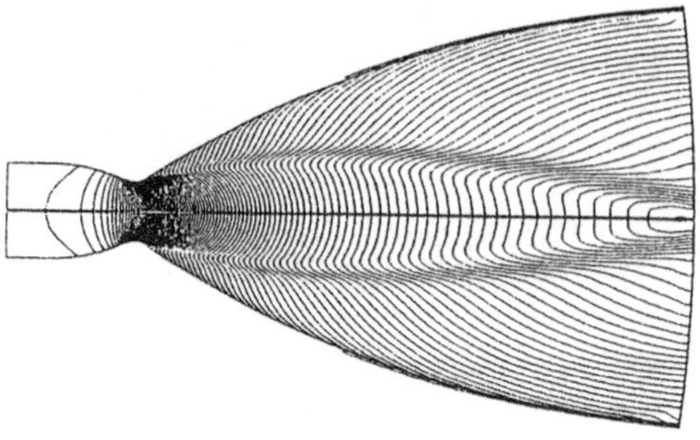

FIG. 9 Mach number contours, laminar
$M_{min} = 0., M_{max} = 5.5, \Delta M = 0.05$

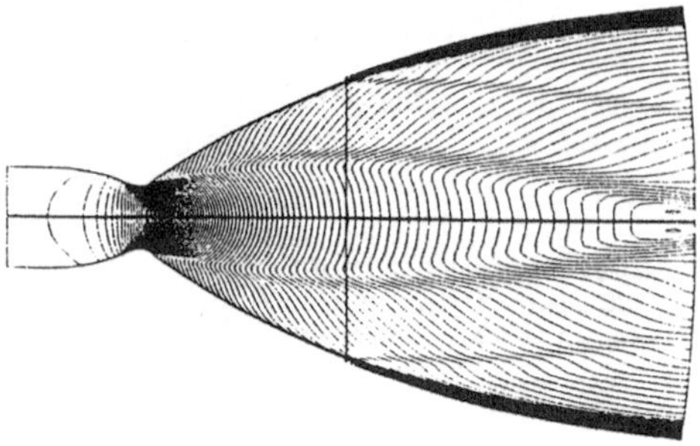

FIG. 10 Mach number contours, turbulent
$M_{min} = 0., M_{max} = 5.5, \Delta M = 0.05$

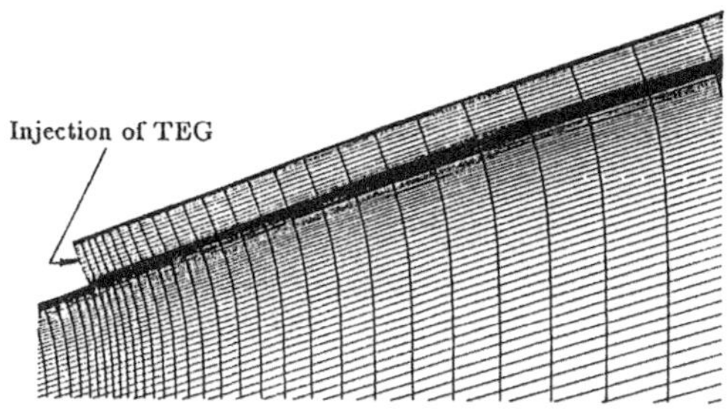

FIG. 11 Detail of computational grid;
2 blocks; 200 * 120 and 79 * 50 cells

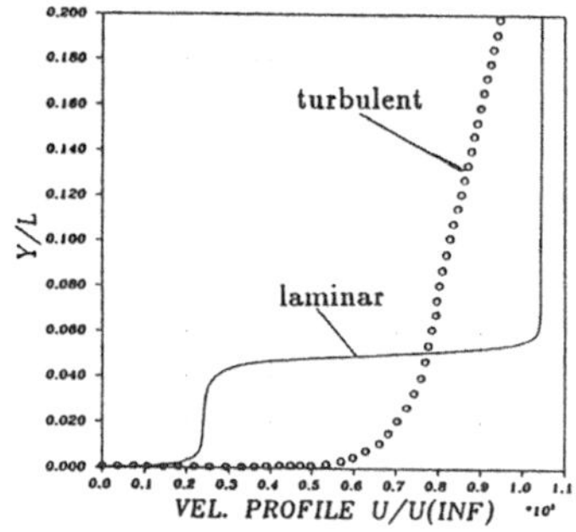

FIG. 12 u/u_∞ profile at nozzle exit cross section
comparison laminar - turbulent

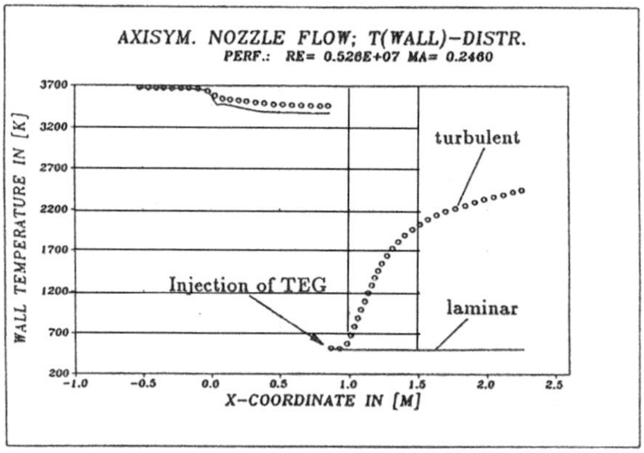

FIG. 13 Temperature distribution at the wall
comparison laminar - turbulent

A GODUNOV-TYPE SCHEME USING STREAMLINE MESHES FOR STEADY SUPERSONIC AERODYNAMICS

J. Y. Yang and C. A. Hsu

Institute of Applied Mechanics, National Taiwan University
Taipei, Taiwan, R.O.C.

1. Introduction

In this paper, motivated by the recent work on new Lagrangian method developed by Hui and Loh [1,2], we describe a streamline Godunov method for solving the steady Euler equations in supersonic flow in the Eulerian formulation. In contrast to the conventional Eulerian formulation, in [1] a stream function and a new Lagrangian time are used as the independent variables and the steady Euler equations can be cast into hyperbolic conservation law form in this new Lagrangian formulation. Thus, many well-established shock capturing schemes such as Godunov method [3] can be employed to solve them. This newly developed Lagrangian method is noticeable for its crisp resolution of the slip lines when compared with results obtained using Eulerian formulation by Glaz and Wardlaw [4]. A flux difference splitting method has also been given by Pandolfi [5] for solving steady supersonic flows. In [4] and [5], regular meshes in the Cartesian coordinates was used. In search of the reason why the Godunov method in Eulerian formulation does not contain the good property of slip line resolution as in the new Lagrangian method, we found that by using streamline mesh system one can share this desirable features of the new Lagrangian method with Eulerian formulation. Our work is akin to [4] but with the streamline and a time-like coordinate as the control volumes. The control volumes or cells which are composed of streamlines and constant x lines are constructed during each marching step and used as computational meshes. These streamline meshes are expected to share the Lagrangian features of the adaptive flow nature and have the advantage of simplifying the implementation of the Godunov scheme. A second order extension using a MUSCL-type approach [6] is also presented. The essentially nonoscillatory (ENO) interpolation of Harten and Osher [7] is used for the top and bottom states at the cell interfaces to achieve higher order representations.

2. Hyperbolic Conservation Equations of Steady Supersonic Flows

The Euler equations of motion of an inviscid, non-heat conducting fluid for steady two-dimensional flows in conservation form can be written as

$$E_x + F_y = 0 \tag{1}$$

where $E = (\rho u, \rho u^2 + p, \rho uv, \rho uH)^T$ and $F = (\rho v, \rho uv, \rho v^2 + p, \rho vH)^T$ are flux vectors. Here, x and y are the physical coordinates, u and v are the corresponding components of the velocity, ρ is the density, p is the pressure, and H is the total enthalpy per unit mass. For a polytropic gas, the pressure is related to other fluid properties by the equation of state $p = \rho(\gamma - 1)(H - \frac{1}{2}q^2)/\gamma$, where $q^2 = u^2 + v^2$ and γ is the ratio of specific heats.

For supersonic flow, the system (1) is hyperbolic and has eigenvalues

$$\lambda_0 = \frac{v}{u} \quad \text{multiplicity 2,} \quad \lambda_\pm = \frac{\mathrm{uv} \pm \mathrm{a}\sqrt{\mathrm{q}^2 - \mathrm{a}^2}}{\mathrm{u}^2 - \mathrm{a}^2} \tag{2}$$

where $a = (\gamma p/\rho)^{1/2}$ is the speed of sound. A characteristic whose slope is λ_0 is a streamline, while one whose slope is $\lambda_\pm$ is a Mach line.

Along Mach lines $C_\pm : \frac{dy}{dx} = \tan(\theta \pm \mu)$

$$\frac{dp}{\rho q^2 \tan\mu} \pm d\theta = 0 \tag{3a,b}$$

Along streamlines $C_0 : \frac{dy}{dx} = \tan\theta$

$$\frac{d\rho}{dp} = \frac{1}{a^2}, \quad \frac{dq^2}{dp} = -\frac{2}{\rho} \tag{3c, d}$$

where θ denotes the angle between the streamlines and the free stream direction x and μ the local Mach angle $\sin^{-1}(1/M)$. The Mach number M is defined by $M = q/a$. For a perfect gas, it can be explicitly integrated to obtain the Riemann invariants in terms of the Prandtl-Meyer function. The simplicity of form of the characteristic equations as expressed in terms of dp and $d\theta$ is to be emphasized. That the quantities p and θ should be considered as the basic variables is related to the fact that it is these quantities which must be matched across a slip line. For a given supersonic flow, if the streamlines can be used as coordinate lines, then the description of flow physics becomes rather simple and natural. One such example is the particularly elegant form of equations in natural coordinates. In the present work, we shall take advantage of this observation and using streamlines and constant coordinate x lines as grid lines for the control volumes or cells used in the Godunov method. This will not only simplify the Godunov algorithm but also enable the excellent resolution of slip lines without smearing.

3. The Riemann Problem of Steady Supersonic Flow

The Riemann problem for two-dimensional steady supersonic flow is specified as the system (1) of hyperbolic conservation laws subject to the initial condition

$$Q(x = x_0, y) = \begin{cases} Q_T, y > y_m \\ Q_B, y \le y_m \end{cases} \tag{4}$$

where the flow states are given at the initial data line $x = x_0$ and Q_T and Q_B denote the top and bottom states.

The solution to the Riemann problem is self-similar in the variable $(y - y_m)/(x - x_0)$ and consists of three types of elementary waves, namely, the oblique shock waves (*sh*), the Prandtl-Meyer expansions (*P-M*) and the slip lines (*sl*) separated by uniform regions in between. Denote the solution as $R[(y - y_m)/(x - x_0); Q_B, Q_T]$. These elementary waves can be used to construct a solution to the Riemann problem for two-dimensional flow. For details, see [4] and [1].

4. A Godunov Method Using Streamline Meshes

The computational mesh is illustrated in Fig. 1. The marching direction is x and the mesh is oriented such that $x_j^n = constant = x^n$ for all j. Here n and j refer to the marching step number and cell number, respectively. The marching distance, $\Delta x^n = x^{n+1} - x^n$, is chosen to satisfy the usual Courant-Friedrichs-Lewy (CFL) linear stability condition. The computational domain in $x - y$ plane is divided into a system of control volumes or cells which in the y-direction are centered at y_j^n and have a height of $\Delta y_j^n = y_{j+1/2}^n - y_{j-1/2}^n$.

The difference equations for the jth zone are obtained by integrating Eq.(1) over the jth cell in Fig. 1 and applying the divergence theorem to get

$$E_j^{n+1} = E_j^n \frac{\Delta y_j^n}{\Delta y_j^{n+1}} - \frac{\Delta x^n}{\Delta y_j^{n+1}}(G_{j+1/2}^N - G_{j-1/2}^N), \tag{5}$$

where the numerical flux vector $G_{j+1/2}^N$ is defined as

$$G_{j+1/2}^N = F_{j+1/2}^{n+1/2} - s_{j+1/2}^n \, E_{j+1/2}^{n+1/2}. \tag{6}$$

Here for any quantity f,

$$f_j^n = \frac{1}{\Delta y_j^n} \int_{y_{j-1/2}^n}^{y_{j+1/2}^n} f(x^n, y)\, dy, \tag{7}$$

is the cell-average of f and

$$f_{j+1/2}^{n+1/2} = \frac{1}{\Delta x^n} \int_{x^n}^{x^{n+1}} f(x, y_{j+1/2}^n + s_{j+1/2}^n \cdot (x - x^n))\, dx, \tag{8}$$

where

$$s_{j+1/2}^n = (y_{j+1/2}^{n+1} - y_{j+1/2}^n)/\Delta x^n. \tag{9}$$

In the first-order Godunov scheme the jth cell average, Q_j^n, at marching step n is considered as constant within that cell and the flux $F_{j+1/2}^{n+1/2} - s_{j+1/2}^n E_{j+1/2}^{n+1/2}$ along the interface between the jth cell and the $(j+1)$th cell from marching step n to $n+1$ is to be obtained from the self-similar solution $R[(y - y_{j+1/2}^n)/(x - x_j^n); Q_j^n, Q_{j+1}^n]$ at $y = y_{j+1/2}^n$ to the Riemann problem formed by two adjacent piecewise constant flow states Q_j^n and Q_{j+1}^n, i.e., with $Q_j^n = Q_B$ and $Q_{j+1}^n = Q_T$ and use the procedure described in Section 3.

The first order Godunov scheme given above can be upgraded to higher order accuracy. Here, we present a new MUSCL-type [6] second order scheme using an essentially nonoscillatory (ENO) interpolation [7] to achieve high order accurate values of the cell interface (see also [8]). The solution algorithm may be divided into a number of steps. These are

Step 1. The "decoding" of E_j^n to obtain Q_j^n. Let

$$\chi^2 = \gamma^2 + (\gamma^2 - 1)(E_3^2 - 2E_1E_4)/E_2^2$$

Then

$$u = \frac{\gamma + \chi}{\gamma + 1}\frac{E_2}{E_1}, \; v = E_3/E_1, \; \rho = E_1/u, \; p = E_2 - uE_1. \tag{10}$$

Step 2. Generation of streamline meshes. To start the calculation, we need grid points at step $n+1$ to build up computational cells. These are carried out by defining

$$x_j^{n+1} = x_j^n + \Delta x^n, \; y_j^{n+1} = y_j^n + \Delta x^n \frac{v_j^n}{u_j^n}, \tag{11}$$

$$s_{j+1/2}^n = (\frac{v_j^n}{u_j^n} + \frac{v_{j+1}^n}{u_{j+1}^n})/2. \tag{12}$$

Step 3. Second order accurate representation of values of cell interfaces. We first use ENO interpolation to get more accurate values at interfaces as

$$E_{j+\frac{1}{2}}^{n-} = E_j^n + \frac{1}{2}\Psi_j^n, \; E_{j-\frac{1}{2}}^{n+} = E_j^n - \frac{1}{2}\Psi_j^n, \tag{13}$$

with

$$\Psi_j^n = m[\Delta_+ E_j^n - \varsigma\overline{m}(\Delta_-\Delta_+ E_j^n, \Delta_+\Delta_+ E_j^n), \Delta_- E_j^n + \varsigma\overline{m}(\Delta_+\Delta_- E_j^n, \Delta_-\Delta_- E_j^n)] \tag{14}$$

where $\Delta_\pm E_j = \pm(E_{j\pm 1} - E_j)$, denote the usual forward and backward difference operators. The limiter functions m and $\overline{m}$ are defined in [7]. Then we advance $E_{j\mp 1/2}^{n\pm}$ to step $n+1/2$ as follows:

$$\overline{E}_{j+\frac{1}{2}}^{-} = E_{j+\frac{1}{2}}^{n-} - \frac{\Delta x}{2\Delta y_j^n}[F(E_{j+\frac{1}{2}}^{n-}) - F(E_{j-\frac{1}{2}}^{n+})], \tag{15a}$$

$$\overline{E}_{j-\frac{1}{2}}^{+} = E_{j-\frac{1}{2}}^{n+} - \frac{\Delta x}{2\Delta y_j^n}[F(E_{j+\frac{1}{2}}^{n-} - F(E_{j-\frac{1}{2}}^{n+})]. \tag{15b}$$

If $\varsigma = 0$, one has a second order TVD scheme [9], and if $\varsigma = 1/2$, one has a uniformly second order ENO scheme [7]. With $\overline{E}^{\mp}_{j\pm 1/2}$, one can decode it and obtain $\overline{Q}^{\mp}_{j\pm 1/2}$.

Step 4. Obtain the self-similar solution $R[(y - y^n_{j+1/2})/(x - x^n_{j+1/2}); \overline{Q}^{+}_{j+1/2}, \overline{Q}^{-}_{j+1/2}]$ by solving the Riemann problem. The solution state is used to evaluate the interface fluxes $E^{n+1/2}_{j+1/2}$ and $F^{n+1/2}_{j+1/2}$ along $(y - y^n_{j+1/2})/(x - x^n_{j+1/2}) = s^n_{j+1/2}$. Finally, the solutions at step $n+1$ can be obtained from Eq. (5).

At this stage the numerical procedure is completed. To march forward further in x, one goes back to *Step 1* and repeats *Step 1-4*.

5. Numerical Results

We first consider an initial value Riemann problem. It is formed by the confluence of two parallel supersonic streams at $y_0 = 0.5$ with different states. The initial conditions are given by

$$Q = \begin{cases} Q_T = (\rho_T, p_T, M_T, \theta_T) = (0.5, 0.25, 4.0, 0.0^\circ), y > y_0, \\ Q_B = (\rho_B, p_B, M_B, \theta_B) = (1.0, 1.0, 2.4, 0.0^\circ), y \le y_0. \end{cases}$$

This is a case considered by Glaz and Wardlaw [4]. Here we use 100 cells and $\Delta y_{init} = 0.01$. Numerical results using Cartesian and streamline meshes for the flow properties at section $A - A'$ ($x = 0.5$) are shown in Fig. 2 along with the exact solution (solid line) which was obtained using the Riemann solver. Both mesh systems indicate good resolutions of the oblique shock, but the streamline mesh system has much better resolution of the slip line. This may be attributed to the streamline meshes, which follow the movement of the fluid particles.

We also consider supersonic flows over two-dimensional profiles. In Fig. 3, the flow fields around a symmetric circular-arc airfoil in a hypersonic flow with $M_\infty = 8.0$ and 0° angle of attack are shown. The leading edge and trailing edge oblique shock waves and the Mach waves emanating from the surface are all crisply resolved. The grid lines (particle traces) clearly indicates that the computational mesh follows exactly the particle movement even when it crosses a shock wave. The initial mesh distribution is stretched with $\Delta y_{min} = 0.001$.

Computed results for supersonic flow past a symmetric double-wedge airfoil with 10° wedge angle are given in Fig. 4 for freestream Mach number $M_\infty = 3.0$ and 0° angle of attack. The leading edge oblique shocks, the Prandtl-Meyer expansion fans and the trailing edge oblique shocks are all crisply represented, and the flow adaptive feature is easily illuminated via the streamlines (grid lines).

Acknowledgement

This research was supported by the National Science Council of the Republic of China under contract NSC 81-0210-E004-49.

References

[1] C. Y. Loh and W. H. Hui, *J. Comput. Phys.*, Vol. 89, (1990), pp. 207-240.
[2] W. H. Hui and C. Y. Loh, A new Lagrangian method for steady supersonic flow computation, II. Slip-line resolution, to appear in *J. Comput. Phys* (1992).
[3] S. K. Godunov, *Mat. Sbor.* (1959) Vol. 47, pp. 271-306.
[4] H. M. Glaz and A. B. Wardlaw, *J. Comp. Phys.* (1985) Vol. 58, pp. 157-187.
[5] M. Pandolfi, *Computers & Fluids*, (1985) Vol. 13, pp. 37-46.
[6] B. van Leer, *J. Comput. Phys.*, (1979) Vol. 32, pp. 101-136.
[7] A. Harten and S. Osher, *SIAM J. Numer. Analy.*, (1987) Vol. 24, No. 2, pp. 279-309.
[8] J. Y. Yang, C. A. Hsu and W. H. Hui, Solution of the steady Euler equations in a generalized Lagrangian formulation, to appear in *AIAA J.*
[9] A. Harten, *J. Compu. Phys.* (1983) Vol. 49, pp. 357-393.

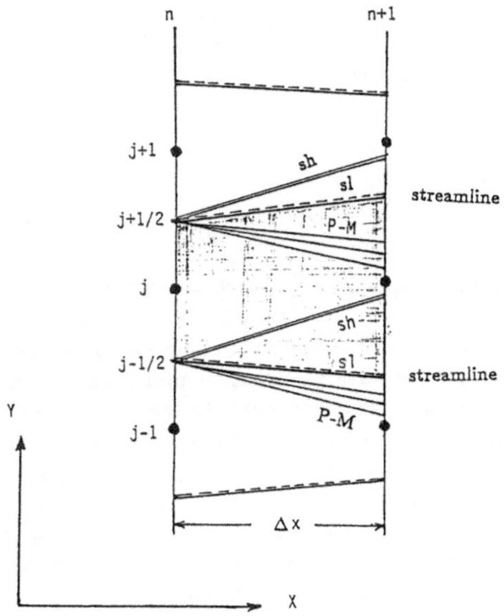

Fig. 1. **Streamline mesh system for Godunov method.**

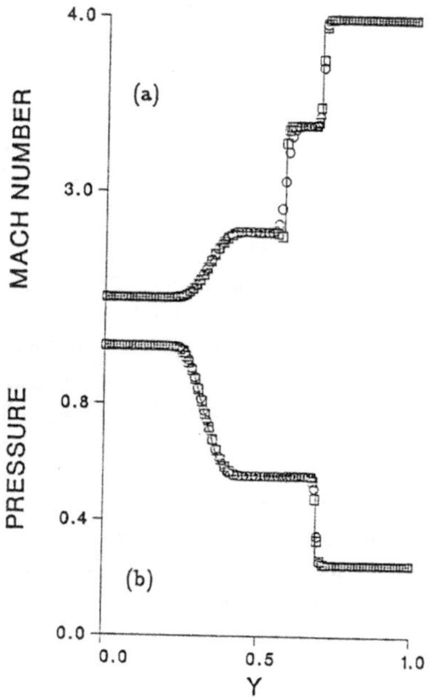

Fig. 2. A comparison of results using streamline mesh (squares) and regular mesh (circles).

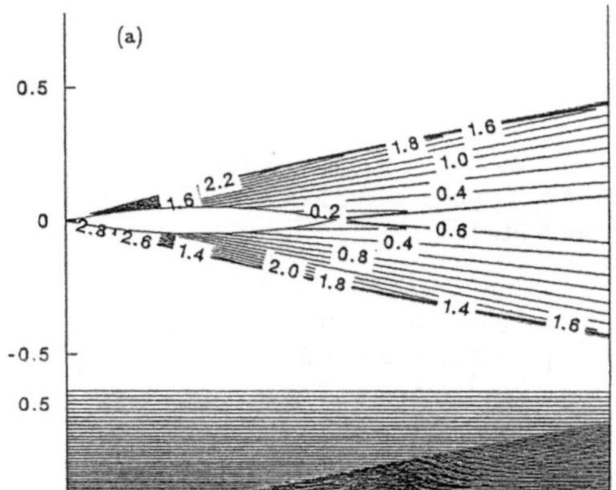

Fig. 3. Steady hypersonic flow past a circular arc airfoil ($M_\infty = 8.0$). (a) pressure contours, (b) grid lines.

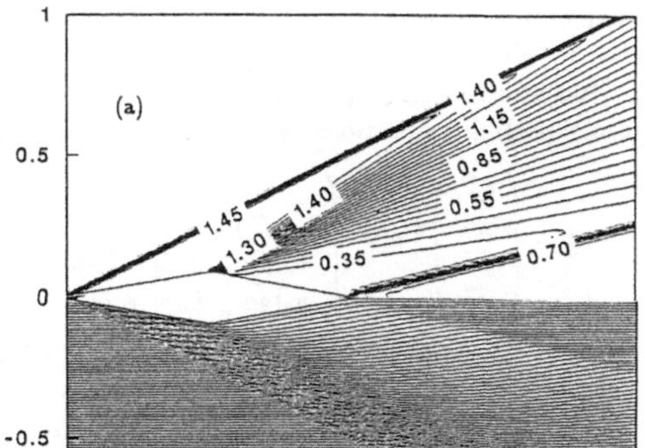

Fig. 4. Steady supersonic flow past a double-wedge airfoil ($M_\infty = 3.0$). (a) pressure contours, (b) grid lines.

A CFD-BASED APPROACH FOR THE SOLUTION OF ACOUSTICS, MAXWELL AND SCHROEDINGER EQUATIONS FOR SCATTERING PROBLEMS [1]

R. K. Agarwal, K. S. Huh and M. Shu
McDonnell Douglas Research Laboratories
St. Louis, Missouri 63166, U.S.A.

INTRODUCTION

It is shown that the acoustics, Maxwell and Schroedinger equations can each be written as a set of first-order partial differential equations in hyperbolic conservation-law form. As a consequence, the well-developed CFD algorithms for the solution of compressible Euler equations can be applied for the numerical solution of these equations. However, for wave propagation problems, the requirements of accuracy are substantially different than that for aerodynamics. In the present work, the governing equations are solved both in time- and frequency-domain using the method of lines which decouples the temporal terms from spatial terms. In frequency domain, a pseudo-time variable is introduced. An explicit node-based finite-volume algorithm is developed wherein the spatial terms are discretized using a fourth-order compact difference operator while the time-integration is carried out using a four-stage explicit/point-implicit Runge-Kutta time-stepping scheme. A sixth-order compact dissipation operator is added to stabilize the algorithm. A novel analytic treatment is developed for the implementation of the far-field radiation boundary condition. It is shown that this boundary condition provides an excellent approximation to exact integral boundary condition.

GOVERNING EQUATIONS

In two dimensions, the acoustics, Maxwell and Schroedinger equations can be expressed in conservation-law form as follow:

$$\frac{\partial \mathbf{Q}}{\partial t} + \frac{\partial \mathbf{F}}{\partial x} + \frac{\partial \mathbf{G}}{\partial y} = \mathbf{S}, \qquad (1)$$

where Q, F, G and S are defined below.

Acoustics Equations

Let (u_0, v_0) denote the steady mean flow velocity components and $\tilde{\rho}$, $\tilde{p}$, $(\tilde{u}, \tilde{v})$ denote the acoustic perturbation density, pressure and velocity components, respectively. Then for the linearized compressible Euler equations (acoustic equations), we can write [1],

$$\mathbf{Q} = \begin{pmatrix} \tilde{\rho} \\ \widetilde{\rho u} \\ \widetilde{\rho v} \\ \widetilde{\rho e} \end{pmatrix}, \mathbf{F} = \begin{pmatrix} \widetilde{\rho u} \\ u_0[2\widetilde{\rho u} - u_0\tilde{\rho}] + \tilde{p} \\ v_0[\widetilde{\rho u} - u_0\tilde{\rho}] + \widetilde{\rho v}u_0 \\ u_0[\widetilde{\rho h} - h_0\tilde{\rho}] + \widetilde{\rho u}h_0 \end{pmatrix}, \mathbf{G} = \begin{pmatrix} \widetilde{\rho v} \\ u_0[\widetilde{\rho v} - v_0\tilde{\rho}] + \widetilde{\rho u}v_0 \\ v_0[2\widetilde{\rho v} - v_0\tilde{\rho}] + \tilde{p} \\ v_0[\widetilde{\rho h} - h_0\tilde{\rho}] + \widetilde{\rho v}h_0 \end{pmatrix}, \mathbf{S} = \mathbf{0}. \qquad (2)$$

A state equation is required for closure and is of the form,

$$\tilde{p} = (\gamma - 1)\left\{\widetilde{\rho e} - \frac{1}{2}\left[u_0(2\widetilde{\rho u} - u_0\tilde{\rho}) + v_0(2\widetilde{\rho v} - v_0\tilde{\rho})\right]\right\}.$$

The variables $\widetilde{\rho u}$, $\widetilde{\rho v}$, $\widetilde{\rho e}$ and $\widetilde{\rho h}$ are defined as,

$$\widetilde{\rho u} = \rho_0\tilde{u} + \tilde{\rho}u_0 \qquad \widetilde{\rho v} = \rho_0\tilde{v} + \tilde{\rho}v_0 \qquad \widetilde{\rho h} = \widetilde{\rho e} + \tilde{p}$$

[1]This research was conducted under the McDonnell Douglas Independent Research and Development Program.

$$\widetilde{\rho e} = \frac{\tilde{p}}{\gamma - 1} + \rho_0 u_0 \tilde{u} + \rho_0 v_0 \tilde{v} + \frac{1}{2}\tilde{\rho}(u_0^2 + v_0^2).$$

Maxwell Equations

Let **D** denote the electric field displacement and **B** the magnetic field induction. For transverse electric (TE) polarization [2],

$$\mathbf{Q} = \begin{pmatrix} B_z \\ D_x \\ D_y \end{pmatrix}, \mathbf{F} = \begin{pmatrix} D_y/\epsilon \\ 0 \\ B_z/\mu \end{pmatrix}, \mathbf{G} = \begin{pmatrix} -D_x/\epsilon \\ -B_z/\mu \\ 0 \end{pmatrix}, \mathbf{S} = \mathbf{0}, \tag{3}$$

where μ and ϵ represent the permeability and permittivity of the homogeneous medium, respectively. Similarly, **Q**, **F**, **G** and **S** can be written for transverse magnetic (TM) polarization.

Schroedinger Equation

For the wave function ψ, the Schroedinger equation can be written as [3], $i\hbar\frac{\partial\psi}{\partial t} = \hat{H}\psi$, where

$\hat{H}$	=	non-relativistic Hamiltonian in Coulomb-variant form			
	=	$\frac{\hat{p}^2}{2m} - \frac{e}{m}(\mathbf{A}\cdot\hat{p}) + \frac{e^2}{2m}\mathbf{A}^2 + U(x,y)$,	e	=	electron charge,
$\hat{p}$	=	momentum operator $= (\hbar/i)\nabla$,	m	=	electron mass,
A	=	vector potential ($\nabla\cdot\mathbf{A} = 0$, homogeneous field),	$\hbar$	=	$h/2\pi$,
U	=	scalar potential,	h	=	Plank's constant.

Letting $\psi = \alpha\exp(i\beta)$, it can be shown after considerable algebraic manipulation that,

$$\mathbf{Q} = \begin{pmatrix} \rho \\ \rho u \\ \rho v \end{pmatrix}, \mathbf{F} = \begin{pmatrix} \rho \\ \rho u^2 \\ \rho v u \end{pmatrix}, \mathbf{G} = \begin{pmatrix} \rho v \\ \rho u v \\ \rho v^2 \end{pmatrix}, \mathbf{S} = -\frac{\rho}{m}\begin{pmatrix} -\frac{e}{\rho}\mathbf{A}\cdot\nabla\rho \\ \hat{U}_x \\ \hat{U}_y \end{pmatrix}, \tag{4}$$

where $\rho = \alpha^2$ =probability density, $\mathbf{V} = (u,v) = (\hbar/m)\nabla\beta$, and $\hat{U} = U - \frac{\hbar^2}{4m\rho}\left[\nabla^2\rho - \frac{(\nabla\rho)^2}{2\rho}\right]$.

Frequency Domain Formulation

Since the governing equations are linear, an assumption that the incident field is harmonic in time with a frequency ω will result in a time-dependent total-field that will also be harmonic with frequency ω. Thus the governing equations can be recast in frequency domain by the use of the single-frequency assumption, $\mathbf{Q} = \Re(\tilde{\mathbf{Q}}(x,y)e^{-i\omega t})$, where tilde denotes a complex quantity and $\Re$ denotes the real part.

By applying the single-frequency assumption, and by recasting the equations in scattered form with pseudo-time difference, the governing equations become,

$$\frac{\partial\tilde{\mathbf{Q}}_s}{\partial t^*} + \frac{\partial\tilde{\mathbf{F}}_s}{\partial x} + \frac{\partial\tilde{\mathbf{G}}_s}{\partial y} - i\omega\tilde{\mathbf{Q}}_s = \tilde{\mathbf{S}} \tag{5}$$

where $\tilde{\mathbf{Q}}$, $\tilde{\mathbf{F}}$ and $\tilde{\mathbf{G}}$ contain the complex coefficients of **Q**, **F** and **G**. The source term $\tilde{\mathbf{S}}$ is defined

$$\tilde{\mathbf{S}} = \mathbf{S} + i\omega\tilde{\mathbf{Q}}_i - \frac{\partial\tilde{\mathbf{F}}_i}{\partial x} - \frac{\partial\tilde{\mathbf{G}}_i}{\partial y}. \tag{6}$$

The total-field is cast as a sum of the known incident value and the scattered value, $\tilde{\mathbf{Q}} = \tilde{\mathbf{Q}}_i + \tilde{\mathbf{Q}}_s$.

Boundary Conditions

For acoustic scattering from a rigid stationary wall, the inviscid flow boundary condition, $\tilde{\mathbf{u}} \cdot \hat{n} = 0$, is applied where $\tilde{\mathbf{u}}$ is composed of both incident and scattered components, and $\hat{n}$ is a unit normal to the surface. In terms of scattered quantities, the boundary conditions becomes $\tilde{\mathbf{u}}_s \cdot \hat{n} = -\tilde{\mathbf{u}}_i \cdot \hat{n}$, where $\tilde{\mathbf{u}}_i$ is the given incident complex velocity.

For electromagnetic scattering from a perfectly conducting surface, the boundary condition is $\hat{n} \times \mathbf{E} = 0$ where $\mathbf{E}$ is the total electric field intensity. In terms of scattered quantities in the frequency domain, the boundary conditions becomes $\hat{n} \times \tilde{\mathbf{E}}_s = -\hat{n} \times \tilde{\mathbf{E}}_i$, where $\tilde{\mathbf{E}}_i$ is the known incident electric field intensity.

For the scattering of a beam of electrons in the spherically symmetric potential field $U(r)$ of an atom, the boundary condition is $\rho(r = 0) = 0$.

Far Field Conditions

A correct mathematical model would extend the domain to infinity and would require all scattered quantities to vanish. This is impractical numerically, and a set of well–posed boundary conditions must be placed on a finite domain.

For acoustic and electromagnetic scattering, the current methodology uses the radiation boundary conditions based on that of Giles [4].

The boundary condition is, in theory, accurate to $O(l/\omega)^4$ where l is the Fourier decomposed wave number in the azimuthal direction; it has been found to be more effective than the commonly used one–dimensional radiation condition or the characteristic condition.

For the Schroedinger equation, the conditions $V_r(r_0, \theta) = \sqrt{2E_k}$, and $V_\theta(r_0, \theta) = 0$ are applied at the outer boundary of the computational domain, where E_k is the kinetic energy of the scattered beam of electrons, and r_0 is the range of the atomic field.

NUMERICAL METHOD

Spatial Discretization

The set of weakly conservative governing equations (1) or (5) is solved numerically using a finite–volume node-based scheme [1]. The physical domain in (x, y) is mapped to the computational domain (ξ, η) to allow for arbitrarily shaped bodies. The semi-discrete form of the method can be written as,

$$\frac{d}{dt} J\mathbf{Q}_{ij} + \frac{\mu_\eta^2 \mu_\xi \delta_\xi}{1 + \delta_\xi^2/6} \mathbf{F}'_{ij} + \mu_\xi^2 \mu_\eta \delta_\eta \mathbf{G}'_{ij} - i\omega J\mathbf{Q}_{ij} = J\mathbf{S}_{ij} + \mathbf{D}_{ij}. \tag{7}$$

where $\mathbf{D}$ is the added dissipation term defined as,

$$\mathbf{D}_{ij} = \nu_6 \delta_\xi (\frac{J}{\Delta \bar{t}} \delta_\xi^5) \mathbf{Q} - \nu_4 \delta_\eta (\frac{J}{\Delta \bar{t}} \delta_\eta^3) \mathbf{Q}$$

and $\bar{t}$ is the time step when the CFL number, σ_i, is 1. $\Delta\xi$ and $\Delta\eta$ are defined to be 1, and hence have been omitted. The vectors $\mathbf{F}'$ and $\mathbf{G}'$ are the curvilinear flux vectors and are defined as, $\mathbf{F}' = \mathbf{F}y_\eta - \mathbf{G}x_\eta$, and $\mathbf{G}' = -\mathbf{F}y_\xi + \mathbf{G}x_\xi$. J is the determinant of the metric tensor, $J = x_\xi y_\eta - x_\eta y_\xi$. δ and μ are the standard difference and averaging operators. The discretized form of equation (1), is the same as (7) except that $-i\omega J\mathbf{Q}_{ij}$ is dropped and $\tilde{\mathbf{S}}$ is replaced by $\mathbf{S}$.

The spatial discretization in η direction is second-order accurate for a smooth grid, the discretization in ξ direction, however, is based on the classical Padé scheme and is fourth–order accurate. In practice, the accuracy of the Padé scheme has been shown to be significantly higher than that of the standard central differencing [1, 2]. The resolution of second–order central differencing with 20 points per wavelength is matched by compact differencing with roughly 8 points per wavelength.

Time Integration

A four stage Runge-Kutta scheme is used to integrate the governing equations (1) in real time and (5) in the pseudo-time plane. The integration method is point implicit for the real–time term $-i\omega\mathbf{Q}$ to alleviate the stiffness for large values of ω [2].

For the current two-dimensional system of equations with dissipation, the time integration is computed as follows,

$$\begin{aligned} \mathbf{Q}_{ij}^{0} &= \mathbf{Q}_{ij}^{n} \\ s_k(\mathbf{Q}_{ij}^{k} - \mathbf{Q}_{ij}^{0}) &= -\alpha^k \frac{\Delta t_{ij}}{J}(\mathbf{R}_{ij}^{k-1} - \mathbf{D}_{ij}^{0}) \qquad k = 1,4 \\ \mathbf{Q}_{ij}^{n+1} &= \mathbf{Q}_{ij}^{4} \end{aligned} \tag{8}$$

where $s_i = 1 + \alpha^i\omega\Delta t$ and $\mathbf{R}_{ij}$ is the discretized form of the sum of the spatial fluxes, real time term and source term. $\mathbf{D}_{ij}$ is the dissipation term described previously. A single evaluation of the dissipative terms is required and the scheme is stable provided that the CFL number, σ_i, does not exceed $2\sqrt{2}$ regardless of the size of ω [2].

RESULTS

First, the accuracy and efficiency of the algorithm was evaluated by computing the acoustic radiation from an oscillating cylinder. It was found that the compact scheme requires eight grid points per wavelength in contrast to the second–order scheme which requires 20 grid points for comparable accuracy. As shown in Fig. 1, the error using second–order central–differencing at roughly 20 points/wavelength ($\log(dx) \approx -3$) is matched by compact differencing at only roughly 8 points/wavelength ($\log(dx) \approx -2$). Furthermore, the explicit/point–implicit algorithm is very efficient. As shown in Fig. 2, in 500 iterations, four orders of magnitude reduction in residual are obtained, versus a single order of magnitude reduction obtained using a four–stage explicit Runge–Kutta scheme. The new farfield boundary treatment reduces the cpu and memory requirements by an order of magnitude. As shown in Fig. 3(a), no numerical reflection was observed in the calculations even by bringing the farfield boundary to a distance of 0.02 diameters from the surface of the cylinder. Accurate solutions were obtained on a 66×10 grid in 24 cpu seconds on a CRAY X-MP with six orders of magnitude reduction in residuals as shown in Fig. 3(b).

Only a few selected results for acoustic and electromagnetic scattering from complex two–dimensional objects are reported here. For the Schroedinger equation, the method has been validated for the case of elastic scattering of a beam of electrons by a hydrogen atom in ground state. For acoustic and EM scattering, incident waves of the form $\tilde{A}e^{ikx}$ were assumed, where $k = \omega/c_0$ and the complex constant $\tilde{A}$ was set to 1. Combined with the time harmonic factor $e^{-i\omega t}$, the incident waves correspond to plane waves which travel in the positive x direction.

Acoustic Scattering by a NACA0012 Airfoil

We consider a NACA0012 airfoil at zero angle of attack in a steady $M_0 = 3.0$ flow. A plane acoustic wave of the form $\tilde{p}_i = e^{i(kx-\omega t)}$ where $k = \omega/[c_0(1 + M_0)]$ impinges on the airfoil. Since the thickness of the airfoil is only 3% of the incident wavelength, the scattering effects were expected to be small. The blunt leading edge accounts for most of the scattering and emulates a compact scatterer radiating sound in all directions. Fig. 4(a) shows contours of the amplitude of acoustic pressure. The region ahead of the airfoil exhibits the interference fringes from the interaction of the incident wave with scattered wave. Fig. 4(b) shows the amplitude of the scattered wave at the farfield using the current method and the integral equation formulation using the Green's function. The agreement is excellent.

TE Scattering by an Ogive

The scattering of a TE electromagnetic wave by a perfectly conducting ogive is computed for $c/\lambda = 8$, and $t/\lambda = 2$, where c and t are the chordlength and thickness of the ogive, respectively. Fig. 5(a) shows lines of constant modulus of the scattered magnetic field intensity. As shown in Fig. 5(b), excellent agreement is obtained for the computed bistatic radar cross section (RCS) between the present method and that obtained using the method of moments (MOM). Again, the farfield boundary is only at a distance of 1.2 times the chord from the center of the ogive.

For several other examples and details, the reader is referred to [1] and [2].

References

[1] Huh, K. S., Agarwal, R. K. and Widnall, S. E., AIAA Paper 90-3920, 1990.

[2] Huh, K. S., Shu, M. and Agarwal, R. K., AIAA Paper 92-0453, 1992.

[3] Landau, L. D. and Lifshitz, E. M., *Quantum Mechanics*, Pergamon Press, 1965.

[4] Giles, M., CFDL-TR-90-1, MIT Department of Aeronautics and Astronautics, 1990.

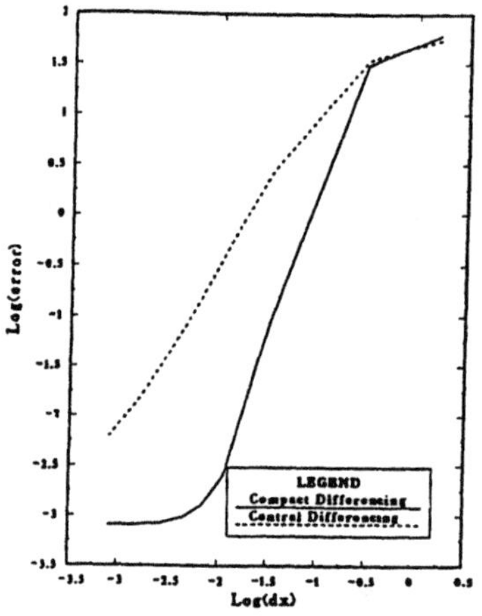

Figure 1: A comparison of the error between second-order and compact fourth-order differencing for the same grid-spacing.

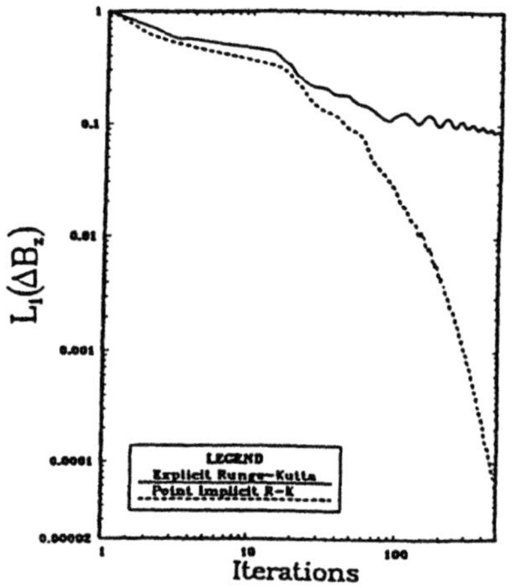

Figure 2: Convergence history for explicit and point-implicit Runge-Kutta time-integration.

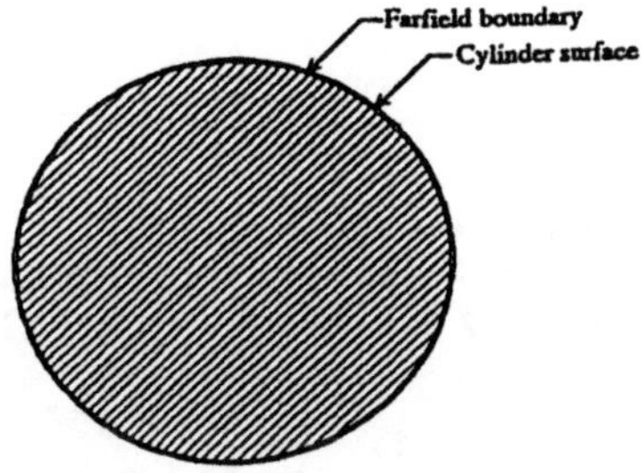

Figure 3: (a) Acoustic radiation from an oscillating cylinder; D=λ, D_0=1.02λ, kD=2π, 65x10 grid.

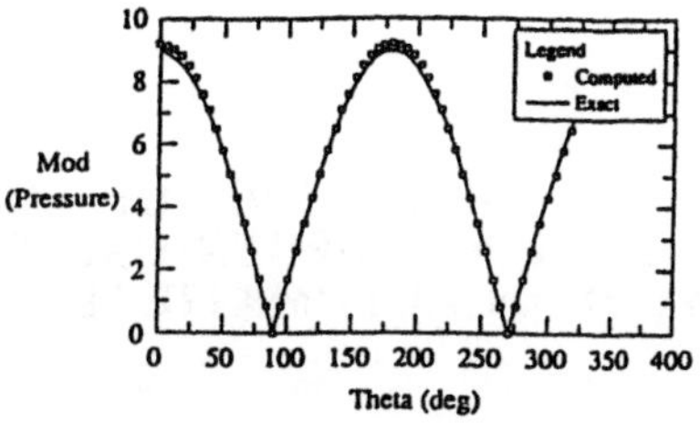

Figure 3: (b) Amplitude of pressure on cylinder, λ=1.

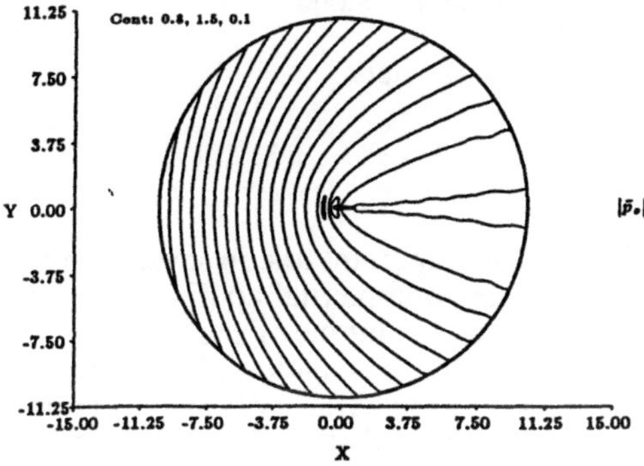

Figure 4: (a) Amplitude of scattered acoustic pressure; ω=2.602, M_0=0.3, 300x150 grid.

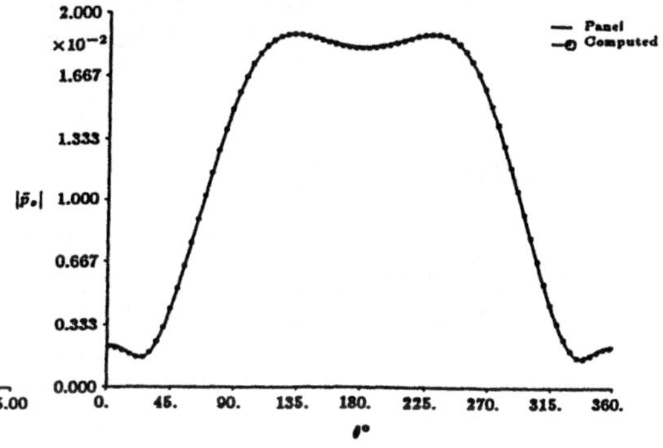

Figure 4: (b) Amplitude of scattered acoustic pressure at far-field; ω=2.602, M_0=0.3, 300x150 grid.

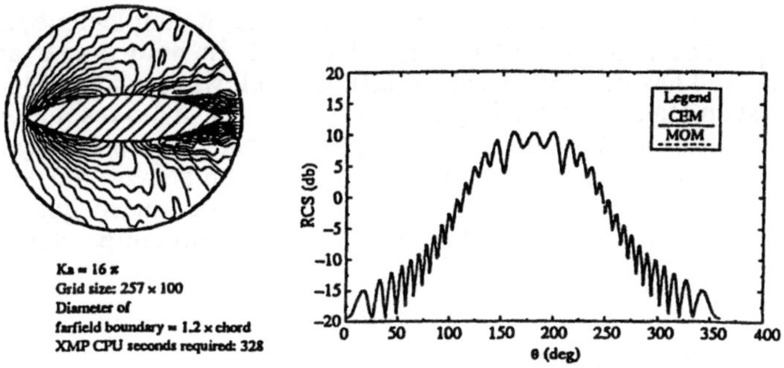

Figure 5: Transverse-Electric electromagnetic scattering by an ogive.

2-D STUDIES OF THE MULTI-WAVELENGTH PLANAR AND SPHERICAL RAYLEIGH-TAYLOR INSTABILITY WITH AN AUTOMATICALLY REZONED LAGRANGIAN CODE

S. Atzeni and A. Guerrieri*

Associazione EURATOM-ENEA sulla Fusione
Centro Ricerche Energia Frascati
C.P. 65 - 00044 Frascati (Roma) - Italy

1. INTRODUCTION

The Rayleigh-Taylor instability (RTI) [1] plays an important role in several physical problems, including explosive stages of stellar evolution and inertial confinement fusion (ICF) [2]. In these cases, understanding the non linear evolution of multi-wavelength perturbation and the induced turbulent mixing [3] is essential. Simulation of such processes requires the capability of following the evolution of multimaterial systems, undergoing strong compression and/or expansion and, of course, being distorted by the instability itself. The substantial volume variation occurring in ICF motivates the preference given to Lagrangian codes for the study of these problems [4], but in multi-dimensional geometry rezoning techniques need being introduced, to deal with the mesh distortion caused by the unstable fluid motion.

In this paper we describe a discrete rezoning technique, its implementation in an ICF code, and its application to the study of the 2-D RTI, in both planar and spherical geometry. The method has been designed to fulfil the following goals: 1) to preserve as much as possible such features of the Lagrangian scheme, as "mass resolution", boundary tracking, low level diffusion; 2) to allow in principle the code to simulate any fluid motion; and 3) to be fully automatic, namely that no user intervention be required while running a problem. The proposed technique has been devised for staggered-mesh, quadrilateral-zone, 2-D Lagrangian codes [5], and has been implemented in a multi-temperature plasma-fluid code for ICF [6].

2. THE REZONING TECHNIQUE

The rezoning technique considered here is a discrete one, i.e. is only performed when the mesh distortion exceeds a given "measure". The criterion for switching on the rezoning is presently as follows. We consider separately each of the two couples of triangles in which a quadrilateral cell is divided by each of the two diagonals. The cell is considered pathological if for at least one of the two couples the ratio y of the areas of the two triangles is outside the range $\varepsilon \leq y \leq 1/\varepsilon$, where ε is a constant smaller than unity (typically $\varepsilon = 0.1$—0.2) When a pathological cell is detected, rezoning is performed in the part of the simulation domain that is affected by some distortion.

---

* ENEA guest, supported by ENEA Contract n. 90/151/96/88.

The new mesh is generated by applying a variational mesh generator separately to each of a pre-determined set of macro-zones, whose boundaries remain Lagrangian (see Fig. 1 and, for details, Ref. 7). Appropriate boundary conditions guarantee continuity between the mesh in adjacent macro-zones. The mesh generator is a modified version [7] of that by Brackbill and Salzmann [8], which yields a mesh, by optimizing between mesh smoothness, volume variation and orthogonality. If a well behaving mesh cannot be generated with the assigned macro-zone boundaries, then the code automatically chooses different macrozone boundaries [7] (see Fig. 1), or couples adjacent macrozones, thus leading sometimes to loss of resolution, but allowing for the simulation of fluid motions with any degree of shear.

The fluid quantities are then mapped from the old mesh to the new one. This is accomplished by using a remapping algorithm, which guarantees conservation of the proper physical quantitites (e.g.: mass for the density remapping). The use of a second-order accurate scheme [9] (with Van Leer's flux limiter) keeps diffusion at very low level and improves substantially upon a previous first order scheme (see [7] and refs. therein). This is particularly important in view of reducing the unphysical mass-diffusion occurring when material interfaces do not coincide with macrozone boundaries. The logics of the remapping process is summarized in Table I, listing the sequence of operations, the mapped quantities and the relevant conserved integral quantities.

TABLE I

Logics of the remapping step of the rezoning process. Here ρ: mass density; $T_{e(i)}$: electron (ion) temperature; T_r: radiation temperature; ρ^*: mass density (on vertices); u_γ: γ-component of the velocity (γ= x,y); E_α: energy density of non-termal alpha-particles; ρ_j: mass density of species j. Primes indicate quantities on the new mesh; the superscrit *prov* indicates a provisional value of the velocity to be corrected to ensure kinetic energy conservation (see main text).

OLD MESH		CONSERVED QUANTITY	NEW MESH	
ρ ↘		mass		↙ ρ'
$T_{e(i)}$ →	$\varepsilon_{e(i)}$	electron (ion) internal energy	$(\varepsilon_{e(i)})'$ →	$(T_{e(i)})'$
T_r		radiation energy		T_r'
ρ^* ↘		mass		↙ $\rho^{*\prime}$
u_γ →	$\rho^* u_\gamma$	$(\text{momentum})_\gamma$	$(\rho^* u_\gamma)'$ →	$(u_\gamma')^{prov}$
E_α		fast α-particle energy		E_α'
ρ_j		$(\text{mass})_j$		ρ_j'

Velocity remapping requires particular care. Because of the staggered-mesh scheme, the use of an auxiliary, vertex-centered mesh is necessary to remap the two momentum components and the kinetic energy. Furthermore, the conflict between energy and momentum conservation is dealt with as follows. First, provisional values of the two components of the velocity on each vertex are computed by remapping the momentum components and by dividing them by the vertex density (in turn obtained by an additional remapping process on the auxiliary mesh). Then, these values are "corrected" so as to ensure kinetic energy conservation, while distributing the relative momentum error in equal quantity between the two velocity components. Temperature remapping is also non trivial, requiring the inversion of the material equation of state (in general, highly non linear)

An illustration of the method and of its quality is given by Fig. 1, showing the mesh and the mass density contours before and after the rezoning, in a typical ICF implosion simulation.

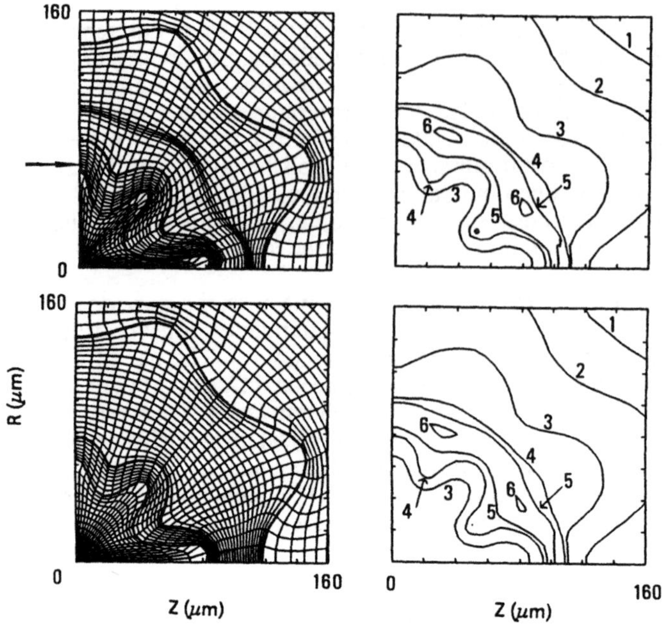

Fig.1: Mesh rezoning during a simulation of the implosion of a gas filled ICF shell. Mesh (left) and iso-density contours (right) are shown before (upper row) and after the rezoning. Thicker curves mark the boundaries of the macrozones (notice that one of these boundaries is changed during the rezoning process; the arrow points to the Lagrangian line which becomes a new macrozone boundary).
The iso-density labels indicate: 1) ρ = 10 g/cm^3; 2) 20 g/cm^3; 3) 40 g/cm^3; 4) 80 g/cm^3; 5) 160 g/cm^3; 6) 320 g/cm^3.

3. RAYLEIGH-TAYLOR INSTABILITY SIMULATIONS

In this section we present applications of the code to the study of an ICF spherical shell implosion, for which a Lagrangian scheme is best suited, and then to a classical, Cartesian RTI problem, showing the capabilities of the code also in problems usually studied by Eulerian codes.

In ICF a thin shell is imploded by the pressure exerted as a consequence of the absorption of laser light, of thermal radiation, or of particle beams (e.g. heavy ion beams). At the collapse, when the shell radius is a small fraction of the initial radius, some inner layer of the shell becomes unstable. Evidence for this process was already given previously [10], referring to single wavelength perturbations. Due to the increased accuracy of the rezoning scheme, the study is now being extended to multiple wavelength perturbations. In fig. 2 we show the case of a gas filled shell, with initial radius r_0 = 2 mm and thickness Δr_0 = 0.2 mm, driven by a pressure pulse, with spatial non-uniformity in the spherical modes l = 4 and l = 12, with peak-to-valley amplitudes C_4 = 1.4% and C_{12} = 0.3%, respectively. The seed for the instability is provided by the mass distribution disturbance associated to the shell distortion caused by the pressure non-uniformity. The faster growth of the l=12 mode is apparent, in agreement with theory.

The complexity of the physical model of the full code, however, does not allow for the use of meshes suitable for studying modes with l > 16. Therefore, mode-coupling processes and turbulent mixing cannot be observed.

Addressing such a problem requires the capability of resolving spatial scales differing by at least one order of magnitude. In particular, it is of interest the case in which the initial perturbation consists of a large number of small amplitude, short wavelength modes (with the possible addition of a few modes with a much longer wavelength). This mandates the use of very large meshes and therefore of a simplified physical model and "experimental set-up". We have then simulated the RTI of superposed ideal gases in planar geometry, considering multi-wavelength and random initial perturbations [11]. Meshes with up to 10^5 mesh points have been used, and instrumental to the success of the simulations has been the control of numerical diffusion at very low levels.

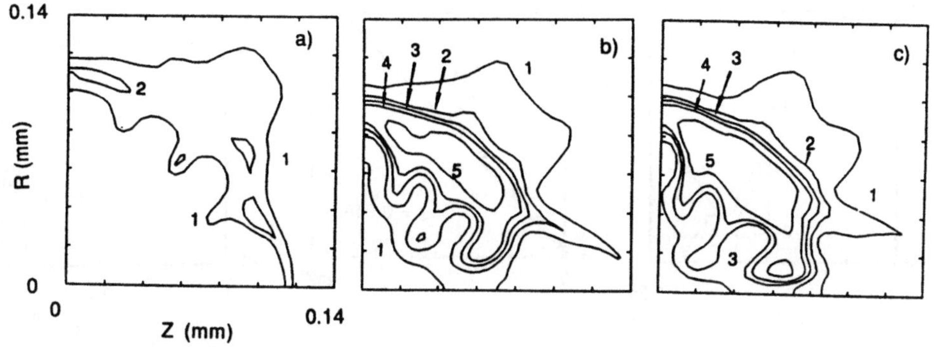

Fig.2. RTI at the collapse of an ICF target, shown by the density contours (ρ) at a sequence of three time levels. Labels on the curves: 1) ρ = 50 g/cm^3; 2) 100 g/cm^3; 3) 150 g/cm^3; 4) 200 g/cm^3; 5) 300 g/cm^3.

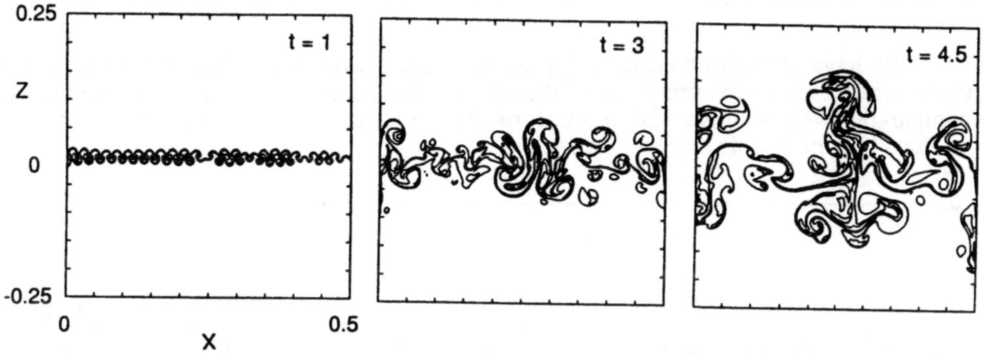

Fig. 3. Multiple wavelength RTI (see main text). Contours of constant volume fractions f (f = 0.25; f = 0.5; f = 0.75) of the ligther fluid are shown (times and lengths scaled as described in the main text).

An example of such a problem is shown in Figs. 3 and 4. Here superposed fluids with density ratio 2:1, in a square box, are considered. The mesh is 200 x 200. The initial perturbation consists in the sum of 19 cosine modes, with mode number n in the range n = 41—59, with random amplitudes, and phase $\psi = 0$ or $\psi = \pi$..[the size of the box is 1/2 the wavelength of the mode n = 1, chosen as the unit length; scales are chosen so as to make equal to unity the linear growth rate of the mode n = 1. The total initial perturbation amplitude is about 10^{-3}]. Fig. 4 shows the occurrence of bubble merging, after saturation of the shortest wavelength modes. This gives rise to larger and larger bubbles, whose maximum size is at any time of the same order of the depth of the mixed layer at the same time. The mixing is also evidenced by the plots of the fluid fractions vs z (the direction parallel to the gravity) at different times, reported in Fig. 4a. Once the shortest wavelength modes saturate, at a time which we take as t = 0, the height of the fastest rising bubbles (of the order of one half of the size of the mixing layer), grows quadratically with time, while the turbulent energy grows with the fourth power of time (see Fig. 4b). It is also found from spectral analysis that the wavelength of the dominant mode grows quadratically with time, in agreement with the observed relationship between bubble size and mixed layer, and with the law for the growth of the mixed layer. Figure 4c shows the pseudo 1-D energy spectra at t = 0 and at t = 4.5, clearly evidencing the emergency of low n modes as a result of non-linear mode coupling.

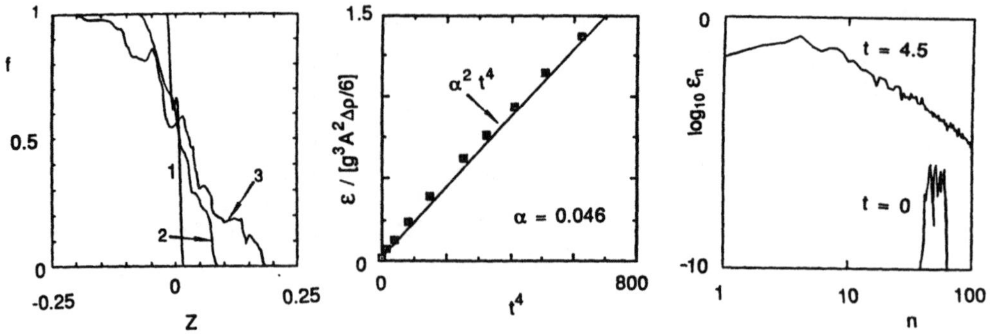

Fig. 4. Same problem as in Fig. 3. a) lighter fluid volume fractions averaged over horizontal layers, f, vs vertical coordinate z, at different times: 1) $t = 1$; 2) $t = 3$; 3) $t = 4.5$; b) time evolution of the turbulent kinetic energy per unit area ε (here A is the Atwood number, g is the gravity, and $\Delta\rho$ is the density difference between the two fluids); c) vertically averaged 1-D kinetic energy spectrum (arbitrary units)

We have also observed (see Fig. 5) that large wavelength, "isolated" modes grow undisturbed even in the presence of short scale turbulence, at least until bubbles of size comparable with a substantial fraction of the wavelength of the large scale mode are produced by the bubble merging process.

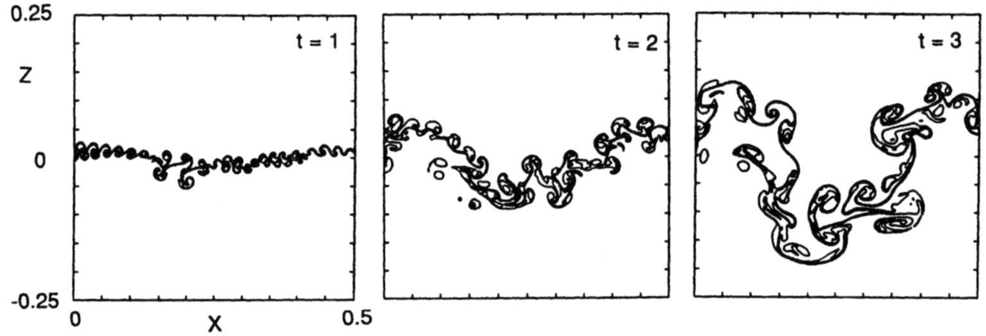

Fig. 5. Same as Fig. 3, but with an additional long wavelength perturbation ($n = 2$; initial amplitude 10^{-2})

REFERENCES

[1] G.I. Taylor, *Proc. Royal Soc. (London)* **201A**, 192 (1950); for a review see D.H. Sharp, *Physica D* **12**, 3 (1984)

[2] for a review see H.J. Kull, *Phys. Rep.* **206**, 197 (1991).

[3] D.L.Youngs, *Physica D* **12**, 19 (1984); *Physica D* **37**, 270 (1989); K.I.Read, *Physica D* **12**, 45, (1984).

[4] S. Atzeni, *Plasma Physics Controll. Fusion* **29**, 1535 (1987).

[5] W.D. Schulz, in *Methods in Computational Physics* (B. Alder et al, Eds.), vol. 3, p.1, Academic Press, New York (1964).

[6] S. Atzeni, *Computer Phys. Commun.* **43**, 107 (1986).

[7] S. Atzeni and A. Guerrieri, *Laser Part. Beams* **9**, 443 (1991).

[8] J.U. Brackbill and J.S. Saltzmann, *J. Comput. Phys.* **46**, 342 (1982).

[9] J.D. Ramshaw, *J. Comput. Phys.* **67**, 214 (1986); J.K. Dukowicz and J.K.Kodis, *SIAM J.Sci.Stat.Comp.*, **8**, 305 (1987).

[10] S. Atzeni, *Laser Part. Beams*, **9**, 233 (1991); *Part. Accel.* **37-38**, 495 (1992).

[11] S. Atzeni and A. Guerrieri, to be published.

MIXED-CONVECTION OF ROTATING FLUIDS IN SPHERICAL ANNULI*

P. Bar-Yoseph, G. Even-Sturlesi, A. Arkadyev, A. Solan, K.G. Roesner†

Faculty of Mechanical Engineering, Technion, Haifa 32000, Israel.
†Institut für Mechanik, Technische Hochschule Darmstadt, 6100 Darmstadt, Germany.

The laminar, axisymmetric flow field of a rotating Boussinesq fluid contained within concentric spherical annuli is investigated for the case where the outer spherical shell is stationary at a given temperature while the inner sphere rotates about a vertical axis in a gravity field and is kept at a lower temperature. The characteristic dimensionless parameters of the problem are: Reynolds number (Re), Rayleigh number (Ra), Froude number (Fr), Eckert number (Ec), Prandtl number (Pr), and the gap width (s). The magnitude of the effect of the density variations on the rotational acceleration terms (Coriolis and centrifugal) is a function of the ratio of the gravitational acceleration to the rotational acceleration of the inner sphere. The ratio of the velocity induced by rotation to that induced thermally is given by Re^2Pr/Ra.

The Galerkin finite element method has been used for the numerical solution of the rotationally symmetric momentum, continuity, and energy equations in terms of the primitive variables. Biquadratic Lagrangian interpolation was employed for the velocity field and bilinear interpolation for the pressure and temperature fields. Several numerical experiments were performed comparing mixed and penalty formulations, each of which confirmed the computational efficiency of the penalty approach over the mixed formulation. Results are therefore reported for the penalty Galerkin finite element formulation. This formulation yields a system of nonlinear algebraic equations which can be written as

$$\mathbf{F}(\mathbf{w},\boldsymbol{\lambda}) = \mathbf{0}, \tag{1}$$

where $\mathbf{w}$ is the N-dimensional vector that consists of the nodal values of the velocity, pressure and temperature fields, $\boldsymbol{\lambda} = (Re,Ra,Fr,Ec,Pr,s,\beta)$ is the parameter set and $\beta = \Omega_o/\Omega_i$ (Ω_o and Ω_i are respectively the angular velocities of the outer and inner spheres). In the present study $\beta=0$, $s=0.15$ or $s=0.5$ and $Ec=0$. The solution of Eq. (1) is a seven-parameter surface in $\mathbf{R}^{N+7}$. Solutions by continuation Newton algorithms are used to trace branches in the parameter space along which lie steady flow states.

A zeroth order continuation in Re and Ra with step size control followed the solution branch from a given unique solution state. At each continuation step, a nonlinear equation set was solved by a Newton iteration, and the approximation to the solution at each iteration was found by a modified vectorized version of the frontal solver for in-core solution of large, sparse nonsymmetric systems of the linearized set of equations. The method was an extension of that used by the authors in previous work [1]. As a test of the present method, solutions were found for mixed-convection in a rotating fluid in cylindrical annuli (finite and infinitely long) and were compared with published results (de Vahl Davis and Hessami [2]), showing very good agreement.

---

*Work supported by the Fund for Basic Research, Israel Academy of Sciences, Deutsche Forschungsgemeinschaft, by the Endowment of the Y. Winograd Chair in Fluid Mechanics and Heat Transfer and by the Center for Absorption in Science, Israel Ministry of Immigrant Absorption.

In the present paper we focus on the formation of various secondary flow patterns in the meridional plane (the flow is assumed to be rotationally symmetric). Such patterns are modifications of the Taylor vortex flow and polar vortex breakdown investigated by the authors in previous work [1,3] or a completely new wide gap instability originated at the south pole. The emphasis is on the combined effect of vertical-gravitational and rotational body forces in a fluid with thermal gradient. The flow regimes were investigated for different Ra and Re numbers ($0 \leq Re \leq 1500, 0 \leq Ra \leq 5 \cdot 10^5$) with the other parameters fixed. The range of investigation covers the subcritical region in which the flow is stable and uniquely determined, as well as the supercritical region where multiple solutions exist. Comparison of the present results with thermal effect with previously published results for incompressible flow shows the following:

(a) For incompressible flow in a narrow gap (s=0.15) at Re=1500 Bartels [4] found three symmetric steady Taylor vortex flow modes near the equator (base flow, one pair of vortices and two pairs of vortices), depending on the acceleration of the inner sphere from rest to the final steady rotation or on an artificial temporary perturbation of equatorial symmetry. Our results for Re=1500 with thermal effects e.g., at $Ra=3.75 \times 10^5$ show, first, that a new non-symmetric flow mode with **three** pairs of vortices can exist. Secondly, transition between different flow modes which can exist at a given value of Re and Ra can be achieved by reaching that value by different paths in the Re-Ra plane (Figs. 1 and 2). In particular, by following path D to its end point Re=1500, Ra=0, we obtain by passing through a thermal perturbation the **one-pure** state which could previously be obtained only by an artifical perturbation

(b) For incompressible flow in a wide gap (s=0.5) our previous results [3] show that a vortex breakdown can occur at the polar axis for Re>3300 while below Re≅3300 the flow is the simple two-cell base recirculation. With thermal effect, vortex breakdown can occur from Re>1500 [5]. This vortex breakdown is a stable and unique solution at each parameter set. The present results with thermal effects show that in a range of Re below that corresponding to vortex breakdown, a flow instability leading to multiple solutions can occur. For example, in the region near Re=100 and $Ra=1.2 \times 10^5$ **five** different flow patterns were found, depending on the path in the Re-Ra plane leading to the point under consideration. These patterns contain **even** or **odd** numbers of recirculation cells, extending over the whole meridional cross-section (Figs. 3 and 4).

The effect of Froude number (Fr) on the flow field has been investigated in the supercritical region for $10^{-3} \leq Fr \leq 10$ and was found to be negligible there.

References

[1] Bar-Yoseph, P., Solan, A., Hillen, R. and Roesner, K.G., Taylor vortex flow between eccentric coaxial rotation spheres, Phys. Fluids A, **2**, 1990, 1564-1573.

[2] de Vahl Davis, G. and Hessami, M.A., Mixed convection in vertical cylindrical annuli, Int. J. Heat and Mass Transfer, **30**, 1987, 151-164.

[3] Bar-Yoseph, P., Roesner, K.G. and Solan, A., Vortex breakdown in the polar region between rotating spheres, Phys. Fluids, A, (to appear).

[4] Bartels, F., Taylor vortices between two concentric rotating spheres, J. Fluid Mech.,**119**, 1982, 1-25.

[5] Arkadyev, A., Bar-Yoseph, P. Solan, A. and Roesner, K.G., Thermal effects on axisymmetric vortex breakdown in spherical gap, Phys. Fluids A, (submitted).

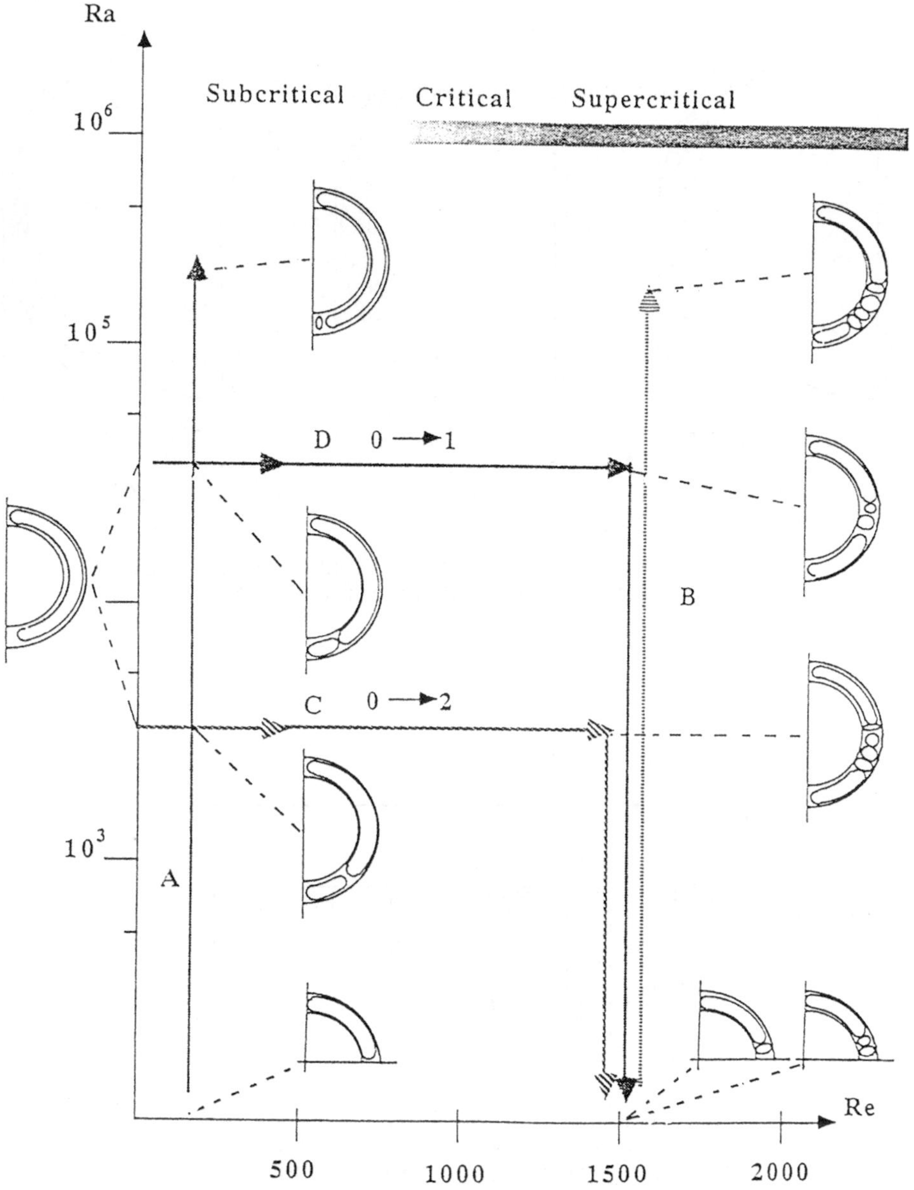

Fig. 1: Flow regimes map in the {Ra;Re} plane.
Path **A**: Re = 200, 0<Ra<10^6. Path **B**: Re=1500, 0<Ra<10^6
Path **C**: Ra=$5 \cdot 10^3$, 0<Re<1500, then Re=1500 Ra → 0
Path **D**: Ra=$5 \cdot 10^4$, 0<Re<1500, then Re=1500 Ra → 0

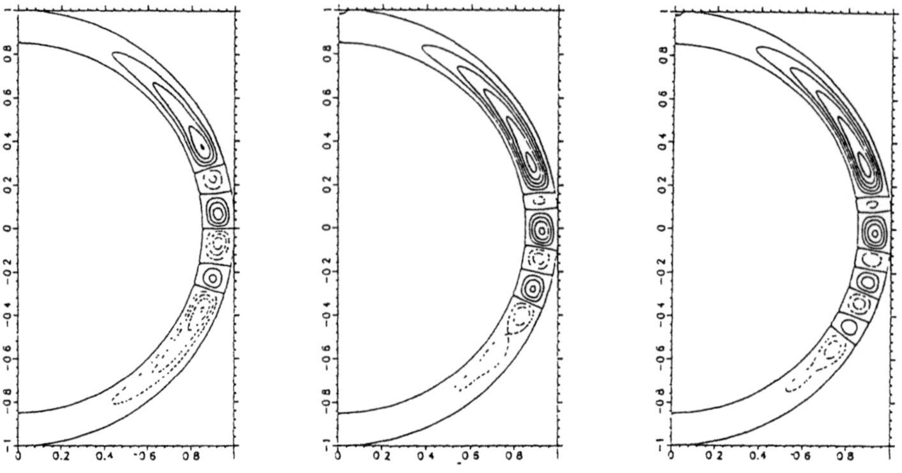

(a) (b) (c)

Fig. 2: Streamlines for path B (Re=1500, Fr=0.01, s=0.15)
(a) Ra=0, (b) Ra=350000 (c) Ra=375000.

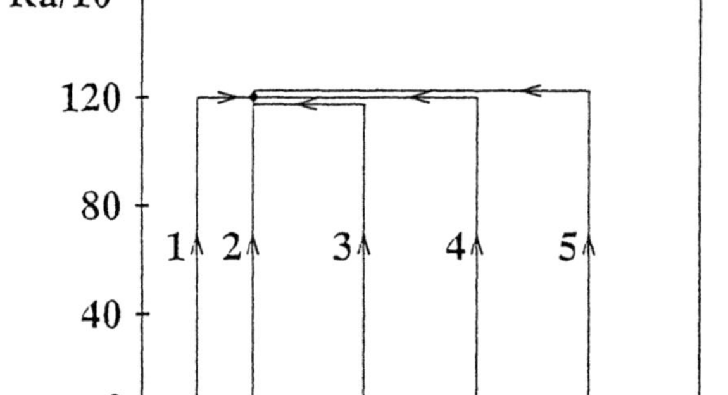

Fig. 3: Computational paths for different flow modes.
Numbers correspond to the numbers of modes in Fig. 4.

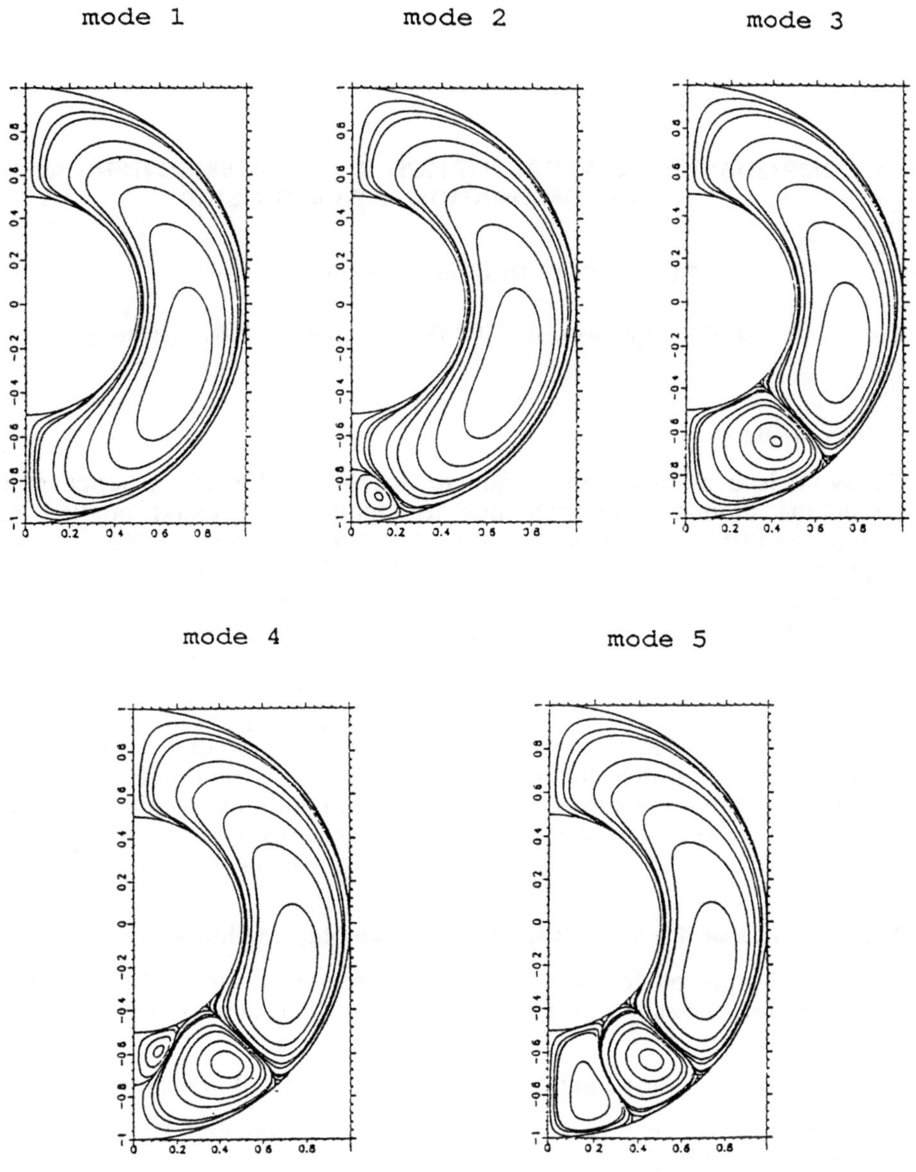

Fig. 4: Streamlines of different modes of secondary flow at the point Re=100, Ra=120000, obtained by the five computational paths shown in Fig. 3.

A COMPARISON OF SEVERAL IMPLICIT RESIDUAL SMOOTHING METHODS IN COMBINATION WITH MULTIGRID

J. Blazek, C.-C. Rossow, N. Kroll, R.C. Swanson *

DLR, Flughafen, W-3300 Braunschweig, F.R. Germany

1. Introduction

Today, it is a standard technique to accelerate the convergence of explicit multistage time-stepping schemes to steady state using implicit residual smoothing (IRS) together with multigrid. In the past, various IRS methods have been developed. They all artificially extend the stability region of the basic explicit time-stepping scheme and thus they permit higher CFL-numbers. Additionally, residual smoothing strongly effects the damping properties of a scheme which are essential for the robustness and fast convergence of multigrid. The different IRS methods can be divided mainly into two categories. The first one contains smoothers with a centrally weighted form of the implicit operator [1, 3]. The second category includes smoothers with an upwind biased form of the operator as developed in [2, 3]. The objective of this paper is to explore the capabilities of different IRS methods in combination with both central [1] as well as upwind [4] spatial discretizations and to compare their efficiency for various 2-D inviscid and viscous flow problems.

2. Description of the Implicit Residual Smoothing Methods

In the following, all the investigated IRS methods will be briefly described. Some important results of a 2-D Fourier analysis code [5] will also be presented.

- Central smoothing based on superposition of 1-D operators (CS1D):

This well known technique introduced by Jameson and Baker [1] reads in I-direction as follows

$$-\varepsilon \vec{\tilde{R}}_{I-1,J} + (1+2\varepsilon)\vec{\tilde{R}}_{I,J} - \varepsilon \vec{\tilde{R}}_{I+1,J} = \vec{R}_{I,J} \, , \tag{1}$$

where $\vec{R}$ represents the unsmoothed residual, $\vec{\tilde{R}}$ the smoothed residual and ε the smoothing coefficient.

- 2-D central smoothing using Laplacian operator (CS2D):

A two-dimensional centrally weighted operator can be defined as

$$-\varepsilon \vec{\tilde{R}}_{I-1,J} - \varepsilon \vec{\tilde{R}}_{I+1,J} - \varepsilon \vec{\tilde{R}}_{I,J-1} - \varepsilon \vec{\tilde{R}}_{I,J+1} + (1+4\varepsilon)\vec{\tilde{R}}_{I,J} = \vec{R}_{I,J} \, . \tag{2}$$

---

* NASA Langley Research Center, Hampton, VA 23665, USA

This technique leads to a sparse matrix which is solved iteratively. Because the coefficient matrix of the system is diagonally dominant, only a few Gauss-Seidel iterations (3-6) are sufficient.

In order to improve the damping and convergence characteristics of multistage time-stepping schemes, particulary in connection with upwind spatial discretizations, new upwind-biased IRS methods have been recently developed by the authors [2, 3]. In contrast to the central IRS, the residuals are smoothed applying upwind operators based on the wave propagation direction. All methods follow the operator splitting approach. The smoothing operators are written here for convenience in I-direction only.

- Simplified upwind smoothing (SUS):

In order to give the CS1D method some kind of upwind character, a slight modification is devised [3]

$$
\begin{aligned}
-\varepsilon\overset{\rightleftarrows}{R}_{I-1,J} + (1+\varepsilon)\overset{\rightleftarrows}{R}_{I,J} \qquad\qquad &= \vec{R}_{I,J} \quad , if\ M_I > 1 \\
-\varepsilon\overset{\rightleftarrows}{R}_{I-1,J} + (1+2\varepsilon)\overset{\rightleftarrows}{R}_{I,J} - \varepsilon\overset{\rightleftarrows}{R}_{I+1,J} &= \vec{R}_{I,J} \quad , if\ |M_I| \le 1 \\
(1+\varepsilon)\overset{\rightleftarrows}{R}_{I,J} - \varepsilon\overset{\rightleftarrows}{R}_{I+1,J} &= \vec{R}_{I,J} \quad , if\ M_I < -1\ .
\end{aligned}
\tag{3}
$$

In this simple procedure the upwind direction is determined by the Machnumber M projected into the coordinate direction.

- Improved upwind smoothing (IUS):

For systems of equations an improved upwind-IRS can be obtained considering the wave propagation direction for each equation separately. This can be accomplished through the splitting of the residual into two parts R^+, R^- corresponding to the positive and negative eigenvalues of the flux Jacobian [3]

$$
\vec{R}_{I,J} = \vec{R}^{+}_{I,J} + \vec{R}^{-}_{I,J}\ . \tag{4}
$$

The smoothing is applied independently to both parts

$$
\begin{aligned}
-\varepsilon\overset{\rightleftarrows}{R}{}^{+}_{I-1,J} + (1+\varepsilon)\overset{\rightleftarrows}{R}{}^{+}_{I,J} \qquad\qquad &= \vec{R}^{+}_{I,J} \\
(1+\varepsilon)\overset{\rightleftarrows}{R}{}^{-}_{I,J} - \varepsilon\overset{\rightleftarrows}{R}{}^{-}_{I+1,J} &= \vec{R}^{-}_{I,J}\ .
\end{aligned}
\tag{5}
$$

After the smoothing, the positive and negative parts of the residual are simply added.

- Full upwind smoothing (FUS):

Within this method the residuals are first transformed from conservative to characteristic variables, smoothed and then transformed back [2, 3] separately for I- and J-direction. The transformed residuals P are smoothed for each equation according to the sign of the corresponding eigenvalue λ. Written for the first component of the residual vector (denoted by '(1)') the operator reads

$$
\begin{aligned}
-\varepsilon\tilde{P}^{(1)}_{I-1,J} + (1+\varepsilon)\tilde{P}^{(1)}_{I,J} \qquad\qquad &= P^{(1)}_{I,J} \quad , if\ \lambda^{(1)} \ge 0 \\
(1+\varepsilon)\tilde{P}^{(1)}_{I,J} - \varepsilon\tilde{P}^{(1)}_{I+1,J} &= P^{(1)}_{I,J} \quad , if\ \lambda^{(1)} < 0\ .
\end{aligned}
\tag{6}
$$

The damping properties of the schemes are exemplary investigated by 2-D Fourier analysis of the two-grid cycle [5]. In contrast to the central smoothers, all upwind smoothers reduce to one single scheme for a scalar advection equation. In **Fig. 1** amplification factors for the central spatial discretization and the three smoother types are displayed. Compared to the CS1D method, the 2-D central IRS shows a better damping of the high-frequency modes. The upwind IRS damps (at this high CFL-number) satisfactorily only in the coordinate directions, revealing the one-dimensionality of the method. Similar behavior is found for the upwind spatial discretization (**Fig. 2**). Therefore, a true 2-D implementation of the upwind IRS is currently under investigation.

3. Numerical Results and Discussion

At first, computations have been carried out for inviscid transonic and supersonic flow past a NACA 0012 airfoil as well as for hypersonic flow around a circular arc. Both central and upwind spatial discretizations have been used. The required multigrid cycles (for residual drop of 5 orders/ 6 orders for the circular arc), the CPU-times (1 processor Cray Y-MP) and the asymptotic convergence rates are summarized in **Tab. 1, 2**. For each test case the most efficient multistage scheme with an optimal CFL-number and smoothing coefficient has been chosen. Additionally, the convergence has been enabled or at least improved (up to factor 2.5 in CPU-time) for super- and hypersonic flows coupling central IRS with an simple upwind prolongation (yet unpublished). One example is shown in **Fig. 3**. In general, central IRS performs best with a central scheme, and upwind IRS (particularly FUS) performs best with an upwind scheme. In accordance with the analysis, CS2D needs fewer cycles than CS1D (except for one transonic case) but the CPU-times are mostly longer. It can be stated, that at higher Mach-numbers some upwind technique (for IRS or the prolongation) is essential for fast convergence.

The last test case presented is a hypersonic viscous flow past a 20° compression ramp [5]. The rendering in **Fig. 4** shows the convergence history of an upwind-TVD scheme smoothed by CS1D or FUS, respectively. By the FUS method the CPU-time could be significantly improved. However, using the semicoarsening technique [5] both smoothers would require approximately the same CPU-time. Further investigations with other IRS methods would be desirable.

4. References

[1] Jameson, A.; Baker, T.J.: *Solution of the Euler Equations for Complex Configurations*. AIAA Paper 83-1929, 1983.

[2] Blazek, J.; Kroll, N.; Radespiel, R.; Rossow, C.-C.: *Upwind Implicit Residual Smoothing Method for Multi-Stage Schemes*. Proceeding AIAA 10th CFD Conference, Hawaii, USA, 1991.

[3] Blazek, J.; Kroll, N.; Rossow, C.-C.: *A Comparison of Several Implicit Residual Smoothing Methods*. ICFD Conference on Numerical Methods for Fluid Dynamics, Reading, UK, 1992.

[4] Kroll, N.; Rossow, C.-C.: *A High Resolution Cell Vertex TVD Scheme for the Solution of the Two- and Three-Dimensional Euler Equations*. Lecture Notes in Physics, Vol. 371, pp. 442-446, Springer Verlag, 1990.

[5] Radespiel, R.; Swanson, R.C.: *Progress with Multigrid Schemes for Hypersonic Flow Problems*. ICASE Report, No. 91-89, 1991.

	IRS method	No. of cycles	Conv. rate	CPU-time (s)
NACA 0012 $M_\infty = 0.8$ $\alpha = 1.25°$ grid 160x48	CS1D	113	0.900	26.5
	CS2D	83	0.870	27.0
	SUS	236	0.952	63.3
	IUS	61	0.827	21.8
	FUS	62	0.830	24.1
NACA 0012 $M_\infty = 3$ $\alpha = 7°$ grid 160x48	CS1D	111	0.841	10.3
	CS2D	107	0.840	12.4
	SUS	157	0.891	17.0
	IUS	138	0.903	19.0
	FUS	123	0.889	17.9
Circular arc $M_\infty = 20$ grid 40x40	CS1D	422	0.969	16.8
	CS2D	293	0.942	17.2
	SUS	400	0.966	17.7
	IUS	236	0.946	12.3
	FUS	422	0.969	25.9

Tab. 1: Convergence data; central spatial discretization

	IRS method	No. of cycles	Conv. rate	CPU-time (s)
NACA 0012 $M_\infty = 0.8$ $\alpha = 1.25°$ grid 160x48	CS1D	413	0.981	84.8
	CS2D	>580	---	$>160.$
	SUS	450	0.982	103.0
	IUS	241	0.953	96.1
	FUS	204	0.945	85.4
NACA 0012 $M_\infty = 3$ $\alpha = 7°$ grid 160x48	CS1D	135	0.912	20.0
	CS2D	95	0.865	25.0
	SUS	91	0.780	15.4
	IUS	181	0.938	41.5
	FUS	70	0.848	16.0
Circular arc $M_\infty = 20$ grid 40x40	CS1D	279	0.929	19.5
	CS2D	189	0.907	19.0
	SUS	310	0.943	23.6
	IUS	218	0.907	15.8
	FUS	219	0.939	14.0

Tab. 2: Convergence data; upwind-TVD spatial discretization

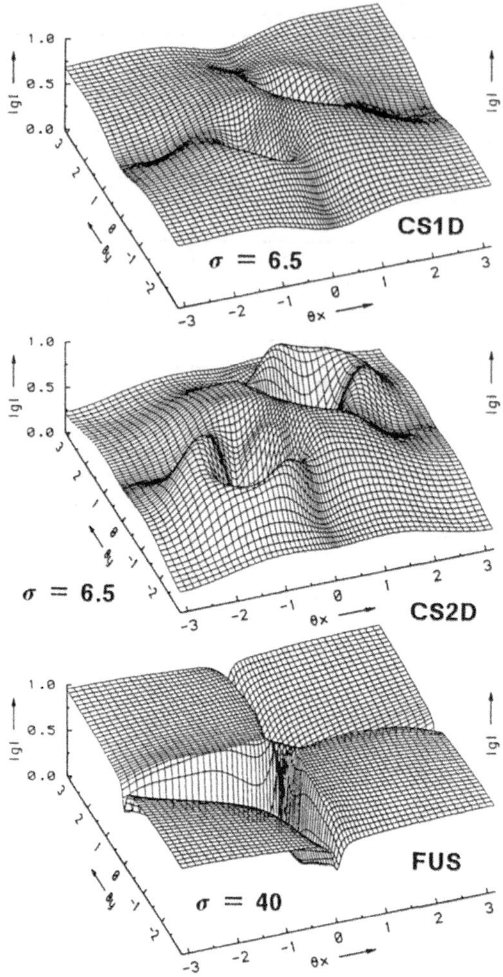

Fig. 1: Amplification factors for 2-D scalar advection problem; central discr., 5-stage scheme, 2 dissipation eval.

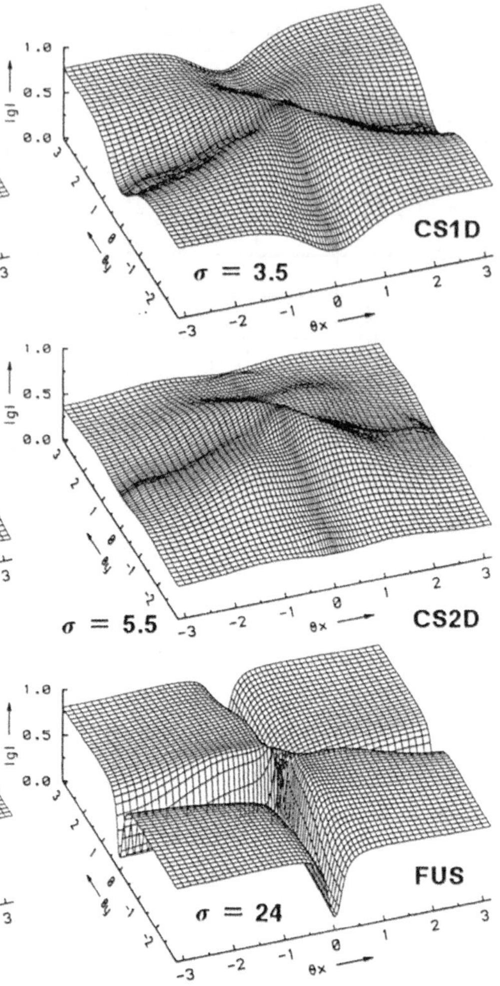

Fig. 2: Amplification factors for 2-D scalar advection problem; 2nd o. upwind, 5-stage scheme, 5 dissipation eval.

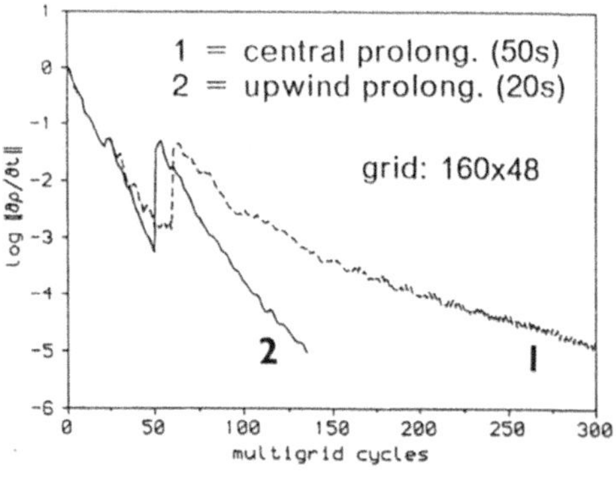

Fig. 3: Convergence histories (multigrid); NACA 0012 airfoil, $M_\infty = 3$, $\alpha = 7°$; inviscid flow, CS1D, upwind-TVD.

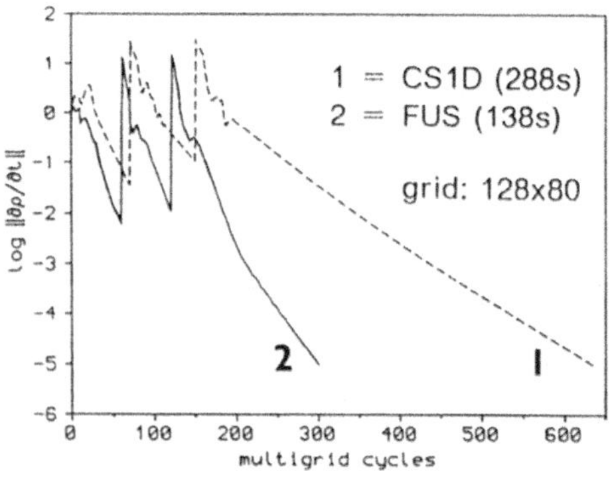

Fig. 4: Convergence histories (multigrid); compression corner, $M_\infty = 10$; laminar flow, upwind-TVD.

THE COMPUTATION OF HIGHLY NONLINEAR FREE SURFACE WAVES WITH A THREE-DIMENSIONAL PANEL METHOD

J. Broeze[1], E.F.G. van Daalen[2] and P. Zandbergen[3]

[1]Delft Hydraulics, P.O. Box 152, 8300 AD Emmeloord, The Netherlands
[2]Maritime Research Institute, P.O.Box 28, 6700 AA Wageningen, The Netherlands
[3]Twente University, P.O.Box 217, 7500 AE Enschede, The Netherlands

In this paper we discuss a higher order three dimensional panel method for nonlinear free surface wave simulations. A high degree of accuracy is obtained by a higher order resolution of the spatial problem and an efficient fourth order time integration method. Due to the accuracy, the algorithms provide stable results in the computation of linear and (highly) nonlinear wave problems.

1. INTRODUCTION

Boundary integral equation methods are very useful for the computation of hydrodynamic (nonlinear) wave problems in domains with arbitrary bottom geometries. They have favourable properties for the application to problems with time-dependent shapes of the free surface. Also their application to problems with fixed or freely floating bodies is very promising.

A number of successful two-dimensional methods have been described in literature (see e.g. [3,4]).

Extending the 2-D methods to three dimensions is possible for most of the methods. However, in practical computations, large problems may occur then. The problems are mainly due to the low order of accuracy of the method, the large amount of CPU time needed for the computations, and the complex algorithms for the geometric description and the evolution of the boundary grid. These are the main reasons why to our knowledge hardly any computations on nonlinear waves with a 3-D boundary integral equation method have been described in literature yet.

In this paper we will discuss some of our experiences in the further development of a higher order three-dimensional panel method for nonlinear free surface gravity waves. The method provides stable and very accurate results for linear, mildly nonlinear and highly nonlinear wave problems.

A 2-D version of the method has also been extended for the computation of the interaction with freely floating bodies. The 3-D code can be applied to problems with submerged bodies. The extension of this method for floating bodies is now in development.

In this paper we will additionally present results of computations on 3-D problems.

2. PROBLEM DEFINITION

The fluid is assumed to be incompressible and inviscid. The motion of the fluid is assumed to be irrotational. Under these circumstances, a potential ϕ for the flow velocity can be introduced:

$$\underline{v} = \nabla\phi \tag{1}$$

This potential satisfies Laplace's equation through the domain:

$$\nabla^2\phi = 0 \qquad (\underline{x}\in\Omega) \tag{2}$$

The elliptic field equation can be solved (providing the velocity field in the whole domain), if boundary conditions are provided on the boundaries: the potential or its normal derivative must be prescribed.

For obtaining the values at a new time level, the boundary conditions for the field equation can be obtained by integrating the time-dependent equations for the boundary conditions. The most important boundary conditions are the dynamic and kinematic boundary conditions that provide expressions for the time derivatives of the potential and the positions of the fluid boundary:

$$\frac{D\phi}{Dt} = -g\cdot z - \frac{1}{2}(\nabla\phi)^2 + \underline{v}\cdot\nabla\phi \tag{3}$$

$$\frac{D\underline{x}}{Dt}\cdot\underline{n} = \frac{\partial\phi}{\partial n} \tag{4}$$

On the bottom and lateral boundaries, some other well-known boundary conditions must be used.

In this paper we will present our special approaches for solving this set of equations in the time domain in three dimensions.

3. SOLUTION METHOD

3.1 The original solution method

Romate [1] has developed an accurate higher order 3-D panel method for the numerical simulation of free surface gravity waves. His program is extremely fast due to the high degree of vectorization (for doing a time step in a problem with 650 panels, he needed about 4 CPU seconds on a CRAY-XMP).

The method is based on Green's third identity for solving Laplace's equation. For the discretization of the boundary integral equation, the boundary of the domain is divided into quadrilateral panels, with one collocation point per panel. For determining the influence coefficients, a linear source distribution and a quadratic dipole distribution is assumed. Tangential derivatives are determined by finite difference approximations over the collocation points of a number of adjacent panels.

For the time integration Romate used the classical fourth order Runge-Kutta method with 'frozen coefficients' (i.e. the influence coefficients were not redetermined at the intermediate time levels).

Romate's method provided stable and very accurate results for linear and mildly nonlinear problems. However, in the computation of highly nonlinear waves, instabilities occur, due to numerically excited cross waves.

In order to improve the results of Romate's method for highly nonlinear wave problems, we have modified the method.

3.2 Evolution of the domain boundary grids

In [2] we presented a variational formulation of the free surface problem, including the motion of lateral boundaries of the domain. We concluded that having a free surface with fixed positions for the vertical boundaries is only allowed when simulating solid walls. This was confirmed in computations on a highly nonlinear wave problem, that showed that instabilities develop near the intersection of the free surface with the inflow and outflow boundaries if their positions is kept fixed.

A well-posed problem is obtained, if all domain boundaries travel along with the fluid particles (a Lagrangian description). In numerical computations on high waves, no instabilities occur with this approach. However, due to variations in the horizontal velocity along the vertical boundaries, in the computation of highly nonlinear waves, after a few wave periods, highly curved inflow and outflow boundaries occur. A very dense grid is needed on these boundaries to prevent numerical instabilities then.

We have also concluded from the variational formulation that also other motions of the lateral boundaries are allowed, if only a Lagrangian description is chosen for the intersection of these boundaries with the free surface. For practical reasons we decided to have the vertical inflow and outflow boundaries travelling in horizontal direction along with the motion of the fluid particles at the intersections of these boundaries with the free surface (without variations in the horizontal velocity of the domain boundaries over the vertical). It should be noted, that we do not prescribe all *fluid particles* at the lateral boundaries to move at this uniform velocity (due to the partially non-Lagrangian description of these boundaries, fluid can flow in or out of the domain). Using this approach, long and stable computations are possible, without instabilities.

3.3 Geometric modelling

In Romate's approach, fixed horizontal positions were used for the inflow and outflow boundaries. Because the position of the grid at the free surface changes in time (also the horizontal positions), a grid redistribution algorithm was used in his method after every time step to preserve equal panel sizes over the surface.

Grid redistributions may reduce the accuracy of the method. That is why we have chosen to replace the algorithms for the geometric modelling and the evolution of the grid. Grid redistributions are no longer necessary due to the basically Lagrangian motion of the surface.

In the present approach, the geometric modelling is based on the positions of the collocation points, and the evolution of the grid in time is determined by the motion of the collocation points. As explained above, for the free surface a Lagrangian description for the motion of the collocation points is chosen. Special algorithms have been developed for the evolution of the grids on the lateral boundaries and the bottom to keep the boundary grids well connected at the intersections.

An adaptive grid motion algorithm has been implemented to ensure that the grid has some desirable properties (such as a dense grid on places with large gradients, and only small curvature of the grid lines over the boundaries).

Because in our method the collocation points are at the centres of the panels, also a new iterative algorithm has been implemented to determine the intersections of adjacent networks.

The geometric data for the panels of each network are determined from a projection of the network on a rectangle in the computational domain. This implies that some boundaries must be split up into a number of networks if constructions are introduced into the fluid domain (multiply connected boundaries). For that aim we have extended the algorithms, so that smooth connections can be determined between the networks, and an arbitrary number of adjacent networks can be connected to one network edge.

3.4 Time integration

Because Romate's time integration method may cause a severe loss of accuracy in the computation of highly nonlinear waves (due to the 'frozen coefficients', which may cause severe errors in situations with rapidly deforming boundaries), we have implemented another time integration method. We have chosen to implement a 2-stage 2-derivative generalized Runge-Kutta method, where the geometric parameters and the influence coefficients are redetermined at the intermediate time level. This method is very accurate, with a minimal increase of CPU time.

4. NUMERICAL RESULTS

4.1 Computation of highly nonlinear periodic waves

Due to the above described improvements, stable and very accurate results are obtained in highly nonlinear wave computations.

As a test case, computations were done on a periodic propagating wave, with amplitude 5m, wave length 60m and wave period 6.55s on 10m water (exact solution is known from Fourier-theory). The height of this wave is over 80% of the theoretical maximum.

The initial solution is prescribed. On the inflow boundary, the exact normal velocities is prescribed (simulating a wave maker). Figure 1 shows the solution obtained after 3 wave periods. It can be seen in this figure, that the errors in the elevation are below 0.1m (which is 2% of the wave height). No growing errors or instabilities are observed if this computation is continued.

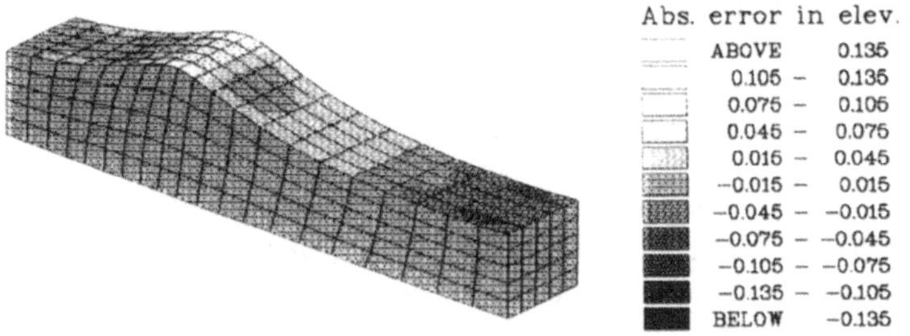

fig.1. Results from computations on a highly nonlinear periodic wave, propagating to the right. Shape of grid and errors in elevation after 3 wave periods.

4.2 Computation of a solitary wave over a construction

Real 3-D effects can be observed from computations on the interaction of a highly nonlinear solitary wave (wave height 3.5m on 5m water) with a smooth construction on the bottom. Fig.2 shows the shape of the free surface and the bottom grid (both including the grid on the lateral boundaries) after the wave has passed over the construction. The results show that the wave becomes vertical and starts to break behind the obstacle.

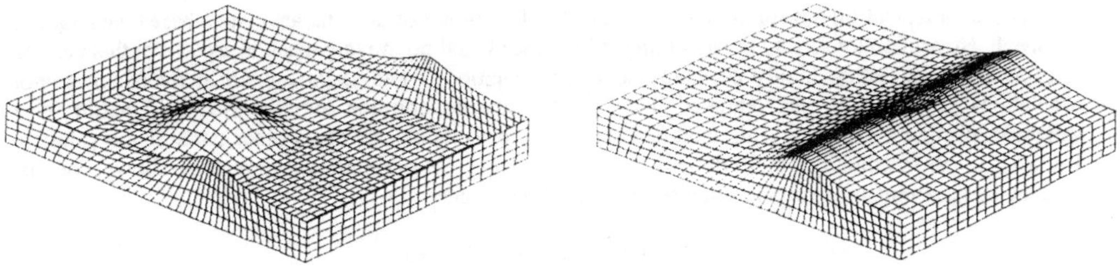

Fig.2. Shape of free surface and bottom grid after interaction of solitary wave with construction.

REFERENCES

1. J.E. Romate: *The numerical simulation of nonlinear gravity waves in three dimensions using a higher order panel method*, PhD-thesis, Enschede, 1989.
2. P.J. Zandbergen, J. Broeze & E.F.G. van Daalen: *A panel method for the simulation of nonlinear gravity waves and ship motions*, in: J.H. Kane et al. (Eds.): Advances in Boundary Element Techniques, Springer-Verlag, Berlin (to appear, 1992).
3. J.W. Dold & D.H. Peregrine: *An efficient boundary-integral method for steep unsteady water waves*, in: K.W. Morton, M.J. Baines (Eds.), Numerical Methods in Fluid Dynamics II, Oxford Univ. Press, 1986, pp.1-10.
4. S.T. Grilli, J. Skourup & I.A. Svendsen: *An efficient boundary element method for nonlinear water waves*, Engineering Analysis with Boundary Element Methods, 6 (1989), pp.97-107.

ACKNOWLEDGEMENTS

These investigations were supported by the Technology Foundation of the Netherlands (STW). Supercomputer usage was financed by the National Computing Facilities Foundation of the Netherlands (NCF).

A BRIEF DESCRIPTION OF A NEW NUMERICAL FRAMEWORK FOR SOLVING CONSERVATION LAWS — THE METHOD OF SPACE-TIME CONSERVATION ELEMENT AND SOLUTION ELEMENT §

S.C. Chang † and W.M. To ‡

† NASA Lewis Research Center, Cleveland, Ohio 44135, U.S.A. (Tel. 216-433-5874)

‡ Sverdrup Technology, Inc., Cleveland, Ohio 44142, U.S.A. (Tel. 216-433-5937)

1. INTRODUCTION

In this paper we shall describe a new numerical method for solving conservation laws. It is much simpler than a typical high resolution method [1]. No flux limiter or any characteristics-based technique is involved. No artificial viscosity or smoothing is introduced, and no moving mesh is used. Yet this method is capable of generating highly accurate shock tube solutions. The slight numerical overshoot and/or oscillations generated can be removed if a simple averaging formula initially used is replaced by a weighted averaging formula. This modification has no discernable effect on other parts of the solution. Because of its simplicity, multi-dimension generalization is straightforward and it allows for the simultaneous treatment of variables in different spatial directions.

2. CONSERVATION LAWS

We consider a dimensionless form of the 1-D unsteady Euler equations for an ideal gas. Let ρ, u, p, and γ, respectively, be the mass density, velocity, static pressure, and constant specific heat ratio. Let

$$q_1 = \rho \quad , \quad q_2 = \rho u \quad , \quad q_3 = p/(\gamma-1) + (1/2)\rho u^2 \tag{2.1}$$

$$f_1 = q_2 \; , \; f_2 = (\gamma-1)q_3 + (1/2)(3-\gamma)(q_2)^2/q_1 \; , \; f_3 = \gamma q_2 q_3/q_1 - (1/2)(\gamma-1)(q_2)^3/(q_1)^2 \tag{2.2}$$

Then the Euler equations can be expressed as

$$\partial q_m/\partial t + \partial f_m/\partial x = 0 \quad , \qquad m = 1, 2, 3 \tag{2.3}$$

Let $x_1 = x$ and $x_2 = t$ be considered as the coordinates of a two-dimensional Euclidean space E_2. The integral form of Eq. (2.3) in the space-time E_2 can be expressed as (see Fig. 1)

$$\oint_{S(V)} \vec{h}_m \cdot d\vec{s} = 0 \quad , \qquad m = 1, 2, 3 \tag{2.4}$$

where *(i)* $S(V)$ is the boundary of an arbitrary space-time volume V in E_2, *(ii)* $\vec{h}_m = (f_m, q_m)$ are space-time current density vectors in E_2, and *(iii)* $d\vec{s} = d\sigma\,\vec{n}$ with $d\sigma$ and $\vec{n}$, respectively, being the area and the outward unit normal of a surface element on $S(V)$. Note that *(i)* $\vec{h}_m \cdot d\vec{s}$ is the space-time flux of $\vec{h}_m$ leaving the volume V through the surface element $d\vec{s}$, and *(ii)* all mathematical operations can be carried out as though E_2 is an ordinary two-dimensional Euclidean space.

3. NUMERICAL METHOD

Let E_2 be divided into nonoverlapping rhombic regions (see Fig. 2) referred to as solution elements (SEs). Each SE is centered at a mesh point (j,n) where $n = 0, 1/2, 1, 3/2, \cdots$, and $j = (n\pm1/2), (n\pm3/2), \cdots$, i.e., j is a half-integer (whole integer) if n is a whole integer (half-integer). In other words, j and n are both whole integers or both half-integers if, as occurring in Fig. 2, a SE is centered at the mesh point $(j,n+1/2)$. Thus, the locations of SEs and their centers are staggered over every half time-step. A SE centered at (j,n), and its interior are denoted by SE(j,n) and SE$'(j,n)$, respectively.

For any $(x,t) \in$ SE$'(j,n)$, $q_m(x,t)$, $f_m(x,t)$, and $\vec{h}_m(x,t)$, respectively, are approximated by $\underset{\sim}{q}_m(x,t;j,n)$, $\underset{\sim}{f}_m(x,t;j,n)$, and $\vec{\underset{\sim}{h}}_m(x,t;j,n)$ which we shall define immediately. Let

$$\underset{\sim}{q}_m(x,t;j,n) = (\sigma_m)_j^n + (\alpha_m)_j^n (x - x_j) + (\beta_m)_j^n (t - t^n) \quad , \qquad m = 1, 2, 3 \tag{3.1}$$

where $(\sigma_m)_j^n$, $(\alpha_m)_j^n$, and $(\beta_m)_j^n$ are constants in SE$'(j,n)$, and (x_j, t^n) are the coordinates of the mesh point (j,n). Note that

§ This work is dedicated to the memory of a teacher, Mr. Nylon Cheng.

$$q_m(x_j,t^n;j,n) = (\sigma_m)_j^n \quad , \quad \partial q_m(x,t;j,n)/\partial x = (\alpha_m)_j^n \quad , \quad \partial q_m(x,t;j,n)/\partial t = (\beta_m)_j^n \tag{3.2}$$

Moreover, if we identify $(\sigma_m)_j^n$, $(\alpha_m)_j^n$, and $(\beta_m)_j^n$, respectively, with the values of q_m, $\partial q_m/\partial x$, and $\partial q_m/\partial t$ at (x_j,t^n), the expression on the right side of Eq. (3.1) becomes the first-order Taylor's expansion of $q_m(x,t)$ at (x_j,t^n). As a result of these considerations, $(\sigma_m)_j^n$, $(\alpha_m)_j^n$, and $(\beta_m)_j^n$ will be considered as the numerical analogues of the values of q_m, $\partial q_m/\partial x$, and $\partial q_m/\partial t$ at (x_j,t^n), respectively.

Let $f_m(x,t;j,n)$, $m = 1, 2, 3$, be defined in terms of $q_m(x,t;j,n)$, $m = 1, 2, 3$, according to Eq. (2.2) with the understanding that f_m and q_m in Eq. (2.2) be replaced by $f_m(x,t;j,n)$ and $q_m(x,t;j,n)$, respectively. Using Eq. (3.1), $f_m(x,t;j,n)$ is expressed as a function of $(x-x_j)$ and $(t-t^n)$, and then expanded as a power series of them. The new method is simplified by truncating the power series after first-order terms. This is consistent with the first-order approximation given in Eq. (3.1). In the new method, only f_m at $x = x_j$ are needed. Let $\sigma_m = (\sigma_m)_j^n$ and $\beta_m = (\beta_m)_j^n$. Then explicitly, we have:

$$f_1(x_j,t;j,n) = \sigma_2 + \beta_2(t-t^n) \tag{3.3}$$

$$\begin{aligned} f_2(x_j,t;j,n) &= (\gamma-1)\sigma_3 + (1/2)(3-\gamma)(\sigma_2)^2/\sigma_1 \\ &+ [(\gamma-1)\beta_3 + (3-\gamma)(\sigma_2\beta_2/\sigma_1) - (1/2)(3-\gamma)(\sigma_2/\sigma_1)^2\beta_1](t-t^n) \end{aligned} \tag{3.4}$$

$$\begin{aligned} f_3(x_j,t;j,n) &= \gamma\sigma_2\sigma_3/\sigma_1 - (1/2)(\gamma-1)(\sigma_2)^3/(\sigma_1)^2 + \{\gamma[(\sigma_2\beta_3+\sigma_3\beta_2)/\sigma_1 \\ &- \sigma_2\sigma_3\beta_1/(\sigma_1)^2] + (1/2)(\gamma-1)[2(\sigma_2/\sigma_1)^3\beta_1 - 3(\sigma_2/\sigma_1)^2\beta_2]\}(t-t^n) \end{aligned} \tag{3.5}$$

Since $\vec{h}_m = (f_m, q_m)$, we define $\vec{h}_m(x,t;j,n) = (f_m(x,t;j,n), q_m(x,t;j,n))$.

Let E_2 be divided into nonoverlapping rectangular regions (see Fig. 2) referred to as conservation elements (CEs). They are also staggered over every half time-step. Let the CE with its upper edge centered at (j,n) be denoted by CE(j,n). Then the current approximation of Eq. (2.4) is

$$\oint_{S(\mathrm{CE}(j,n))} \vec{h}_m \cdot d\vec{s} = 0 \qquad (\text{all possible } m \text{ and } (j,n)) \tag{3.6}$$

Because the entire boundary (except for three isolated points) of a CE is located within the interiors of three neighboring SEs, $\vec{h}_m$ is continuous across any interface separating two neighboring CEs. Thus Eq. (3.6) will remain valid if CE(j,n) is replaced by the union of any combination of CEs.

Because each $S(\mathrm{CE}(j,n))$ is a simple closed curve in E_2 (see Fig. 1), the surface integration form Eq. (3.6) can be converted into a line integration form [2, p.14], i.e.,

$$\oint_{S(\mathrm{CE}(j,n))} \vec{g}_m \cdot d\vec{r} = 0 \qquad (\text{all possible } m \text{ and } (j,n)) \tag{3.7}$$

where $\vec{g}_m = (-q_m, f_m)$ and $d\vec{r} = (dx,dt)$.

For each SE(j,n), let

$$s_1(j,n) = (\Delta x/8)\alpha_1 + (1/2)(\Delta t/\Delta x)[\sigma_2 + (\Delta t/4)\beta_2] \tag{3.8}$$

$$\begin{aligned} s_2(j,n) &= (\Delta x/8)\alpha_2 + (1/2)(\Delta t/\Delta x)[(\gamma-1)\sigma_3 + (1/2)(3-\gamma)(\sigma_2)^2/\sigma_1] \\ &+ (1/8)[(\Delta t)^2/\Delta x]\{(\gamma-1)\beta_3 + (3-\gamma)[\sigma_2\beta_2/\sigma_1 - (1/2)(\sigma_2/\sigma_1)^2\beta_1]\} \end{aligned} \tag{3.9}$$

$$\begin{aligned} s_3(j,n) &= (\Delta x/8)\alpha_3 + (1/2)(\Delta t/\Delta x)[\gamma\sigma_2\sigma_3/\sigma_1 - (1/2)(\gamma-1)(\sigma_2)^3/(\sigma_1)^2] + (1/8)[(\Delta t)^2/\Delta x] \\ &\times \{(\gamma/\sigma_1)(\sigma_2\beta_3 + \sigma_3\beta_2 - \sigma_2\sigma_3\beta_1/\sigma_1) + (\gamma-1)[(\sigma_2/\sigma_1)^3\beta_1 - (3/2)(\sigma_2/\sigma_1)^2\beta_2]\} \end{aligned} \tag{3.10}$$

where $\sigma_m = (\sigma_m)_j^n$, $\alpha_m = (\alpha_m)_j^n$, and $\beta_m = (\beta_m)_j^n$. Then Eq. (3.7) implies that, for each SE$(j,n+1/2)$,

$$(\sigma_m)_j^{n+1/2} = (1/2)[(\sigma_m)_{j-1/2}^n + (\sigma_m)_{j+1/2}^n] + s_m(j-1/2,n) - s_m(j+1/2,n) \quad , \quad m = 1, 2, 3 \tag{3.11}$$

i.e., $(\sigma_m)_j^{n+1/2}$ is determined in terms of the numerical variables associated with SE$(j-1/2,n)$ and SE$(j+1/2,n)$. Similar formulae for $(\alpha_m)_j^{n+1/2}$ and $(\beta_m)_j^{n+1/2}$ will be given next.

Let A_+, A, and A_- (see Fig. 2) denote $(x_{j+1/2},t^{n+1/2})$, $(x_j,t^{n+1/2})$, and $(x_{j-1/2},t^{n+1/2})$, respectively. Let

$$q_{m\pm} = q_m(x_{j\pm1/2},t^{n+1/2};j\pm1/2,n) \tag{3.12}$$

Because $A_\pm$ do not belong to SE$'(j\pm1/2,n)$, the expression on the right side of Eq. (3.12) is to be evaluated at the two points immediately below them. A central-difference formula for evaluating $(\alpha_m)_j^{n+1/2}$, the numerical analogue of $\partial q_m/\partial x$ at point A, is

$$(\alpha_m)_j^{n+1/2} = (\underline{q}_{m+} - \underline{q}_{m-}) / \Delta x \tag{3.13}$$

This formula is valid as long as no discontinuity of q_m (or its derivatives) occurs between A_- and A_+. In the following discussion, we develop an alternate which is valid even in the presence of discontinuity.

Let

$$(\alpha_{m\pm})_j^{n+1/2} = \pm [\underline{q}_{m\pm} - (\sigma_m)_j^{n+1/2}] / (\Delta x/2) \tag{3.14}$$

where $(\sigma_m)_j^{n+1/2}$ has just been obtained using Eq. (3.11). Because $\underline{q}_{m+}$, $(\sigma_m)_j^{n+1/2}$, and $\underline{q}_{m-}$ are the numerical analogues of q_m at A_+, A, and A_-, respectively, $(\alpha_{m+})_j^{n+1/2}$ and $(\alpha_{m-})_j^{n-1/2}$ are two numerical analogues of $\partial q_m(x_j, t^{n+1/2}) / \partial x$ with one being evaluated from the right and another from the left. Note that $(\alpha_m)_j^{n+1/2}$ defined by Eq. (3.13) is equal to the average of $(\alpha_{m+})_j^{n+1/2}$ and $(\alpha_{m-})_j^{n+1/2}$.

In case that a discontinuity occurs between A and A_+ but not between A and A_-, one would expect that $|(\alpha_{m+})_j^{n+1/2}| \gg |(\alpha_{m-})_j^{n+1/2}|$. Moreover, because A and A_- are on the same side of the discontinuity while A and A_+ are on the opposite sides, $(\alpha_m)_j^{n+1/2}$ should be closer to $(\alpha_{m-})_j^{n+1/2}$ than $(\alpha_{m+})_j^{n+1/2}$. This observation suggests that $(\alpha_m)_j^{n+1/2}$ should be a weighted average of $(\alpha_{m+})_j^{n+1/2}$ and $(\alpha_{m-})_j^{n+1/2}$ biased toward the one with the smaller magnitude.

As a result of the above and other considerations [3], Eq. (3.13) will be generalized by

$$(\alpha_m)_j^{n+1/2} = F((\alpha_{m-})_j^{n+1/2}, (\alpha_{m+})_j^{n+1/2}; c) \tag{3.15}$$

Here $c \geq 0$ is an adjustable constant and the function F is defined by *(i)* $F(0,0;c) = 0$ and *(ii)*

$$F(\alpha_-, \alpha_+; c) = (|\alpha_+|^c \alpha_- + |\alpha_-|^c \alpha_+) / (|\alpha_+|^c + |\alpha_-|^c) \quad , \quad (|\alpha_+| + |\alpha_-| > 0) \tag{3.16}$$

where α_- and α_+ are any two real variables. Note that $F(\alpha_-, \alpha_+; c) = (\alpha_- + \alpha_+)/2$ if $c = 0$ or $|\alpha_-| = |\alpha_+|$, i.e., Eq. (3.15) is reduced to Eq. (3.13) if $c = 0$ or $|\alpha_-| = |\alpha_+|$. Also the expression on the right side of Eq. (3.16) represents a weighted average of α_- and α_+ with the weight factors $|\alpha_+|^c / (|\alpha_+|^c + |\alpha_-|^c)$ and $|\alpha_-|^c / (|\alpha_+|^c + |\alpha_-|^c)$. For $c > 0$, this average is biased toward the one among α_- and α_+ with the smaller magnitude. For the same values of $|\alpha_+|$ and $|\alpha_-|$, the bias increases as c increases.

Substituting Eq. (2.2) into Eq. (2.3), one obtains three equations in which $\partial q_m / \partial t$, $m = 1, 2, 3$, are expressed in terms of q_m and $\partial q_m / \partial x$, $m = 1, 2, 3$. Let q_m and their derivatives be replaced by the corresponding numerical analogues at the mesh point $(j, n+1/2)$, one obtains that

$$(\beta_1)_j^{n+1/2} = -\alpha_2 \tag{3.17}$$

$$(\beta_2)_j^{n+1/2} = (1/2)(3-\gamma)(\sigma_2/\sigma_1)^2 \alpha_1 - (3-\gamma)(\sigma_2/\sigma_1)\alpha_2 - (\gamma-1)\alpha_3 \tag{3.18}$$

$$\begin{aligned}(\beta_3)_j^{n+1/2} = {} & [\gamma\sigma_2\sigma_3/(\sigma_1)^2 - (\gamma-1)(\sigma_2/\sigma_1)^3]\,\alpha_1 \\ & + [(3/2)(\gamma-1)(\sigma_2/\sigma_1)^2 - \gamma\sigma_3/\sigma_1]\,\alpha_2 - \gamma(\sigma_2/\sigma_1)\,\alpha_3\end{aligned} \tag{3.19}$$

where $\sigma_m = (\sigma_m)_j^{n+1/2}$ and $\alpha_m = (\alpha_m)_j^{n+1/2}$.

With the aid of Eqs. (3.11), (3.15), and (3.17) – (3.19), $(\sigma_m)_j^n$, $(\alpha_m)_j^n$, and $(\beta_m)_j^n$ can be determined in terms of the initial values $(\sigma_m)_{\pm 1/2}^0, (\sigma_m)_{\pm 3/2}^0, \cdots$, and $(\alpha_m)_{\pm 1/2}^0, (\alpha_m)_{\pm 3/2}^0, \cdots$.

4. NUMERICAL RESULTS

We consider a shock tube problem used by Sod [4]. Let $\gamma = 1.4$. At $t = 0$, let *(i)* $(\rho, u, p) = (1,0,1)$, i.e., $(q_1, q_2, q_3) = (1,0,2.5)$, if $x < 0$, and *(ii)* $(\rho, u, p) = (0.125,0,0.1)$, i.e., $(q_1, q_2, q_3) = (0.125,0,0.25)$, if $x > 0$. Thus

$$(i) \qquad ((\sigma_1)_j^0, (\sigma_2)_j^0, (\sigma_3)_j^0) = \begin{cases} (1,0,2.5) & \text{if } j = -1/2, -3/2, \cdots \\ (0.125,0,0.25) & \text{if } j = 1/2, 3/2, \cdots \end{cases} \tag{4.1}$$

and *(ii)* $(\alpha_m)_j^0 = 0$, $j = \pm 1/2, \pm 3/2, \cdots$. Eqs. (3.17) – (3.19) imply that $(\beta_m)_j^0 = 0$, $j = \pm 1/2, \pm 3/2, \cdots$.

The above initial conditions, and Eqs. (3.11), (3.15), and (3.17) – (3.19), imply that $(\sigma_m)_j^n$, $(\alpha_m)_j^n$, and $(\beta_m)_j^n$ are constant in two separate regions which, respectively, are defined by $j \leq -(n+1/2)$ and $j \geq (n+1/2)$. Thus one needs to evaluate the above variables only if $(n+1/2) > |j|$.

The current scheme is stable if $CFL = max\,(|u| + |a|)\Delta t / \Delta x \leq 1$ [3]. Here a = local sound speed. In the current computations, $\Delta x = 0.01$, $\Delta t = 0.004$, and $CFL \doteq 0.88$. Numerical results (dots) at $t = 0.4$ are compared with the exact solutions (solid lines) in Fig. 3. Since each marching step advances the solution from t to $t+\Delta t/2$, these results are obtained after 200 steps. Note that *(i)* shock discontinuity is resolved almost within one mesh interval, and *(ii)* the slight numerical overshoot and oscillations generated when $c = 0$ essentially disappear when $c = 1$ is used.

5. CONCLUSIONS AND DISCUSSIONS

The current scheme has a stencil containing only two points. This minimization of stencil has the effect of reducing numerical diffusion [5]. It is achieved by including $(\alpha_m)_j^n$ and $(\beta_m)_j^n$ as numerical variables. The fluxes at an interface separating two CEs are evaluated with no interpolation or extrapolation. Accuracy of flux evaluation is enhanced by requiring that the solution given in Eq. (3.1) satisfies the Euler equations at the center of every SE. This makes the use of characteristics-based techniques less necessary. The above key features all contribute to the simplicity, generality, and accuracy of the current scheme. They all owe their existence to the use of staggered SEs and CEs.

In the current method, flux evaluation within each SE(j,n) is required only at a subset of SE(j,n), i.e., a horizontal line segment centered at (j,n) and a vertical line segment starting upward from (j,n) (see Fig. 2). As a result, we may redefine SE(j,n) to be this subset. This new definition is used in the following sketch of an extension of the the current scheme to a three-dimensional Euclidean space E_3 ($x_1 = x$, $x_2 = y$, and $x_3 = t$).

A SE contains three mutually perpendicular rectangles (see Fig. 4a). The point of intersection is referred to as the center of this SE. The SEs are staggered in both $x-$ and $y-$ directions over every half time-step. The CEs are rectangular boxes (see Fig. 4b) also staggered in both $x-$ and $y-$ directions over every half time-step. From Fig. 4b, it is seen that the boundary of a CE can be divided into five parts which, respectively, belong to five neighboring SEs. As a result, the solution procedure described in Section 3 can be easily extended to E_3.

By replacing Eq. (3.1) with the second-order approximation, other versions of the current scheme with higher-order accuracy were also developed. The details will be reported in [3].

REFERENCES

1. R.J. LeVeque, *Numerical Methods for Conservation Laws*, Birkhäuser Verlag, 1990.
2. S.C. Chang and W.M. To, *A New Numerical Framework for Solving Conservation Laws — The Method of Space-Time Conservation Element and Solution Element*, NASA TM 104495.
3. S.C. Chang and W.M. To, *A New Development in the Method of Space-Time Conservation Element and Solution Element — Application to Shock Tube Problem*, in preparation.
4. G.A. Sod, J. Computational Phys. **27** (1978), 1.
5. S.C. Chang, *On An Origin of Numerical Diffusion: Violation of Invariance under Space-Time Inversion*, to appear in the proceeding of 23rd Conference on Modeling and Simulation, April 30 – May 1, 1992, Pittsburgh, PA, USA.

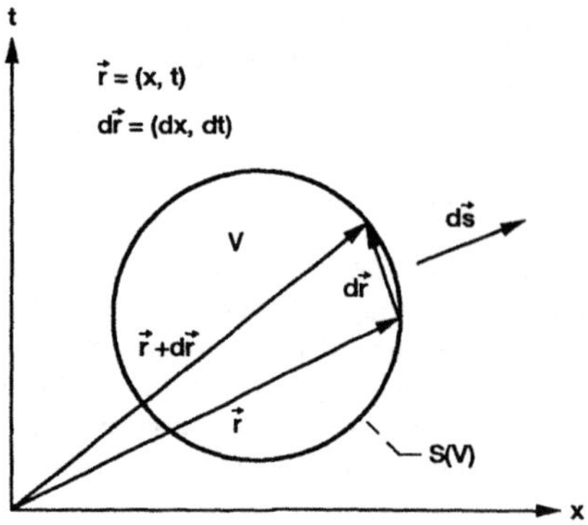

Figure 1.—A surface element $d\vec{s}$ and a line segment $d\vec{r}$ on the boundary S(V) of a volume V in a space-time E_2.

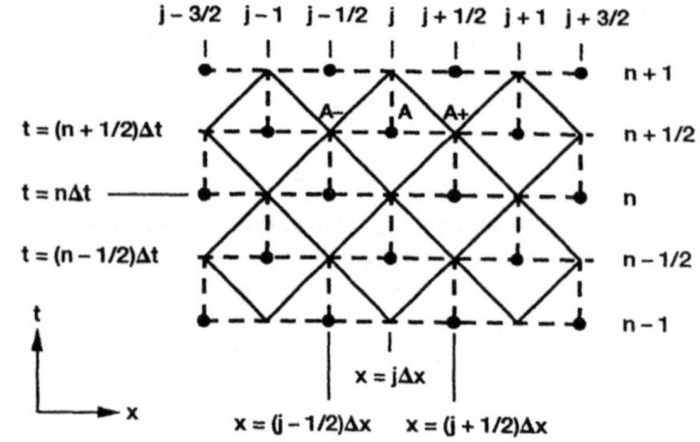

Figure 2.—The SEs (the rhombuses formed by solid lines) and the CEs (the rectangles formed by dashed lines). The dots represent the centers of SEs.

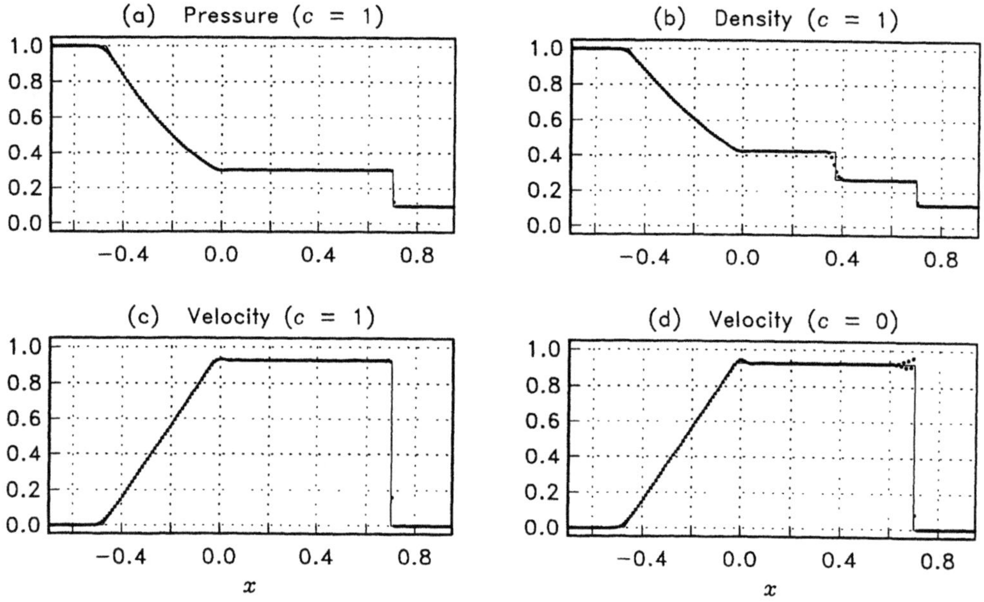

Figure 3.—Shock tube solution at t = 0.4.

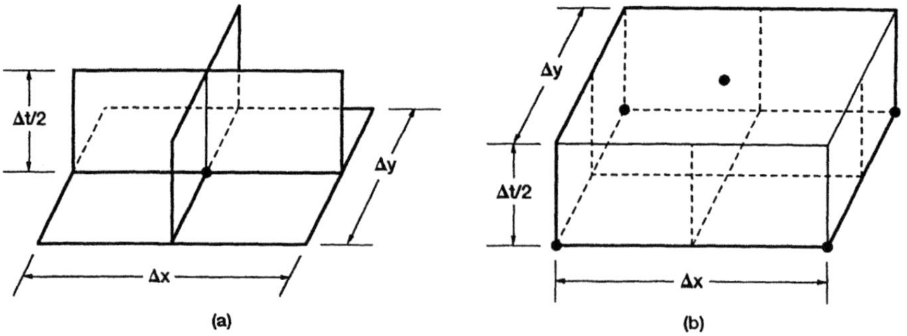

Figure 4.—A SE and a CE in E_3. The dots represent the centers of SEs.

COMPUTATION OF AIRFOIL AND CASCADE FLOWS USING THE CLEBSCH DECOMPOSITION METHOD

S.S. Chu

Department of Aeronautical Engineering
Airforce Academy, Taiwan, R.O.C.

INTRODUCTION

The Conventional Euler equations can be recast into another form by using the Hamilton's principle [1]. The new set of Euler equations contains one elliptic-like equation, and several convection equations. The velocity is decomposed into several terms, one of which is a gradient of the potential function, $\nabla\phi$. One advantage with the formulation, due to Clebsch [2], is the continuity in modeling the physics. For example, the velocity will be defined only by $\nabla\phi$, while all other terms vanish for the case of potential flow. This implies that an Euler code can be built from a full potential code by supplying convection solvers and adding source terms due to the convective quantities. In the meantime, it becomes easier to control the modeling of physics locally by setting the behavior of any convective quantities a priori, and therefore some possible sources of numerical errors are eliminated.

Another merit with the formulation is that each variable can be integrated by following its own characteristics, which distinguishes the present approach from the conventional Euler technology [3,4,5,6]. The variables in the steady Euler equations are divided into elliptic and convective quantities. Elliptic quantities are integrated with a relaxation procedure, while the convective quantities are integrated using upwind differencing. In conventional Euler technologies, the convective quantities, such as entropy, enthalpy, and swirl, are contaminated due to finite truncations and numerical diffusion, since these quantities are not primary variables in the integration.

Computations are performed for various two-dimensional, transonic, inviscid flows over single airfoil and in cascades.

GOVERNING EQUATIONS

For rotational, non-isentropic, inviscid flows, Clebsch defined the velocity vector as:

$$\vec{q} = \nabla\phi + \mu\nabla\lambda + s\nabla\eta \qquad (1)$$

where ϕ is the velocity potential, λ and η are called the Clebsch potentials, and μ and s are the Clebsch densities. Using a vector identity, the velocity decomposition yields the vorticity as:

$$\vec{\xi} = \nabla\mu\times\nabla\lambda + \nabla s\times\nabla\eta \qquad (2)$$

In this equation, the first term on the right hand side is the contribution of pure inviscid vorticity, such as that existing initially in the flow or that created to accommodate lift across the inviscid wake downstream of the trailing edges. The second term is the vorticity due to the presence of entropy gradients, as in the case of shocks.

With constraints on mass conservation, energy conservation, and kinematic relations, the variational principle recast the conventional Euler equations into the following set of equations which describe inviscid Euler physics:

$$\nabla\cdot(\rho\nabla\phi) = -\nabla\cdot(\rho\mu\nabla\lambda + \rho s\nabla\eta) \qquad (3)$$
$$\rho\vec{q}\cdot\nabla\mu = 0 \qquad (4a)$$
$$\rho\vec{q}\cdot\nabla\lambda = 0 \qquad (4b)$$
$$\rho\vec{q}\cdot\nabla s = 0 \qquad (4c)$$
$$\rho\vec{q}\cdot\nabla\eta = -p \qquad (4d)$$

The first equation is of an elliptic-like form for the potential function, and the others are convection equations for convective variables. The right hand side of equation (3) represents the contribution of convective quantities to mass conservation. The above five equations construct a complete system for the five variables(ϕ,μ,λ,s,η), along with the velocity definition, the Bernoulli equation, and the equation of state

$$p = k\rho^{\gamma}\exp(s/c_v) \qquad (5)$$

Here, c_v is the specific heat at constant volume, γ is the ratio of specific heats, and k is a constant.

INITIAL AND BOUNDARY CONDITIONS

Equation (3) is a boundary value problem which requires boundary conditions at all boundaries. At solid boundaries, the flow tangency condition is enforced using the Neumann boundary condition for the potential function. The velocity potential function and density at the inlet boundary of the computational mesh are hold fixed at the initial free-stream values. At the exit boundary the condition of no mass flux change is specified. For cascade flows, periodicity must be satisfied on the upstream boundary of the leading edge and the downstream boundary of the trailing edge.

For lifting cases, the Kutta condition should be satisfied in order to make static pressure continuous at both sides of the trailing edge. For single airfoil flows, the outer boundary points have the specified circulation consistent with that of a compressible vortex and updated at the end of each iteration step.

The entropy jump across a shock wave can easily be determined in terms of the local upstream Mach number normal to the shock by the Rankine-Hugoniot relations.

For solving convection equations only initial conditions are necessary at the upstream side. For example, in the case of no vorticity initially existing in a flow, the initial conditions for s and η can be constructed as :

$$s = 0 \qquad (6a)$$
$$\eta = -(x\cos\alpha + y\sin\alpha)a_\infty^2/\gamma \qquad (6b)$$

where a_∞ is the freestream sonic speed, and α is the angle of attack.

NUMERICAL COMPUTATION

As in conventional full potential codes, the elliptic-like equation is solved using central differencing, upwind differencing is implemented at supersonic points by adding artificial viscosity terms to provide an upwind bias. A flux conservative finite volume method [7] is used to discretize the equation, and a line successive over-relaxation method is adopted for solving the difference equation when it is elliptic.

The convection process is an initial value problem and convective variables are transported following streamlines. The appropriate way of solving the convection equations is therefore by upwind differencing according to the local flow direction. A line implicit method is used for integration, which yields a convergence rate of about than two orders of magnitude in each sweep.

The solution procedure is as follows. With initial estimates of the velocity potential and convective variables, the elliptic-like continuity equation is relaxed for new potential values, which provide the velocity field. Type-dependent operators are used as in the full potential equation. Next, the convective quantities are transported downstream via a convection operator, which in turn influence the potential distribution. The physical production of the convective quantities is accounted for depending on local flow conditions and boundary conditions. Convergent solutions are obtained through an iteration cycle. Clebsch variables are defined at nodes, and all other flow quantities and metrics are defined at the cell centers of the computational mesh. All numerical schemes for the Clebsch Euler equations are implicit and second-order accurate.

TYPICAL RESULTS

Several transonic cases were tested, including flows around single airfoil and through airfoil cascade, to demonstrate the applicability of the proposed technology in obtaining Euler solutions. All the test problems are for steady, inviscid,adiabatic,transonic flows of a perfect gas with a uniform freestream.

A transonic flow around a NACA 0012 airfoil was first considered at freestream Mach number of 0.85 and zero angle of attack. The resulting surface pressure distributions are compared with the predictions using the method of Jameson et al. [3] in Fig.1. Shock positions captured by the two Euler technologies are closer than that from potential equation, which is located farther downstream. Shock strengths and shock positions are functions of rotational physics that allow the entropy change. The effect of flow rotation behind the shock can be observed from the Mach contours in Fig.2. Entropy contours from the proposed technology are shown in Fig.5. Note that the non-smooth shock shape is attributed to the grid size.

The second problem is a transonic lifting case for a NACA 0012 airfoil at freestream Mach number of 0.75 and angle of attack of 2 degrees. Close agreements between the Clebsch Euler and the conventional Euler technologies are obtained in pressure distributions and Mach contours. Figure 3 compares the pressure distributions on the airfoil surface. Mach contours and entropy contours are given in Figs. 4 and 6 respectively.

A nonlifting case, NACA 0012 cascade with a pitch-to-chord ratio of 3.6, was next considered. The flow conditions are inflow Mach number of 0.8, zero staggered angle, zero inflow and outflow angles. The resulting surface pressure distributions are shown in Fig.7. Comparison of the resulting Mach contours shown in Fig.8 shows the differences between potential and non-isentropic rotational flows in this cascade problem. Entropy contours from the present method are shown in Fig.11.

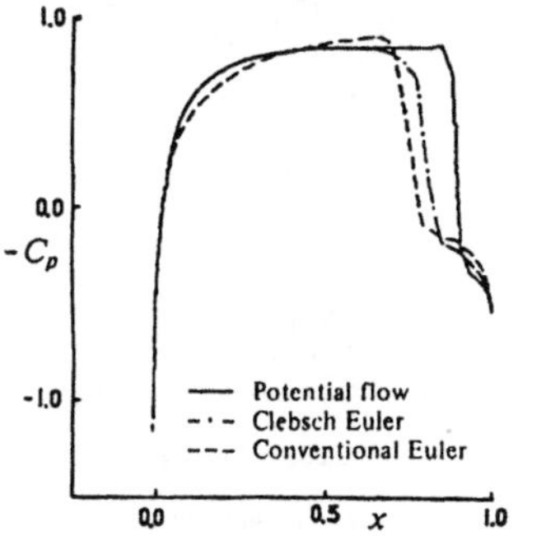

Fig. 1 Pressure distribution on airfoil at $M_\infty = .85$, $\alpha = 0°$.

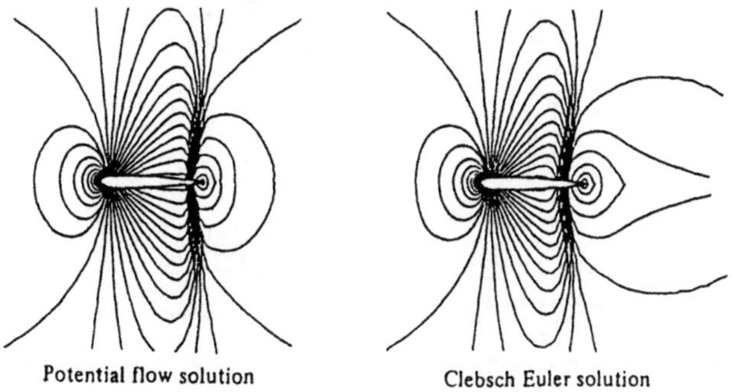

Potential flow solution

Clebsch Euler solution

Fig. 2 Mach contour for flow around airfoil at $M_\infty = .85$, $\alpha = 0°$

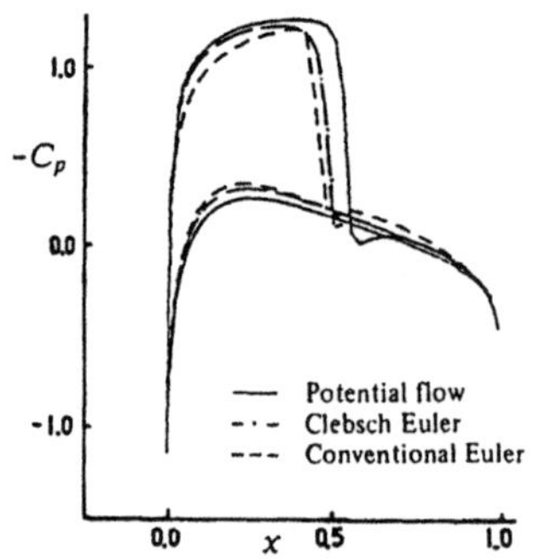

Fig.3 Pressure distribution on airfoil for the lifting case

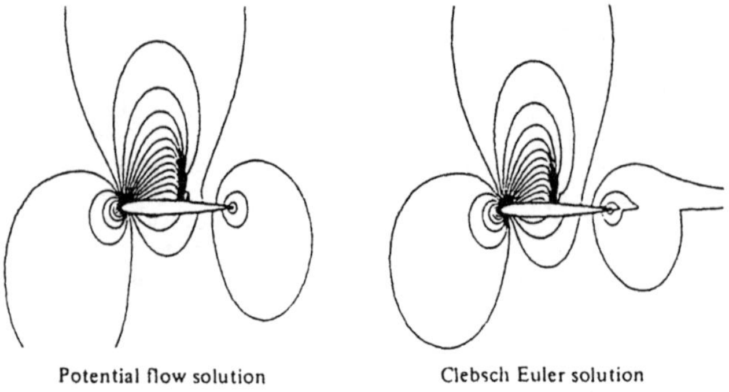

Fig.4 Mach contours for the lifting airfoil case

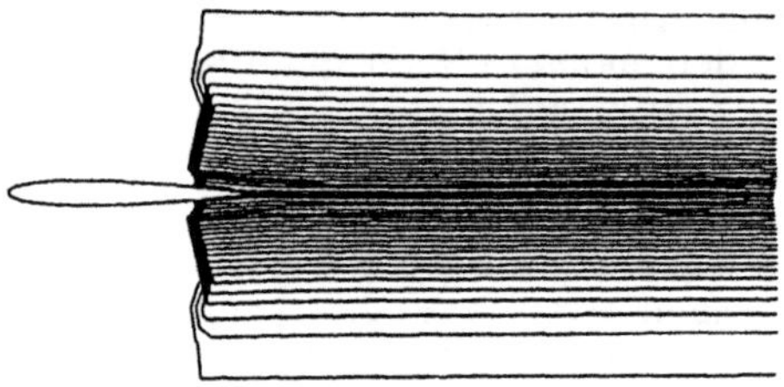

Fig.5 Entropy contours for the nonlifting airfoil case

Fig.6 Entropy contours for the lifting airfoil case

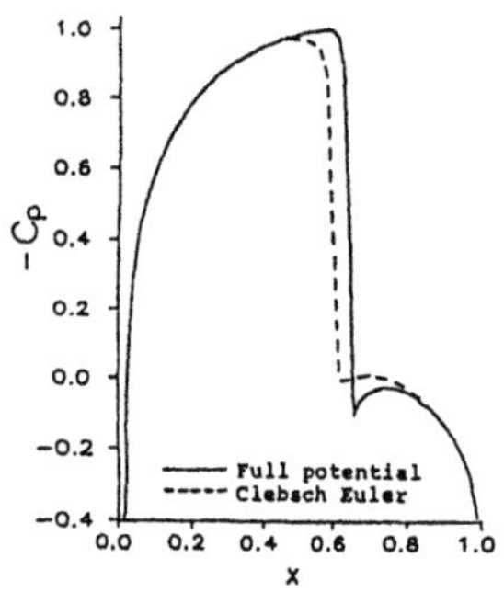

Fig.7 Surface pressure distribution, the nonlifting cascade

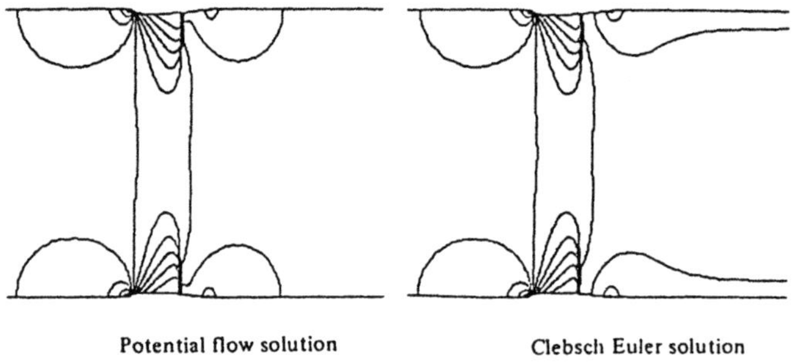

Fig.8 Mach number contours for the nonlifting cascade

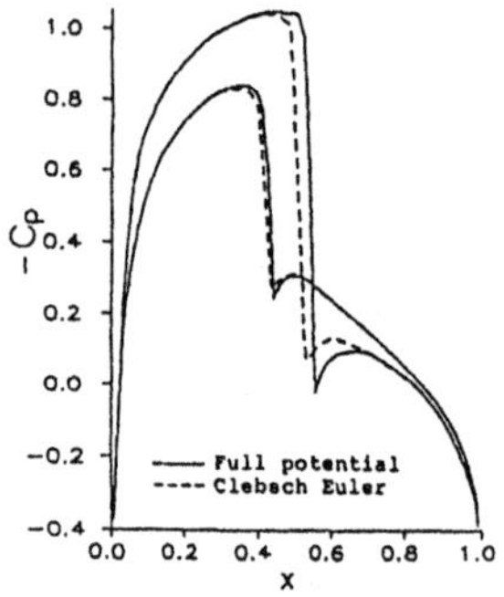

Fig.9 Surface pressure distribution, the lifting cascade

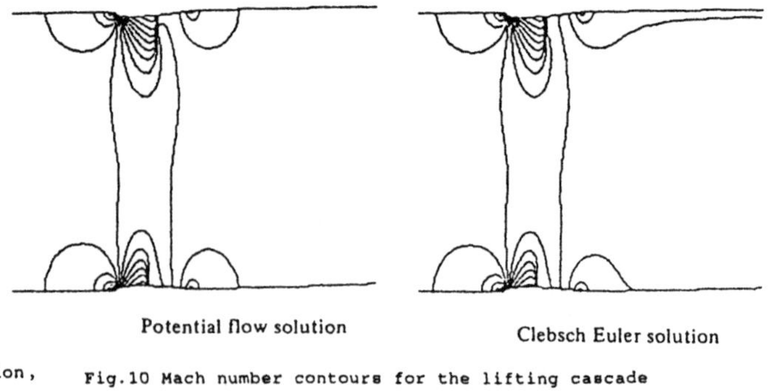

Fig.10 Mach number contours for the lifting cascade

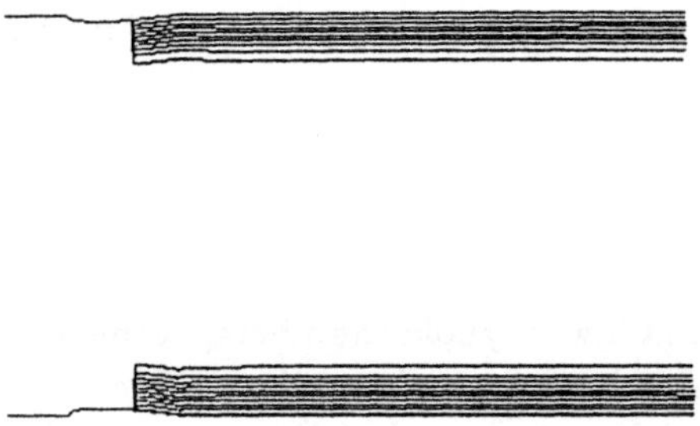

Fig.11 Entropy contours for the nonlifting cascade

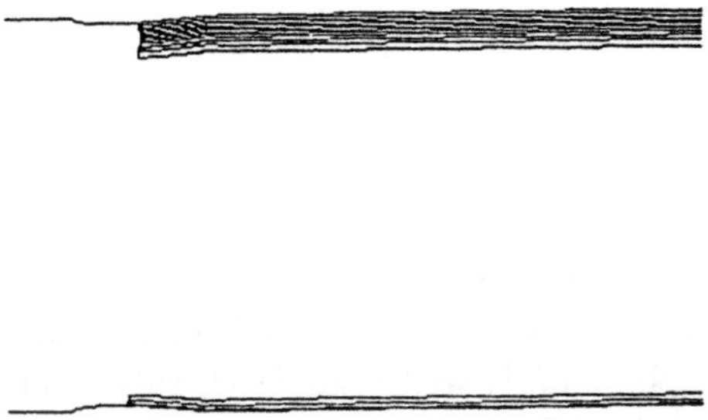

Fig.12 Entropy contours for the lifting cascade

The last test case is of a lifting cascade. Figure 9 presents the pressure distributions over the suction and pressure sides of a 1 degree staggered cascade of NACA 0012 airfoils with pitch-to-chord ratio of 3.6. The flow conditions are inflow Mach number of 0.78 and inflow angle of zero degree. Mach number contours and entropy contours are shown in Fig. 10 and 12 respectively.

CONCLUSIONS

An alternative method for the solutions of compressible, rotational, transonic flows has been presented. The method can also be regarded as a natural zonal decomposition technique without switching the formulation between zones of different physics. In the approach, local physics are accommodated by the control of convective quantities, without drawing zonal boundaries for different physics.

The four test cases verified the applicability of the proposed method in two-dimensions, and compared to the conventional Euler and full potential technologies. The Clebsch Euler formulation solves for each variable by following individual characteristics, which distinguishes the present approach from the conventional Euler method. The proposed technology is directly extensible to three dimensional flows. Further research is desirable to extend the method into viscous flow problems.

REFERENCES

1. Serrin, J.,"Mathematical Principles of Classical Fluid Mechanics," Encyclopedia of Physics, Volume VIII/1,1959.
2. Clebsch, A.,"Uber die Integration der Hydrodynamischen Gleichungen," J. Reine Angew. Math., Vol.57, pp. 1-10, 1859.
3. Jameson, A. Schmidt,W. and Turkel,E., "Numerical Solution of the Euler Equations by Finite Volume Methods using Runge Kutta Time Stepping Schemes," AIAA Paper 81-1259.
4. Steger, J. L., and Warming, R. F.,"Flux Vector Splitting of the Inviscid Gas Dynamics Equations with Applications to Finite Difference Methods,"J. Comp. Physics, Vol.40, pp.263-293, 1981.
5. Verhoff, A. and O'Neil, P.J.,"A Natural Formulation for Numerical Solution of the Euler Equations," AIAA Paper 84-0163.
6. Moretti, G.,"The λ - Scheme," Computers and Fluids, Vol.7, No.3, pp.191-205, 1979.
7. Jameson, A. and Caughey, D.A.,"A Finite-Volume Method for Transonic Potential Flow Calculations," Proc. of AIAA 3rd CFD Conference, pp. 35-54, June 1977.

Solutions of the Vorticity Transport Equation at High Reynolds Numbers

A. Dagan and R. Arieli
Rafael, P.O.Box 2250, Haifa 31021, Israel.

Introduction

The flow field about a two dimensional section at low Reynolds numbers, of the order of $Re \approx 60$ is well organized, with a pair of steady symmetric vortices[1]. Increasing the flow Reynolds number within the range of $10^2 < Re < 5 \cdot 10^5$, causes the flow to become unstable to asymmetric modes, resulting into a periodic vortex shedding behind the configuration. This proccess is the well known as the von Karman vortex street. For higher Reynolds numbers ($Re > 5{\cdot}10^5$), this flow pattern disappears, and a chaotic wake structure arises. The chaotic structure is essentially a turbulent flow regime, in which the stretching mechanism of the line vortices plays an important role. Thus, 2-D laminar model is unable to correctly predict the flow field.

The approximate factorization technique (AF), has been widely used in solving the compressible N.S. equation. The main advantage in the AF approach, is that the technique permits the use of a relatively large time step and the efficiency in their solving the resulting tri-diagonal block structure of the equations. However, for the incompressible case this method seems to be not sufficiently effective due to the absence of compressibility effects in the continuity equation. A steady state solution of the vorticity stream function system, can be achieved by adding a pseudo derivative term of ψ_t to the Poisson equation[2]. The resultant set of equations is easily solved using the AF-1 scheme, in a similar fashion to that used for the compressible case.

Another point of interest and importance is the out-flow boundary conditions (OFBC). The OFBC plays a crucial role in the success of any numerical scheme. This condition must prevent the reflection of error waves from the computational boundary, especially when the solution contains developing wakes. The usual way to handle the OFBC is to imply vanishingly small normal derivatives for both the vorticity and the stream function at the outflow computational boundaries[2]. Beher et. al.[2] has shown that the traction-free condition performs very well for most cases and allows to place the outflow BC at a distance of about 5 to 6 diameters away, for a $Re = 10^2$ case.

Mathematical formulation

A major concern in any numerical discretization scheme is the presence of numerical diffusion and dispersion. Central differencing of the advection terms is known to be dispersive while a first order upwind scheme based upon the vector flux splitting is diffusive. In the present study, central differencing schemes are used and they contribute truncation errors, that should be considered as a dispersion term in the equations. In the case of high Reynolds number flows, or in the presence of large gradients, additional time scales appear which cannot be resolved by the numerics. The lower frequencies associated with these time scales, do not cause any special problems. The high frequencies, on the other hand, cannot be ignored and must be accounted for in the algorithm. The most convenient way to eliminate the difficulties associated with the high frequencies, is to add to the algorithm some form of numerical dissipation. This dissipation introduces an error that does not exceed the truncation error and thus, does not affect the accuracy of any one of the physical viscous effects. In the present investigation a second order artificial diffusion term is introduced into the discretized form of the vorticity transport equation.

The governing equation namely, the 2-D vorticity equation and the stream function Poisson equation, are solved coupled together using a block ADI method[3,4]. The simultaneous solution of the two equations eliminate the difficulty emerging from the coupling between the vorticity wall condition and the Neuman boundary condition for the stream function. The discretization of the two equations in an orthogonal body fitted coordinates[5] is written:

$$\frac{1}{\Delta t}\delta_t(J_0\omega) - \delta_\eta\psi\delta_\xi\omega + \delta_\xi\psi\delta_\eta\omega = \frac{1}{Re}\{\delta_\xi(\alpha\delta_\xi) + \delta_\eta(\beta\delta_\eta)\}\omega + D_2\omega$$

$$\{\delta_\xi(\alpha\delta_\xi) + \delta_\eta(\beta\delta_\eta)\}\psi = J_0\omega \qquad (1)$$

where $D_2 = \delta_\xi(\epsilon_\xi\delta_\xi) + \delta_\eta(\epsilon_\eta\delta_\eta)$ while $\epsilon_\xi = \epsilon_\xi^{(2)}|\delta_\eta\psi|\sqrt{|\delta_\xi^2\omega/(\mu_\xi^2\omega)|}$, and a similar expression is valid for ϵ_η. $J_0 = x_\xi y_\eta - y_\xi x_\eta$ is the Jacobian, and α, β are the transformation metric quantities. The δ, μ are the central differencing and the averaging operators respectively.

Artificial diffusion is widely applied in any solution of compressible flow. Usually, this diffusion is composed of a second order term and a fourth order term. However, when an AF technique is used, it requires a block penta-diagonal solver for an implicit form of the fourth order diffusion term, or a tri-diagonal block solver for an explicit version of this term. Both methods require considerable computer resorces, when the explicit form is used, the required time step is smaller than that needed with the implicit form of the fourth order diffusion term. Moreover, past experience indicates that the solution of eq. (1) with a second order diffusion term such as in ref [6], might lead to numerical instability. Therefore, it is essential to construct the second order artificial diffusion term that ensures numerical stability. This form of artificial viscosity is of the same order of magnitude as the truncation error of equation (1) i.e. $O(\Delta x^2, \Delta y^2)$.

Numerical Results

Two test cases are considered in the present paper: a circular cylinder and a thick diamond shape profile. In order to initiate the asymmetric mode, the cylinder has been slightly distorted by generating an "elliptical bump" with a maximum height of 1.0% of the cylinder radius. This "bump" is located in a circumferential position at an angle of 45^0 to the flow direction. The flow impulsively starts from rest at $T = 0.0$ where the nondimensional time T is defined as $T = (tU_\infty)/D$. Considering the incompressible flow around this "distorted" circular cylinder, at a Reynolds number (based on free stream conditions and the cylinder diameter) of $Re_D = 10^4$, the computed velocity vectors in the vicinity of the solid surface are presented in fig. 1. On the same figure, though without sufficient resolution, are presented also the location of the two primary separation points and the location of the secondary separation as computed at a time level of $T = 10$. The location of the two primary separation points oscilates in the range of $90° \pm 10°$. The non-dimensional frequency of these oscilationas closely resembles the Strouhal number $S = nD/U_{\infty}$ of the von Karman vortex street.

A color map of equal vorticity regions in the flow around a circular cylinder is presented in fig. 2. The vorticity pattern of the flow at $Re = 10^2$ and at $Re = 10^3$ are shown in fig. 2(a) and 2(b) respectively. The numerical diffusion in the two cases is identical. Therefore, the major differences between the two figures can be entirely attributed to real viscous effects. For the case of $Re = 10^3$, the presently computed drag coefficient $C_d = 1.1$ compares well with the value [7] of $C_d = 1.0$. Similarly, the computed value of the Strouhal number is $S = 0.20$ while that of ref. [7] is $S = 0.21$. These results considerably differ from previous numerical solutions[8,9] of the 2-D N.S. equations, where computed values of the drag coefficient of the order of 1.7 to 1.8 were reported at Re numbers of $3 \cdot 10^3$, 10^4 and 10^5. These values overoredict by about 50% the results from avialable experimental data.

A sequence of eight different pictures presented in fig. 3(a) to 3(h) reveals the time history of the flow about the same "slightly distorted" circular cylinder at $Re = 10^4$. The vorticity levels in the field are ploted at $T = 1$, $T = 2$, $T = 3$, $T = 4$, $T = 5$, $T = 7$, $T = 9$ and $T = 10$, in figures 3(a) to 3(h) respectively. At the early stages of the flow, ($T = 1$, $T = 2$) the feeding sheet rolls up until two concentrated vortices appear. The roll up of the feeding sheet is clear and sharp thus demonstrating the relatively low artificial viscosity of the present solution. Due to the asymmetry, the inner vortex is slightly advected towards the feeding sheet of its counterpart ($T = 3$, $T = 4$), resulting in a pile up of the feeding sheet and a formation of a new vortex. This process is continued, until a complete von-Karman street wake is established ($T = 9$, $T = 10$). In this process, an annihilation takes place between the roled up feeding sheet and the counterpart vortex ($T = 7$) and it is found to play a minor role in the establishment of the wake formation. In addition, computed values of the sectional drag coefficient and the Strouhal number compares very well with available data from the literature. At $Re = 10^4$ the computed drag coefficient is $C_d = 1.25$ whilst the experimental result is found to be $C_d = 1.20$, the computed Strouhal number of 0.19 compares well with the experimental value of 0.20 from ref. [7].

The vorticity field as computed at $Re = 10^4$ around a diamond shape section, at flow conditions where the flow has a 45° inclination, is presented in figs. 4(a) to 4(d). The formation of the primary vortex as well as the secondary are clear in the early stage of the impulsively start flow. At $T = 1$, (fig. 4(a)) the separation and the reattachment of the sheed vorticity is clearly observed. However, at later stages, $T = 2$, $T = 3$ and $T = 7$, it is evident that the mechanism that governs the formation of the wake is the annihilation process. Such a process is observed in figs. 4(c) and 4(d), between the confined vortex generated at the central corner and the feeding sheet of leading edge.

References

1 Ta Phuoc Loc; "Numerical Analysis of Unsteady Secondary Vortices Generated by an Impulsively Started Circular Cylinder", J. Fluid Mech., 100, 111-128 (1980)

2 M. Behr. J. Liou. R. Shih and T.E. Tezduyar; "Vorticity- Stream function Formulation of Unsteady Incompressible Flow Past a Cylinder: Sensitivity Of the Computed Flow Field to the Location of the Outflow Boundary" Int Journal Nume. Meth. in Fluids. Vol 12,pp 323-342 1991.

3 M. Napolitano; "Efficient Solution of Two-Dimensional Steady Separated Flows" Computers Fluids Vol 20. No 3. pp 213-222, 1991.

4 A. Dagan and D. Almosnino; "Vorticity Equation Solutions of Slender Wings at High Incidence". AIAA Journal Vol. 29, no. 4, pp.497-504 April 1991.

5 J.L. Steger Y.M. Rizk; "Generation of Three -Dimensional Body-Fitted Coordinates Using Hyperbolic Partial Differential Equations", NASA TM 86753, June 1985.

6 T. H. Pulliam; "Euler and Thin Layer Navier-Stokes Codes: ARC2D, ARC3D" Notes for Computational Fluid Dynamics user's Workshop, the University of Tennessee Space Institute Tullahoma, Tennessee, March 1984.

7 X. Wang, and C. Dalton; "Numerical Solutions for Impulsively Started and Decelerated Viscous Flow Past a Circular Cylinder", Int Journal Nume. Meth. in Fluids. Vol 12 pp 383-400, 1991.

8 T. Tamura, K. Tsuboi and K. Kuwahara; "Numerical Simulation of Unsteady Flow Patterns Around a Vibrating Cylinder", AIAA Paper 88-0128, Reno NV, 1988

9 K. Tsuboi, T. Tamura and K. Kuwahara; "Numerical Study for Vortex Induced Vibration of a Circular Cylinder in High Reynolds Number Flow",AIAA Paper 89-0294, Reno, NV, 1988.

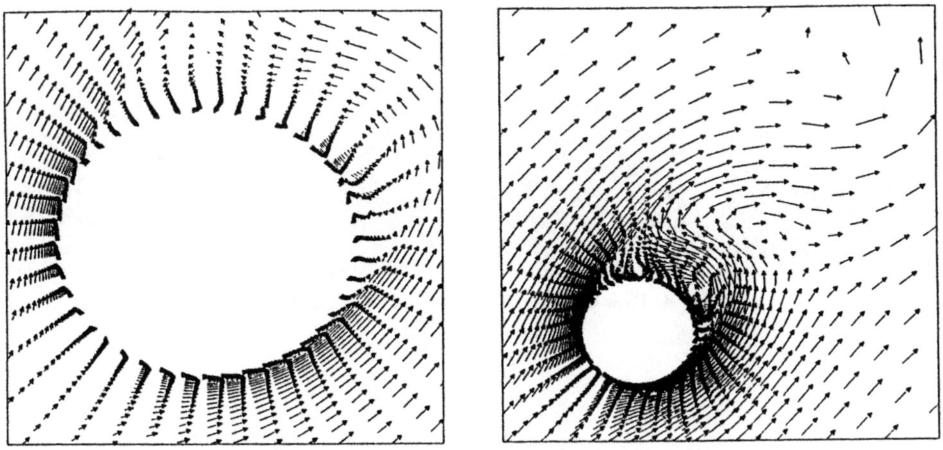

Fig. 1: Velocity vectorplot of Flow around a circular cylinder $Re = 10^4$ $T = 10.0$.

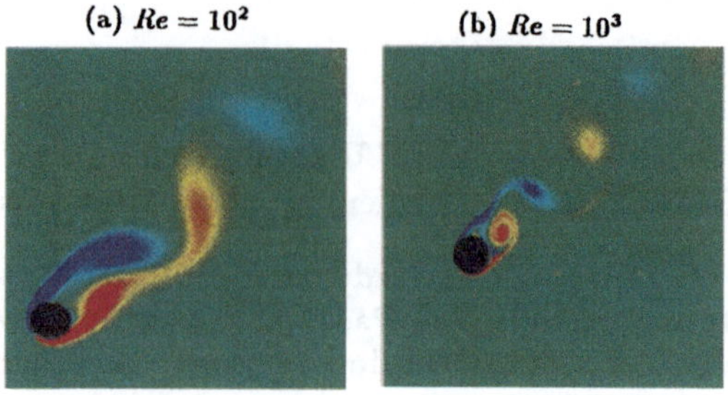

Fig. 2: Flow around a circular cylinder.

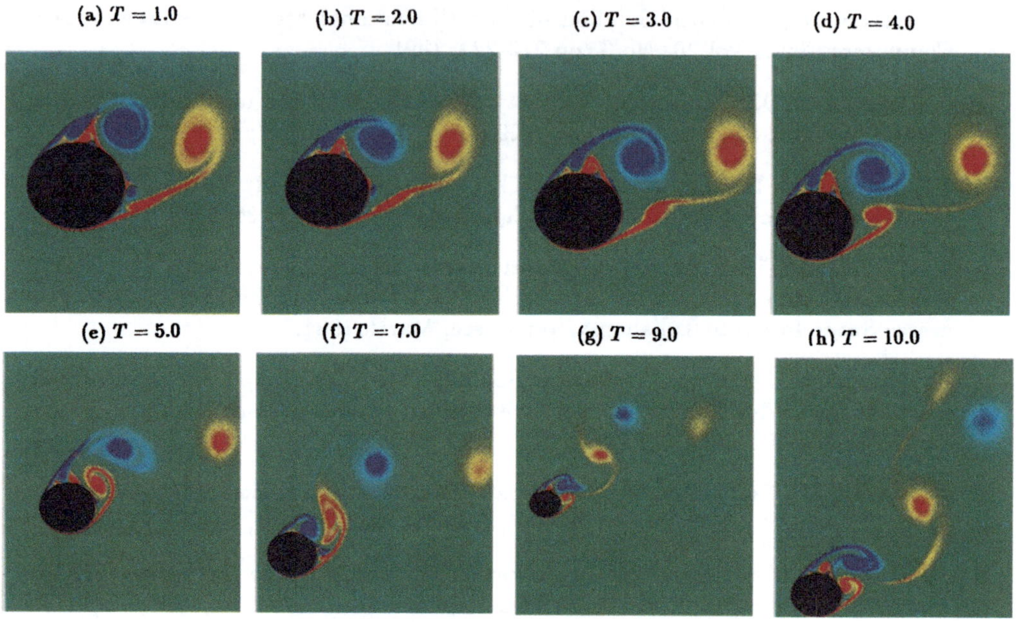

Fig. 3: Flow around a circular cylinder at $Re = 10^4$.

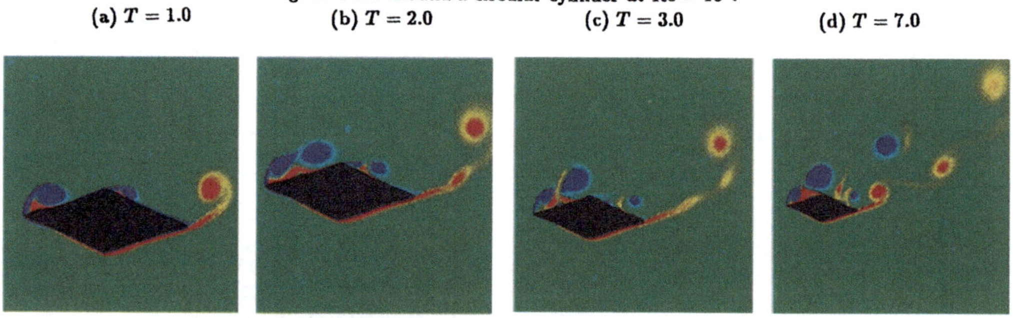

Fig. 4: Flow around a 2:1 "diamond" section at $Re = 10^4$.

MAGNETICALLY DRIVEN FLOW IN AN ANNULUS

G. de Vahl Davis[1], E. Leonardi[1], Y. Meinhardt[2] O. Pade[2] and M. Wolfshtein[3]

[1]University of New South Wales, Kensington, Australia. [2]Scientific Department, Ministry of Defense, Haifa, Israel. [3]Technion-Israel Institute of Technology, Haifa.

1 Introduction

This paper describes the study of the MHD flow which is produced in an annular region containing a weakly ionised gas by the simultaneous application of a radial electric current and an axial magnetic field. The resultant body force, and hence the base flow, is in a circumferential direction.

A rotating flow can be generated in ways other than by MHD forces: for example, by the rotation of one or both of the cylindrical walls of an annulus (the Taylor problem) or by a pressure gradient imposed on a fluid in a curved duct (the Dean problem). In these and other cases, the base flow may become unstable at some value of the relevant driving force, leading to the generation of secondary flow, e.g. Taylor vortices. The same is true here. It has been found that when the product of the imposed electric current and magnetic force exceeds a value which depends upon the properties of the fluid and the geometry of the annulus, a secondary flow in the form of toroidal vortices is established. Our first purpose in this work has been to explore and describe these vortices and the conditions for their formation, and to relate these conditions to those applying in other, similar, flows. A further consequence of the passage of an electric current through a fluid is the generation of heat (Joule heating). The overall rate of heat transfer, and its distribution along the annulus walls, is affected by the secondary flow. A study of this effect has been our secondary aim in this work.

2 A Description of the Problem

A weakly ionised plasma at low pressure is contained in an annular cavity of length L and inner and outer radii R_i and R_o. The gas is ionised by the application of an electric potential between the cylindrical walls; hence a radial electric current **I** passes through the fluid. In addition, a uniform axial magnetic field **B** is imposed.

All walls of the annulus are at rest. The curved walls are held at a constant and uniform temperature T_w, while the plane end walls are adiabatic. We simplify by assuming the fluid to be incompressible with constant properties. Buoyancy forces can be shown to be small in comparison with the electromagnetic body forces and are neglected.

3. Mathematical Formulation

3.1 The flow equations The flow is axi-symmetric, incompressible and laminar, with zero dissipation and flow work. The governing equations are the conservation equations for mass, axial, radial and circumferential momentum and energy. The interesting feature of the equations in this context is the MHD body force. We have sought their steady solution in (r, ϕ, z) coordinates as $t \to \infty$.

3.2 The electromagnetic interaction The complete body force in a conducting fluid depends upon the current **I** (or, which is equivalent, the current flux **J**), the magnetic field **B** and the electric field **E**. Contributions from variations in the relative permittivity and relative magnetic permeability are negligible in MHD.

The body force is $\mathbf{F} = \mathbf{J} \times \mathbf{B}$; and the heat generation per unit volume by the electric current is $S_J = \mathbf{J}\cdot\mathbf{J}/\sigma_0$ where σ_o is the electrical conductivity of the fluid. The vectors **J** and **B** are governed by the Maxwell equations. In the present context, when we are considering a single component, weakly ionised plasma, we may make the following further assumptions:

(i) constant thermodynamic, transport, magnetic and electric properties;
(ii) $\mathbf{B} = (0, 0, B_z)$ and $\mathbf{E} = (E_r, 0, 0,)$ so that **B** and **E** are orthogonal; also B_z is uniform and E_r is not a function of z; and $\mathbf{I} = (I_r, 0, 0)$;
(iii) the magnetic Reynolds number $Re_m = \sigma_0 \mu_m \nu << 1$, where $\nu = \mu/\rho$ is the kinematic viscosity and μ_m is the magnetic permeability; thus the induced magnetic field is small compared with the applied magnetic field;
(iv) the Hartmann number $Ha = B_z\, a(\sigma_o / \mu)^{1/2} < 1$, where $a = R_o - R_i$;
(v) end effects of the magnetic field are neglected;
(vi) the Hall parameter of the electrons β_e is of order unity;
(vii) the Hall parameter of the ions $\beta_i << \beta_e$ and can be neglected;
(viii) the temperature and velocity of the neutral particles are similar to those of the ions, so that the fluid can be treated as a single-component continuum;
(ix) the temperature of the electrons is much higher than that of all the other constituents;
(x) the ions are singly ionised;
(xi) the bremsstrahlung radiation of the electrons is negligible.

With these assumptions, the body force and Joule heating terms can be shown to be

$$F_r = J_\phi B_z = (\beta_e J_r - \sigma_0\, u\, B_z) \qquad F_\phi = -J_r B_z = -\, I_r B_z / 2\pi r L$$

$$F_z = 0 \qquad S_J = (J_r^2 + (\beta_e J_r - \sigma_0\, u\, B_z)^2) / \sigma_0$$

4 Stream function/vorticity formulation

The governing equations were transformed from the primitive variable formulation (u, v, w and p) to one in terms of the stream function ψ, vorticity ζ and swirl Γ. The equations were made dimensionless using the quantities $a = R_o - R_i$, ν/a, a^2/ν, T_w and

$\rho\nu^2/a^2$ as reference values for length, velocity, time, temperature and pressure respectively.

A stream function ψ, vorticity ζ and swirl Γ were introduced by the definitions

$$u = \frac{1}{r}\frac{\partial\psi}{\partial z} \qquad w = \frac{1}{r}\frac{\partial\psi}{\partial r} \qquad \Gamma = rv \qquad \zeta = \frac{\partial u}{\partial z} - \frac{\partial w}{\partial r}$$

The following parameters are used:

$A = L/a,$ aspect ratio

$\gamma = C_p/C_v$ specific heat ratio;

$Br = \rho\nu^3/kT_w a^2$ Brinkman number;

$Ha = aB_z(\sigma_0/\mu)^{1/2}$ Hartmann number;

$Pr = C_p\mu/k$ Prandtl number;

$C = I_r B_z a/2\pi\rho\nu^2$

The vorticity transport equation is

$$\frac{\partial\zeta}{\partial t}+\frac{1}{r}\frac{\partial}{\partial r}(ru\zeta)+\frac{\partial}{\partial z}(w\zeta)-\frac{2\Gamma}{r^3}\frac{\partial\Gamma}{\partial z}-\frac{u\zeta}{r}=-Ha^2\frac{\partial u}{\partial z}+\frac{\partial^2\zeta}{\partial r^2}+\frac{1}{r}\frac{\partial\zeta}{\partial r}+\frac{\partial^2\zeta}{\partial z^2}-\frac{\zeta}{r^2}$$

The stream function - vorticity relationship is

$$\zeta = \frac{1}{r}\left\{\frac{1}{r}\frac{\partial\psi}{\partial r}+\frac{\partial^2\psi}{\partial r^2}+\frac{\partial^2\psi}{\partial z^2}\right\}$$

and the circumferential momentum equation may be written, in terms of the swirl,

$$\frac{\partial\Gamma}{\partial t}+\frac{1}{r}(ru\Gamma)+\frac{\partial}{\partial z}(w\Gamma)=-\frac{C}{A}+\frac{\partial^2\Gamma}{\partial r^2}+\frac{1}{r}\frac{\partial\Gamma}{\partial z}+\frac{\partial^2\Gamma}{\partial z^2}$$

The parameter C is more conveniently expressed as an MHD Dean number

$$De = \frac{V_m a}{\nu}\left(\frac{a}{R_o}\right)^{1/2}$$

where $V_m = R_O \dfrac{I_r B_z}{4\pi L\mu}\dfrac{(1-\eta^2)^2-4\eta^2(2\eta\ell n\eta)^2}{4(1-\eta)(1-\eta^2)}$ in which $\eta = R_i/R_o$.

The energy equation is

$$\frac{\partial\theta}{\partial t}+u\frac{\partial\theta}{\partial x}+w\frac{\partial\theta}{\partial z}=\frac{\gamma}{\Pr}\left(\frac{\partial^2\theta}{\partial r^2}+\frac{1}{r}\frac{\partial\theta}{\partial r}+\frac{\partial^2\theta}{\partial z^2}\right)+\frac{\gamma}{\Pr}\left\{\frac{C^2\,Br}{A^2\,Ha^2\,r^2}+\left(\frac{\beta_c\,C^{1/2}\,Br^{1/2}}{A\,Ha\,r}-Br^{1/2}\,Ha\,u\right)^2\right\}$$

5 Boundary conditions

The normal velocity components are zero on the boundaries. Accordingly, the stream function is a constant there; we have set the constant to be zero. The boundary condition for vorticity is derived from the vorticity transport equation. Temperature is uniform at T_w on the curved boundaries, so that $\theta = 1$ there, and the end walls are adiabatic.

6 Solution Procedure

The equations were solved by finite differences, using forward differences in time and central differences for all spatial derivatives. The resulting FDAs were solved by the Samarskii-Andreyev ADI scheme using the false transient technique. Convergence was tested by checking the norm of the relative change of all variables (ψ , ζ , Γ and θ) over the entire solution region, as well as by computing the norm of the residuals of all equations (in their stationary form) over the solution region.

7 Results

Solutions have been computed for a range of values of radius ratio β and Dean number *De*. All results presented here are for $A = 15$; $Pr = 0.7$, $\gamma = 1.63$, $Br = 1.5944 \times 10^{-8}$, $\beta_e = 1.767$, $Ha = 0.1679$. The mesh size used was 11×181; a test of 21×181 showed a change in solution characteristics of 2% or less and the extra computing cost associated with the finer mesh was not felt to be warranted.

For a given radius ratio, three classes of solution were obtained, depending upon *De*. At low values of *De*, the flow consisted of a simple circumferential motion modified only by the no-slip condition at the walls. The stream function showed two weak cells at each end of the annulus, and the radial and axial motions were essentially zero. The temperature θ and circumferential velocity v increased monotonically from their wall values (1 and 0 respectively) to a maximum in the centre of the solution region. Above a certain critical Dean number De_l, however, a multi-cellular motion in the *r-z* plane was formed, superimposed on the circumferential motion, and reminiscent of the Taylor vortex phenomenon. The distributions of θ ,ψ and v shown in Figure 1, show this flow.

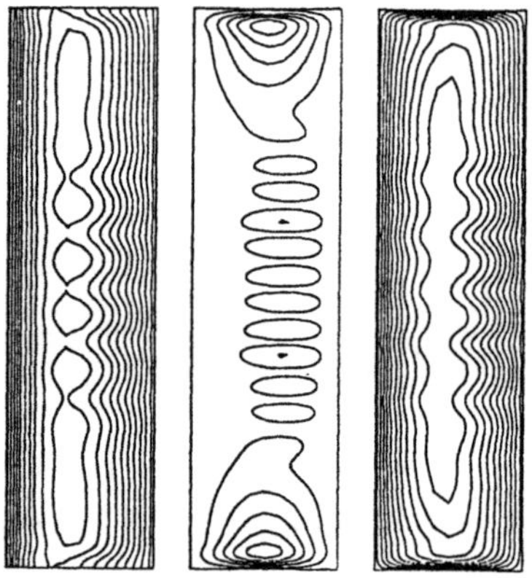

(a) (b) (c)

Fig. 1: Contours of (a) θ (b) ψ (c) v

The steady, cellular motion persisted up to a further critical value, De_u . Beyond this second critical value, the solution became unsteady and oscillated periodically. Although the false transient solution procedure does not permit an accurate tracking of these oscillations, computational experience shows that it enables the limit between the steady and unsteady multi-cellular regions to be established. As a function of radius ratio η, the critical Dean numbers are - for the fixed values of the other parameters as listed above,

$$De_l = 29.8 + 7.69 / \eta$$
$$De_u = 34.8 + 7.69 / \eta$$

Figure 2 shows the three regions. The region of steady cells is quite narrow. In the unsteady region, cells are also observed at the values of *De* investigated. However, the present code does not allow the three-dimensionality, which is likely to arise at a sufficient value of *De*, or the true transient behaviour of the flow, to be computed.

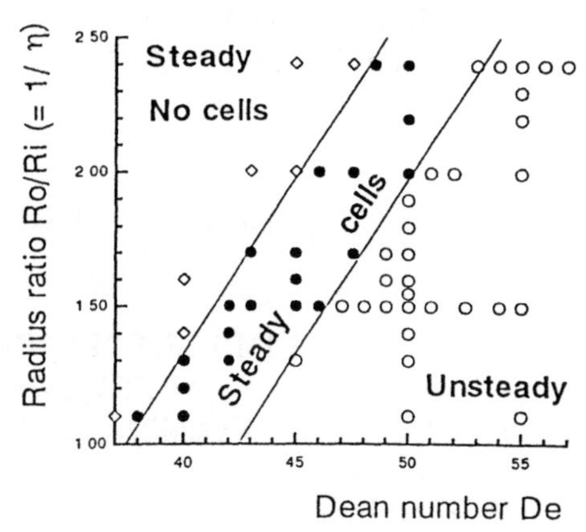

Figure 2: Stability limits

The secondary flow affects the rate of heat transfer, especially at the outer surface of the annulus, as shown in Figure 3

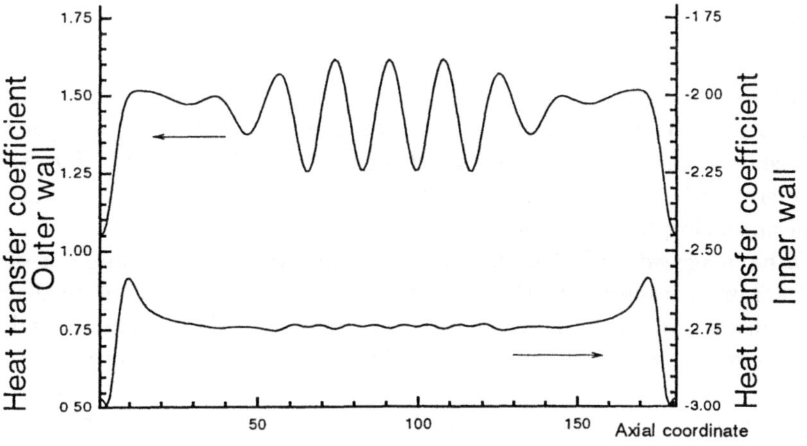

Figure 3: Local heat transfer coefficients

8. Conclusion

It has been shown that, at sufficiently high Dean number, a multi-cellular secondary flow is imposed on the circumferential base flow, with a consequential effect also on the axial distribution of the heat transfer. The similarities to the Taylor and Dean instabilities are remarkable.

MULTIBLOCK NAVIER STOKES METHOD TO SIMULATE TURBULENT TRANSONIC FLOW OVER A DELTA-WING

C. Eldem[1], J. Vos[1], A. Rizzi[2] & J. Oppelstrup[3]

[1] IMHEF/EPFL, CH-1015 Lausanne, Switzerland
[2] FFA, S-16111 Bromma, Sweden
[3] C2M2/KTH, S-10044 Stockholm, Sweden

1 Introduction

Even though the geometry of a delta wing is reasonably simple, it still is a challenging problem to simulate with quantitative accuracy the vortex-dominated flow over the wing at high angle of attack. The reasons for this are: that multiple vortices (primary, secondary, and even tertiary vortices) interact in complex ways, that to obtain the correct separation of the boundary layers which create these vortices is difficult, that shock waves may interact with the vortices, and finally that the flow is turbulent and needs to be modelled. For these reasons this problem has become something of a standard for validation of Navier-Stokes solvers, e.g. [1].

The wing we have chosen to study here is the delta with 65° sweep angle that NASA Langley is measuring in its NTF cryogenic windtunnel. What makes this case unique for validation is the wide range in Reynolds number at which experimental data have been obtained, approximately between 10 and 150 millions. Here the challenge is even greater because the boundary layer and separating shear layer are thinner and the mesh spacing has to decrease to resolve them properly.

We have agreed with Langley to exchange our data for a number of cases that are of common interest. In this report we concentrate on a numerical study of one flow condition, $M_\infty = 0.85$ and $\alpha = 12°$, at two Reynolds numbers, 9 and 72 million, in both laminar and turbulent flow. This paper reports the progress of our work including the Reynolds number effect on our computations and the difficulties we have encountered in computing high Reynolds number vortex flows.

2 Numerical Approach

Our past work [2] has studied vortices that develop over blunt delta wings in laminar hypersonic flow for relatively low Reynolds number. This was carried out with a single-block Navier Stokes code. A multiblock Euler code has also been developed for inviscid hypersonic flow [3]. These two efforts have led now to the multiblock Navier-Stokes solver NSMB, and we present first results here for turbulent transonic flow.

We briefly outline the by-now rather standard finite volume approach we take , to solve the Reynolds-averaged Navier-Stokes equations written in Cartesian coordinates over a control volume $\mathcal{V}$ with boundary $\partial\mathcal{V}$

$$\frac{\partial}{\partial t}\int\int\int_{\mathcal{V}} \mathbf{u}\, d\mathcal{V} + \int\int_{\partial\mathcal{V}} \mathcal{F}(\mathbf{u})\cdot\mathbf{n}\, dS = 0 \tag{1}$$

where the vector of state variables $\mathbf{u} = (\rho \;\; \rho u \;\; \rho v \;\; \rho w \;\; \rho E)^T$ contains density ρ, x, y, and z components of velocity u, v, and w, and energy per unit mass E. The x direction is aligned with the body axis. The flux vector $\mathcal{F}$ is composed of inviscid and viscous parts

$$\mathcal{F} = (\mathbf{F}_I - \mathbf{F}_V)\mathbf{e}_x + (\mathbf{G}_I - \mathbf{G}_V)\mathbf{e}_y + (\mathbf{H}_I - \mathbf{H}_V)\mathbf{e}_z \tag{2}$$

in the x, y, and z coordinate directions, respectively. The Baldwin-Lomax model with the Degani-Schiff modification completes the system for the turbulent viscosity and heat conduction coefficients.

The solver uses a cell- centered discretization in space with explicit Runge- Kutta time stepping. The convective flux at a cell face is computed as the flux of the weighted average of the two state variables to the left and right of the face. The gradients of the velocities and the temperature in the viscous terms are computed at the cell vertices using a staggered grid. The gradients at the cell faces are obtained by averaging of the cell vertex gradients leading to a compact stencil for the second derivatives.

To stabilize the numerical discretization, an artificial smoothing term consisting of a nonlinear blend of second and fourth differences is added to the numerical scheme. The smoothing terms are scaled by the local eigenvalue in each coordinate direction for the Navier-Stokes solutions to better handle the high aspect ratios of the mesh in the boundary layer. It is very important to treat the numerical smoothing terms correctly near the boundaries in order to maintain good convergence and to keep the variations in entropy small.

The spatially discretized equations are integrated in time with a standard explicit four stage Runge-Kutta scheme that uses the usual local time step scaling. The time integration is augmented by an optional implicit residual smoothing procedure.

The usual no-slip boundary conditions are enforced on the wing. The farfield is assumed to be inviscid and is treated either by setting or extrapolating the locally one dimensional Riemann invariants. The condition for pressure on the wing surface is set by second order extrapolation from the field values.

The use of multiple patched blocks is novel. In principle the method permits the use of an arbitrary number of patched blocks. To reduce memory requirements, the allocation of space for the variables is done block by block using a Dynamic Memory Manager (DMM) [4]. Temporary storage is allocated only for the largest block. The solver uses the MEM-COM data base system [4] as data structure. The grid system and calculated results are stored in MEM-COM block by block, and can be accessed independent of each other. The main advantage of using a data base system is that all data required for a flow simulation is stored in a single data base file. Exchange of information between the different processes (grid generation, solving the equations, flow visualization) is done via the data base file.

3 Computed Results

We focus on the NASA-Langley NTF Experiment of laminar as well as fully turbulent flow over a 65° swept delta wing with a round leading edge and adiabatic wall conditions at 12° angle of attack, Mach number of 0.85, and two *Re* numbers: 9.0×10^6 and 7.2×10^7 based on the root wing chord. These are our first results with this code for turbulent transonic flow, and these two cases are initial verification of the code. We experienced more difficulty to reach convergence at the higher Reynolds number than at the lower one. Becker [5] also reported difficulty in studying this wing.

3.1 Mesh

With the mesh spacing normal to the wall set to resolve the higher Reynolds number case, we choose a C-O type of mesh around the delta wing that envelopes the wing back to the trailing edge, and continues to contain the wake downstream of the wing (Fig 1). At the finest resolution the two-block C-O type grid contains $97 \times 57 \times 81$ points. The grid in the streamwise direction extends downstream 3 chord lengths. Solutions were obtained by successively refining the mesh from a coarse to a medium and then to a fine resolution numbering approximately 448 K points after 6000 iterations.

Case 1: $R_e = 9 \times 10^6$

The particle traces in Fig 2 indicate the vortex flow above the wing. The isobar contours on the upper surface (Fig 3) also show the footprint of the vortex as a pressure trough. Figure 3 also presents isobars in the plane at the 40% chord station, indicating a secondary vortex. The comparison of surface C_p along the span (Fig 4) shows that the position of the vortex is somewhat more inboard in the computations than in the measurements. That there is little difference between the laminar and turbulent computations is unreasonable, and suggests that our turbulence model provides too little turbulent viscosity.

Case 2: $R_e = 7.2 \times 10^7$

To reach convergence for this case we had to increase the artificial viscosity very substantially. The results must be viewed with this in mind. This may explain why the pressure trough (Fig 5) is not as deep as in Fig 3 and the absence of a secondary vortex in the isobars in the plane at the 40% chord station. The comparison of surface C_p along the span (Fig 6) shows a lower suction level in the computed pressure than in the measurements, which is also expected. At this time we have no good explanation of what causes our convergence problems, and we can offer no remedies. It is however significant to point out the very substantial differences between the experimental C_p values for the low and high Reynolds number cases (Figs 4 and 6).

This suggests that there is in fact a dominant effect on the flow with increasing Reynolds number. How to treat this effect numerically in order to obtain an accurate solution remains for future work.

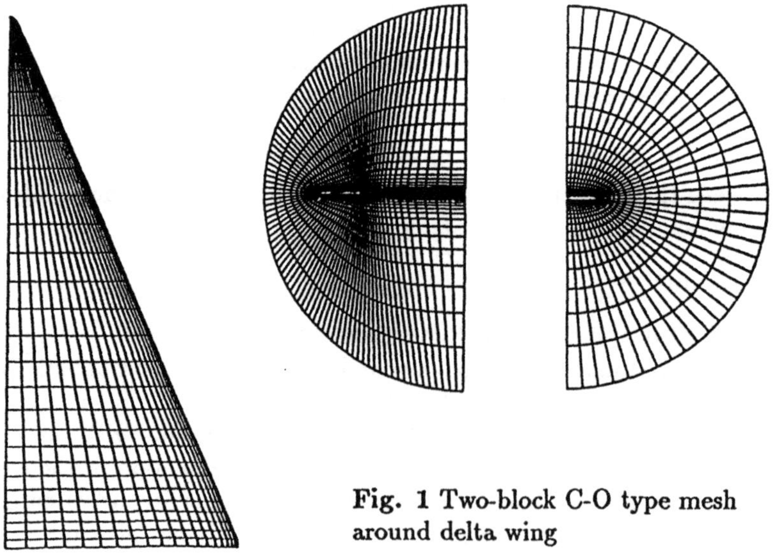

Fig. 1 Two-block C-O type mesh around delta wing

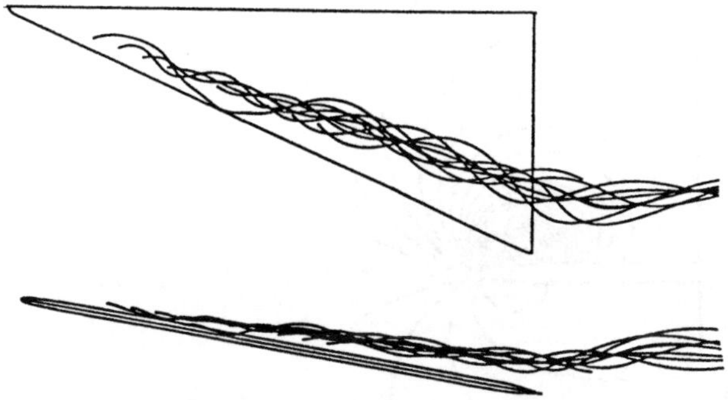

Fig. 2 Streamlines over delta wing showing vortex

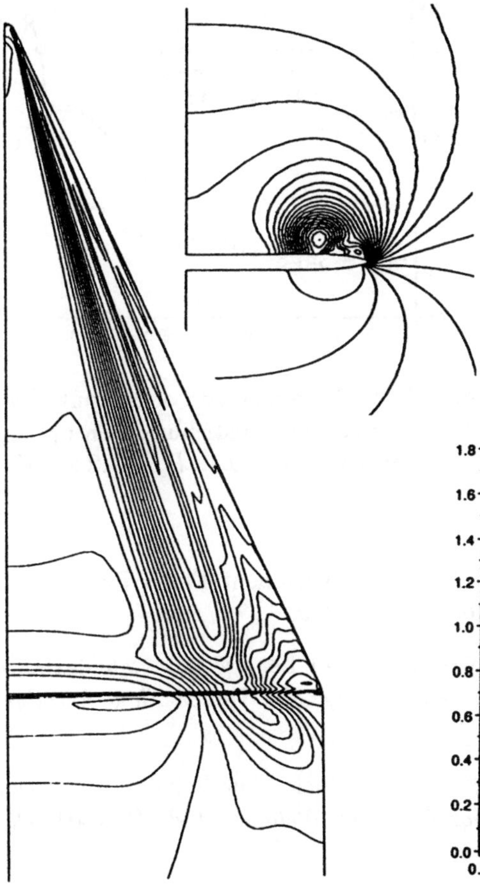

Fig. 3 Isobars on upper surface of wing and in 40% chord section, $R_e = 9.0 \times 10^6$

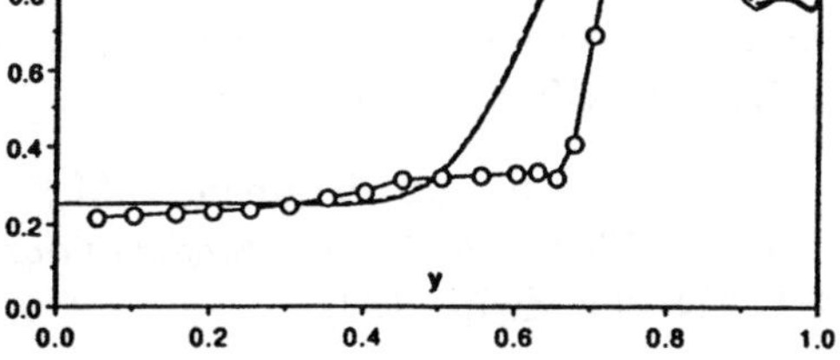

Fig. 4 Comparison of computed and measured C_p distribution on upper surface, $x/c = 0.4$, $R_e = 9.0 \times 10^6$

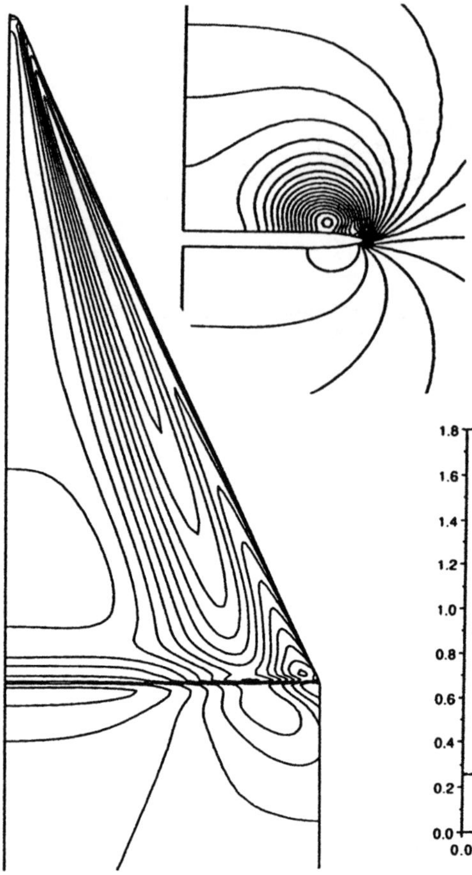

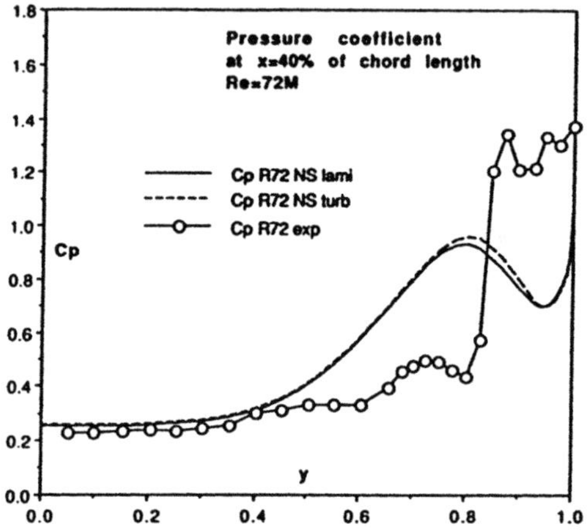

Fig. 5 Isobars on upper surface of wing and in 40% chord section, $R_e = 7.2 \times 10^7$

Fig. 6 Comparison of computed and measured C_p distribution on upper surface, $x/c = 0.4$, $R_e = 7.2 \times 10^7$

References

1. Hilgenstock, A.: Validation of Transonic Turbulent Flows Past Delta Wing Configurations, **Aero J.**, Aug/Sept 1991, pp.219-230.

2. Rizzi, A., Murman, E.M., Eliasson, P., and Lee, M.K.: Calculation of Hypersonic Leeside Vortices Over Blunt Delta Wings, in *Vortex Flow Aerodynamics*, **AGARD CP-494**, Paris 1990.

3. Vos, J.B. and Bergman, C.M.: A MultiBlock Flow Solver For Inviscid Hypersonic Flows, in *Aerothermodynamics for Space Vehicles*, (ed) B. Battrick, **ESA SP-318** Noordwijk, 1991.

4. MEM-COM An Integrated Memory and Data Management System, MEM-COM User Manual Ver 5.6, **SMR TR-5056**, SMR Corp., Box 41, CH-2500 Bienne, 1988.

5. Becker, T.M.: Hybrid 3D Euler/Navier Stokes Calculations of Transonic Vortex Flows over the NTF Delta Wing, Masters Thesis, CFDL-TR-89-7, MIT, Cambridge, 1989.

Quasigasdynamic equations and computer simulation of viscous gas flows

Elizarova T., Chetverushkin B., Sheretov Yu.

Institute for Mathematical Simulation Russian Akademy of Science,
Moscow 125047, Miusskaya sq.,4
June 25, 1992

1. Quasigasdynamic model.

Quasigasdynamic (QGD) equations may be constructed using the next assumption. Assume that kinetic model for the behaviour of the distribution function $f(\mathbf{x}, \xi, t)$ is a cyclically recurring process of collision - free scattering of gas molecules, followed by instantaneous Maxwellization. So distribution function in a new time step t^{j+1} may be defined with the help of the expression

$$f^{j+1}(\mathbf{x}, \xi, t) = f_0{}^j(\mathbf{x} - \xi\tau, \xi, t) \tag{1}$$

where $f_0{}^j$ is the locally - Maxwellian on the time layer t^j, $\tau = t^{j+1} - t^j$ is a characteristic mean collision time.

Expanding (1) in a Taylor series in the parameter $\xi\tau$ we will have a model kinetical equation [1]. In standard notations it looks in the next way [3]

$$f_t + (\xi\nabla)f - (\xi\nabla)\tau(\xi\nabla)f = \mathcal{J}(f, f') \tag{2}$$

Then we successively multiply (2) by the summation invariants $\varphi(\xi) = 1, \xi, 0.5\xi^2$ and integrate over all molecular velocities ξ. Then we will have the system of differential equations for the macroparameters - QGD model [1], [2] . In invariant form it may be written in the invariant form [3]

$$\rho_t + div\rho\mathbf{u} = div\tau(div\rho\mathbf{u} \otimes \mathbf{u} + gradp) \tag{3}$$

$$\begin{gathered}(\rho\mathbf{u})_t + div(\rho\mathbf{u} \otimes \mathbf{u}) + gradp = \\ = div\tau div\rho\mathbf{u} \otimes \mathbf{u} \otimes \mathbf{u} + \nabla\tau(\nabla \otimes p\mathbf{u}) + grad\tau divp\mathbf{u} + \nabla\tau(\nabla \otimes p\mathbf{u})\end{gathered} \tag{4}$$

$$E_t + div((E + p)\mathbf{u}) = div\tau(div(E + 2p)\mathbf{u} \otimes \mathbf{u} + gradp/\rho(E + p)) \tag{5}$$

By augmenting (3) - (5) with the equations of state of an ideal gas

$$E = \rho(\mathbf{u}^2/2 + \varepsilon), \varepsilon = p/\rho(\gamma - 1) \tag{6}$$

and, also, the initial and boundary conditions, we obtain a closed system of equations which describes the space-time evolution of the macroscopic parameters of the gas: $\mathbf{u}$ - velocity, ρ - density, p - pressure, E - energy. Where $\tau = \lambda/c$ - characteristic relaxation time for gas molecules, c - sound velocity , λ - molecule free-pass.

2. QGD and Navier-Stokes system correlation.

Correlation QGD and Navier-Stokes equations is analysed [3]. It is shown that in the case of one particle Boltzmann gas the asymptotic of QGD equations when $\lambda \to 0$ are Navier-Stokes equations in the form (λ assumed to be a small parameter).

$$\rho_t + div\rho\mathbf{u} = 0 \tag{7}$$

$$(\rho\mathbf{u})_t + div(\rho\mathbf{u} \otimes \mathbf{u}) + gradp = div\sigma \tag{8}$$

$$E_t + div((E+p)\mathbf{u}) + divq = div\mathbf{u}\sigma \tag{9}$$

where

$$\sigma = \mu((\nabla \otimes \mathbf{u}) + (\nabla \otimes \mathbf{u})^T - 2/3Idiv\mathbf{u})$$

$$\mathbf{q} = -\kappa\nabla T$$

Viscosity and heat-conductivity coefficients correspond to one-particle gas model and are represented in the form $\mu = p\tau$, $\kappa = c_p p\tau$ and Prandtl number $Pr = 1$.

Particularly in boundary layer approach QGD equations degenerate to Prandtl equations [2].

QGD system requires the formulation of additional boundary conditions compared with Euler or Navier-Stokes equations. It is the sequence of the second order space derivatives in the right hand of the equation (3). This additional boundary condition for pressure may be written in the form $\partial p/\partial n = 0$. It ensures that there is no mass flow through the solid wall.

3.Entropy equation.

For the QGE system the equation for entropy $S = c_v \ln p/\rho^\gamma + S_0$ was constructed in the form:

$$(\rho S)_t + div\rho\mathbf{u}S = div\tau div\rho S\mathbf{u}\otimes\mathbf{u} + div\tau Sgradp + div\kappa(\nabla T)/T + \Phi_{QG}/T + \kappa(\nabla T/T)^2 \tag{10}$$

Here we have used the dissipative function Φ_{QG} that was build up for QGD equations. This function is rigorous positive and is defined as the sum of Navier-Stokes dissipative function Φ_{NS} and the left hands of stationary Euler equations in square with positive coefficients [3]

$$\Phi_{QG} = \Phi_{NS} + \tau p/\rho^2(div\rho\mathbf{u})^2 + \tau/\rho(\rho(\mathbf{u}\nabla)\mathbf{u} + \nabla p)^2 + \tau/(\rho\varepsilon)(\rho(\mathbf{u}\nabla)\varepsilon + pdiv\mathbf{u})^2 \tag{11}$$

From (10) it turns out the suggestion about a rise of the full thermodynamic entropy function for adiabatically isolated system in the form

$$\frac{d}{dt}\int(\rho S)dx \geq 0$$

The equal sign takes place at the equilibrium state. We supposed that the domain boundary is piecewise smooth.

It is known that similar results were obtained for Navier-Stokes system. Recently the similar results were obtained for hydrodynamic equations with thermal diffusion.

4. Shock-wave structure problem.

Here we consider the problem about the structure of a static shock wave. It is known that the solution of that problem based on Navier-Stokes equations for Mach numbers $M \geq 2$ is disagreed with the experimental data.

The problem is examined basing on QGD system in plane one-dimensional form

$$\frac{\partial \rho}{\partial t} + \frac{\partial \rho u}{\partial x} = \frac{\partial}{\partial x}\tau\frac{\partial}{\partial x}(\rho u^2 + p) \tag{12}$$

$$\frac{\partial \rho u}{\partial t} + \frac{\partial}{\partial x}(\rho u^2 + p) = \frac{\partial}{\partial x}\tau\frac{\partial}{\partial x}(\rho u^3 + 3pu) \tag{13}$$

$$\frac{\partial E}{\partial t} + \frac{\partial}{\partial x}u(E+p) = \frac{\partial}{\partial x}\tau\frac{\partial}{\partial x}u^2(E+5/2p) + \frac{\gamma}{\gamma-1}\frac{\partial}{\partial x}\frac{p}{\rho}\tau\frac{\partial p}{\partial x} + \frac{\gamma Pr^{-1}}{\gamma-1}\frac{\partial}{\partial x}p\tau\frac{\partial p/\rho}{\partial x} \tag{14}$$

Here parameter τ is taken in accordance with

$$\tau = (p_0\tau_0)(T/T_0)^s/p \tag{15}$$

and is represented as characteristic relaxation time for gas molecules. For (12) - (14) we have used the spatial discretization of the second order. The resulting equations are solved by time-relaxation method. As initial data we choose parameters of the flow according with the Hugoniot conditions. We take into consideration ideal polytropical one-particle gas $\gamma = 5/3, Pr = 2/3, S = 0.72$. The results of the computations are represented in Fig.1.

The normalized shock thickness is plotted against the Mach number for moderate-strength shock range. For comparison experiments, Navier-Stokes and Mott-Smith solution are plotted together. These are reproduced from [6]. It is seen that QGD solution correlates much better with the experimental value than Navier-Stokes one. This investigations are now being continued.

5.Kinetical-Consistent Finite-Difference Schemes.

Based on the kinetical model mentioned at the beginning of the paper kinetical-consistent finite-different (KCFD) schemes were constructed in [1]. For viscous heat-conducting flows KCFD schemes were obtained in [2]. The same result was obtained by averaging finite-difference scheme for model kinetic equation (2). The system (3) (5) can be used for the KCFD scheme construction in the next manner.

We note the superficial similarity between a number of the terms occurring in the QGD system and the corresponding terms in the Navier-Stokes equations. So it is possible to replace this terms by the Navier-Stokes terms and to consider them as the natural viscosity and thermal conductivity. (Mainly it is shifted viscosity terms).

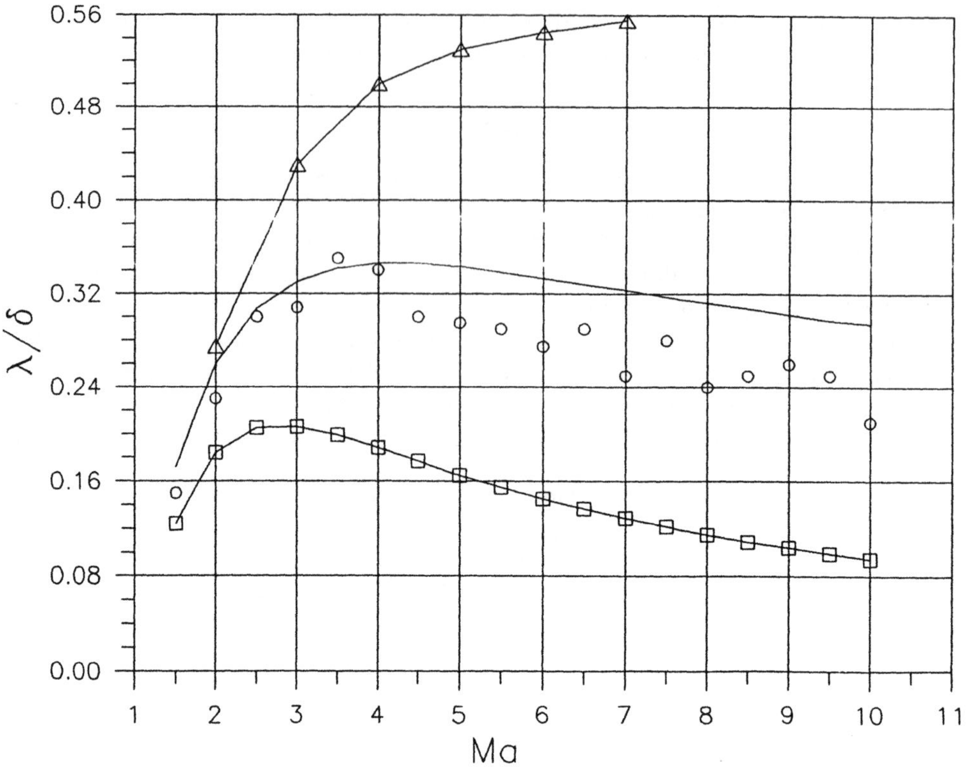

Figure 1: Shock thickness λ_1/δ versus M, solid line - QGD solution, $\triangle$ - Navier-Stokes solution, $\square$ - Mott-Smith solution, $\circ$ - experiment

In fact those terms have the maximum order of magnitude in the boundary layer approximation. Other terms in right hands (3) - (5) are transformed by replacing the characteristic mean free pass λ by step size of the spatial mesh h in the way $\tau \sim h/c$. So in this terms τ is supposed to be a characteristic time for gas-particle to pass a computational cell h. Mentioned terms are considered as mesh regularizers which play the role of the artificial viscosity. Moreover, this terms are small compared with natural viscosity close to solid walls. This manner of choosing τ in QGD equations provides for KCFD schemes an integral transition from simulation a nonviscous parts of the flow to those regions, where function of viscosity and heat conductivity is important. KCFD schemes are shown in [2], [4] and [5].

Here is an example of KCFD scheme for momentum equation for plane 3-D flow

$$(\rho u)_t + (\rho u^2 + p)_{\dot{x}} + (\rho uv)_{\dot{y}} + (\rho uw)_{\dot{z}} =$$
$$(\tau^x(\rho u^3 + 3pu)_{\bar{x}})_x + (\tau^y(\rho uv^2)_{\bar{y}})_y + (\tau^z(\rho uw^2)_{\bar{z}})_z +$$
$$M_\infty/Re_\infty((\mu u_{\bar{y}})_y + (\mu u_{\bar{z}})_z)$$

Based on KCFD schemes a number of supersonic flow problems both stationary and non-stationary types have been carried out. Particularly it was computer simulation of supersonic flows in boundary layers, in the vicinity of cavities and steps, flows near airfoils. Also we investigated pulsating regimes accompanying supersonic flows round a hollow cylinder and a spiked bodies.

KCFD schemes in explicit form are suitable for implementation on powerful MIMD-type computers, particularly on transputer systems. The computations of 3-D separation flows were made on massively transputer system [5]. The investigations in this field are supported by Russian Transputer Association.

References

[1] Elizarova T.G., Chetverushkin B.N. Kinetic algorithms for calculating gas dynamic flows. U.S.S.R. Comput. Math. Phys. Vol.25, No 5, pp.164-169, 1985.

[2] Elizarova T.G., Chetverushkin B.N. Kinetically coordinated difference schemes for modelling flows of a viscous heat- conducting gas. U.S.S.R. Comput. Math. Phys. Vol. 28, No 6, pp. 64-75, 1988.

[3] Elizarova T.G., Sheretov Yu.V. Invariant form and asymptotic features of extended quasigasdynamic equations. J. Comput. Mathem. and Math. Phys. 1991, v. 31, N 7, p.1042-1050 (in Russian).

[4] Antonov A.N., Elizarova T.G., Chetverushkin B.N., Sheretov Yu.V. Numerical modelling of pulsating regimes accompanying supersonic flow round a hollow cylinder. U.S.S.R. Comput. Math. Phys. Vol. 30, No 2, pp. 139-144, 1990.

[5] Dujsekulov A.E., Elizarova T.G., Aspnas M. An Implementation of gasdynamic problems on multiprocessor systems. Report on computer science and mathematics, Ser. A, No 123, 1991, Abo-Akademy, SF-20500, Abo, Finland.

[6] Honma H.,Xu D., Oguchi H. Kinetic model approach to shock structure problem revisited. 17-th Int. Symp. on Rarefied Gas Dynamics., 1990, Aachen.

EFFECTS OF ANGLE OF ATTACK ON THE LEADING EDGE ATTACHMENT LINE

Jolen Flores

Fluid Dynamics Division
NASA Ames Research Center
Moffett Field, CA 94035, U.S.A.

Introduction

A major design parameter for future transport aircraft is fuel efficiency. The aerodynamic efficiency of aircraft configurations can be improved through increases in lift and/or decreases in total drag. Skin friction contributes a large portion to the drag of an aircraft. In fact, turbulent skin friction can contribute up to 50% of the total drag for a subsonic transport. A method to reduce the turbulent skin friction is through natural laminar flow (NLF) or laminar flow control (LFC). The purpose of the current work is to analyze transition mechanisms (e.g. attachment line location and crossflow velocities) near a swept wing leading edge. Flow visualization is used to study the influence of angle-of-attack on the aforementioned transition mechanisms. A three-dimensional Navier-Stokes code is used to computationally simulate the flow about a highly swept wing-fuselage configuration. The flow solver used in this study is the Transonic Navier-Stokes (TNS) code. This code uses a zonal approach to solve complex configurations. Zones used in the TNS code are overlapped in order to exchange zonal boundary conditions. The TNS code has been previously used to compute the flow about different types of wings[1] and wing-fuselage geometries[2]. Recently, the numerical simulation of flow about the complete F-16A for transonic conditions and at yaw[3] was accomplished.

Results and Discussion

The first numerical simulation was conducted with flow conditions approximately matching the flight conditions (i.e., $M_{inf} = 1.6$, $\alpha = 2.0^o$ and a $Re_L = 116$ million). The Reynolds number is based on the fuselage length. These flow conditions are hereafter referred to as the base conditions. Nineteen zones were used for the computation with a total of one million grid points. This required approximately 2500 iterations to drop the initial L2-norm in each zone by three orders of magnitude. On the NASA supercomputer this required approximately 13 hours of cpu time.

Figure 1 shows a planform view of the surface grid used in the computations. Not shown, but modeled, are the inlet, diverter and environmental control system on the underside of the geometry. Instrumented on the actual flight configuration is a fitted glove on the upper surface of the wing. The glove is located on the left wing inboard of the wing-break. The inboard (beginning of the glove) location is at a semi-span value of approximately 72 inches and the outboard (end of the glove) location is at a semi-span value of 114 inches. Figure 2 illustrates the pressure contours on the symmetry plane. Shocks can be seen at the nose, canopy and lower lip of the inlet on the geometry. The smoothness of the contours, even though they are traversing different zones, is also noted. An expansion wave at the top of the canopy, as well as a recompression shock at the back of the canopy, can also be seen. These regions cause adverse pressure gradients which can cause the flow to separate.

Sensitivity Study

The following results represent a sensitivity study on the attachment line location and boundary layer velocity profiles for $\alpha = 0.0^o$, 2.0^o, and 3.67^o. Figure 3 illustrates the attachment line along the span, between the inboard and outboard portion of the glove. The experimental data points exist only at the inboard and outboard stations. The vertical axis indicates the position of the attachment point, either on the upper surface (positive x) or on the lower surface(negative x). The leading edge itself is at x = 0.0. There are twenty equally-spaced interpolated span stations between the inboard and outboard stations. The stagnation point, at each span station, was determined by finding the grid point corresponding to the maximum pressure coefficient. The experimental stagnation positions at the inboard and outboard stations were also determined in a similar manner. Therefore the true stagnation points for the computations (and experiment) may actually occur between grid points (and pressure taps). This explains the jagged appearance of the computational attachment line.

At the inboard station, increasing the angle of attack from 0.0^o to 3.67^o moves the computational stagnation point from the upper surface of the wing to the lower surface of the wing. A similar trend occurs for the outboard station as well. Notice that for the 0.0 and 3.67^o cases, the outboard stagnation point is further downstream than the inboard points. This is due to the ``effective angle of attack" at the outboard station which is larger in magnitude than the angle of attack at the inboard station. That is, each span station creates a line vortex which changes the effective angle of attack for the adjacent span station outboard. A change in angle-of-attack is therefore expected to have a greater effect on the outboard attachment point. Experimentally, the location of the inboard and outboard attachment points indicate the same trends.

Figure 4 illustrates the flow about the leading edge in a cross-flow plane. The geometry of the leading edge is outlined in black, with a black horizontal line indicating the leading edge point. At the inboard station, the flow about the leading edge is illustrated and the stagnation point can be seen to be slightly above the leading edge. At the 82 inch span station, the flow near the leading edge experiences a positive angle-of-attack. Again, this is due to upwash effects from the adjacent inboard stations which increases the effective angle-of-attack. At the 114 inch station, it appears that the effective angle-of-attack has decreased relative to the 82 inch span station. The attachment line, being extremely close to the leading edge, is very sensitive to small changes due to upwash effects and results in the small perturbations of the attachment line location about the wing leading edge.

Simulated oil flow patterns are conducted near the leading edge of the wing. The flow patterns should reflect what has been discussed in Figs. 3 and 4. Figure 5 shows three different sections of the wing leading edge. The location of the leading edge of the wing is indicated by the leading edge grid line arrow. The attachment line is the dividing streamline where the particles go onto the upper and lower portion of the wing. Examining the flow at the inboard part of the wing (Fig. 5a) it can be readily seen that the attachment line is located on the upper portion of the wing. The flow at the middle portion (Fig. 5b) of the wing now indicates that the attachment line has moved to the lower surface of the wing. Finally, as the outboard part of the wing (Fig. 5c) is approached, the attachment line is right at the leading edge of the wing. As the flow proceeds down the wing the attachment line can be seen to move to the upper surface.

The effect of different angles-of-attack on the crossflow profiles is illustrated in Fig. 6a at x/c = .7%. As the angle-of-attack is increased from 0.0^o to 3.67^o, the maximum crossflow velocity decreases at a given location. As the angle of attack is increased, the stagnation point is pushed further underneath the leading edge to the lower surface. The crossflow velocity profiles are also rotated towards the stagnation point yielding a lower maximum crossflow value at a fixed chord station. Since the location of a given crossflow velocity value moves forward, one would expect that increasing alpha will move transition due to crossflow instability upstream. Decreasing the angle of attack appears to lower the height of the inflection point.

Figures 6b and 6c show the crossflow profiles at two more stations downstream. The same trend continues, i.e., decrease in angle of attack increases the maximum crossflow velocity at a given location. The location of the inflection point becomes more difficult to ascertain as the profiles have less curvature. But qualitatively, Figs. 6a-6c illustrate the effects of different angles-of-attack on the maximum crossflow component and the associated inflection point location.

Summary

The computational results presented show good agreement with existing flight test data. A parametric study on the sensitivity of attachment line location and crossflow profiles to alpha was conducted. For the 2° case, the attachment line is very close to the leading edge. However, for $\alpha = 0.0^o$ and 3.67^o, the outboard attachment point moves much further downstream relative to the inboard attachment point. At different angles-of-attack the boundary layer thickness and stream-wise velocity profiles do not change significantly. There is, however, a significant change in the crossflow velocity profiles with different angles-of-attack. The trends exhibited in the crossflow velocity profiles indicate that as the angle of attack is decreased from 3.67^o to 0.0^o, the maximum crossflow velocity at a fixed location increases.

References

[1]Flores, J., Holst, T.L., Kaynak, U., and Thomas, S.D., ``Transonic Navier-Stokes Wing Solutions Using A Zonal Approach: Part 1. Solution Methodology and Code Validation," AGARD Paper 30A, April 1986.

[2]Flores, J., Reznick, S.G., Holst, T.L., and Gundy, K., ``Navier-Stokes Solutions for a Fighter-Like Configurations," AIAA Paper 87-0032, Jan. 1987.

[3]Flores, J. and Chaderjian, N.M., ``Zonal Navier-Stokes Methodology for Flow Simulation about a Complete Aircraft", Journal of Aircraft, Vol. 27, No. 7, July 1990, pp.583-590.

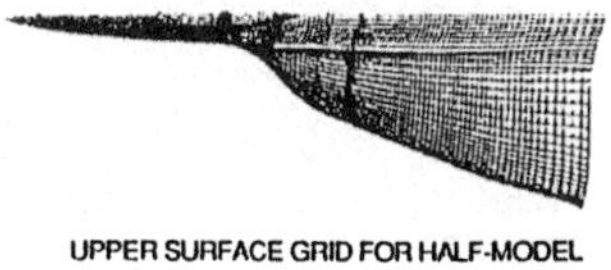

Fig. 1 Top perspective of the surface grid.

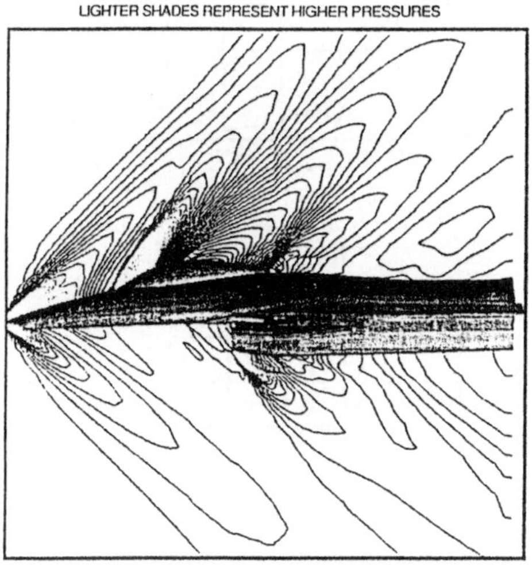

Fig. 2 Symmetry plane pressure contours. $M_{inf} = 1.6$, $\alpha = 2.0^{0}$ and $Re_L = 116$ million.

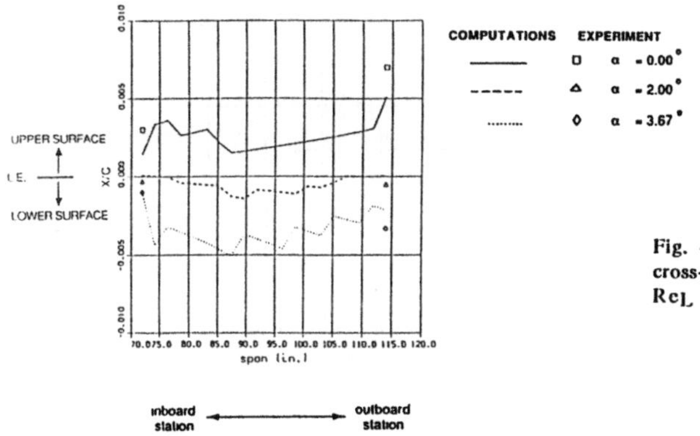

Fig. 3 Sensitivity of attachment line location to α. $M_{inf} = 1.6$ and $Re_L = 116$ million.

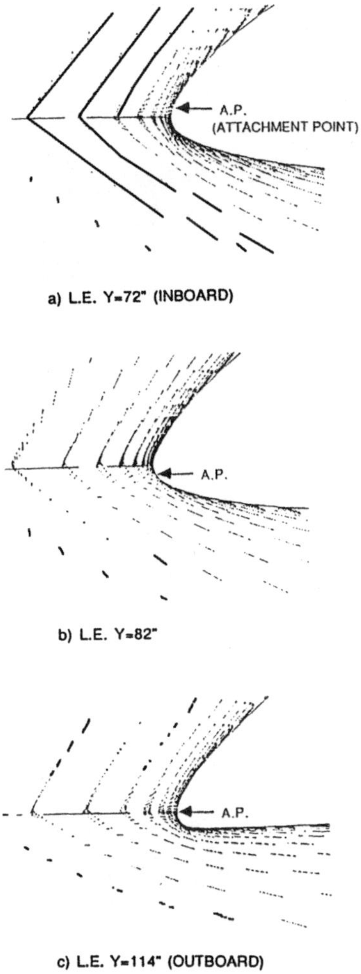

Fig. 4 Flow about the leading edge at selected cross-flow stations. $M_{inf} = 1.6$, $\alpha = 2.0^{0}$ and $Re_L = 116$ million. (Leading-edge span indicated)

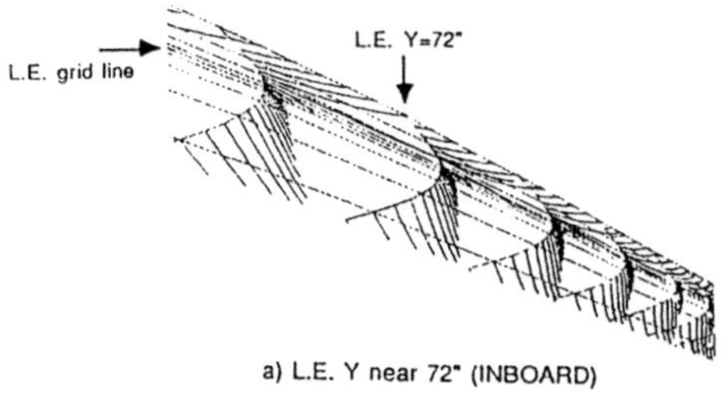

a) L.E. Y near 72" (INBOARD)

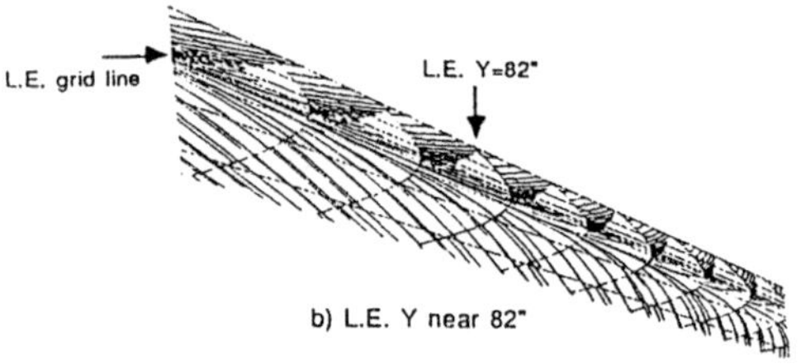

b) L.E. Y near 82"

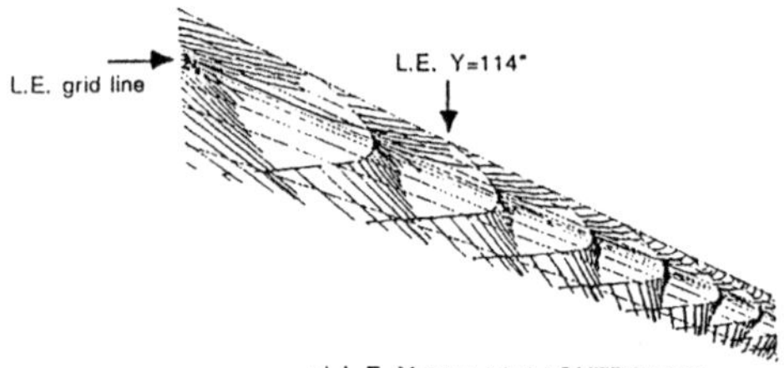

c) L.E. Y near 114" (OUTBOARD)

Fig. 5 Perspective view of simulated oil-flow patterns about the leading edge. M_{inf} = 1.6, α = 2.0° and Re_L = 116 million.

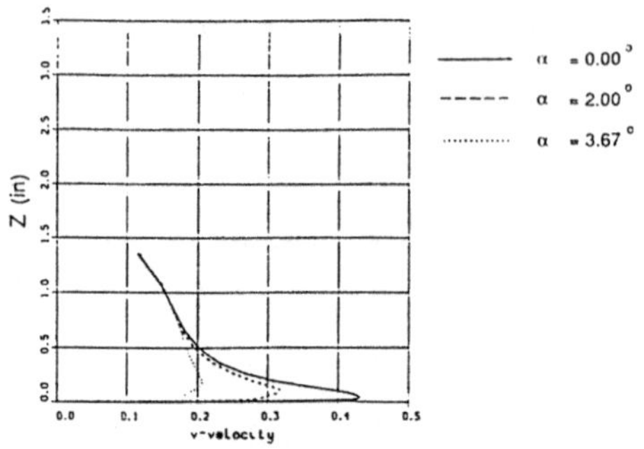

a) x/c = 0.7 percent.

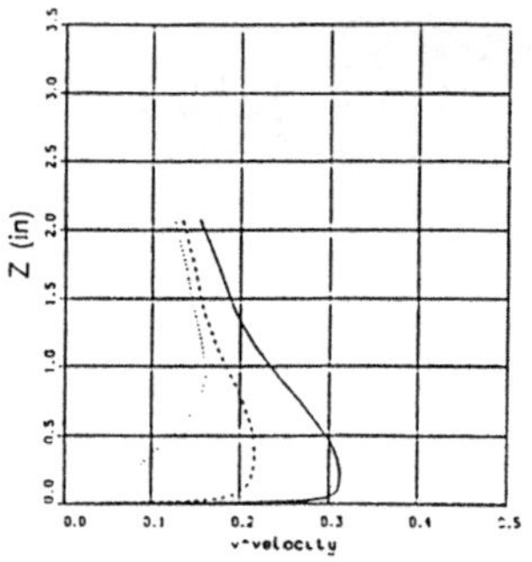

b) x/c = 2.2 percent.

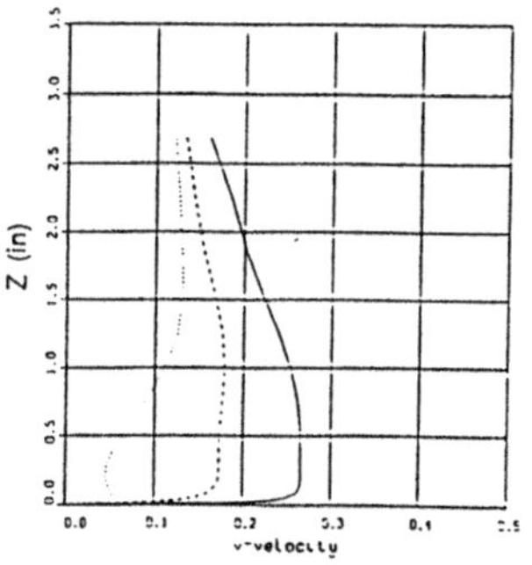

c) x/c = 3.2 percent.

Fig. 6 Sensitivity of inboard station crossflow velocity profile to α at various chordwise locations. M_{inf} = 1.6 and Re_L = 116 million.

FINITE ELEMENT SIMULATION OF TRANSONIC FLOWS WITH SHOCK WAVES

M. Hafez and D. Kinney
University of California, Davis, CA 95616

Introduction

Accurate and efficient tools are still in demand for analysis and design of transonic aerodynamic configurations. In this paper, the artificial viscosity method is adopted to solve the transonic potential flow equation. Several forms are studied. Finite elements techniques are employed to allow for handling complex geometries using unstructured grids. The advantage of using higher order elements even in the presence of strong shocks is explored. Comparisons of linear and quadratic elements on triangles as well as bilinear and biquadratic interpolations on four sided elements demonstrate the new capabilities of the proposed formulation. Application of the standard Galerkin procedure results in a nonlinear system of algebraic equations which are solved using both Newton's method with direct banded Gaussian elimination at each iteration as well as with a preconditioned conjugate gradient type algorithm. It is shown that the latter is more efficient even for a moderate size problem. Finally, possible extensions to three dimensional problems are outlined.

Governing Equations

For steady inviscid and adiabatic flows over a body, the fluid motion can be described by the conservation laws of mass, momentum and energy. For smooth subsonic or supersonic flows (with no shocks), the total derivatives of both entropy and total enthalpy vanish, hence if the incoming flow is uniform, the flow remains isoenergetic as well as isentropic. It follows, from Crocco's relation that the vorticity is identically zero and a potential function exists such that

$$\vec{q} = \nabla\phi \tag{1}$$

where $\vec{q}$ is the velocity vector.

The governing equations become

$$\nabla \cdot \rho\nabla\phi = 0 \tag{2}$$

where

$$\rho = \left(1 - \frac{\gamma-1}{2} M_\infty^2 \left(|\nabla\phi|^2 - 1\right)\right)^{\frac{1}{\gamma-1}} \tag{3}$$

In the above formula, ρ is the normalized density and M_∞ is the free stream Mach number.

The pressure coefficient is given by

$$C_p = 2\left(\rho^{\gamma} - 1\right) / \left(\gamma M_{\infty}^2\right) \tag{4}$$

As $M_{\infty}^2 \to 0$, $\rho \to 1$ and $C_p \to \left(1 - |\nabla\phi|^2\right)$ which is consistent with Bernoulli's law for incompressible flows.

For transonic flows with shock waves, the potential flow model consists of conservation of mass and energy. The momentum equations are replaced by the irrotationality and isentropic conditions and the unbalance of momentum produces the wave drag. For more details see ref (1).

Numerical Method

The potential as well as Euler equations admit both expansion and compression shocks. To exclude the nonphysical solution (negative drag), and to guarantee numerical stability in the supersonic flow regions, artificial viscosity is introduced in the above formulations.

One choice is to modify the density relation to read

$$\rho = \left(1 - \frac{\gamma - 1}{2} M_{\infty}^2 \left(|\nabla\phi|^2 - 1\right)\right)^{\frac{1}{\gamma - 1}} - \mu \frac{\partial \rho}{\partial s} \tag{3'}$$

where $\frac{\partial}{\partial s}$ is the derivative along the particle path and μ is a parameter function of Mach number.

Alternatively, the continuity equation can be augmented with derivatives of the density in the form

$$\nabla \cdot \rho \nabla \phi = \nabla \cdot \varepsilon \nabla \rho \tag{2'}$$

where ε is the artificial viscosity coefficient.

Another modification, consistent with the potential flow model conserving mass and entropy, can be derived by integrating the momentum equation including the artificial viscosity terms

$$\rho \frac{D\vec{q}}{Dt} = -\nabla p + \varepsilon \rho \left(\nabla^2 \vec{q} + \frac{1}{3} \nabla \left(\nabla \cdot \vec{q}\right)\right) \tag{5}$$

Assuming the flow is isentropic and taking the curl of equation (5) yields

$$\frac{D\vec{\omega}}{Dt} + \vec{\omega}\left(\nabla \cdot \vec{q}\right) - \vec{\omega} \cdot \nabla \vec{q} = \varepsilon \nabla^2 \vec{\omega} \tag{6}$$

Equation (6) admits a zero solution, hence the flow can be irrotational even in the presence of artificial viscosity. Moreover, equation (5) can be integrated to give a modified Bernoulli's law including the artificial viscosity effects as

$$\rho = \left(1 - \frac{\gamma - 1}{2} M_\infty^2 \left(|\nabla\phi|^2 - 2\varepsilon\nabla^2\phi - 1\right)\right)^{\frac{1}{\gamma-1}} \tag{3''}$$

Notice, the total enthalpy H, is no longer constant, instead: $H - \varepsilon\nabla^2\phi = H_\infty$.

In the present work, the above forms of artificial viscosity have been tested using a standard Galerkin finite element method. For more details see ref (2). A combination of (2') and (3") is recommended.

Summary of Results

Three issues are addressed briefly:

i) Flexibility and quality of unstructured grids.
ii) Accuracy of the solution in the presence of shock waves.
iii) Efficiency of calculations based on a preconditioned conjugate gradient type procedure.

To demonstrate the capability of unstructured grids, triangular elements are used to discretize the flow over a cylinder in a tunnel. The distributions of the nodes along the tunnel walls and the cylinder surface are prescribed and the grid size is smoothly varied in the domain as shown in figure (I-a). An automatically refined mesh is produced in figure (I-b). Both linear and quadratic interpolations have been considered with triangular elements. For four sided elements, the results based on bilinear and biquadratic interpolations are compared for the test problem proposed by a GAMM workshop for transonic flows over a parabolic arc in a tunnel. In figure (II-a) the surface pressure distributions are shown for $M_\infty = 0.85$. The density contours are given in figures (II-b) and (II-c). The calculations based on the higher elements produce basically the same sharp shock at the same place. As expected, fewer higher order elements are needed in the smooth part of the flow. The efficiency of the calculations for a moderate size problem (17 x 117) is shown in figures (III-a) and (III-b). Quadratic convergence is obtained with Newton's method if a direct solver is used at each iteration. The direct solver can be replaced, however, by a conjugate gradient algorithm for nonsymmetric systems, with SSOR as a preconditioner (see ref. (3)). The total number of the CGSTAB steps, when the solution converges to machine accuracy at each Newton's iteration, is 2474. With adaptive convergence strategy, the above number is reduced to 872. In terms of CPU time as well as storage, CGSTAB is superior.

Currently, the CGSTAB algorithm is applied to three dimensional flow problems using linear tetrahedral as well as trilinear elements. With higher order interpolations, the density can be eliminated and only one equation for ϕ, including the artificial viscosity terms, is solved. Preliminary results indicate that such a strategy is promising.

References

1. M. Hafez, "Numerical Algorithms For Transonic Inviscid Flow Calculations" in Advances in Computational Transonics, Pineridge Press, 1985.

2. D. Kinney, "Finite Element Simulation of Compressible Inviscid and Viscous Flows" M.S. Thesis, UCD, June 1992.

3. Vorst and Sonneveld, Delft University Report #90-50, May 1990.

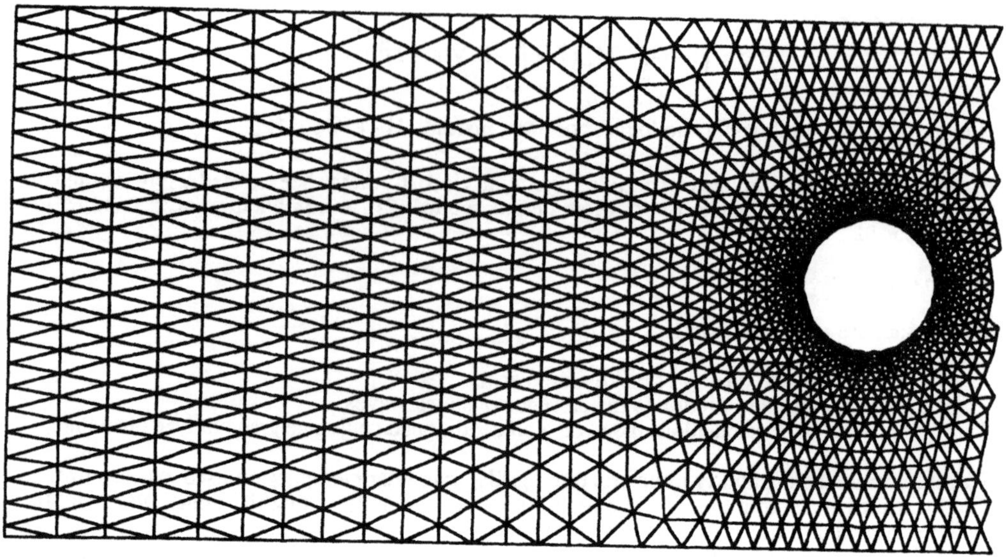

Fig. (I-a) Unstructured Grid For a Cylinder in a Tunnel
Nodes 1848, Elements 3482

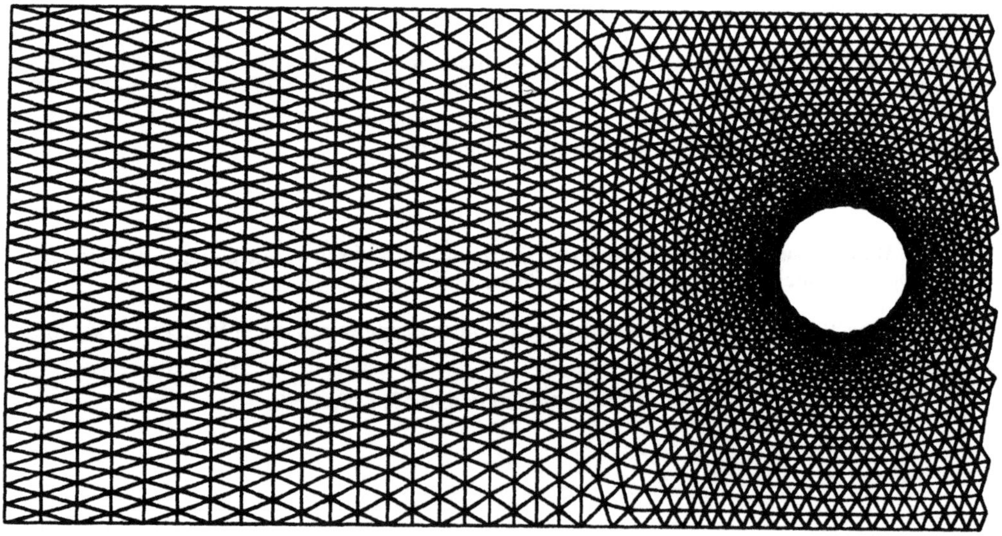

Fig. (I-b) Mesh Refinement: Nodes 2812, Elements 5372

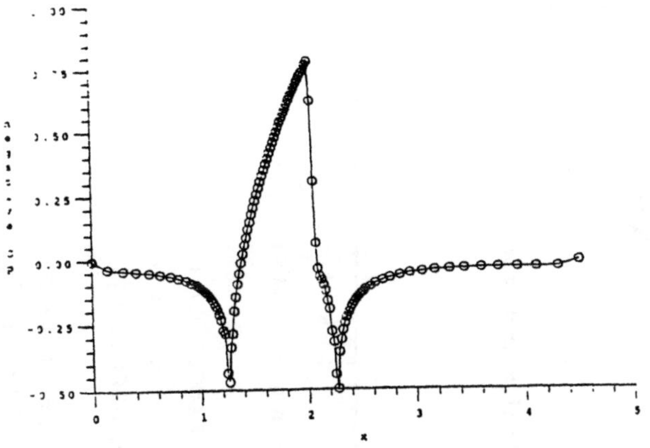

Fig. (II-a) Surface Pressure Distribution Over a Parabolic Arc in a Duct
$M_\infty = 0.85$, Bilinear (circles), Biquadratic (line)

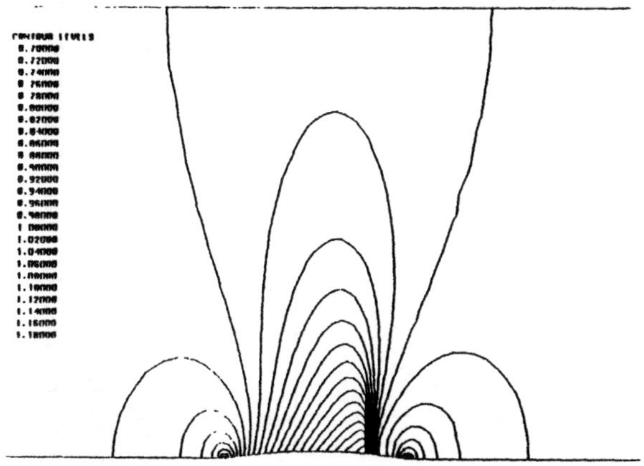

Fig. (II-b) Density Contours, Bilenar Elements: Grid 17 x 117

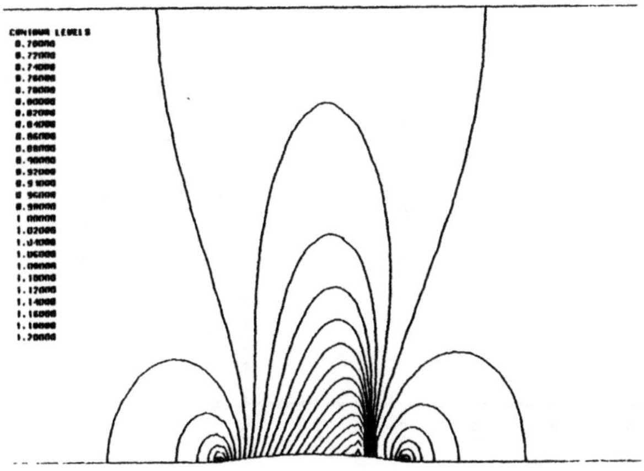

Fig. (II-c) Density Contours, Biquadratic Elements: Grid 17 x 117

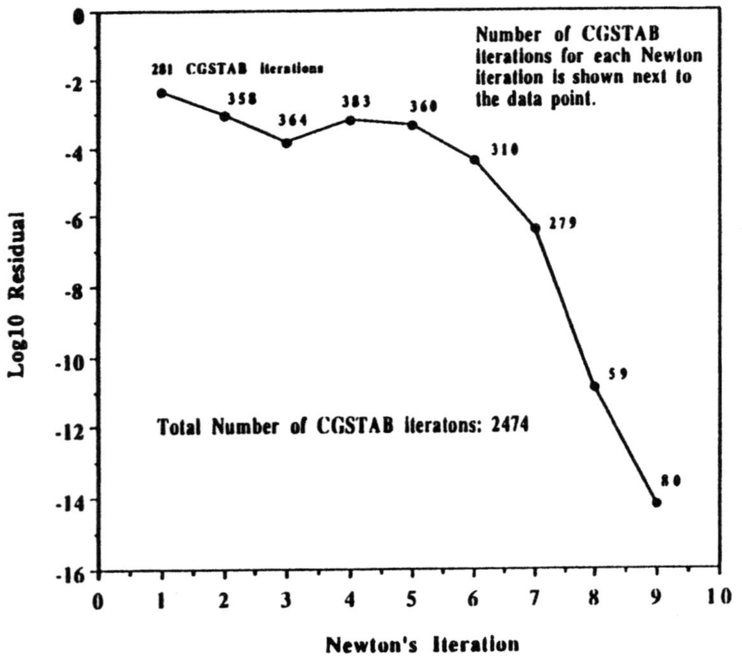

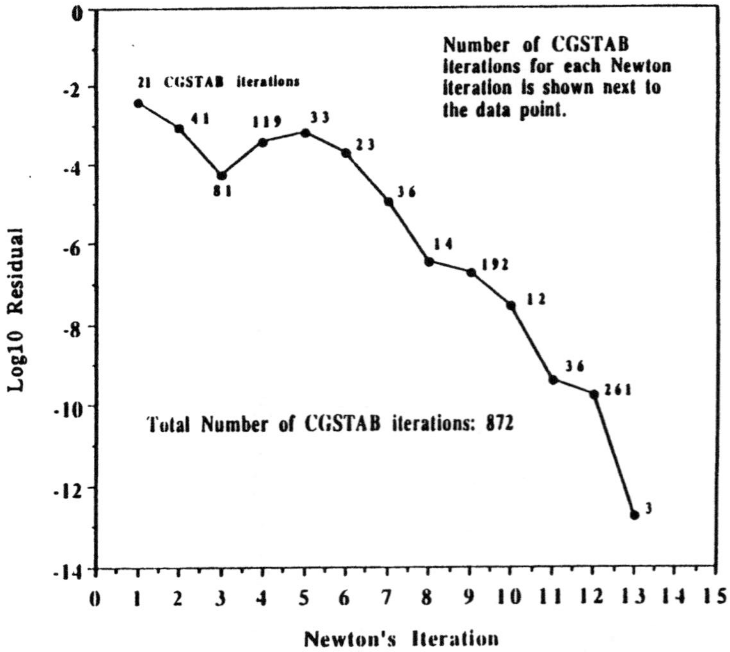

Fig. (III-a) Convergence History of Newton's Method using a direct solver and a preconditioned conjugate gradient method. (At each Newton's iteration, solution is converged to machine accuracy)

Fig. (III-b) Adaptive Convergence of a Preconditioned Conjugate Gradient method (At each Newton's step, solution converged two order of magnitude higher than the solution obtained at the previous step)

Navier-Stokes Analysis and Experimental Data Comparison of Compressible Flow Within Ducts

G. J. Harloff,[*] B. A. Reichert,[**] J. R. Sirbaugh,[*] S. R. Wellborn[+]

[*]Sverdrup Technology, Inc., LeRC Group, Brook Park, Ohio, [**]NASA Lewis Research Center, Cleveland, Ohio, [+]Iowa State University, Ames, Iowa

Introduction

Many aircraft employ ducts with centerline curvature or changing cross-sectional shape to join the engine with inlet and exhaust components. S-ducts convey air to the engine compressor and circular-to-rectangular transition ducts connect the engine to the rectangular nozzle. Changes in duct centerline curvature or cross-sectional shape give rise to streamline curvature which cause cross stream pressure gradients. Secondary flows can be caused by deflection of the transverse vorticity component of the boundary layer. This vortex tilting results in counter-rotating vortices. Additionally the adverse streamwise pressure gradient caused by increasing cross-sectional area can lead to flow separation. Vortex pairs, due to secondary flows induced by pressure gradients resulting from streamline curvature, occur in the exit planes of both S-duct and transition duct flows. The purpose of the present study is to predict the measured flow field in a diffusing S-duct and a circular-to-rectangular transition duct with a full Navier-Stokes computer program, PARC3D, and to compare the prediction with new detailed experimental measurements. The work extends previous S-duct studies[1,2], provides additional CFD validation data, and compares results for both Baldwin-Lomax and k-ϵ turbulence models. For the transition duct both straight and swirling inflow conditions were considered.

Experimental and Numerical Methods

The geometry of the circular diffusing S-duct is shown in Fig. 1. The duct centerline is defined by two circular arcs in the xz-plane with identical radii, R, and subtended angle Θ of 30°. The ratio of the centerline curvature to the inlet radius is R/R_{inlet} = 10. The axial position, s/D_1, is along the duct centerline from the inlet, normalized by the inlet diameter. The polar angle, ϕ, is measured from the vertical in a positive clockwise direction. The duct offset is $1.34D_1$ and the length is l/D_1 = 5.23. The diameter increases with s/D_1 and the area ratio is A_2/A_1 = 1.52. Upstream of the inlet and downstream of the exit are straight pipes.

The circular-to-rectangular transition duct is shown in Fig. 2. The x-axis is along the duct centerline and the y- and z- axes are parallel to the major and minor axes of the duct exit. Axial locations are described in terms of x/D_1. The duct is symmetric with respect to reflections across the horizontal xy- and vertical xz-planes. The duct cross section in the yz-plane changes from circular to nearly rectangular in the region $1.0 \leq x/D_1 \leq 2.5$. The duct aspect ratio (ratio of major to minor axis lengths at $x/D_1 \geq 2.5$) is 3.0. The cross-sectional areas at $x/D_1 \leq 1.0$ and $x/D_1 \geq 2.5$ are equal. In the region $1.0 \leq x/D_1 \leq 2.5$ the cross-sectional area increases to 1.15 times the area at $x/D_1 \leq 1.0$.

Experiment

The inlet diameter of both the S-duct and transition duct was 10.24 cm. The inlet conditions for both ducts are summarized in Table 1. For the transition duct experiment[1,2] with swirling flow, the maximum swirl angle was 15.6°. A swirl generator was used to produce solid body rotation. Additional information about the experiment is contained in Refs. 3 and 4.

Detailed measurements of velocity, total and static pressure were made at cross-stream planes inside both ducts. Data were acquired at about 530 locations near the S-duct inlet (at s/D_I = -0.5) and at nearly 1220 locations near the S-duct exit (at s/D_I = 5.73). In the transition duct, 480 measurements were made in four cross-stream planes located at x/D_I = 1.49, 1.99, 2.55, and 3.93. Surface static pressure measurements were obtained for both ducts.

Computation

The PARC3D[5] computer program solves the full, three-dimensional Reynolds-averaged Navier-Stokes equations in strong conservation form with the Beam and Warming approximate factorization algorithm. The implicit scheme uses central differencing for a curvilinear set of coordinates. A Baldwin-Lomax[6] algebraic turbulence model was used. The model was modified to use only vorticity in the local boundary layer[7] and for the swirling flow transition duct computation the constant level of axial vorticity was removed from the vorticity in the boundary layer. For the S-duct computation a low-Reynolds number k-ϵ turbulence model of Speziale *et al.*[8] was also used as implemented by Nichols[9].

The S-duct grid was composed of three blocks with grid distributions of 32 x 71 x 53, 69 x 71 x 53, and 32 x 71 x 53 and an H-grid of 129 x 11 x 15 was used in the center. The first point off the wall had an average y^+ of less than 1. The boundary conditions were no slip at the walls, inlet total pressure and temperature specified, static pressure specified at the exit plane, and symmetry about the *xz*-plane.

For the nonswirling transition duct flow the grid modeled one quadrant and contained 97 x 51 x 53 grid points. The inlet total temperature and pressure were specified locally. At the duct exit, a static pressure was specified and density and velocity were extrapolated from the interior. The swirling flow transition duct computation was performed with a 97 x 51 x 97 grid that modeled two quadrants, and the inlet boundary condition used the measured inlet flow angles.

Results and Discussion

The pressure coefficients are defined by Eq.(1) where total pressures is p_o and p is static pressure. The reference variables, cl (centerline) or wall, were evaluated at one-half inlet diameter upstream of either the S-duct or transition duct inlet.

$$Cp_o=p_o-p_{wall}/p_{o,cl}-p_{wall}, \quad Cp=p_o-p_{wall}/p_{o,cl}-p_{wall} \qquad (1)$$

Diffusing S-Duct Results

The computed and experimental surface static pressure distributions are shown in Fig. 3. Both the algebraic and k - ϵ turbulence model results agree with the experimental values upstream of the experimentally determined separation (shaded region in Figs. 3). In the separation zone, the k - ϵ model agrees better with experimental values, and downstream of separation both models begin to agree with the experimental data. The measured region of flow separation is $2.02 \leq s/D_I \leq 4.13$ and the k - ϵ and the algebraic turbulence models predict separation between $2.59 \leq s/D_I \leq 4.25$ and $2.69 \leq s/D_I \leq 4.25$, respectively.

Computational and experimental total pressure contours near the S-duct exit

(s/D_I = 5.73) are compared in Fig. 4. The predicted region of diminished total pressure is smaller than the measured region for both turbulence models. This discrepancy is possibly due to a turbulence modeling deficiency to account for with strong cross flow (three-dimensional) effects or a modeling deficiency of artificial viscosity in the boundary layer separation region. The grid resolution should be adequate as the distance of the first grid point off the wall is approximately y^+ of 1.

The predicted velocity profiles for both turbulence models near the S-duct exit are shown in wall coordinates in Fig. 5. Near the S-duct exit the vortices have convected low velocity fluid away from the wall and the boundary layer in this region, particularly at ϕ = 170°, departs significantly from this law.

Circular-to-Rectangular Transition Duct Results

The numerical and experimental values of surface static pressure for both the nonswirling and swirling flow cases are plotted in Fig. 6. The prediction is very good indicating that the aerodynamic blockage is correctly predicted. Experimental and computed contours of the total pressure coefficient at x/D_I = 2.55 are shown in Fig. 7 (a) for flow without inlet swirl. The concentric contours extending inward from the duct side walls are regions of diminished total pressure that result from side wall vortices convecting low total pressure fluid away from the duct surface. The CFD results predict the same flow structure at x/D_I = 2.55, but the region of diminished total pressure is not as large as the experimental results reveal. Total pressure contours at x/D_I = 3.93, Fig. 7 (b) also show that the predicted region of diminished total pressure does not extend as far from the side walls as was experimentally observed. Comparing experimental and CFD surface flow visualization[7] indicates that the point of formation of the vortices is correctly predicted. This suggests that the turbulence model, restricted to wall bounded shear flow, can not account for turbulent mixing in the vortex region. The agreement outside the region affected by the side wall vortices at x/D_I = 2.55 and 3.93 is excellent.

The total pressure at x/D_I = 2.55 and 3.93 for flow with inlet swirl are shown in Figs. 8 (a) and (b). With swirl the flow is symmetric with respect to 180° rotations about the x-axis. Regions of diminished total pressure near the duct side walls (that were produced in the nonswirling flow by the side wall vortices) are absent. However, surface oil film visualization[7], indicates cross flow near the duct corners in the down-stream region of the duct. The effect of the cross flow on the total pressure coefficients is visible, particularly in the upper left (experiment) and lower right (computation) quadrant at x/D_I = 2.55. In general, the agreement between the experiment and computation is better for the swirling case (no side wall vortices) than the nonswirling case (side wall vortices). This is most apparent at x/D_I = 3.93 where the comparison in Fig. 8 (b) is noticeably better than Fig. 7 (b).

Conclusion

The PARC3D Navier-Stokes code has been used to compute flow through a diffusing S-duct and a circular-to-rectangular transition duct with and without inlet swirl. The S-duct computed flow fields are generally in good agreement with the experimental data. However, both turbulence models underpredict the length and angular extent of the boundary layer separation, and the predicted separation occurs one half s/D_I late. Neither algebraic nor k - ϵ turbulence model adequately accounts for strong secondary flows with separation. The transition duct flow field matched the experimental total and surface static pressure coefficients well. The agreement appeared better for flow with inlet swirl, where the pairs of counter-rotating vortices at the duct exit were absent. For attached flow, PARC3D has demonstrated reasonable accuracy. However, for strong cross flow or separated boundary layers, the modeled turbulence and or artificial viscosity should be improved.

Acknowledgment

Part of this effort was supported by NASA Lewis, contract No. NAS3-25266.

References

1. Smith, C. F., Bruns, J. E., Harloff, G. J., and DeBonis, J. R., "Three-Dimensional Compressible Turbulent Computations for a Diffusing S-Duct," Sverdrup Technology, Inc. NASA CR 4392, 1992.
2. Harloff, G. J., Smith, C. F., Bruns, J. E., and DeBonis, J. R., "Navier-Stokes Analysis of Three-Dimensional S-Ducts," Journal of Aircraft, in press.
3. Reichert, B. A., Hingst, W. R., and Okiishi, T. H., "An Experimental Comparison of Nonswirling and Swirling Flow in a Circular-to-Rectangular Transition Duct," AIAA 91-0342, 1991. (also NASA TM 104359).
4. Wellborn, S. R., Reichert, B. A., and Okiishi, T. H., "Aerodynamic Measurement of the Subsonic Flow Through a Diffusing S-Duct," AIAA Paper 92-3622, 1992.
5. Cooper, G. K., "The PARC Code: Theory and Usage," A.E.D.C. Tech. Rep. TR-87-24, Oct. 1987.
6. Baldwin, B. S. and Lomax, H., "Thin Layer Approximation and Algebraic Model for Separated Turbulent Flows," AIAA 78-257, 1978.
7. Sirbaugh, J. R. and Reichert, B. A., "Computation of a Circular-to-Rectangular Transition Duct Flow Field," AIAA 91-1741, 1991.
8. Speziale, C. G., Ridha, A., and Anderson, E. C., "A Critical Evaluation of Two-Equation models for Near Wall Turbulence," ICASE Tech. Rep. 90-46, June 1990.
9. Nichols, R. H., "A Two-Equation Model for Compressible Flows," AIAA 90-0494, 1990.

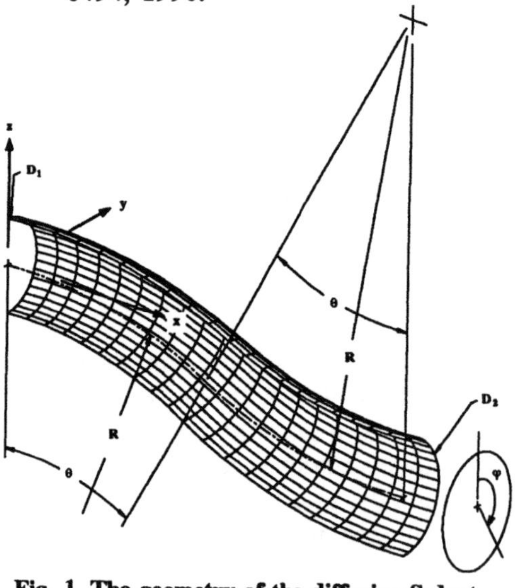

Fig. 1 The geometry of the diffusing S-duct.

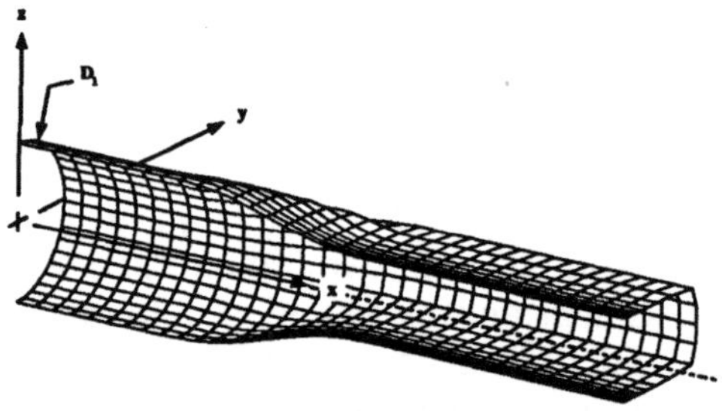

Fig. 2 The geometry of the circular-to-rectangular transition duct.

Table 1 Experimental inlet conditions.

	S-duct	Transition duct	
		No swirl	Swirl
$M_{centerline}$	0.60	0.50	0.35
$Re_{D_1} \times 10^{-6}$	2.20	2.09	1.37
$\delta_{0.95}/D_1 \times 100$	3.04	3.76	10.24

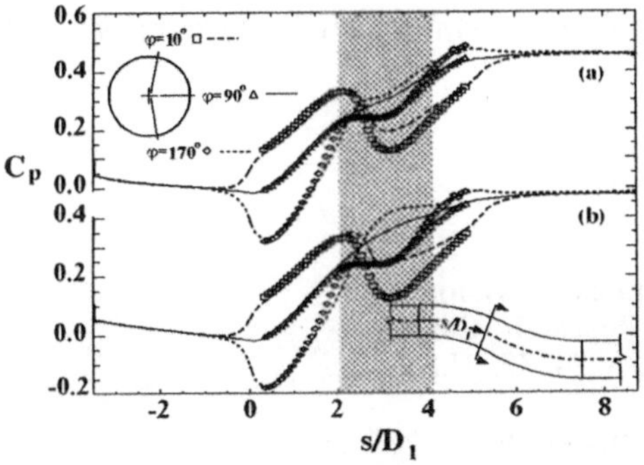

Fig. 3 S-duct surface pressure coefficient, (a) $k-\epsilon$, (b) algebraic turbulence models.

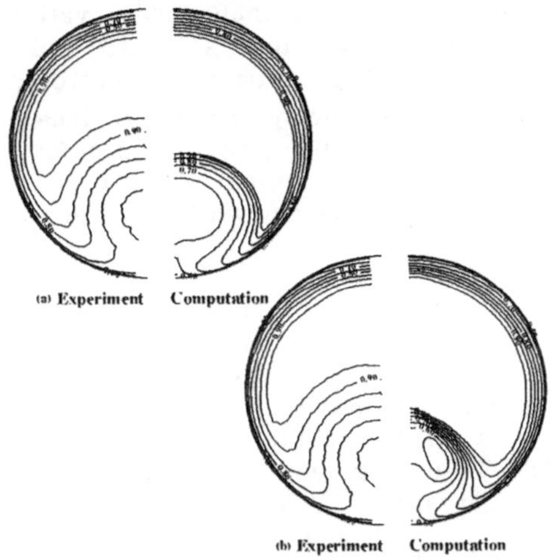

Fig. 4 S-duct exit plane total pressure coefficient, (a) $k-\epsilon$, (b) algebraic turbulence models.

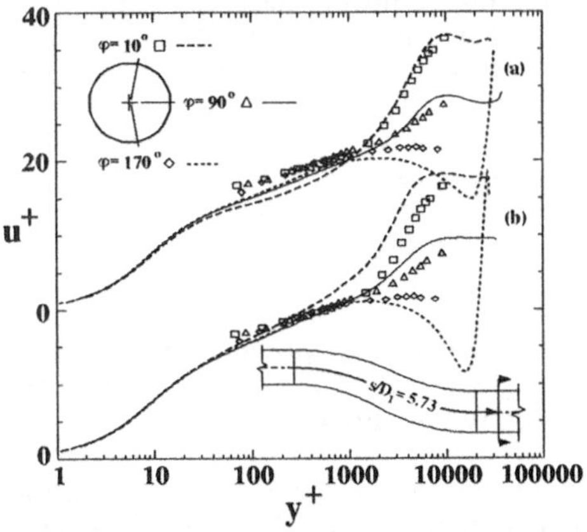

Fig. 5 S-duct exit law of the wall velocity profiles, (a) $k-\epsilon$, (b) algebraic turbulence models.

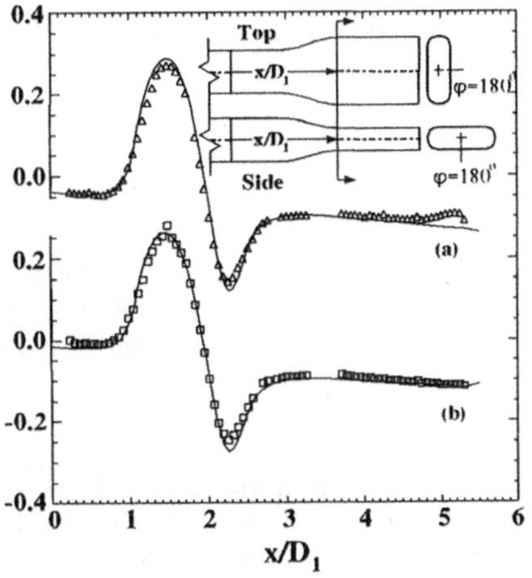

Fig. 6 Transition duct surface static pressure coefficient, (a) without, (b) with inlet swirl.

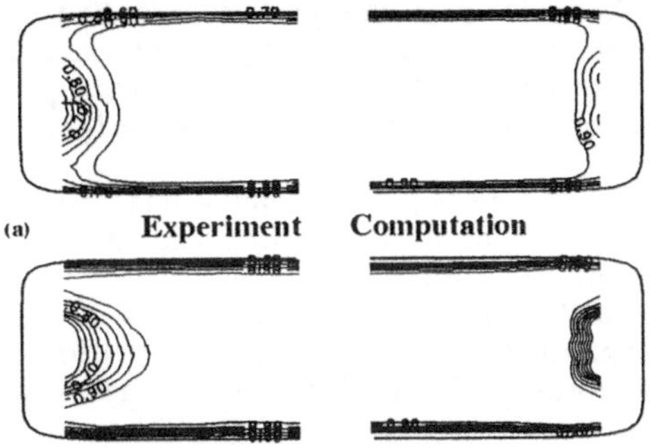

Fig. 7 Total pressure coefficient for transition duct flow without inlet swirl, (a) at $x/D_1 = 2.55$, (b) at $x/D_1 = 3.93$.

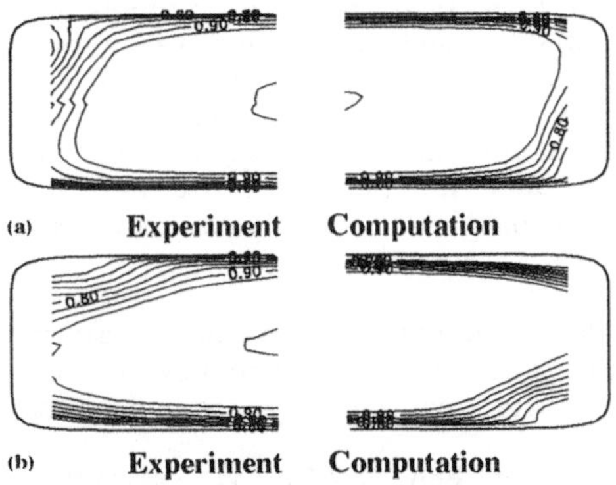

Fig. 8 Total pressure coefficient for transition duct flow with inlet swirl, (a) at $x/D_1 = 2.55$, (b) at $x/D_1 = 3.93$.

MULTIDISCIPLINARY COMPUTATIONAL AEROSCIENCES

Paul Kutler

Fluid Dynamics Division
NASA Ames Research Center
Moffett Field, CA 94035, U.S.A.

Introduction

The use of computational fluid dynamics (CFD) in the aerospace business has steadily increased during the past decade and especially during the last several years. The primary reasons for this usage are the enhanced, validated applications software, available computer power to run that software and the past successful application of CFD to "real world" design problems. According to Holst et. al.,[1] computer hardware execution speed has increased by a factor of about 15 over the past decade and by over 200 during the past two decades. It can safely be said that the increased use of computational simulations has improved the efficiency of aerospace vehicle performance while at the same time reducing their cost to design.

The discipline of computational fluid dynamics has proven its value to many in the aerospace profession, and it is routinely being used to simulate flows about vehicles and components that in some instances don't yet fully represent "real" aerospace configurations. The overarching goal for computational fluid dynamics or more appropriately "multidisciplinary computational aerosciences" (MCAS) in the aerospace field is to simulate the actual flow field about a computer optimized, realistic aerospace vehicle at true flight conditions in a reasonable amount of time on the computer. Phenomena such as shock and expansion waves, vortices, shear layers, separation, reattachment and unsteadiness are part of the actual flow field that must be predicted. A realistic configuration consists of a structure that is geometrically accurate, flexes under aerodynamic loads, has deflecting control surfaces that perturb the flow field, possesses propulsion systems that interact with the airframe and radiates acoustic, electromagnetic and infrared signatures. True flight conditions consist of the actual speed, attitude and altitude, all of which can vary with time. A reasonable amount of time on the computer depends on the need for the data. Use for design studies requires a quick turn-around time, probably on the order of one hour. Use for design validation, that might supplement or compliment an experiment or flight test, would not require such a short time.

The simulation of an actual aircraft is truly multidisciplinary in nature and offers numerous challenges to the CFD scientist. The dimensions of the computational challenge are depicted graphically in Figure 1.

To that end, NASA is currently involved in a national research program entitled High-Performance Computing and Communication (HPCCP). One part of the NASA element involves the solution of computational aeroscience grand challenge problems, that is, multiple disciplinary problems, on massively parallel computers. The primary objective of the computational aeroscience applications portion of HPCCP is to develop robust computational methods and the associated pilot application computer programs that will enable integrated multidisciplinary analysis and design of advanced aerospace systems on massively parallel computers.

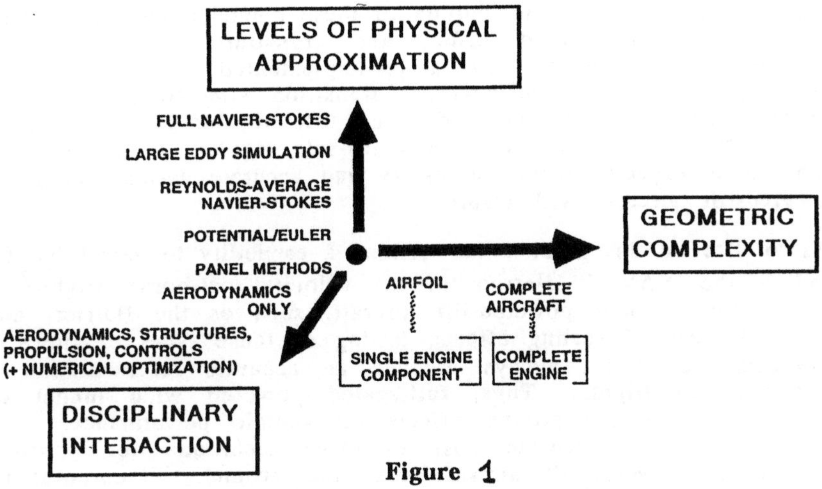

Figure 1

One goal of the program is to make available computational power 1000 times greater than that existing today to solve the computational grand challenge problems. Research for this element of the program is focusing on the development of algorithms for multidisciplinary equation set coupling, algorithm/architecture mapping of the multidisciplinary models, multidisciplinary optimization techniques, artificial intelligence procedures for multidisciplinary simulation and analysis, and application of these technologies to airframes, propulsion systems, and vehicle systems.

Challenges offered the CFD scientist include coupling the equations governing the necessary disciplines, developing the appropriate algorithms to solve that all inclusive set of equations and efficiently programming those procedures on the computer. Solution of such a set of equations is estimated to require a teraFLOPS (10^{12}) computer and between 10 and 100 hours per solution. Examples of MCAS are discussed in the next section.

Multidisciplinary Example

An example is discussed below that demonstrates the MCAS program being pursued at the NASA Ames Research Center. Other examples being pursued at Ames include high-speed civil transport configurations and hypersonic vehicles. These problems were selected because of the following reasons: 1) development of technology for the numerical simulation of these vehicles will help the U.S. aerospace industry, 2) their simulation involves strongly-coupled multidisciplinary phenomena, and 3) their solution, it is believed, is tractable for advanced parallel computers. In the example discussed below, the plans for the project under NASA's HPCCP are first discussed followed by the current status and latest results of the project.

High-Performance Aircraft

Multidisciplinary numerical simulations about high-performance aircraft are planned under HPCCP that include the disciplines of aerodynamics, engine stability, and aircraft controls.[1] These disciplines will be integrated for two different but related problems; a powered lift aircraft undergoing a transition maneuver from hover to forward flight and a multi-role fighter undergoing a low-speed, high-g turn. Both of these simulations involve unsteady maneuvers and as such will require advanced computational resources, i.e., massively parallel computers. The propulsion challenges involved in this element are concentrated on simulating the behavior of the engine operating with distorted flow at its inlet. In addition to these computations, others are planned involving hover in ground effect and transonic cruise. A classical aerodynamic control system with a jet-implemented reaction control system will be employed for the powered-lift transition simulation while a classical aerodynamic control system with advanced forebody devices and thrust vector control system will be used for the multi-role fighter simulation. Following validation, it is expected that a necessary and accurate design and analysis tool for powered-lift aircraft will result.

Progress to date on this problem has yielded a capability to model the flow field surrounding the YAV-8B Harrier in low-velocity, jet-borne flight.[2] Ground effect flow fields about powered-lift aircraft, such as the Harrier, are highly complex. Because of scaling effects, testing of these types of flows in small wind tunnels does not always produce an accurate representation of what would happen in flight. Thus, full-scaled powered wind tunnel tests are required to quantify the ground effects on vehicle performance. Numerical simulation offers an accurate cost effective technique for predicting the performance of powered-lift aircraft near the ground. Incorporated in the simulation is an engine component deck which models the response of an engine, given the throttle position and inlet temperature and pressure; it returns values for the mass flow and nozzle exit temperature.

Elements Pacing MCAS

Requirements for performing such simulations include multidisciplinary application software, systems software and computers powerful enough to run that software. The successful development of all three of these technologies is mandatory for a positive outcome of MCAS. There are a number of elements for MCAS that are pacing the progress of multidisciplinary problems. They include physical modeling (e.g., turbulence and transition modeling, chemistry modeling, other physics such as electromagnetics and structures), solution methodology (e.g., equation set coupling and algorithms), computer power, and multidisciplinary validation data. Secondary pacing technologies include pre- and post-data processing (e.g., surface, internal and structural definition, grid generation and scientific visualization). The primary pacing elements are discussed below in more detail.

Solution Methodology

With the potential for enhanced computer power offered by machines with parallel architectures, will come a search for algorithms that run effectively on those machines. Not only will existing algorithms have to be programmed and evaluated, but also "old" algorithms will have to be revisited and tried to determine their effectiveness. In addition, the development of new algorithms specifically designed to take advantage of these architectures will have to be researched. These new algorithms should be those that can efficiently solve such equation sets as those governing gas-dynamics, structural dynamics, controls and optimization as well as some combination of those equation sets.

Problems governed by multiple equation sets of different physical disciplines will require research as to the most optimum procedure for couping those equation sets. Both indirect and direct procedures will have to be investigated. The indirect approach links the physical dependence of one discipline on another by sensitivity matrices.

The direct approach mathematically links the governing equation sets. They can be linked in either a strong or weak form. For example, hypersonic problems involve the solution of both the gas dynamic and finite-rate chemistry equations. In the strongly coupled approach, the two equation sets are solved simultaneously by a large matrix inversion process for an implicit method. In the weakly coupled approach, each equation set is solved independently at each iteration, and the information from each equation set is then used to update the other equation set for the next iteration. The strongly coupled approach takes considerably more computer time than the weakly coupled approach. For hypersonic problems the weakly coupled approach has proved successful.

Physical Modeling

The development of suitable turbulence models for the RANS equations to date remains highly problem dependent. They are dependent on such elements of the problem as the free-stream conditions (for example, Mach and Reynolds number), configuration geometry (for example, deflected control surfaces and wing-body junctures) and flow field behavior (for example, shock boundary layer interaction). Accurate and numerically efficient turbulence models for numerical simulations, especially those types of models that can be used to obtain engineering answers for separated flow problems, must be developed.

One multidisciplinary area of application requiring advanced turbulence models is associated with combustion. In this area, turbulence and chemistry models must be developed to appropriately account for chemistry/turbulence interactions. Transition models, that predict the flow characteristics leading up to fully developed turbulent flows are also important, especially for laminar flow control (LFC) applications, e.g., those associated with the high-speed civil transport, and high-speed, high-altitude flows, e.g., those associated with the National Aero-Space Plane (NASP).

To improve the turbulence-model development process, it should be considered at the early stage of code development according to Marvin[3]. The first step in such a process is to identify those flows pacing the development of the aerodynamic computations. The second step is to develop models through a phased approach of building-block studies that combine theory, experiment and computations. The final step is to provide verification and/or limits of the modeling through benchmark experiments over a practical range of Reynolds and Mach numbers.

Turbulence models that complete the above process must then be installed in the application software. This requires close collaboration between the model developer and software author. Issues associated with overall code stability must be addressed

Computer Technology

Conventional serial computer architectures will probably not be capable of fulfilling the requirement of multidisciplinary computational aeroscience challenges in a reasonable amount of time. For that matter, some single discipline problems, such as direct simulation of turbulence, will also demand computer power beyond that offered by serial machines. Those machines are limited by the speed of light. To alleviate that problem, computers based on parallel architectures are being developed. Massively parallel computer architectures offer another avenue for improvement in computing speed. To obtain a numerical simulation on such a computer, the problem is divided among hundreds or even thousands of processors. It is important that the work be balanced across the processors and the communication between processors is minimized. By doing this, the time to compute a solution is dramatically reduced.

The goal for the CAS element within the HPCCP is a sustained execution speed of one teraFLOPS. This will require a peak speed of more than one teraFLOPS or approximately 10 teraFLOPS. The CAS performance goal was designed to push the current state of computing technology. The HPCCP, it is hoped, will lead to the commercial development of fully scaled systems complete with system software.

Parallel computers currently available at NASA Ames for performing calculations include the Connection Machine (CM-2) with its 32,768, 7 megahertz (Mhz) processors and the Intel iPSC/860 Gamma with 128 nodes based on 40 Mhz i860 superscalar RISC chips. The CM-2 has 128 kilobytes (KB) of memory per processor for a total of 4 gigabytes (GB) while the Intel has 8 MB of memory per processor for a total of 1GB. Scientists at Ames also have available to them the Intel iPSC/860 Delta machine at the California Institute of Technology which has 528 processors.

These machines are currently being used by Ames scientists to solve such problems as 1) chimera grid generation for ARC3D on the CM-2, 2) implicit algorithm efficiency on the Gamma, 3) three-dimensional unstructured grids on the Gamma, 4) turbulent channel flow on the Gamma, 5) compressible homogeneous turbulence on the Gamma, 6) delta wing and high-performance aircraft wing on both the CM-2 and Gamma, 7) strake-wing configuration for aeroelastic studies on the Gamma, 8) high-speed civil transport on the Gamma, and 9) hypersonic blunt body with chemistry using the Monte Carlo approach on the Gamma. One problem at Ames that has been quite successfully solved on a parallel machine from an efficiency point of view is a very large-scale turbulent-flow simulation.

Turbulent flow numerical simulations require extremely large data bases and long run times. To solve problems such as these on the Intel machines, Wray [4] first had to implement Vectoral (a highly optimized compiler), a very efficient FFT and an efficient inter-processor communication scheme. Once this software was written, then the applications software could be easily ported to the machines.

The first code ported simulates homogeneous turbulence in a three-dimensional box. For this code on the 128 processor Intel Gamma, it runs more than 11 times faster than on a single Cray Y-MP processor. However, that is about 25% its theoretical speed. If it were not for limitations imposed by the speed of memory accesses and of interprocessor communication, the code would run at 80% of the absolute upper bound of performance (80 MFLOPS per processor in single precision). The performance on the Delta, using 512 processors, is about 30 times a Y-MP processor.

The results of these calculations indicate that highly parallel, distributed memory computers are well suited to large-scale turbulence simulations, memory speed degrades the i860 processor performance, memory sizes for the speed available are inadequate, and the interprocessor communication speeds are out of balance with the processor speeds.

Multidisciplinary Validation

Validation of multidisciplinary application software will be a key pacing item. It takes a long time for design engineers to gain confidence in applications software. That confidence is in part gained by comparison with experimental or flight test data. It will be important to conceive and perform, in parallel with the application software development, companion experiments designed to validate the multidisciplinary computer programs being created. It is also important that these experiments be designed in concert with the MCAS scientists but at an early stage so that the data produced can impact the software development program in a timely fashion. In the past such attempts resulted in a missed opportunity because the experiments required such a long time to produce the validation data.

A new aeroelastic experiment is being planned at NASA Ames to validate software being developed there. In it, the unsteady aerodynamic and structural responses of a flexible delta-wing configuration representative of a high-speed civil transport wing will be studied. Simple semi-span models will be tested in the High-Reynolds Channel-2. Unsteady pressures will be measured using a new technology called pressure-sensitive luminescent coatings. This reduces the cost of the model. The data produced from this experiment will be used to validate the ENSAERO program developed by Guruswamy. [5] That program couples the Euler or Navier-Stokes equations with the structural equations of motion to predict the aeroelastic response of simple wings and wing-body configurations.

Conclusions

Solution of the multidisciplinary computational aerosciences problems described above will go a long ways towards helping the aerospace industry fulfill its basic two objectives, that is, (1) to reduce the development cost and (2) to increase the performance, safety and environmental compatibility of aerospace vehicles and their components. The tasks are challenging but the CFD scientific community is poised to meet that challenge. Accomplishment of those lofty goals will require computers orders of magnitude larger and faster than those currently available. It will also require systems software that permits easy use of those computers. Massively parallel computers offer a means to meeting the multidisciplinary computational aeroscience objectives. The High- Performance Computing and Communications Program designed by NASA provides a framework and the resources necessary for systematically developing the requisite elements needed to successfully accomplish multidisciplinary simulations.

References

[1] Holst, T. L., Salas, M.D. and Claus, R. W., "The NASA Computational Aerosciences Program - Toward Teraflops Computing," AIAA-92-0558, AIAA 30th Aerospace Sciences Meeting & Exhibit, Reno, Nevada, January 6-9, 1992.

[2] Smith, M. H., Chawla, K., Van Dalsem, W. R., "Numerical Simulation of a Complete STOVL Aircraft in Ground Effect," AIAA Paper No. 91-3293-CP, AIAA 9th Applied Aerodynamics Meeting, Baltimore, Maryland, September 23-25, 1991.

[3] Marvin, J. G., "Future Requirements of Wind Tunnels for Computational Fluid Dynamics Code Verifications," AIAA Paper 86-0752-CP, 1986.

[4] Wray, A. A. and Rogallo, R. S., "Simulation of Turbulence on the Intel Gamma and Delta," NASA Technical Memorandum, April 1992.

[5] Guruswamy, G.P., "Vortical Flow Computations on a Flexible Blended Wing-Body Configuration," AIAA Paper 91-1013, 32nd Structures, Structural Dynamics and Materials Conference, April 1991, Baltimore, Maryland.

NUMERICAL METHODS FOR SIMULATING UNSTEADY INCOMPRESSIBLE FLOWS

D. Kwak[1], C. Kiris[1], N. Wiltberger[1], S. Rogers[1], and M. Rosenfeld[2]

[1] NASA Ames Research Center, Moffett Field, CA 94035, U.S.A.
[2] Tel Aviv University, Tel Aviv 69978, Israel

I. Introduction

Unsteady calculations of incompressible flows are especially time consuming due to the elliptic nature of the governing equations. Therefore, it is particularly desirable to develop computationally efficient methods. In the present paper, a brief description of the unsteady methods recently developed by the present authors is given followed by a discussion on some computed results. Since our main interest is in 3-D applications, methods presented here are limited to the primitive variable formulation, which is considered to be the most flexible for complex 3-D geometries.

II. Solution Methods

The incompressible Navier-Stokes equations for 3-D flows can be written in curvilinear coordinates, (ξ, η, ζ), as

$$\frac{\partial}{\partial t}\hat{u} = -\frac{\partial}{\partial \xi_i}(\hat{e}_i - \hat{e}_{vi}) \tag{2.1}$$

$$\nabla \cdot \hat{u} = 0 \tag{2.2}$$

Here, t is the time, $\xi_i = \xi$, η or ζ for $i = 1, 2,$ or 3, $\hat{u}$ the velocity vector normalized by the Jacobian of the transformation, $\hat{e}_i$ the convective flux vectors and $\hat{e}_{vi}$ the flux vectors for the viscous terms. All variables are nondimensionalized by a reference velocity and length scale. In this formulation, satisfying the mass conservation equation in each time step is the primary issue for obtaining time accuracy. Two methods are discussed below.

Pseudocompressibility (PC) Method:

The first unsteady method was developed using pseudocompressibility approach [1]. In this approach the time derivatives in the momentum equations are differenced first. To solve the resulting difference equations for a divergence free velocity at the *n+1* time level, a pseudotime level is introduced and is denoted by a superscript m. The equations are iteratively solved such that $\hat{u}^{n+1,m+1}$ approaches the new velocity $\hat{u}^{n+1}$ as the divergence of $\hat{u}^{n+1,m+1}$ approaches zero. To drive the divergence of this velocity to zero, the following pseudocompressibility relation is introduced:

$$\frac{p^{n+1,m+1} - p^{n+1,m}}{\Delta\tau} = -\beta\nabla \cdot \hat{u}^{n+1,m+1} \tag{2.3}$$

where τ denotes pseudotime and β is a pseudocompressibility parameter.

Since, in this formulation, the governing equations are changed into hyperbolic-parabolic type, the upwind differencing scheme derived from Roe [2] was used to discretize the convective flux terms. The viscous fluxes are approximated using central differences. Then the set of numerical equations is solved using an unfactored line relaxation scheme similar to that employed by MacCormack [3].

Fractional Step (FS) Method:

The second method was developed using a fractional step approach [4]. The dependent variables are the pressure, defined at the center of the primary cells, and the volume fluxes defined on the faces of the primary cells. First the momentum equations are integrated for intermediate velocities. This first step velocity correction is obtained by evaluating the pressure gradient term at time level n, thus the continuity is not satisfied at this intermediate level.

In the second step, the velocity is updated such that the continuity equation is satisfied at the next time level. This is achieved by a Poisson type equation as below:

$$D_{iv}\left(\frac{R_l(\phi)}{V}\right) = f \tag{2.4}$$

where $R_l(\phi)$ represents discrete gradient-like terms of the modified pressure, ϕ, and V, the volume of a computational cell. The right hand side, f, is a function of velocities at the previous and intermediate time level. Here, D_{iv} is a discrete form of divergence operator. Then, the variables at the new time level $n+1$ are computed from the intermediate velocities and ϕ. The most CPU-time consuming part of the FS method is the solution of a discrete Poisson-like equation. To achieve computational efficiency, multi-grid techniques have been employed for accelerating the convergence rate of the Poisson solver [5]. The total computational time required for solving the Poisson equation was reduced by an order of magnitude, while the overall computational time of the flow solver was reduced by a factor of 3-4.

III. Computed Results

Computed examples using the flow solvers discussed above are presented next.

Flow in a Constricted Channel

The two methods discussed above achieve time-accuracy using different numerical algorithms. The PC method recovers the divergence free velocity field by iterating at the pseudo time level. During the subiteration in pseudo time, the pressure wave maps the entire flow field until the velocity field becomes divergence free. For fast convergence, the speed of the pressure wave has to be made large by using a large value of the pseudocompressibility parameter, β. On the other hand, the FS method achieves time accuracy by solving for the pressure through a Poisson equation such that the flow field is divergence free at the next time level.

To test the time accuracy and the iteration process, the flow in a constricted channel with time dependent inflow was computed and compared with the experimental measurements by Park [6]. In Figure 1, the geometry and the inflow velocity profile are given. In Figure 2, computed results from the two methods are compared. The results of the two methods compare well with the experimental data. In Figure 3, pressure contours during an iteration process are compared. To illustrate the pressure wave propagation during subiteration in PC method, a small value of β was chosen at this particular time step. A much larger value of β was used for actual calculations so that the incompressibility condition is satisfied within 10 to 15 subiterations. The cost per iteration for PC method is higher than the FS method. However, since a large value of β can be used, the total number of iterations required for the PC method is much smaller than the FS method making computational efficiency of the two methods competitive [7].

Artificial Heart Flow

Artificial heart flow offers a wealth of CFD-related problems as well as physical phenomena, involving moving boundaries and time-dependent flow. The flow through the Penn State artificial heart is numerically simulated by using the time-accurate mode of the pseudocompressibility method [8]. As shown in Figure 4, the artificial heart model is composed of a cylindrical chamber with two tube extensions. The inflow (mitral) and outflow (aortic) tubes contain concave tilting disks which open and close to act as valves. The pumping action is provided by a pusher plate whose velocity is sinusoidal in time. To handle the geometric complexity and the moving-domain problem, a zonal method and an overlapped grid-embedding scheme is combined as shown in Figure 4.

Computationally, the flow is started at rest, and four cycles of pusher plate motion were completed to obtain cyclic-flow solutions. During most of the cycle, 10-20 subiterations at each physical time step were required. In Figure 5, unsteady particle traces are illustrated at a non-dimensional time of $t/T = 0.45$, where t and T denote the physical time and a period of the pusher plate motion, respectively. In light of the difficulties associated with measuring flow quantities in the region where a boundary is moving rapidly, the present unsteady flow capability will be a valuable alternative to obtain flow data in those areas.

Turbopump Flow Simulation

A crucial element for liquid rocket propulsion systems is the design of an efficient and reliable turbopump, which requires a good understanding of the complex internal flow through the inducer. To develop and validate a computational procedure for turbomachinery applications, the flow through an inducer, which was developed by Rocketdyne Division of Rockwell International Corporation, was studied [9]. The design flow is 2236 GPM with a design speed of 3600 RPM. The geometry is illustrated in Figure 6. In Figure 7, the planes are shown where experimental data were taken. The Reynolds number was 191,000 based on a reference length of 1 inch. The computation was performed using the time-accurate mode of the pseudocompressibility method to obtain steady solutions in a rotational frame of reference. In Figure 8, the computed relative total velocities on four different measuring planes are compared with experimental data. The computed surface pressure is shown in Figure 9. In the present study, an algebraic turbulence model was used. Considering the simplicity of the turbulence model, the results are quite satisfactory.

IV. Concluding Remarks

In the present paper, two unsteady methods for the incompressible Navier-Stokes equations are discussed. The main emphasis has been placed on achieving computational efficiency in three-dimensional applications. Implementation of an implicit upwinding scheme to the pseudocompressibility method and the multigrid acceleration scheme to the fractional step method are a part of such an effort. Overall, the two solution procedures produce comparable results.

REFERENCES

1. Rogers, S. E., Kwak, D. and Kiris, C., AIAA J. vol. 29, No. 4, April 1991, pp603-610.
2. Roe, P. L., J. Comput. Phys., vol. 43, p. 357, 1981.
3. MacCormack, R. W., AIAA Paper 85-0032, 1985.
4. Rosenfeld, M., Kwak, D. and Vinokur, M., J. Comp. Phys. vol. 94, 1991, pp102-137.
5. Rosenfeld, M., and Kwak, D., AIAA Paper 92-0185, 1992.
6. Park, D.K., M.Sc. Thesis, Lehigh University, Bethlehem, Pennsylvania, 1989.
7. Wiltberger, N.L., Rogers, S.E. and Kwak, D., NASA TM , to appear, 1992.
8. Kiris, C., Rogers, S., Kwak, D. and Chang, I-D, Proc. International Symposium on Biofluiddynamics, July 6-12, 1991.
9. Kiris, C., Kwak, D. and Rogers, S.E., 4th International Symposium on Transport Phenomena and Dynamics of Rotating Machinery (ISROMAC-4) April 5-8, 1992.

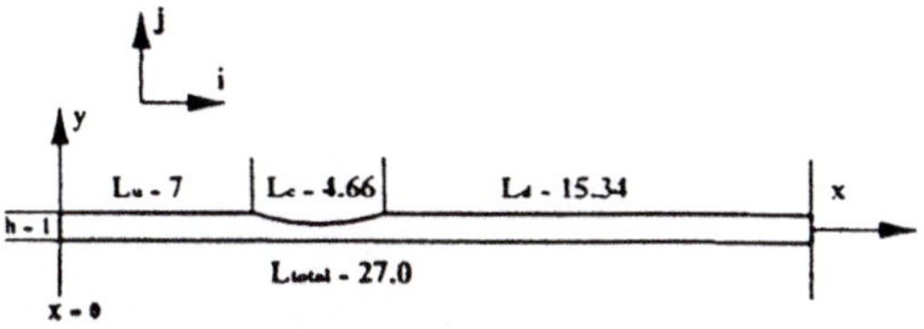

(a)

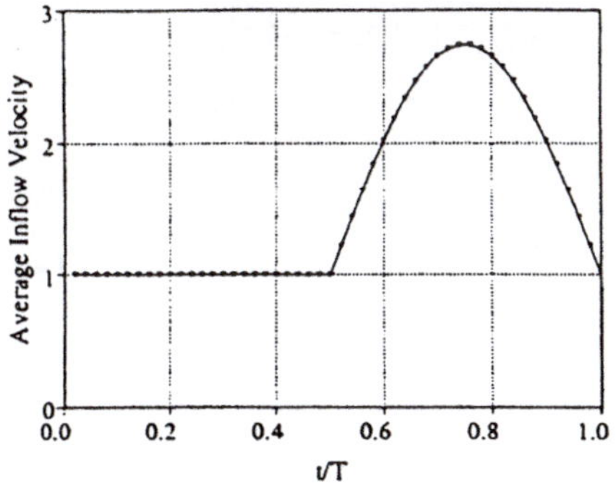

(b)

Fig. 1. - Pulsatile flow through a constricted channel: (a) schematic of geometry, (b) average inflow velocity for one period

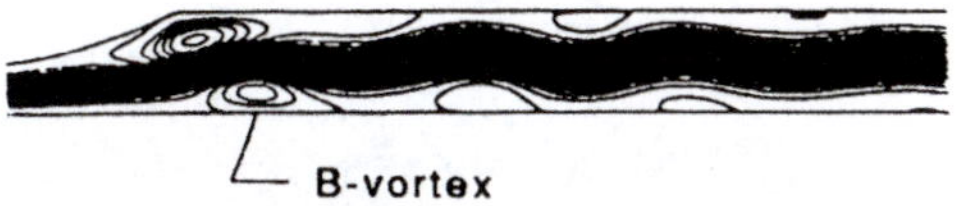

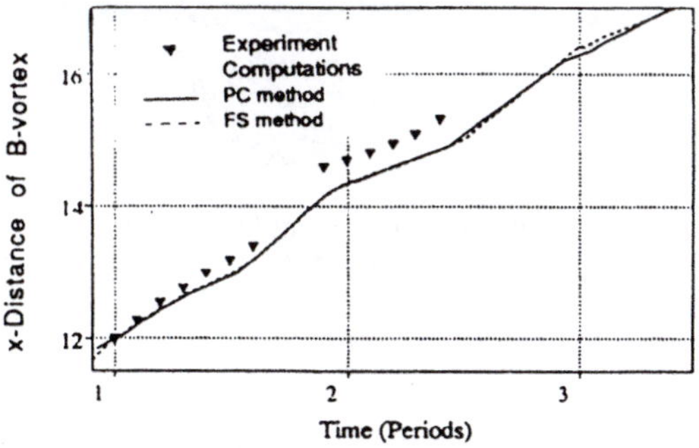

Fig. 2. - Time evolution of B-vortex at Re=132.

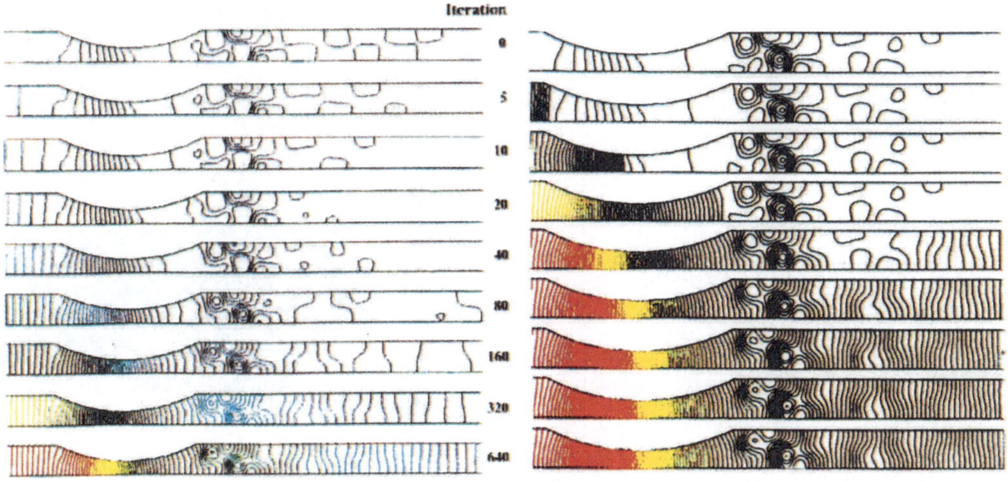

(a) (b)

Fig. 3. - Evolution of pressure contour during iteration within one time step at t/T=0.5 using Δt/T=0.01 : (a) PC method with β=100, (b) FS method.

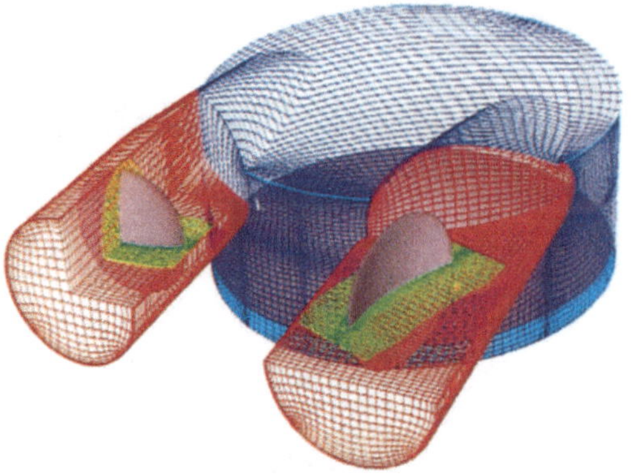

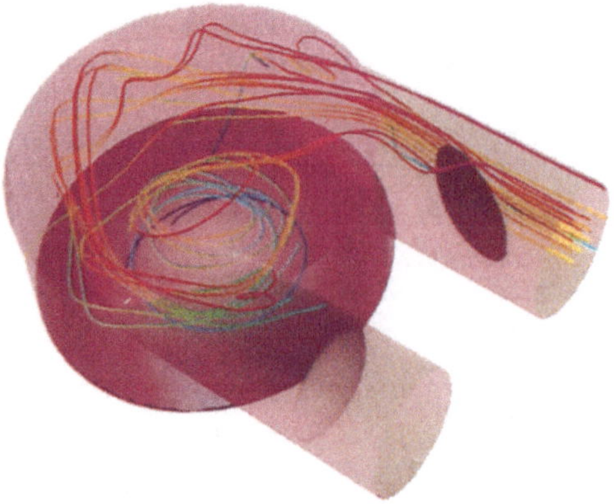

Fig. 4. - Computational model for the Penn State artificial heart

Fig. 5. - Computed particle traces at t/T=0.45

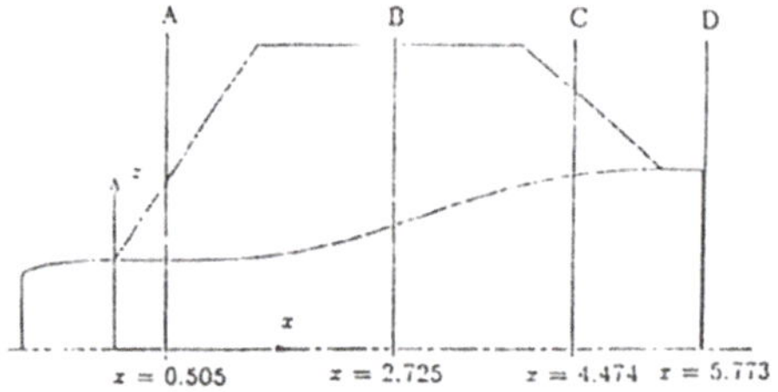

Fig. 6. - Experimental inducer geometry

Fig. 7. - Schematic of the experimental measurement planes

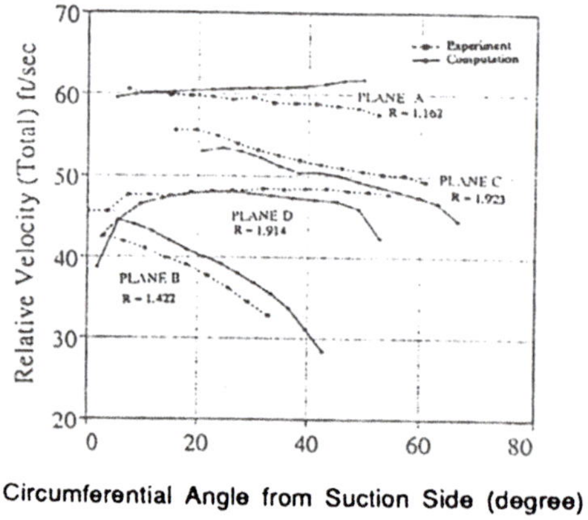

Fig. 8. - Comparison of relative total velocities

Fig. 9. - Computed surface pressure for inducer at Re=191,000

TIME-MARCHING COMPUTATIONS IN THE COMPLETE MACH/REYNOLDS NUMBER DOMAIN

C. L. Merkle, S. Venkateswaran and Y-H. Choi

Department of Mechanical Engineering
The Pennsylvania State University
University Park, PA 16802

Sverdrup Technology, Inc.
NASA Lewis Research Center
Brookpark, OH 44142

INTRODUCTION

Time-marching density-based methods become stiff at low Mach numbers and low Reynolds numbers because of the disparate time scales in these flowfields. Preconditioning the time-derivative to alter the acoustic speeds is an effective method of controlling these disparities [1]. At low Mach numbers, the acoustic time scales are altered to be of the same order as the fluid convective time scale, while at low Reynolds numbers, the acoustic time scales are made the same order as the viscous time scale. Thereby, convergence rates that are independent of Mach number and Reynolds numbers may be obtained.

A problem closely related to solving viscous flows is the stiffness induced by grid refinement and grid stretching. In localized regions of the flowfield, such as the near-wall region in high Reynolds number flows, very strong grid stretching is often used. Such regions typically cause slowdown in convergence because of reduced propagation speeds of disturbances. In this paper, we describe a new time-step definition that enables virtually grid-independent convergence.

We also describe an extension of the preconditioning procedure to include chemical species transport. Clearly, the approach may be generalized to include other sets of transport equations as well. We apply the preconditioned method to a variety of reacting and non-reacting flows to demonstrate its capability. These problems include fluid dynamics in a combustor and nozzle, flow over a backward facing step and mixing and combustion in a shear co-axial injector.

THEORETICAL DEVELOPMENT

We start with the non-conservative governing equations, with $\tilde{Q} = (\rho,\ u,\ v,\ s,\ Y_i)^T$ as the set of primary dependent variables. For the time-derivative term, this vector $\tilde{Q}$ is transformed in terms of an alternate set of dependent variables, $Q_v = (P,\ u,\ v,\ T,\ Y_i)^T$. The resulting equation equation is written as:

$$\frac{\partial \tilde{Q}}{\partial Q_v}\frac{\partial Q_v}{\partial t} + \tilde{A}\frac{\partial \tilde{Q}}{\partial x} + \tilde{B}\frac{\partial \tilde{Q}}{\partial y} = \tilde{H} + \tilde{L}(Q_v) \tag{1}$$

The matrix multiplying the time derivative is replaced by a new preconditioning matrix (Γ_v). These matrices are given as:

$$\frac{\partial \tilde{Q}}{\partial Q_v} = \begin{pmatrix} 1/RT & 0 & 0 & -\rho/T & -\rho\frac{R_i - R_N}{R} \\ 0 & \rho & 0 & 0 & 0 \\ 0 & 0 & \rho & 0 & 0 \\ -1 & 0 & 0 & \rho C_p & \rho(h_i - h_N) \\ 0 & 0 & 0 & 0 & \rho \end{pmatrix} \quad \Gamma_v = \begin{pmatrix} 1/\beta M_r^2 & 0 & 0 & 0 & 0 \\ 0 & \rho & 0 & 0 & 0 \\ 0 & 0 & \rho & 0 & 0 \\ -1 & 0 & 0 & \rho C_p & \rho(h_i - h_N) \\ 0 & 0 & 0 & 0 & \rho \end{pmatrix}$$

Note that the new matrix is derived by replacing the coefficient of the pressure time-derivative term. Here, $\beta = k\gamma RT$ (where $\gamma = C_p/C_v$) and M_r is a reference Mach number. This replaces the acoustic speeds of the system with pseudo-acoustic speeds that are of the same order as the fluid velocity. The rescaling of the pressure derivative renders the preconditioned system 'pressure-based' at low Mach numbers. Pressure-based methods [2], developed originally for incompressible flows, resemble the preconditioning method in philosophy. These methods solve a separate equation for the pressure instead of the continuity equation. In this manner, the propagation of acoustic waves are treated independent of the other characteristics of the sytem.

In addition, some of the off-diagonal terms are also dropped in deriving Γ_v. These terms do not alter the eigenvalues of the system, which remain the same as for the non-reacting equations [1] and, hence, retain the same convergence properties. However, dropping the $\rho(h_i - h_N)$ term, gives the same eigenvalues but an incomplete set of eigenvectors. Therefore, care should be taken in selecting the preconditioning matrix.

Preconditioning to maintain well-conditioned eigenvalues is appropriate when viscous effects are negligible because the system is hyperbolic. On the other hand, when viscous and diffusion effects become dominant, the system becomes parabolic and convergence is controlled by the viscous time step. However, the propagation of acoustic waves remains hyperbolic to first order. Therefore, at low Reynolds numbers, the preconditioning matrix is used to optimize both the acoustic time scale and the viscous time step. This is done by scaling the parameter k in the definition of β and is similar to that outlined in Ref. 1.

The final step in deriving the preconditioning matrix is to transform the equation to conservative form. This is achieved by premultiplying the equation by the Jacobian $\frac{\partial Q}{\partial \hat{Q}}$, where Q is the standard conservative flux vector. The final form of the preconditioned system is given by

$$\Gamma \frac{\partial Q_v}{\partial t} + \frac{\partial E}{\partial x} + \frac{\partial F}{\partial y} = H + L(Q_v) \tag{2}$$

Equation (2) is solved by an Euler Implicit algorithm using Douglas-Gunn approximate factorization. Traditionally, the time-step size is selected by using a fixed CFL (around 5 to 10) and the maximum eigenvalue at each location. For high aspect ratio grid cells, this results in reduced wave propagation speeds, resulting in poor convergence rates. Stability analysis reveals that the optimum CFL for high aspect grids can be much higher than for regular grids. Accordingly, we use the following empirical relationship to vary the local CFL based on the grid aspect ratio (AR):

$$\Delta t = Min\ (\frac{CFL_x \Delta x}{\lambda_x},\ \frac{CFL_y \Delta y}{\lambda_y}) \tag{3}$$

where

$$CFL_x = CFL, \qquad CFL_y = CFL\sqrt{(AR)\ \frac{\lambda_y}{\lambda_x}}$$

This definition appears to give nearly grid-independent convergence in the presence of strong wall stretching. A similar expression has been used for grid-stretching in the axial direction and has been found to give similar convergence behaviour.

RESULTS

The grid geometry used for the combustor/nozzle computations is shown in Fig. 1(a). The grid size is 200 X 50 with wall stretching to resolve the boundary layer

($Re = 1 \times 10^5$). Convergence from three different runs is shown in Fig. 1(b). Curve I shows convergence without preconditioning. The poor convergence is due to the low Mach numbers encountered in the combustor section. Curve II shows convergence with preconditioning and time-step based on maximum eigenvalue ($CFL = 4$). Though, this is an improvement over the previous case, convergence is still quite poor. The aspect ratio of the cells in the near wall regions is about 100 and this is most likely the reason for the poor convergence. Curve III shows the convergence with preconditioning and the new time-step definition based on Eqn. (3). Much better performance is obtained with the residuals going down about ten orders of magnitude in 2000 steps. Figure 1(c) shows the converged Mach number contours for this case.

Figure 2(a) shows the convergence with preconditioning for flow over a backward facing step. The length and width of the domain are 10 and 1 cm respectively and the step height is 0.5 cm. A uniform grid of size 101 X 61 is used giving an aspect ratio of 6. Convergence is shown for both time definitions. Once again, the new definition shows a significant improvement in convergence for this problem. Figure 2(b) shows the converged velocity contours for four Reynolds number conditions and Fig. 2(c) compares the predicted recirculation lengths to experimental data [3]. Very good agreement is evident except for a slight discrepancy at Re = 400, at which Re number, the experiments start to show three-dimensionality.

Figure 3(a) shows the grid used for the shear co-axial injector. Oxygen gas flows in the core while hydrogen flows in the surrounding annular passage. This flowfield is complex, involving mixing, recirculation and combustion. For these initial studies, a simplified chemistry model is used and only laminar Reynolds numbers are considered. Figure 3(b) shows convergence without and with reactions. At the higher Reynolds number (Re = 2000), excellent convergence is obtained for both cases. However, at the lower Re, convergence with reactions deteriorates. This is because the stronger combustion source terms at the lower flow speed require a reduction in the time-step size. The effect of source terms is an important factor, especially for preconditioned schemes because they typically operate at larger time step sizes than conventional methods. Future work will be targetted on such source terms and how they may be better tackled.

ACKNOWLEDGMENTS

This work is supported by NASA/MSFC Grant NASA/NAS8-38861, and NASA USERC Grant NAGW-1356. Partial computational support was provided by the NAS Program.

REFERENCES

[1] Choi, Y.-H. and Merkle, C. L., "Time-Derivative Preconditioning for Viscous Flows," AIAA-91-1652, AIAA 22nd Fluid Dynamics, Plasma Dynamics and Lasers Conference, June, 1991, Honolulu, HW.

[2] Merkle, C. L., Venkateswaran, S. and Buelow, P. E. O., "The Relationship Between Pressure-Based and Density-Based Algorithms," AIAA 92-0425, 30th Aerospace Sciences Meeting, January, 1992, Reno, NV.

[3] Armaly, B. F., Durst, F., Pereira, J. C. F. and Schonung, B., "Experimental and Theoretical Investigation of Backward-Facing Step Flow," *Journal of Fluid Mechanics*, Vol. 127, 1983, pp. 473-496.

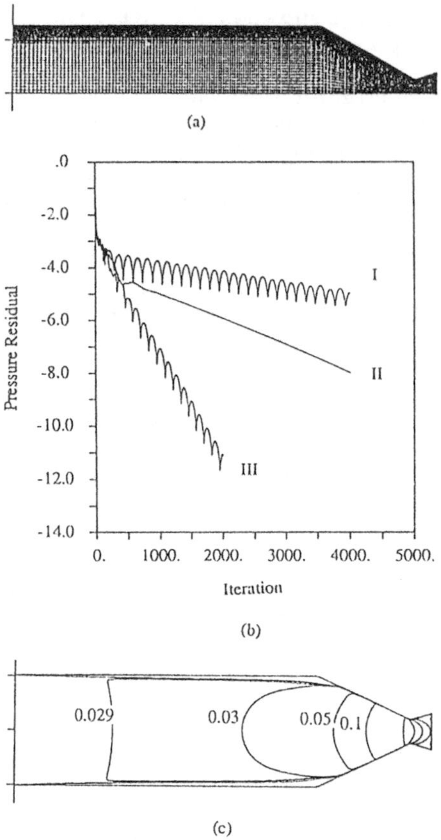

Figure 1 Combustor nozzle flowfield:
(a) Computational grid.
(b) Convergence.
(c) Mach number contours.

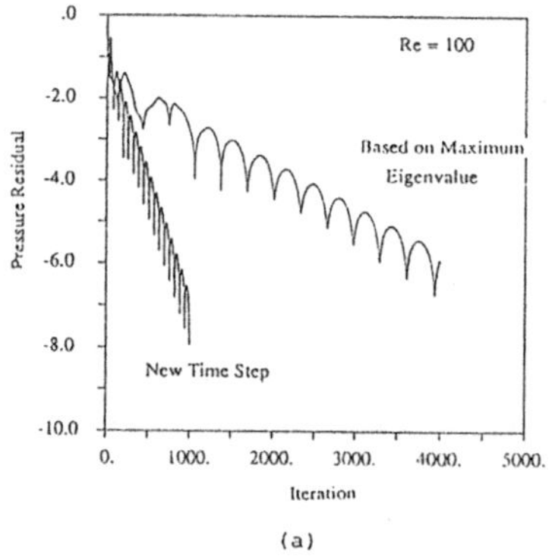

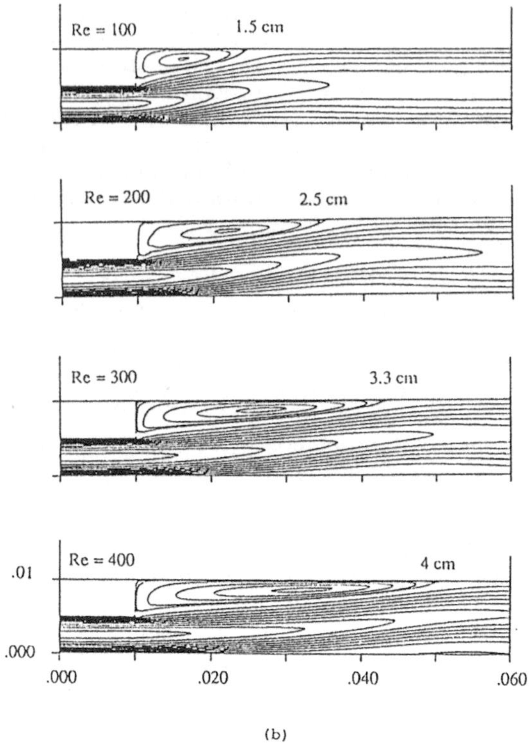

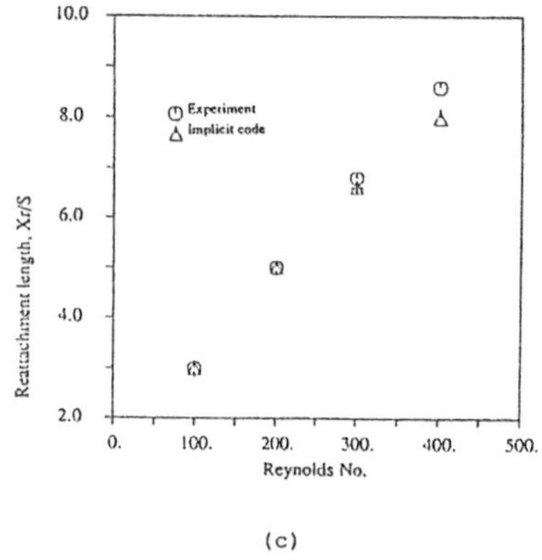

Figure 2 Backward facing step flowfield.
(a) Convergence with preconditioning.
(b) Velocity Contours for four Re's.
(c) Comparison of recirculation length with experiment.

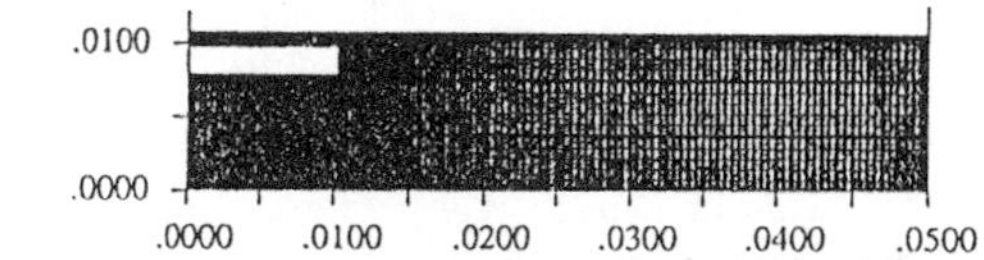

(a)

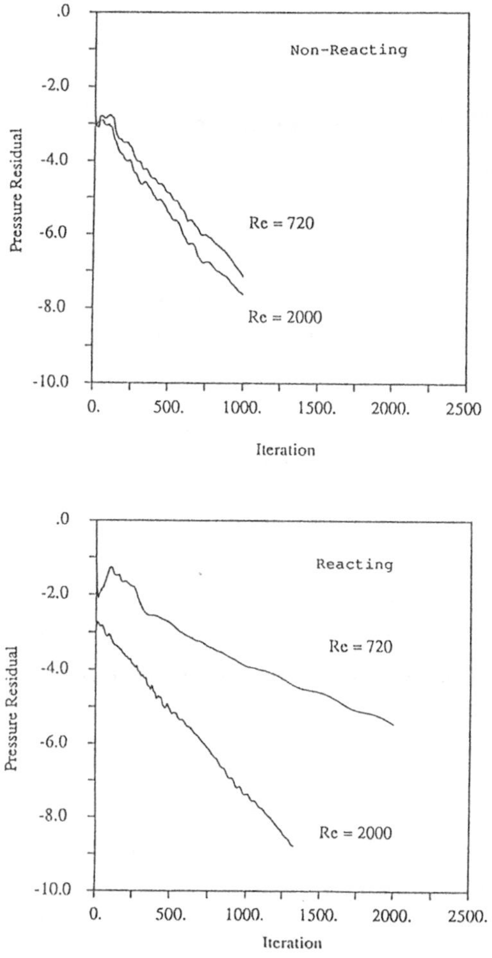

(b)

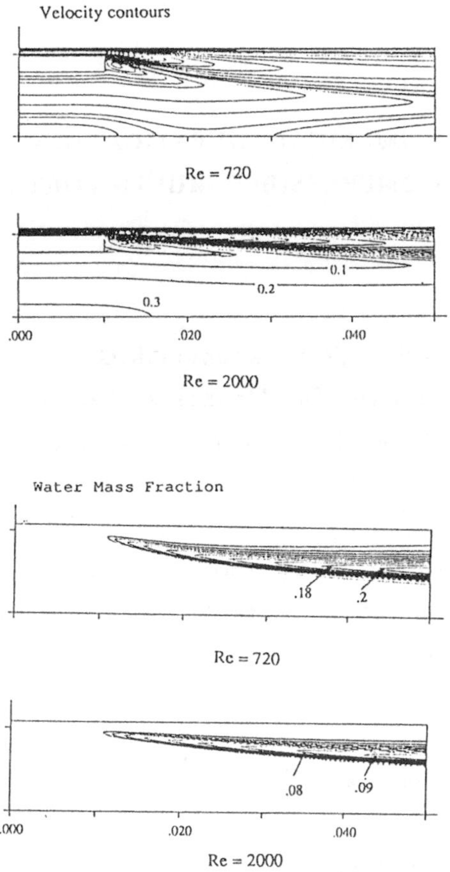

(c)

Figure 3 Shear co-axial injector.
(a) Grid geometry.
(b) Convergence without and with reactions.
(c) Velocity and water mass fraction contours.

NUMERICAL INVESTIGATION OF A DOUBLE-PERIODIC COMPRESSIBLE MULTI-VORTEX-FIELD

Müller,K.J., Roesner,K.G.
Institut für Mechanik, Technische Hochschule Darmstadt
Hochschulstr.1, 6100 Darmstadt, Germany

In this numerical study stable, laminar solutions and the transition to unsteady, unstable situations for a three-dimensional double- or triple-periodic Vortex-field are calculated. These vortices are proposed for application to gas-gas-separation like centrifuges. The parameters and mechanisms of flow-field-instability are investigated for a viscous gas.

Problems:

The vortices are induced by corotating disks on top and bottom of a infinitely extended box. The distribution of the disks in the infinite x-y-plane is shown in figure 1. The height of the box is chosen equal to the width and the diameter of the rotating disks of an elementary vortex cell. Geometrical parameters are of no interest in this investigation. Because of the periodicity only one cell has to be considered. The boundary conditions are given by symmetry to the surrounding cells. On top and bottom no-slip conditions are assumed on the rotating disks. The outer part of top and bottom shows in the first case tangential flow due to a assumed periodic continuation in z-direction. Otherwise a wall part at rest is assumed. No turbulence model is used. In the first case a spin-up was calculated. For the second problem the initial flow field is given by an interpolation from the known distribution on the walls and a Taylor-Green-flow [1],[5] in the middle part of the elementary cell.

Numerical approach:

In order to solve the time dependent Navier-Stokes-equations a Chebyshev-Collocation-Method [2] with Ralston-integration [3] of fourth order in time is used. The collocation points are Gauß-Lobatto-points which fits very well to the estimated and calculated flow properties. Due to the explicit time discretization a fast 3D Chebyshev-transformation was needed. The usual reduction to a 3D FFT needs a lot of pre- and postprocessing and limits the order of discretization. Furthermore it is very difficult to use symmetries of the solution with readily implemented FFT's. Therefore a direct transformation basing on the principles of the FFT was developed and a very important part of this issue was the adaptation to the vector-computer architecture. Aliasing errors are reduced by a padding technique and the viscous terms are frozen during the four stages. As the boundary and initial conditions have several internal symmetry properties the number of used grid points and Chebyshev-coefficients can be reduced to 1/4 of the complete cube.
The calculations were made for air and a Re-Number range from 1600 to $8 \cdot 10^5$. The solution was achieved on grids of 41^3 points to grids with 121^3 points for maximum accuracy.
A 2D Taylor-Green-vortex was used to check the code and to get an estimation for the numerical dissipation of the scheme.

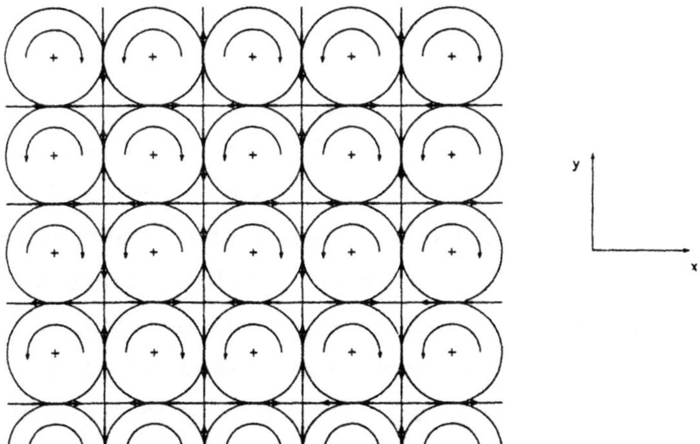

Fig. 1: Distribution of rotating disks in the infinite x-y-plane

Results:

In the first case top and bottom of the cells have impermeable walls which allow slip. The solution is assumed to be periodic in z-direction too. It was possible to find laminar solutions in a small region of the Re-Ma-plane. Figure 2 shows for Re=1600 and Ma=0.5 the stationary flow field in the middle plane z=0. In general it's a kind of Ekman-layer flow with sucking disks. The vertical motion is concentrated along the edges of the cube.

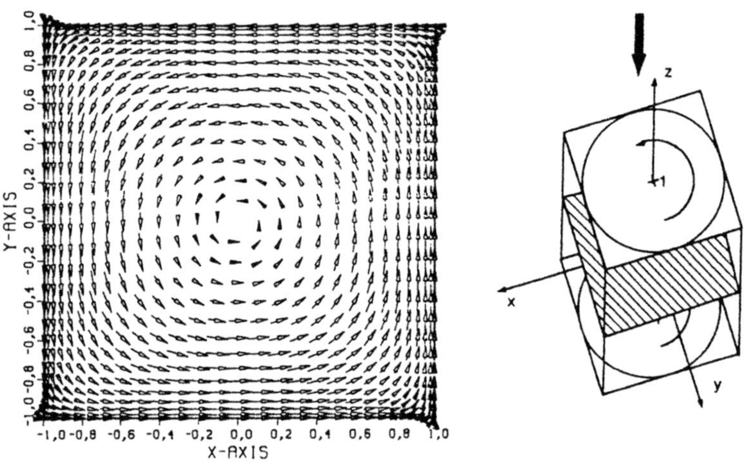

Fig. 2: Flow field for Re=1600, Ma=0.5 in the plane z=0

It turned out that for higher Re- and Mach-numbers the flowfield becomes unstable due to Taylor-Vortices emanating from the stagnation line region. The circumferential velocity plotted over the diagonal shows similarity to a unstable Taylor-Couette- or Görtler-flow. A kind of critical 'Taylor-number'-value has been detected. This critical 'Taylor-number' can be expressed as the product of Mach-number and the square-root of the Reynolds-number. Stability corresponds to a 'Taylor-number' less than 19. When this parameter reaches a value of 38 the the vortices increases in strength and frequency. They destabilize the Ekman-Layer of the rotating disks. Finally the flow seperates from the disks and turns into turbulence. The numerical results in the three different stability-regions are marked with squares, circles and triangles in figure 3.

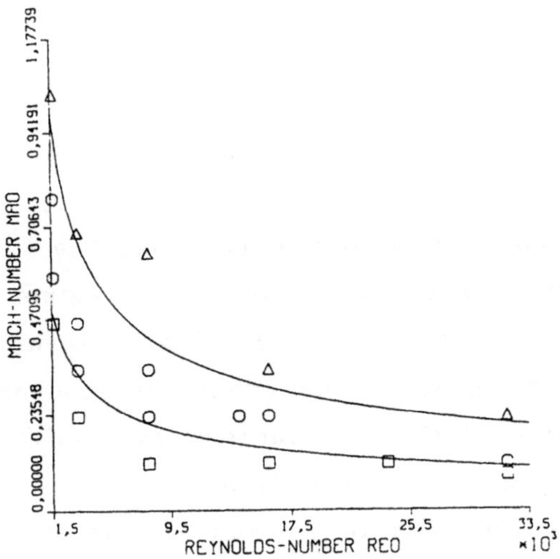

Fig. 3: Stability-regions in the Re-Ma-plane

In the second case top and bottom of the cells are solid walls with no-slip conditions. It turns out out that the boundary layer above is unconditionally unstable [4]. The external flow is decelerated towards the stagnation line along the edge. The pressure increases significantly und the boundary layer forms a separation bubble. The separation turns into turbulence for Ekman-numbers less than 1.6E-3. Figure 4 gives a cut through the recirculation zone.

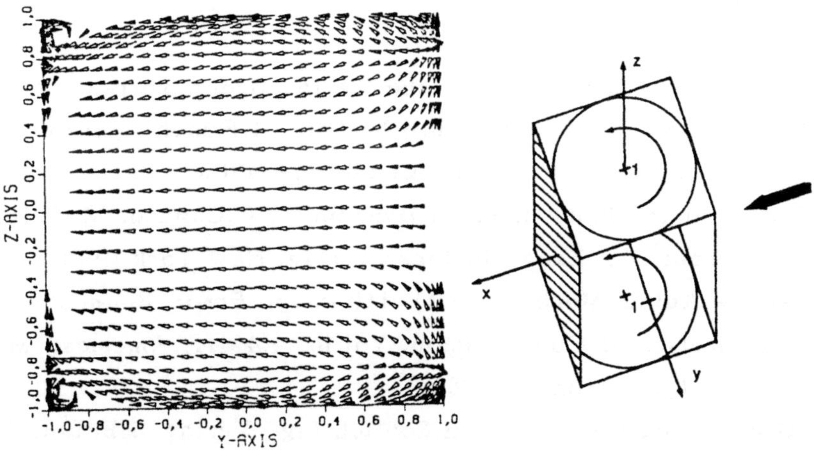

Fig. 4 : Flow-field in the side plane showing separation of the boundary layer for Re=8000, Ma=0.25

Conclusions:

A three-dimensional time-dependent Chebyshev-Collocation-scheme was developed and tested. It was applied to a compressible double- or triple-periodic Vortex-field.

In the first triple-periodic case without boundary-layers the flowfield shows a kind of Taylor-Görtler-instability. The stable regime of the Reynolds-Mach-number-plane was determined. For higher Reynolds- or Mach-numbers secondary vortices along the edges are stochastically genererated. Further increasing of the two parameters results in destabilized Ekman-layers and a turbulent flow-field.

In the second problem with solid walls at top and bottom of the vortex-cells the boundary-layer above predominates the flowfield. An unsteady three-dimensional separation-bubble evolves in front of the stagnation lines. For higher Reynolds- or vice versa small Ekman-numbers the recirculation zone becomes turbulent and influences the complete vortex-cell.

The proposed two-dimensional array of vortices has no technical use for gas-gas-separation due to the prescribed instabilities. The stable Reynolds- and Mach-Numbers are orders of magnitude smaller than the corresponding values for real centrifuges.

References:

[1] "Small-scale structure of the Taylor-Green vortex", M.E.Brachet, D.I.Meiron, S.A.Orszag, B.G.Nickel, U.Frisch, Journal of Fluid Mechanics, Vol.130, p.411ff, Cambridge, 1983

[2] "Spectral Methods in Fluid Dynamics", C.Canuto, M.Y.Hussaini, A. Quarteroni, T.A.Zang, Springer Verlag, New York, 1988

[3] "Runge-Kutta Methods with Minimum Error Bounds", A.Ralston, Mathematics of Computation, Vol.16, p.431ff, American Mathematical Society, Baltimore, 1962

[4] "Grenzschicht-Theorie", H.Schlichting, Verlag G.Braun, Karlsruhe, 1982

[5] "On the decay of vortices in a viscous fluid", G.I.Taylor, Philosophical Magazine, Vol.46, p.671ff, 1923

A DIRECT IMPLICIT METHODOLOGY FOR THE SIMULATION OF THE RAPID PITCH-UP OF AN AIRFOIL WITH AND WITHOUT LEADING-EDGE FLOW CONTROL*

G.A. Osswald, K.N. Ghia and U. Ghia†

Department of Aerospace Engineering and Engineering Mechanics
†Department of Mechanical, Industrial, and Nuclear Engineering
Computational Fluid Dynamics Research Laboratory
University of Cincinnati, Cincinnati, Ohio 45221 USA

During the last decade, a number of experimental and analytical/computational studies have examined the large-amplitude rapid pitching motion associated with high angle-of-attack maneuvers. Such motion typically leads to the generation of a dynamic stall vortex whose evolution results in large transient lift, drag and moment variation that can, for short periods of time, produce loadings significantly larger than those expected during either steady, or quasi-steady flight. Recently, Osswald, K. Ghia and U. Ghia (1991) and K. Ghia, Yang, Osswald and U. Ghia (1991) have analyzed this flow using direct numerical simulation and a vorticity-stream function (ω, ψ) formulation of the unsteady incompressible Navier-Stokes(NS) equations. In the present paper, this formulation is extended to include viscous circulation at infinity, $\Gamma = \oint_\infty \bar{V} \cdot d\bar{l}$, as an additional unknown resulting in the development of an (ω, ψ, Γ) formulation of the 2-D unsteady NS equations. The inclusion of circulation as a primary unknown closes the problem associated with the creation of surface vorticity. By demonstrating that all involved matrices are non-singular and well-conditioned, the present implicit algorithm shows that only one distribution of surface vorticity and only one value of circulation will result in a continuous single-valued pressure distribution along the body surface, while remaining consistent with the NS equations for vorticity transport and stream function within the interior of the flow domain.

The problem of a 2-D body, currently a NACA 0015 airfoil, subjected to an arbitrary three-degree-of-freedom maneuver is sketched in Fig. 1. The present analysis is developed in terms of a generalized body-fixed non-inertial coordinate system (ξ^1, ξ^2), whose covariant base vectors $\bar{e}_1$ and $\bar{e}_2$ are also sketched in Fig. 1. This coordinate generality is used to introduce the (461,129) body-fixed C-grid distribution whose near-airfoil region is shown in Fig. 2. Through analytical stretching transformations, the far-field boundary is placed at infinity, with semi- infinite cells occurring adjacent to this boundary in the physical plane.

The (ω, ψ, Γ) formulation of the 2-D unsteady incompressible NS equations in non-inertial body-fixed generalized coordinates is given as follows:

Vorticity Transport Equation:

$$\frac{\partial \bar{\omega}_I}{\partial t} + \nabla \times \left(\bar{\omega}_I \times \bar{V}_A\right) + \frac{1}{Re}\left(\nabla \times \nabla \times \bar{\omega}_I\right) = 0 \tag{1}$$

Deviational Stream Function Equation:

$$\frac{\partial}{\partial \xi^1}\left(\frac{g_{22}}{\sqrt{g}}\frac{\partial \psi_D}{\partial \xi^1}\right) + \frac{\partial}{\partial \xi^2}\left(\frac{g_{11}}{\sqrt{g}}\frac{\partial \psi_D}{\partial \xi^2}\right) = -\sqrt{g}\,\omega_I \tag{2}$$

*This research is supported,in part, by AFOSR grant No. F49620-92-J-0292 and Ohio Supercomputer Center grant No. PES080.

where the apparent velocity $\bar{V}_A$ is given as $\bar{V}_A = \bar{V}_I - \bar{V}_{B/I} - \bar{\Omega}_B \times \bar{r}$. Here, $\bar{V}_{B/I}$ and $\bar{\Omega}_B$ are, respectively, the translational and of angular velocities of the body, $\bar{r}$ is the relative position vector as shown in Fig. 1 and V_I is the inertial velocity obtained in terms of an inertial stream function ψ_I given as

$$\psi_I = \{x^2 \cos\theta(t) - x^1 \sin\theta(t)\} + \frac{\Gamma}{2\pi} ln\left(\left|\frac{\bar{r}-d_o}{a}\right|\right) + \frac{S}{2\pi} arg\left(\frac{\bar{r}-d_o}{a}\right) + \psi_D.$$

Here, ψ_D is the deviational stream function governed by Eq. (2). The far-field boundary conditions are $\omega_I = 0$ and $\psi_D = 0$ at infinity. For a maneuvering airfoil with surface suction/injection flow control, the body-surface boundary conditions are $\bar{V}_A = T\hat{t} + N\hat{n}$; $S = \oint N\sqrt{g_{11}}d\xi^1$;

$$\psi_D = -\left[x^2(\cos\theta(t) - V^1_{B/I}(t)) - x^1(\sin\theta(t) - V^2_{B/I}(t)) + \frac{\Omega_{B/I}(t)}{2}\{(x^1)^2 + (x^2)^2\}\right.$$
$$\left. + \frac{\Gamma}{2\pi} ln\left(\left|\frac{\bar{r}_B-d_o}{a}\right|\right) + \frac{S}{2\pi} arg\left(\frac{\bar{r}_B-d_o}{a}\right) + \oint_{\xi^1_{TEL}}^{\xi^1} N(\xi^1)\sqrt{g_{11}}d\xi^1 - S\right]$$

and $\bar{\omega}_I = \bar{\nabla} \times \bar{V}_I$ subject to the constraints that

$$\left(\frac{\partial\psi_D}{\partial\xi^2}\right) = -\left[\frac{\partial}{\partial\xi^2}\{x^2(\cos\theta(t) - V^1_{B/I}(t))\} - \frac{\partial}{\partial\xi^2}\{x^1(\sin\theta(t) - V^2_{B/I}(t))\}\right.$$
$$+ \frac{\Omega_{B/I}(t)}{2}\frac{\partial}{\partial\xi^2}\{(x^1)^2 + (x^2)^2\} + \frac{\Gamma}{2\pi}\frac{\partial}{\partial\xi^2}\left\{ln\left(\left|\frac{\bar{r}_B-d_o}{a}\right|\right)\right\}$$
$$\left. + \frac{S}{2\pi}\frac{\partial}{\partial\xi^2}\left\{arg\left(\frac{\bar{r}_B-d_o}{a}\right)\right\} - T(\xi^1)\sqrt{g_{11}}\right]$$

and that pressure is single-valued on the body surface so that $\oint(\nabla p)\cdot d\bar{l} = 0$. For maneuvering bodies,

$$p = q - \left(\frac{\bar{V}_A\cdot\bar{V}_A}{2}\right) - \bar{a}_B\cdot\bar{r} + \frac{|\bar{\Omega}_B\times\bar{r}|^2}{2},$$
$$\nabla q = -\left\{\frac{\partial\bar{V}_A}{\partial t} + (\bar{\omega}_I \times \bar{V}_A) + (\bar{\alpha}_B \times \bar{r}) + \frac{1}{Re}(\nabla \times \bar{\omega}_I)\right\}$$

where $\bar{a}_B$ and $\bar{\alpha}_B$ are the translational and angular accelerations, respectively, of the body, and Re the Reynolds number.

The numerical technique employed is that of direct implicit solution. The finite-difference discretization is second-order accurate in time and space, except for vorticity convection, which is fourth-order accurate in slow regions ($| V_A | \leq 0.1$) and third-order accurate in fast regions ($| V_A | \leq 0.1$). The vorticity transport equation (1) is solved using a Douglas-Gunn type ADI scheme, while the elliptic stream function equation (2) is solved using direct block-Gaussian elimination. The advanced levels of wall vorticity, circulation and interior stream function are determined simultaneously, so as to guarantee pressure continuity along the body surface. The creation of surface vorticity is the critical driving mechanism associated, with the evolution of these incompressible viscous flows. The algorithm both vectorizes and parallelizes. It currently achieves $\simeq$ 130 M flops per processor, with parallel speed-up of $\simeq$5 on an 8-processor Cray Y-MP 8/864.

Figure 3 presents a comparison of vorticity contours for the simulation of a single pitch-up event and the ensemble-averaged experimental data of Ramaprian (1992). The maneuver is a slow constant-rate pitch-up with a nondimensional pitch rate $\alpha^+ = \left(\frac{c\dot{\theta}}{U_\infty}\right) = 0.072$ and Re = 52,000. The initial condition chosen corresponds to a C_L decreasing zero-lift state of the asymptotic shedding-wake solution at zero degree angle of attack.

Figure 4 presents the effectiveness of leading-edge suction as a flow control mechanism for a rapid pitch-up maneuver with $\alpha^+ = 0.2$ at Re = 45,000. Contours of vorticity are presented for both the controlled and uncontrolled cases for angles of attack $\alpha \simeq 20°, 26°$ and $30°$. For the uncontrolled case, the dynamic-stall vortex forms at $\alpha \simeq 18°$. Leading-edge suction over the first 15% of chord at a volumetric suction rate of S = -0.0074 was initiated at $\alpha = 17.2°$. Control authority through $\alpha = 30°$ is clearly evident from Fig. 4. Figure 5 displays stream-function and vorticity contours for the $\alpha^+ = 0.072$, Re = 52,000 case corresponding to the experiment of Ramaprian (1992). At this slower pitch rate, a leading-edge disturbance is seen to reach the region of the dynamic-stall vortex and trigger its earlier formation at $\alpha \simeq 16°$. However, its subsequent growth appears to be a leading-edge driven phenomenon. In both cases, the video animation shows the dynamic stall vortex to be an amalgamation of strong, discretely shed vortices from the leading edge.

In summary, an unsteady NS analysis is developed that permits arbitrary three-degree-of-freedom maneuvering of an arbitrarily shaped body. A generalized Schwarz-Christoffel transformation technique is used to place a clustered C-grid about a NACA 0015 airfoil. Precise grid-point control is achieved through the use of analytical 1-D clustering transformations. Far-field boundaries are analytically mapped to infinity. Results are presented for Re = 45,000 and 52,000, both with and without flow control, and compared with experimental data. Also, video animation sequences detailing the vortex dynamics of the complete dynamic stall event, both with and without control, are available.

References

Ghia, K.N., Yang, J., Osswald, G.A. and Ghia, U. (1991), *AIAA Paper* No. 91-0546.

Osswald, G.A., Ghia, K.N., and Ghia, U. (1991), *Computer Physics Communications*, Vol. 65, pp. 209-218.

Ramaprian (1992), *AFOSR Workshop on Supermaneuverability: Physics of Unsteady Separated Flows at High Angle-of-Attack*, Lehigh University, pp. 119-129.

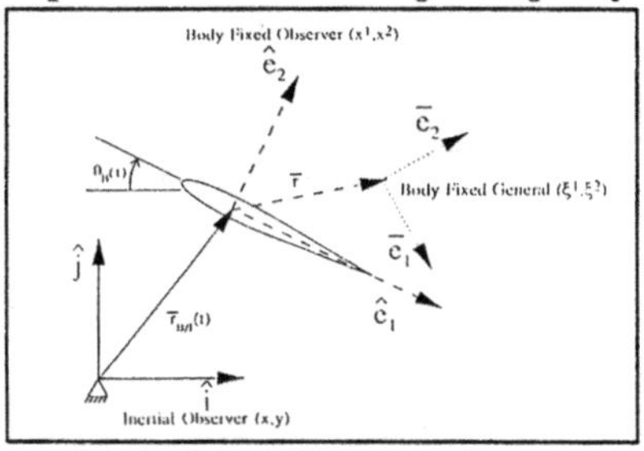

Fig. 1. Inertial/Body Fixed Coordinate Systems for Arbitrary Maneuvering Body.

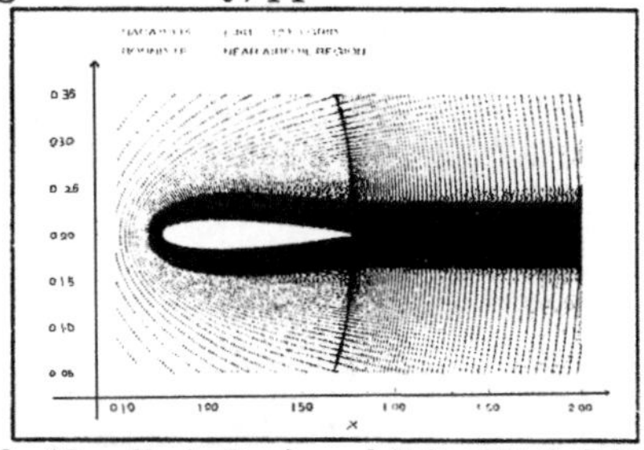

Fig. 2. Near Body Region of (461,129) C-Grid Distribution. Far-Field Boundary is Analytically Mapped to Infinity.

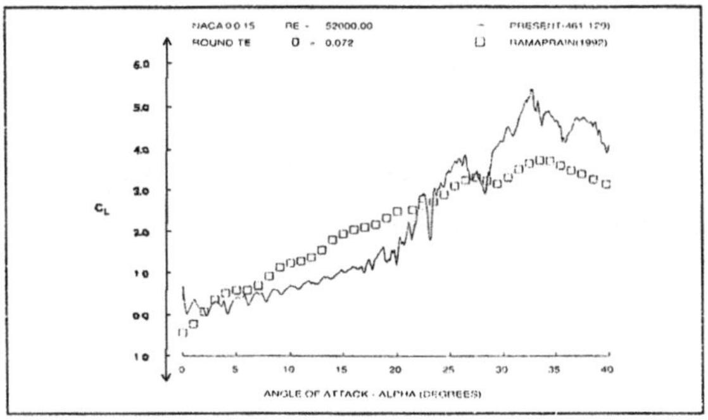

Fig. 3. Comparison of a Single Pitch-Up Instantaneous Lift History with the Ensemble Averaged Experimental Data of Ramaprian (1992).

Re = 52,000 , $\alpha^+ = 0.072$.

NO FLOW CONTROL

UNIFORMLY DISTRIBUTED LEADING-EDGE SUCTION 0 - .15 CHORD, S = -0.0074

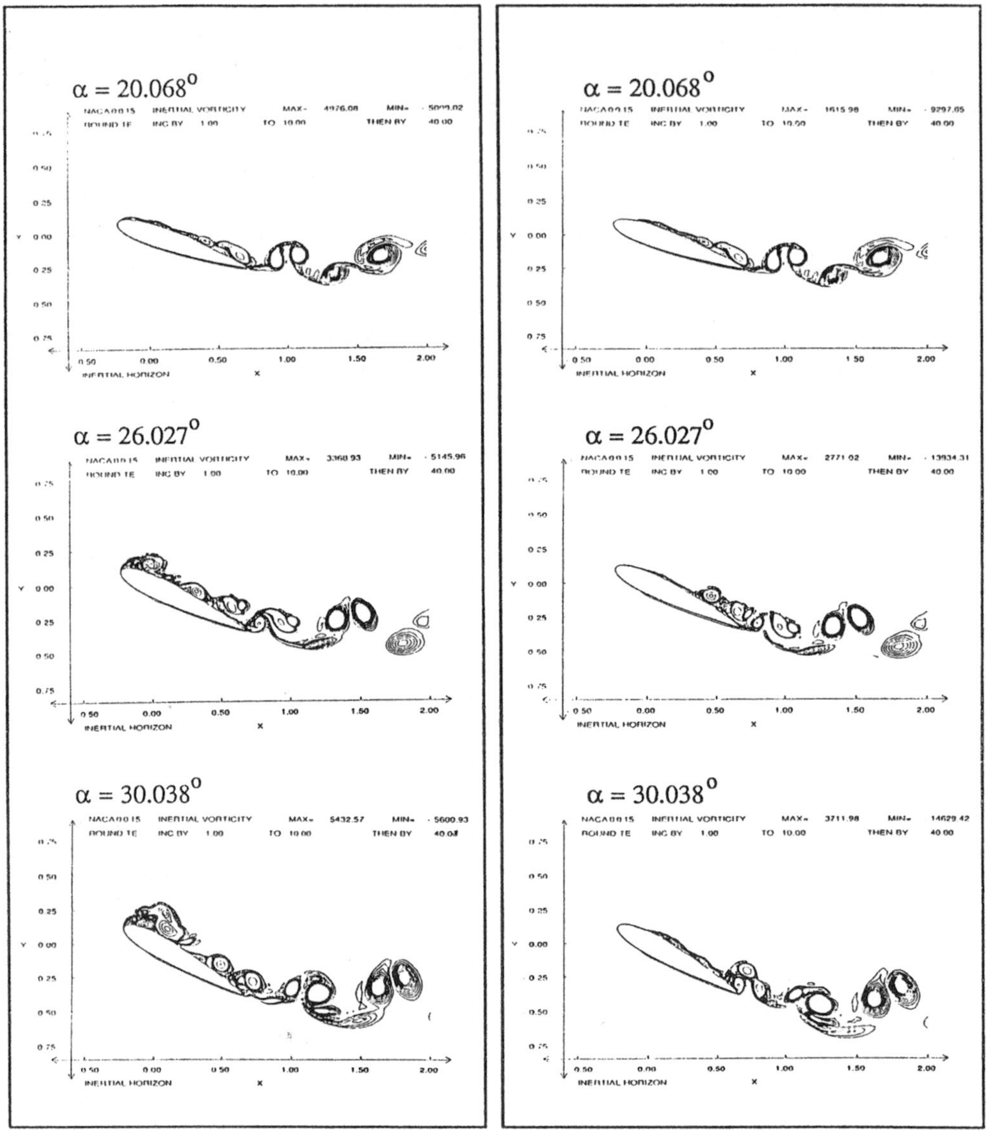

Fig. 4. Comparison of Instantaneous Vorticity Contours for Pitch-up Maneuver with and without Flow Control. Re = 45,000 $\alpha^{+} = 0.2$

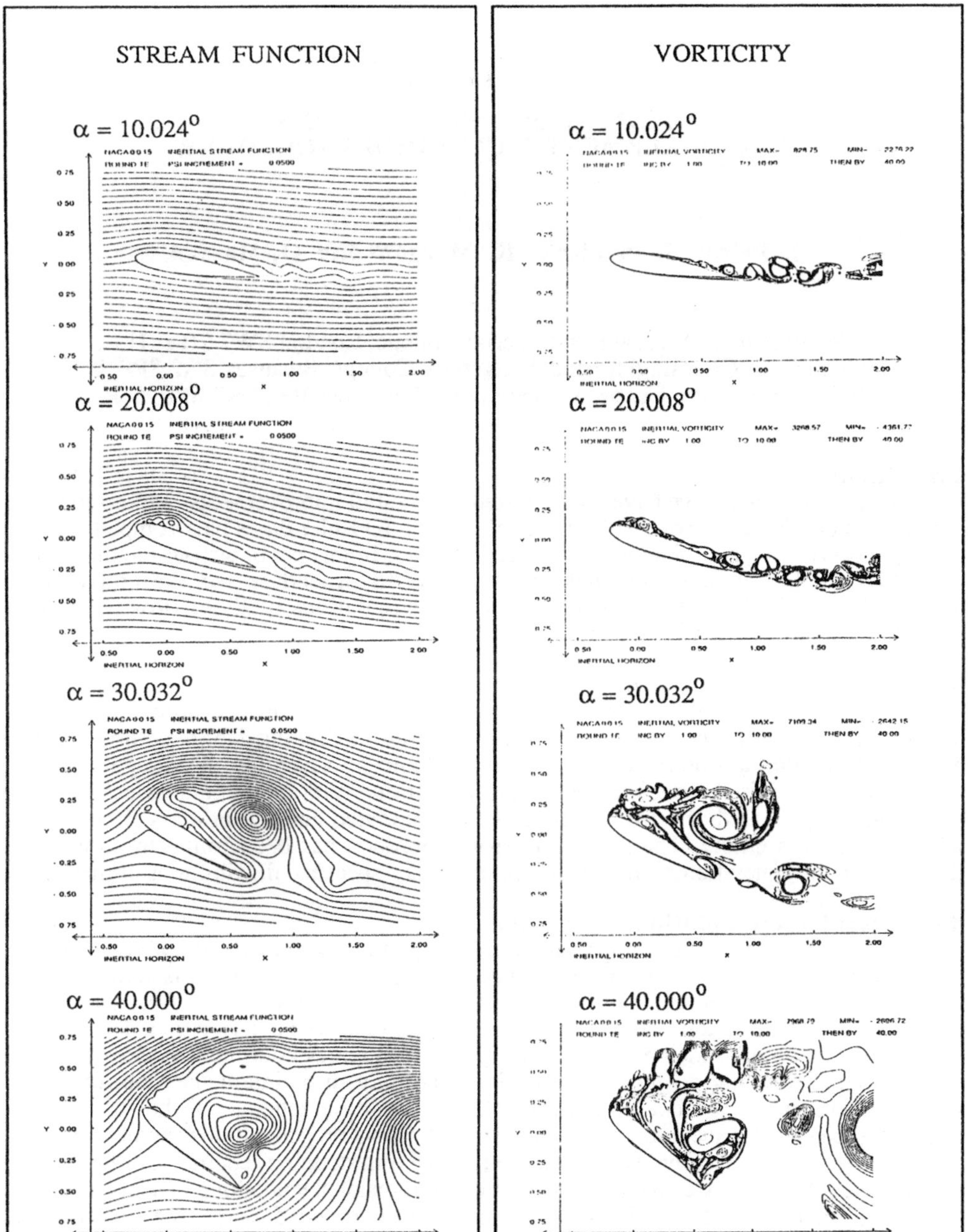

Fig. 5. Instantaneous Stream Function and Vorticity Contours for Slow Pitch-up Maneuver. Re = 52,000 $\alpha^+ = 0.072$

THE SOLUTION OF 3D INVISCID FLOW USING A FINITE ELEMENT MULTIGRID PROCEDURE

J. Peiró[1], J. Peraire[1], K. Morgan[2] and O. Hassan[3]

[1] Department of Aeronautics, Imperial College, London SW7 2BY, U.K.
[2] Department of Civil Engineering, University College, Swansea SA2 8PP, U.K.
[3] CDR, Innovation Centre, University College, Swansea SA2 8PP, U.K.

Introduction

Over the past four years, we have incorporated substantial improvements into the approach which we had originally proposed[1] for the solution of 3D Euler flows past complete aircraft configurations using completely unstructured meshes. In this paper we outline our current capability in this area, which includes a multigrid procedure for the convergence acceleration of the basic flow solver. Numerical simulations of Euler flows past some realistic configurations are presented.

Mesh Generation Approach

The subdivision of a 3D computational domain, into an unstructured assembly of teterahedral elements, is achieved by means of the advancing front method[2]. In this procedure, the domain boundary is first divided into an assembly of triangular facets, which forms the initial front for the 3D discretisation. With elements and points being simultaneously created, the front is continuously updated and the process is completed when the front is empty. The advancing front approach has been found to allow great flexibility in the generation of meshes which meet user-specified distributions of mesh size.

Single Grid Flow Solution Algorithm

Steady state solutions of the compressible Euler equations are obtained by advancing the transient form of the equations to steady state. The spatial discretisation is achieved via a standard Galerkin procedure, leading to a centered approximation of the flux derivatives. Stabilisation of the solution is achieved by the addition of artificial dissipation. The artificial dissipation, which is of matrix form[3], is designed to suppress hourglass modes in smooth regions of the flow and to produce sufficient damping to maintain the stability of the scheme in the presence of discontinuities[4,5]. The resulting scheme can be shown to be formally second order accurate in space for smooth flows on general unstructured meshes. Time discretisation is accomplished via a multi-stage timestepping scheme[6] and the stability range of the scheme is extended by the use of explicit residual smoothing. The memory and CPU requirements of this scheme are considerably reduced when the unstructured tetrahedral mesh is represented in terms of an edge based data structure.

Multigrid Acceleration

The convergence of the basic flow algorithm has been accelerated by the addition of an unstructured multigrid procedure[4,7]. With the geometry of the computational domain defined, a sequence of totally unstructured grids, of increasing fineness, is automatically constructed by the advancing front method. The objective of the multigrid procedure is then to compute, on the coarser grids, corrections to the fine mesh solution. This requires the identification of suitable methods for the transfer of unknowns, corrections and residuals

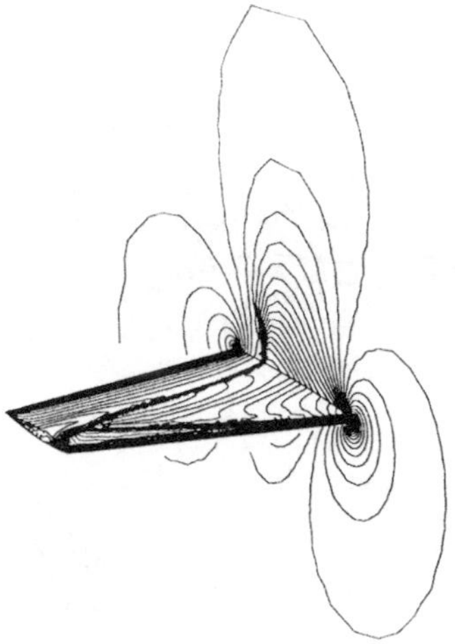

Figure 1
ONERA M6 wing; $M_\infty = 0.84$ and $\alpha = 3.06$ degrees. Pressure contour distribution on the wing upper surface and on the symmetry plane.

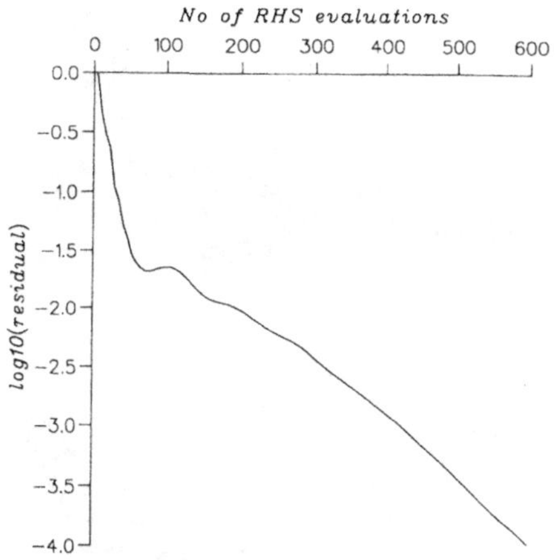

Figure 2
ONERA M6 wing; $M_\infty = 0.84$ and $\alpha = 3.06$ degrees. Convergence history of the multigrid scheme using a sequence of six grids.

between general tetrahedral meshes[8]. The algorithmic implementation which has been employed to date uses the single grid scheme outlined above on the finest mesh and a more economical first order scheme on the coarser meshes. The multigrid cycle starts on the finest mesh and one relaxation sweep is performed on each mesh before restriction to the next coarser mesh. The corrections computed on all the coarser grids are then simply added to the fine mesh solution. This procedure is such that the converged fine mesh solution is unaltered by the multigrid process.

Numerical Examples

Transonic Flow Over an ONERA M6 Wing. The first example consists of the simulation of the flow over an ONERA M6 wing at a free stream Mach number, M_∞, of 0.84 and at an angle of attack, α, of 3.06 degrees. A sequence of six grids was employed, with the finest mesh consisting of approximately 200000 points. A five stage version of the timestepping scheme is employed on each grid. Figure 1 shows the distribution of the pressure contours on the upper surface of the wing and the symmetry plane. The convergence history is shown in figure 2 and it may be observed that the residual is reduced by four orders of magnitude in only 100 multigrid cycles.

Flow About an Installed Aero-Engine Nacelle. The second example involves the simulation of the flow over a Rolls-Royce model of a twin engined civil transport with long cowl nacelle powerplants[9]. The mass inflow to the engine was prescribed and the engine outlet conditions were determined under the assumption that the separate core and fan streams were completely mixed before exhausting through the nozzle. To investigate the effect of varying the free stream conditions, three different flow simulations have been computed. In Case (i), $M_\infty = 0.77$ and $\alpha = 2.743$ degrees; in Case (ii), $M_\infty = 0.801$ and $\alpha = 2.738$ degrees; in Case (iii), $M_\infty = 0.84$ and $\alpha = 2.739$ degrees. For each flow condition, the same sequence of three meshes, consisting of 787568, 262664 and 172066 elements, was used for the multigrid procedure. The corresponding surface discretisations are shown in figure 3. The computed pressure contour distributions on the surface, obtained after 300 multigrid cycles, for the three flow cases are also shown in this figure. A comparison between the convergence behaviour of the single grid and the multigrid schemes for this example is shown in figure 4. For Case (ii), a comparison between the results of experiment and of computation for typical wing and nacelle sections is shown in figure 5. The overprediction of the pressure on the upper surface of the wing is attributed to

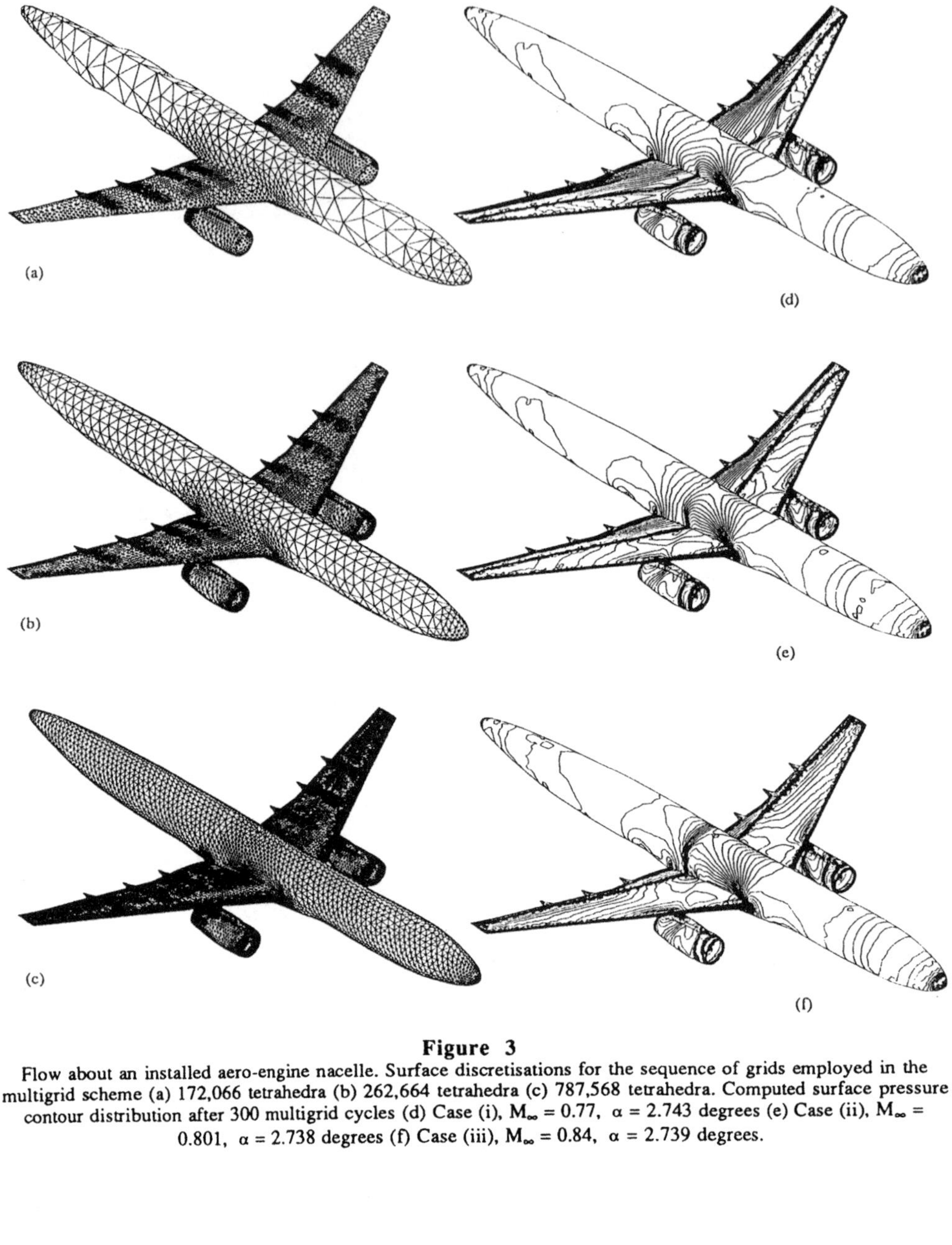

Figure 3

Flow about an installed aero-engine nacelle. Surface discretisations for the sequence of grids employed in the multigrid scheme (a) 172,066 tetrahedra (b) 262,664 tetrahedra (c) 787,568 tetrahedra. Computed surface pressure contour distribution after 300 multigrid cycles (d) Case (i), $M_\infty = 0.77$, $\alpha = 2.743$ degrees (e) Case (ii), $M_\infty = 0.801$, $\alpha = 2.738$ degrees (f) Case (iii), $M_\infty = 0.84$, $\alpha = 2.739$ degrees.

the inviscid flow model which has been employed. On an IBM RS6000/530H workstation, the grid generation for this example required approximately seven cpu hours, while each of flow cases was completed in approximately eight cpu hours. The flow simulation required less than fifteen megawords of memory.

Conclusions

It has been demonstrated that the use of an inviscid flow solver and a fully unstructured mesh approach leads to an efficient computational technique for rapidly analysing the effects of design changes to aerodynamic configurations. Current efforts are aimed at providing an equivalent viscous capability.

Acknowledgements

This work has been partially supported by the Aerothermal Loads Branch, NASA Langley Research Center, under NASA Research Contract NAGW-1809, and by Rolls-Royce plc, Derby. Computational resources were provided by IBM UK.

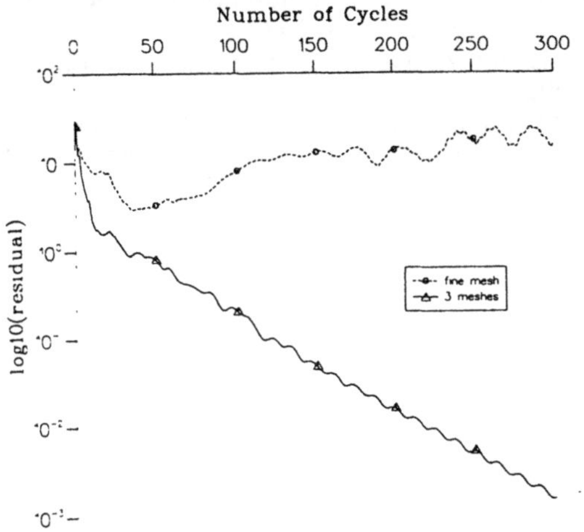

Figure 4
Flow about an installed aero-engine nacelle. Comparison of the convergence behaviour of the single grid and the multigrid schemes.

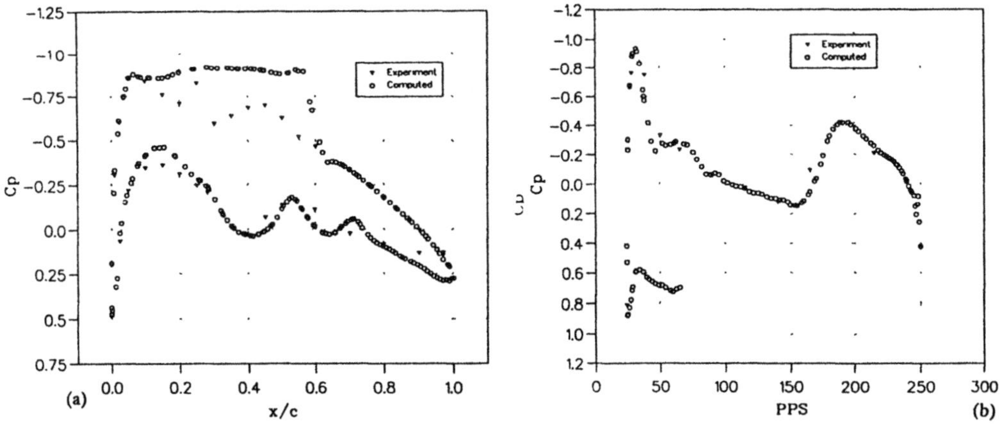

Figure 5
Flow about an installed aero-engine nacelle; Case (ii), $M_\infty = 0.801$, $\alpha = 2.738$ degrees. Comparison between the results of experiment and computation at (a) a wing section (b) a nacelle section.

References

[1] J. Peraire, J. Peiró, L. Formaggia, K. Morgan and O.C. Zienkiewicz, 'Finite element Euler computations in three dimensions', *AIAA Paper 88-0032*, 1988

[2] J. Peraire, K. Morgan and J. Peiró, 'Unstructured finite element mesh generation and adaptive procedures for CFD', in *Application of Mesh Generation to Complex 3-D Configurations, AGARD Report 464*, Paris, 18.1-18.12, 1990

[3] R.C. Swanson and E. Turkel, 'On central difference and upwind schemes', *ICASE Report 90-44*, 1990

[4] J. Peraire, J. Peiró and K. Morgan, 'A 3D finite element multigrid solver for the Euler equations', *AIAA Paper 92-0449*, 1992

[5] J. Peraire, K. Morgan, M. Vahdati and J. Peiró, 'The construction and behaviour of some unstructured grid algorithms for compressible flows', to appear in *Proc. ICFD Conference on Numerical Methods for Fluid Dynamics*, Oxford University Press, 1992

[6] A. Jameson, 'Transonic flow calculations', *Princeton University Report MAE 1751*, 1984

[7] D. Mavriplis and A. Jameson, 'Multigrid solution of the Euler equations on unstructured and adaptive meshes', *ICASE Report 87-53*, 1987

[8] J. Peraire, J. Peiró, Y-J. Li and K. Morgan, 'Unstructured multigrid solution of transonic flows', *in preparation*, 1992

[9] J. Peiró, J. Peraire, K. Morgan, O. Hassan and N. Birch, 'The numerical simulation of flow about installed aero-engine nacelles using a finite element Euler solver on unstructured meshes', *accepted for publication in Aero. J.*, 1992

Eulerian-Eulerian and Eulerian-Lagrangian Methods in Two Phase Flow

S. Sivier*, E. Loth†, J. Baum††, R. Löhner**

* Graduate Research Assistant, Department of Aeronautical and Astronautical Engineering, University of Illinois at Urbana-Champaign, Urbana, IL 61801

† Assistant Professor, Department of Aeronautical and Astronautical Engineering, University of Illinois at Urbana-Champaign, Urbana, IL 61801

†† Senior Research Scientist, Science Applications International Corporation, McLean, VA 22102

** Research Professor, CMEE, School of Engineering and Applied Science, The George Washington University, Washington, D.C. 20052

This paper describes a Finite Element Method - Flux Corrected Transport (FEM-FCT) approach with an unstructured adaptive grid scheme to the simulation of two-phase flow. The gas equations are computed using an Eulerian frame of reference, while the particle transport is computed using both Eulerian and Lagrangian frames of reference. One-dimensional shock wave attenuation was investigated to evaluate the performance of these methodologies and the gas-solid interphase transport models.

1. INTRODUCTION

When solving the two-phase flow equations, the gas, as a continuum, is best represented by an Eulerian description, that is, the gas characteristics are calculated at fixed points within the flow. However, as the particles may be relatively sparse in the flowfield, they can be modeled by either the Eulerian description (in the same manner as the gas flow) or Lagrangian description (where individual particles or particle groups are monitored and tracked in the flow). Crowe[1] presented a review of numerical models for two-phase flow. Both Eulerian and Lagrangian methods were discussed and compared. Eulerian methods allow particle diffusion to be incorporated into the model. Lagrangian methods require less memory overhead when particles of different sizes are considered and when particles rebound off of boundary surfaces. These two approaches were both employed in the present study to allow a direct comparison.

Sommerfeld[2] performed an experiment to study the attenuation of a shock as it propagates into a gas-particle mixture. A vertical shock tube was used to provide a homogeneous gas-particle mixture. Air was used as the gas and glass spheres were used for the particles. The shocks were generated in pure air and propagated into the dusty flow. A number of different tests were run at different initial Mach numbers (M_0) and particle mass fraction (η) values. Numerical experiments to examine the dusty shock attenuation were conducted in the same study using the random-choice method and by adding the particle phase in an Eulerian manner.

Igra and Ben-Dor[3] conducted a numerical study of the relaxation zone behind a normal shock in a dusty gas. The particle phase was added in an Eulerian fashion and various representations of the drag coefficient, C_D, and the Nusselt number, Nu, were examined. For the Mach numbers considered in the current study, compressibility effects in the C_D formulation and radiative heat transfer in the energy equation were found to be negligible.

Additional numerical studies based on Sommerfeld's experimental data were conducted by Olim, et al.[4] using a one-dimensional Eulerian two-phase formulation in conjunction with a finite difference FCT method. The study was used to show that the attenuation of the shock wave takes the form of an exponential decay curve. The choice of FCT showed marked improvement over Sommerfeld's random-choice method.

2. NUMERICAL METHOD

2.A. Gas Equations

The gas equations were integrated based on an Eulerian frame of reference. The governing equations for this flow are written in conservation form as

$$\frac{\partial \mathbf{U}}{\partial t} + \frac{\partial \mathbf{F}_j}{\partial x_j} = \mathbf{S} \tag{1}$$

where the summation convention is used and

$$\mathbf{U} = \begin{bmatrix} \rho \\ \rho u_i \\ \rho e \end{bmatrix}, \quad \mathbf{F}_j = \begin{bmatrix} \rho u_j \\ \rho u_i u_j + p\delta_{ij} \\ u_j(\rho e + p) \end{bmatrix}, \quad \mathbf{S} = \begin{bmatrix} 0 \\ -D_i \\ -Q - u_{pi}D_i \end{bmatrix} \tag{2}$$

where D_i is the component of drag force per unit volume of the gas in the i direction on the particles, and Q is the heat transferred from the gas to the particles. The state equations are

$$p = (\gamma - 1)\rho[e - \frac{1}{2}u_j u_j], \quad T = [e - \frac{1}{2}u_j u_j]/C_v \tag{3}$$

where ρ, p, e, T, k, γ, and C_v are density, pressure, specific total energy, temperature, thermal conductivity of the fluid, ratio of specific heats, and specific heat at constant volume, respectively, and u_i is the component of the fluid velocity in the i direction of a Cartesian coordinate system. Quantities with a subscript p indicate a discrete phase quantity. These equations were integrated in time by limiting a high order solution (which included interphase source contributions) with a low order contribution as described in Löhner, et al.[5], for the FEM-FCT method. A two-step second-order Taylor-Galerkin algorithm was used to produce the high order solution. Grid adaption was employed to optimize the distribution of grid points by refining areas with high gradients of density and coarsening areas of low gradients of density.

2.B. Particle Description: Eulerian

For the two dimensional case, five particle equations[4] were added to solve for the five particle unknowns: spatial density (σ), particle u velocity, particle v velocity, particle energy, and number density (n). The two densities are defined as:

$$\sigma = \frac{\text{mass of particles}}{\text{unit volume}}, \qquad n = \frac{\text{number of particles}}{\text{unit volume}}.$$

The two densities are required to allow for a variable particle diameter, such as occurs in combustion or evaporation.The particle equations then become:

$$\frac{\partial \mathbf{U}_p}{\partial t} + \frac{\partial \mathbf{F}_{pj}}{\partial x_j} = \mathbf{S}_p \tag{4}$$

where:

$$\mathbf{U_p} = \begin{bmatrix} \sigma \\ \sigma u_{pi} \\ \sigma e_p \\ n \end{bmatrix}, \quad \mathbf{F_{pj}} = \begin{bmatrix} \sigma u_{pj} \\ \sigma u_{pi} u_{pj} \\ u_{pj} \sigma e_p \\ n u_{pj} \end{bmatrix}, \quad \mathbf{S_p} = \begin{bmatrix} 0 \\ D_i \\ Q + u_{pi} D_i \\ 0 \end{bmatrix} \tag{5}$$

where u_p and e_p are the particle velocity and specific energy. FEM-FCT was used to advance these equations similar to that used for the gas equations, except that the limiter was modified to prevent non-monotonic behavior in the non-conservative velocities.

The particles affect the gas through the interphase coupling terms D_i and Q. The drag force per unit volume, D_i, is:

$$D_i = \frac{\pi}{8} n \rho C_D \left| u_i - u_{pi} \right| (u_i - u_{pi}) d^2 \tag{6}$$

where summation is not carried out over the indices, d is the particle diameter, and C_D is the coefficient of drag. The C_D used was given by Clift, et al.[6] as:

$$C_D = \frac{24}{Re} (1 + 0.15 Re^{0.687}) \tag{7}$$

where this C_D is valid for $Re < 800$, and Re is the flow Reynolds number based on slip velocity defined by:

$$Re = \frac{\rho \left| \mathbf{u} - \mathbf{u_p} \right| d}{\mu} \tag{8}$$

The heat transferred from the gas to the particles is described by[4]:

$$Q = A \cdot n \cdot [h \cdot (T - T_p) + \sigma^* \cdot (T^4 - T_p^4)] \tag{9}$$

where A is the area of the particle (πd^2), h is the coefficient of heat convection, σ^* is the Stefan-Boltzmann constant, and T is the temperature. Since the particles are inert and M_0 is small, the radiation term was ignored. The coefficient of heat convection is:

$$h = \frac{Nu \cdot k}{d} \tag{10}$$

where Nu is the Nusselt number, and the thermal conductivity of the gas, k, is taken as a constant, $1.787 \cdot 10^3$ g/cm·s. Nu was formulated with:

$$Nu = 2 + 0.459 \cdot Pr^{0.333} \cdot Re^{0.55} \tag{11}$$

where the Prandtl number, Pr, is taken as a constant, 0.75.

2.C. Particle Description: Lagrangian

The Lagrangian description was implemented by tracking individual groups of particles, referred to as parcels[7]. The previous timestep's gas characteristics are interpolated to each parcel to determine the interphase coupling terms, using the same equations as the Eulerian coupling terms. The parcel shape functions are then used to scatter each parcel's contributions to the host element's nodes. The Lagrangian parcel equations[7] are then solved directly for each parcel to update the parcel unknowns:

$$\frac{dx_{pi}}{dt} = u_{pi} \tag{12}$$

$$\frac{du_{pi}}{dt} = \frac{3\rho}{4\rho_p d} C_D(u_i - u_{pi})|u_i - u_{pi}| \tag{13}$$

$$\frac{dT_p}{dt} = 6\frac{k}{C\rho_p d^2} \mathrm{Nu}\,(T - T_p) \tag{14}$$

where C is the specific heat capacity of the parcels. A large part of the coding change required to implement the Lagrangian description involved determining which computational element contained a given parcel after the parcel had moved or when the unstructured mesh was refined/coarsened. The use of a vectorized successive neighbor search[8] was found to provide an elegant solution to this problem.

3. DISCUSSION

Figure 1 shows the numerical domain used in the study. The particles consisted of glass spheres with an average diameter, $d = 27\ \mu m$, a particle density $\rho_p = 2.5\ g/cm^3$ and a specific heat $C = 7.66 \cdot 10^6\ cm^2/s^2 \cdot K$. The shock was initiated upstream of the particle region in the single phase domain.

Figure 2 shows a comparison of the Sommerfeld[2], Olim et al. and FEM-FCT (with both Eulerian and Lagrangian particles) data for the case of a shock with initial Mach $M_0 = 1.25$ propagating into a dusty gas with loading η ($= \sigma/\rho$) $= 0.63$. All predictions tend to overestimate the attenuation. At four meters, data from Olim et al. give better predictions than the FEM-FCT data or the Sommerfeld data. Both Eulerian and Lagrangian versions of FEM-FCT gave similar results. This was the only case Olim et al. computed.

Figure 3 shows a comparison of the Sommerfeld[2] and FEM-FCT data (with only Eulerian particles) for the case of a shock with initial Mach $M_0 = 1.49$ propagating into a dusty gas with loading $\eta = 0.63$. Both the Sommerfeld predictions and the FEM-FCT predictions gave reasonable agreement.

Figure 4 shows a comparison of the Sommerfeld[2], FEM-FCT data (with only Eulerian particles) for the case of a shock with initial Mach $M_0 = 1.48$ propagating into a dusty gas with loading $\eta = 1.25$. FEM-FCT gave excellent agreement with experiment, especially during the first meter. Data from Sommerfeld tend to underpredict the attenuation at the beginning of the computation and overpredict near the end. Such a discrepency may be consistent with the drag coefficient chosen by Sommerfeld.

4. CONCLUSIONS

The developed FEM-FCT has been shown to give good agreement with experimental results. Lagrangian and Eulerian implementations give similar results. However, the current Lagrangian version requires much more memory and CPU time due to the need to saturate elements with parcels in order to guarantee that a reasonably uniform parcel distribution will exist after mesh refinement. Similar parcel refinement is required to reduce the number of parcels that must be maintained.

5. REFERENCES

1. Crowe, C.T., "REVIEW - Numerical Models for Dilute Gas-Particle Flows," Journal of Fluids Engineering, **104**, Sept. 1982, 297-303.

2. Sommerfeld, M., "The Unsteadiness of Shock Waves Propagating Through Gas-Particle Mixtures," Experiments in Fluids, **3**, 197-206 (1985).
3. Igra, O., and Ben-Dor, G., "Parameters Affecting the Relaxation Zone behind Normal Shock Waves in a Dusty Gas," Israel Journal of Technology, **18**, 1980, 159-168.
4. Olim, Igra, Mond, and Ben-Dor, "A General Attenuation Law of Planar Shock Waves Propagating into Dusty Gases," Proceedings of the 16th International Symposium on Shock Waves And Shock Tubes, 217-225 (1987).
5. Löhner, R., Morgan, K., Peraire, J., and Vahdati, M., "Finite Element Flux Corrected Transport (FEM-FCT) for the Euler and Navier-Stokes Equations," International Journal for Numerical Methods in Fluids, **7**, 1987, 1093-1109.
6. Clift, R., Grace, J.R., and Weber, M.E., Bubbles, Drops, and Particles, Academic Press, 1978.
7. Sommerfeld, M., "Numerical Simulation of Supersonic Two-Phase Gas-Particle Flows," Proceedings of the 16th International Symposium on Shock Tubes and Waves, 235-241 (1987).
8. Löhner, R., and Ambrosiano, J., "A Vectorized Particle Tracer for Unstructured Grids," Journal of Computational Physics, **91**, No. 1, November 1990, 22-31.

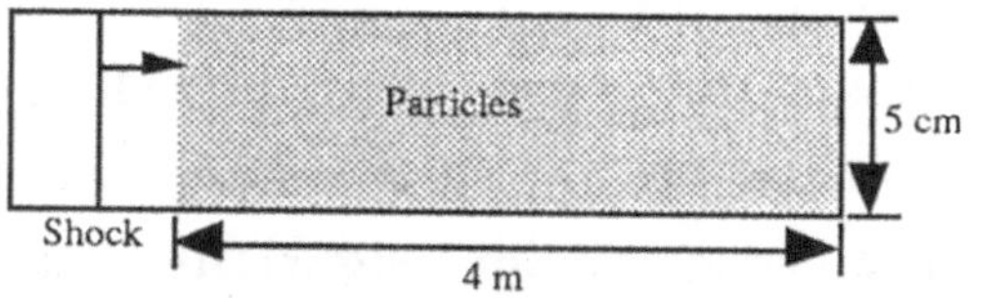

Fig. 1: Numerical Domain

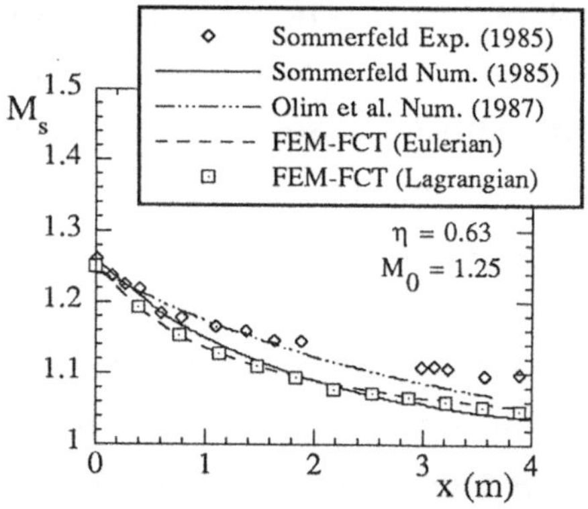

Fig. 2: Comparison between numerical and experimental data for shock attenuation with $M_0 = 1.25$, $\eta = 0.63$

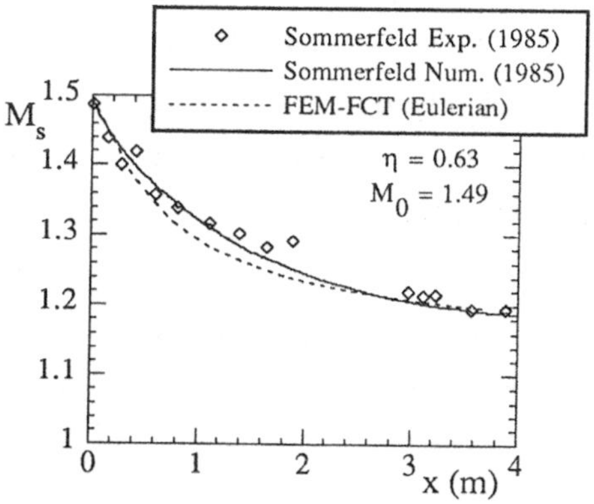

Fig. 3: Comparison between numerical and experimental data for shock attenuation with $M_0 = 1.49$, $\eta = 0.63$

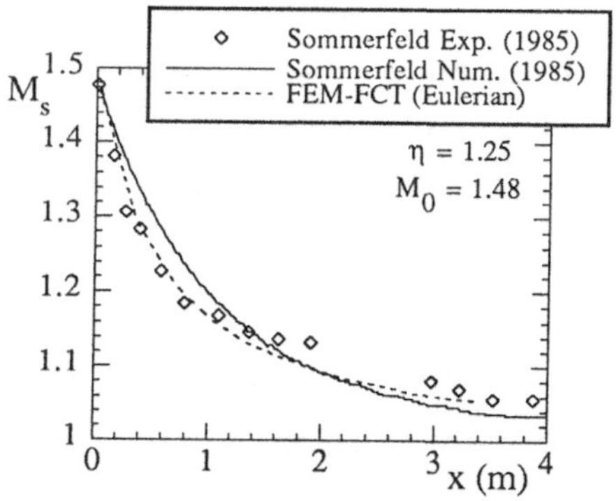

Fig. 4: Comparison between numerical and experimental data for shock attenuation with $M_0 = 1.48$, $\eta = 1.25$

FAST-CONVERGENT METHOD FOR FINDING STEADY-STATE SOLUTIONS OF THE EULER EQUATIONS

I.L.Sofronov

Keldysh Institute of Applied Mathematics, Moscow

Introduction

The method based on the explicit Lax-Wendroff-type difference scheme and special procedure accelerating the process of temporal build-up is suggested. Acceleration procedure uses the standard fast Fourier transform algorithm to modify the residuals on each step of the temporal integration process. Numerical experiments show that this procedure is not worse in efficiency than a multigrid one. Computational costs are also economized by decomposition of the given domain into subdomains with embedded finer grids. Using the acceleration algorithm and the decomposition procedure of flow into subdomains with simple and complicated flow structure one may reduce the computational cost till 5÷20 times depending on a problem.

This paper contains only the description of the convergence acceleration idea (method in details and numerical results were published in [1]). Some generalization of this idea is also given here.

1. Spectral analysis and idea of acceleration

Consider in the square $x \in [-\pi,\pi]$, $y \in [-\pi,\pi]$ the specimen equation

(1) $$u_t + u_x + u_y + f(x,y) = 0$$

with time-independent periodic boundary conditions in both directions.

Let u_0 be the steady state solution of this equation. Starting from some function $u(x,y)$ given at zero temporal point and using the Lax-Wendroff-type scheme we build the discrete process of finding u_0 . According to this scheme the value of $\bar{u}$, that is the value of u on upper time level, is calculated as

$$\bar{u} = u + \tau u_t + \frac{\tau^2}{2} u_{tt} ,$$

where τ is the time step. Expressing u_t and u_{tt} from (1) and introducing additional parameter a which influences the stability of the process, we obtain

$$\text{(2)} \qquad \bar{u} = u - \tau(u_x+u_y+f) + a\frac{\tau^2}{2} (u_{xx}+2u_{xy}+u_{yy}+f_x+f_y)$$

The evaluation of deviation $v \equiv u - u_0$ is obviously described as

$$\text{(3)} \qquad \bar{v} = v - \tau(v_x+v_y) + a\frac{\tau^2}{2} (v_{xx}+2v_{xy}+v_{yy})$$

In order to analyze the behavior of v let us introduce the discrete set $I = \{0, \pm h, \pm 2h, \ldots, \pm 1\}$, $h = 1/N$, N is natural, and represent v as

$$\text{(4)} \qquad v = \sum_{\xi,\eta\in I}\sum a_{\xi\eta} e^{-i(\xi x+\eta y)/h}$$

We obtain from (3) and (4) that the amplificator $\lambda_{\xi\eta}$ of the transition to the next time level for some harmonica $a_{\xi\eta}e^{-i(\xi x+\eta y)/h}$ equals to

$$\lambda_{\xi\eta} = 1 - ir(\xi + \eta) - a\frac{r^2}{2} (\xi + \eta)^2 \quad , \quad r = \tau/h$$

Consequently,

$$\text{(5)} \qquad |\lambda_{\xi\eta}|^2 = 1 - (a-1)r^2(\xi + \eta)^2 + \left[a\frac{r^2}{2} (\xi + \eta)^2\right]^2$$

Let us introduce the operators M and M^2 acting on an arbitrary function of kind (4), as follows

$$Mv = \sum_{\xi,\eta\in I}\sum a_{\xi\eta} \frac{e^{-i(\xi x + \eta y)/h}}{\sqrt{\xi^2 + \eta^2}} \quad , \qquad M^2 = MM$$

Parallel with (2) consider the process

$$\text{(6)} \qquad \bar{u} = u - \tau M(u_x + u_y + f) + a\frac{\tau^2}{2} M^2(u_{xx}+2u_{xy}+u_{yy}+f_x+f_y)$$

Analogously to (5) we get the following expression

$$(7) \qquad |\lambda_{\xi\eta}|^2 = 1 - (a-1)r^2 \frac{(\xi + \eta)^2}{\xi^2 + \eta^2} + \left[a\frac{r^2}{2} \frac{(\xi + \eta)^2}{\xi^2 + \eta^2} \right]^2$$

for the amplificator $\lambda_{\xi\eta}$ corresponding to (6).

To compare formulae (5) and (7) let us fix numbers $a > 1$, and r, $0 < r < [2(a-1)]^{0.5}/a$ (in this scale $|\lambda_{\xi\eta}| < 1$) and find frequencies (ξ,η) yielding $|\lambda_{\xi\eta}| > 1-\varepsilon$, where $\varepsilon > 0$ is some small number. Figs. *1,a* and *1,b*

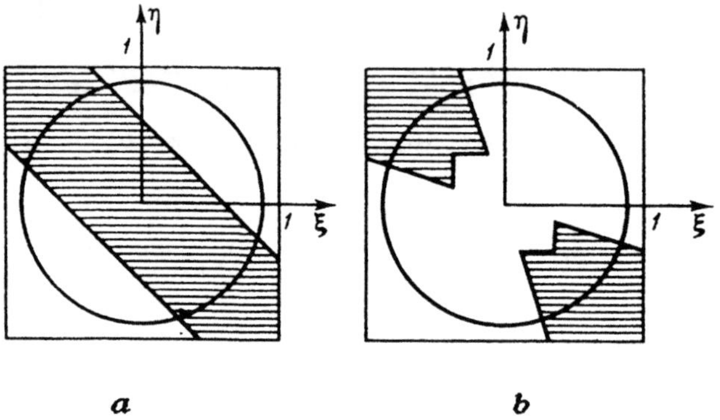

Fig. 1

show the regions of corresponding (ξ,η). The boundaries of these regions are described by the equations

$$\eta = \begin{cases} -\xi \pm \sqrt{\delta} & \text{for (5)} \\ (-1 \pm \sqrt{2\delta})\xi, \quad |\eta| = |\xi| = h/\sqrt{2\delta} & \text{for (7)} \end{cases}$$

where $\delta \cong 2\varepsilon/[r^2(a-1)]$ for sufficiently small ε.

From the viewpoint of convergence, the process (6) is more preferable than (2) since it has better dissipation properties. Fig. *1,b* shows that in the first and the third quadrants the rate of drop in residual does not depend on (ξ,η) ,i.e. on frequency of harmonica. Indeed, it follows from (7) that

$$|\lambda_{\xi\eta}|^2 < \max\{1 - (a-1)r^2 + 0.25a^2r^4, \; 1 - 2(a-1)r^2 + a^2r^4\}$$

for ξ,η of the same signs. For example, if $a = 2$, $r = 0.5$ then $|\lambda_{\xi\eta}| < 0.9$.

Moreover, for all low frequencies, i.e. for $|\eta|,|\xi| < nh$, $n = 1, 2, \ldots, \ll h^{-1}$, the value $|\lambda_{\xi\eta}|$ in (7) is also restricted:

$$|\lambda_{\xi\eta}| < 1 - 0.125(a-1)r^2/n^2$$

The above distinction in properties of (2) and (6) leads to the convergence acceleration of the process (6) in comparison with (2) . Numerical experiments for the real problems show that this reduction in a temporal steps number may reach 2÷12 times.

2. *Difference scheme and method.*

In order to obtain solution at the upper time level using the Lax–Wendroff scheme one adds two terms – of the first and the second order in time – to the solution at the lower time level:

$$X(t+\tau) \approx X(t) + \tau X_t + \frac{\tau^2}{2} X_{tt} ,$$

where $X = (\rho, \rho u, \rho v, e)^T$ is the vector of gas dynamics variables; τ is time step. When the original continual problem is replaced by a difference one, the quadrilateral body-fitted grid in the computational domain is introduced at first. Then the Euler equations are being approximated on this grid by the scheme of Lax–Wendroff type obtained by using the finite volumes method. Artificial viscosity term is also added to this scheme. The resulting residual ΔX to next time level in each grid node is the following:

(8) $$\Delta X = \Delta X^{(1)} + \Delta X^{(2)} + \Delta X^{(\nu)} ,$$

where $\Delta X^{(1)}$ is residual of term τX_t , $\Delta X^{(2)}$ is residual of term $\frac{\tau^2}{2} X_{tt}$, $\Delta X^{(\nu)}$ is residual of artificial viscosity.

In order to accelerate convergence using the idea which was considered in *Sect.1.* we change formula (8) by introducing two additional linear operators $M^{(1)}$ and $M^{(2)}$:

$$\Delta X = M^{(1)}\left[\Delta X^{(1)} + \Delta X^{(\nu)}\right] + M^{(2)}\Delta X^{(2)} .$$

Application of these operators is based on the sine FFT algorithm. The cost of one temporal step increases by 70%

due to this procedure but the total number of steps decreases over and over.

The scheme obtained has the second order of accuracy on smooth steady-state solutions and resolves well shocks in transonic case.

3. Application for other difference schemes

As we have seen the suggested acceleration procedure requires implementing the residual by sum of terms corresponding to derivatives in time of the first and second order. The Lax-Wendroff scheme gives those terms directly. In order to apply this procedure for other schemes one can calculate those terms by numerical differentiation, namely:

$$\Delta X^{(1)} = (X_{n+1} - X_{n-1})/2, \quad \Delta X^{(2)} = (X_{n+1} - 2X_n + X_{n-1})/2,$$

where X_{n-1} , X_n , X_{n+1} are sequential values of X calculated without acceleration.

Temporal step with acceleration is performed as

$$\tilde{X}_{n+1} = X_n + M^{(1)}\Delta X^{(1)} + M^{(2)}\Delta X^{(2)}$$

Then one makes two next steps without acceleration: X_{n+2}, X_{n+3} ; calculates $\Delta X^{(1)}$, $\Delta X^{(2)}$ anew using $\tilde{X}_{n+1}$, X_{n+2}, X_{n+3} ; makes step with acceleration to obtain $\tilde{X}_{n+3}$ and so on.

Note that the original difference scheme must be really of the second order in time since the acceleration procedure will not be stable in other case.

1. *Sofronov I.L.* A fast-convergent method for solving Euler equations, *Journal of Computational Math. and Math. Physics. 1991. V.31. #4. P. 575–591* (in Russian, the English version of this Journal is *U.S.S.R. Computational Math. and Math. Physics*)

NUMERICAL STUDY OF THE CLOSE-COUPLED CANARD-WING-BODY AERODYNAMIC INTERACTION

E. L. Tu *

NASA Ames Research Center, Moffett Field, California 94035, USA

Introduction

Many modern aircraft, both operational and experimental, utilize canards for improved aerodynamic performance. The influence of canards can often result in increased maximum lift and decreased trim drag. For close-coupled canards, aerodynamic performance is a function of the aerodynamic interaction between the canard and wing. However, depending upon geometry and flow parameters, this interaction can be either favorable or unfavorable. Proper utilization of canards requires an accurate understanding of their influence on the structure of the flow about the wing.

The flow structure of highly-swept or delta canard-wing configurations is characterized by a canard downwash which modifies the wing flowfield and an interaction between the canard and wing vortex systems. A schematic of the canard-wing vortex interaction is given in Fig. 1. The downwash of the canard modifies the flowfield of the wing within the canard-tip span-line by decreasing the effective angle of attack of the wing. Beyond the canard tip, upwash from the canard increases the wing's effective angle of attack. The downwash and upwash effects of the canard have a significant influence on the formation of the wing leading-edge vortex. The canard downwash can weaken or delay the formation of the wing vortex, thus affecting its position over the wing surface.

Detailed experimental and computational studies have been performed on canard configurations and are given in Ref. 1. However, limited computational work has been performed utilizing the Navier-Stokes equations, which are required to capture some of the important flow features of the canard-wing interaction. With the emergence of faster computers and increased memory capacities, the Navier-Stokes equations can now be utilized.

Using an extension of the NASA Ames Research Center's Transonic Navier-Stokes (TNS) code,[2] the thin-layer Navier-Stokes equations are solved for the flow about a highly-swept coplanar canard-wing-body configuration. The results of the computations are used in the investigation of the aerodynamic characteristics of the canard configuration, including the canard-wing leading-edge vortex interaction and its effects on wing vortex breakdown. The current application of the TNS code expands the capability for analysis of the complex aerodynamics of canard configurations.

Computational Modeling

Numerical Procedure

The TNS code is a Reynolds-averaged thin-layer Navier-Stokes flow solver with structured zoning capability and has been applied to a wide range of wing and aircraft configurations. Since the TNS code solves the Reynolds- (or time-) averaged equations, the Baldwin-Lomax algebraic eddy-viscosity model[3] is chosen to compute the effects of turbulence on the flow. Due to the vortex-dominated flow structures of the highly-swept sharp-leading-edge canard and wing, a modification to the original Baldwin-Lomax formulation is required. For this study, the Degani-Schiff modification,[4] as originally developed for crossflow type separations, is employed. Without such a modification, the overpredicted eddy-viscosity levels tend to dominate the vortical flowfield. Further details about the TNS code, algorithm, zonal approach, and general performance are given in Ref. 2.

Geometry Modeling and Grid Generation

The geometry in this study is based on the wind-tunnel model used by Gloss and Washburn[5] and is illustrated in Fig. 2. Using the S3D surface geometry and grid generation code,[6] the canard, wing and body component surface geometries are modeled from their original analytical definitions. Further details of the S3D code and its applications are given in Ref. 6.

---

* Research Scientist

The flowfield grid is generated using the 3DGRAPE program[7] and is illustrated in Fig. 3. The 3DGRAPE program is a block-type general-purpose elliptic grid generator which allows for user-specified orthogonality and normal grid spacing conditions. The flowfield grid is H-O topology, which is ideal for this type of configuration because of the natural clustering of grid points in certain regions of the grid.

Results and Discussion

Experimental Comparisons

To investigate the influence of the canard, computations are conducted on the wing-body geometry with and without a canard. All computational results are for fully turbulent flow at a transonic Mach number (M_∞) of 0.90, a Reynolds number based on mean wing aerodynamic chord ($Re_{\bar{c}}$) of 1.52 million, and nominal angles of attack (α) ranging from 0 to 12 degrees. To validate the computational modeling, comparisons between the computed results and the force balance and wing surface pressure measurements are made.

A comparison of computed wing surface pressure coefficients (C_p) with experimental data[5] at $\alpha \approx 12°$ is illustrated in Fig. 4. Comparisons with the experimental data are given at representative wing span-stations of 25% and 45% as measured from the symmetry plane.

For the canard-off case, the leading-edge vortex is indicated by the suction peak in the C_p distribution. At this angle of attack, the major canard influence on wing surface pressures is the canard downwash effect. Significant differences in wing surface pressure between the canard-on and canard-off cases are observed. The effective local angle of attack (α_{eff}) of the wing is reduced and the wing leading-edge vortex is considerably weakened in the canard-on case.

Although detailed comparisons of flow quantities such as surface pressure are better measures of the computational accuracy, integrated force quantities such as lift, drag, and moments are often used to assess the overall aerodynamic performance characteristics of a given configuration. Figure 5 presents the comparison of canard-on and canard-off integrated force quantities for the canard-wing-body configuration. The canard-on computations are also compared with experimental data.[5] Figure 5 shows that the nonlinearity of the lift coefficient curve for the canard-on case is captured well by the computations. For angles of attack less than 6°, the computed lift coefficients for the canard-off and canard-on cases are comparable. At these low angles of attack, the reduction in wing lift due to the canard downwash is balanced by the additional canard lift.

The drag polar in Fig. 5 indicates comparable levels of drag coefficient for both the canard-on and canard-off cases. The "cross-over" of the two drag curves shows the potential of the canard configuration for reduced drag at a given lift.

Due to the relative location of the canard and the moment center, the pitching moment curve in Fig. 5 illustrates the typical nose-up pitching moment which is characteristic of many canard configurations. At higher angles of attack, the computations overpredict the nose-up moment of the canard-on case. This overprediction is further studied by examining the moment curves of the canard and wing regions separately.

The curves of pitching moment coefficients for different regions of the canard-wing-body configuration are given in Fig. 6. The forward and aft regions, including the body, (henceforth designated as the canard and wing regions) are chosen to correspond with regions measured in the experiment. The pitching moment curves show that the canard's influence on the wing region pitching moment is nominal. The nose-up pitching moment of the canard-on configuration is almost entirely due to the canard region. Good comparisons with experimental data indicate an accurate distribution of pitching moment in the computed results. Figure 6 also shows that the overpredicted pitch-up moment for the complete configuration, noted in Fig. 5, is approximately evenly distributed between the canard and wing regions.

The effects of grid refinement and a more detailed description of the comparisons between computations and experiment are given in Ref. 1.

Visualization of the Canard-Wing Interaction

For the current configuration, the primary mechanism for the canard-wing interaction is the canard's influence on the wing leading-edge vortex. Both the canard downwash and the canard leading-edge vortex have pronounced effects on the formation and subsequent trajectory of the wing vortex. At higher angles of attack, the canard's influence may extend to delaying or eliminating wing vortex breakdown.

The dominating effect of the canard downwash on the inboard wing has been illustrated in the surface pressures of Fig. 4. Outboard of the canard-tip span-line, the canard leading-edge vortex is the primary mechanism for the canard's influence on the wing flowfield. Crossflow-plane vortex visualizations in the form of normalized total pressure contours at two constant streamwise stations are given in Fig. 7, and show the effect of the canard-vortex induced flow on the wing. Although the computed canard vortex is considerably weaker than the corresponding wing vortex, its influence on the wing flowfield is still significant. For the two co-rotating vortices, each with a counter-clockwise rotation and position as shown in Fig. 7, the wing vortex flowfield induces a relative downward and inward motion of the canard vortex while the canard vortex induces an upward and outward movement of the wing vortex. In the absence of such interaction, the canard and wing vortex trajectories would be expected to follow an upward and outward path which would be dependent upon the angle of attack and respective sweep angles.

The potential for a canard to delay or eliminate wing vortex breakdown is of significant interest and has been the topic of numerous experimental and computational studies, some of which are listed in Ref. 1. By comparing vortex lift theory with experimental results, Gloss and Washburn[5] found that wing vortex burst occurs at $\alpha \approx 13°$ for the current canard-off case. For the canard-on case, their study indicated no evidence of wing vortex burst for angles of attack up to 20°.

Figure 8 shows off-surface particle traces for the canard-off case at three angles of attack. At the higher angles of attack, increased vortex strength and a relative shift of the core location is observed. Evidence of vortex burst is observed at $\alpha = 12.38°$. Crossflow-plane visualization of scaled axial velocity contours in Fig. 9 serves to confirm the presence of vortex breakdown over the wing for the canard-off case. Reversed axial flow in the core of the primary vortex is identified as a qualitative indication of vortex breakdown, and is observed in Fig. 9. From Fig. 8, the wing vortex burst location appears to be near the trailing edge of the wing and indicates that the computed angle of attack for vortex burst is approximately 12°, which is within reasonable agreement with the experimental observations.[5]

Computed particle traces for the canard-on case at $\alpha = 12.38°$ are given in Fig. 10. The lower leading-edge-sweep angle of the canard results in a canard vortex burst further upstream of the canard trailing edge compared to the wing vortex burst relative to its trailing edge shown earlier in Fig. 8. Figure 10 shows a stable wing vortex with no evidence of wing vortex breakdown in the presence of the canard.

Concluding Remarks

A numerical investigation of the canard-wing aerodynamic interaction has been performed using the Navier-Stokes equations. Favorable comparisons with experimental data verify that the present method is capable of predicting the flow about such configurations.

On the inboard wing, the canard-induced downwash was found to weaken the wing leading-edge vortex at $\alpha \approx 12°$. The wing vortex on the outboard wing was shown to be influenced by both the canard vortex and the aforementioned canard downwash. The presence of the canard was also shown to eliminate the wing vortex breakdown which was evident in the canard-off case at $\alpha \approx 12°$. These computational results confirm the canard's potential for delaying wing vortex breakdown which has been documented in numerous experimental studies.

References

1. Tu, E.L., "Navier-Stokes Simulation of a Close-Coupled Canard-Wing-Body Configuration," AIAA Paper 91-0070, January 1991.
2. Holst, T.L., Kaynak, U., Gundy, K.L., Thomas, S.D. and Flores, J., "Numerical Solution of Transonic Wing Flows Using an Euler/Navier-Stokes Zonal Approach," *Journal of Aircraft*, Vol. 24, No. 1, January 1987, pp. 17-24.
3. Baldwin, B.S. and Lomax, H., "Thin Layer Approximation and Algebraic Model for Separated Turbulent Flows," AIAA Paper 78-257, January 1978.
4. Degani, D. and Schiff, L.B., "Computation of Turbulent Supersonic Flows Around Pointed Bodies Having Crossflow Separation," *Journal of Computational Physics*, 66, 1986, pp. 173-196.
5. Gloss, B.B. and Washburn, K.E., "Load Distribution on a Close-Coupled Wing Canard at Transonic Speeds," *Journal of Aircraft*, Vol. 15, No. 4, April 1978, pp. 234-239.
6. Luh, R.C., Pierce, L. and Yip, D., "Interactive Surface Grid Generation," AIAA Paper 91-0796, January 1991.
7. Sorenson, R.L., "The 3DGRAPE Book: Theory, Users' Manual, Examples," NASA TM 102224, July 1989.

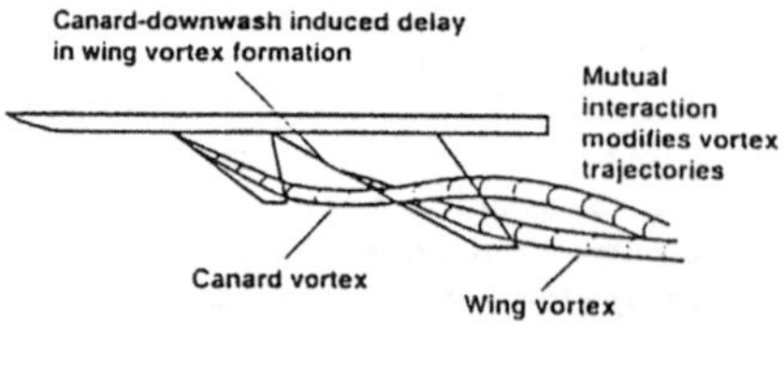

Fig. 1 Schematic of the canard-wing vortex interaction.

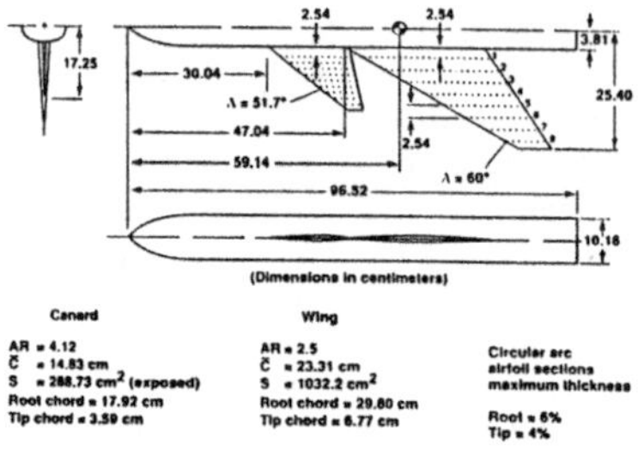

Fig. 2 Coplanar canard-wing-body geometry.

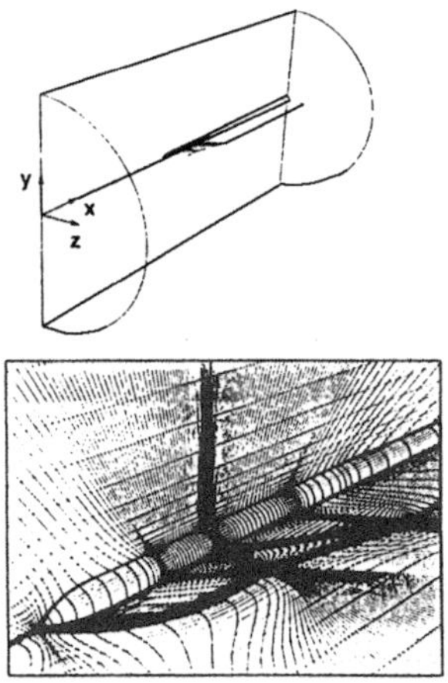

Fig. 3 Canard-wing-body flowfield grid topology.

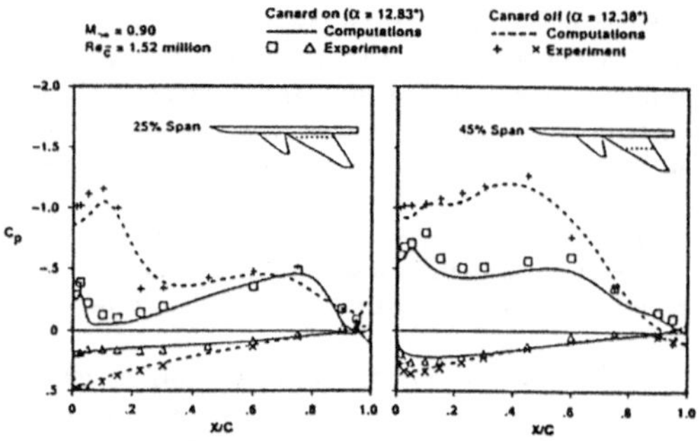

Fig. 4 Comparison of computed and experimental surface pressure coefficients with and without canard at $\alpha \approx 12°$.

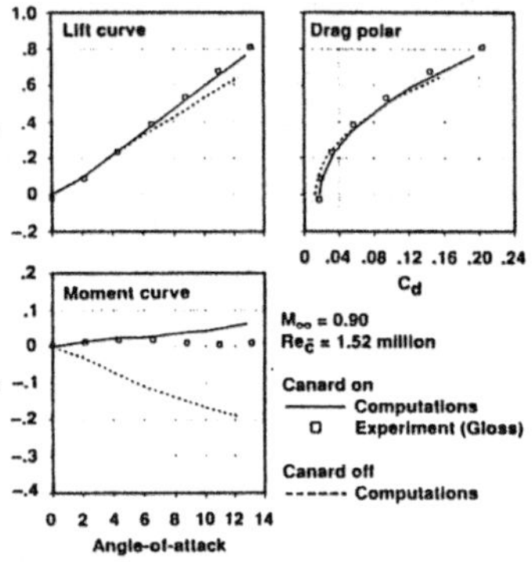

Fig. 5 Comparison of computed and experimental force coefficients with and without canard.

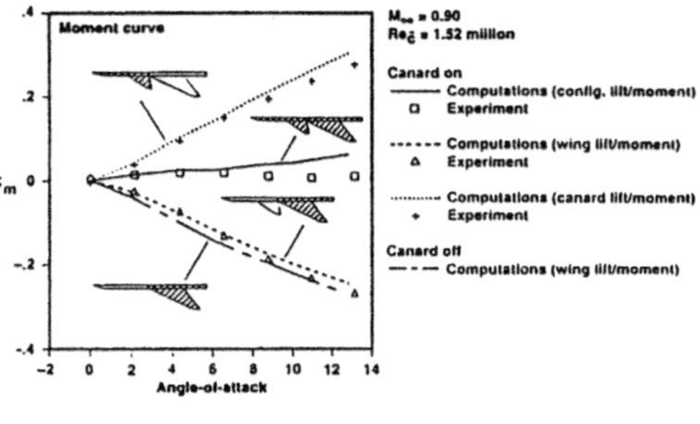

Fig. 6 Comparison of component moment coefficients with and without canard. (Moment curves are given for shaded regions of the geometry).

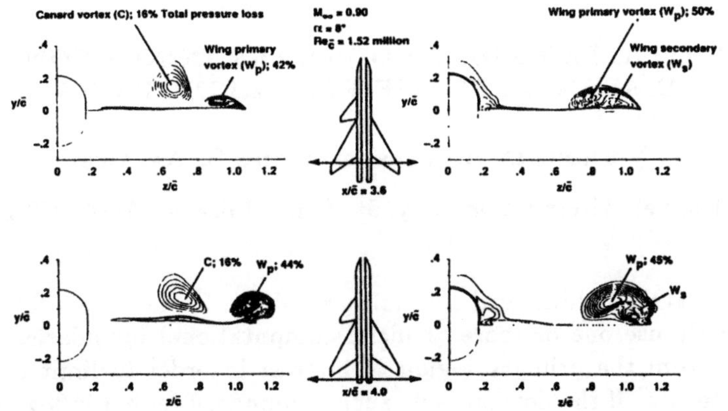

Fig. 7 Crossflow plane visualization of normalized total pressure contours of the canard and wing leading-edge vortices at $\alpha \approx 8°$.

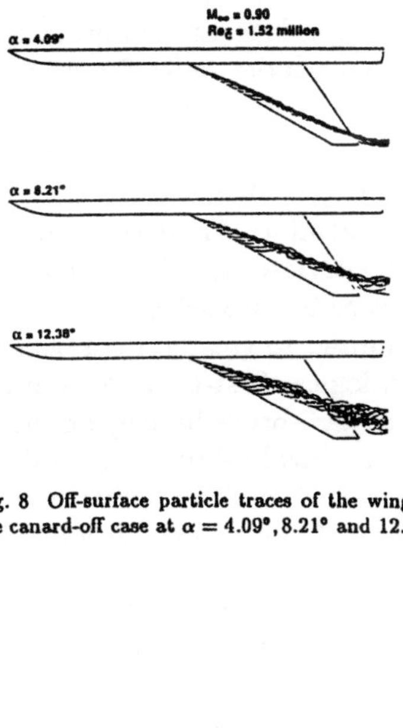

Fig. 8 Off-surface particle traces of the wing vortex for the canard-off case at $\alpha = 4.09°, 8.21°$ and $12.38°$.

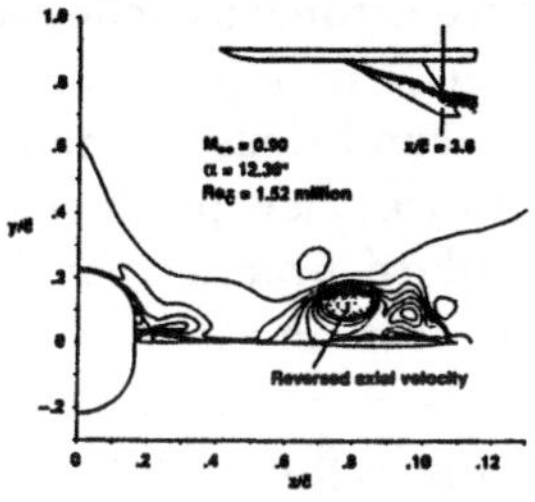

Fig. 9 Crossflow plane visualization of scaled axial velocity contours for the canard-off case at $\alpha = 12.38°$ and $x/\bar{c} = 3.6$.

Fig. 10 Off-surface particle traces of the canard and wing vortices for the canard-on case at $\alpha = 12.38°$.

ASYMPTOTIC FAR-FIELD BOUNDARY CONDITIONS FOR NUMERICAL FLOWFIELD PREDICTIONS

A. Verhoff, D. Stookesberry and S. Agrawal

McDonnell Aircraft Company, St. Louis, Missouri 63166 USA

INTRODUCTION - Numerical solution procedures for nonlinear fluid dynamic equations usually use one or more artificial computational boundaries located at some distance from the primary region of interest in order to limit the physical domain to finite size. If the flow crossing such a boundary (either inflow or outflow) is subsonic, then some type of computational boundary conditions must be imposed which simulate the influence of the true far-field conditions at infinity. These boundary conditions must be such that waves propagating outward across the boundary do not produce erroneous reflections back into the computational field to degrade the calculations. It is generally acknowledged that simply imposing free stream conditions (or conditions at infinity) at computational boundaries is usually inappropriate because they produce spurious reflections back into the computational domain. Standard practice consists of locating the boundaries quite far from the region of interest in an attempt to simplify the boundary condition models and minimize any effects of inconsistent modeling. The net effect is a significant increase in the number of grid points required for an accurate flowfield calculation.

Boundary modeling procedures for two-dimensional internal and external flows have been developed in References [1] and [2] which alleviate the difficulties mentioned above and allow the computational boundaries to be located much closer to the nonlinear region of interest. The procedure is limited to steady, inviscid flow, although the flow can be rotational. It represents a logical first-order asymptotic extension of the so-called characteristic (referred to as zero-order herein) boundary conditions commonly used with inviscid or viscous numerical solution methods. It also illustrates a consistent method for coupling linearized analytic solutions with nonlinear numerical solutions by means of computational boundary conditions. The procedure is derived from the Euler equations. Therefore, it is applicable to flows which have strong entropy producing effects (e.g., shock waves) within the computational region. Such effects can produce large variations in density and Mach number in the far-field in the direction normal to streamlines which persist to infinity and which cannot be treated as small perturbations. A potential flow description of such a far-field containing rotational flow is inappropriate.

Many far-field boundary condition models are derived by linearizing the steady or unsteady fluid dynamic equations (usually Euler) about constant far-field conditions and solving the resulting system assuming a generalized wave form. However, their utility is compromised if strong rotational effects are produced in the computational domain, because flow variables other than pressure and flow angle are not constant in the far field if the flow is treated as inviscid and subsonic. In that case linearization in terms of primitive or conservative variables yields perturbation quantities which do not vanish at infinity. Moreover, linearization about average far-field conditions becomes questionable if rotational effects are strong. It was shown in Reference [1] that conditions are achieved readily where a shock wave produces far-field Mach numbers which vary normal to streamlines by a factor of two. Streamline-normal gradient variations in some cases can be even more pronounced which cause strong interactions to persist in the far field. It was also shown in Reference [1] that a linearization with decaying perturbations is possible, but the linear system has non-constant coefficients.

The present boundary condition analysis is based on a Riemann variable formulation of the Euler equations given in Reference [3]. This represents a natural starting point because the zero-order (or characteristic) boundary conditions for subsonic flow mentioned above are expressed in terms of Riemann variables. The equations are linearized about a constant pressure, rectilinear flow condition, which truly represents conditions at infinity. These linearized equations are assumed applicable in the far-field region beyond a computational boundary. The first-order linearized Euler equations are solved using integral transform and Fourier analysis techniques coupled with an iterative procedure. Coupling of these analytic solutions with a nonlinear numerical solution is accomplished by the boundary conditions in an unambiguous manner. The analytic solutions provide a smooth transition across the computational boundary to the true far-field conditions at infinity. The first-order boundary conditions are in the form of distributions of flow quantities to be imposed along the boundary, not constant conditions. The additional computational effort required to impose the first-order boundary conditions is modest. Furthermore, the boundary analysis can be coupled with any inviscid numerical solution method. It can also be coupled with a viscous method by expressing a wake as a vorticity distribution and convecting this distribution downstream via the Euler analytic model.

ANALYSIS - The system of two-dimensional, steady, linearized Euler equations which describe first-order perturbations from a constant pressure state is derived from the Riemann variable formulation of Reference [3]. This formulation is used because of its close relationship with zero-order (or characteristic) boundary conditions commonly used in nonlinear numerical solution methods. These equations in natural streamline coordinates are

$$\frac{\partial Q}{\partial t}+(q+a)\frac{\partial Q}{\partial s} = -\frac{\gamma-1}{2}qaS\frac{\partial \theta}{\partial n} - \frac{\gamma-1}{2}a(S-\frac{2}{\gamma-1})[\frac{\partial}{\partial s}(q-\frac{2}{\gamma-1}a)]$$

$$\frac{\partial R}{\partial t}+(q-a)\frac{\partial R}{\partial s} = +\frac{\gamma-1}{2}qaS\frac{\partial \theta}{\partial n} + \frac{\gamma-1}{2}a(S-\frac{2}{\gamma-1})[\frac{\partial}{\partial s}(q+\frac{2}{\gamma-1}a)]$$

$$\frac{\partial \theta}{\partial t}+q\frac{\partial \theta}{\partial s} = -\frac{a^2}{\gamma q}\frac{\partial P}{\partial n}, \qquad \frac{\partial S}{\partial t}+q\frac{\partial S}{\partial s}=0$$

The Riemann variables Q and R are defined as $Q \equiv q + aS$ and $R \equiv q - aS$, S is entropy, and θ is the flow angle. Velocity magnitude and speed of sound are denoted by q and a, respectively, and P is the logarithm of pressure.

For steady flow the analysis can be greatly simplified by defining a new dependent variable $T \equiv Q - R$. In far-field regions of a flowfield where asymptotic conditions prevail, the flow can be treated as a perturbation to a constant pressure, rectilinear subsonic flow. Such regions occur near and beyond far-field computational boundaries. The dependent variables in the Euler formulation can then be expanded in asymptotic series

$$T = T_\infty + T_1 + T_2 + \cdots, \quad \theta = \theta_\infty + \theta_1 + \theta_2 + \cdots, \quad S = S_\infty + S_1 + S_2 + \cdots$$

The flow direction at infinity is assumed constant and denoted by θ_∞ ; the perturbation quantities T_1, θ_1 and S_1 vanish at infinity. Entropy variation is not excluded so that the flow can be rotational in which case S_∞ and T_∞ are not constant, but vary normal to the streamline direction. Furthermore, far-field entropy variations can be strong (i.e., not small perturbations). Analytic solutions of the first-order perturbed Euler equations were developed in References [1] and [2] for various computational far-field boundary shapes associated with internal and external flows. The first-order boundary conditions were then developed from these analytic solutions.

APPLICATIONS - Euler airfoil calculations were carried out for a NACA 0012 airfoil using the ET12D and the FLO672D codes. These two-dimensional codes are derivatives of their three-dimensional counterparts (References [4] and [5]) and are based on different numerical solution algorithms.

Reference Euler results were calculated using a baseline C-grid whose far-field boundaries were located at a large distance (approximately 40 chord lengths) from the airfoil. The calculations were repeated on a much smaller core grid using both the zero- and first-order boundary conditions. The small core grid shown in Fig. 1 had 209 points distributed along the C-curves and 41 points in the outward direction. It extended one chord length upstream of the airfoil leading edge and one chord length downstream of the trailing edge. The large baseline grid had dimensions 273x65 and was constructed by simply adding parabolic C-lines outside of the core grid inflow boundary and vertical lines downstream of the outflow boundary. The small grid had 52 percent fewer grid points than the baseline grid. The relative locations of the outer boundaries of the two grids are shown in Fig. 2.

Isentropic Flow: For a free stream Mach number of 0.50 and angle of attack of 4 degrees, boundary condition accuracy can be quantified by comparing computed lift. Figure 3 shows the variation in lift coefficient predicted by the FLO672D code using the zero- and first-order boundary conditions as the downstream boundary location is moved upstream toward the airfoil. With the first-order boundary conditions, the lift coefficient remains essentially constant. With the zero-order boundary conditions, accuracy degrades rapidly as the downstream boundary distance is decreased. Consistency of the coupling between the nonlinear numerical solution and the analytic solution is demonstrated by the pressure contours shown in Fig. 4. Near-field contours of the ET12D computation on the core grid match smoothly with far-field contours of the analytic solution. Only a modest increase in computational effort (approximately five percent) was required for the first-order boundary conditions. An additional efficiency gain was also provided in that fewer iterations were typically required for solution convergence because of the closer proximity of the far-field boundaries.

Non-Isentropic Flow: With a free stream Mach number of 0.85 and angle of attack of zero, shock waves occur on the upper and lower surfaces of the airfoil. This case was calculated using the ET12D code with the downstream boundary located one chord from the airfoil trailing edge and the overall boundary at ($\overline{BCDB}$).

Pressure contours are shown in Figs. 5 and 6 which compare results using zero- and first-order boundary conditions, respectively, with the reference solution. Very accurate results are obtained on the small grid when the first-order conditions are used. Surface Mach numbers predicted by FLO672D using the core grid ($\overline{GHIG}$) are presented in Fig. 7.

Viscous Flow: Viscous flow results were calculated for a NACA 0012 airfoil using the CFL2D code (Reference [6]). The small core grid was modified by clustering points in the boundary layer and increasing its dimensions to 209x61. The downstream boundary was located one chord from the airfoil trailing edge at $\overline{BD}$ (see Fig. 2) to demonstrate the applicability of the first-order boundary conditions to viscous flow. The free stream Mach number for these calculations was 0.80, angle of attack was 0 degree, and the Reynolds number was 1 million. A Baldwin-Lomax turbulence model was used. Surface pressures calculated using zero- and first-order boundary conditions on the downstream boundary ($\overline{BD}$) are compared with the reference solution in Fig. 8. Figure 9 shows the effect of the boundary conditions on the velocity boundary layer.

Inviscid Duct Flow: Steady converging/diverging duct flow results were calculated using the ET12D code. The duct geometry and grid are shown in Fig. 10. The computational grid for the sinusoidal portion of the duct had dimensions 41 x 21. Additional rectangular grid cells were added within the constant area sections of the duct. Results for isentropic, subsonic duct flow are shown in Fig. 11. Only one column of grid cells was used in the upstream and downstream constant-area portions of the duct. The linearized far-field analytic solution provides a smooth transition across the computational boundary to the true far-field conditions at infinity. Non-isentropic duct flow results are shown in Fig. 12. Five columns of grid cells were used in the constant area sections of the duct.

SUMMARY - Far-field computational boundary conditions have been developed for two-dimensional internal and external flow problems. These first-order boundary conditions are derived from analytic solutions of the linearized Euler equations and represent a logical extension of the zero-order (or characteristic) boundary conditions commonly used in the numerical solution of nonlinear fluid dynamic equations. The first-order boundary conditions allow the far-field boundaries to be located much closer thereby reducing the number of grid points needed for the numerical solution and also the number of iterations for solution convergence. This allows a significant reduction in the amount of computational effort required for the nonlinear numerical solution because the additional calculations required for the first-order boundary conditions are modest. Results are shown where the number of grid points was reduced by approximately 50 percent.

REFERENCES -

1. Verhoff, A. and Stookesberry, D., Naval Postgraduate School, Monterey, CA, NPS67-88-001CR, Sept. 1988.
2. Verhoff, A., Stookesberry, D., and Agrawal S., AIAA – 91-0630, Jan. 1991.
3. Verhoff, A. and O'Neil, P. J., AIAA – 84-0163, Jan. 1984.
4. Agrawal, S., Vermeland, R. E., et al., AIAA – 88-0009, Jan. 1988.
5. Jameson, A., ASME AMD, Vol. 78, Dec. 1986, pp. 45-73.
6. Rumsey, C. L., Taylor, S., et al., AIAA – 87-0413, Jan. 1987.

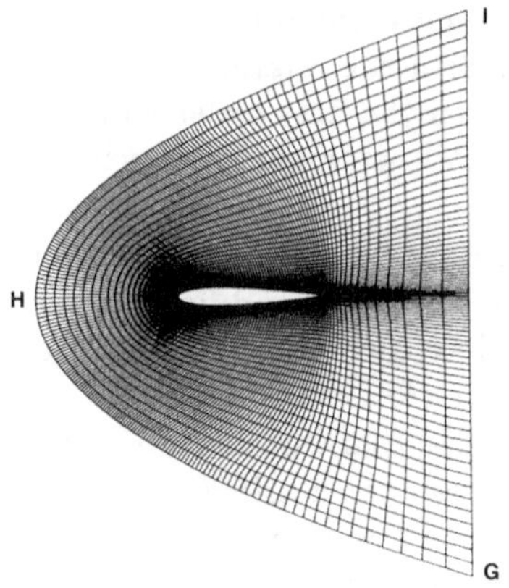

Fig. 1 Core Grid for NACA 0012 Airfoil

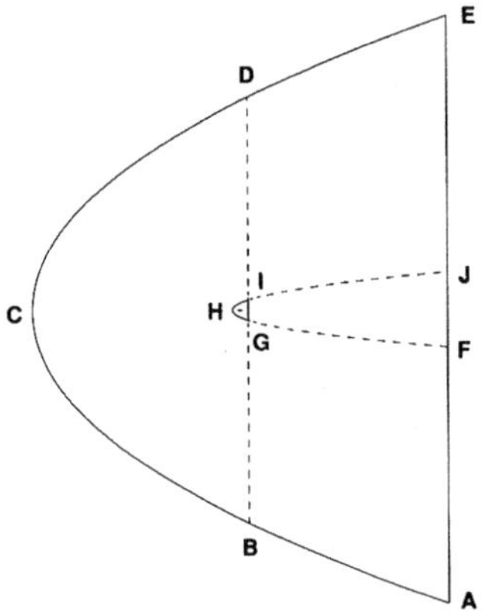

Fig. 2 Relative Boundary Locations of Baseline and Core Grids

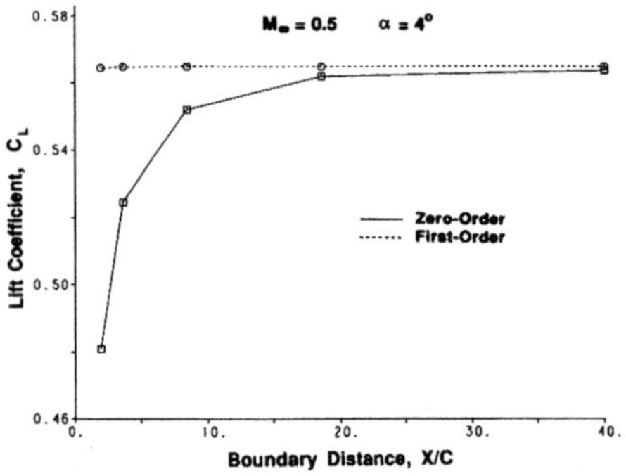

Fig. 3 Effect of Downstream Boundary Location on Lift Coefficient (FLO672D)

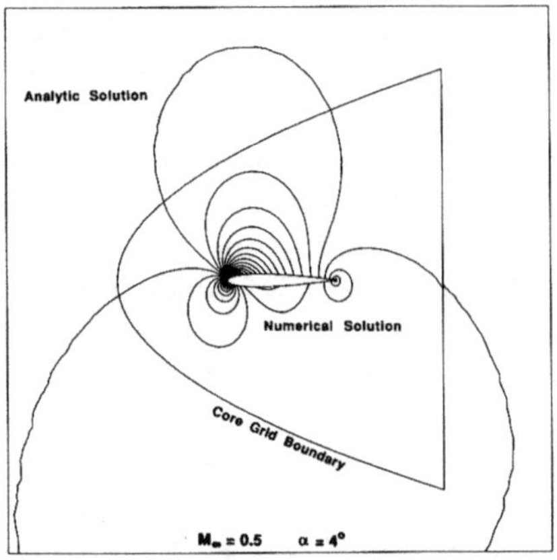

Fig. 4 Pressure Contours Predicted by ET12D With Far-Field Analytic Solution Added

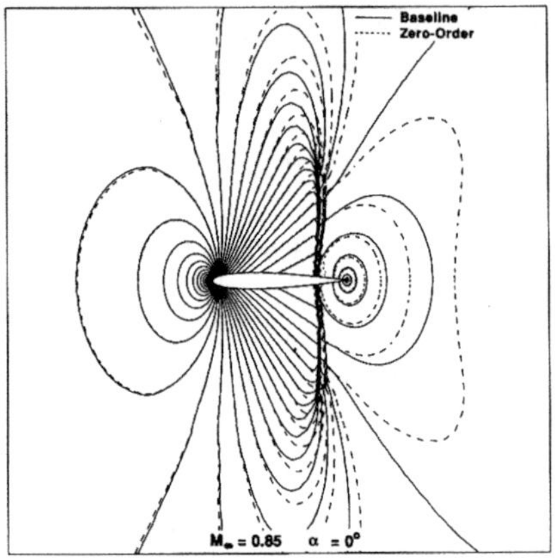

Fig. 5 Pressure Contours Predicted by ET12D Using Zero-Order Boundary Conditions

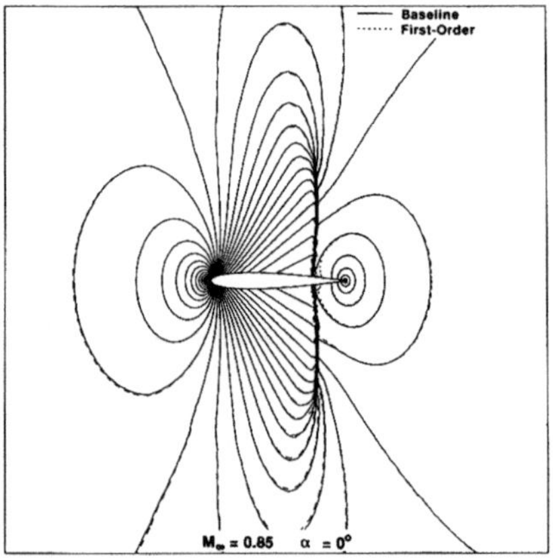

Fig. 6 Pressure Contours Predicted by ET12D Using First-Order Boundary Conditions

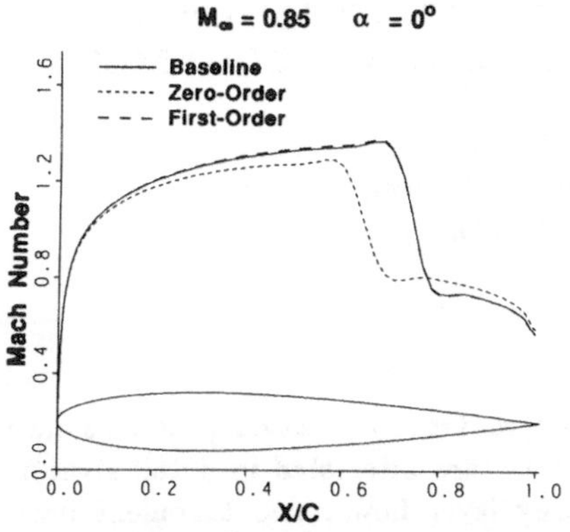

Fig. 7 FLO672D Surface Mach Number Predictions with Outer Boundary at $\overline{GHIG}$

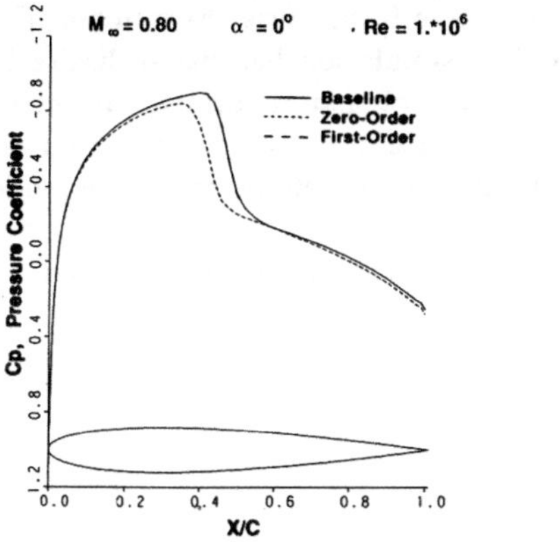

Fig. 8 CFL2D Surface Pressure Predictions with Downstream Boundary at $\overline{BD}$

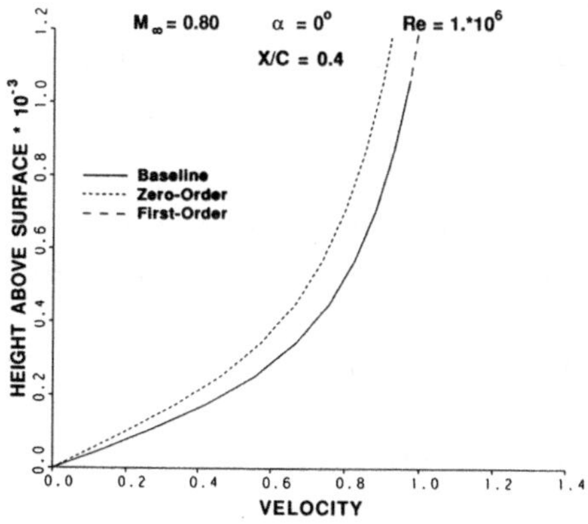

Fig. 9 Velocity Boundary Layer Profile Predicted by CFL2D

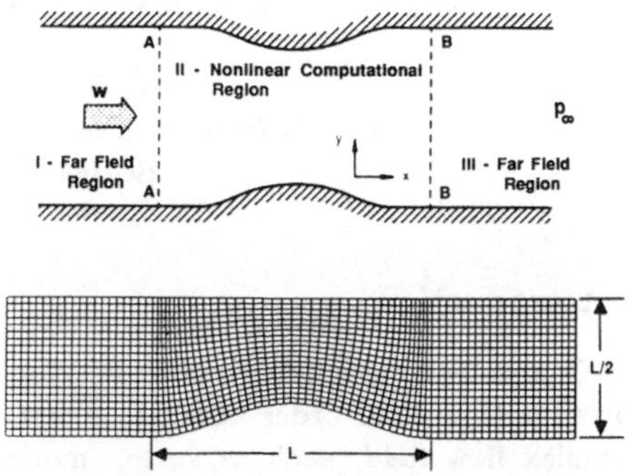

Fig. 10 Duct/Nozzle Geometry and Grid

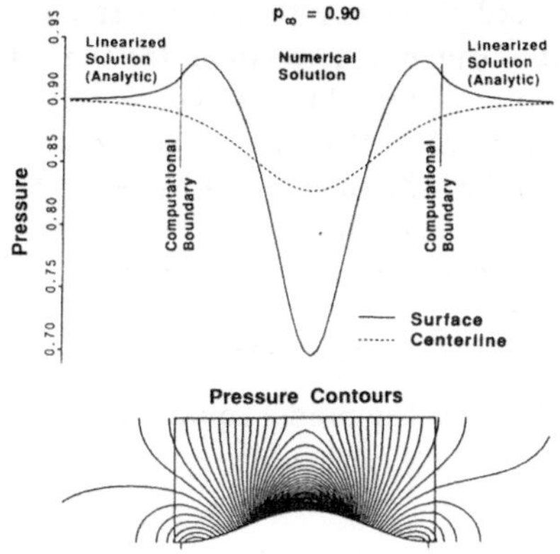

Fig. 11 Pressure Distributions and Contours First-Order Boundary Conditions

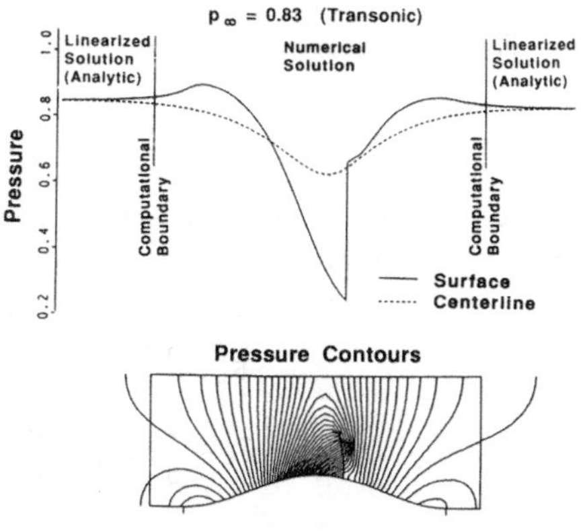

Fig. 12 Pressure Distributions and Contours First-Order Boundary Conditions

NUMERICAL SIMULATION OF INVISCID AND VISCOUS FLOW WITH HIGH ORDER ACCURATE SCHEMES

Ma Yanwen and Fu Dexun
(LNM,Institute of Mechanics, Chinese Academy of Sciences, Beijing, 100080, P.R. China)

1. Introduction

Computational fluid dynamics develop very fast, and many practical problems have been solved with second order accurate schemes. Now we are interested in detail structures of complex flow field, such as vortex motion, mixing layer flow , and turbulent flow. For solving this kind of problems the harmonics with moderate and high wavenumbers must be simulated well.

The R–K method can be used in time integration, and only space discretization is discussed. One way to improve the accuracy of solution is to increase the number of mesh points. But it is impossible for the complex flow field simulation because of limitation of computer power. The another way is to increase the order of spatial approximation. It was shown in ref.[1] that high order accurate difference approximations for the first and second spatial derivatives can give better approximation of effective wavenumber for the moderate and high wave number.

Before using high order accurate schemes several questions have to be answered.

2. Difference Approximations

Consider the model equation

$$\frac{\partial u}{\partial t} + \frac{\partial f}{\partial x} = 0, \qquad f = cu \tag{1}$$

and it's semi–discrete approximation

$$\frac{\partial u_j}{\partial t} + \frac{1}{\Delta x} F_j = 0 \tag{2}$$

where F_j is a approximation of $\Delta x \frac{\partial f}{\partial x}$, and it can be obtained from the equation

$$\sum_k \alpha_k F_{j+k} = \sum_k \beta_k (f_{j+k+1} - f_{j+k}) \tag{3}$$

Some relations must be satisfied between α_k and β_k. The 4th order accurate compact difference and the 3rd order accurate upwind compact difference for $c > 0$ are listed below

$$\frac{1}{6}F_{j+1} + \frac{2}{3}F_j + \frac{1}{6}F_{j-1} = \frac{1}{2}(f_{j+1} - f_{j-1})$$

$$\frac{2}{3}F_j + \frac{1}{3}F_{j-1} = \frac{5}{6}(f_j - f_{j-1}) + \frac{1}{6}(f_{j+1} - f_j)$$

3. Diffusion Analogy

Using the modified equation for kth order accurate difference scheme (2) we can get the following diffusion analogy equation [2]

This work is supported by National Natural Science Foundation of China

$$\frac{\partial u}{\partial t}+\frac{\partial f}{\partial x}=\Delta x\frac{\partial}{\partial x}\mu\frac{\partial u}{\partial x} \tag{4}$$

where μ is called diffusion analogy coefficient and it is defined as follows

$$\mu=\sigma\Delta x^{k-1}\Big(\frac{\partial^k f}{\partial x^k}/\frac{\partial f}{\partial x}\Big)c \tag{5}$$

The behavior of the solution for difference scheme near the shock (continuous solution with steep gradient) depends on the concrete scheme, the order of approximation, and the ratio $\frac{\partial^k f}{\partial x^k}/\frac{\partial f}{\partial x}$.

4. Restriction on the Mesh Reynolds Number $Re_{\Delta x}$ and Stability Analysis

Consider the following model equation and it's difference approximation

$$\frac{\partial u}{\partial t}+\frac{\partial f}{\partial x}=\frac{\partial}{\partial x}\tilde{\mu}\frac{\partial u}{\partial x} \tag{6}$$

$$\frac{\partial u_j}{\partial t}+\frac{1}{\Delta x}F_j=\frac{\tilde{\mu}}{\Delta x^2}S_j \tag{7}$$

where $\tilde{\mu}$ is the physical diffusion coefficient , S_j is a approximation of $\Delta x^2\frac{\partial^2 u}{\partial x^2}$. Taking leading term of the modified equation, the following equation can be obtained

$$\frac{\partial u}{\partial t}+\frac{\partial f}{\partial x}=\frac{\partial}{\partial x}\Big[(\tilde{\mu}+\Delta x\mu)\frac{\partial u}{\partial x}\Big] \tag{8}$$

It is reasonable to require $|\Delta x\mu|<\tilde{\mu}$, or

$$Re_{\Delta x}<\Big|\frac{\partial f}{\partial x}/\Big(\sigma\Delta x^{k-1}\frac{\partial^k f}{\partial x^k}\Big)\Big|,\qquad Re_{\Delta}x=\Delta x\frac{c}{\tilde{\mu}} \tag{9}$$

The restriction on $Re_{\Delta}x$ depends on the accuracy of the scheme, and behavior of the solution. Generally speaking, for the high order accurate schemes with smooth solutions the restriction on $Re_{\Delta x}$ can be relaxed.

Consider equation (1),(2) and (4). Putting a elementary solution $u=e^{\alpha t+imx}$ into equation(5) for the case $k+1=2n$ where n is a integer, we can obtain

$$\mu=\sigma c(m\Delta x)^{2(n-1)}(-1)^{n-1} \tag{10}$$

For the coefficient to be positive the following inequality for stability is required

$$\sigma c(-1)^{n-1}>0 \tag{11}$$

This is the same requirement as obtained in Ref.[3]. Satisfication of condition (11) does not guarantee to get solution without nonphysical oscillations. Careful study of the exact solution of the nonlinear Burgers' equation shows:
(a) for the case $k=2$ the oscillations may be produced only in one side of the shock;
(b) for the case $k=2n(n>1)$ the oscillations may be produced in both side of the shock, but the oscillation with the largest amplitude exists in one side of the shock;
(c) for the case $k=2n+1$ the oscillations may be produced in the both side of the shock, but with satisfication of (11) the oscillation with the largest amplitude is removed. This is the reason why the upwind biased scheme can give better solutions.

5.High Wave Number Effect

A. Large Diffusivity of High Order Upwind Scheme

Suppose condition (11) is satisfied. From expression (10) it can be seen the diffusion analogy coefficient increases with increasing the wave number. For the small and moderate wave number the diffusion analogy coefficient decreases rapidly with increasing the order of accuracy of approximation. Effect of large diffusivity for the high wave number can be improved with increasing mesh points.

B. Severe Deformation of Harmonic with High Wave Number for the Symmetric Difference

For the elementary solution $e^{\alpha t}\sin(mx)$ we have

$$\mu = (-1)^n c\sigma (m\Delta x)^{2n-1} \mathrm{tg}(mx) \tag{12}$$

From expression (12) it can be seen that μ oscillates across zero with large amplitude for the high wave number, and the harmonic is severely deformed. The deformation is smaller for the low and moderate wave number for the high order accurate scheme.

C. Restriction on $Re_{\Delta x}$ for the Harmonic with High Wave Number

Putting $e^{\alpha t}\sin(mx)$ into equation (9) we can obtain

$$Re_{\Delta x} < \left|\frac{1}{\sigma(m\Delta x)^{2n}}\right|, \quad \text{for } k = 2n+1, \quad \text{(dissipative difference)}$$
$$Re_{\Delta x} < \left|\frac{\cos(mx)}{\sigma(m\Delta x)^{2n-1}\sin(mx)}\right|, \quad \text{for } k = 2n, \quad \text{(symmetric difference)}$$

It can be seen that the restriction on the mesh Reynolds number is smaller for the dissipative scheme. It also can be seen that for the high wave number the restriction on $Re_{\Delta x}$ is severe, but for the low and moderate wave number the restriction is small.

From above discussion we see the high order accurate scheme can give better numerical solutions. The high wave number effect can be improved with increasing mesh points.

6. Construction of Scheme with High Resolution of the Shock

In order to get high resolution of the shock the diffusion analogy coefficient must be reconstructed.

Three methods of reconstructing μ for the kth order accurate difference schemes are suggested: (a)Taking absolute value; (b)Using simple minmod function; (c)Using logarithm–minmod function. For example, for the fourth order accurate difference scheme we have

$$\mu = \sigma\Delta x^3 \left(\frac{\partial^4 f}{\partial x^4} / \frac{\partial^3 f}{\partial x^3}\right)\left(\frac{\partial^3 f}{\partial x^3} / \frac{\partial^2 f}{\partial x^2}\right)\left(\frac{\partial^2 f}{\partial x^2} / \frac{\partial f}{\partial x}\right) c$$

One of it's difference approximation is

$$\mu_{j+\frac{1}{2}} = \sigma(1-\gamma^{1,-}_{j+\frac{1}{2}})(1-\gamma^{2,-}_{j+\frac{1}{2}})(1-\gamma^{3,-}_{j+\frac{1}{2}})c$$

where

$$\gamma^{1,\pm}_{j+\frac{1}{2}} = \frac{\delta^+_x f_{j\pm1}}{\delta^+_x f_j}, \qquad \gamma^{2,\pm}_{j+\frac{1}{2}} = \left(\frac{\delta^2_x f_{j+1}}{\delta^2_x f_j}\right)^{\pm1}, \qquad \gamma^{3,\pm}_{j+\frac{1}{2}} = \frac{\delta^3_x f_{j+\frac{1}{2},\pm1}}{\delta^3_x f_{j+\frac{1}{2}}}$$

For making $\mu_{j+\frac{1}{2}}$ positive the minmod function can be used for $\gamma^{n,\pm}_{j+\frac{1}{2}}$.

7. Numerical Examples

A. Driven cavity flow [4]. 2–D incompressible N–S equations are approximated with 3rd order accurate upwind compact difference for the convection terms. With increasing the Reynolds number periodic motion is found. The stream lines on different time steps for case $Re = 10000$ are given in Fig.1

B. 2–D shock reflection. The fourth order accurate compact differences are used to approximate the convection terms in the compressible N–S equations. For getting high resolution the diffusion analogy coefficient is modified with three above mentioned methods. The incident shock angle θ is 29° and free Mach number M_∞ is 2.9. The computed results are given in Fig.2.

REFERENCES

[1] Lele, S.K. AIAA 89–0374 (1989)

[2] Fu Dexun, Ma Yanwen, 4th International Symposium on Computational Fluid Dynamics, Davis, 1991

[3] Worming.R.F. and Hyett . B.J., J.Comp.Phys. Vol.14,p159–179

[4] Liu Hong, Fu Dexun, Ma Yanwen, Science in China, to appear.

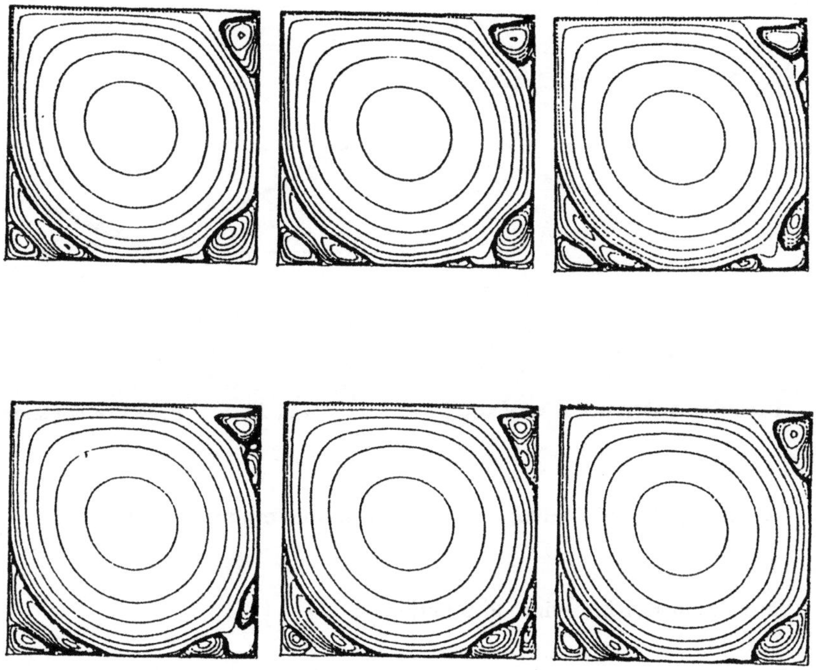

Fig.1 Stream line pattern at different time in a period

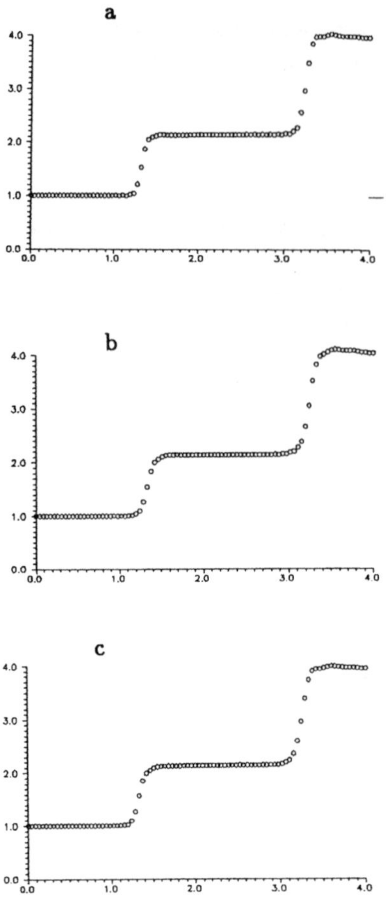

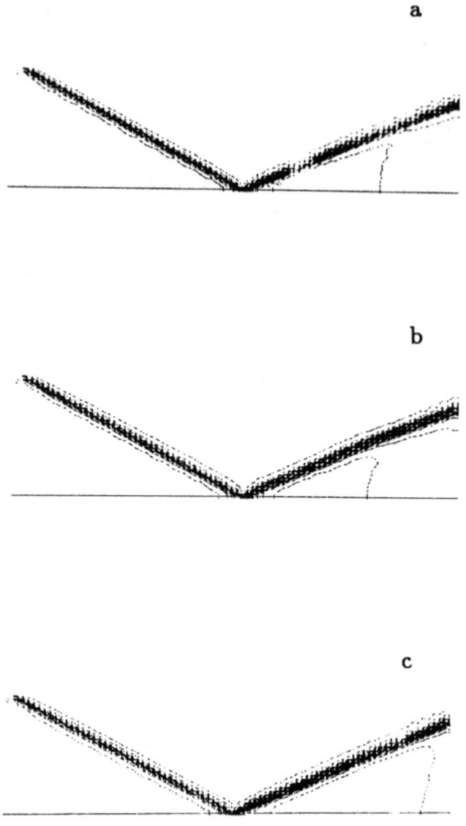

Fig.2a **Pressure distribution at $y = 0.5$** **Fig.2b** **Pressure contours**

a: taking absolute value; b: using simple minmod function

c: using logarithm minmod function

SOLUTION OF THREE-DIMENSIONAL NAVIER-STOKES EQUATIONS USING AN IMPLICIT GAUSS-SEIDEL SCHEME

S. Yoon

MCAT Institute, NASA Ames Research Center
MS 258-1, Moffett Field, California 94035, U.S.A.

I. Introduction

Although unstructured grid methods have been used successfully in solving the Euler equations for complex geometries, structured zonal grid solvers still remain the most useful for the Navier-Stokes equations because of their natural advantages in dealing with the highly clustered meshes in the viscous boundary layers. Zonal structured grid methods not only handle reasonably complex geometries using multiple blocks, but also offer a hybrid grid scheme to alleviate difficulties which unstructured grid methods have encountered. Recent developments in structured grid solvers have been focused on the efficiency as well as the accuracy since existing three-dimensional Navier-Stokes codes are not efficient enough to be used routinely for aerodynamic design.

The author[1] has introduced an implicit algorithm based on a lower-upper factorization and symmetric Gauss-Seidel relaxation. The scheme has been used successfully in computing chemically reacting flows due in part to the algorithm's property which reduces the size of the left hand side matrix for nonequilibrium flows with finite rate chemistry.[2,3] More recently, a study[4] suggests that the three-dimensional extension of the method is one of the most efficient ways to solve the Navier-Stokes equations. Consequently, a new three-dimensional Navier-Stokes code named CENS3D was produced. CENS3D requires less computational work per iteration than most existing codes on a Cray YMP supercomputer and in addition converges reasonably fast. The performance of the code is demonstrated for a viscous transonic flow past an ONERA M6 wing.

II. Numerical Methods

Let t be time; $\widehat{Q}$ the vector of conserved variables; $\widehat{E}$, $\widehat{F}$, and $\widehat{G}$ the convective flux vectors; and $\widehat{E}_v$, $\widehat{F}_v$, and $\widehat{G}_v$ the flux vectors for the viscous terms. Then the three-dimensional Navier-Stokes equations in generalized curvilinear coordinates $(\xi,\ \eta,\ \zeta)$ can be written as

$$\partial_t \hat{Q} + \partial_\xi(\hat{E} - \hat{E}_v) + \partial_\eta(\hat{F} - \hat{F}_v) + \partial_\zeta(\hat{G} - \hat{G}_v) = 0 \tag{1}$$

where the flux vectors are found in Ref. 4.

An unfactored implicit scheme can be obtained from a nonlinear implicit scheme by linearizing the flux vectors about the previous time step and dropping terms of the second and higher order.

$$[I + \alpha\Delta t(D_\xi \hat{A} + D_\eta \hat{B} + D_\zeta \hat{C})]\delta\hat{Q} = -\Delta t \hat{R} \tag{2}$$

where $\hat{R}$ is the residual

$$\hat{R} = D_\xi(\hat{E} - \hat{E}_v) + D_\eta(\hat{F} - \hat{F}_v) + D_\zeta(\hat{G} - \hat{G}_v) \tag{3}$$

and I is the identity matrix. $\delta\hat{Q}$ is the correction $\hat{Q}^{n+1}$ - $\hat{Q}^n$, where n denotes the time level. D_ξ, D_η, and D_ζ are difference operators that approximate ∂_ξ, ∂_η, and ∂_ζ. $\hat{A}$, $\hat{B}$, and $\hat{C}$ are the Jacobian matrices of the convective flux vectors.

An efficient implicit scheme can be derived by combining the advantages of LU factorization and Gauss-Seidel relaxation.

$$LD^{-1}U\delta\hat{Q} = -\Delta t \hat{R} \tag{4}$$

Here,

$$\begin{aligned}
L &= I + \alpha\Delta t(D_\xi^- \hat{A}^+ + D_\eta^- \hat{B}^+ + D_\zeta^- \hat{C}^+ - \hat{A}^- - \hat{B}^- - \hat{C}^-) \\
D &= I + \alpha\Delta t(\hat{A}^+ - \hat{A}^- + \hat{B}^+ - \hat{B}^- + \hat{C}^+ - \hat{C}^-) \\
U &= I + \alpha\Delta t(D_\xi^+ \hat{A}^- + D_\eta^+ \hat{B}^- + D_\zeta^+ \hat{C}^- + \hat{A}^+ + \hat{B}^+ + \hat{C}^+)
\end{aligned} \tag{5}$$

where D_ξ^-, D_η^-, and D_ζ^- are backward difference operators, while D_ξ^+, D_η^+, and D_ζ^+ are forward difference operators.

In the framework of the LU-SGS algorithm, a variety of schemes can be developed by different choices of numerical dissipation models and Jacobian matrices of the flux vectors. Jacobian matrices leading to diagonal dominance are constructed so that " + " matrices have nonnegative eigenvalues while " - " matrices have nonpositive eigenvalues. For example,

$$\begin{aligned}\widehat{A}^{\pm} &= \widehat{T}_{\xi}\Lambda_{\xi}^{\pm}\widehat{T}_{\xi}^{-1}\\ \widehat{B}^{\pm} &= \widehat{T}_{\eta}\Lambda_{\eta}^{\pm}\widehat{T}_{\eta}^{-1}\\ \widehat{C}^{\pm} &= \widehat{T}_{\zeta}\Lambda_{\zeta}^{\pm}\widehat{T}_{\zeta}^{-1}\end{aligned} \tag{6}$$

where $\widehat{T}_{\xi}$ and $\widehat{T}_{\xi}^{-1}$ are similarity transformation matrices of the eigenvectors of $\widehat{A}$. Another possibility is to construct Jacobian matrices of the flux vectors approximately to yield diagonal dominance.

$$\begin{aligned}\widehat{A}^{\pm} &= \frac{1}{2}[\widehat{A} \pm \tilde{\rho}(\widehat{A})I]\\ \widehat{B}^{\pm} &= \frac{1}{2}[\widehat{B} \pm \tilde{\rho}(\widehat{B})I]\\ \widehat{C}^{\pm} &= \frac{1}{2}[\widehat{C} \pm \tilde{\rho}(\widehat{C})I]\end{aligned} \tag{7}$$

where

$$\tilde{\rho}(\widehat{A}) = \kappa \; max[|\; \lambda(\widehat{A}) \;|] \tag{8}$$

for example. Here $\lambda(\widehat{A})$ represent eigenvalues of the Jacobian matrix $\widehat{A}$ and κ is a constant that is greater than or equal to 1. Stability and convergence are controlled by adjusting κ either manually or automatically as the flowfield develops.

It is interesting to note that the need for block inversions along the diagonals can be eliminated if we use the approximate Jacobian matrices of Eq. (7). Setting $\alpha = 1$ and $\Delta t = \infty$ yields a Newton-like iteration. Although a quadratic convergence of the Newton method cannot be achieved because of the approximate factorization, a linear convergence can be demonstrated. The use of Newton-like iteration offers a practical advantage in that one does not have to find an optimal Courant number or time step to reduce the overall computer time.

The cell-centered finite-volume method[4] is augmented by a numerical dissipation model with a minmod flux limiter. The coefficients of the dissipative terms are the directionally scaled spectral radii of Jacobian matrices.

III. Results

In order to demonstrate the performance of the CENS3D code, transonic flow calculations have been carried out for an ONERA M6 wing. A 289 x 50 x 44 C-H mesh (635,800 points) is used as a fine grid. The distance of the first grid point from the wing surface is 1.0 x 10^{-5} times the chord length at the root section. The freestream conditions are at a Mach number of 0.8395, Reynolds number of 1.5 x 10^7, and a 3.06° angle of attack. This is an unseparated flow case. The algebraic turbulence model by Baldwin and Lomax is employed for mathematical closure of the Reynolds-averaged Navier-Stokes equations. The root-mean-squared residuals drop 3 orders of magnitude in about 380 iterations or 38 minutes of CPU time on the fine grid. In the present implementation, the implicit left hand side viscous terms are not included which decreases the computational work per iteration. To investigate the effect of this left hand side compromise on the convergence rate, a grid-convergence study has been performed using a 171 x 25 x 44 (188,100 points) coarse grid. Although the number of radial grid points to resolve the viscous boundary layer is doubled in the fine grid case, the fine grid convergence is slowed by only twenty percent. Fig. 1 and Fig. 2 show a good agreement between experimental data[5] and the pressure coefficients at 44% and 65% semi-span stations computed on the fine grid. This comparison validates the present code CENS3D.

The CENS3D code requires only 9 μsec per grid-point per iteration for the thin-layer Navier-Stokes equations with an algebraic turbulence model on a single Cray YMP processor at the sustained rate of 175 Mflops. It is interesting to note that the LU-SGS implicit scheme requires less computational work per iteration than a Runge-Kutta explicit scheme.

Conclusions

Good performance of a three-dimensional Navier-Stokes solver CENS3D based on an implicit lower-upper Gauss-Seidel scheme is demonstrated for nonseparated transonic flow past a wing. In addition to its reasonabe convergence rate, the code requires very low computational time per iteration. The three-dimensional Navier-Stokes solution of a high Reynolds number flow using 636K grid points is obtained in 38 minutes. The computational results compare well with available experimental data.

References

1. Yoon, S. and Jameson, A., "Lower-Upper Symmetric-Gauss-Seidel Method for the Euler and Navier-Stokes Equations," *AIAA Journal*, Vol. 26, Sep. 1988, pp. 1025-1026.

2. Shuen, J.S. and Yoon, S., "A Numerical Study of Chemically Reacting Flows Using a Lower-Upper Symmetric Successive Overrelaxation Scheme," *AIAA Journal*, Vol. 27, Dec. 1989, pp. 1752-1760.

3. Park, C. and Yoon, S., "A Fully-Coupled Implicit Method for Thermo-Chemical Nonequilibrium Air at Sub-Orbital Flight Speeds," *Journal of Spacecraft and Rockets*, Vol. 28, No. 1, Jan.-Feb. 1991, pp. 31-39.

4. Yoon, S. and Kwak, D., "An Implicit Three-Dimensional Navier-Stokes Solver For Compressible Flows," AIAA Paper 91-1555, June 1991.

5. Schmitt, V. and Charpin, F., " Pressure Distributions on the ONERA M6 Wing at Transonic Mach Numbers," AGARD AR-138-B1, 1979.

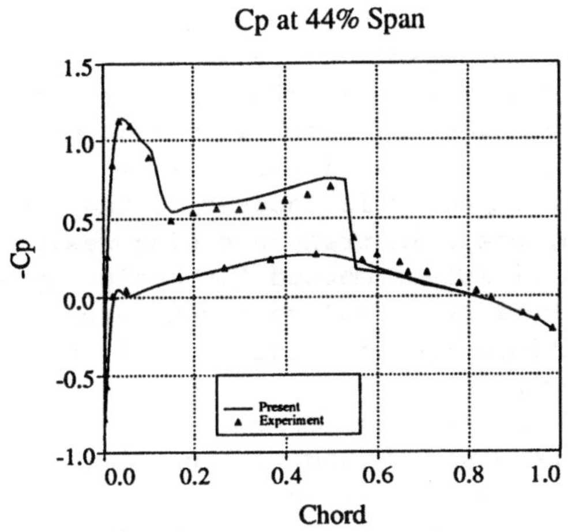

Figure 1.

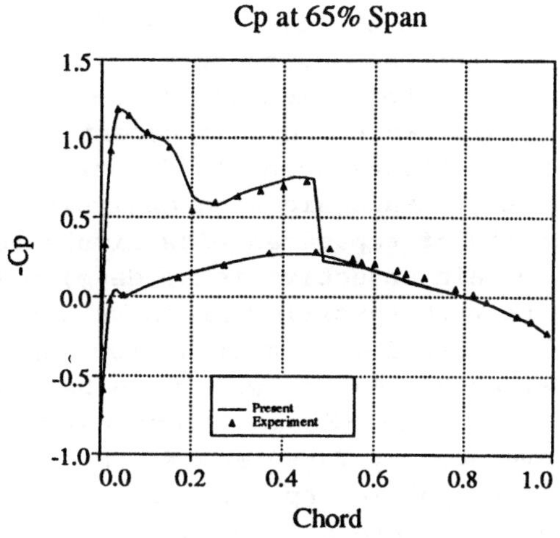

Figure 2.

TRANSONIC WING ANALYSIS AND DESIGN USING NAVIER-STOKES EQUATIONS

N. J. Yu

Boeing Commercial Airplane Group
Seattle, Washington

Introduction

Recent advances in computational methods have made Navier-Stokes solvers practical and useful engineering tools for wing or wing-body analysis [1]. With improvements in numerical accuracy and advances in turbulence models [2-6], results obtained from Navier-Stokes codes have shown good test/theory correlations for a number of configurations at a range of flow conditions [7,8].

Although accurate and reliable analysis codes provide useful tools for airplane design, it would be even more valuable to have a design capability within the Navier-Stokes codes. There are a few design methods in the literature [9-13] that have been incorporated into Navier-Stokes codes to provide a design capability. However, applications of these methods to 3-D wing design need further development.

This paper presents recent developments and applications of a Navier-Stokes code [1] for transonic wing analysis and design. In the area of transonic wing analysis, the emphasis is focused on the capability of separated flow simulation, where in transonic wing design, the main objective is to develop an efficient method for designing a wing with specified pressure distributions. Problems encountered in the airfoil design with flow separations are discussed in this paper.

Navier-Stokes Analysis Code Development

The Navier-Stokes algorithm used in the present study was developed by Vatsa and Wedan [1]. The basic method follows Jameson's finite volume approach [14]. The unsteady thin-layer Navier-Stokes equations are solved using a multistage Runge-Kutta time-stepping scheme. To speed up convergence, a V-cycle full multigrid method is implemented in the solver.

Several different turbulence models were implemented in the code [1,7], including Baldwin-Lomax [2], Johnson-King [3], and Baldwin-Barth [5] models. Through a number of numerical experiments, it was found that the Johnson-King model provides good correlation with test data for a range of flow conditions, including flows with mild shock-induced and trailing edge separations. This model is used in all results presented in this paper.

To solve the flowfield around a wing or wing-body configuration with viscous effects along the wing surface, a high-quality Navier-

Stokes grid is needed. The elliptic grid generation method explored earlier [15] has been proved effective in generating useful Navier-Stokes grids and is used in the present study. The grid spacing normal to the wing surface is of the order of 10^{-7} to 10^{-5} x/C, depending on the Reynolds number of the flow to be analyzed. As a general rule, the first grid spacing normal to the wall, in terms of the wall unit y^+, should be of the order one so that the viscous effects can be realistically simulated.

Computed results for a research wing at an off-design condition is shown in Figure 1. The trailing edge region of the upper surface showed small separation, as can be seen from the surface streamline plot. The computed wing pressures correlated well with test data, including the trailing edge recovery region. The same code has also been used for the analysis of conventional wings, such as the ONERA wing M6 and the Boeing 747-200 wing. Results also correlated quite well with test data.

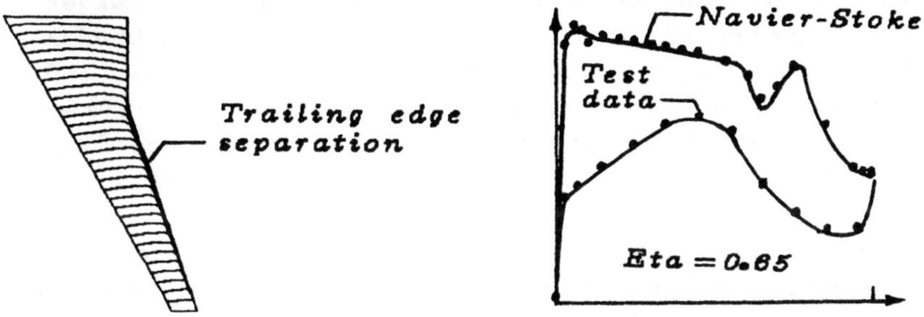

a. Upper surface streamlines b. Pressure distributions

Fig. 1. Test/theory comparisons for an advanced wing at off-design condition (flow separation at trailing edge)

Design Method Development

The design method explored in this paper uses an iterative approach, where the target pressures of the wing to be designed are compared with the analysis pressures of an initial input wing. The differences between target and analysis pressures are used to drive the geometry change of the wing through a streamline curvature method [16-17]. The updated geometry is then analyzed using the same Navier-Stokes solver. The process repeats until the final pressures match the target pressures within a prescribed tolerance.

The design driver, developed by Campbell and Smith [12-13], relates pressure differences to geometry changes. Two different algorithms are used to determine the geometry changes required to produce the desired change in surface pressures, depending on the local Mach number of the flowfield. In subsonic regions, the change in surface curvature is related to the change in pressure through a surface curvature method, where in supersonic regions, the change in surface slope is related to the change in pressure based on the supersonic thin-airfoil theory. The change in curvature or slope is integrated along the streamwise direction to produce the geometry correction. More details on the basic method are discussed in references 13 and 18.

Several issues involving wing design need to be addressed, especially the trailing edge closure and the smoothness of the designed geometry. During the iterative design process, the geometry correction added to the original surface may result in a crossover or an open trailing edge wing section. To alleviate this problem, the newly designed geometry is rotated with respect to the leading edge point (or the first design point) so that the trailing edge returns to its original location. To ensure smoothness of the redesigned geometry, the wing surface coordinates are smoothed in both chordwise and spanwise directions after each design cycle. Typically, a third-order polynomial with a least square fit is used in the geometry smoothing process.

To demonstrate this design concept, we used the pressure distributions of an ONERA wing M6 at Mach=0.84, α=3.06°, and Reynolds number=21.66 million per unit length (semispan), as target pressures. The input geometry was the same wing planform, with the NACA 0012 airfoil section. Figure 2 compares the initial, final, and target design pressures at two span stations, and Figure 3 compares the corresponding geometries at the same stations for the final converged solution. The complete design process takes 28 iterative design cycles. Notice that both the target pressures and the target geometry were recovered accurately. In fact, at the 15th design cycle, the design solution was reasonably well converged. The last few design cycles primarily provided fine adjustments of the surface pressures and geometry in order to recover the targets accurately.

The same design process has also been successfully applied to 2-D airfoil design. An interesting case is the design of an airfoil at a

Fig. 2. Comparisons of initial, target, and final design pressures for a wing at M=0.84, α=3.06°, Re=21.66 million/semispan

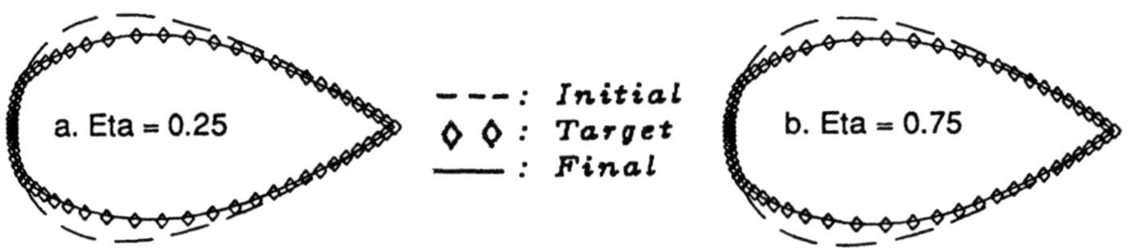

Fig. 3. Comparisons of initial, target, and final design geometries for a wing at M=0.84, α=3.06°, Re=21.66 million/semispan

flow condition with shock-induced separation. Here an RAE2822 airfoil at Mach=0.75, α=2.81°, and Re=6.2 million/C is used as the target. This is a well-known "case 10" solution, which shows a moderate amount of flow separation downstream of the shock (Figure 4). The NACA 0012 airfoil at the same flow conditions is used as the initial input. After 30 design cycles, the surface pressures approach the target pressures reasonably well, with a slight mismatch in shock location and in pressure level just aft of the shock. The geometry obtained from the design code shows a correspondingly good match to the target airfoil, with a slight bulge in the upper surface ordinates just aft of the maximum thickness location. Apparently in this case, the design method produces a geometry that encloses the separation bubble downstream of the shock, as one of the possible solutions. The present iterative design method does not guarantee recovering the target geometry if substantial flow separations are present in the analysis.

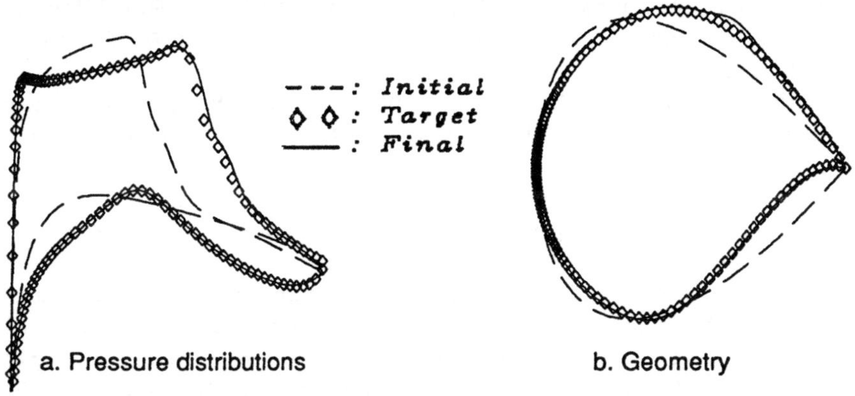

Fig. 4. Airfoil design—RAE2822 at M=0.75, α=2.81°, Re=6.2 million/C

Conclusions

The thin-layer Navier-Stokes code can be used as an effective tool for simple wing or wing-body analysis. Results show good test/theory correlations for various configurations at a range of flow conditions. The simple nonequilibrium turbulence model of Johnson and King [3] provides improvements in solution quality for the Navier-Stokes analysis. Further developments in turbulence modeling are needed to simulate flow with substantial separations reliably and accurately.

The design method using the direct iterative surface curvature approach explored by Campbell and Smith [12-13] has been extended to 3-D wing design with prescribed surface pressures. Preliminary results are encouraging. Further developments of the design method to include additional geometry as well as aerodynamic constraints are needed to make the method a useful tool.

The author would like to acknowledge Drs. H. C. Chen and W. H. Jou of The Boeing Company and Dr. R. L. Campbell of NASA Langley Research Center for many valuable discussions on design methods.

References

1. Vatsa, V. N., and Wedan, B. W., "Development of an Efficient Multigrid Code for 3-D Navier-Stokes Equations," AIAA-89-1791, 1989.

2. Baldwin, B. S., and Lomax, H., "Thin Layer Approximation and Algebraic Model for Separated Turbulence Flows," AIAA 78-257, 1978.

3. Johnson, D. A., and King, L. S., "A Mathematically Simple Turbulence Closure Model for Attached and Separated Turbulent Boundary Layers," AIAA J., Vol. 23, No. 11, Nov. 1985, pp. 1684-92.

4. Johnson, D. A., and Coakley, T. J., "Improvements to a Nonequilibrium Algebraic Turbulence Model," AIAA J., Vol. 28, No. 11, Nov. 1990.

5. Baldwin, B. S., and Barth, T. J., "A One Equation Turbulent Transport Model for High Reynolds Number Wall Bounded Flows," AIAA 91-0610, 1991.

6. Spalart, P. R., and Allmaras, S. R., "A One-Equation Turbulence Model for Aerodynamic Flows," AIAA 92-0439, 1992.

7. Yu, N. J., Allmaras, S. R., and Moschetti, K. G., "Navier-Stokes Calculations for Attached and Separated Flows Using Different Turbulence Models," AIAA-91-1791, 1991.

8. Garner, P., Meredith, P., and Stoner, R., "Areas of Future CFD Development as Illustrated by Transport Aircraft Applications," AIAA-91-1527, 1991.

9. Malone, J. B., Narramore, J. C., and Sankar, L. N., "Airfoil Design Method Using the Navier-Stokes Equations," J. Aircraft, Vol. 28, No. 3, March 1991.

10. Malone, J. B., and Swanson, R. C., "Inverse Airfoil Design Procedure Using a Multigrid Navier-Stokes Method," Third International Conference on Inverse Design Concepts and Optimization in Engineering Sciences, Ed: G. S. Dulikravich, Washington, D. C., Oct. 23-25, 1991.

11. Birckelbaw, L., "Inverse Airfoil Design Using the Navier-Stokes Equations," AIAA Paper 89-2202, 1989.

12. Campbell, R. L., and Smith L. A., "A Hybrid Algorithm for Transonic Airfoil and Wing Design," AIAA Paper 87-2552, 1987.

13. Smith, L. A., and Campbell, R. L., "A Method for the Design of Transonic Flexible Wings," NASA Technical Paper 3045, December 1990.

14. Jameson, A., "Multigrid Algorithms for Compressible Flow Calculations," 2nd European Conference on Multigrid Methods, Cologne, Oct. 1985, also MAE Report 1743, Princeton University.

15. Yu, N. J., Kusunose, K., Chen, H. C., and Sommerfield, D. M., " Flow Simulations for a Complex Airplane Configuration Using Euler Equations," AIAA Paper 87-0454, 1987.

16. Barger, R. L., and Brooks, C. W., Jr., "A Streamline Curvature Method for Design of Supercritical and Subcritical Airfoils," NASA TN D-7770.

17. Davis, W. H., Jr., "Technique for Developing Design Tools From the Analysis Methods of Computational Aerodynamics," AIAA Paper 79-1529, 1979.

18. Yu, N. J., and Campbell, R. L., "Transonic Airfoil and Wing Design Using Navier-Stokes Codes," AIAA-92-2651, 1992.

A SECOND ORDER METHOD FOR THE FINITE ELEMENT SOLUTION OF THE EULER AND NAVIER–STOKES EQUATIONS

G.S. Baruzzi[1], W.G. Habashi[1] and M. Hafez[2]

[1] Dep't of Mechanical Engineering, Concordia University, Montreal, Canada
[2] Dep't of Mechanical Engineering, University of California, Davis, California

1. Introduction

Most of the numerical methods for the solution of the inviscid Euler equations require the use of artificial viscosity. For the Navier-Stokes equations at high Reynolds numbers, the viscous terms are dominant only in a thin layer outside which the flow is nearly inviscid. An artificial viscosity is still needed for numerical stability in the nearly inviscid region but must be eliminated in the viscous layer, otherwise the numerical solution is contaminated with excessive dissipation. For first order methods the amount of artificial viscosity necessary for stable solutions is proportional to the mesh size. Its detrimental effects, therefore, can be reduced by refining the mesh. A uniformly fine mesh throughout the solution domain, however, is impractical and the grid must reflect the disparate characteristic lengths of the viscous and inviscid regions. The practical alternative is to adopt a higher order artificial viscosity.

In finite elements, Hughes et al. [1] developed the Petrov-Galerkin and streamline upwinding approaches and Morgan et al. [2] adopted a flux-corrected transport strategy. In previous work [3] the authors have presented a method for the solution of the Euler equations, based on an artificial viscosity introduced in the continuity and momentum equations in the form of Laplacians of the pressure and the velocity components, respectively. Equal order interpolation for pressure and velocity is allowed. The same technique has also been applied to subsonic Navier-Stokes flows [4].

To further enhance the quality of the finite element solutions for the Euler and Navier-Stokes equations, a higher order artificial viscosity is proposed. The artificial viscosity terms of the first order solver are balanced with correction terms obtained from the governing equations. The balancing terms are evaluated using a standard Galerkin method and are lagged, hence the Jacobian matrix is unaltered by the modifications. The residual, however, is of higher order accuracy and the solution is not contaminated with the effects of the excessive viscosity usually needed for numerical stability of the first order method.

2. Second Order Scheme

To demonstrate the second order scheme, the modified governing equations of viscous, steady compressible flow can be written as:

$$\nabla \bullet (\rho \vec{V}) - \varepsilon_1 \left[\nabla^2 p - \varepsilon_2 \nabla \bullet (\nabla \hat{p}) \right] = 0 \tag{1a}$$

$$\nabla \bullet (\rho \vec{V} \vec{V}) + \nabla p - \nabla \bullet \tau - \varepsilon_1 \left[\nabla^2 \vec{V} - \varepsilon_2 (\nabla s + \nabla \times \omega) \right] = 0 \tag{1b}$$

$$H_\infty = \frac{\gamma}{\gamma - 1} \frac{p}{\rho} + \frac{\vec{V} \bullet \vec{V}}{2} \tag{1c}$$

$$\frac{\mu}{\mu_\infty} = \left(\frac{T_\infty + 110^\circ k}{T + 110^\circ k} \right) \left(\frac{T}{T_\infty} \right)^{3/2} \tag{1d}$$

$$\nabla \bullet \tau = \frac{1}{Re} \left[-\frac{2}{3} \nabla (\mu \nabla \bullet \vec{V}) + \nabla \times \mu (\nabla \times \vec{V}) + 2 (\nabla \bullet \mu \nabla) \vec{V} \right]$$

For convenience, the assumption of constant total enthalpy is used instead of the full energy equation. This approximation is acceptable in the transonic flow regime. In two dimensions the system reduces to three coupled equations in terms of the variables $\{u, v, p\}$, where the density in Eqs. (1a) and (1b) is replaced using Eq. (1c).

The Laplacian of pressure in Eq. (1a) is balanced by an additional term that is a function of the velocity components and their derivatives. This term is obtained by taking the divergence of the momentum equations, as follows:

$$\nabla \bullet (\nabla \hat{p}) = \nabla \bullet \left(g_1 \vec{i} + g_2 \vec{j} \right) \tag{2}$$

where, in Cartesian coordinates:

$$g_1 = -\left\{ \left[\rho u^2 - \frac{2\mu}{3Re} (2u_x - v_y) \right]_x + \left[\rho u v - \frac{\mu}{Re} (u_y + v_x) \right]_y \right\}$$

$$g_2 = -\left\{ \left[\rho u v - \frac{\mu}{Re} (u_y + v_x) \right]_x + \left[\rho v^2 - \frac{2\mu}{3Re} (2v_y - u_x) \right]_y \right\}$$

The artificial term of Eq. (1a) does not vanish identically, even in the steady-state when the discrete momentum equations are satisfied, since the divergence of the momentum equations is not necessarily zero in the discrete sense. Nevertheless, conservation of mass improves at least an order of magnitude over the first order method.

For the momentum equations the Laplacians of the velocity components are balanced according to the following vector identity:

$$\nabla \bullet \left(\rho \vec{V}\vec{V}\right) + \nabla p - \nabla \bullet \tau - \varepsilon_1\left[\nabla^2 \vec{V} - \varepsilon_2(\nabla s + \nabla \times \omega)\right] = 0 \qquad (3)$$

where $s = \nabla \bullet \vec{V}$ and $\omega = \nabla \times \vec{V}$. The artificial viscosity coefficient ε_1 is a constant and ε_2 vanishes in the shock region. The correction terms g_1, g_2, s and ω are evaluated at the nodes in a standard Galerkin way. For s, for example:

$$\iint_A W_i\left(s - \nabla \bullet \vec{V}\right)dA = 0 \qquad (4)$$

The finite element discretization is based on isoparametric bilinear elements. The standard Galerkin weighted residual method in weak form is applied to each equation and the system is linearized by Newton's method. For the Euler and Navier-Stokes equations proper boundary conditions are imposed, in genuine finite element style, as demonstrated in previous work [3,4]. At each iteration, a fully coupled, sparse linear algebraic system of the form [*J*] {ΔU} = – {*Res*}, where ΔU = {$\Delta u, \Delta v, \Delta p$} and {*Res*} is the residual, is solved by a direct solver. The added artificial viscosity balancing terms are lagged through the iteration, hence the Jacobian matrix [*J*] is unaltered with respect to the first order approach. The residual, however, attains higher order accuracy.

3. Numerical Results

An inviscid shock reflection problem is considered: at the inlet Mach=2 and the upper domain boundary conditions are set to produce a 40° incident shock. No outflow boundary conditions are imposed. The Mach number contours for the second order method are shown in Fig. 1a and the Mach number distribution at the midsection of the channel is shown in Fig. 1b.

The second case is inviscid flow around a NACA0012 at Mach=0.8 and α=1.25°. The Mach number contours for the second order method are shown in Fig. 2a and the surface Mach number is compared to the results of Pulliam and Barton [5] in Fig. 2b.

The last case is viscous compressible flow around a NACA0012 at Mach=0.8 and Re=500 at α=10°. Fig. 3a shows the Mach number contours obtained with the second order scheme and Fig. 3b shows the surface Cp distribution compared to the results of Hollanders and Ravalason [6].

References

1. Tezduyar, T.E. and Hughes, T.J.R., "Finite Element Formulation for Convection Dominated Flows with Particular Emphasis on the Compressible Euler Equations," AIAA Paper 83-0125.
2. Hassan, O., Morgan, K. and Peraire, J., "An Implicit Finite Element Method for High Speed Flows," AIAA Paper 90-0402.

3. Baruzzi, G.S., Habashi, W.G. and Hafez, M.M., "Finite Element Solutions of the Euler Equations for Transonic External Flows", AIAA Journal, Vol. 29, No. 11, November 1991, pp. 1886-1893.
4. Peeters, M.F., Habashi, W.G., Nguyen, B.Q. and Kotiuga, P.L., "Finite Element Solutions of the Navier-Stokes Equations for Compressible Internal Flows", AIAA Journal of Propulsion and Power, Vol. 8, No. 1, January 1992, pp. 192-198.
5. Pulliam, T.H. and Barton, J.T., "Euler Computations of AGARD Working Group 07 Airfoil Test Cases", AIAA Paper 85-0018.
6. H. Hollanders and W. Ravalason, "Résolution des Equations de Navier-Stokes en Fluide Compressible par Méthode Implicite", La Recherche Aérospatiale, No. 1, Jan.-Feb. 1986, pp. 23-46.

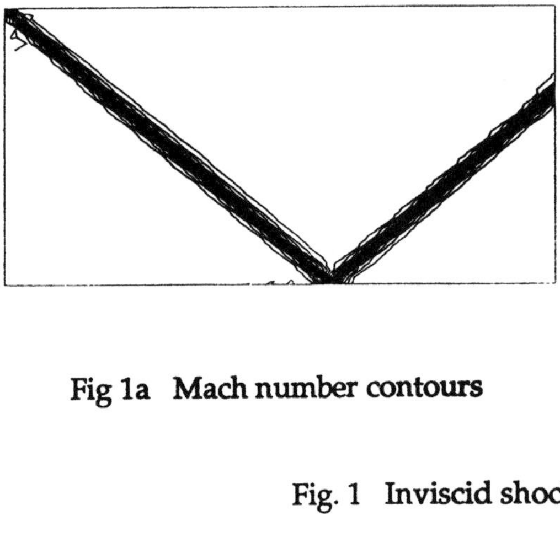

Fig 1a Mach number contours

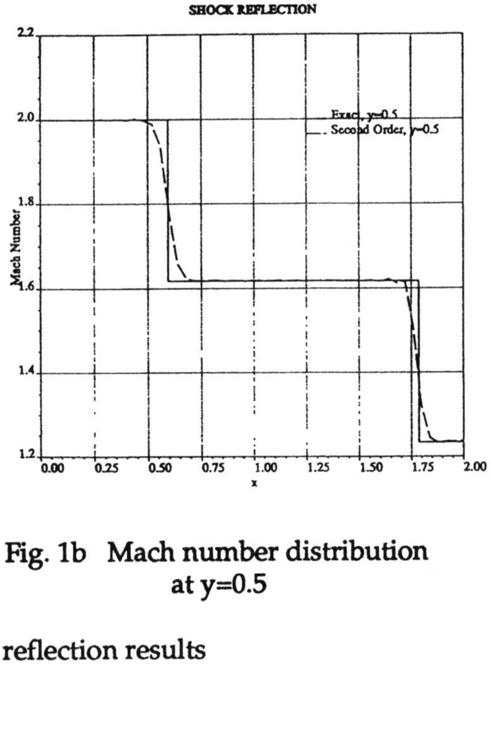

Fig. 1b Mach number distribution at y=0.5

Fig. 1 Inviscid shock reflection results

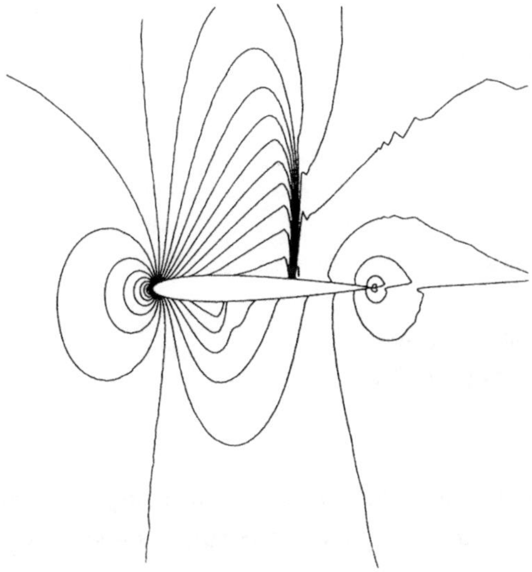

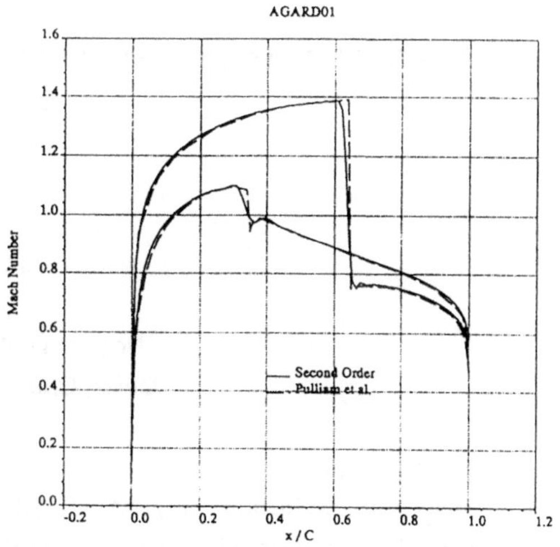

Fig 2a Mach number contours

Fig. 2b Surface Mach number distribution

Fig. 2 Inviscid flow around a NACA0012 at M=0.8 and α=1.25°

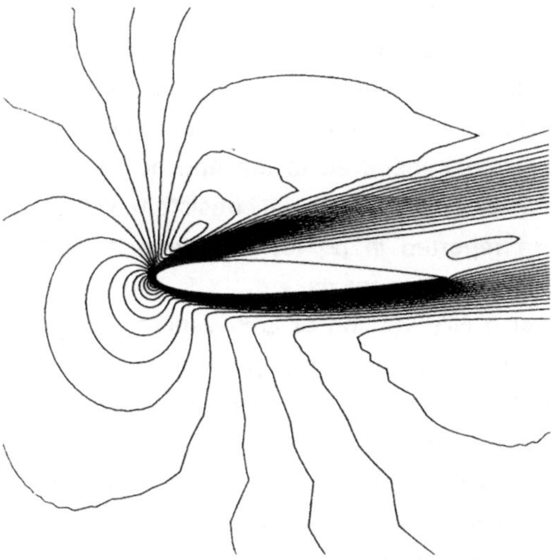

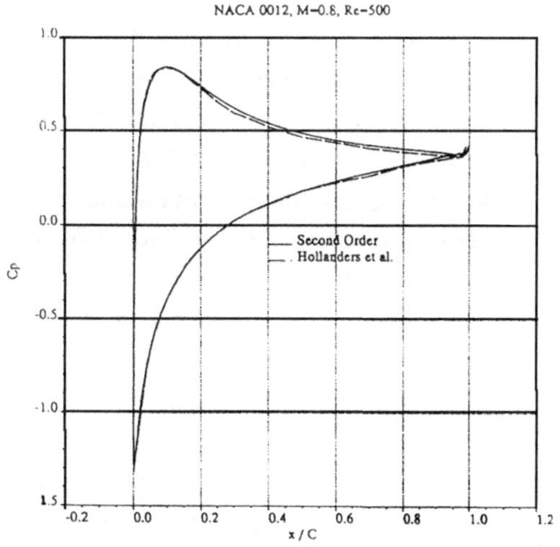

Fig 3a Mach number contours

Fig. 3b Surface Cp distribution

Fig. 3 Viscous flow around a NACA0012 at M=0.8, Re=500 and α=10°

IMPLEMENTATION OF CELL-VERTEX SCHEMES ON A MASSIVELY PARALLEL COMPUTER[1]

J.J. Chattot

University of California Davis
Davis, California 95616, U.S.A.

1.Introduction

Cell-vertex schemes are particularly well suited for computation using domain decomposition on a massively parallel computer, since they have a compact stencil, each cell being used as control volume for the conservation laws, and a minimum of information needs to be exchanged among the subdomains.

A cell-vertex or "box" scheme was introduced initially by Keller in 1970 [1], for the discretization of the boundary-layer equations. For first order systems of partial differential equations it seems natural to discretize the first derivatives using a two-point scheme in 1-D and a four-point scheme on quadrilaterals in 2-D. This allows the time derivatives to be evaluated at the center of the cell. The question then arises as to how the time derivatives (at the cell center) is to be distributed among the nodes (at the corner of each cell). For the quasi one-dimensional Euler equations it has been shown that the time derivatives in the "box" [i-1,i] can be transferred to the end points i-1 or i, depending on the characteristic directions, in a way which is consistent with the physical properties of the flow. Results were reported in [2] for various test cases, including the shock tube problem and the flow in a slender nozzle.

Extension to 2-D has been investigated using Ni's version of Lax-Wendroff scheme [3], and developing a new, characteristic based scheme [4]. Both approaches have been applied to a 2-D linear hyperbolic model problem, governing a low speed flow past slender profiles. It has been shown that Ni's scheme is a box scheme. Ni's "distribution formulae" distribute the contributions of the derivatives from the center of the cell to the nodes, taking into account consistency and accuracy requirements (second order scheme). The characteristic based scheme makes use of characteristic information to distribute the changes from the cell center to the nodes.

The special feature of a box scheme is that the steady conservation equations are

[1] Funds for the support of this study have been allocated by the NASA-Ames Research Center, Moffett Field, California, under Interchange No: NCA2-627

satisfied exactly, in the discrete sense, in each cell. This is believed to be a stronger enforcement of the conservation laws than with non-box-type schemes, for which the equations are satisfied on staggered control volumes requiring interpolation of the flow variables.

Results were compared with the exact solution for the flow past a wavy wall and the flow past a parabolic profile at zero incidence, using uniform meshes. Interestingly, the two scheme exhibit second order accuracy as expected for the first problem, but only first order accuracy for the latter, due to a weak singularity at the leading and trailing edges, fig.1.

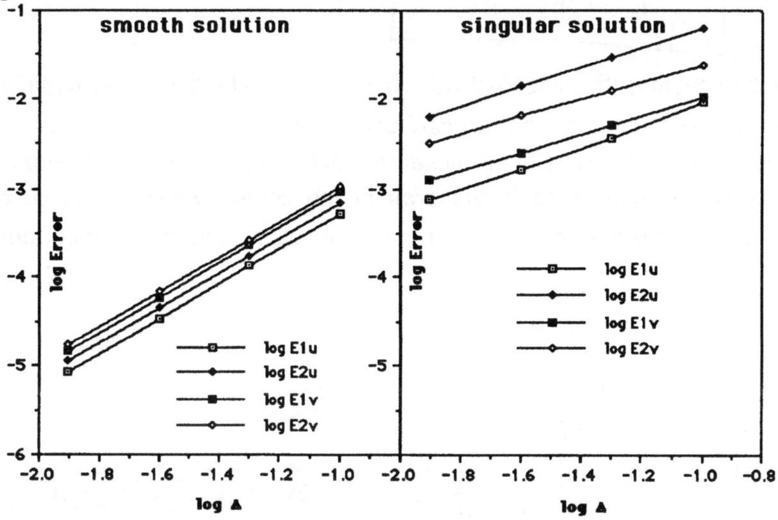

figure 1-Error norms vs. mesh size, for the linear hyperbolic problem

2.Comparison of two schemes for the Euler Equations

The Ringleb flow [5] is one of the few exact solutions of the Euler Equations, which can be found in a report by G. Chiocchia [6]. Ni's scheme is compared to a version of Lax-Wendroff scheme due to Lerat [7]. The computational features of both schemes are presented below, as implemented by the author in the codes.

Cell and control volume are defined on the sketch below. The conservation equations are written in compact form, with the usual notation, as:

$$w_{,t} + f_{,x} + g_{,y} = 0$$

In Ni's scheme, the time derivatives, $w_{,t}=-(f_{,x}+g_{,y})$, are evaluated at the cell centers, say G, using a cell-vertex formula for the divergence (linear variation along the edges). Then these are distributed to the nodes, say A, in proportion to the area ratio $a_1/(a_1+a_2+a_3+a_4)$. The second order contributions, $w_{,tt}=-(f_{,w} \cdot w_{,t})_{,x}-(g_{,w} \cdot w_{,t})_{,y}$, are evaluated on the control volume using the same formula.

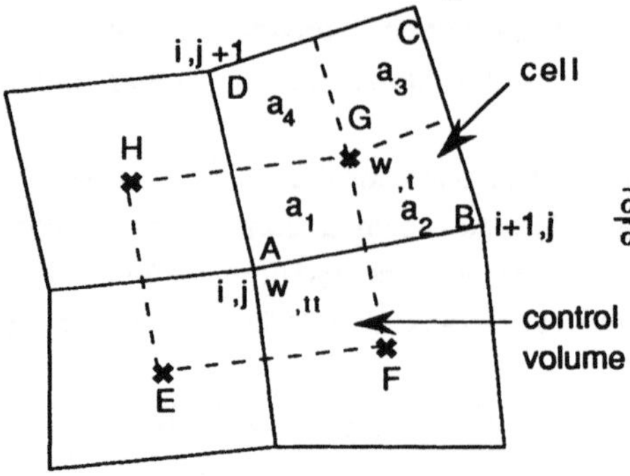

$$\frac{\partial f}{\partial x} = \frac{(f_C - f_A)(y_D - y_B) - (f_D - f_B)(y_C - f_A)}{(x_C - x_A)(y_D - y_B) - (x_D - x_B)(y_C - y_A)}$$

In Lerat's scheme, the time derivatives, $w_{,t}$, are evaluated at the mid-edge, say I, using a cell vertex formula on the auxiliary control volume AFBG, the values at F and G being averaged from the cells. Then these are sent to the nodes, say A, in proportion to the area a_1/Area, where Area is the area of the control volume. The second order contributions, $w_{,tt}$, are evaluated using the control volume and a mid-edge formula.

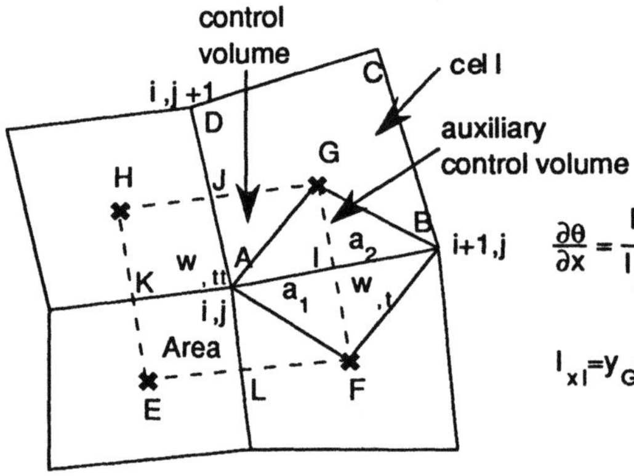

$$\frac{\partial \theta}{\partial x} = \frac{l_{xI}\theta_I + l_{xJ}\theta_J + l_{xK}\theta_K + l_{xL}\theta_L}{l_{xI}x_I + l_{xJ}x_J + l_{xK}x_K + l_{xL}x_L}$$

$$l_{xI} = y_G - y_F \ , \quad x_I = \frac{1}{2}(x_F + x_G)$$

For both schemes, following Ni's idea, these expressions are obtained by summing up contributions computed as the domain is swept, on a cell by cell basis. Lerat's scheme is not a box scheme, because of the averaging which is done at the cell centers. This may be the reason for the more dissipative properties of this scheme.

Comparisons, using Ringleb flow, indicate that both schemes have the same order of accuracy (close to 2) as the mesh is refined, fig.2. However, the level of error is somewhat higher in Lerat's scheme, probably in relation to the dissipation built into the scheme. Another feature that made these schemes differ, is the stability requirement to run the two explicit versions of Lax-Wendroff scheme: It has been found, from the test case, that Ni's scheme could be run with a time step which is twice that needed by Lerat's version to maintain stability of the computation.

The use of the cell-vertex and mid-edge formulae allow to solve the cartesian form of the Euler equations on a curvilinear mesh system, without the need to introduce a transformation and metric coefficients. This has been found to be extremely flexible. Only nodal coordinates are needed.

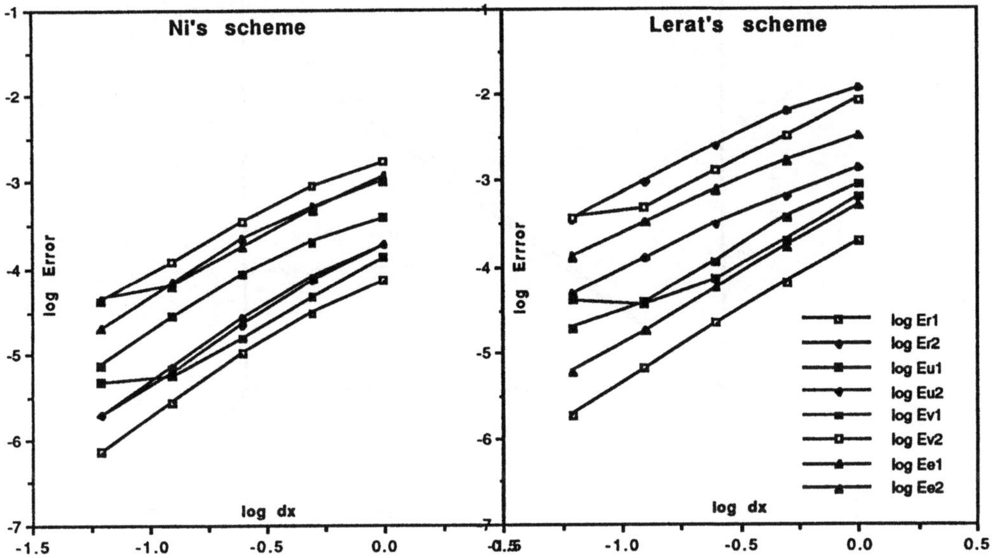

figure 2-Error norms vs. mesh size, for the Euler test case

3.Implementation on the iPSC/860

The Intel iPSC/860 'hypercube' is a massively parallel computer. Each processor carries its own program. This is a MIMD architecture. The machine installed at NASA/Ames has 128 processors (nodes), each of which has 8 Mbytes of memory and an Intel i860 chip capable of 80 Mflops ideally. The processors are interconnected using a hypercube network topology. Taken as a whole, this machine has a peak speed of 10 Gflops and a memory of 128 Mwords (64 bits) which puts it (potentially) among the largest and fastest computers in the world.

The box scheme described above was used to solve the Euler test case. The finest mesh used was 1281x81. The problem was divided among the processors geographically. The domain was divided into regions having equal number of points. Each region was assigned to a single processor. Communication between neighboring regions was required once per iteration. This was not found to be a significant cost. The flow solution was obtained using

1,4,16,64 and 128 processors. As shown in fig. 3, a nearly linear speedup was obtained when more processors were used. The slight degradation is due to communication costs. These could be further reduced by using asynchronous I/O or by choosing a larger problem. With the latest compiler the speed has been evaluated at 233 Mflops.

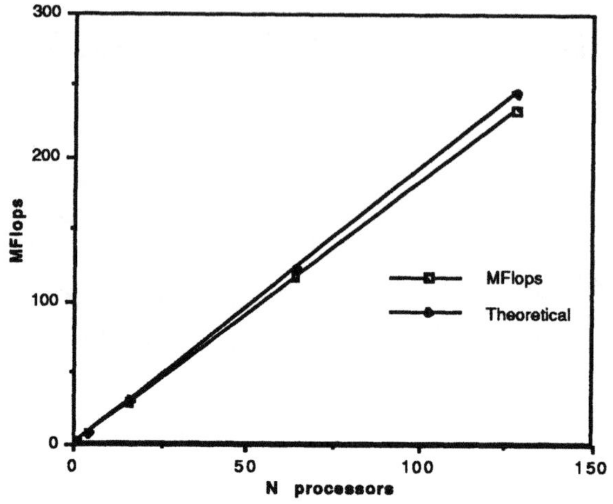

figure 3-Speed-up (MFlops) vs. number of processors of the iPSC/860

References

[1] Keller, H.B., A New Difference Scheme for Parabolic Problems, Numerical Solutions of Partial Differential Equations, vol. 2, (J. Bramble, ed.), Academic Press, New York, 1970.

[2] Chattot, J.J., and Malet, S., A Box Scheme for the Euler Equations, Proceedings of Non Linear Hyperbolic Problems, (C. Carasso, P.A. Raviart, D. Serre, eds.), Lecture Notes in Mathematics, No 1270, Springer-Verlag, 1987.

[3] Ni, R.H., A Multiple-Grid Scheme for Solving the Euler Equations, AIAA Paper 81-1025. Also: AIAA Journal, 20, 1565-1571, 1982.

[4] Chattot, J.J., Box Schemes for First Order Partial Differential Equations, Advances in Computational Fluid Mechanics, (W. Habashi and M. Hafez, eds.), to appear, 1992.

[5] Ringleb, F., Exacte Lösungen der Differentialgleichungen einer adiabatischen Gasströmung, ZAMM, vol. 20, pt 4, 1940, pp 185-198.

[6] Chiocchia, G., Exact Solutions to Transonic and Supersonic Flows, in Test Cases for Inviscid Flow Fields Methods, AGARD-AR-211, 1985.

[7] Lerat, A. and Sides,J., Efficient Solution of the Steady Euler Equations with a Centered Implicit Method, Numerical Methods for Fluid Dynamics III, K.W. Morton & M.J. Barnes, Oxford University Press, 1988.

A CONSISTENT CELL VERTEX FINITE VOLUME METHOD FOR THE COMPRESSIBLE NAVIER-STOKES EQUATIONS

P.I. Crumpton, J.A. Mackenzie and K.W. Morton

ICFD, Oxford University Computing Laboratory, 11 Keble Road, Oxford

1 Introduction

Two objectives guided the development of the cell vertex algorithm which is described here and in more detail in [1]: firstly, the algorithm should give a practical means of calculating external aerodynamical flows, both laminar and turbulent, on a multiblock mesh in two and three dimensions; secondly, there should be a sufficient mathematical basis to the method to yield efficient a posteriori error criteria to be used in mesh adaption, as well as strict a priori error bounds for some relevant model problems.

We have so far had to achieve these aims by means of a compromise. The algorithm is built on the definition of a cell residual $\mathbf{R}_\alpha(\mathbf{W})$, in which the inviscid and viscous fluxes are consistently approximated on the boundaries of a quadrilateral cell Ω_α of the primary mesh. The error analysis is based on cases where $\mathbf{W}$ can be found such that $\mathbf{R}_\alpha(\mathbf{W}) = \mathbf{0}\ \forall\alpha$. On the other hand, in order to solve the Navier-Stokes equations we have had to introduce nodal residuals $\mathbf{N}_j(\mathbf{W})$, obtained by combining the cell residuals for the cells with common node $\mathbf{x}_j$; $\mathbf{W}$ is then determined by setting $\mathbf{N}_j(\mathbf{W}) = \mathbf{0}\ \forall j$. In one dimensional problems and simple two-dimensional problems, $\mathbf{R}_\alpha$ and $\mathbf{N}_j$ can be defined so that they are driven to zero together — see [7], [8], [3]; but this is not generally true and hence leads to the compromise.

In the present paper we report on progress in the overall aim of ensuring that $\mathbf{N}_j(\mathbf{W}) = \mathbf{0}\ \forall j \Rightarrow \mathbf{R}_\alpha(\mathbf{W}) = \mathbf{0}\ \forall\alpha$ for the Navier-Stokes equations. Plots of the flux budget balance for each type of residual are used to show how the design of the artificial viscosity terms and the distribution matrices which are used in the definition of $\mathbf{N}_j$ affect the attainment of this goal, see Fig.1. Our view is that artificial viscosity is an overused expedient, which we aim to relegate to a minor rôle.

2 The basic algorithm and some variants

In the cell vertex algorithm, the conserved unknown variables $\mathbf{W}_j$ are associated with the vertices $\mathbf{x}_j$ of the primary mesh, which we take to be a body-fitted multiblock mesh, composed of quadrilaterals in 2D and hexadedra in 3D. The equations are written in conservation law form, in 2D cartesian co-ordinates

$$\mathrm{div}(\mathbf{f},\mathbf{g}) \equiv \frac{\partial \mathbf{f}(\mathbf{w},\nabla\mathbf{w})}{\partial x} + \frac{\partial \mathbf{g}(\mathbf{w},\nabla\mathbf{w})}{\partial y} = \mathbf{0}. \tag{1}$$

Integration of (1) over a quadrilateral cell Ω_α of measure V_α (with vertices $\mathbf{x}_1, \mathbf{x}_2, \mathbf{x}_3, \mathbf{x}_4$, ordered anticlockwise), and application of Gauss' theorem, gives

$$\frac{1}{V_\alpha}\int_{\Omega_\alpha} \mathrm{div}(\mathbf{f},\mathbf{g})\mathrm{d}\Omega = \frac{1}{V_\alpha}\oint_{\partial\Omega_\alpha} \mathbf{f}\mathrm{d}y - \mathbf{g}\mathrm{d}x; \tag{2}$$

and approximation of the edge integrals by the trapezoidal rule yields the cell residual,

$$\begin{aligned}\mathbf{R}_\alpha(\mathbf{W}) := \frac{1}{2V_\alpha}[&(\mathbf{F}_1 - \mathbf{F}_3)(y_2 - y_4) + (\mathbf{F}_2 - \mathbf{F}_4)(y_3 - y_1) \\ &-(\mathbf{G}_1 - \mathbf{G}_3)(x_2 - x_4) - (\mathbf{G}_2 - \mathbf{G}_4)(x_3 - x_1)].\end{aligned} \tag{3}$$

Here we have used the notation $\mathbf{w}$ for the exact solution of (1) and $\mathbf{W}$ for the approximation, with $\mathbf{F}_j = \mathbf{f}(\mathbf{W}, \nabla\mathbf{W})$ evaluated at the vertex $\mathbf{x}_j$.

Note here the distinctive feature of this scheme: $\nabla\mathbf{W}$ is recovered at each vertex from the approximation over the cells that have the vertex in common; hence all the inviscid and viscous flux terms, for a laminar or turbulent flow model, are approximated in a consistent manner at each vertex of the mesh. The recovery of $(\nabla\mathbf{W})_j$ can be achieved in various ways. In the simplest technique, which is used in the basic algorithm, $\partial\mathbf{W}/\partial x$ is written as $\mathrm{div}(\mathbf{W}, \mathbf{0})$ and integrated over a subsidiary cell whose boundary is formed by primary cell diagonals enclosing the vertex $\mathbf{x}_j$. Then the integrals are approximated as in (2) and (3). Recovery at boundary nodes is accomplished by extrapolation from the interior.

On the mesh away from inter-block boundaries, $\mathbf{R}_\alpha(\mathbf{W})$ thus involves twelve mesh points, the four that comprise the vertices of Ω_α plus the further eight vertices of the cells which have edges in common with Ω_α. For the most part, one would then like to determine $\mathbf{W}$ by setting $\mathbf{R}_\alpha(\mathbf{W}) = \mathbf{0}$ for every primary cell Ω_α of the mesh and satisfying the boundary conditions. However, without a direct association of the unknowns $\mathbf{W}_j$ at the vertices $\mathbf{x}_j$ with the residual equations at the cells Ω_α this may not be possible; also, in the neighbourhood of shocks it may not be desirable without shock fitting; and, in any case, one needs an iterative procedure which uses discrepancies in the $\mathbf{R}_\alpha$ to compute nodal updates $\delta\mathbf{W}_j$.

Hence in the basic algorithm the nodal residuals $\mathbf{N}_j(\mathbf{W})$ are formed at each vertex by using distribution matrices to combine the surrounding cell residuals; artificial viscosity (AVIS) terms may also be added at this stage to give

$$\mathbf{N}_j(\mathbf{W}) := \frac{\sum_{(\alpha)} V_\alpha (D_{\alpha,j}\mathbf{R}_\alpha + \mathbf{A}_{\alpha,j})}{\sum_{(\alpha)} V_\alpha}. \tag{4}$$

The distribution matrices are based on a generalised Lax-Wendroff procedure, using the flux Jacobian matrices for the inviscid fluxes only, $A := \partial\mathbf{f}^I/\partial\mathbf{w}, B := \partial\mathbf{g}^I/\partial\mathbf{w}$. Integrating the second order update term $\mathrm{div}(A\mathbf{R}, B\mathbf{R})$ over the same subsidiary cell used in recovering $\nabla\mathbf{W}$, we obtain for example

$$D_{\alpha,1} := I - \nu_C \frac{\Delta t_\alpha}{V_\alpha}[(y_4 - y_2)A_\alpha - (x_4 - x_2)B_\alpha], \tag{5}$$

where ν_C is a global cell CFL number and Δt_α is the maximum local timestep for stability of the standard Lax-Wendroff scheme. Then the nodal update is given by

$$\mathbf{W}_j^{n+1} = \mathbf{W}_j^n - \nu_N \Delta t_j \mathbf{N}_j(\mathbf{W}^n) \tag{6}$$

where ν_N is a global nodal CFL number and Δt_j is the minimum of the Δt_α for the surrounding cells. Convergence requires

$$\nu_N \leq \nu_C \quad \nu_N \nu_C < 1 \tag{7}$$

and is accelerated by a FAS multigrid procedure. Boundary conditions are applied in a standard way, by performing a Lax-Wendroff update using dummy exterior cells with zero residuals and then imposing either solid body or characteristic far-field boundary conditions.

Clearly at convergence we have $\mathbf{N}_j(\mathbf{W}) = \mathbf{0}$, but what happens to the cell residuals $\mathbf{R}_\alpha(\mathbf{W})$ of our target scheme? In [7] a detailed study has been made of the situation in one dimension. For the Euler equations the optimal distribution matrix is then given by

$$D_\alpha^{\pm} := I \pm \text{sign}(A_\alpha), \tag{8}$$

where $A_\alpha = L_\alpha^{-1}\Lambda_\alpha L_\alpha$ diagonalises A_α and $\text{sign}(A_\alpha) = L_\alpha^{-1}\,\text{sign}(\Lambda_\alpha)L_\alpha$. That is, a matrix CFL parameter is used so that each residual component is fully upwinded. This automatically takes care of counting problems, ensuring that the number of equations and boundary conditions matches the number of unknowns so that in general $\mathbf{R}_\alpha(\mathbf{W}) = \mathbf{0}$. Exceptions occur at shocks and sonic points; at a shock two residual components, corresponding to the switching characteristic, are summed to maintain conservation and give a very local error without the use of artificial viscosity; but a cell containing a sonic point should ideally be split to give two residuals, though the use of fourth order AVIS can be almost as effective. This study goes on to show how the choice of $\nu_C > 1$ in (5) can approximate the optimal choice (8), what the rôle of the AVIS is at sonic points and shock points and how the efficiency of the update procedure (6) can be greatly improved.

In two dimensions the only problem that has been studied in the same detail is the scalar convection-diffusion problem. In [2] it is shown how extraordinarily good results are obtained over the whole range of mesh Péclet numbers, and with no adjustable AVIS parameters, by setting $\mathbf{R}_\alpha(\mathbf{W}) = \mathbf{0}$ except at outflow Dirichlet boundaries. These are just the nodes where extra boundary conditions are imposed when diffusion is added to convection; and, as with the change from the Euler equations to the Navier-Stokes equations, the use of a conventional Lax-Wendroff iteration does not lead to the uncoupling and hence the setting to zero of the cell residuals. However, if the distribution matrices like (5) are replaced by products of those like (8), that is in the scalar case

$$D_\alpha^{\pm\pm} := [1 \pm \text{sign}(a_\alpha)][1 \pm \text{sign}(b_\alpha)], \tag{9}$$

we obtain the same results as in [2] and set all the selected cell residuals to zero.

3 Recent developments and numerical studies

The results presented in [1] and [4] for the Navier-Stokes equations were obtained with the basic cell vertex algorithm, that is, using distribution matrices and updates based on the two scalar CFL numbers ν_C and ν_N. This is now seen to be quite restrictive: if the effective ν_C for each component is to be larger than unity, in order to obtain a smooth decay of errors away from key flow features, the scalar ν_C has to be very large and hence, from (7), ν_N has to be very small. Even with multigrid the resulting slow convergence cannot be tolerated. Hence second order and fourth order AVIS is brought in to replace the use of $\nu_C >> \nu_N$. The studies presented here aim to show how the deleterious effects of this compromise can be minimised.

Loss of accuracy, especially with entropy variables, is particularly serious if the second order AVIS is applied too freely. It is needed by the cell vertex scheme only near shocks. Hence we have used the techniques of shock detection (see [5], [6], [4]) to eliminate its introduction in regions of high gradients, such as near the leading edge, where it is not needed; and we have scaled the fourth order AVIS by the local Mach number. The benefits are shown in Fig.2 for a transonic turbulent flow computation. Moreover, AVIS is needed only for the residual component corresponding to the switching characteristic; the result of using a more discriminating matrix-based AVIS is shown in Fig.3, which when coupled with the shock detection procedure ensures that the computational overhead is minimal.

References

[1] P.I. Crumpton, J.A. Mackenzie, and K.W. Morton. Cell vertex algorithms for the compressible Navier-Stokes equations. OUCL Report NA91/12.

[2] J.A. Mackenzie and K.W. Morton. Finite volume solutions of convection-diffusion test problems. To appear in Mathematics of Computation.

[3] K.W. Morton. Cell-node mappings for the cell vertex method. To appear in the Proc. of the Conf. on Numerical Methods for Fluid Dynamics, Reading, Apr, 1992.

[4] K.W. Morton, P.I. Crumpton, and J.A. Mackenzie. Cell vertex methods for inviscid and viscous flows. To appear in Computers and Fluids, 1992.

[5] K.W. Morton and M.F. Paisley. A finite volume scheme with shock fitting for the steady Euler equations. *Journal of Computational Physics*, 80:168–203, 1989.

[6] K.W. Morton and M.A. Rudgyard. Shock recovery and the cell vertex scheme for the steady Euler equations. *Proc. of 11th Int. Conf. on Numerical Methods in Fluid Dynamics*, pages 424–428, 1989.

[7] K.W. Morton, M.A. Rudgyard, and G.J. Shaw. Upwind iteration methods for the cell vertex scheme in one dimension. OUCL Report NA91/09.

[8] K.W. Morton and M. Stynes. An analysis of the cell vertex method. OUCL Report NA91/07.

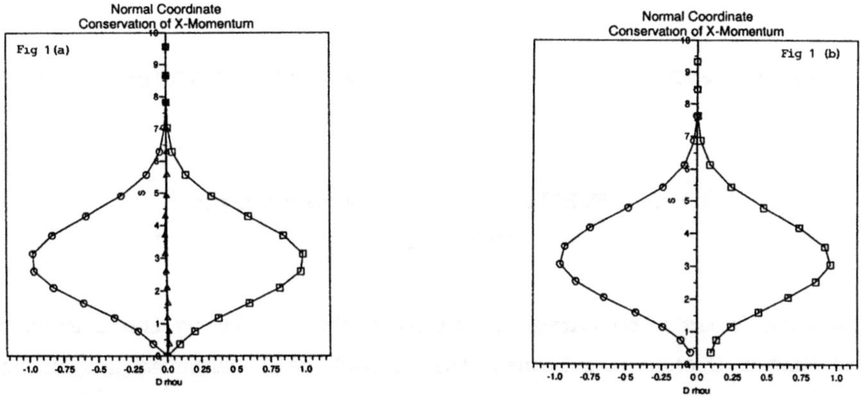

Figure 1: Equation budget of x-momentum for flow over a flat plate with $M_\infty = 0.5$ and $Re = 5000$: (a) flux balance of nodal residual and (b) flux balance of the cell residual; □ inviscid, ○ viscous and △ artificial dissipation.

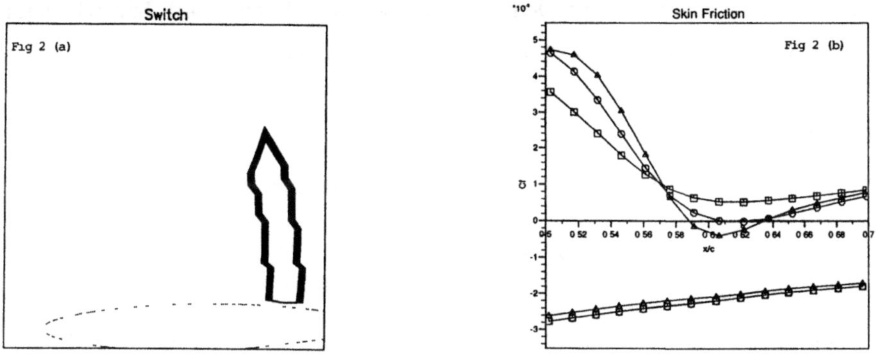

Figure 2: RAE 2822 aerofoil $M_\infty = 0.73$, $\alpha = 2.79$, $Re = 6.5 \times 10^6$, (257×65) grid: (a) region of shock and (b) improvement of skin friction coefficient in shocked region; □ standard, ○ with shock detection, △ Mach number scaling and shock detection.

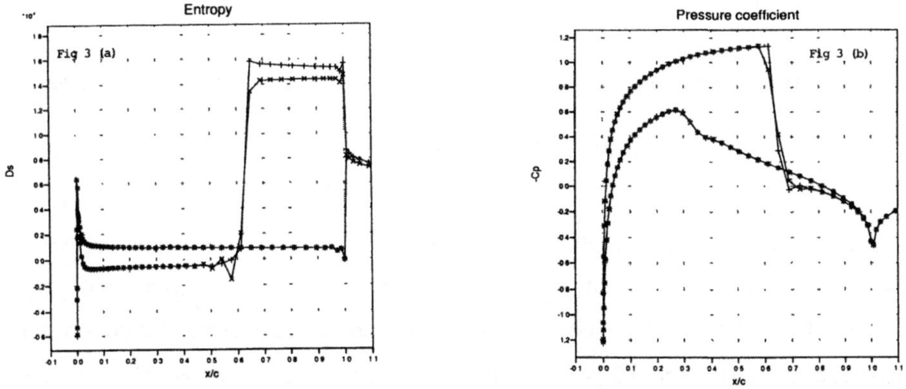

Figure 3: Comparison of (a) entropy deviation and (b) pressure coefficient for Euler flow over a NACA0012 aerofoil $M_\infty = 0.85$ and $\alpha = 1^o$: $+$ matrix based capturing AVIS; $\times$ scalar AVIS.

REMARK ON NUMERICAL SIMULATION OF 2D UNSTEADY TRANSONIC FLOWS

J.Fořt, T.Hůlek, K.Kozel, M.Vavřincová
TU Prague

The work deals with three numerical methods solving the system of Euler or Navier-Stokes equations. Mac Cormack cell centered and Ni cell vertex finite volume schemes were used for simulation of inviscid unsteady solution of transonic flows through a 2D cascade. Unsteady motion is caused by a periodic change of downstream pressure. The Runge-Kutta multistage cell centered finite volume scheme has been used for viscous laminar steady and unsteady transonic flows over NACA 0012.

I. Numerical solution

The system of Euler equations in conservative form,

$$W_t + F_x + G_y = 0 \qquad (1)$$

$W = \text{col}\,\|\,\rho,\ \rho u,\ \rho v,\ e\,\|$, $F = \text{col}\,\|\,\rho u,\ \rho u^2 + p,\ \rho uv,\ (e+p)p\,\|$,
$G = \text{col}\,\|\,\rho v,\ \rho uv,\ \rho v^2 + p,\ (e+p)v\,\|$, $p = (\varkappa - 1)[e - 0{,}5\,\rho(u^2+v^2)]$,*

is rewritten in integral form

$$\frac{\partial}{\partial t}\iint_{D} W\,dxdy + \oint_{\partial D} F\,dy - G\,dx = 0 \qquad (2)$$

and used to find a new value of numerical solution in (n+1)-time level

$$W_{ij}^{n+1} = W_{ij}^{n} - \frac{1}{\mu_{ij}} \oint_{\partial D_{ij}} \mathcal{F} dy - \mathcal{G} dx, \quad \mu_{ij} = \iint_{D_{ij}} dxdy\ , \qquad (3)$$

W_{ij} is mean value of W in quadrilateral cell D_{ij} (cell centered scheme) or value of W in a vertex of D_{ij} (cell vertex form); $\mathcal{F}, \mathcal{G}$ are some approximations of F, G in D_{ij}. Mac Cormack and Ni scheme are considered as some approximation of (3), see [1],[2]. Relation (3) is also possible to rewrite in residual form

---

* ρ,u,v,e are density, velocity components, total energy per unit volume; p is pressure

$$W_{ij}^{n+1} = W_{ij}^{n} - \Delta t \,.\, \mathrm{Rez}\, W_{ij}^{n} \qquad (4)$$

that defines steady residual Rez W_{ij}^{n} in cell D_{ij} after n-iterations steps. The system of Navier-Stokes equations in conservative form

$$W_t + F_x + G_y = \frac{1}{Re}(R_x + S_y) \qquad (5)$$

$$R = \mathrm{col}\,||0, \tau_{xx}, \tau_{xy}, u\tau_{xx} + v\tau_{xy} + kT_x||$$

$$S = \mathrm{col}\,||0, \tau_{xy}, \tau_{yy}, u\tau_{xy} + v\tau_{yy} + kT_y||. \quad ^{*)}$$

In the case of Navier-Stokes equations the following relations correspond (2)

$$\frac{\partial}{\partial t}\iint_D W\,dxdy + \oint_{\partial D}\left(F - \frac{1}{Re}R\right)dy - \left(G - \frac{1}{Re}S\right)dx = 0 \qquad (6)$$

and when one can approximate line integral along boundary ∂D_{ij} of each computational cell we have the system of ordinary differential equations

$$W_t|_{ij}^{n} = -\frac{1}{\mu_{ij}}\oint_{\partial D_{ij}}\left(\mathcal{F} - \frac{1}{Re}\mathcal{R}\right)dy - \left(\mathcal{G} - \frac{1}{Re}\mathcal{S}\right)dx = \widetilde{RW}_{ij}^{n} \qquad (7)$$

that is solved by multistage Runge-Kutta method

$$W_{ij}^{n} = W_{ij}^{(0)}$$

$$W_{ij}^{(k)} = W_{ij}^{n} - \alpha_k \cdot \Delta t.\, \widetilde{RW}_{ij}^{(k-1)}, \quad (k = 1,\dots,m) \qquad (8)$$

$$W_{ij}^{m+1} = W_{ij}^{(n)}.$$

We used two method: a) $\alpha_1 = \alpha_2 = \frac{1}{2}$, $\alpha_3 = 1$

b) $\alpha_1 = \alpha_2 = 0,6$; $\alpha_3 = 1$.

The first method showed the better numerical results [3]. Artificial dissippation terms were used in each method. The forms of the terms are published in [1],[2],[3],[4]. The same is true for boundary conditions used in all computed cases.

II. Unsteady solution through SE 1050 cascade

Consider unsteady solution caused by known change $p_2 = p_2(t)$ as a periodical function of time t:

---

*) T is temperature and $\tau_{xx} = (\frac{4}{3}u_x - \frac{2}{3}v_y)\mu$, $\tau_{yy} = \mu(\frac{4}{3}v_y - \frac{2}{3}u_x)$, $\tau_{xy} = \mu(u_y + v_x)$, Re is Reynolds number.

$$p_2 = (k_1 + k_2 \sin (k_3 t))p_1 \ , \qquad (9)$$

$t = \tilde{t} a_\infty / c$, $\tilde{t}$ is physical time, a_∞ is upstream sonic velocity, c is lenth of chord of the profile.

We considered the following cases:

a) $k_1 = 0.6$, $k_2 = 0.15$, $k_3 = \frac{1}{3}$, 1,3,6,9 - influence of frequency

b) $k_1 = 0.6$, $k_2 = 0.15$, $k_3 = 1$ and three different possibilies for a location of BB´(see fig.1) α) volume $F_2BB'F_2'$ given by steady state solution

ß) BB´ is moved to the right (compared to case b α)

γ) BB´ is moved to the left (compared to case bα).

In all cases mentioned here we computed results with Δt given by stability limitation and also with $\Delta t/2$ and we achieved the same numerical results. Fig 2 shows steady state solution (fig 2a), unsteady solution for t = 3,66 and volume given by (bγ) - 2b or t = 3,66 and volume given by (bß) - 2c and the same frequency (k_3=1). Fig.3 shows results of unsteady solution for t = 3,66, volume (bα) and k_3=1 (fig 3a) or k_3=3 (fig.3b). We also investigated time period T needed to achieve the periodic motion. Fig.4 shows comparison of computed function $\alpha_2(M_2)$ for $p_2 = p_2(t)$ a) near initial conditions and b) in one period during periodical motion for the case of Mac Cormack scheme using three considered downstream volumes bα), bß), bγ).

Conclusion: a) value of downstream volume influenced the qualitative properties of unsteady flowfield, not time needed for stabilized periodic motion

b) frequency dominantly influenced time needed to stabilize periodic motion.

The same results as by Mac Cormack scheme were achieved by Ni scheme.

III. Unsteady viscous solution over NACA 0012

GAMM Workshop 87 showed that case M_∞=0,85, α= 0°, Re = 10^4 is regime where stable steady solution should not exist. We wanted to test our method and simulate flow behavior in this case. In the first case symmetric initial conditions were used. We observed symmetric behavior of the results but not steady. The same small disturbance in initial velocity near trailing edge (order of magnitude - 2) was considered and we can expect periodical behavior near trailing edge and in a wake. When we continue with first case (symmetric initial conditions), for long-run computation (20 000 - 25 000 time steps) the round-off errors effect was strong to break symmetry of the flow. Fig.5 shows isomachlines of the flow with $M_\infty = 0{,}85$, $\alpha = 0^o$, $Re = 10^4$ (case 2) and nonsymmetric initial conditions after 5100 time steps and details near trailing edge. Fig.6 shows isomachlines of the same flowfield with symmetric initial conditions after 20,500 time steps.

References:

[1] Kozel K.,Nhac N.,Vavřincová M.: Notes in Numerical Fluid Mechanics, Vol.24, 1988.

[2] Ni R.H.: AIAA Journal, Vol.20, 1982.

[3] Fořt J.,Kozel K.,Vavřincová M.: Proceedings of IMACS´91 (Dublin 1991), Editors: R.Vichnevetsky, .J.H.Miller, pp.489-490.

[4] Svanson R.S.,Turkel E.: ICASE Report No 84-62, 1985.

[5] Kozel K.(Ed): Report K 201 TU Prague, No 201-3-91-111, Prague 1991.

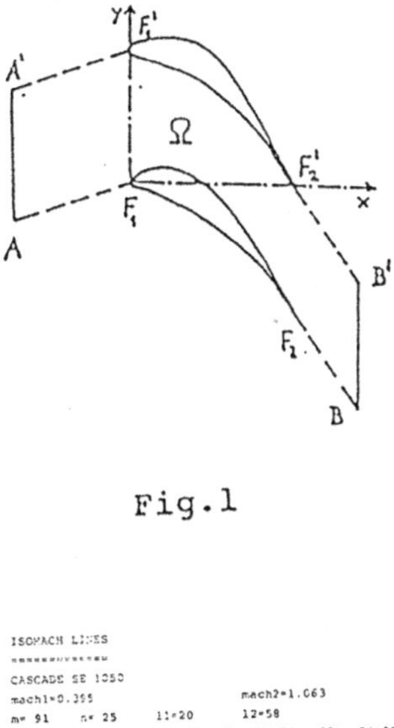

Fig.1

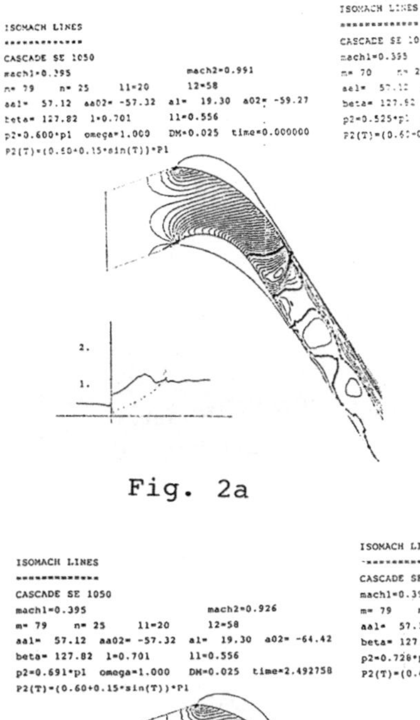

Fig. 2a

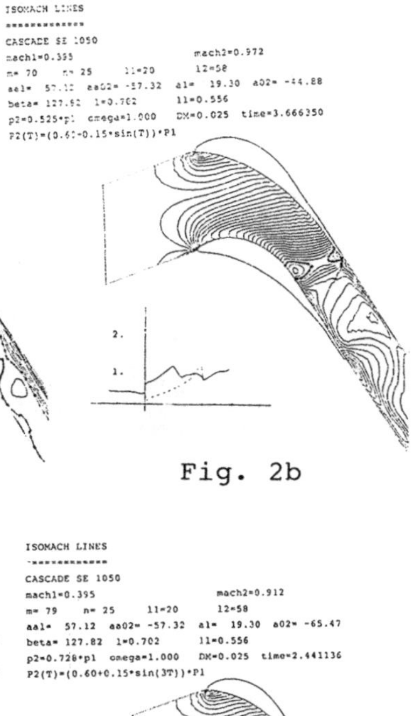

Fig. 2b

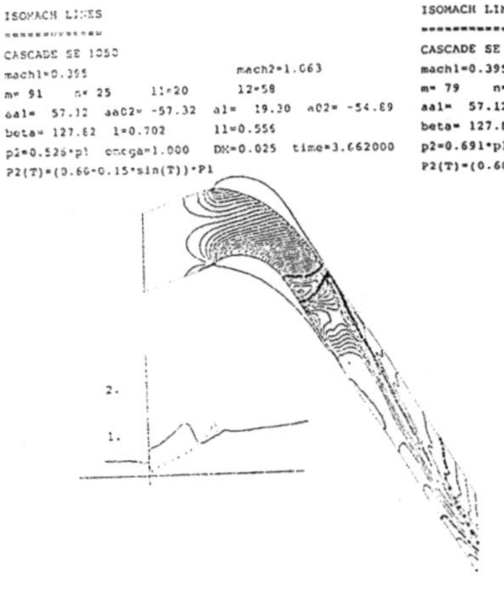

Fig. 2c

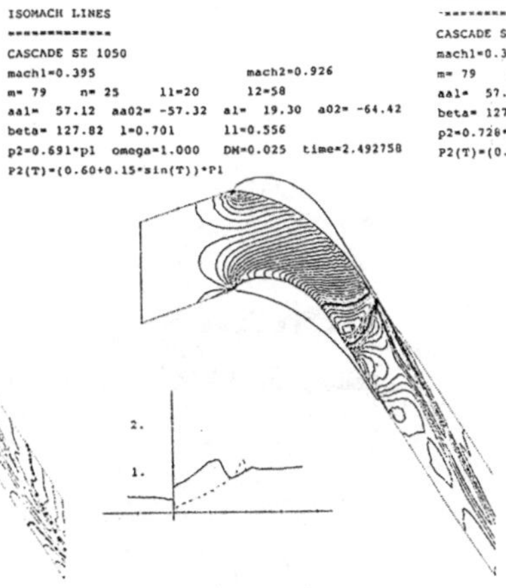

Fig. 3a

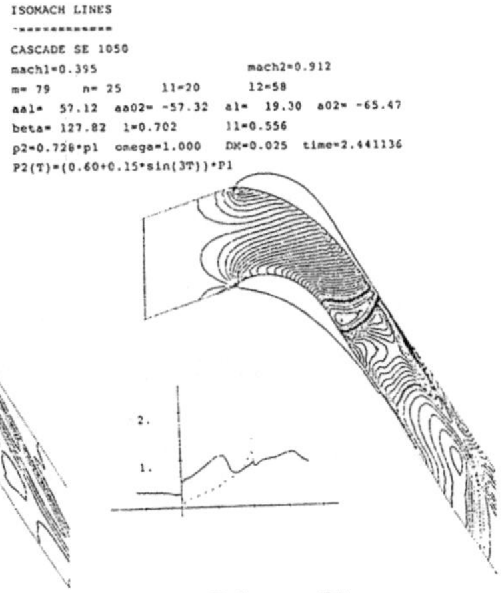

Fig. 3b

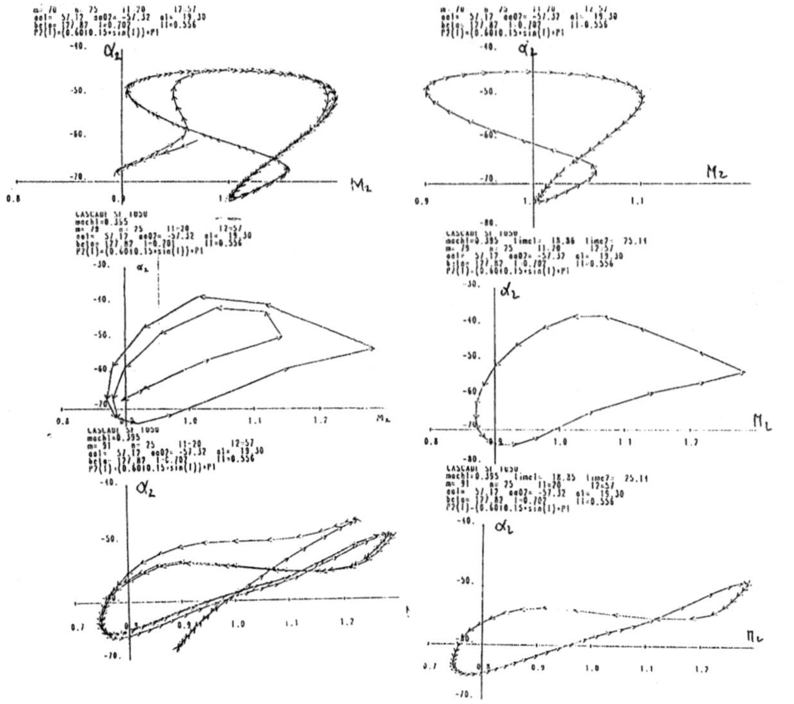

Fig. 4

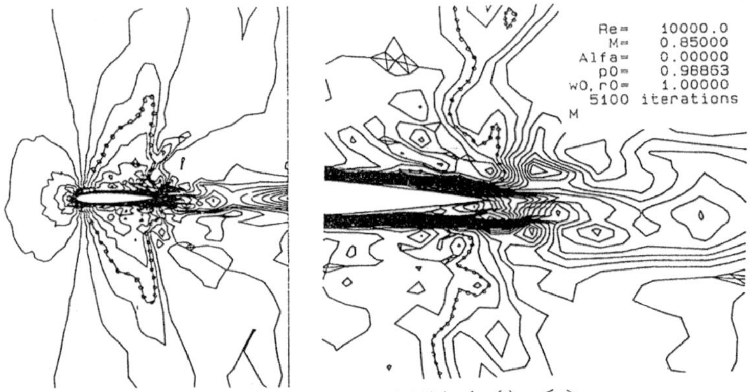

Fig. 5

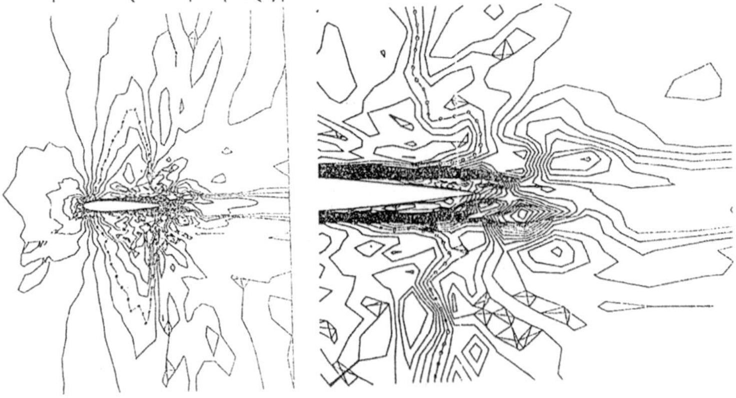

Fig. 6

DEVELOPMENT AND APPLICATIONS OF A 3D COMPRESSIBLE NAVIER-STOKES SOLVER

H. Kuerten, B. Geurts, J. van der Burg, B. Vreman, P. Zandbergen

Department of Applied Mathematics, University of Twente,
P.O. Box 217, 7500 AE Enschede, The Netherlands

1 Introduction

As a part of the Dutch ISNaS project our group and NLR jointly develop a flow solver for compressible, turbulent flow. This flow solver is especially aimed at applications on the industrial level: the multi-element airfoil and wing/body combination, both at transonic flow conditions. The flow solver is based on the Reynolds-averaged Navier-Stokes equations, in which presently the algebraic Baldwin-Lomax turbulence model is adopted. In reference [1] the first results, for laminar and turbulent flow around a single airfoil and over a finite flat plate have been shown. In the present paper recent developments in the solver are discussed.

In section 2 the numerical method used in the ISNaS solver is briefly described. Section 3 discusses the role of the numerical, or artificial dissipation in relation to the physical dissipation. In section 4 numerical aspects of the extension of the monoblock solver to a multiblock solver are described. The numerical method used in the ISNaS solver serves as a basis for many CFD programs used in our group. These programs are not only intended for the two applications mentioned above, but also for more fundamental studies of turbulence (with the help of large eddy simulation (LES) and direct numerical simulation (DNS)) and for the simulation of viscous water waves. In section 5 of this paper the use of the numerical method in large eddy simulation is discussed.

2 Numerical Method

The formulation of the numerical method starts from the integral formulation of the unsteady Reynolds-averaged Navier-Stokes equations, in which the densities of the conserved quantities are used as dependent variables. This integral formulation leads in a natural way to a finite volume approach in physical space, preserving the conservation property. A cell-vertex method with overlapping control volumes is used, since it remains accurate in the neighbourhood of grid distortions, which are unavoidable in the relevant applications. In figure 1 the control volumes used for the convective and viscous fluxes are sketched.

For the spatial discretization central differencing is applied with additional second- and fourth order nonlinear artificial dissipation. The formulation of the artificial dissipation is based on the work of Jameson [2], and is described in more detail in the next section. At a solid surface the velocity components are set equal to zero, while the density and energy density are obtained by solving their conservation equations, assuming an adiabatic wall.

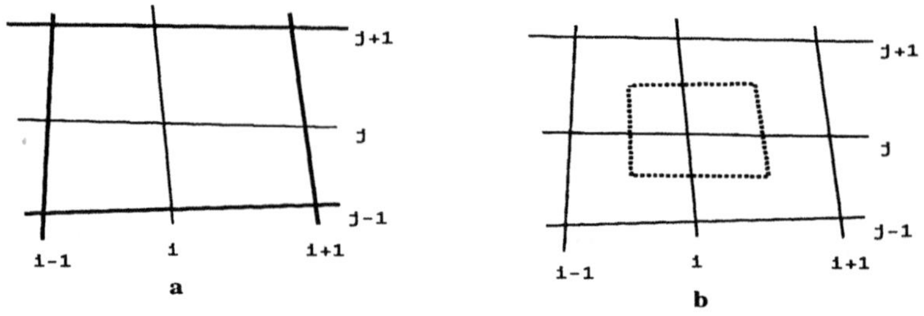

Figure 1: Control volumes used for the convective terms (a) and viscous terms (b)

The treatment of the far-field boundaries is based on approximate Riemann invariants, with the optional inclusion of a circulation correction [3].

The system of ordinary differential equations which results from the spatial discretization is integrated in time with a time-explicit multistage Runge-Kutta method, where local time stepping is used. The calculation time can drastically be reduced by the use of a multigrid technique and implicit residual averaging. Both for two-dimensional turbulent flow around an airfoil and three-dimensional turbulent flow around a wing the gain in calculation time, using these acceleration techniques is of the order of a factor of 10.

For calculations on smooth grids instead of the cell-vertex method, sketched in figure 1, a vertex-based method can be used as an alternative. This method differs in the calculation of the convective fluxes, for which the same control volume is used as for the viscous fluxes. This and the fact that the control volume on which the artificial dissipation method is based is the same, increase the consistency of the method.

3 Artificial Dissipation

In refs. [1], [4] it has been shown that the numerical method described in the previous section captures the physics accurately for subsonic, laminar flow over a flat plate. For transonic, turbulent flow at high Reynolds number, however, two disadvantages of the usual formulation of artificial dissipation [2], [5] become apparent. In the first place, shock waves are not accurately captured, and in the second place, the artificial dissipation in the boundary layer is rather high, leading to too high drag coefficients. In this section it is shown that these drawbacks can be met by several modifications in the formulation.

After spatial discretization the Navier-Stokes equations can schematically be written as $dw/dt = f_c + f_v + f_a$, where w is the vector of dependent variables in all grid points, f_c is the convective flux, f_v the viscous flux and f_a the artificial dissipation. The latter consists of contributions from every spatial direction, each of which has the form $d_{i+1/2} - d_{i-1/2}$, where

$$d_{i+1/2} = S_{i+1/2}[\epsilon^{(2)}_{i+1/2}\Delta_{i+1/2}w - \epsilon^{(4)}_{i+1/2}\Delta^3_{i+1/2}w].$$

Here, $S_{i+1/2}$ is a scaling factor, $\epsilon^{(2)}_{i+1/2}$ and $\epsilon^{(4)}_{i+1/2}$ are functions of a shock sensor, and $\Delta_{i+1/2}$ and $\Delta^3_{i+1/2}$ are first and third order difference operators.

An easy way to reduce the artificial dissipation in the boundary layer, where the viscous dissipation should be dominant, is a multiplication of $S_{i+1/2}$ with a function of the local Mach number. Application of this scaling in the direction normal to a solid wall with a linear function leads to a significant reduction of the drag coefficient.

For two-dimensional calculations on meshes with grid cells of high aspect ratio the scaling factor $S_{i+1/2}$ in the i-direction is usually taken as $S_i = [1 + (\lambda_j/\lambda_i)^{2/3}]\lambda_i$, where λ_i is the maximum eigenvalue of the flux Jacobian matrix in the i-direction [5]. The term between the brackets increases the artificial dissipation in the streamwise direction in the boundary layer. It appears that in the vertex-based discretization $S_i = \lambda_i$ can be chosen, at the cost of a small reduction in multigrid efficiency. However, the decrease of artificial dissipation leads to a better caption of the physical phenomena in the solution, which would otherwise require a finer grid.

Shock waves can be captured more accurately, if the first order difference terms in the artificial dissipation, which are triggered by a shock sensor, are replaced by upwind differences [6]. This approach works both for inviscid and turbulent, viscous flow problems (see figure 2).

4 Numerical Aspects of Multiblock Solver

For the multi-element airfoil application a multiblock solver will be constructed. The total computational domain will be divided into blocks in such a way that in each block a structured, boundary-conforming grid can be adopted. This block structure opens the possibility to solve the Euler equations in those blocks which are situated in the 'outer' regions of the flow, and hence save calculation time. The numerical aspects of the multiblock solver have been investigated, using a multiblock structure around a single airfoil [7].

To this end each of the blocks is taken out of the total domain separately and 'dressed' with two rows of dummy grid points. The blocks can be updated over one or more time steps independently, whilst keeping the variables in the dummy grid points frozen. After each block has been treated in this way the variables at the block interfaces are averaged.

It appears that the steady-state solution and convergence behaviour are unchanged, irrespective of the number of blocks and location of the interfaces, if the number of time steps over which the dummy variables are kept frozen is not too large [7]. Further, the solution is unchanged, if outside the boundary layer the Euler equations are solved instead of the Navier-Stokes equations (see figure 3), even if the grid is distorted near the interface.

5 Large Eddy Simulation

Especially for complex flows, with large separation regions and shock-boundary layer interaction, the results of flow simulation based on the Reynolds-averaged Navier-Stokes equations inadequately describe related physical experiments. This is mainly caused by the turbulence model. Information on improvements of turbulence models can be obtained through comparison with results from large eddy simulations of flow in simpler geometries. It has been shown that large eddy simulations of compressible flow are possible within a finite-volume approach, if the discretizations of the convective and viscous fluxes are performed on the same control volume [8]. If the Simpson rule is used for the integration of both fluxes, the theoretically expected velocity correlation spectrum is obtained in the case of homogeneous, isotropic, decaying turbulence in a 3D box. With the cell-vertex method direct numerical simulations of turbulent flow in a compressible mixing layer have been performed. The pairing of vortices and the correlation between pressure and vorticity are in agreement with literature.

Acknowledgements

The authors are greatly indebted to D. Dijkstra and F.J. Brandsma for many stimulating discussions. This work is partly subsidized by the Dutch Ministries of Education and Sciences and of Transport and Public Works. Part of the necessary supercomputer time was

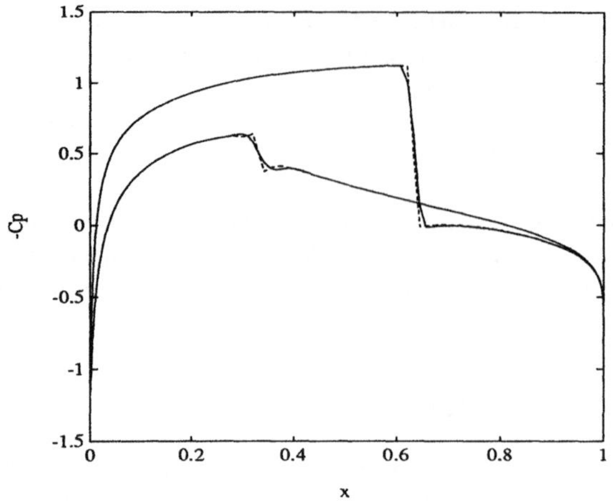

Figure 2: Pressure coefficient on the airfoil for inviscid flow around a NACA0012 profile at $M_\infty = 0.8$ and $\alpha = 1.25°$; central differences in artificial dissipation (solid) and upwind differences (dashed).

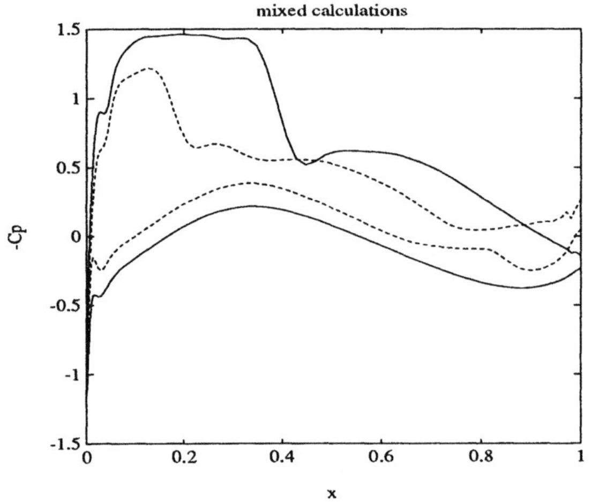

Figure 3: Pressure coefficient on the airfoil for turbulent flow around a CAST7 profile at $M_\infty = 0.7$, $Re = 4 \times 10^6$ and $\alpha = 2°$; only Navier-Stokes (solid), Euler-Navier-Stokes interface just outside boundary layer (dotted, coincides with solid) and just inside boundary layer (dashed).

sponsored by the National Computing Facilities Foundation, NCF, with financial support from the Netherlands Organisation for Scientific Research, N.W.O.

References

[1] F.J. Brandsma, J.G.M. Kuerten: The ISNaS compressible Navier-Stokes solver; first results for single airfoils, in: K.W. Morton (Ed.), Twelfth International Conference on Numerical Methods in Fluid Dynamics, Oxford, U.K., July 9-13, 1990, Springer Verlag, pp. 152-156.

[2] A. Jameson, W. Schmidt, E. Turkel: Numerical Solutions of the Euler Equations by Finite Volume Methods Using Runge-Kutta Time-Stepping-Schemes, AIAA Paper 81-1259 (1981).

[3] J.G.M. Kuerten: Numerical Definition Document for the Time-explicit ISNaS Solver, ISNaS 88.10.031 (1988).

[4] D. Dijkstra, J.G.M. Kuerten: An easy test-case for a Navier-Stokes solver, in: Proceedings of the First European Computational Fluid Dynamics Conference, Brussels, Belgium, September 7-11, 1992.

[5] V.N. Vatsa, B.W. Wedan: Development of an efficient multigrid code for 3-d Navier-Stokes equations, AIAA Paper 89-1791 (1989).

[6] J.W. van der Burg, J.G.M. Kuerten, P.J. Zandbergen: Improved shock capturing of Jameson's scheme for the Euler equations, to appear in Int. J. Num. Meth. Fluids.

[7] B.J. Geurts, J.G.M. Kuerten: Numerical aspects of a block structured compressible flow solver, to appear in J. Engg. Math.

[8] A.W. Vreman, B.J. Geurts, J.G.M. Kuerten, P.J. Zandbergen: A finite volume approach to large eddy simulation of compressible, homogeneous, isotropic, decaying turbulence, to appear in Int. J. Num. Meth. Fluids.

NUMERICAL SIMULATION OF THREE-DIMENSIONAL VORTEX FLOWS OVER DELTA WING CONFIGURATIONS

L. Martinelli [1], A. Jameson [1], E. Malfa [2]

[1]Department of M.A.E, Princeton University, Princeton N.J. 08544, USA
[2]Aermacchi Spa, Via Sanvito 80, Varese, Italy

1 Introduction

The simulation of three-dimensional vortex flows like the ones established by delta wing configurations presents several challenges. First of all, there is the selection of an appropriate mathematical model. It is well established that if the separation line of the primary vortex is known *a priori* (on geometries with sharp leading edges, for example), then a useful prediction of the integral aerodynamic characteristics can be obtained by solving the non-linear Euler equations (Murman and Rizzi 1986). However, when the separation lines are not known (e.g., on delta wings with round leading edges), or when a more detailed prediction of the aerodynamic loads is needed, the mathematical model used to represent the flow must account for the effects of viscosity. This work is a numerical study of vortex flows using a hierarchy of mathematical models of increasing complexity and completeness, ranging from the inviscid Euler equations to the full three-dimensional Navier-Stokes equations. Since vortex flows contain strong gradients in both the normal (to the wall), and cross-flow (spanwise) directions, the use of a thin-layer approximation is not justifiable. Thus the full three-dimensional form of the viscous equations was selected. In the turbulent regime, the flow was modeled by the full three-dimensional mass averaged Reynolds equations, and a turbulence model was used for closure.

2 Numerical Scheme

A very efficient finite-volume multigrid time-stepping algorithm was used to integrate the Euler equations. Since some of the geometries studied have sharp leading edges, which inevitably cause sharp corners between grid lines, a cell vertex formulation was chosen. The formulation of the equations and the basic discretization scheme is given elsewhere (Jameson (1986), Volpe, Siclari, and Jameson (1987)). The scheme has been extensively validated for the simulation of vortical flows (Malfa, Guarino and Visentini (1991)) and allows for convenient extension of the algorithm to treat viscous flows.

2.1 Discretization of the Viscous Fluxes

The three-dimensional viscous flow solver employed in this study follows guidelines originally proposed by the first two authors(Martinelli (1987), Martinelli and Jameson (1988)) for the simulation of two-dimensional viscous flows. A finite-volume discretization of the viscous fluxes using a compact support which avoids odd-even decoupling modes was devised as follows. The components of the stress tensor and of the heat-flux are computed at the centers of the computational cells with the aid of Gauss's formula. Then, the viscous fluxes are computed by making use of an auxiliary cell bounded by faces lying on the planes containing the centers of the cells surrounding a given vertex and the mid-lines of the cell faces. This scheme has been shown to to maintain good local accuracy on grids with kinks (Liu and Jameson (1992)).

2.2 Artificial Dissipation

Since the discretization of the convective operator reduces to central differences on a regular Cartesian grid, it is necessary to introduce a dissipative operator to avoid aliasing at odd and even points, and to allow for a clean capture of shock waves. This is accomplished by the introduction of blended second and fourth differences. However, in order to maintain accuracy on highly stretched grids, an appropriate rescaling of the artificial dissipation terms (Martinelli (1987)) was implemented.

2.3 Convergence Acceleration

Time integration is carried out by making use of a five-stage scheme which requires re-evaluation of the dissipative operators only at alternate stages (Martinelli (1987)). This scheme couples the desirable feature of a wide stability region along both the imaginary and the real axis with good high frequency damping. The efficiency of the scheme was enhanced by using an implicit residual averaging scheme with variable coefficients, and an effective multigrid strategy based on a W-cycle. Throughout the course of this study, the proposed algorithm has been found to be both accurate and robust. Moreover, its high efficiency has made the solution of viscous flow problems on meshes containing up to 1.5 million points routinely feasible, even on a mini-supercomputer.

3 Simulation of A Low Reynolds Number Flow

A validation of the numerical scheme for a flat-plate laminar boundary layer has been carried out elsewhere (Liu and Jameson 1992).

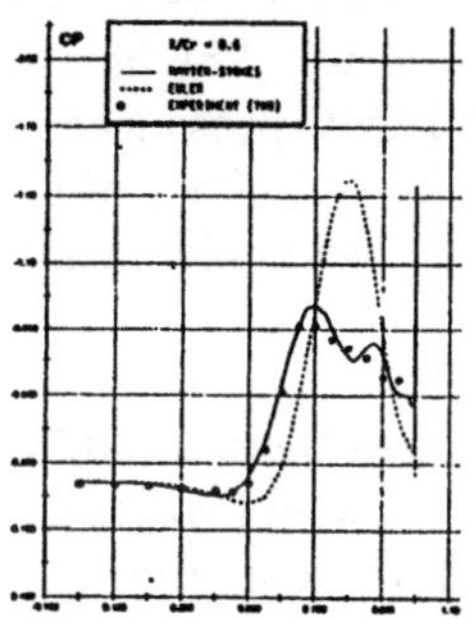

Fig. 1.a. Pressure Coefficient at $\frac{x}{c_r} = .6$

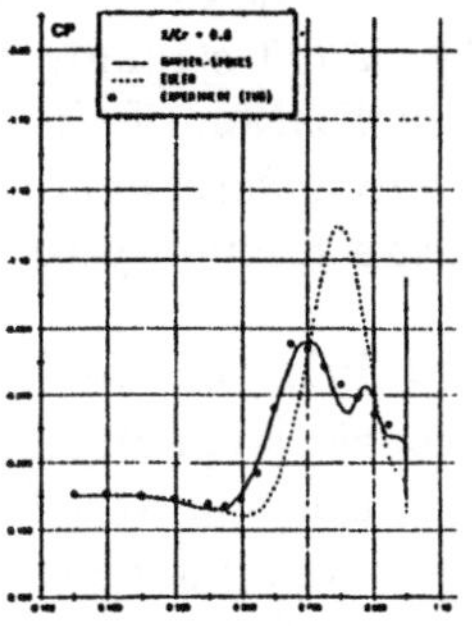

Fig. 1.b. Pressure Coefficient at $\frac{x}{c_r} = .8$

The set of results presented here is aimed at demonstrating the accuracy achieved by the numerical scheme for vortex-flow applications. We conducted a numerical simulation of the flow over a cropped delta wing with a sharp leading edge. Both the geometry and the flow conditions corresponds to those tested by Hummel (Bergmann and Hummel (1990)). The relatively low Reynolds number (based on the root chord) of the experiments ($Re = 440,000$), allowed the calculation of the viscous flow to be successfully computed by using the pure Navier-Stokes formulation. By avoiding any uncertainty attributable to turbulence modeling, it provided us with an ideal test-case for establishing the accuracy of the basic numerical scheme for vortex-dominated flows. A $161 \times 49 \times 49$ C-H grid was employed. For the sake of brevity the results for only two cross-sections of the wing are reported here. Figures 1.a and 1.b show a comparison of the pressure coefficient on two cross-sectional cuts of the wing computed by modeling the flow using the Navier-Stokes equations and the Euler equations.

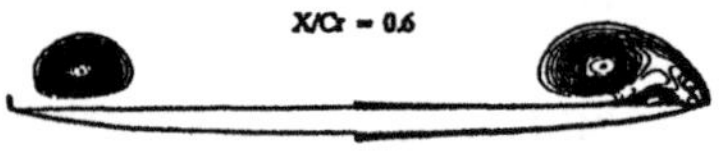

Fig. 2.a. Cross-flow at $\frac{x}{c_r} = .6$
Euler simulation (left)
Laminar simulation (right)

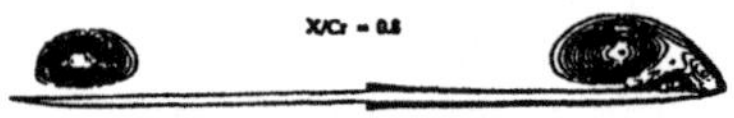

Fig. 2.b. Cross-flow at $\frac{x}{c_r} = .8$
Euler simulation (left)
Laminar simulation (right)

It can be seen that the two mathematical models are in good agreement in the inboard section of the wing, whereas they differ substantially in the outboard region. Moreover, the position of the primary vortex, for both the Euler and the Navier-Stokes solutions, compares reasonably well with experiments, and the results of the viscous simulation are also in good agreement with the measurements in the region of the secondary vortex. This seems to confirm the expectation that the computed evolution of the primary vortex is not strongly affected by the effects of viscosity. This is also confirmed by the visualization of the streamwise evolution of the field presented in Figures 2.a and 2.b. From those figures, we can also verify that the main effect of the viscosity is to generate a secondary vortex, which is absent in the inviscid calculation.

4 Simulation of High Reynolds Number Flows

Several turbulence models, including a Baldwin & Lomax (Baldwin and Lomax (1978), a Johnson & King (Johnson and King (1984)), and an algebraic RNG based (Martinelli and Yakhot (1989)) were considered to account for turbulent transport. Although such models are well established for simulating turbulent wall flows, their application to three-dimensional vortex flows presents some new challenges.

4.1 Implementation and Selection of Scales

The three turbulence models considered in this study require the selection of an outer length scale. The main advantage of the Baldwin & Lomax over previous algebraic models is in the determination of the outer length scale, which is accomplished by computing the location of the maximum of the function $F(y) = |\omega| y[1 - exp(-\frac{y^+}{A^+})]$. Here $|\omega|$ is the magnitude of the vorticity vector, y is the distance from the wall in the normal to the wall direction, y^+ is the normal distance from the wall in wall coordinate, and A^+ a damping constant. The convenience of determining an outer length scale based on the maximum of this function and the relative ease of implementation provided the rationale for adopting the Baldwin & Lomax formulation as the equilibrium model in the modified Johnson & King model implemented in this study (Radespiel (1989)). In particular, following Radespiel, we use $\delta = 1.9 * y_{max}$ as an estimate of the boundary layer thickness. For consistency, such an estimate has also been employed in the present study in the implementation of the RNG based algebraic model. For a turbulent wall flow it can be verified that such a function possess a unique and well defined maximum: however this is far from being the case for the vortical flows of interest in this work. Depending on the location, the function $F(y)$ can exhibit multiple local maxima leading to the need of establishing a selection rule. For example, the choice of an *erroneous* maximum in the region of the core of the primary vortex can cause the overprediction of the eddy-viscosity which, in turn, will damp out the main features of the flow. This fact has been recognized in the literature (Degani and Schiff (1986)). The modification of the Baldwin & Lomax model proposed by Degani and Schiff, although effective in the region of the primary vortex, showed a tendency to overpredict the pressure coefficient in the region where the secondary separation takes place. There the flow is dominated by the effect of viscosity and turbulence transport, and the increased strength of the secondary vortex may be attributable to an underprediction of the eddy viscosity. In the region of the secondary separation the boundary layer and the vortex sheet which originates from the secondary vortex merge, and the criteria proposed by Degani and Schiff for the selection of the maximum of $F(y)$ seems to be inadequate. For this reason an alternative procedure has been devised, in which we compute the location of all the local maxima and we select the one with a maximum in curvature. In this way, a larger length scale is selected in the region of the secondary vortex and the calculated pressure distribution reproduces the experimentally observed plateau (Elsenaar and Hoeijmakers (1991)).

4.2 Numerical Results

An extensive series of simulations has been carried out to assess the behaviour of the three turbulent models as well as their implementation.

A $193 \times 65 \times 49$ C-H grid was employed for all the calculations reported here over a 65° swept cropped delta wing with a sharp leading edge. The wing was set at 10° angle of attack into a free stream flowing at $M = .85$. The Reynolds number based on the root chord was set to $9,000,000$. Figure 3 shows a comparison of the pressure distribution computed using each of the three turbulence models with the experimental data of Elsenaar and Hoeijmakers. It can be seen the the results compare reasonably well in the inboard section, while there are minor differences in the outboard region. A more detailed analysis of the cross-flow, which is not presented here because of space constraints, shows that the pressure distributions computed with the Johnson and King and the RNG based models agree quite well, and that the Baldwin and Lomax model produces a slightly more dissipative solution with a less developed secondary vortex.

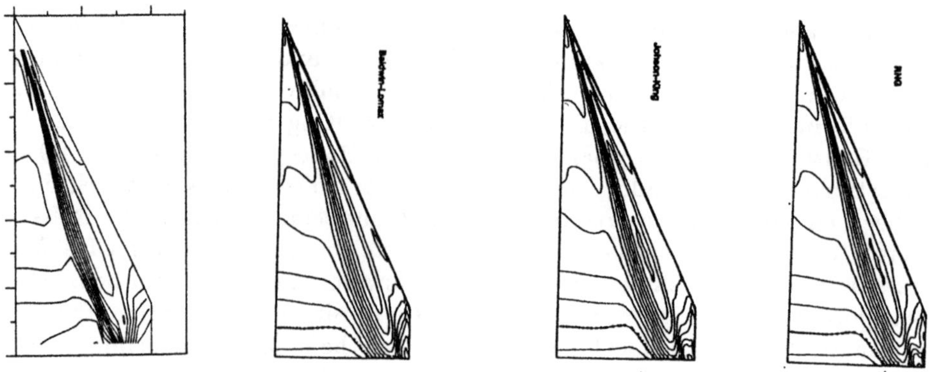

Fig. 3. Measured and Computed Isolines of the Pressure Coefficient. From left to right: Experimental Data, Baldwin & Lomax Model, Johnson & King Model, RNG Model.

References

Murman E., A. Rizzi (1986): "Applications of Euler Equations to Sharp Edge Delta Wings With Leading Edge Vortices", In Applications of Computational Fluid Dynamics in Aeronautics, AGARD-CP-412.

A. Jameson (1986): "A Vertex Based Multigrid Algorithm for Three Dimensional Compressible Flow Calculations", In Proceedings of the ASME Symposium on Numerical Methods for Compressible Flow, Anaheim.

G. Volpe, M.J. Siclari, and A. Jameson (1987): "A New Multigrid Euler Method for Fighter-Type Configuartions", AIAA-87-1160-CP

E. Malfa, L. Guarino, and L. Visentini (1991): "Application of Euler Equations to Computation of Vortex Flow on Wing-Body and Close-Coupled Wing-Body-Canard Configuration" AIAA-91-3306-CP

L. Martinelli (1987): "Calculations of Viscous Flows with a Multigrid Method", Princeton University, MAE Dept., Ph.D Thesis 1754T.

L. Martinelli, and A. Jameson (1988): "Validation of a Multigrid Method for the Reynolds Averaged Equations", AIAA-88-0414.

B.S. Baldwin, and H. Lomax (1978): "Thin Layer Approximation and Algebraic Model for Separated Turbulent Flows", AIAA-78-257.

D.A. Johnson, and L.S. King (1984): "A New Turbulence Closure Model for Boundary Layer Flows with Strong Adverse Pressure Gradients and Separation", AIAA-84-0175.

L. Martinelli, and V. Yakhot (1989): "RNG-Based Turbulent Transport Approximations with Applications to Transonic Flows", AIAA-89-1950.

F. Liu, and A. Jameson (1992): "Multigrid Navier-Stokes Calculations For Three-dimensional Cascades", AIAA-92-0190.

D. Degani, and L. Shiff (1986): "Computation of Turbulent Supersonic Flows around Pointed Bodies Having Crossflow Separation", Journal of Computational Physics, **66**, 173-196.

R. Radespiel (1989): "A Cell-Vertex Multigrid Method for the Navier-Stokes Equations", NASA-TM-101557.

A. Bergmann, and D. Hummel (1990): "Pressure Distributions on the Flat Surface Wing WI1 at $\alpha = 10^o$", TU Braunschweig. (private communication).

A. Elsenaar , and H.W.M. Hoeijmakers (1991): "An Experimental Study of the Flow Over a Sharp-Edged Cropped Delta Wing at Subsonic, Transonic and Supersonic Speed", NLR TP91117L

Index

Lecture Notes in Physics

For information about Vols. 1–374
please contact your bookseller or Springer-Verlag

Vol. 375: C. Bartocci, U. Bruzzo, R. Cianci (Eds.), Differential Geometric Methods in Theoretical Physics. Proceedings, 1990. XIX, 401 pages. 1991.

Vol. 376: D. Berényi, G. Hock (Eds.), High-Energy Ion-Atom Collisions. Proceedings, 1990. IX, 364 pages. 1991.

Vol. 377: W. J. Duschl, S.J. Wagner, M. Camenzind (Eds.), Variability of Active Galaxies. Proceedings, 1990. XII, 312 pages. 1991.

Vol. 378: C. Bendjaballah, 0. Hirota, S. Reynaud (Eds.), Quantum Aspects of Optical Communications. Proceedings 1990. VII, 389 pages. 1991.

Vol. 379: J. D. Hennig, W. Lücke, J. Tolar (Eds.), Differential Geometry, Group Representations, and Quantization. XI, 280 pages. 1991.

Vol. 380: I. Tuominen, D. Moss, G. Rüdiger (Eds.), The Sun and Cool Stars: activity, magnetism, dynamos. Proceedings, 1990. X, 530 pages. 1991.

Vol. 381: J. Casas-Vazquez, D. Jou (Eds.), Rheological Modelling: Thermodynamical and Statistical Approaches. Proceedings, 1990. VII, 378 pages. 1991.

Vol. 382: V.V. Dodonov, V. I. Man'ko (Eds.), Group Theoretical Methods in Physics. Proceedings, 1990. XVII, 601 pages. 1991.

Vol. 384: M. D. Smooke (Ed.), Reduced Kinetic Mechanisms and Asymptotic Approximations for Methane-Air Flames. V, 245 pages. 1991.

Vol. 385: A. Treves, G. C. Perola, L. Stella (Eds.), Iron Line Diagnostics in X-Ray Sources. Proceedings, Como, Italy 1990. IX, 312 pages. 1991.

Vol. 386: G. Pétré, A. Sanfeld (Eds.), Capillarity Today. Proceedings, Belgium 1990. XI, 384 pages. 1991.

Vol. 387: Y. Uchida, R. C. Canfield, T. Watanabe, E. Hiei (Eds.), Flare Physics in Solar Activity Maximum 22. Proceedings, 1990. X, 360 pages. 1991.

Vol. 388: D. Gough, J. Toomre (Eds.), Challenges to Theories of the Structure of Moderate-Mass Stars. Proceedings, 1990. VII, 414 pages. 1991.

Vol. 389: J. C. Miller, R. F. Haglund (Eds.), Laser Ablation-Mechanisms and Applications. Proceedings. IX, 362 pages, 1991.

Vol. 390: J. Heidmann, M. J. Klein (Eds.), Bioastronomy - The Search for Extraterrestrial Life. Proceedings, 1990. XVII, 413 pages. 1991.

Vol. 391: A. Zdziarski, M. Sikora (Eds.), Ralativistic Hadrons in Cosmic Compact Objects. Proceedings, 1990. XII, 182 pages. 1991.

Vol. 392: J.-D. Fournier, P.-L. Sulem (Eds.), Large-Scale Structures in Nonlinear Physics. Proceedings. VIII, 353 pages. 1991.

Vol. 393: M. Remoissenet, M.Peyrard (Eds.), Nonlinear Coherent Structures in Physics and Biology. Proceedings. XII, 398 pages. 1991.

Vol. 394: M. R. J. Hoch, R. H. Lemmer (Eds.), Low Temperature Physics. Proceedings. X,374 pages. 1991.

Vol. 395: H. E. Trease, M. J. Fritts, W. P. Crowley (Eds.), Advances in the Free-Lagrange Method. Proceedings, 1990. XI, 327 pages. 1991.

Vol. 396: H. Mitter, H. Gausterer (Eds.), Recent Aspects of Quantum Fields. Proceedings. XIII, 332 pages. 1991.

Vol. 398: T. M. M. Verheggen (Ed.), Numerical Methods for the Simulation of Multi-Phase and Complex Flow. Proceedings, 1990. VI, 153 pages. 1992.

Vol. 399: Z. Švestka, B. V. Jackson, M. E. Machedo (Eds.), Eruptive Solar Flares. Proceedings, 1991. XIV, 409 pages. 1992.

Vol. 400: M. Dienes, M. Month, S. Turner (Eds.), Frontiers of Particle Beams: Intensity Limitations. Proceedings, 1990. IX, 610 pages. 1992.

Vol. 401: U. Heber, C. S. Jeffery (Eds.), The Atmospheres of Early-Type Stars. Proceedings, 1991. XIX, 450 pages. 1992.

Vol. 402: L. Boi, D. Flament, J.-M. Salanskis (Eds.), 1830-1930: A Century of Geometry. VIII, 304 pages. 1992.

Vol. 403: E. Balslev (Ed.), Schrödinger Operators. Proceedings, 1991. VIII, 264 pages. 1992.

Vol. 404: R. Schmidt, H. O. Lutz, R. Dreizler (Eds.), Nuclear Physics Concepts in the Study of Atomic Cluster Physics. Proceedings, 1991. XVIII, 363 pages. 1992.

Vol. 405: W. Hollik, R. Rückl, J. Wess (Eds.), Phenomenological Aspects of Supersymmetry. VII, 329 pages. 1992.

Vol. 406: R. Kayser, T. Schramm, L. Nieser (Eds.), Gravitational Lenses. Proceedings, 1991. XXII, 399 pages. 1992.

Vol. 407: P. L. Smith, W. L. Wiese (Eds.), Atomic and Molecular Data for Space Astronomy. VII, 158 pages. 1992.

Vol. 408: V. J. Martínez, M. Portilla, D. Sàez (Eds.), New Insights into the Universe. Proceedings, 1991. XI, 298 pages. 1992.

Vol. 409: H. Gausterer, C. B. Lang (Eds.), Computational Methods in Field Theory. Proceedings, 1992. XII, 274 pages. 1992.

Vol. 410: J. Ehlers, G. Schäfer (Eds.), Relativistic Gravity Research. Proceedings, VIII, 409 pages. 1992.

Vol. 411: W. Dieter Heiss (Ed.), Chaos and Quantum Chaos. Proceedings, XIV, 330 pages. 1992.

Vol. 413: Aa. Sandqvist, T. P. Ray (Eds.); Central Activity in Galaxies. From Observational Data to Astrophysical Diagnostics. XIII, 235 pages. 1993.

Vol. 414: M. Napolitano, F. Sabetta (Eds.), Thirteenth International Conference on Numerical Methods in Fluid Dynamics. Proceedings, 1992. XIV, 541 pages. 1993.

Vol. 415: L. Garrido (Ed.), Complex Fluids. Proceedings, 1992. XIII, 413 pages. 1993.

New Series m: Monographs

Vol. m 1: H. Hora, Plasmas at High Temperature and Density. VIII, 442 pages. 1991.

Vol. m 2: P. Busch, P. J. Lahti, P. Mittelstaedt, The Quantum Theory of Measurement. XIII, 165 pages. 1991.

Vol. m 3: A. Heck, J. M. Perdang (Eds.), Applying Fractals in Astronomy. IX, 210 pages. 1991.

Vol. m 4: R. K. Zeytounian, Mécanique des fluides fondamentale. XV, 615 pages, 1991.

Vol. m 5: R. K. Zeytounian, Meteorological Fluid Dynamics. XI, 346 pages. 1991.

Vol. m 6: N. M. J. Woodhouse, Special Relativity. VIII, 86 pages. 1992.

Vol. m 7: G. Morandi, The Role of Topology in Classical and Quantum Physics. XIII, 239 pages. 1992.

Vol. m 8: D. Funaro, Polynomial Approximation of Differential Equations. X, 305 pages. 1992.

Vol. m 9: M. Namiki, Stochastic Quantization. X, 217 pages. 1992.

Vol. m 10: J. Hoppe, Lectures on Integrable Systems. VII, 111 pages. 1992.

Vol. m 11: A. D. Yaghjian, Relativistic Dynamics of a Charged Sphere. XII, 115 pages. 1992.

Vol. m 12: G. Esposito, Quantum Gravity, Quantum Cosmology and Lorentzian Geometries. XVI, 326 pages. 1992.

Vol. m 13: M. Klein, A. Knauf, Classical Planar Scattering by Coulombic Potentials. V, 142 pages. 1992.

Vol. m 14: A. Lerda, Anyons. XI, 138 pages. 1992.

Vol. m 15: N. Peters, B. Rogg (Eds.), Reduced Kinetic Mechanisms for Applications in Combustion Systems. X, 360 pages. 1993.

GPSR Compliance

The European Union's (EU) General Product Safety Regulation (GPSR) is a set of rules that requires consumer products to be safe and our obligations to ensure this.

If you have any concerns about our products, you can contact us on ProductSafety@springernature.com

In case Publisher is established outside the EU, the EU authorized representative is:

Springer Nature Customer Service Center GmbH
Europaplatz 3
69115 Heidelberg, Germany

Batch number: 10838062

Printed by Printforce, the Netherlands